Methods in Enzymology

Volume 125
BIOMEMBRANES
Part M
Transport in Bacteria, Mitochondria, and Chloroplasts:
General Approaches and Transport Systems

METHODS IN ENZYMOLOGY

EDITORS-IN-CHIEF

Sidney P. Colowick   Nathan O. Kaplan

Methods in Enzymology

Volume 125

Biomembranes

Part M
Transport in Bacteria, Mitochondria, and
Chloroplasts:
General Approaches and Transport Systems

EDITED BY

Sidney Fleischer
Becca Fleischer

DEPARTMENT OF MOLECULAR BIOLOGY
VANDERBILT UNIVERSITY
NASHVILLE, TENNESSEE

1986

ACADEMIC PRESS, INC.
Harcourt Brace Jovanovich, Publishers
Orlando   San Diego   New York   Austin
London   Montreal   Sydney   Tokyo   Toronto

ACADEMIC PRESS, INC.
Orlando, Florida 32887

United Kingdom Edition published by
ACADEMIC PRESS INC. (LONDON) LTD.
24–28 Oval Road, London NW1 7DX

LIBRARY OF CONGRESS CATALOG CARD NUMBER: 54-9110

ISBN 0–12–182025–4

PRINTED IN THE UNITED STATES OF AMERICA

86 87 88 89   9 8 7 6 5 4 3 2 1

Table of Contents

Section I. General Approaches

A. Membrane Characteristics and Chemistry

v

B. Membrane Assembly, Mutations, and Cloning Strategy

Section II. Bacterial Transport

Section III. Transport in Mitochondria and Chloroplasts

A. Mitochondria

B. Chloroplasts

Contributors to Volume 125

Article numbers are in parentheses following the names of contributors.
Affiliations listed are current.

SURESH V. AMBUDKAR (43), *Department of Physiology, The Johns Hopkins University School of Medicine, Baltimore, Maryland 21205*

HEINRICH AQUILA (51), *Institut für Physikalische Biochemie, Universität München, 8000 München 2, Federal Republic of Germany*

ANGELO AZZI (7), *Medizinisch-chemisches Institut, Universität Bern, CH-3000 Bern 9, Switzerland*

ULI BENBOW (9), *1810 Brownell Street, Kansas City, Missouri 64124*

KLAUS BEYER (49), *Institut für Physikalische Biochemie, Universität München, 8000 München 2, Federal Republic of Germany*

MARC R. BLOCK (50, 52), *Laboratoire de Biochimie, Département de Recherche Fondamentale, Centre d'Etudes Nucléaires, 85 X, 38041 Grenoble Cedex, France*

WERNER BOGNER (51), *Institut für Physikalische Biochemie, Universität München, 8000 München 2, Federal Republic of Germany*

FRANÇOIS BOULAY (50, 52), *Laboratorie de Biochimie, Département de Recherche Fondamentale, Centre d'Etudes Nucléaires, 85 X, 38041 Grenoble Cedex, France*

GÉRARD BRANDOLIN (50, 52), *Laboratoire de Biochimie, Département de Recherche Fondamentale, Centre d'Etudes Nucléaires, 85 X, 38041 Grenoble Cedex, France*

JOHANN M. BRASS (23), *Universität Konstanz, Facultät für Biologie, D-7750 Konstanz, Federal Republic of Germany*

BARRY S. BRAUN (9), *Department of Biochemistry and Biophysics, University of Pennsylvania, Philadelphia, Pennsylvania 19104*

J. MALCOLM BRUCE (8, 9), *Department of Chemistry, University of Manchester, Manchester M13 9PL, England*

WOLFGANG BUCKEL (42), *Lehrstuhl Biochemie I, Universität Regensburg, D-8400 Regensburg, Federal Republic of Germany*

NANCY CARRASCO (33), *Department of Biochemistry, Roche Institute of Molecular Biology, Roche Research Center, Nutley, New Jersey 07110*

ROBERT P. CASEY[1] (7), *Department of Medical Chemistry, University of Helsinki, SF-00170 Helsinki 17, Finland*

S. CASTLEMAN (3), *College of Osteopathic Medicine, Michigan State University, East Lansing, Michigan 48824*

BRAD CHAZOTTE (4), *Laboratories for Cell Biology, Department of Anatomy, School of Medicine, University of North Carolina at Chapel Hill, Chapel Hill, North Carolina 27514*

VINCENT P. CIRILLO (20), *Department of Biochemistry, State University of New York at Stony Brook, Stony Brook, New York 11794*

PETER S. COLEMAN (53), *Laboratory of Biochemistry, Department of Biology, New York University, New York, New York 10003*

ANNE COLLINS (55), *Department of Cell Physiology, Boston Biomedical Research Institute, Boston, Massachusetts 02114*

WALEED DANHO (33), *Department of Molecular Genetics, Hoffmann-La Roche Inc., Roche Research Center, Nutley, New Jersey 07110*

PETER DIMROTH (40, 41), *Institut für Physiologische Chemie, Technische Universität München, 8000 München 40, Federal Republic of Germany*

[1] Deceased.

YVES DUPONT (50), *Laboratoire de Biologie Moléculaire et Cellulaire, Département de Recherche Fondamentale, Centre d'Etudes Nucléaires, 85 X, 38041 Grenoble Cedex, France*

P. LESLIE DUTTON (8, 9), *Department of Biochemistry and Biophysics, University of Pennsylvania, Philadelphia, Pennsylvania 19104*

GUNNAR FALK (18), *The Arrhenius Laboratory, Department of Biochemistry, University of Stockholm, Stockholm, Sweden*

SHELAGH M. FERGUSON-MILLER (2, 3), *Department of Biochemistry, Michigan State University, East Lansing, Michigan 48824-1319*

U. I. FLÜGGE (56, 57), *Institut für Biochemie der Pflanze, Universität Gottingen, 3400 Gottingen, Federal Republic of Germany*

DAVID L. FOSTER (32), *Donner Laboratory, University of California, Berkeley, California 94720*

S. FOXALL-VANAKEN (3), *Ford Motor Co., Research Division, Dearborn, Michigan 48121*

CLEMENT E. FURLONG (22), *Departments of Genetics and Medicine, Division of Medical Genetics, Center for Inherited Diseases, University of Washington, Seattle, Washington 98195*

R. MICHAEL GARAVITO (25), *Biozentrum, University of Basel, CH-4056 Basel, Switzerland*

MARGARET M. GIBSON (29), *Department of Biochemistry, University of Dundee, Dundee DD1 4HN, Scotland*

MARIE A. GILLES-GONZALEZ (16), *Department of Biology, Massachusetts Institute of Technology, Cambridge, Massachusetts 02139*

A. A. GUFFANTI (28), *Department of Biochemistry, Mount Sinai School of Medicine, New York, New York 10029*

CHARLES R. HACKENBROCK (4), *Laboratories for Cell Biology, Department of Anatomy, School of Medicine, University of North Carolina at Chapel Hill, Chapel Hill, North Carolina 27514*

NEIL R. HACKETT (16), *Departments of Biology and Chemistry, Massachusetts Institute of Technology, Cambridge, Massachusetts 02139*

H. W. HELDT (56, 57), *Institut für Biochemie der Pflanze, Universität Goettingen, 3400 Goettingen, Federal Republic of Germany*

PETER J. F. HENDERSON (13, 31), *Department of Biochemistry, University of Cambridge, Cambridge CB2 1QW, England*

DORIS HERZLINGER (33), *Department of Medicine, College of Physicians and Surgeons, Columbia University, New York, New York 10032*

CHRISTOPHER F. HIGGINS (29), *Department of Biochemistry, University of Dundee, Dundee DD1 4HN, Scotland*

WILHELM HILPERT (41), *Institut für Physiologische Chemie, Technische Universität München, 8000 München 40, Federal Republic of Germany*

H. HIRATA (36), *Department of Biochemistry, Jichi Medical School, Tochigi, Japan 329-04*

JEN-SHIANG HONG (14, 24), *Department of Cell Physiology, Boston Biomedical Research Institute, Boston, Massachusetts 02114*

ARTHUR G. HUNT (24), *Department of Agronomy, University of Kentucky, Lexington, Kentucky 40506*

YASUO IMAE (45), *Institute of Molecular Biology, Faculty of Science, Nagoya University, Nagoya 464, Japan*

MASAYORI INOUYE (11), *Department of Biochemistry, State University of New York at Stony Brook, Stony Brook, New York 11794*

SHIGERU ITOH (6), *National Institute for Basic Biology, Okazaki 444, Japan*

SIMON J. JONES (16), *Genetics Institute, 225 Longwood Avenue, Boston, Massachusetts 02115*

MAURICE C. JONES-MORTIMER (13), *Department of Biochemistry, University of*

Cambridge, Cambridge CB2 1QW, England

H. RONALD KABACK (17, 32, 33), Department of Biochemistry, Roche Institute of Molecular Biology, Roche Research Center, Nutley, New Jersey 07110

NAOKI KAMO (5), Department of Biophysics, Faculty of Pharmaceutical Sciences, Hokkaido University, Sapporo 060, Japan

RONALD S. KAPLAN (53), Department of Biological Chemistry, The Johns Hopkins University School of Medicine, Baltimore, Maryland 21205

H. GOBIND KHORANA (16), Departments of Biology and Chemistry, Massachusetts Institute of Technology, Cambridge, Massachusetts 02139

MARTIN KLINGENBERG (48), Institut für Physikalische Biochemie, Universität München, 8000 München 2, Federal Republic of Germany

YONOSUKE KOBATAKE (5), Department of Biophysics, Faculty of Pharmaceutical Sciences, Hokkaido University, Sapporo 060, Japan

SYOYU KOBAYASHI (45), Institute of Molecular Biology, Faculty of Science, Nagoya University, Nagoya 464, Japan

HANNO V. J. KOLBE (55), Institute for Monoclonal Antibodies, D-3000 Hannover 61, Federal Republic of Germany

WIL N. KONINGS (37), Microbiology, University of Groningen, 9750 AA Haren, The Netherlands

REINHARD KRÄMER (47, 54), Institute für Physikalische Chemie und Physikalische Biochemie, Universität München, 8000 München 2, Federal Republic of Germany

T. A. KRULWICH (28), Department of Biochemistry, Mount Sinai School of Medicine, New York, New York 10029

JANOS K. LANYI (38), Department of Physiology and Biophysics, University of California, Irvine, California 92717

GUY J. M. LAUQUIN (50, 52), Laboratoire de Physiologie Cellulaire, Département de Biologie, Université de Luminy, 13288 Marseille Cedex 9, France

DAE-SIL LEE (16), Research Center of Genetic Engineering, KAIST, Dongdaemoon, Seoul, Korea

JOSEPH W. LENGELER (35), Fachbereich Biologie/Chemie, University of Osnabrück, D-4500 Osnabrück, Federal Republic of Germany

E. C. C. LIN (34), Department of Microbiology and Molecular Genetics, Harvard Medical School, Boston, Massachusetts 02115

PAUL LLOYD-WILLIAMS (8, 9), Department of Chemistry, University of Manchester, Manchester M13 9PL, England

KIN-MING LO (16), Damon Biotechnology, Needham Heights, Massachusetts 02194

CHARLES A. LUNN (11), Department of Biochemistry, State University of New York at Stony Brook, Stony Brook, New York 11794

ROBERT M. MACNAB (44), Department of Molecular Biophysics and Biochemistry, Yale University, New Haven, Connecticut 06511-8112

ANDREW J. S. MACPHERSON (31), Department of Biochemistry, University of Cambridge, Cambridge CB2 1QW, England

PETER C. MALONEY (43), Department of Physiology, The Johns Hopkins University School of Medicine, Baltimore, Maryland 21205

CARMEN A. MANNELLA (46), Wadsworth Center for Laboratories and Research, New York State Department of Health, Albany, New York 12201

HIROSHI MATSUKURA (45), Institute of Molecular Biology, Faculty of Science, Nagoya University, Nagoya 464, Japan

WARREN MCCOMAS (17), Department of Molecular Genetics, Hoffmann-La Roche Inc., Roche Research Center, Nutley, New Jersey 07110

JOHN M. MCCOY (16), Genetics Institute, 225 Longwood Avenue, Boston, Massachusetts 02115

ANTON MUNDING (49), Institut für Physikalische Biochemie, Universität München,

8000 München 2, Federal Republic of Germany

MACIEJ J. NAŁĘCZ (7), *Nencki Institute of Experimental Biology, Polish Academy of Sciences, Department of Cellular Biochemistry, PL 02-093 Warsaw, Poland*

MICHAEL J. NEWMAN (32), *Graduate Department of Biochemistry, Brandeis University, Waltham, Massachusetts 02254*

HIROSHI NIKAIDO (21), *Department of Microbiology and Immunology, University of California, Berkeley, California 94720*

MITSUO NISHIMURA (6), *Department of Biology, Faculty of Science, Kyushu University, Fukuoka 812, Japan*

ETANA PADAN (17, 27), *Institute of Life Sciences, Division of Microbial and Molecular Ecology, The Hebrew University–Hadassah Medical School, Jerusalem 91010, Israel*

FERDINANDO PALMIERI (54), *Department of Pharmaco-Biology, Laboratory of Biochemistry, University of Bari and CNR Unit for the Study of Mitochondria and Bioenergetics, 70126 Bari, Italy*

RISA A. PARLO (53), *Laboratory of Biochemistry, Department of Biology, New York University, New York, New York 10003*

CHARLES A. PLATE (15), *Abbott Diagnostics Division, Cancer Research, Abbott Laboratories, North Chicago, Illinois 60064*

MOHINDAR S. POONIAN (17), *Department of Molecular Genetics, Hoffmann-La Roche Inc., Roche Research Center, Nutley, New Jersey 07110*

GIROLAMO PREZIOSO (54), *Department of Pharmaco-Biology, Laboratory of Biochemistry, University of Bari and CNR Unit for the Study of Mitochondria and Bioenergetics, 70126 Bari, Italy*

ROGER C. PRINCE (8), *Exxon Research & Engineering Company, Annandale, New Jersey 08801*

LINDA L. RANDALL (10), *Biochemistry/Biophysics Program, Washington State University, Pullman, Washington 99164*

BARRY P. ROSEN (26), *Department of Biological Chemistry, University of Maryland School of Medicine, Baltimore, Maryland 21201*

JÜRG P. ROSENBUSCH (25), *Biozentrum, University of Basel, CH-4056 Basel, Switzerland, and European Molecular Biology Laboratory, D-6900 Heidelberg, Federal Republic of Germany*

SHLOMO ROTTEM (20), *Department of Membranes and Ultrastructure Research, The Hebrew University–Hadassah Medical School, Jerusalem 91010, Israel*

HAGAI ROTTENBERG (1), *Department of Pathology, Hahnemann University School of Medicine, Philadelphia, Pennsylvania 19102*

HEMANTA K. SARKAR (17), *Department of Biochemistry, Roche Institute of Molecular Biology, Roche Research Center, Nutley, New Jersey 07110*

BRIGITTE SCHOBERT (38), *Department of Physiology and Biophysics, University of California, Irvine, California 92717*

SHIMON SCHULDINER (27), *Department of Molecular Biology, Institute of Microbiology, The Hebrew University–Hadassah Medical School, Jerusalem 91010, Israel*

HOWARD A. SHUMAN (12), *Department of Microbiology, College of Physicians and Surgeons, Columbia University, New York, New York 10032*

ITALO STIPANI (54), *Department of Pharmaco-Biology, Laboratory of Biochemistry, University of Bari and CNR Unit for the Study of Mitochondria and Bioenergetics, 70126 Bari, Italy*

CLARENCE H. SUELTER (2), *Department of Biochemistry, Michigan State University, East Lansing, Michigan 48824*

MASAYASU TAKAHARA (11), *Biochemical Research Laboratory, Toyo Jozo Company Ltd., Tagata-gun, Shizouka-ken 410-23, Japan*

BART TEN BRINK (37), *Microbiology, TNO–CIVO, 3700 AJ Zeist, The Netherlands*

HAJIME TOKUDA (39), *Department of Membrane Biochemistry, Research Institute* for Chemobiodynamics, Chiba University, Chiba 280, Japan

PETER P. TOTH (2), Department of Biochemistry, Michigan State University, East Lansing, Michigan 48824

WILLIAM R. TRUMBLE (17), Battelle Laboratories, Northwest Biotechnical Operations, Richland, Washington 99352

TOMOFUSA TSUCHIYA (30), Department of Microbiology, Faculty of Pharmaceutical Sciences, Okayama University, Okayama 700, Japan

VICTOR L. J. TYBULEWICZ (18), MRC Laboratory of Molecular Biology, Cambridge CB2 2QH, England

T. VANAKEN (3), BonSecours Hospital, Grosse Pointe, Michigan 48230

PIERRE V. VIGNAIS (50, 52), Laboratoire de Biochimie, Département de Recherche Fondamentale, Centre d'Etudes Nucléaires, 85 X, 38041 Grenoble Cedex, France

PAUL VIITANEN (17, 32), Plant Sciences, du Pont Experimental Station, Wilmington, Delaware 19898

JOHN E. WALKER (18), MRC Laboratory of Molecular Biology, Cambridge CB2 2QH, England

DOROTHY M. WILSON (30), Department of Physiology and Biophysics, Harvard Medical School, Boston, Massachusetts 02115

T. HASTINGS WILSON (30, 32), Department of Physiology and Biophysics, Harvard Medical School, Boston, Massachusetts 02115

HERBERT H. WINKLER (19), Department of Microbiology and Immunology, University of South Alabama College of Medicine, Mobile, Alabama 36688

HARTMUT WOHLRAB (55), Department of Cell Physiology, Boston Biomedical Research Institute, Boston, Massachusetts 02114

Preface

Volumes 125 and 126 of *Methods in Enzymology* initiate the transport volumes of the Biomembranes series. Biological transport represents a continuation of methodology for the study of membrane function, Volumes 96–98 having dealt with membrane biogenesis, assembly, targeting, and recycling.

This is a particularly good time to cover the topic of biological membrane transport because a strong conceptual basis for its understanding now exists. Membrane transport has been divided into five topics. Topic 1 is covered in Volumes 125 and 126. The remaining four topics will be covered in subsequent volumes of the Biomembranes series.

1. Transport in Bacteria, Mitochondria, and Chloroplasts
2. ATP-Driven Pumps and Related Transport
3. General Methodology of Cellular and Subcellular Transport
4. Cellular and Subcellular Transport: Eukaryotic (Nonepithelial) Cells
5. Cellular and Subcellular Transport: Epithelial Cells

We are fortunate to have the advice and good counsel of our Advisory Board. Additional valuable input to these volumes was obtained from many individuals. Special thanks go to Giovanna Ames, Angelo Azzi, Ernesto Carafoli, Hans Heldt, Lars Ernster, Peter Pedersen, Youssef Hatefi, Dieter Oesterhelt, Saul Roseman, and Thomas Wilson. The enthusiasm and cooperation of the participants have enriched and made these volumes possible. The friendly cooperation of the staff of Academic Press is gratefully acknowledged.

These volumes are dedicated to Professor Sidney Colowick, a dear friend and colleague, who died in 1985. We shall miss his wise counsel, encouragement, and friendship.

SIDNEY FLEISCHER
BECCA FLEISCHER

METHODS IN ENZYMOLOGY

EDITED BY

Sidney P. Colowick and Nathan O. Kaplan

VANDERBILT UNIVERSITY
SCHOOL OF MEDICINE
NASHVILLE, TENNESSEE

DEPARTMENT OF CHEMISTRY
UNIVERSITY OF CALIFORNIA
AT SAN DIEGO
LA JOLLA, CALIFORNIA

I. Preparation and Assay of Enzymes
II. Preparation and Assay of Enzymes
III. Preparation and Assay of Substrates
IV. Special Techniques for the Enzymologist
V. Preparation and Assay of Enzymes
VI. Preparation and Assay of Enzymes (*Continued*)
Preparation and Assay of Substrates
Special Techniques
VII. Cumulative Subject Index

METHODS IN ENZYMOLOGY

EDITORS-IN-CHIEF

Sidney P. Colowick and Nathan O. Kaplan

VOLUME XXXII. Biomembranes (Part B)
Edited by SIDNEY FLEISCHER AND LESTER PACKER

VOLUME XXXIII. Cumulative Subject Index Volumes I–XXX
Edited by MARTHA G. DENNIS AND EDWARD A. DENNIS

VOLUME XXXIV. Affinity Techniques (Enzyme Purification: Part B)
Edited by WILLIAM B. JAKOBY AND MEIR WILCHEK

VOLUME XXXV. Lipids (Part B)
Edited by JOHN M. LOWENSTEIN

VOLUME XXXVI. Hormone Action (Part A: Steroid Hormones)
Edited by BERT W. O'MALLEY AND JOEL G. HARDMAN

VOLUME XXXVII. Hormone Action (Part B: Peptide Hormones)
Edited by BERT W. O'MALLEY AND JOEL G. HARDMAN

VOLUME XXXVIII. Hormone Action (Part C: Cyclic Nucleotides)
Edited by JOEL G. HARDMAN AND BERT W. O'MALLEY

VOLUME XXXIX. Hormone Action (Part D: Isolated Cells, Tissues, and Organ Systems)
Edited by JOEL G. HARDMAN AND BERT W. O'MALLEY

VOLUME XL. Hormone Action (Part E: Nuclear Structure and Function)
Edited by BERT W. O'MALLEY AND JOEL G. HARDMAN

VOLUME XLI. Carbohydrate Metabolism (Part B)
Edited by W. A. WOOD

VOLUME XLII. Carbohydrate Metabolism (Part C)
Edited by W. A. WOOD

VOLUME XLIII. Antibiotics
Edited by JOHN H. HASH

VOLUME XLIV. Immobilized Enzymes
Edited by KLAUS MOSBACH

VOLUME XLV. Proteolytic Enzymes (Part B)
Edited by LASZLO LORAND

Section I

General Approaches

Article 1

A. Membrane Characteristics and Chemistry
Articles 2 through 9

B. Membrane Assembly, Mutations, and Cloning Strategy
Articles 10 through 18

[1] Energetics of Proton Transport and Secondary Transport

By Hagai Rottenberg

Proton pumps and their associated proton-coupled transport systems are the most common biological systems of energy conversion. Most organisms depend on proton pumps for chemical (e.g., oxidative phosphorylation) and photochemical (e.g., photophosphorylation) energy conversion. In addition, in many microorganisms and in the organelles of both animals and plant cells, the proton electrochemical potential, which is generated by a proton pump, is the direct driving force for specific transport systems of ions and substrates. In describing the energetics of these systems, it is useful to distinguish between primary active transport systems or pumps, in which transport of an ion is directly coupled to a chemical (or photochemical) reaction, and secondary active transport systems, in which the vectorial transport of ions is coupled to the transport of other ions and substrates. The existence of several primary and secondary active transport systems which share common ions and substrates in the same membrane leads to complex energy conversion pathways. For example, oxidative phosphorylation in mitochondria is the sum of energy conversion by several types of proton pumps and proton-coupled transport systems.[1,2] The energetics of such complex, high-order coupling is beyond the scope of this chapter.[3] Here, we shall only discuss the energetics of individual transport systems. However, since it is not always possible to study the energetics of a totally isolated system (except perhaps in a reconstituted system) the existence and the activity of other systems in the same membrane must be accounted for in a quantitative description of the energetics of primary and secondary transport.

Basic Concepts

The first law of thermodynamics defines and describes the properties of energy (ΔE), and the second law defines and describes the properties of entropy (ΔS). In classical energetics these two laws are combined in the

---

[1] P. Mitchell, in "Membrane and Ion Transport" (E. E. Bittar, ed.), Vol. 1, pp. 192. Wiley (Interscience), London, 1970.

[2] P. Mitchell, *Science* **206,** 1148 (1979).

[3] H. Rottenberg, *Biochim. Biophys. Acta* **549,** 225 (1979).

METHODS IN ENZYMOLOGY, VOL. 125

definition of Gibbs free energy (ΔG)[4]

$$\Delta G = \Delta H - T\Delta S \leq 0 \tag{1}$$

(at constant pressure and temperature where ΔH, the enthalpy, is given by $\Delta E + P\Delta V$). This fundamental statement means that in a spontaneous change of state the Gibbs free energy always decreases. For systems of both chemical and osmotic (transport) processes the Gibbs free energy of each chemical species is referred to as the chemical potential. We can define the chemical potential of each chemical species μ_i as the sum of a standard potential (at 1 M concentration) μ_i° and a concentration-dependent term; hence,

$$\mu_i = \mu_i^\circ + RT \ln C \tag{2}$$

(here we use concentrations as an approximation for the activity, a). To find the free energy change, ΔG, for a chemical reaction,

$$a\mathrm{A} + b\mathrm{B} + \cdots \leftrightharpoons m\mathrm{M} + n\mathrm{N} + \cdots \tag{3}$$

we add up the potentials of the products and subtract the potentials of the substrates while taking into account the reaction stoichiometry, thus

$$\Delta G = \Delta G^\circ + RT \ln[(\mathrm{M})^m(\mathrm{N})^n/(\mathrm{A})^a(\mathrm{B})^b] \tag{4}$$

where ΔG° is the reaction standard and free energy (at 1 M concentration). Since, at equilibrium, $\Delta G = 0$, it follows that

$$\Delta G^\circ = -RT \ln K_{\mathrm{eq}} \tag{5}$$

The free energy change for the flow of a solute across the membrane from high to low concentration is given by the difference in chemical potential between the two compartments. Hence,

$$\Delta G = \Delta \mu = RT \ln(C_{\mathrm{in}}/C_{\mathrm{out}}) \tag{6}$$

For ions the flow is also dependent on the electrical potential difference between the two compartments,[4a] thus the ions' electrochemical potential is defined as

$$\tilde{\mu}_i = \tilde{\mu}_i^\circ + RT \ln C_i + Fz\psi \tag{7}$$

---

[4] I. M. Klotz, "Energy Changes in Biochemical Reactions." Academic Press, New York, 1967.

[4a] In a more rigorous treatment both the electrical potential gradients and the concentration gradients must be considered in three dimensions. Here we consider vesicular systems such as bacterial cells or organelles for which only two values of the potential (and concentrations) must be considered—inside the vesicle and outside the vesicle.

and the change of free energy for ion flow is

$$\Delta G = \Delta \bar{\mu}_i = RT \ln(C_{in}/C_{out}) + Fz\Delta\psi \qquad (8)$$

Energetics of Primary Active Transport

The second law of thermodynamics states that spontaneous processes are associated with negative free energy change as the systems approach equilibrium, and the energetics of an isolated system always obey this principle. However, it is often observed that deviations from this rule do exist when it is applied to individual processes. This deviation arises whenever coupling between processes exist. For example, a diluted solution can be pumped by hydrostatic pressure into a concentrated solution forcing the solute to flow against its concentration gradient. In this case the solvent flow, driven by hydrostatic pressure, is coupled to the solute flow. However, the sum of free energy change of the coupled process always obeys Eq. (3). Similarly, many biological active transport systems have evolved in which the flow of a spontaneous energy yielding chemical process is coupled to the flow of an ion which proceeds against its own electrochemical potential difference. In general, only a fraction of the energy dissipated by the spontaneous process is captured and utilized by the driven process. However, many membrane pumps and carriers are coupled very tightly. It is easy to describe energy conversion in these processes quantitatively, within the boundaries of classical energetics, if we can consider the two processes as completely coupled, effectively treating them as a single process.[1] Hence, for electrogenic ion pumps such as an ATPase proton pump we may combine the description of the driving chemical reaction with the driven proton flow into a single scheme as follows:

$$ATP + nH_{in} \leftrightharpoons ADP + P_i + nH_{out} \qquad (9)$$

It follows from Eqs. (4), (8), and (9) that the change of free energy in this process is given by

$$\Delta G_p = \Delta G_p^\circ + RT \ln([ATP]/[ADP][P_i]) + nF\Delta\psi + nRT2.3\Delta pH \quad (10)$$

It is convenient to express those relations in electrical potential units as in Eq. (10a),

$$\Delta G_p/F = \Delta G_p^\circ/F + Z \log([ATP])/[ADP][P_i]) + n\Delta\psi + nZ\Delta pH \quad (10a)$$

where Z is $2.3RT/F$ (which is approximately 59 mV at 25°). By calculating the value of ΔG for any given initial condition it is possible to determine the direction of the reaction. Indeed, proton ATPases, as well as other cation pumps, can be reversed and induced to synthesize ATP if the ions electrochemical potential is increased and the phosphate potential ([ATP]/[ADP][P$_i$]) is reduced.[5]

Stoichiometry of Proton Pumps

One of the most useful applications of Eq. (10) is the determination of pump stoichiometry. Traditionally, the stoichiometry of a chemical process is determined from the relative rate of change in reactant concentrations during the course of the reaction. For proton pumps this procedure is equivalent to measuring the ratio between the rate of proton pumping and the rate of the driving reaction. This method is referred to as the flow ratio method.[3] However, several difficulties arise in such procedures because of the considerable back flow of protons through passive diffusion and through various proton-coupled transport systems. Appropriate experimental design requires prevention of the buildup of the proton electrochemical potential and inhibition of all H$^+$-coupled proton transport.[6,7] However, these efforts do not always lead to satisfactory results.

The energetic approach to the estimation of stoichiometry is to allow the pump to reach equilibrium and to measure the value of each of the variable terms in Eq. (10a). Since at equilibrium $\Delta G = 0$, then

$$\Delta G_p^\circ + Z \log([ATP]/[ADP][P_i]) = -[n\Delta\psi + nZ\Delta pH] \qquad (11)$$

or

$$\Delta G_p/-\Delta\bar{\mu}_H = n \qquad (12)$$

This method is referred to as the force ratio method.[3]

Similarly, for a redox proton pump driven by electron transport

$$\Delta G_R + Z \log(A^{red}D^{ox}/A^{ox}D^{red}) = -[n\Delta\psi + nZ\Delta pH] \qquad (13)$$

or

$$\Delta G_R/-\Delta\bar{\mu}_H = n \qquad (14)$$

The stoichiometry of various proton pumps has been evaluated by both the force ratio and flow ratio method (see Table I). Experimentally the force method requires the simultaneous determination of membrane potential and ΔpH as described in a previous volume (Volume LV, Section III) and the determination of free energy of the driving reactions. A

[5] P. C. Maloney, *J. Membr. Biol.* **67**, 1 (1982).

[6] P. Mitchell and J. Moyle, *Eur. J. Biochem.* **4**, 530 (1968).

[7] M. D. Brand, B. Reynafarge, and A. L. Lehninger, *Proc. Natl. Acad. Sci. U.S.A.* **73**, 431 (1976).

TABLE I
STOICHIOMETRY OF THE PROTON ATPASE: COMPARISON OF ESTIMATES
FROM FORCE RATIO AND FLOW RATIO IN DIFFERENT SYSTEMS[a]

Experimental system	$\Delta G_{ATP}/\Delta\bar{\mu}_{H^+}$ force ratio	Ref.	H^+/ATP flow ratio	Ref.
Mitochondria	2.5–3.5	19–21	2.0–3.3	22–24
Submitochondrial particles	2.4–3.0	25, 26	2.0	27
Chloroplasts	3.0	28	3.0	28
Chromatophores	2.6	29		

[a] The estimation from force ratio assumes full equilibrium and complete coupling while the estimation from flow ratio assumes no back flow of protons. The values obtained in mitochondria were not corrected for the proton-driven transport of phosphate and the potential driven exchange of ATP for ADP. If these two systems are also equilibrated, one H^+ must be allowed to account for transport of each ATP synthesized.

reliable application of this method calls for extreme caution as discussed below.

1. In most cells and organelles the driving reaction, for example, ATP hydrolysis, occurs on one side of the membrane either internally or externally. Since the reactants often participate in secondary active transport the reaction free energy is often different in the two compartments. To determine the stoichiometry of the proton ATPase the reaction free energy must be evaluated in the compartment in which the reaction occurs. For example, in mitochondria, ATP is hydrolyzed (or synthesized) in the matrix. Routinely, ΔG_p is determined in the external medium and the ratio $\Delta G_p/\Delta\bar{\mu}_H$ is evaluated (see Table I). However, because ATP, ADP, and P_i participate in secondary active transport this ratio does not represent the stoichiometry of the pump itself but the combined stoichiometry of the two transport systems (ADP/ATP exchange and P_i–H^+ cotransport) and the ATPase. To obtain the stoichiometry it is necessary to correct $\Delta G_p/\Delta\bar{\mu}_H$ for the contribution of secondary transport. Alternatively, the stoichiometry can be evaluated in submitochondrial particles, which are inverted inner membrane vesicles in which the reactions occur outside the vesicle. An even more complicated situation is encountered in redox proton pumps since very often the product and the substrate of a particular pump react on opposite sides of the membrane. This situation has lead to controversy concerning the correct evaluation of the reaction free energy change.[8,9] However, a meticulous application of the laws of equilib-

[8] N. G. Forman and D. F. Wilson, *J. Biol. Chem.* **257**, 12908 (1982).
[9] J. J. Lemasters, R. Grunwald, and R. K. Emaus, *J. Biol. Chem.* **259**, 3058 (1984).

rium thermodynamics[10] indicates the free energy change of redox reactions must be evaluated separately for each compartment if these are not equilibrated. Hence the reaction free energy change must be evaluated either externally or internally without mixing values from the two compartments.

2. Perhaps the most difficult task is to ascertain that the process is at equilibrium. In reality, the system never attains a true equilibrium. In most natural membranes the chemical reactants often participate in other reactions which buffer the change of their concentration, thus inhibiting or even preventing the approach to equilibrium. Adenylate kinase, creatine kinase, and uncoupled ATPases are examples of reactions that often delay and prevent the approach of the ATPase to full equilibrium. Moreover, protons are transported passively through the membrane (leak), and also by other pumps and by various proton-coupled transport systems, all of which tend to lower $\Delta\bar{\mu}_H$ and prevent the approach of the pump to true equilibrium. While the proton flows in many of these processes can be inhibited considerable leaks persist which prevent the establishment of a true equilibrium. Indeed, there is considerable doubt whether any pump (or any coupled system for that matter) can be considered completely coupled. Real systems are always irreversible which means that "slips" probably occur in the most tightly coupled pumps. Nonequilibrium thermodynamics provides the means to deal with such partially coupled system (see below). An alternative solution, still within the framework of classical energetics, is described here.

Conceding the existence of leaks, slips, and interference from other reactions which prevent the establishment of true equilibrium, we may evaluate the approach to equilibrium from both directions. The establishment of $\Delta\bar{\mu}_H$ driven by ATP hydrolysis proceed toward equilibrium in such a way that $\Delta G_p \geq n\Delta\bar{\mu}_H$, while the synthesis of ATP driven by $\Delta\bar{\mu}_H$ approaches equilibrium in such a way that $\Delta G_p \leq n\Delta\bar{\mu}_H$. Hence,

$$(\Delta G_p/\Delta\bar{\mu}_H)_{ATPase} > n > (\Delta G_p/\Delta\bar{\mu}_H)_{synthase} \qquad (15)$$

Thus, evaluating the force ratio from both directions provides an upper and lower limit to the value of n.

It should be realized that proton leaks do not affect the results when equilibrium is approached from $\Delta\bar{\mu}_H$-driven ATP synthesis (as in state 4 of oxidative phosphorylation), while interference from other enzymes does not affect the results when equilibrium is approached from ATP hydrolysis. In general, proton leaks are a more severe problem and, hence, the determination of the stoichiometry from the force ratio at state 4 appears to be more reliable.

[10] D. Walz, *Biochim. Biophys. Acta* **456,** 1 (1974).

Energetics of Secondary Active Transport

Energy conversion due to coupling by cotransport and exchange carriers is quite efficient. It is therefore justified, as in the preceding evaluation of proton pumps, to assume that these proteins catalyze a fully coupled process. It is useful to distinguish between two types of secondary transport: (1) cotransport carriers (symport) in which protons are cotransported with the substrate in the same direction and (2) exchange-transport carriers (antiport) in which protons are transported in one direction in exchange for transport of a substrate or other ion in the reverse direction. There also exists electrogenic secondary transport systems that do not transport protons at all but result in net charge transfer. Since the transport is driven by the membrane electrical potential, which is generated by the proton pump, these systems are also ultimately dependent on the proton pump even though no protons are transported.

Both the charge stoichiometry (electrogenicity) and the proton stoichiometry can be determined from a measurement of the force ratio close to equilibrium.[1,3]

For cotransport of neutral substrates the reaction scheme is

$$S_{out} + nH_{out}^+ \rightleftharpoons S_{in} + nH_{in}^+ \tag{16}$$

At equilibrium, from Eqs. (8) and (16), we get

$$Z \log(S_{in}/S_{out}) = n(Z\Delta pH + \Delta\psi) \tag{17}$$

For cotransport of ions of charge z (either positive or negative) the reaction scheme is

$$S_{out}^z + nH_{out}^+ \rightleftharpoons S_{in}^z + nH_{in}^+ \tag{18}$$

At equilibrium, from Eqs. (8) and (18), we get

$$Z \log(S_{in}^z/S_{out}^z) = nZ\Delta pH + (n + z)\Delta\psi \tag{19}$$

As can be seen from Eq. (19), if the charge of a transported anion is balanced by the charge of the protons (i.e., $n + z = 0$) the transport is electroneutral and is driven solely by ΔpH. The substrate ion may exist in several ionization forms, as for example

$$S^{-1} \xrightarrow{K_1} SH \xrightarrow{K_2} SH_2^+ \tag{20}$$

It is important in calculating the ratio in Eq. (19) to take into account the pH in each compartment and to calculate the ratio for the transported species using the appropriate ionization constants. Often, the identity of the transported species is unknown, though in principle it may be evaluated from kinetic experiments.[11] It should also be noted that the stoichiometry of transport need not be constant under all conditions. Many carriers appear to have a pH-dependent stoichiometry, apparently due to the ionization of the carrier itself.[12] For example, several bacterial cotransport system for anions are driven by ΔpH alone at low external pH but also by membrane potential at high external pH[30] (see Table II). Thus at low pH the overall process is

$$S_{\text{out}}^- + H_{\text{out}}^+ \leftrightharpoons S_{\text{in}}^- + H_{\text{in}}^+ \tag{21}$$

while at high pH the overall scheme is

$$S_{\text{out}}^- + 2H_{\text{out}}^+ \leftrightharpoons S_{\text{in}}^- + 2H_{\text{in}}^+ \tag{22}$$

For an exchange (transport) carrier of the general type,

$$S_{\text{in}}^z + nH_{\text{out}}^+ \leftrightharpoons S_{\text{out}}^z + nH_{\text{in}}^+ \tag{23}$$

the force ratio at equilibrium is

$$Z \log(S_{\text{out}}^z / S_{\text{in}}^z) = nZ\Delta\text{pH} + (n - z)\Delta\psi \tag{24}$$

Hence, for an exchange transport carrier in which the substrate is positively charged, and when $n = z$, the exchange is electroneutral while if $n \neq z$ it will be electrogenic. Exchange transport carriers may also exhibit pH-dependent stoichiometry as was observed for the Na^+/H^+ carrier in *Escherichia coli*.[13]

Electrogenic carriers which do not transport H^+ but a single ion (uniport) obey the following relationship (at equilibrium)

$$Z \log(S_{\text{in}}^z / S_{\text{out}}^z) = z\Delta\psi \tag{25}$$

Electrogenic exchange carriers of the type

$$S_{\text{in}}^z + P_{\text{out}}^y \leftrightharpoons S_{\text{out}}^z + P_{\text{in}}^y \tag{26}$$

will equilibrate according to the relationship

$$Z \log S_{\text{in}}^z / S_{\text{out}}^z = Z \log P_{\text{in}}^y / P_{\text{out}}^y + (z - y)\Delta\psi \tag{27}$$

Table II gives several examples for systems in which the stoichiometry and/or the electrogenicity was determined by the force ratio method.

Most of the precautions discussed in the preceding sections apply also to the measurement of force ratios in secondary transport. Again, the systems may never reach true equilibrium and it is important to evaluate

[11] R. G. Johnson, S. E. Carty, and A. Scarpa, *J. Biol. Chem.* **256,** 5773 (1981).
[12] H. Rottenberg, *FEBS Lett.* **66,** 159 (1976).
[13] S. Schuldiner and H. Fishkes, *Biochemistry* **17,** 706 (1978).

TABLE II
EXAMPLES OF STOICHIOMETRY DETERMINATIONS BY THE FORCE RATIO METHOD OF
SECONDARY TRANSPORT SYSTEMS

System or carrier	Force ratio	Stoichiometry	Ref.
H$^+$ cotransport systems (symport)			
Lactose (*E. coli*)	$\Delta\bar{\mu}_{lac}/\Delta\bar{\mu}_H$	1–2	30, 31
Proline (*E. coli*)	$\Delta\bar{\mu}_{pro}/\Delta\bar{\mu}_H$	1	30
Glucose-6-P$^-$ (*E. coli*, low pH)	$\Delta\bar{\mu}_{glu-6P}/\Delta pH$	1	30
Glucose-6-P$^-$ (*E. coli*, high pH)	$\bar{\mu}_{glu-6P}/-2\Delta pH + \Delta\psi$	2	30
phosphate–H$^+$	$\mu_{HPC^{2-}}/\Delta pH$	1	32, 14
H$^+$ exchange systems (antiport)			
Na$^+$/H$^+$ (*Halobacterium halobium*)	$\Delta\mu_{Na}/\Delta\mu_H$	2	33
Catecholamine/H$^+$ (chromaffin granules)	$\Delta\bar{\mu}_{cat}/\Delta\psi - 2Z\Delta pH$	0–2[a]	11
Electrogenic carriers			
ADP^{3-}/ATP^{4-} (mitochondria)	$\bar{\mu}_{ADP^{3-}}/\bar{\mu}_{ATP^{4-}}$	1	34
Ca^{2+} (mitochondria)	$\bar{\mu}_{Ca^{2+}}/\Delta\psi$	2	35

[a] The stoichiometry depends on the identity of the transported species which must be
established by kinetic measurements.

the effects of the various factors that prevent true equilibration. In general, and particularly in the case of organic substrates, the proton leak is a
more serious problem than substrate leak or metabolism and the force
ratio is therefore best evaluated under conditions in which $\Delta\bar{\mu}_H$ is generated by a proton pump and allowed to drive the substrate to a steady
maximal accumulation ratio ("static head," see below). With most secondary systems, the approach to equilibrium by substrate driven $\Delta\bar{\mu}_H$ (or
ΔpH) is not feasible because the activity of the carrier is negligible compared to the total activity of all other H$^+$-linked systems. However, some
carriers, for example, the phosphate translocator of mitochondria, can
generate ΔpH in equilibrium with the driving substrate gradient.[14] The
metabolism of substrates must be inhibited for reliable estimates of force
ratio. Many substrates and ions, both in bacterial cells and organelles, are
transported by more than one system and it is necessary to suppress or, if
possible, to inhibit completely the alternative carriers before the force
ratio can be evaluated with confidence. Specific inhibitors or specific
mutations often help to overcome these difficulties but extreme caution
should be exercised if the alternative transport pathways are not well
characterized.

[14] S. Ogawa, H. Rottenberg, T. R. Brown, R. G. Shulman, C. L. Castillo, and P. Glynn,
Proc. Natl. Acad. Sci. U.S.A. **75,** 1796 (1978).

Nonequilibrium Thermodynamics of Active Transport

Basic Principles. The inherent irreversibility of all machines, including ion pumps and carriers, and the complicated interactions between several processes in natural membrane vesicles complicate the reversible approach to true equilibrium which is assumed in classical energetics. It is, therefore, desirable to develop an alternative approach which provides a better description of the energetics of real processes. The fundamental consideration of nonequilibrium thermodynamics is the *rate* of change of the system's free energy whereas the fundamental consideration of classical energetics is the *extent* of change of the system's free energy.[15,16] The rate of change of free energy is described by the dissipation function

$$\phi = \sum_i J_i X_i \tag{28}$$

That is, the rate of the dissipation of free energy is given by the sum of the products of the flows (J_i) and their conjugated force (X_i).

The second law of thermodynamics requires that ϕ is always positive,

$$\phi \geq 0 \tag{29}$$

For our purpose it is sufficient to consider two types of flows: (1) a transport process in which J_i represents the rate of transport of species i and X_i is its conjugated driving force, the electrochemical potential difference μ_i; (2) a chemical reaction in which J_i represents the rate of the reaction while X_i, its conjugated driving force is the reaction free energy $-\Delta G$. (More precisely, the driving force is the reaction affinity A, which is a stoichiometric function of $-\Delta G$. However, by proper choice of the reaction rate, $A = -\Delta G$.) We shall restrict our discussion to two types of systems. A system composed of a chemical reaction and an ion flow is sufficient to describe a proton pump such as the proton ATPase. In this case from Eqs. (28) and (29)

$$\phi = -J_{\text{ATP}}\Delta G_\text{p} + J_\text{H}\Delta\bar{\mu}_\text{H} \geq 0 \tag{30}$$

Note that only the sum of the terms of the dissipation function must be positive. If the two processes are coupled one of the terms can be negative, i.e., the flow may proceed against its conjugated driving force.

[15] A. Katchalsky and P. F. Curran, "Nonequilibrium Thermodynamics in Biophysics." Harvard Univ. Press, Cambridge, Massachusetts, 1965.
[16] R. S. Caplan and A. Essig, "Bioenergetics and Linear Non-Equilibrium Thermodynamics." Harvard Univ. Press, Cambridge, Massachusetts, 1983.

Similarly for secondary transport we may consider a two flow system such as

$$\phi = J_S \Delta \bar{\mu}_S + J_H \Delta \bar{\mu}_H \geq 0 \qquad (31)$$

Describing the flows as a function of the forces we write the phenomenological equations (assuming, as a first approximation, linearity of flow and forces). Thus, for example, an ATPase proton pump [Eq. (30)] is described by the phenomenological equations

$$J_{ATP} = L_p \Delta G_{ATP} + L_{pH} \Delta \bar{\mu}_H \qquad (32a)$$
$$J_H = L_{HP} \Delta G_{ATP} + L_H \Delta \bar{\mu}_H \qquad (32b)$$

A secondary transport system [as in Eq. (31)] is described by

$$J_S = L_S \Delta \bar{\mu}_S + L_{SH} \Delta \bar{\mu}_H \qquad (33a)$$
$$J_H = L_{HS} \Delta \bar{\mu}_S + L_H \Delta \bar{\mu}_H \qquad (33b)$$

The cross-coefficients (L_{pH}, L_{HP} and L_{SH}, L_{HS}) describe the coupling between the two flows and their value in relation to the straight coefficient expresses the tightness of the coupling. Under some conditions,[16] reciprocity is observed; in which case $L_{12} = L_{21}$.

For a two flow system one can define the degree of coupling,[17] q, as

$$q = L_{12}/\sqrt{L_{11}L_{22}} \qquad (34)$$

This is a dimensionless quantity and takes values between -1 and $+1$. For completely coupled systems $q = 1$ and for completely uncoupled systems $q = 0$. For a cotransport system q is positive, since the coupled flows move in the same direction while for an exchange system q is negative.

When a coupled system is allowed to change its state spontaneously it would gravitate into a state in which ϕ takes a minimal value. In this state the driving force reaches its maximal value while the driven flow vanishes. Such a state is called static head.[17] For example, when a proton cotransport system is allowed to reach static head, the substrate transport rate may vanish but proton transport may continue. The force ratio at static head will be greater, the higher the degree of coupling, q. It is easy to see from Eq. (33a) that when $J_S = 0$, $\Delta \bar{\mu}_S/\Delta \bar{\mu}_H = -L_{SH}/L_S$, if we define Z as $\sqrt{L_H/L_S}$ then

$$(\Delta \bar{\mu}_S/\Delta \bar{\mu}_H) = -qZ \qquad (35)$$

Thus for a given value of Z, the higher the degree of coupling the larger the force ratio. Moreover, Z is closely related to the stoichiometry of the

[17] O. Kedem and R. S. Caplan, *Trans. Faraday Soc.* **61**, 1897 (1965).

process since it expresses the relative rate of flow per unit force. In fact, for a completely coupled process $Z = n$. Therefore, for systems with a high degree of coupling we may assume, as a first approximation that

$$(\Delta\bar{\mu}_S/\Delta\bar{\mu}_H) = -qn \qquad (36)$$

similarly for proton pumps

$$(\Delta G_p/\Delta\bar{\mu}_H) = qn \qquad (37)$$

Thus, according to the nonequilibrium approach the estimation of stoichiometry from the force ratio has to be corrected by the degree of coupling q. For that purpose it is necessary to determine the value of q. An exact determination of q requires an experimental determination of the coefficients in Eqs. (32a,b) or (33a,b). However, for strictly linear systems q may be determined from the ratio of the rate of the driving reaction at static head (J_{SS}) and level flow (J_{LV}) according to the following relation.[17]

$$q^2 = 1 - J_{SS}/J_{LV} \qquad (38)$$

For example, for proton pumps this value can be estimated from the stimulation of the rate of the driving reaction (ATP hydrolysis, electron transport, etc.) by uncouplers, which abolish $\Delta\bar{\mu}_H$. It must be realized, however, that these linear relations hold only approximately and under restricted conditions.[16] In most pumps for which the degree of coupling was estimated, the values are higher,[3] usually larger than 0.9. This approximate figure can be used to correct the stoichiometry estimated by the force ratio method (Table I).

The Efficiency of Energy Conversion in Active Transport

Since every process is associated with net reduction of the system free energy the efficiency of energy conversion in primary and secondary active transport is always less than 1.0. In general, thermodynamic efficiency, η, is defined as the ratio of free energy gained by a driven process (output) to the free energy expended by the driving process (input). Thus, in a two flow system, Eqs. (30) and (31), the efficiency is

$$\eta = J_1 X_1/J_2 X_2 = 1 - \phi/J_2 X_2 \qquad (39)$$

It must be realized that the efficiency depends on the state of the system, namely, the magnitude of the flows and the forces. For instance, at level flow, since $X_1 = 0$, $\eta = 0$. Similarly at static head, since $J_1 = 0$ ($J_2 \neq 0$), $\eta = 0$. It follows that the efficiency must reach a maximal value in a particular state between level flow and static head. In general the maximal efficiency depends only on the degree of coupling and is obtained close to level flow for low degree of coupling and close to static head for high-degree of coupling.[17] For completely coupled systems ($q = 1$) a maximal efficiency of 1.0 is approached at equilibrium. A proton pump with a degree of coupling of 0.95 will show maximal efficiency in a state in which the value of $\Delta G_p / \Delta \bar{\mu}_H$ is 80% of the value obtained at steady state. For instance, in oxidative phosphorylation, *in situ,* $\Delta G_p / \Delta \bar{\mu}_H$ at state 3, i.e., during net phosphorylation, is approximately 2.5 while the value at static head (i.e., state 4) is about 3 suggesting that this process occurs close to its maximal efficiency.[17,18] It is, however, possible that biological energy converters have evolved to maximize parameters other than efficiency. For instance, it is possible that the system is controlled to maintain a state of maximum output flow, or alternatively a maximum output force. These states also depend on the degree of coupling. When the degree of coupling is low (<0.6) these states coincide. With higher q a maximum output flow is obtained at a force ratio considerably lower than the state of maximal efficiency, while a maximum output force is obtained at force ratio higher than the state of maximal efficiency. The value of $\Delta \bar{\mu}_H$ reported for many cells and organelles, *in situ,* is sufficiently close to the maximal $\Delta \bar{\mu}_H$ obtained at static head, in the isolated system, to suggest that proton pumps work close to their maximal efficiency, but are possibly regulated to maintain maximal output force (i.e., $\Delta \bar{\mu}_H$).

[18] H. Rottenberg, *Prog. Surf. Membr. Sci.* **12**, 245 (1979).
[19] D. G. Nichols and V. S. M. Bernson, *Eur. J. Biochem.* **75**, 601 (1977).
[20] G. F. Azzone, T. Pozzan, S. Massari, and L. Pregnolato, *Biochim. Biophys. Acta* **501**, 307 (1978).
[21] H. Woelders, W. J. Van der Zande, A. A. F. Colen, R. J. A. Wanders, and K. Van Dam, *FEBS Lett.* **179**, 278 (1984).
[22] P. Mitchell and J. Moyle, *Eur. J. Biochem.* **4**, 530 (1968).
[23] A. Alexander, B. Reynatarje, and A. L. Lehninger, *Proc. Natl. Acad. Sci. U.S.A.* **75**, 5296 (1978).
[24] F. Sholtz, I. A. Gorskaya, and A. V. Kotelnikova, *Eur. J. Biochem.* **136**, 129 (1983).
[25] M. C. Sorgato, S. J. Ferguson, D. B. Kell, and P. John, *Biochem. J.* **174**, 237 (1978).
[26] E. A. Berry and P. C. Hinkle, *J. Biol. Chem.* **258**, 1474 (1983).
[27] W. S. Thayer and P. C. Hinkle, *J. Biol. Chem.* **248**, 5395 (1973).
[28] R. E. McCarty, *Curr. Top. Bioenerg.* **7**, 245 (1978).
[29] M. Leiser and Z. Gromet-Elhanan, *Arch. Biochem. Biophys.* **178**, 79 (1977).
[30] S. Ramos and H. R. Kaback, *Biochemistry* **16**, 4271 (1977).
[31] D. Zilberstein, S. Schuldiner, and E. Padan, *Biochemistry* **18**, 669 (1979).
[32] F. Palmieri, E. Quagliariello, and M. Klingenberg, *Eur. J. Biochem.* **17**, 230 (1970).
[33] J. K. Lanyi and M. P. Silverman, *J. Biol. Chem.* **25**, 4750 (1979).
[34] M. Klingenberg and H. Rottenberg, *Eur. J. Biochem.* **73**, 125 (1977).
[35] H. Rottenberg and A. Scarpa, *Biochemistry* **13**, 4811 (1974).

[2] Isolation of Highly Coupled Heart Mitochondria in High Yield Using a Bacterial Collagenase

By PETER P. TOTH, SHELAGH M. FERGUSON-MILLER, and CLARENCE H. SUELTER

Introduction

Numerous methods have been developed for the isolation of mitochondria from a variety of tissues.[1,2] These procedures usually involve homogenization of the tissue by mechanical means, and separation of the mitochondria containing fraction by differential centrifugation. The consistency of myocardium often necessitates the use of a nonspecific proteinase such as Nagarse to facilitate the disruption process.[3-5] However, as demonstrated by Pande and Blanchaer,[6] mitochondria isolated in the presence of Nagarse lose all long-chain fatty-acyl-CoA synthase (EC 2.3.1.86) activity, and likely lose other outer membrane-associated enzymes as well. Heart mitochondria isolated according to currently available methods also show significant rates of state 4 respiration. Although oligomycin[7,8] and an enzymatic ADP sink[9] have been successfully used to reduce the rate of oxygen consumption in state 4, it is recognized that state 4 respiration is phenomenologically complex[10-15] and difficult to suppress or eliminate.

Because collagen plays a major role in the organization of cells into

[1] J. Nedergaard and B. Cannon, this series, Vol. 55, p. 5.
[2] L. Ernster and G. Schatz, *J. Cell Biol.* **91,** 227s (1981).
[3] B. Chance and B. Hagihara, *Proc. Int. Congr. Biochem., 5th, 1961,* **5,** 3 (1963).
[4] J. W. Palmer, B. Tandler, and C. L. Hoppel, *J. Biol. Chem.* **252,** 8731 (1977).
[5] L. Mela and S. Seitz, this series, Vol. 55, p. 39.
[6] S. V. Pande and M. C. Blanchaer, *Biochim. Biophys. Acta* **202,** 43 (1970).
[7] A. Masini, D. Ceccarelli-Stanzani, and U. Muscatello, *FEBS Lett.* **160,** 137 (1983).
[8] A. Masini, D. Ceccarelli-Stanzani, and U. Muscatello, *Biochim. Biophys. Acta* **767,** 130 (1984).
[9] P. D. Bishop and D. E. Atkinson, *Arch. Biochem. Biophys.* **230,** 335 (1984).
[10] J. W. Stucki and E. A. Ineichen, *Eur. J. Biochem.* **48,** 365 (1974).
[11] D. Nicholls, *Eur. J. Biochem.* **50,** 305 (1974).
[12] J. W. Stucki, *Eur. J. Biochem.* **68,** 551 (1976).
[13] J. J. Lemasters and C. B. Hackenbrock, *J. Biol. Chem.* **255,** 5674 (1980).
[14] A. K. Groen, R. J. A. Wanders, H. V. Westerhoff, R. van der Meer, and J. M. Tager, *J. Biol. Chem.* **257,** 2754 (1982).
[15] A. Masini, D. Ceccarelli-Stanzani, and U. Muscatello, *Biochim. Biophys. Acta* **724,** 251 (1983).

tissues,[16] and heart is known to have a highly developed collagen infrastructure,[17] a collagenase (clostridiopeptidase A, EC 3.4.24.3) was selected to facilitate degradation of intercellular connections. The use of collagenase in place of a nonspecific proteinase has the advantage of being specific toward collagen, with no activity toward other proteins.[18] One potential obstacle to the use of collagenase for the disruption of heart muscle is that the enzyme has an absolute requirement for Ca(II),[19] and addition of this divalent cation might be expected to damage the mitochondria.[20-22] This difficulty is circumvented by allowing the collagenase to be activated by endogenous Ca(II) released from the myocardium during homogenization.

In this chapter we detail a collagenase-facilitated isolation of highly coupled heart mitochondria in high yield with morphologically intact inner and outer membranes.

Experimental Procedures

Materials

Water. In order to isolate highly coupled mitochondria, it is critical that the water used be as pure as possible. For the preparations described herein, water is glass distilled, deionized by passage over a column of Dowex MR-3 (Sigma Chemical Co., St. Louis, MO) mixed bed ion-exchange resin, and then redistilled from alkaline permanganate. The last step of this procedure may be essential for removing organic contaminants.

Collagenase. Collagenase (Sigma type VII, lot 33F-6819) is reconstituted prior to each experiment in 0.225 M mannitol/0.075 M sucrose.

Reagents. The following substances were reagent grade or better, used without further purification, and obtained from the sources noted: sucrose (RNase free), coenzyme A (type III-L), fatty acid free bovine serum albumin (BSA; fraction V), ethylene glycol bis(β-aminoethyl ether)

[16] E. D. Hay, *J. Cell Biol.* **91,** 205s (1981).
[17] J. B. Caulfield and T. K. Borg, *Lab. Invest.* **40,** 364 (1979).
[18] S. Seifter and E. Harper, *in* "The Enzymes" (P. D. Boyer, ed.), 3rd Ed., Vol. 3, p. 649. Academic Press, New York, 1971.
[19] M. D. Bond and H. E. Van Wart, *Biochemistry* **23,** 3085 (1984).
[20] A. L. Lehninger, E. Carafoli, and C. S. Rossi, *Adv. Enzymol.* **29,** 259 (1967).
[21] A. L. Lehninger, *Biochem. J.* **119,** 129 (1970).
[22] A. M. Stadhouders, *in* "Mitochondria and Muscular Diseases" (H. F. M. Busch, F. G. I. Jennekens, and H. R. Scholte, eds.), p. 77. Mefar b.v. Beetstwerg, The Netherlands, 1981.

N,N,N',N'-tetraacetic acid (EGTA), 2-(4-iodophenyl)-3-(4-nitrophenyl)-5-phenyltetrazolium chloride (INT), EDTA, rotenone, the sodium salts of NADH, ADP, and ATP, and the sodium salts of pyruvic, L-malic, L-glutamic, DL-3-hydroxybutyric, 2-oxoglutaric, and succinic acids (Sigma); Tris (Boehringer Mannheim Biochemicals, Indianapolis, IN); mannitol, potassium permanganate, and phosphoric acid (Mallinckrodt, Paris, KY); palmitic acid (Nutritional Biochemical Corp., Cleveland, OH); and L-(−)-carnitine (gift of L. Bieber, Michigan State University). 7-Deoxycholic acid (Sigma, grade III) was recrystallized three times from 80% acetone. Cytochrome c (horse heart, Sigma type VI) was purified and reduced as described by Thompson and Ferguson-Miller.[23]

Animals. Fertile single comb white leghorn eggs were purchased from a local farm and allowed to hatch in incubators at Michigan State University's Department of Animal Science. Chicks were maintained on a diet of Chick G0125 feed (Kent Feeds, Inc., Muscatine, IA) and were not starved prior to sacrifice.

Glassware. All glassware must be clean and maintained as described by Mela and Seitz.[5]

Enzyme Assays

Cytochrome c oxidase (cyt aa_3, EC 1.9.3.1) concentrations are quantitated from difference spectra of solubilized mitochondria (dithionite reduced minus ferricyanide oxidized) obtained with a Perkin-Elmer 559-A spectrophotometer. Mitochondria (~2 mg/ml) are solubilized in a buffer containing 40 mM sodium phosphate, pH 7.8, 1% (w/v) deoxycholate. A $\Delta\varepsilon(605-630\ \text{nm}) = 24\ \text{m}M^{-1}\ \text{cm}^{-1}$ is used.[24]

Succinate dehydrogenase (succinate:INT reductase, EC 1.3.99.1)[24a] is quantitated by the endpoint assay of Pennington[25] after 7–10 min incubations at 37°. The percentage yield of mitochondria is estimated from the fraction of succinate:INT reductase activity recovered from the crude homogenate. One unit of succinate:INT reductase catalyzes the synthesis of 1 μmol of INT-formazan per min at 37°.

Hexokinase (EC 2.7.1.1) is assayed according to the method of Polakis and Wilson.[26] The hexokinase activity remaining in the mitochondrial fraction after five rinses with 0.225 M mannitol/0.075 M sucrose is solubilized by incubation in 0.1 M sodium phosphate, pH 6.5, 1% Triton X-100

[23] D. A. Thompson and S. Ferguson-Miller, *Biochemistry* **22**, 3178 (1983).
[24] G. von Jagow and M. Klingenberg, *FEBS Lett.* **24**, 278 (1972).
[24a] INT, 2-(4-Iodophenyl)-3-(4-nitrophenyl)-5-phenyltetrazolium chloride.
[25] R. J. Pennington, *Biochem. J.* **80**, 649 (1961).
[26] P. G. Polakis and J. E. Wilson, *Biochem. Biophys. Res. Commun.* **107**, 937 (1982).

for 20 min. One unit of hexokinase catalyzes the synthesis of one μmole of glucose 6-phosphate per min at 30°.[27]

Long-Chain Fatty-Acyl-CoA Synthetase. To test for the presence of this enzyme, mitochondria ($\sim$0.78 nmol cyt aa_3) are incubated at 30.5° in a 0.225 M mannitol/0.075 M sucrose solution (1.75 ml) containing 450 μM coenzyme A, 5 mM ATP, 5 mM magnesium acetate, 50 μM palmitate, 2 mM L-($-$)-carnitine, and 10 mM Tris-buffered P_i, pH 7.4. In addition, 0.2% (w/v) fatty acid free BSA is included in the assay medium to control the concentration of free fatty acid in the mitochondrial suspension by creating an equilibrium between the free and bound forms,[28] thereby safeguarding against the uncoupling effect of fatty acid,[29] the detergent effect of acylcarnitine derivatives (L. Bieber, personal communication), and inhibition of the adenine nucleotide translocase.[30] β-Oxidation of palmitoylcarnitine is monitored polarographically.

Other Assays

ADP. A 2- to 5-μl aliquot of an ADP solution is added to 1.0 ml of an assay mixture containing 25 mM MOPS [3-(N-morpholino)propanesulfonic acid], pH 6.8, 5 mM magnesium acetate, 1 mM EDTA, 1.8 mM phosphoenolpyruvate, 0.32 mM NADH, 44 IU lactate dehydrogenase, and 12 IU pyruvate kinase. The oxidation of NADH is monitored at 340 nm using a Beckman DU spectrophotometer thermostatted at 30°. ADP concentrations are calculated by assuming a molar extinction coefficient for NADH of $\Delta\varepsilon(340 \text{ nm}) = 6.23 \times 10^3 \ M^{-1} \text{ cm}^{-1}$.

Mitochondrial protein is determined by the modified Lowry method of Markwell *et al.*,[31] using bovine serum albumin (BSA) as a standard.

Isolation of Mitochondria

Mitochondria are isolated from ventricular myocardium according to the scheme shown in Fig. 1. Hearts are removed from three 14- to 20-day-old chickens. The pericardium, major vessels, and atria are excised, and the ventricles of each heart are immediately minced with scissors into 5 ml of fresh ice-cold osmotic support medium containing 0.225 M mannitol, 0.075 M sucrose, and 0.2% (w/v) fatty acid free BSA (MSB medium).

[27] A. C. Chou and J. E. Wilson, *Arch. Biochem. Biophys.* **151,** 48 (1972).
[28] A. A. Spector, J. E. Fletcher, and J. D. Ashbrook, *Biochemistry* **10,** 3229 (1971).
[29] S. G. Van den Bergh, this series, Vol. 10, p. 749.
[30] G. Woldegiorgis, S. Y. K. Yousufzai, and E. Shrago, *J. Biol. Chem.* **257,** 14783 (1982).
[31] M. A. K. Markwell, S. M. Haas, N. E. Tolbert, and L. L. Bieber, this series, Vol. 72, p. 296.

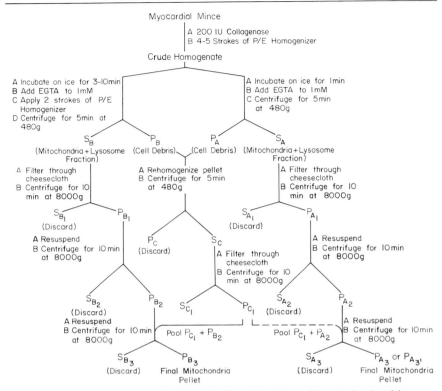

FIG. 1. Flow chart for collagenase-facilitated isolation of heart mitochondria.

In order to maintain optimal mitochondrial respiratory control, it is critical that the heart tissue from each animal be isolated and minced within 30 sec after decapitation and that all subsequent operations be performed at 0–4°. The combined mince is washed with three, 30 ml volumes of MSB medium, resuspended in approximately 10 ml of MSB per gram of heart, and 200 IU of collagenase is added. Homogenization is performed in a glass vessel with four to five up and down strokes of a motor driven Potter-Elvehjem (P/E) pestle set at 500 rpm with a clearance of 0.117 mm (caliper measurement). The homogenate is allowed to incubate on ice for 1 min, and EGTA is added to a final concentration of 1 mM to halt collagenolysis and prevent mitochondrial uptake of Ca(II). The volume of the homogenate is measured and a sample for yield determination is withdrawn.

The homogenate is centrifuged at 480 g for 5 min in 15-ml Corex tubes with a Sorvall SS-34 head in order to sediment cellular debris. The supernatant liquid is filtered through cheesecloth and centrifuged at 8000 g for 10 min to sediment the mitochondria. The pelleted mitochondria are gently resuspended with a glass stirring rod in MSB medium that is 1 mM EGTA (MSBE medium). The mitochondria are washed twice by centrifuging at 8000 g for 10 min and resuspended with a Pipetman (Gilson Medical Instruments, Middleton, WI) at a final concentration of 15–20 mg mitochondrial protein per ml of MSBE medium. The above protocol should take no more than 1 hr to complete. The resulting mitochondria pellet is designated P_{A_3}.

If enhanced yields are needed, the following protocol should be followed. For a 10–15% increase in yield, the first low-speed pellet is rehomogenized and centrifuged at 480 g. The new supernatant liquid is subjected to a high speed spin, the resulting mitochondria pellet (P_{C_1}) is washed once and pooled with P_{A_2}. The pooled pellets are washed and resuspended with a Pipetman in approximately 1 ml of MSBE medium. This mitochondrial suspension corresponds to $P_{A_{3'}}$.

For maximal yield, the crude homogenate is allowed to incubate on ice for 3 min in the presence of collagenase, made 1 mM EGTA, further homogenized with two strokes of the Potter-Elvehjem pestle, and centrifuged at 480 g for 5 min. The mitochondria in the supernatant liquid are pelleted and washed (P_{B_2}) and then pooled with mitochondria recovered from the crude pelleted material (P_{C_2}). The combined pellet is washed and resuspended as above in MSBE medium (P_{B_3}).

Oxygen Consumption Assays

Mitochondrial respiratory control parameters are assessed at 30.5° from data obtained with a Gilson model K-IC oxygen polarograph (Gilson Medical Electronics) equipped with a Yellow Springs Instruments (Yellow Springs, OH) Clarke oxygen electrode and a glass-stoppered, 1.75 ml, water-jacketed reaction chamber. Constant temperature is maintained with a Haake circulating water bath. Chicken heart mitochondria are assayed for oxygen consumption in a medium consisting of 0.225 M mannitol, 0.075 M sucrose, and 10 or 20 mM P_i, pH 7.4. State 3 respiration is induced with approximately 400 nmol of ADP in the presence of 5 mM pyruvate and 2.5 mM malate. Respiratory control ratios are calculated as previously described.[32,33] ADP:O ratios are calculated using "total" oxygen as defined by Lemasters.[34] The concentration of oxygen atoms in the

---

[32] B. Chance and G. R. Williams, *Adv. Enzymol.* **17**, 65 (1956).
[33] R. W. Estabrook, this series, Vol. 10, p. 41.
[34] J. J. Lemasters, *J. Biol. Chem.* **259**, 13123 (1984).

TABLE I
YIELD OF HEART MITOCHONDRIA ISOLATED WITH OR WITHOUT
COLLAGENASE

Mitochondria pellet[a]	Collagenase	Yield (%)	Number of preparations
P_{A_3}	−	17 ± 4	7
P_{A_3}	+	31 ± 8	29
$P_{A_{3'}}$	+	41 ± 3	5
P_{B_3}	+	70 ± 8	4

[a] These designations are shown in Fig. 1.

air saturated assay medium at 30.5° is assumed to be 440 μM.[35] Mitochondria and all reaction components are added to the assay medium with Hamilton glass syringes (Hamilton Co., Reno, NE).

Results

Yield of Mitochondria. Percentage recoveries of the mitochondrial marker enzyme succinate:INT reductase from the crude homogenate shown in Table I clearly demonstrate that collagenase facilitates the isolation of mitochondria from myocardium. A 1 min incubation of the crude homogenate with collagenase enhances the yield of mitochondria nearly twofold. Rehomogenizing and recentrifuging the pelleted crude material gives a 10–15% increase in the yield. The highest yield is obtained after a 3 min incubation of the crude homogenate with collagenase. Subjecting the crude homogenate to two additional strokes of the Potter-Elvehjem pestle and pooling pellets P_{B_2} and P_{C_1} results in a yield as high as 76% of the total mitochondria contained in the starting tissue. Significantly, the mitochondria in both pellets P_{A_3} and P_{B_3} have comparable respiratory parameters.

Mitochondrial Respiratory Control. The respiratory control ratio (RCR) of chicken heart mitochondria is strikingly dependent on cyt aa_3 concentration (Fig. 2). While state 3 respiration is a linear function of cyt aa_3 concentration, the relationship of state 4 respiration to cyt aa_3 concentration is more complex (inset, Fig. 2). State 4 rates initially increase with increasing concentration of mitochondria until, at an optimum concentration that is very similar for each preparation, the rates become very low or frequently zero. Beyond the optimum mitochondrial concentration, state 4 rates again increase. This rate minimum yields the peak in RCR values shown in Fig. 2.

[35] J. B. Chappell, *Biochem. J.* **90**, 225 (1963).

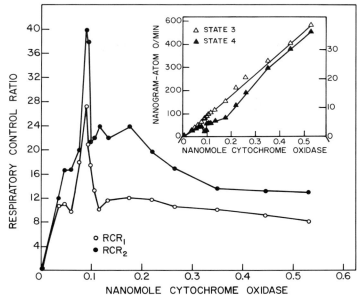

FIG. 2. Respiratory control ratios plotted as a function of increasing concentration of mitochondria (expressed as nmol cyt aa_3). RCR_2 values were calculated from the state 3 (left ordinate) and state 4 (right ordinate) rates of respiration that are presented in the inset. RCR_1 values are calculated from state 3 and state 4 rates of respiration subsequent to ADP_1 (not shown). Respiration rates were measured as described under Experimental Procedures using 5 mM pyruvate/2.5 mM malate, 20 mM P_i, pH 7.4, at 30.5°.

Figure 3A is an example of an oxygen electrode tracing obtained with an optimal concentration of chicken heart mitochondria showing very low rates of state 4 respiration for five phosphorylation cycles. These mitochondria typically have a significant state 4 rate of respiration after the first phosphorylation cycle (ADP_1), but subsequent cycles (ADP_2, ADP_3) frequently show zero state 4 rates. Although zero rates of state 4 respiration mathematically imply infinite respiratory control, an RCR of 40 was assumed as an arbitrary maximum (Fig. 2). The respiratory parameters presented in Table II are based on data derived from ADP_2. Controversy exists with regard to the precise stoichiometries at each coupling site of the respiratory chain[34,36]; however, the ADP:O ratios greater than 3 for pyruvate/malate indicate that electron transport is highly coupled to phosphorylation and that pyruvate is probably oxidized past the level of 2-

[36] T. A. Scholes and P. C. Hinkle, *Biochemistry* **23**, 3341 (1984).

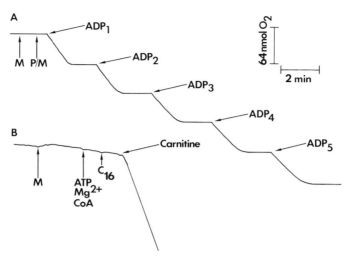

FIG. 3. Oxygen electrode tracings of chicken heart mitochondria respiration. (A) State 3 to state 4 transitions at an optimal concentration of mitochondria under conditions described in the legend to Fig. 2. M, Addition of mitochondria equivalent to 0.085 nmol cyt aa_3; P/M, addition of pyruvate and malate; ADP_{1-5}, sequential additions of 418 nmol of ADP. A systematic investigation indicates that above 40% oxygen saturation, the range in which the respiration induced by ADP_{1-3} is measured, back-diffusion of oxygen does not artificially depress state 4 respiration. (B) Long-chain fatty-acyl-CoA synthetase activity measured under conditions described under Experimental Procedures. M, Addition of mitochondria; C_{16}, addition of palmitic acid.

TABLE II

RESPIRATORY PARAMETERS OF CHICKEN HEART MITOCHONDRIA ISOLATED WITH COLLAGENASE[a]

[P_i] (mM)	Respiratory rate[b] ng-atom O min^{-1} (nmol cyt aa_3)$^{-1}$		ADP:O[c]	RCR[d]	cyt aa_3[e] (nmol)	Number of preparations
	State 3	State 4				
10	721 ± 122	20.7 ± 11.2	3.17 ± 0.12	33.8 ± 7.5	0.137 ± 0.020	19
20	843 ± 120	7.4 ± 9.1	3.37 ± 0.12	39.6 ± 1.6	0.089 ± 0.011	16

[a] All values are expressed as a mean ± SE for the second phosphorylation cycle of the indicated number of preparations.
[b] These rates of respiration are converted to ng-atom O min^{-1} mg^{-1} mitochondrial protein by using the factor 0.41 ± 0.07 ($n = 23$) nmol cyt aa_3 mg^{-1}.
[c] Values represent the ratio of nmol of ADP phosphorylated per ng-atom of oxygen consumed.
[d] These RCR values are obtained at the optimum concentration of cyt aa_3 using 40 as the arbitrary maximum value. Consequently, the RCR values calculated from the average state 3/state 4 ratios are not identical to the more conservative estimates of RCR shown.
[e] Values represent the amount of cyt aa_3 in the 1.75 ml assay vessel that gives the highest respiratory control ratio.

TABLE III
RECOVERY OF HEXOKINASE IN HEART MITOCHONDRIA ISOLATED WITH
COLLAGENASE

Trial	Crude homogenate[a] (units)	Mitochondria[a] (units)	Recovery[b] (%)	Specific activity mitochondrial hexokinase (units mg^{-1})
1	11.1	6.0	54.1	0.092
2	8.6	5.0	58.1	0.10

[a] The values are units of hexokinase in the crude homogenate or in the mitochondrial fraction, corrected in the latter case for yield of mitochondria.

[b] 100% recovery of hexokinase activity is not expected due to the metabolic regulation of this enzyme's ambiquitous binding to the mitochondrial outer membrane [J. E. Wilson, *Curr. Top. Cell. Regul.* **16**, 2 (1980)].

oxoglutarate dehydrogenase as shown in rabbit heart mitochondria by Hansford and Johnson.[37]

The biochemical basis of state 4 respiratory behavior as a function of mitochondrial concentration is at present unknown. In each experiment, the concentration of mitochondria at which the lowest state 4 rate is observed must be found by trial and error. As indicated in Table II, the optimum concentration of mitochondria also depends on the concentration of phosphate in the assay medium: at 10 mM P$_i$, the peak concentration of cyt aa_3 equals 0.137 ± 0.02 ($n = 19$) nmol; at 20 mM P$_i$, the optimum concentration equals 0.089 ± 0.011 ($n = 16$) nmol.

Membrane Intactness. Mitochondria isolated according to the procedure described herein have intact inner and outer membranes as evidenced by the inability of exogenous 1 mM NADH to support the phosphorylation of ADP and by the lack of respiratory stimulation in the presence of 2 mM reduced cytochrome c.

Outer Membrane Enzymes. As shown in Fig. 3B, significant long-chain fatty-acyl-CoA synthetase activity remains in the mitochondrial fraction. Addition of carnitine initiates state 3 respiration, indicating that the free fatty acid has been converted to the CoA thioester and is thus available for transport into the mitochondrial matrix for β-oxidation. ADP is provided by the adenylate kinase (EC 2.7.4.3) transphosphorylation of AMP formed during the activation of fatty acid. Table III details the hexokinase activity associated with mitochondria isolated with collagen-

[37] R. G. Hansford and R. N. Johnson, *J. Biol. Chem.* **250**, 8361 (1975).

ase. The IU of hexokinase per mg of mitochondrial protein is in good agreement with studies on rabbit heart mitochondria.[38]

Comments. At the optimum concentration of mitochondria, a zero or very low rate of state 4 respiration subsequent to ADP_2 is observed between 22 and 33°. Above 33°, state 4 rates increase as a function of temperature.

The respiratory control parameters of heart mitochondria isolated according to this method are extremely sensitive to micromolar concentrations of Mg(II), Mn(II), and Ca(II). State 4 respiration is increased and ADP:O ratios are decreased by these cations as a function of concentration. Mn(II) and Ca(II) also decrease the rate of state 3 respiration, whereas Mg(II) has no effect on this parameter even at 5 mM. Consequently, these divalent cations should not be constituents of the assay medium if ideal respiratory parameters are desired. Concentrations of EDTA in excess of 100 μM decrease RCR values. EGTA concentrations up to 5 mM exert no effect on mitochondrial respiration.

All solutions must be kept rigorously free of microbial contamination. For optimum respiratory control, mannitol/sucrose solutions should be freshly made for each mitochondria preparation and stock solutions of P_i and substrate should be made every 14 days. Of all the respiratory substrates studied (5 mM pyruvate/2.5 mM malate, 5 mM glutamate/2.5 mM malate, 5 mM 2-oxoglutarate/2.5 mM malate, 5 mM 3-hydroxybutyrate, and 5 mM succinate/5 μM rotenone), the combination of pyruvate and malate supported the highest state 3 rates and the lowest state 4 rates.

These heart mitochondria remain optimally coupled for 3 to 3.5 hr. At longer time periods the mitochondria develop an increasing rate of state 4 respiration. Chickens over 20 days old should not be used because the quality of respiratory parameters progressively decreases as a function of age.

Collagenase can also be used to facilitate the isolation of highly coupled heart mitochondria from 14- to 20-day-old Sprague–Dawley rats. For optimal respiratory control, 0.12 M KCl should be used as the osmotic support substance in both the isolation and assay media.

Conclusions. The results of this investigation indicate that collagenase facilitates the release of mitochondria from myocardium. The resulting mitochondria are highly coupled, have enzymatically intact outer membranes, morphologically intact inner and outer membranes, and can be isolated in high yield.

---

[38] E. Aubert-Foucher, B. Font, and D. C. Gautheron, *Arch. Biochem. Biophys.* **232,** 391 (1984).

Acknowledgments

This work was supported by National Institutes of Health Grants GM20716 (to CHS) and GM26916 (to SFM), an MSU Biomedical Research Support Grant and an All University Research Initiation Grant, and by the Michigan State Agricultural Experiment Station. The authors wish to thank Dr. Leena Mela (Oregon Health Sciences University) and Dr. Richard Bukoski (Michigan State University) for valuable discussions during the early stages of this work; Dr. Loran L. Bieber (Michigan State University), Dr. Richard W. von Korff (Michigan Molecular Institute, Midland, MI), and Dr. John E. Wilson (Michigan State University) for helpful advice; Mr. Randy Shoemaker for his assistance in animal care; Mrs. Helen J. Farr for her careful typing of the manuscript; and the Department of Animal Sciences, Michigan State University, particularly Mr. Bruce Buckmaster, Dr. Robert K. Ringer, and Dr. Donald Polin for their assistance in handling and hatching eggs.

[3] Alkyl Glycoside Detergents: Synthesis and Applications to the Study of Membrane Proteins

By T. VanAken, S. Foxall-VanAken, S. Castleman, and S. Ferguson-Miller

In 1975, Baron and Thompson[1] introduced the nonionic detergent octyl glucoside, designed to be easily removed by dialysis to facilitate reconstitution of membrane proteins into phospholipid vesicles. Our interest was in studying and reconstituting cytochrome oxidase, an enzyme that is highly sensitive to the detergent environment and not satisfactorily activated and dispersed by any commercially available detergents. Octyl glucoside was not effective in activating and dispersing with this protein, but since pure, homogeneous analogs could be synthesized using different sugars and alkyl tails, we decided to examine some of these detergents in the hope of finding a version that retained the advantageous qualities of octyl glucoside while supporting optimal activity of cytochrome oxidase. To accomplish this we devised a synthesis[2] for alkyl glycosides that facilitates large scale production by eliminating some of the more complicated procedures and employing a one-step purification on Dowex 1 (hydroxide form). A variety of detergents prepared by this method were investigated for their ability to disperse and activate cytochrome oxidase. Lauryl-maltoside (dodecyl-β-D-maltopyranoside) was found to be the most ef-

---

[1] C. Baron and T. E. Thompson, *Biochim. Biophys. Acta* **382**, 276 (1975).
[2] P. Rosevear, T. VanAken, J. Baxter, and S. Ferguson-Miller, *Biochemistry* **19**, 4108 (1980).

fective.[2-4] Other investigators have found it to be the detergent of choice for studying rhodopsin[5] and photosynthetic reaction centers,[6] suggesting that this detergent may have generally useful properties for studies on a variety of intrinsic membrane proteins. Other analogs may prove to be equally valuable as agents for solubilizing, reconstituting, and crystallizing proteins with different specific requirements.

Synthetic Procedure

The method outlined is applicable to the preparation of all the detergents shown in the table, but will be discussed in terms of laurylmaltoside synthesis. It is scaled up 10-fold over the previously published procedure,[2] and uses higher levels of dodecanol to increase the yield. In cases where the acetobromo derivative of the desired sugar is not commercially available, the procedure for making it has been detailed in the original publication.[2]

Materials

The following materials were obtained from the sources indicated: acetobromomaltose, and Dowex 1 (Cl), 2% cross-linked, 200–400 mesh (Sigma Chemical Co.); dodecanol, dichloromethane, silver carbonate (Aldrich Chemical Co. Milwaukee, WI); silica gel G plates, 250 μm thick (Analtech). All other solvents and reagents were the highest grade available.

Methods

Thin-Layer Chromatographic Systems for Analyzing the Reaction Progress and Products. Solvent system I, ethyl acetate/hexane (1 : 1 v/v); solvent system II, ethyl acetate/methanol (4 : 1 v/v). The TLC plates are developed by spraying with 2 N sulfuric acid and heating in a 90° oven.

Alkyl Glycoside Formation from Acetobromo Sugar and Alcohol (Large Scale). In a foil-covered, 4-liter round bottom flask, 280 mmol (200 g) of acetobromomaltose is stirred with 2800 ml dichloromethane. To this is added 660 ml dodecanol (previously stirred with Drierite for 24 hr at 60° and stored over Drierite), 60 g silver carbonate (dried for 24 hr), 5 g

[3] S. Ferguson-Miller, T. VanAken, and P. Rosevear, *in* "Electron Transfer and Oxygen Utilization" (H. Chien, ed.), p. 297. Elsevier Biomedical, Amsterdam. 1982.

[4] D. A. Thompson and S. Ferguson-Miller, *Biochemistry* **22**, 3178 (1983).

[5] P. Knudsen and W. L. Hubbell, *Membr. Biochem.* **1**, 297 (1978).

[6] M. W. Kendall-Tobias and M. Seibert, *Arch. Biochem. Biophys.* **216**, 255 (1982).

SUMMARY OF PROPERTIES OF ALKYL GLYCOSIDE DETERGENTS

Alkyl glycoside	Critical micelle concentration[a] (mM)	Critical micelle temperature[b] (Krafft point)	Solubility at 25°	Micelle size[c] (Da), Stokes radius (Å)	Effect on cytochrome oxidase
Octyl-β-D-glucoside	23.4	0°	(++)	8,000 (15 Å)	Aggregates
Octyl-α-D-glucoside	—	40°	(−)	—	—
Octyl-β-D-lactoside	—	43°	(−)	—	—
Lauryl-β-D-lactoside	—	73°	(−)	—	—
Lauryl-β-D-cellobioside	—	48°	(−)	—	—
Octyl-β-D-maltoside	23.4	0°	(++)	10,000 (15.5 Å) 23,000 (21 Å)	Aggregates
Lauryl-β-D-maltoside	0.165	0°	(++)	50,000 (29 Å) 76,000[d]	Activates and disperses
Lauryl-α-D-maltoside	0.156	0°	(++)	46,000 (27.5 Å)	Activates and disperses
Oleoyl-β-D-maltoside	—	—	gel (0–90°)	—	
Oleoyl-β-D-maltotrioside	<0.005	0°	(++)	125,000 (42 Å)	Stabilizes dilute enzyme

[a] Critical micelle concentrations were measured using Cibacron blue F3GA and 2-p-toluidinylnaphthalene 6-sulfonate (TNS) as spectral and fluorescent indicators of micelle formation, following a titration procedure similar to that described by Mast and Haynes.[30] An increase in absorbance at 324 nm (compared to 362 nm) was followed with Cibacron blue, and an increase in fluorescence intensity (excitation: 360 nm; emission, 460 nm) was followed with TNS.

[b] Critical micelle temperatures were measured as described in the legend to Fig. 1.

[c] The micelle size was measured by gel filtration on LKB-Ultrogel 34 or 54 as described in Rosevear et al.[2]

[d] This value of the micelle size was determined by sedimentation equilibrium analysis as described in Suarez et al.[11]

iodine, and 200 g Drierite, in that order. The mixture is stirred for 12 hr with a propeller-type stirrer inserted through the neck of the vessel. (Magnetic stirrers have been found unsuccessful with this mass of material.) The reaction mixture is checked by TLC in solvent system I and should show two main spots: laurylmaltoside peracetate (R_f 0.70) and maltose peracetate (R_f 0.38). The reaction mixture is filtered through a pad of celite on a 2-liter scintered glass funnel (small pore) and washed with 1000 ml dichloromethane. The filtrate is concentrated to a syrup by rotary evaporation using a 50° water bath.

Hydrolysis of Orthoester, Deacetylation, and Removal of Sugar By-products. Orthoester side product[7] is hydrolyzed by dissolving the syrup in 3 liters of 0.01 N H_2SO_4 in 90% aqueous acetone, allowing it to stand for 30 min, and then neutralizing with pyridine until a slight cloudiness appears (pH 6 with pH paper). The mixture is again concentrated to a syrup. Deacetylation is performed by dissolving in 2000 ml methanol : triethylamine : water (2:1:1) and letting stand overnight. Two main spots are then apparent in solvent system II: lauryl-β-D-maltoside (R_f 0.43) and maltose (R_f 0.13). Dodecanol runs at the solvent front. The deacetylated mixture is concentrated until the methanol and triethylamine are removed but the water remains (mixture begins to foam), and then allowed to separate into two phases in a separatory funnel. The lower phase (brown, aqueous) contains the unreacted sugar and by-products, and can be discarded. The upper phase contains dodecanol and detergent, and can be washed several times with water in the separatory funnel until most of the brown color is removed. If it is then allowed to stand for several days, a crude precipitate of laurylmaltoside forms, from which the excess dodecanol can be decanted (but the dodecanol should be checked by TLC to determine whether a significant amount of laurylmaltoside remains).

Purification by Dowex 1 (OH) Chromatography. Aliquots of the laurylmaltoside precipitate or the dodecanol–detergent mixture are dissolved in a minimum amount of anhydrous methanol and applied in 50 to 100 ml portions to a 3.0 × 150 cm column of Dowex 1 (OH) previously converted from the chloride form by extensive washing with 2 N NaOH (20 liters, or until the silver nitrate test for chloride is negative), followed by distilled H_2O until no longer alkaline, and 3 liters of anhydrous methanol. (This same procedure is used to regenerate the column when it has turned from brown to black, indicating accumulation of sugar products.[2]) The column is eluted with methanol at a rate of 10 ml/hr. After collecting about 1200 ml in bulk, fractions of 10 to 20 ml are collected, and the alcohol and

[7] N. K. Kochetkov, A. J. Khorlin, and A. F. Bochkov, *Tetrahedron* **23**, 693 (1967).

detergent peaks are located by adding 1 ml 5% aqueous phenol to 100-μl aliquots of each tube (cloudiness indicates dodecanol is present) followed by 5 ml concentrated sulfuric acid (orange-brown color indicates detergent). The alcohol runs ahead of the detergent and is usually found in the initial 1200 ml. Any lauryl-α-D-maltoside moves ahead of the β-form, and can be detected in solvent system II, the α being the slower moving spot (R_f 0.38). The α- and β-laurylmaltoside fractions are pooled separately and any overlapping region can be concentrated and added to the next sample to be chromatographed. When a large (10-fold) excess of dodecanol is used as described in this preparation, no α-form is observed, possibly because it is more soluble in dodecanol and moves ahead with that peak. The methanol is removed by rotary evaporation and the detergent is taken up in water, lyophilized, and stored in a desiccator. Any stock solutions of the detergents in water are kept frozen to avoid any possibility of hydrolysis. A yield of 100 g of detergent can be obtained from the reaction described (approximately 70% yield, from acetobromomaltose).

Properties of Alkyl Glycoside Detergents

The characteristics of the detergents studied so far are summarized in the table. It was found[2,3] that only the α 1,4-linked disaccharides made suitable head groups, since the β-linked sugars (cellobiose, lactose) formed insoluble glycosides, possibly because of dimer formation between sugar head groups.

Octylmaltoside was synthesized to determine if the head group would affect any of the micelle properties, hopefully creating a more stable, better dispersing micelle than octyl glucoside. However, no difference was observed in its physical properties or in its effects on cytochrome oxidase compared to octyl glucoside.

In contrast to the striking differences between the α and β forms of octyl glucoside[8,9] (see the table), α- and β-laurylmaltoside behaved in an almost identical manner, though the tendency of the α-laurylmaltoside to be contaminated with dodecanol (because it runs closer to the alcohol peak on the column) originally led us to believe that it had different properties, including the ability to denature cytochrome oxidase. This deleterious effect was not observed when the α-form was repurified. The β-form does have the distinguishing feature that when it is contaminated

[8] H. Schindler and J. P. Rosenbusch, *Proc. Natl. Acad. Sci. U.S.A.* **78**, 2302 (1981).
[9] D. L. Dorset and J. P. Rosenbusch, *Chem. Phys. Lipids* **29**, 299 (1981).

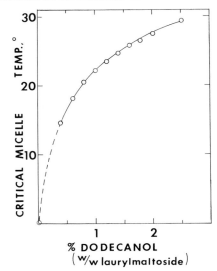

FIG. 1. Effect of dodecanol on the critical micelle temperature of laurylmaltoside. To a solution of pure laurylmaltoside in water (10 g/100 ml) was added various amounts of dodecanol to give effective contamination levels of 0.5 to 2.5% with respect to the weight of laurylmaltoside (1% dodecanol = 0.1 g dodecanol added per 10 g laurylmaltoside). The critical micelle temperature was measured in a capillary tube in a controlled temperature bath, as the temperature at which the detergent/dodecanol solution redissolved after precipitating out at lower temperatures.

with even small amounts of dodecanol it will form a white precipitate at 4° in aqueous solution. Figure 1 illustrates the effect of dodecanol contamination on the solubility of β-laurylmaltoside. α-Laurylmaltoside remains soluble in the presence of substantial dodecanol contamination.

Oleoylmaltotrioside is highly soluble and produces large discrete micelles, but does not disperse or activate cytochrome oxidase to the same extent as laurylmaltoside. It is, however, more effective as a stabilizing agent for lipid-depleted enzyme preparations in dilute solution, as evidenced by a linear rate of oxygen consumption by cytochrome oxidase in an assay system containing 50 mM potassium phosphate, 0.06% oleoylmaltotrioside, pH 6.5, conditions under which laurylmaltoside supports higher initial rates but the activity drops off rapidly.[10]

---

[10] D. A. Thompson, Ph.D. thesis, Michigan State University, 1984.

Applications of Alkyl Glycoside Detergents

Studies on Cytochrome Oxidase

We have found that laurylmaltoside is able to both disperse and activate purified cytochrome oxidase better than any other detergent tested. Although considerable variability in activity is found among different beef heart oxidase preparations, a range of molecular activities from 200 to 700 electrons per second per heme aa_3 has been observed in laurylmaltoside. Cytochrome oxidase prepared in cholate and dissolved in laurylmaltoside behaves consistently on gel filtration columns at medium ionic strength (100 ml KCl, 10 mM Tris–Cl, pH 7.8) revealing a major homogeneous species with an apparent molecular weight of 320,000 to 370,000. All other detergents tested (Tween 20, Tween 80, Triton X-100, deoxycholate, octyl glucoside, oleoylmaltotrioside, octylmaltoside) gave a predominance of a much larger molecular weight species (>1,000,000) under these conditions. More rigorous analysis of beef heart cytochrome oxidase by sedimentation equilibrium[11] shows it to be a monomer (2 heme a, 2 copper atoms) in laurylmaltoside with a protein moiety of 194,000 Da and associated detergent equal to 108,000 Da (~200 molecules of laurylmaltoside). Although laurylmaltoside does not dialyze out rapidly (half-time of dialysis ~8 hr),[12] when the enzyme is dialyzed for 20 hr with cholate-solubilized phospholipids, it is readily incorporated into vesicles that exhibit high levels of respiratory controls,[4] indicating that little detergent remains.

Purification of cytochrome oxidase from whole mitochondria by a one-step cytochrome c affinity chromatography procedure is greatly facilitated with respect to yield, purity, and activity by use of laurylmaltoside as the solubilizing agent, apparently because of its superior dispersing ability compared to Triton X-100, the only other detergent shown to be effective for this purpose.[4,13,14]

Purification and Stabilization of Other Membrane Proteins

The unusual effectiveness of laurylmaltoside compared to several ionic detergents as a protein stabilizing agent was first observed by Knud-

[11] M. D. Suarez, A. Revzin, R. Narlock, E. S. Kempner, D. A. Thompson, and S. Ferguson-Miller, *J. Biol. Chem.* **259,** 13791 (1984).
[12] W. J. DeGrip and P. H. M. Bovee-Geurts, *Chem. Phys. Lipids* **23,** 321 (1979).
[13] H. Weiss and B. Juchs, *Eur. J. Biochem.* **88,** 17 (1978).
[14] K. Bill, C. Broger, and A. Azzi, *Biochim. Biophys. Acta* **679,** 28 (1982).

sen and Hubbell[5] studying rhodopsin. They concluded that large micelle size and high packing density of the nonionic head groups were responsible for preventing access of water and protein denaturation. Evidence from ^{13}C NMR studies on laurylmaltoside micelles in aqueous solution (T. VanAken and S. Ferguson-Miller, unpublished) support the idea that the maltose head groups are indeed relatively immobilized, compared to glucose in an octyl glucoside micelle.

More recently, laurylmaltoside has been reported to increase the stability of photosynthetic reaction centers that were purified in lauryldimethylamine oxide (LDAO) from *Rhodopseudomonas sphaeroides*.[6] Similarly, a final sucrose-gradient purification step in laurylmaltoside of photosystem II from thermophilic blue–green algae results in a significant increase in purity, activity, and stability of the complex.[15]

Octyl glucoside has been used successfully in numerous purification procedures[16–18] and has led to the development of some novel purification schemes.[19–21]

Preparation of Lipid Vesicles and Enzyme Reconstitution

The high critical micelle concentration of octylglucoside has proved useful not only in facilitating reconstitution of enzymes into artificial vesicles,[1,22,23] but also in the preparation of specific types of phospholipid vesicles.[23,24] In addition, alkyl glucosides of different chain lengths have been used to produce vesicles of various sizes in relatively homogeneous populations.[25]

Crystallization of Membrane Proteins

An important recent breakthrough in the study of membrane proteins is the demonstration that three-dimensional crystals can be produced that

[15] J. Bowes, A. C. Stewart, and D. S. Bendall, *Biochim. Biophys. Acta* **725**, 210 (1983).
[16] G. W. Stubbs, H. Smith, Jr., and B. J. Litman, *Biochim. Biophys. Acta* **425**, 46 (1976).
[17] J. T. Lin, S. Reidel, and R. Kinne, *Biochim. Biophys. Acta* **577**, 179 (1979).
[18] B. Wittenberger, D. Raben, M. A. Lieberman, and L. Glaser, *Proc. Natl. Acad. Sci. U.S.A.* **75**, 5457 (1978).
[19] P. Felgner, J. Messer, and J. E. Wilson, *J. Biol. Chem.* **254**, 4946 (1979).
[20] M. Stadt, P. A. Banks, and R. D. Kobes, *Arch. Biochem. Biophys.* **214**, 223 (1982).
[21] J. E. Wilson, J. L. Messer, and P. L. Felgner, this series, Vol. 97, p. 469.
[22] E. Racker, B. Violand, S. O'Neal, M. Alfonzo, and J. Telford, *Arch. Biochem. Biophys.* **198**, 470 (1979).
[23] L. T. Mimms, G. Zampighi, Y. Nozaki, C. Tanford, and J. A. Reynolds, *Biochemistry* **20**, 833 (1981).
[24] O. Zumbuehl and H. G. Weder, *Biochim. Biophys. Acta* **640**, 252 (1981).
[25] R. A. Schwendener, A. Asanger, and H. G. Weder, *Biochem. Biophys. Res. Commun.* **100**, 1055 (1981).

are suitable for high resolution X-ray analysis. In the first successful crystallizations reported[26,27] octyl glucoside was used as the solubilizing agent. More recently other small detergents also have been found to be effective[28,29] and it appears that different proteins will have different specific requirements for a detergent that can be incorporated into the crystal lattice and yet also maintain a stable homogeneous form of the enzyme.[29] A variety of pure alkyl glycosides may prove useful for this purpose.

Acknowledgment

This work was supported by National Institutes of Health Grant GM 26916 to SFM.

[26] H. Michel and D. Oesterhelt, *Proc. Natl. Acad. Sci. U.S.A.* **77,** 1283 (1980).
[27] R. M. Garavito and J. P. Rosenbusch, *J. Cell Biol.* **86,** 327 (1980).
[28] H. Michel, *J. Mol. Biol.* **158,** 567 (1982).
[29] H. Michel, *Trends Biochem. Sci.* **8,** 56 (1983).
[30] R. C. Mast and L. V. Haynes, *J. Colloid Interface Sci.* **53,** 35 (1975).

[4] Lipid Enrichment and Fusion of Mitochondrial Inner Membranes

By CHARLES R. HACKENBROCK and BRAD CHAZOTTE

Introduction

Membrane engineering, reconstitution, and modification are important approaches in facilitating the analysis of the activities and interactions of catalytic membrane components. It is the purpose of this chapter to present the details of methods developed in our laboratory over the past few years which permit a significant incorporation or bulk enrichment of the mitochondrial inner membrane with a variety of membrane lipids.[1,2] Related techniques which permit the conversion of the typically small mitochondrial inner membranes to ultralarge inner membranes through a fusion process are also presented. Significantly, one method provides osmotically active, ultralarge (>200 μm) individual, spherical inner mem-

[1] H. Schneider, J. J. Lemasters, M. Höchli, and C. R. Hackenbrock, *Proc. Natl. Acad. Sci. U.S.A.* **77,** 442 (1980).
[2] H. Schneider, J. J. Lemasters, M. Höchli, and C. R. Hackenbrock, *J. Biol. Chem.* **255,** 3748 (1980).

branes. The lipid-enrichment procedure gives enriched inner membranes in bulk suspension whereas the fusion procedures are best suited for microanalysis of individual, enlarged inner membranes on microscope slides.

Lipid-enriched inner membranes and fused inner membranes have been used in studies on the nature of the interaction of oxidation–reduction components utilizing routine reaction–kinetic approaches,[2] fluorescence recovery after photobleaching,[3-6] and flash-induced absorption anisotropy.[7] Such studies reveal that the major oxidation–reduction components of the inner membrane diffuse laterally and independently in the membrane plane, as well as rotate on an axis normal to the membrane plane. The ultralarge, osmotically active inner membranes are amenable to microelectrode studies. We have used these lipid-enrichment and fusion methods successfully on other membrane systems, such as microsomes and red blood cell ghosts and believe most membrane systems can be treated similarly.

Lipid Enrichment

Preparation of Inner Membranes

The lipid-enrichment procedures outlined here are routinely carried out using rat liver mitochondrial inner membranes. Liver mitochondria are isolated from male Sprague–Dawley rats using a 300 mosM isolation medium composed of 70 mM sucrose, 220 mM mannitol, 2 mM HEPES [4-(2-hydroxyethyl)-1-piperazineethanesulfonic acid], 0.5 mg bovine serum albumin/ml, with KOH added to pH 7.4.[8] This solution is designated H300 medium. The outer mitochondrial membrane is removed by a controlled incubation using digitonin.[9] The resulting mitoplasts (inner membrane-matrix particles) in H300 medium contain a complex membrane topography which is converted to a simple spherical configuration by washing and resuspending in a 7.5 times diluted, bovine albumin free, H300 medium. The diluted solution (40 mosM) is designated H40 medium.

[3] S. S. Gupte, E.-S. Wu, L. Höchli, M. Höchli, K. A. Jacobson, A. E. Sowers, and C. R. Hackenbrock, *Proc. Natl. Acad. Sci. U.S.A.* **81**, 2606 (1984).

[4] S. S. Gupte, K. Jacobson, L. Höchli, and C. R. Hackenbrock, *Biophys. J.* **41**, 371a (1983).

[5] B. Chazotte, E.-S. Wu, and C. R. Hackenbrock, *Biochem. Soc. Trans.* **12**, 463 (1984).

[6] B. Chazotte and C. R. Hackenbrock, *Biophys. J.* **47**, 197a (1985).

[7] S. Kawato, C. Lehner, M. Müller, and R. J. Cherry, *J. Biol. Chem.* **257**, 6470 (1982).

[8] J. W. Greenawalt, this series, Vol. 31, p. 310.

[9] C. Schnaitman and J. W. Greenwalt, *J. Cell Biol.* **38**, 158 (1968).

Preparation of Liposomes

Small unilamellar vesicles (SUV) can be prepared using soybean phospholipid (asolectin); asolectin enriched with specific lipoidal membrane components, such as cholesterol or ubiquinone; asolectin enriched with specific phospholipids, mixtures of specific phospholipids; and mixtures of specific phospholipids containing any number of lipoidal or lipid soluble probes and other agents. The phospholipids used should contain some negatively charged headgroups. SUV should be prepared just before use for membrane enrichment.

The most commonly used SUV are prepared from asolectin, which contains negatively charged phospholipids.[1,2] Dry asolectin (stored at −20° under nitrogen), 0.5 to 2.5 g (typically 1 g) is suspended in 7.5 ml of H40 medium, pH 7.4, hydrated a minimum of 1 hr, and sonicated at 0° using the microtip probe of a Branson sonifier (model W185) at 42 W for three 10–15 min durations. After each of the three sonications, the pH of the mixture is readjusted to 7.4 with 0.2 N KOH. Sonication in the presence or absence of nitrogen gives the same results.

For SUV of asolectin enriched with various specific lipids the following have proved to be reliable. Asolectin–cardiolipin SUV are prepared by adding 900 mg of dry asolectin to 100 mg of cardiolipin in 71 ml of ethanol and vacuum evaporated to near dryness in a Büchler rotary evaporator at 30°. The dried mixture is redissolved in 10 ml of chloroform and evaporated. SUV containing different quantities of cardiolipin are prepared by varying the ratio of dry asolectin to cardiolipin. The dry asolectin–cardiolipin mixture is suspended in 7.5 ml of H40 medium and sonicated as outlined above for asolectin SUV. Sonication is always carried out at a temperature to ensure the liquid crystalline state of the lipid mixture.

For SUV composed of asolectin–cholesterol,[10] 2.5 g of dry asolectin containing 26 to 240 mg of cholesterol/g of asolectin is dissolved in 8 ml of chloroform and evaporated to dryness under high vacuum for 2 hr at 30° in a Büchler rotary evaporator. Of the dry asolectin–cholesterol mixture, 1.5 g is added to 7.5 ml of H40 medium, hydrated for 1 hr and sonicated at 0° as above for asolectin SUV.

For ubiquinone-containing SUV,[11] 2.5 g dry asolectin containing 3.6 mg of ubiquinone–10/g of asolectin or 2.6 mg of ubiquinone–6/g of asolectin is dissolved in 7 ml of chloroform and evaporated to dryness as above. To vary the content of ubiquinone in the SUV, different ratios of ubiquinone to asolectin are used in the mixture. After drying, 1.5 g of the

[10] H. Schneider, M. Höchli, and C. R. Hackenbrock, *J. Cell Biol.* **94**, 387 (1982).
[11] H. Schneider, J. J. Lemasters, and C. R. Hackenbrock, *J. Biol. Chem.* **257**, 10789 (1982).

asolectin–ubiquinone mixture is dissolved in 7.5 ml of H40 medium and sonicated as outlined above for asolectin SUV.

Induction of Lipid Enrichment

Spherical inner membranes are enriched with SUV–lipid mixtures by a low pH method.[1] Of the stock inner membranes in H40 medium, 7.5 ml (100 mg protein) is placed into a 50-ml glass beaker, kept at 30°, and stirred continuously on a magnetic stirrer. To this, 2.6 ml of freshly sonicated SUV is added. The pH is monitored continuously, is adjusted dropwise to 6.35–6.5 with 0.01 N HCl immediately following the addition of SUV, and is maintained at the selected pH dropwise with 0.01 N HCl. Lower pH values tend to increase the rate of enrichment but also increase the occurrence of irreversible aggregation of the inner membranes. Higher pHs tend to slow the enrichment process with no significant enrichment occurring at pH 7.0. After 15 min a second aliquot of 2.6 ml of SUV is added and pH adjusted dropwise again. A third and last addition of 2.6 ml of SUV is added after another 15 min and the pH adjusted. After a total incubation time of 45 min or more, the pH is adjusted dropwise to 7.4 with 0.01 N KOH and the inner membrane-SUV mixture is placed on ice.

Purification of Membrane Populations

After the lipid enrichment of inner membranes, the resulting membranes can be separated by discontinuous sucrose density gradient centrifugation into populations, depending on the degree of their lipid incorporation. Routinely the gradients selected are 0.6, 0.75, 1.0, and 1.25 M sucrose in H40 medium containing 2 mM HEPES, having densities at 2° of 1.0803, 1.0987, 1.1322, and 1.1674 g/cm^3, respectively. The density of nonenriched inner membranes is 1.188. The enriched membrane–SUV mixture is loaded in 4.5-ml aliquots (16 to 18 mg protein) onto 33 ml of the ice cold sucrose gradient in 38.5 ml centrifuge tubes. The tubes are then centrifuged at 70,000 g (at r_{max}) in a swinging bucket rotor for 14 to 16 hr at 4°. Four relatively sharp density distinct inner membrane populations sediment into the four density gradients according to the degree of their lipid enrichment; these populations have been designated Band 1, Band 2, Band 3, and pellet membranes, from least to most dense, respectively. Residual SUV remain above the gradient. The percentage yield of membrane protein per membrane band varies with the SUV–membrane incubation time at low pH. After 45 min incubation, the distribution of membrane protein in the four bands is 6–9, 9–11, 15–20, and 60–70% for Bands 1, 2, 3, and pellet, respectively. Longer incubations and negatively charged phospholipids shift the yields in all bands progressively toward

COMPOSITIONAL ANALYSIS OF LIPID-ENRICHED MITOCHONDRIAL
INNER MEMBRANE

Fraction	μmol lipid phosphorus/ mg protein	nmol heme a/ mg protein	mol lipid phosphorus/ mol heme a ($\times 10^3$)	Increase in lipid bilayer surface area (%)
Control	0.24 ± 0.01	0.27 ± 0.01	0.9	—
Pellet	0.61 ± 0.05	0.51 ± 0.02	1.2	30
Band 3	0.90 ± 0.09	0.57 ± 0.03	1.6	80
Band 2	1.65 ± 0.13	0.54 ± 0.03	3.1	240
Band 1	3.76 ± 0.29	0.52 ± 0.02	7.2	700

Band 1. After the upper layer of SUV is removed, each band or fraction is removed by gentle aspiration from the gradient tubes kept on ice. Each band is then washed with H40 medium and centrifuged for 40 min at 11,000 g twice. The pellets are resuspended to the desired volume.

Assessment of Lipid Enrichment

Assessment of the lipid enrichment, i.e., the amount of exogenous lipid incorporated into the bilayer of the inner membrane, can be carried out by compositional, ultrastructural, and functional analysis.

Compositional Analysis. Lipid phosphorus, total membrane protein, and hemes a, b, c_1, and c are quantitated. Lipid phosphorus can be determined by the Bartlett method[12] after acid hydrolysis with perchloric acid, total protein by the Lowry method,[13] and hemes by the simultaneous method of Williams[14] using the coefficients of Schneider *et al.*[2] Alternatively, heme a can be determined by measuring the ΔA (605–630) after dithionite reduction using an extinction coefficient of 13.1 mM^{-1} cm^{-1}.[15] An increase in the lipid phosphorus/heme a ratio from pellet to Band 1 indicates that exogenous lipid is associated with the inner membranes, while a constant heme a/protein ratio in all four membrane bands indicates that integral membrane proteins are not lost during the lipid-enrichment process. Cytochrome c, a peripheral membrane protein, is lost during the lipid-enrichment process and should be added for functional assays. The degree of lipid enrichment during a 45 min incubation and calculated increase in bilayer surface area is shown in the table. When

[12] G. R. Bartlett, *J. Biol. Chem.* **234**, 466 (1959).
[13] O. H. Lowry, N. J. Rosebrough, A. L. Farr, and R. J. Randall, *J. Biol. Chem.* **193**, 265 (1951).
[14] J. N. Williams, *Arch. Biochem. Biophys.* **107**, 537 (1964).
[15] W. H. Vanneste, *Biochim. Biophys. Acta* **113**, 175 (1966).

enriching with cholesterol, quantification may be determined colorimetrically[16] after membrane extraction.[17] Ubiquinone enrichment may be assessed after membrane extraction by methanol/ethanol.[18]

Ultrastructural Analysis. Although compositional analysis will reveal that exogenous lipid or SUV associate with the inner membrane, freeze fracture electron microscopy is capable of showing that the associated lipid has been incorporated into the inner membrane bilayer. Unfixed lipid enriched inner membranes, as well as nonenriched, control inner membranes, are mixed with 30% glycerol in H40 medium, rapidly frozen in liquid Freon 22, and freeze fractured by routine methods.[19] Freeze fractured, lipid-enriched, spherical inner membranes show an average increase in membrane surface area. More significantly, the enriched membranes show a progressive decrease from pellet to Band 1, in the density distribution of integral proteins (i.e., intramembrane particles per membrane unit surface area), in the membrane plane (Fig. 1). Thus incorporation of exogenous lipid into the inner membrane bilayer results in an average increase in the distance between integral proteins proportional to the degree of lipid enrichment. The integral proteins maintain a random distribution as they are diluted in the membrane plane by the newly incorporated lipid, and their relative distribution in each half of the inner membrane bilayer remains unchanged (Fig. 2). When using cholesterol enrichment, it is important to note that 20 mol% or more of cholesterol in the inner membrane results in clustering and aggregation of integral proteins.[10]

Functional Analysis. Since electron transfer appears to be coupled to the lateral diffusion rate and collision frequency bewteen the major oxidation–reduction (redox) components in the inner membrane,[1] it follows that the greater the lipid enrichment and two-dimensional dilution of the redox components, the slower the electron transfer rate.[2] Electron transfer rates between specific redox components are determined by routine assays in the lipid-enriched inner membranes for cytochrome oxidase, succinate oxidase, NADH oxidase, succinate dehydrogenase, NADH dehydrogenase, ubiquinol-cytochrome c reductase, duroquinol oxidase, NADH ubiquinone reductase, and succinate ubiquinone reductase. It is to be noted that lipid enrichment of the inner membrane usually results in a permeability for NADH and loss in ATP synthesis[1] as well as a loss in osmotic sensitivity.

[16] A. J. Couchrine, W. H. Miller, and D. B. Stein, Jr., *Clin. Chem.* **5**, 609 (1959).
[17] M. Kates, *in* "Laboratory Techniques in Biochemistry and Molecular Biology" (T. S. Work and E. Work, eds.), Vol. 3, Pt. II, p. 269. North Holland Publ., Amsterdam, 1972.
[18] A. Kröger and M. Klingenberg, *Eur. J. Biochem.* **34**, 358 (1966).
[19] K. Fisher and D. Branton, this series, Vol. 32, p. 35.

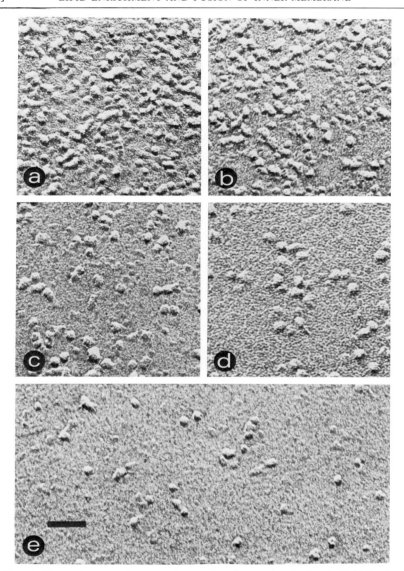

FIG. 1. Freeze fracture electron micrographs of control and phospholipid (asolectin)-enriched mitochondrial inner membranes at ×250,000. Bar = 0.04 μm. (a) Control, non-enriched; (b) pellet, 30% phospholipid enriched; (c) Band 3, 80% phospholipid enriched; (d) Band 2, 240% phospholipid enriched; (e) Band 1, 700% phospholipid enriched. Two-dimensional density of integral proteins decreases proportional to phospholipid enrichments.

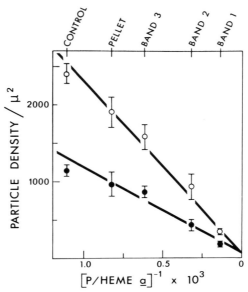

FIG. 2. Intramembrane particle (integral protein) density distribution on the convex (○) and concave (●) freeze fracture faces of phospholipid-enriched inner membranes as a function of phospholipid to heme *a* mole ratios. The higher density of integral proteins found on the convex compared to concave fracture face of the control inner membrane is maintained with enrichment, and the two-dimensional density of integral proteins decreases proportional to phospholipid enrichment.

Decreases in electron transfer between specific redox components that occur after lipid enrichment relate to the degree of enrichment (Fig. 3). Redox reactions not dependent on lateral diffusion and collision between redox components will not show decreases in reaction rates such as cytochrome oxidase when assayed at low ionic strength and NADH dehydrogenase.

Membrane Fusion

Preparation of Inner Membranes

The fusion of inner membranes is carried out using spherical inner membranes in H40 medium prepared from rat liver mitochondria as previously outlined.

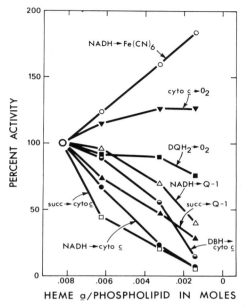

Fig. 3. Normalized electron transfer activities for phospholipid-enriched inner membranes as a function of heme *a* to phospholipid mole ratios. Electron transfer rates between redox partners requiring collisions decrease proportional to phospholipid enrichment. NADH hydrogenase [NADH → Fe(CN)$_6$]; cytochrome *c* oxidase (cyto *c* → O$_2$); duroquinol oxidase (DQH$_2$ → O$_2$); NADH-ubiquitone reductase (NADH → Q-1); succinate-ubiquinone reductase (succ → Q-1); ubiquinol-cytochrome *c* reductase (DBH → cyto *c*); succinate-cytochrome *c* reductase (succ → cyto *c*).

Induction of Inner Membrane Fusion

The fusion of spherical inner membranes to one another to obtain ultralarge, i.e., in excess of 10 μm, diameter spherical inner membranes is carried out on control spherical inner membranes attached to a glass slide. A glass coverslip (18 mm, #1), is pressed onto filter paper wetted with 0.5% gelatin solution, dried, then mounted on a glass microscope slide with double stick tape as shims between the two ends of the coverslip and the slide. The resulting chamber between the coverslip and glass slide is filled with several drops (~40 μl) of spherical inner membranes in H40 medium (12 to 20 mg protein/ml). Inner membranes which are not attached are washed out from the chamber by pulling dropwise 0.1 ml of 10 m*M* KP$_i$, pH 7.4 through the chamber with filter paper. The remaining, attached inner membranes are induced to fuse to one another by pulling into the chamber 10 mM Ca^{2+} in 5 mM KP$_i$, pH 6.5 and incubating for 5 to 10 min at 30° on a thermostatically controlled heating block. Fusion is stopped by pulling 0.1 ml, 10 mM KP$_i$, pH 7.4 followed by 0.1 ml, 10 mM EDTA in 10 mM KP$_i$, pH 7.4 through the chamber. This procedure will result in glass attached spherical inner membranes with diameters larger than 10 μm and many with diameters of 50 μm and more. Generally, larger inner membranes are produced by increasing the concentration of control inner membranes added initially. Generally this method produces enlarged but not osmotically active inner membranes.

Fusion of Lipid-Enriched Inner Membranes

A similar method to that outlined above for fusing control spherical inner membranes can be used to fuse inner membranes which have been previously enriched with lipid. Lipid enrichment and purification of density distinct inner membranes are carried out as previously described. To induce fusion of lipid enriched inner membranes, volume for volume microliter quantities of the membranes in H40 medium and 20 mM CaCl$_2$ in 5 mM HEPES buffered to pH 6.5 with KOH, are mixed thoroughly, then a 40 μl sample is transferred by micropipette to the surface of a microscope slide. For dilute stock solutions of enriched inner membranes, e.g., Band 1, the volume ratios of membranes to calcium solution may be varied, provided the final CaCl$_2$ concentration remains at 10 mM. Two small strips of weighing paper are placed along side the membrane sample and act as a support for a 22-mm-square cover glass which is placed on top of the sample. The two edges of the coverslip by the weighing paper are sealed with paraffin with the other edges sealed partially. The sample is incubated for 15 min at 37° in a humidified environment on a thermostatically controlled heating block, then cooled. Washing out the calcium solution is carried out as gently as possible with a desired osmotically balanced solution through the partially sealed edges of the glass chamber as outlined above. The majority of fused membranes range from 20 to 50 μm with many in excess of 200 μm. Most of the ultralarge fused membranes are attached to aggregated, nonfused membranes which are attached to the glass surfaces. This method can also be used for control, nonlipid-enriched spherical inner membranes and whether enriched or not, produces osmotically active ultralarge inner membranes.

Assessment of Membrane Fusion

The occurrence and degree of inner membrane fusion are routinely determined by phase contrast microscopy preferably with a dry objective

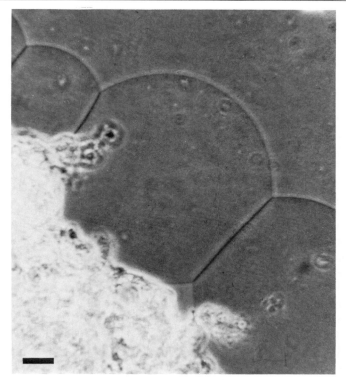

FIG. 4. Phase contrast image of calcium fused, 30% phospholipid (asolectin)-enriched mitochondrial inner membranes. The fusion produces aggregation of the small enriched inner membranes seen in the lower left and the fusion of such small membranes into ultra-large osmotically active, intact inner membranes (center). Bar = 5 μm.

lens (Fig. 4). Freeze fracture of fused, lipid-enriched membranes reveals that the density of integral proteins in these inner membranes corresponds to the protein density distribution in nonfused lipid-enriched membranes.

Acknowledgments

The authors express their gratitude to Drs. Heinz Schneider, Matthius Höchli, En-Shinn Wu, John J. Lemasters, and Sharmila S. Gupte for contributions to the procedures described herein. These procedures were developed in research funded under Grants NIH GM 28704 and NSF PCM79-10968 and PCM84-02569 to CRH.

[5] Changes of Surface and Membrane Potentials in Biomembranes

By Naoki Kamo and Yonosuke Kobatake

According to the chemiosmotic theory developed by Mitchell,[1,2] the difference in the electrochemical potential of proton across the membrane is an intermediate for ATP synthesis and is the motive force for active transport. The difference expressed in the unit of volt is called protonmotive force. It is composed of two terms, the membrane potential and pH difference. Measurement of these quantities is important for detailed analysis in the process concerned with oxidative- or photophosphorylation and active transport of the cells. The changes in these quantities are not limited to energy-converting membranes, and many articles have been published regarding membrane functions associated closely with the change in those quantities.[3]

The biological membrane is usually negatively charged at neutral pH. In the electrolyte solution which is contiguous to the charged membrane, an electric double layer forms, where an electric field extends from the membrane surface to the bulk solution.[4,5] The surface potential is defined as the electrical potential at the membrane surface with reference to the potential in the bulk aqueous phase. Details about the potential profile near the membrane surface and the exact definition of surface potential can be found elsewhere.[5–7] It is reported that the changes in the surface potential are associated with many functions of biomembranes.[7–12] In the present chapter, we describe the methods used to measure the membrane potential and the surface potential in biological membranes.

[1] P. Mitchell, *Biol. Rev.* **41**, 445 (1966).

[2] P. Mitchell, *Trans. Biochem. Soc.* **4**, 399 (1976).

[3] D. Epel, *Curr. Top. Dev. Biol.* **12**, 185 (1978).

[4] H. A. Abramson, L. S. Moyer, and M. H. Gorin, "Electrophoresis of Proteins and the Chemistry of Cell Surface." Reinhold, New York, 1942.

[5] J. T. Davies and E. K. Rideal, "Interfacial Phenomena." Academic Press, New York, 1963.

[6] S. Mclaughlin, *Curr. Top. Membr. Transp.* **9**, 71 (1977).

[7] D. L. Gilbert, *in* "Biophysics and Physiology of Excitable Membrane" (W. J. Adelman, Jr., ed.). Van Nostrand-Reinhold, New York, 1971.

[8] G. F. W. Searle and J. Barber, *Biochim. Biophys. Acta* **545**, 508 (1979).

[9] L. Wojtczack and M. J. Nalecz, *Eur. J. Biochem.* **94**, 99 (1979).

[10] S. Itoh, *Biochim. Biophys. Acta* **591**, 346 (1980).

[11] W. K. Chandler, A. L. Hodgkin, and H. Meves, *J. Physiol.* (*London*) **180**, 821 (1965).

[12] K. Kurihara, N. Kamo, and Y. Kobatake, *Adv. Biophys.* **10**, 27 (1978).

METHODS IN ENZYMOLOGY, VOL. 125

Measurement of the Membrane Potential of Small Cells or Vesicles

For measuring the membrane potential of small cells or vesicles which are too small for microelectrode techniques, we should employ probe methods. Two methods have been developed so far: (1) potential-dependent fluorescent dye[13] such as cyanine or oxonol and (2) distribution of lipophilic ions[14] between internal space of cells and external medium.

Since many articles describing these methods have already been published,[15–19] we will describe the ambiguity in method (2) and attempts to remove it. The basic principle of the method can be described as follows. Since lipophilic ions permeate freely through the membrane, the probe ions distribute between intra- and extracellular spaces of cells according to the Nernst equation:

$$\Delta\varphi = (RT/zF)\,\ln(C_{out}/C_{in}) \tag{1}$$

where $\Delta\varphi$ is the membrane potential with respect to the outside of cells, C_{out} is the concentration of the probe ion in the external solution, C_{in} is the concentration of the probe ion inside the cell, and z is the valency of the probe. R, T, and F have their usual thermodynamic significances. Here, the activity coefficients of probe ions are assumed to be unity. Typical lipophilic ions are valinomycin + K^+ (Rb^+), triphenylmethylphosphonium ($TPMP^+$), tetraphenylphosphonium (TPP^+), tetraphenylboron (TPB^-), and SCN^-. Several methods for the determination of C_{in} and C_{out} have been proposed: (1) filtration method,[20,21] (2) flow-dialysis,[18] and (3) lipophilic ion-selective electrode.[22] The first and second methods employ radioactive isotope-labeled probes while the third does not. For the second and third methods, C_{in} is calculated from the decrease of the probe concentration in the external medium, C_{out} with use of mass conservation of the probe.

When probe molecules bind to the membrane and/or intracellular constituents, all these methods lead to an erroneous estimation of the mem-

[13] P. J. Sims, A. S. Waggoner, C. H. Wang, and J. F. Hoffman, *Biochemistry* **13**, 3315 (1974).

[14] E. A. Liberman and V. P. Skulachev, *Biochim. Biophys. Acta* **216**, 30 (1970).

[15] H. Rottenberg, this series, Vol. 55, 547.

[16] C. L. Bashford and J. C. Smith, this series, Vol. 55, 569.

[17] V. P. Skulachev, this series, Vol. 55, 586.

[18] S. Ramos, S. Schuldiner, and H. R. Kaback, this series, Vol. 55, 680.

[19] A. S. Waggoner, this series, Vol. 55, 689.

[20] H. Hirata, K. Altendorf, and F. M. Harold, *Proc. Natl. Acad. Sci. U.S.A.* **70**, 1804 (1973).

[21] P. C. Maloney, E. R. Kashket, and T. H. Wilson, *Methods Membr. Biol.* **5**, 1 (1975).

[22] N. Kamo, M. Muratsugu, R. Hongoh, and Y. Kobatake, *J. Membr. Biol.* **49**, 105 (1979).

brane potential.[23–27] We will consider this problem at some length. For the case where probe binds or adsorbs only to the intracellular constituents, the problem is relatively easy. If the amounts of binding is proportional to C_{in} with the proportionality constant k, the total amount (bound plus free) of probe inside the cell is represented by $(k + v)C_{in}$, where v is the intracellular volume. Then, the apparent intracellular concentration is given by $[(k/v) + 1]C_{in}$. This indicates that the overestimation caused by the binding may be constant irrespective of the membrane potential. For intact cells, this assumption seems applicable because cells contain a lot of intracellular hydrophobic proteins to which probes bind much more strongly than to the membranes. However, invalidity of this assumption for intact cells of *Halobacterium halobium* was demonstrated,[28] implying that the correction for binding to cytoplasmic membrane is necessary to estimate the true membrane potential of cells.

To bypass the complexity caused by the binding of the probe to intracellular constituents, we used envelope vesicles of *H. halobium,* where the binding of the probe is confined only to the membrane. The membrane of *H. halobium* contains a light-driven ion pump and illumination generates the interior negative membrane potential.[29] The phosphonium probes used were a homologous series of $(Phe)_3\text{-}P^+\text{-}(CH_2)_n\text{-}CH_3$ and TPP^+. The abbreviations used for probes are given in the legend to Fig. 2. The uptake of phosphonium probes was measured with use of a selective electrode[22,29] (Fig. 1).

Figure 2 shows the amount of binding of various phosphonium probes to the membrane in the dark, where no membrane potential is generated. These are basic data for the evaluation of the membrane potential and binding in the presence of membrane potential. The amount of binding increases with the increase in hydrophobicity of the probes. Binding is not proportional to the concentration of the free probe but approaches a plateau level with an increase in free probe concentration. This suggests that the saturating binding model such as the Langmuir adsorption isotherm is applicable.[30]

Upon actinic illumination, an interior negative membrane potential,

[23] A. Zaritsky, M. Kihara, and R. H. Macnab, *J. Membr. Biol.* **63**, 215 (1981).
[24] R. Casadio, G. Venuroli, and B. A. Melandri, *Photobiochem. Photobiophys.* **2**, 245 (1982).
[25] E. P. Bakker, *Biochim. Biophys. Acta* **681**, 474 (1982).
[26] J. S. Lolkema, K. J. Hellingwerf, and W. N. Konnings, *Biochim. Biophys. Acta* **681**, 85 (1982).
[27] J. S. Lolkema, A. Abbing, K. J. Hellingwerf, and W. N. Konnings, *Eur. J. Biochem.* **130**, 287 (1983).
[28] M. Demura, N. Kamo, and Y. Kobatake, *Biochim. Biophys. Acta* **820**, 207 (1985).
[29] N. Kamo, T. Racanelli, and L. Packer, this series, Vol. 88, 356.
[30] M. Demura, N. Kamo, and Y. Kobatake, *Biochim. Biophys. Acta* **812**, 377 (1985).

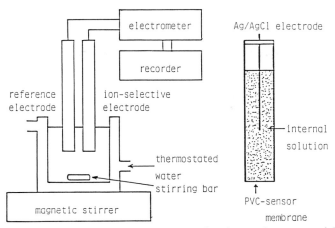

FIG. 1. Schematic illustration of setup for measuring the membrane potential with a lipophilic cation-sensitive electrode. The sensor membrane is made of polyvinyl chloride film containing tetraphenyl borate as an ion exchanger. The film was prepared as by Kamo *et al.*[29] and Demura *et al.*[30] The diameter of the electrode is 5 mm and the volume of the cuvette is 1.0 ml. The reference electrode is a calomel electrode with a salt bridge. The potential difference between these two electrodes is measured with a high imput-impedance electrometer (Takeda Riken TR8651 or laboratory-made amplifier using FET operational amplifiers). The response of the electrode follows the Nernst equation until the concentration of lipophilic cations decreases to 10^{-6} or $5 \times 10^{-7} M$ (even in $4 M$ NaCl). A chart-stripping recorder is connected to the electrometer to record a trace of the potential change. Data are also able to be stored in a microcomputer with a A/D converter. When necessary, the medium is aerated with water-saturating air.

$\Delta\varphi$ is generated, which causes the uptake of phosphonium probes. The discussion is focused on the amounts of binding in the presence of $\Delta\varphi$, $U_b(\Delta\varphi)$. It is now accepted that the potential energy of hydrophobic ions has deep minimum near the membrane–solution interfaces of lipid bilayer membranes. The phosphonium ions within the membrane are mainly located at these binding sites. The binding model according to this concept has been adopted by several authors,[24,26,31] especially by Cafiso and Hubbel.[32] Then, taking the facts that adsorption follows the Langmuir equation into consideration as described above, we can formulate $U_b(\Delta\varphi)$ as follows:

$$U_b(\Delta\varphi) = \frac{A_{out}C_{out}}{C_{out} + K} + \frac{A_{in}C_{in}}{C_{in} + K}$$

$$= \frac{fAC_{out}}{C_{out} + K} + \frac{(1 - f)AC_{out}}{C_{out}\exp(-F\Delta\varphi/RT) + K]} \tag{2}$$

[31] M. Muratsugu, N. Kamo, Y. Kobatake, and K. Kimura, *Bioelectrochem. Bioenerg.* **6,** 477 (1979).

[32] S. J. Ferguson, W. J. Lloyd, and G. K. Radda, *Biochim. Biophys. Acta* **243,** 174 (1976).

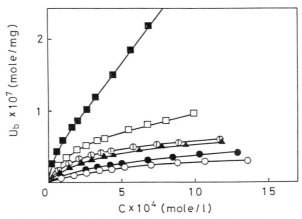

FIG. 2. Plot of amounts of binding to membrane, U_b, against the free concentration of the probe, C.[30] The experiments were performed under the condition that $\Delta\varphi = 0$. The medium was 4 M NaCl buffered with 10 mM HEPES [4-(2-hydroxyethyl)-1-piperazineethanesulfonic acid] at pH 7.0. (○) Triphenylmethylphosphonium (TPMP$^+$); (●) triphenylethylphosphonium (TPEP$^+$); (▲) triphenylbutylphosphonium (TPBP$^+$); (◑) tetraphenylphosphonium (TPP$^+$); (□) triphenylamylphosphonium (TPAP$^+$); (■) triphenylhexylphosphonium.

where A_{out} and A_{in} stand for the maximum amount of binding at outer and inner binding sites. A and f are defined as $A = A_{out} + A_{in}$ and $f = A_{out}/A$. K is the dissociation constant, and we assume that the values of K at inner and outer binding sites are equal. We found that K may depend slightly on the membrane potential in *H. halobium* vesicles, and neglecting K did not affect the estimated values significantly (M. Demura *et al.*, unpublished). Cafiso and Hubbel[31] also showed that the K value of a spin-labeled phosphonium ion to the liposome is independent of $\Delta\varphi$. For the sake of simplicity, we assume that K is independent of $\Delta\varphi$.

In the dark where $\Delta\varphi = 0$, the amount of binding, U_b is expressed by the Langmuir equation as described above and values of A and K are determined as usual.

The total amounts of uptake by vesicle, U, is the sum of internal free and membrane-bound probes. Hence we obtain

$$U = vC_{in} + U_b(\Delta\varphi)$$
$$= v_{in}C_{out} \exp(-F\Delta\varphi/RT) + U_b(\Delta\varphi) \tag{3}$$

where $U_b(\Delta\varphi)$ is given in Eq. (2). In this equation, the values of U, v, C_{out}, A, and K are obtained from experiments. It is not easy to solve Eqs. (2) and (3) to obtain the explicit expression for $\Delta\varphi$. Hence we employed an iterative procedure with the aid of a microcomputer. First, we estimate

$\Delta\varphi$ without consideration of the binding. Then $U_b(\Delta\varphi)$ is calculated from Eq. (2) with this first estimate of $\Delta\varphi$, and C_{in} is calculated according to $[U - U_b(\Delta\varphi)]/v$ to give $\Delta\varphi$, which is taken as the second estimate. Again, $U_b(\Delta\varphi)$ is estimated with the second estimate and the third estimate is obtained as described above. This procedure is repeated until two successive estimates agree with each other with the desired accuracy.

Unfortunately, f is not known. Therefore, we varied f from 0 to 1.0, using intervals of 0.05, and calculated $\Delta\varphi$ by the iterative method described above. The criterion for collect correction of binding is that the estimated values of $\Delta\varphi$ be equal irrespective of the probe used. We found that $f = 0.85$ fulfills this criterion. The results obtained are shown in the table, illustrating that any probe gives almost the same estimated value for the same experimental conditions.

The problems to be solved are (1) development of the method of determination of f. The present method requires the use of at least two kinds of probes because $\Delta\varphi$ and f are unknown. If f can be determined from another experiment, such as kinetics of uptake, we can evaluate $\Delta\varphi$ with only one probe. (2) Does the Langmuir equation hold for every membrane? Adsorption of lipophilic ions changes the surface charge density of the membrane which may alter the binding constant. When, as in the case of the *H. halobium* envelope vesicle, the medium is 4 *M* NaCl, then this effect may be small due to high ionic strength. We should find the binding equation incorporating this effect. (3) Usually, salt composi-

MEMBRANE POTENTIAL AFTER BINDING CORRECTION ESTIMATED WITH VARIOUS PROBES[a]

Light intensity (%)	TPMP+	TPEP+	TPBP+	TPP+	TPAP+
100.0	−107.2	−103.8	−108.6	−103.8	−118.9
81.7	−101.9	−98.5	−103.2	nd[b]	nd
64.3	−94.1	−91.9	−97.0	−95.6	nd
53.4	−90.1	−87.1	−89.7	nd	nd
41.5	−83.4	−81.7	−85.4	−84.9	−76.9
31.4	−76.9	−73.5	−75.1	−79.2	nd
17.9	−59.4	−56.6	−57.6	−62.7	−51.0
13.9	−50.3	−51.8	−52.2	nd	nd
7.3	−30.4	−35.1	−36.6	−44.5	−33.1

[a] KH-10 vesicles (1.8 mg/ml) were suspended in 4 *M* NaCl buffered with 10 m*M* Hepes/NaOH. Membrane potential is defined with reference to outside (mV). Temperature was 30°; 100% of the actinic light intensity was 1120 W/m².

[b] nd, not determined.

tion inside and outside cells is different, and the respective K values may be different from each other. We should incorporate this effect into the calculation of $U_b(\Delta\varphi)$.

Measurement of the Surface Potential

Surface potential can be estimated[5] either from (1) partition of charged probes between the membrane surface and the bulk solution or from (2) the electrophoretic mobilities of cells or vesicles.

Partition of Charged Probes. 8-Anilino-1-naphthalene sulfonate (Ans) and its derivatives are the typical fluorescent probes for this purpose.[22,33] This dye has relatively low quantum yield in water and the quantum yield increases in hydrophobic region. Thus, in a membrane suspension, the fluorescence intensity of this dye stems from the dye bound on the membrane. Merocyanine is also used for this purpose.[34]

For varying concentrations of dye ($<100\ \mu M$), fluorescence intensity, f, is measured. (The calculation of f is referred to later.) Analysis in accordance with the Scatchard or double reciprocal plot gives the binding (or dissociation) constant. Binding is governed by electric and nonelectric interactions with membrane and then the binding constant of the dye, K is expressed by

$$K = K_0 \exp(-zF\psi/RT)$$

where K_0, ψ, and z are intrinsic binding constant (nonelectrical interaction), the surface potential with respect to the bulk solution, and the valency of the probe (for Ans, $z = -1$), respectively. Under the assumption that K_0 and maximum amount of binding are constant, the fluorescence change is caused by the change in the binding constant, K, and then the fluorescence monitors the surface potential change. To evaluate the absolute value of the surface potential, however, we should estimate K_0 separately. K_0 is estimated by (1) binding constant obtained in the solution of high ionic strength where $\psi = 0$[35] or by (2) use of structural analogs without a charged group (N-phenyl-1-naphthylamine for Ans[36]). The former assumes the invariance of K_0 by ionic composition of the milieu, which may not be the case. The ambiguity of the latter method is whether the binding site of Ans and the analog is the same.

[33] T. Aiuchi, N. Kamo, K. Kurihara, and Y. Kobatake, *Biochemistry* **16**, 1626 (1977).
[34] T. Aiuchi and Y. Kobatake, *J. Membr. Biol.* **45**, 233 (1979).
[35] D. E. Robertson and H. Rottenberg, *J. Biol. Chem.* **258**, 11039 (1983).
[36] H. Tanabe, N. Kamo, and Y. Kobatake, *Biochim. Biophys. Acta* **805**, 345 (1984).

The dye is available commercially. Fluorescence was measured with an ordinary fluorescence spectrophotometer, equipped with a thermostated water jacket and a stirring apparatus. The latter apparatus is necessary for large cells which sink to the bottom of the cuvette during measurements. Excitation and emission wavelengths are 373 and 470 nm, respectively. The following procedure is an example for measurement of Ans fluorescence change in intact mitochondria during energization.[33] (1) Rotenone, inhibitor of oxidation of endogenous substrates (final concentration, 1 μM), was added to mitochondrial suspension (0.3 mg of protein/ml) in the buffer solution; (2) 3 ml of the suspension was pipetted into a cuvette and two identical samples were prepared. A given volume of concentrated Ans was added to one cuvette and an equal volume of the buffer solution was added to another cuvette. Fluorescences of both cuvettes were measured. (3) ATP or succinate was added for energization to measure the intensity. (4) In order to determine the free concentration of Ans, the mitochondrial suspension prepared separately which are the same as those of fluorescence measurements was centrifuged in a microfuge and an aliquot of supernatant was separated. The optical density at 350 nm was measured to determine the free concentration of Ans. (5) The fluorescence intensities (see later) obtained under varying concentration of free Ans were analyzed by means of the Scatchard or double reciprocal plot.

We should always take into consideration that the emission intensity may come from the dye and other substances which are contained in the sample, and that the intensity is influenced by the turbidity of the sample.[37] For mitochondria, such a substance is NADH and it is well known that mitochondria swell or shrink, depending on the energy state. Thus, we should measure the fluorescence intensity with and without the dye. The observed fluorescence intensity depends on the instrumental setup, since it is affected by factors such as the magnitude of amplification and Xe lamp intensity. In order to compare the intensity of emission obtained under various conditions, the fluorescence intensity of quinine sulfate (2 mg/liter in 0.1 N H_2SO_4) was used as a reference. The data are calculated from Eq. (4)

$$f = (f_a - f_m)/f_s \tag{4}$$

where f_a, f_m, and f_s stand for the fluorescence intensities of the mitochondria–Ans suspension, the mitochondrial suspension, and quinine solution, respectively.

[37] A. Azzi, this series, Vol. 32, 240.

The spin-labeled quaternary amine (abbreviated CAT) has been introduced to measure the surface potential.[38] (The probe is available from Molecular Probe, Inc.) Two populations of spin probes, i.e., bound and free states, are identified from the difference of resonance magnetic field or from the decrease of peak height caused by the binding. The partition is also governed by Eq. (4). This method has been applied to purple membrane,[39] mitochondria,[40] chloroplast,[41] and leukocytes.[42]

Jasaitis et al.[43] proposed that Ans fluorescence is interpreted as an indication of the membrane potential change. This notion is based on the assumption that Ans is permeating the membrane as lipophilic ion. In other words, the binding of Ans depends on the membrane potential as described in the section of lipophilic membrane potential probes. A report, however, has been published, stating the lack of correlation between Ans fluorescence and the membrane potential determined by impaling a microelectrode.[44] The spin-labeled probe, CAT, is also inferred to be permeable through mitochondrial membranes[45] and through H. halobium membrane vesicles.[28] But, the rate of transport is slow, at least, in the case of H. halobium vesicles. Kinetic measurement may resolve the binding process and transport through membrane. The binding process is governed by the surface potential.

If we have some means for detecting free or bound nonpermeating ions, these ions may be used for a probe for surface potential. In fact, polarography was used for detection for the partitioning of some dyes and the surface potential change was measured.[46] The phosphorescence of terbium ion was introduced.[47]

It is noted that for the interpretation of data obtained with use of charged probes, we should pay attention that binding of charged probes modulates the surface charge, hence the surface potential, especially for the membrane whose surface charge density is small.

[38] J. D. Castle and W. L. Hubbell, Biochemistry 15, 4818 (1976).
[39] C. Carmeli, A. T. Quintahilha, and L. Packer, Proc. Natl. Acad. Sci. U.S.A. 77, 4707 (1980).
[40] A. T. Quintahilha and L. Packer, FEBS Lett. 78, 161 (1977).
[41] A. T. Quintahilha and L. Packer, Arch. Biochem. Biophys. 190, 206 (1978).
[42] L. A. Boxer, R. A. Haak, H. H. Yang, J. B. Wolach, J. A. Whitcomb, C. J. Butterick, and R. L. Bachner, J. Clin. Invest. 70, 1047 (1982).
[43] A. A. Jasaitis, V. V. Kuliene, and V. P. Skulachev, Biochim. Biophys. Acta 234, 177 (1971).
[44] M. Miyake, A. Mekomiya, and K. Kurihara, Brain Res. 301, 73 (1984).
[45] K. Hashimoto, P. Angiolillo, and H. Rottenberg, Biochim. Biophys. Acta 764, 55 (1984).
[46] D. B. Kell and A. M. Griffiths, Photobiochem. Photobiophys. 2, 105 (1981).
[47] K. Hashimoto and H. Rottenberg, Biochemistry 22, 5738 (1983).

Electrophoretic Mobility. The surface potential can be estimated from electrophoretic mobility measurement.[4,5] It is noted that this method gives the potential at the slipping plane with respect to the bulk solution, the zeta-potential, ζ. Although ζ-potential is not generally equal to the surface potential, it is often used as an experimental approximation of the surface potential. ζ-Potential is calculated from the electrophoretic mobility, u, with the aid of the Helmholtz–Smoluchowski equation.[4,5]

$$\zeta = 4\pi\eta u/D \tag{5}$$

Here, η and D are the viscosity and dielectric constant of the dispersion medium, respectively. The viscosity was measured by an Ostwald viscometer. Usually D is taken as 80. Equation (5) is derived under the condition that the size of particle is much larger than the thickness of ionic atmosphere (the Debye length) of the electrolyte solution. (It is ~ 10 nm in 1 mM of uni-uni valent salt and it shortens as concentration increases.) The equation which is applicable to small particles whose diameter is comparable to the Debye length has been derived.[48]

The electrophoretic mobilities were measured by a microelectrophoretic apparatus: the rate of movement of particle due to the externally imposed electric field is observed directly with a microscope (see Fig. 3). The following description is based on a Carl Zeiss cytopherometer. The eyepieces contain a special reticule which serves both as eyepiece micrometer and as object finder. The measuring cuvette is placed in a chamber whose temperature is controlled by circulating thermostated water. The cuvette is connected on both sides to a pair of unpolarizable electrodes with use of sintered glass filter tee. The time required for a particle to travel the distinct distance is measured when the electrical field is applied externally and u is calculated as $u = $ (velocity)/(electric field strength). The electric field is created by passing an electric current, which is supplied from a constant current supplier, through the cuvette. The electric field strength is computed from the specific resistance of the suspension and electric current strength:

Electric field strength $= (I\rho)/S$

where I is the current (amp), ρ is the specific resistance of the medium (Ω-cm), and S is the cross-sectional area of the cuvette, respectively. For Carl Zeiss cytopherometer, the cuvette is rectangular and the area is $(1.4 \pm 0.05) \times (0.07 \pm 0.01)$ cm^2. The observed velocity should be proportional to the electric field.

---

[48] J. T. G. Overbeek, *Kolloid Beih.* **54**, 287 (1943).

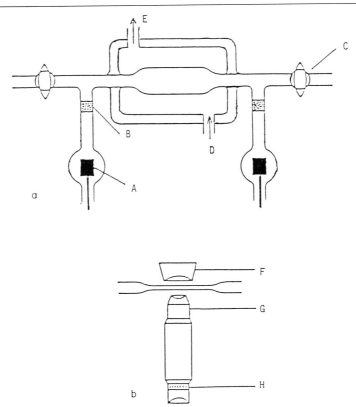

Fig. 3. (a) Schematic illustration for microelectrophoretic apparatus. A, unpolarizable electrode; B, sintered glass filter tee; C, stopcock to enclose the fluid into the electrophoretic system composed of the cuvette and electrode compartment; D, thermostated water inlet; E, thermostated water outlet. (b) Top view of the arrangement. F, condensor lens; G, objective lens; H, eyepiece micrometer.

Because the quartz surface of the cuvette is negatively charged, an externally imposed electrical field moves the electolyte solution (so-called electroosmosis). But, there exist positions near the cuvette surface where the liquid does not flow, and this is called the "stationary layer."[4,5] The electrophoretic mobility should be measured only for the particle positioned at this layer. The focus of a microscope must be adjusted to this position. This position depends on the shape of cuvette and should be referred to a manual of the apparatus. Note that the position of the stationary layer is obtained under the condition that the electrophoretic system composed of the cuvette and the electrode compartment is closed completely and that no net fluid flow occurs. Then, the tight enclosement of measuring fluid into the system is essential and is done with well-ground stopcock or pinched cock. One must make sure that in the absence of external electric field, particles do not move in one direction.

The density of the particle in the test solution is suitable when one can see about 10–30 particles in the optical field. For mitochondria, 0.1 mg/ml or less is suitable. As described above, the measuring cuvette is enclosed tightly and then the medium becomes anaerobic during the measurement. The medium should be frequently exchanged. The period from aerobic to anaerobic condition is easily checked with use of methylene blue.[49] Twenty-five or more particles were timed in each direction of the electric field to eliminate the polarization of electrodes and the averaged value was taken as the velocity of particles. If the averaged value in each direction does not agree, the particle may move in one direction in the absence of the externally applied electric field due perhaps to bad enclosement. In this case, it is recommended that the experiment be performed again after the enclosement of the electrophoretic system. If a repeat experiment is not possible, the true period for traveling (denoted as T_0) is calculated as follows. When the period required for each direction is denoted as T_1 and T_2 ($T_2 > T_1$), and the arithmetic mean value is denoted as T, the following equation holds:

$$(T - T_0)/T_0 = (1 - a)^2/4a$$

where $a = T_2/T_1$.

The change in surface potential of mitochondria during energization was reported.[39,49–51] Mitochondria were isolated from rat liver as usual and suspended in 250 mM sucrose containing 10 mM potassium phosphate (pH 7.0), 20 mM KCl, and 0.1 mM EDTA. Carl Zeiss cytopherometer was used and operated at 25°. The averaged period for traveling between 4 spaces of the eyepiece micrometer which is equivalent to 57.0 μm was 15.8 sec. Then,

$$\text{Velocity} = 57.0 \times 10^{-4}/15.8 = 3.61 \times 10^{-4} \text{ cm/sec}$$

Since $I = 0.88$ mA and $\rho = 0.368 \times 10^3$ Ω-cm,

$$\text{Electric field strength} = (0.88 \times 10^{-3} \times 0.368 \times 10^3)/(1.4 \times 0.07)$$
$$= 3.30 \text{ V/cm}$$

[49] G. P. R. Archbold, C. L. Farrington, S. A. Lappin, A. M. McKay, and F. H. Malpres, *Biochem. Int.* **1**, 422 (1980).

[50] N. Kamo, M. L. Muratsugu, K. Kurihara, and Y. Kobatake, *FEBS Lett.* **72**, 247 (1976).

[51] S. I. Yeremenko, V. G. Budker, and Zh. M. Bekker, *Biofizika* **25**, 294 (1980).

Therefore,

$$u = 3.61 \times 10^{-4}/3.30 = 1.09 \times 10^{-4} \text{ cm}^2 \text{ V}^{-1} \text{ sec}^{-1}$$

For calculating the ζ-potential, we should notice the factor of 1/300 between practical volt units and cgs esu. Then, ζ-potential in volts is

$$\zeta = 4\pi\eta(300)^2 u/80$$

The dielectric constant of the medium is taken as 80. The viscosity of the solution was 1.05×10^{-2} P. Then,

$$\begin{aligned}\zeta &= 4 \times 3.14 \times 1.05 \times 10^{-2} \times (300)^2 \times 1.09 \times 10^{-4}/80 \\ &= 0.0162 \text{ V} = 16.2 \text{ mV}\end{aligned}$$

The apparatus is available from Carl Zeiss (West Germany), Rank Brothers (U.K.) and Mitamura (Japan). The rectangular and vertical electrophoretic cuvette is preferable. The flat surface of the cuvette makes the observation easier than a cylindrical one. When large particles which sink during measurements are used, they will be missed if the chamber is not vertical. In other words, the stationary layer is easier to locate vertically.

For small particles or membrane fragments which are not visible with a microscope, this method is not applicable. Laser Doppler scattering apparatus can be used for this purpose and several experiments have been performed.[52]

[52] A. W. Preece and D. Sabolovic (eds.), "Cell Electrophoresis: Clinical Application and Methodology," Elsevier, Amsterdam, 1979.

[6] Rate of Redox Reactions Related to Surface Potential and Other Surface-Related Parameters in Biological Membranes

By SHIGERU ITOH and MITSUO NISHIMURA

Ion Concentration and Electrical Potential at the Membrane Surface

The generation and use of the electrochemical free energy of protons or other ions play a crucial role in energy coupling between electron transfer, ATP synthesis, ion transport, etc., performed on the membranes according to the chemiosmotic theory.[1] The intensive variable expressing free energy of the ith ion in the aqueous phase is the electrochemical potential $(\bar{\mu}_i)$,[1a]

$$\bar{\mu}_i = \bar{\mu}_{i0} + RT \ln a_i + Z_i F \varphi \tag{1}$$

where $\bar{\mu}_{i0}$ is the potential in the arbitrarily defined reference state, a_i and Z_i are the activity and the valence of the species, i, and φ is the electrical potential with respect to the reference state. Other symbols have their usual meanings. In the aqueous phase on each side of the vesicular membrane, $\bar{\mu}_i$ is uniform from the membrane surface to the bulk phase far from the surface. However, the latter two terms in Eq. (1) (chemical and electrical potential terms) are interchangeable and each of them is not necessarily uniform throughout the aqueous phase even at the equilibrium.

The surfaces of biological membranes are generally at a negative electrical potential with respect to the bulk aqueous phase in the physiological pH range due to the charges immobilized on lipid and protein components, according to the Gouy–Chapman diffuse double layer theory (Fig. 1a).[2] They generally show isoelectric points between pH 4 and 5 as studied by particle electrophoresis.[3] This electrical potential difference is compensated by a change in the chemical potential (the activity term) of each ion to maintain an equilibrium of free energy. At the negative membrane surface, concentrations of cations/anions are higher/lower than those in the bulk aqueous phase (Fig. 1b). The interchange between the electrical and chemical terms introduces significant effects on the reactions of the membrane components introducing electrical and chemical environments different from those in the bulk phase. The change in the

---

[1] P. Mitchell, "Chemiosmotic Coupling in Oxidative and Photosynthetic Phosphorylation." Glynn Research, Bodmin, Cornwall, 1969.

[1a] Abbreviations: a, activity; C_b, concentration in the bulk phase; CCCP, carbonylcyanide m-chlorophenylhydrazone; DCMU, 3-(3,4-dichlorophenyl)-1,1-dimethylurea; E_0, standard redox potential; E_m, midpoint redox potential; F, Faraday constant; k, apparent rate constant; k^0, actual rate constant; K_b, dissociation constant of surface charge groups defined on the basis of the bulk pH; K_s, dissociation constant of surface charge groups defined on the basis of the surface pH; pH_b, pH in the bulk aqueous phase; pH_s, pH at the membrane surface; q, net surface charge density; R, gas constant; T, absolute temperature; Z, valence of ion; γ_b, activity coefficient in the bulk phase; ε, dielectric constant of water; κ, Debye–Hückel parameter; $\bar{\mu}$, electrochemical potential; μC, microcoulomb; φ, electrostatic potential; $\Delta\varphi_{s-s}$, intramembrane electrical potential difference (surface-to-surface potential difference); $\Delta\varphi_{b-b}$, membrane potential (potential difference between bulk aqueous phases); Ψ_0, surface potential; Ψ_x, electrical potential at distance x from surface.

[2] J. Th. G. Overbeek, in "Colloid Science" (H. R. Kruyt, ed.), Vol. I, p. 115. Elsevier, Amsterdam, 1950.

[3] See N. Kamo and Y. Kobatake, this volume [5].

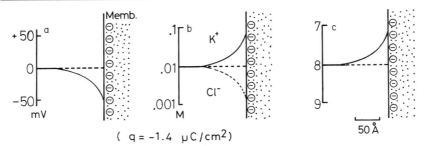

FIG. 1. Schematic profiles of (a) electrical potential, (b) ion concentrations, and (c) pH on the membrane surface. The values were calculated for the case with a net surface charge density (q) of -1.4 $\mu C/cm^2$, estimated for the outer surface of chloroplast thylakoid membrane.

chemical term at the surface results in the change of the apparent reactivity of the membrane component to the ionic reactant added in the aqueous phase, since the reactivity depends on the surface activity but not on the bulk activity of the reactant. *In vivo* and *in vitro* regulations of various membrane reactions, especially those through changes of pH or ion concentrations, can thus be understood more adequately by considering "surface activities" of ions.[4] Surface charges also affect the surface proton concentration, redox state of the membrane components, transmembrane electrical potential profile, intramembrane movements of charges, arrangement of protein complexes on the membrane, membrane–membrane interaction, etc. Effects of these phenomena on the energy transducing reactions are briefly discussed in this chapter. See previous reviews[5–8] for the detailed discussion and background of the phenomena.

Theoretical Background: The Gouy–Chapman Diffuse Double Layer Theory

Many aspects of colloidal and electrochemical phenomena of membranes have been extensively studied in colloid chemistry and many papers and textbooks have dealt with these phenomena. However, the simple classical theory of Gouy and Chapman,[2] which takes into account the

[4] S. Itoh, *Biochim. Biophys. Acta* **504**, 324 (1978).
[5] J. Barber, *Biochim. Biophys. Acta* **594**, 253 (1981).
[6] U. Siggel, *Bioelectrochem. Bioenerg.* **8**, 327 (1981).
[7] H. Traüble, *in* "Structure of Biological Membranes" (S. Abrahamsson and I. Pascher, eds.), p. 509. Plenum, New York, 1977.
[8] S. McLaughlin, *Curr. Top. Membr. Transp.* **9**, 71 (1977).

fact that the counterions at a charged surface will form a diffuse layer in an aqueous phase close to the surface, seems to be adequate to describe the electrical phenomena at the surface of rather complex biological membranes.[4-8]

Immobilized charges on the membrane surface give rise to a difference in electrical potential at the surface with respect to the bulk aqueous phase. According to the theory, the relation between the net surface charge density, q, and the electrical potential of the surface, Ψ_0, is given as follows,

$$q = \pm \left| \frac{RT\varepsilon}{2\pi} \sum_i C_{ib} \left[\exp(-Z_i F \Psi_0 / RT) - 1 \right] \right|^{1/2} \tag{2}$$

or

$$\begin{aligned} q &= 2A\, C_b^{1/2}\, \sinh(Z_i F \Psi_0 / 2RT) \\ &= 11.74\, C_b^{1/2}\, \sinh(Z_i \Psi_0 / 59) \quad (\Psi_0, \text{ in mV}) \end{aligned} \tag{3}$$

where $A = (RT\varepsilon/2\pi)^{1/2}$, after numerical substitutions at 25° in the case of symmetrical ion salt. q is expressed in μC/cm^2 (1 electronic charge per 1000 Å^2 corresponds to 1.6 μC/cm^2). In the equation, Z_i and C_{ib} represent valence and the bulk concentration of ions (in M), and ε is the dielectric constant of water.

For low values of Ψ_0 ($\Psi_0 < 50/Z_i$ mV) in the case of a $Z_i : Z_i$ symmetrical salt, Eq. (2) reduces to

$$\Psi_0 = \left(\frac{2\pi RT}{F^2 \varepsilon C_b} \right)^{1/2} q/z_i \tag{4}$$

where C_b is the bulk concentration of the salt.

For the mixture of 1 : 1 and 2 : 2 symmetrical salts (at concentrations of $C_b^{(1)}$ and $C_b^{(2)}$, respectively), Eq. (2) reduces to[5]

$$2C_b^{(2)} \cosh^2 \left(\frac{F\Psi_0}{RT} \right) + C_b^{(1)} \cosh \left(\frac{F\Psi_0}{RT} \right) - \left(2C_b^{(2)} + C_b^{(1)} + \frac{q^2}{2A^2} \right) = 0 \tag{5}$$

and for the mixture of 1 : 1 and 3 : 3 [at $C_b^{(1)}$ and $C_b^{(3)}$],

$$4C_b^{(3)} \cosh^3 \left(\frac{F\Psi_0}{RT} \right) + (C_b^{(1)} - 3C_b^{(3)}) \cosh \left(\frac{F\Psi_0}{RT} \right)$$

$$- \left(C_b^{(1)} + C_b^{(3)} + \frac{q^2}{2A^2} \right) = 0 \tag{6}$$

Activity of the ith ion at the membrane surface, a_{is}, with respect to that in the bulk phase, a_{ib}, can be obtained when the Boltzmann distribu-

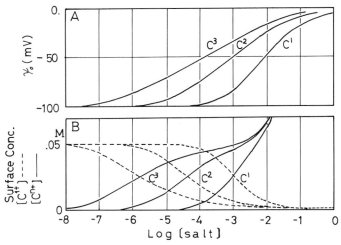

FIG. 2. (A) Dependence of the value of surface potential on bulk concentrations of symmetrical salts added to the medium which originally contains 0.001 M salt of monovalent ions. C^1, C^2, and C^3 indicate cases with salts of ions with 1:1, 2:2, and 3:3 symmetrical valences, respectively. (B) Dependence of the surface concentrations of added cations (solid lines) and preexisting monovalent cation (broken lines) on the bulk concentrations of added symmetrical ion salts. The surface charge density was assumed to be -1.4 $\mu C/cm^2$. The relation between the surface potential and the bulk concentrations of mixture of salts with different valences were calculated by Eqs. (5) and (6).

tion of the ion is assumed [i.e., a constant $\bar{\mu}_i$ value in Eq. (1) is assumed at the bulk phase and at the surface].

$$a_{is} = a_{ib} \exp(-Z_i F \Psi_0 / RT) \tag{7}$$

Thus, the higher valence cations/anions (with larger Z_i) are expected to be more concentrated at the negatively/positively charged membrane surface and to reduce the Ψ_0 value even at low bulk concentrations as shown in Fig. 2. On the other hand, surface concentrations of the higher valence anions are expected to be very low at the negative membrane surface [Eq. (7)] and to increase with the decrease of negative surface potential. It is also to be noted that even at extremely low bulk ion concentrations, surface concentrations of preexisting cations are very high due to the large negative surface potential (Fig. 2B, dashed lines).

The influence of surface charges on the electrical potential in the aqueous phase can be given by the equation below with the Debye–Hückel parameter, $\kappa = (8\pi Z_i^2 F^2 C_b / RT)^{1/2} = 0.33 \ I^{1/2}$ Å^{-1} at 25° (I is the ionic strength):

$$\Psi_x = \Psi_0 \exp(-\kappa x) \tag{8}$$

where electrical potential Ψ_x at a distance x from the surface is related with the surface potential Ψ_0. Ψ_x value becomes $1/e$ at $x = 1/\kappa$. The distance is 100, 30, and 10 Å at 0.001, 0.01, and 0.1 M monovalent ion salt, respectively.

Surface potential value is related to the value of zeta-potential, which is the electrical potential at the hydrodynamic plane of shear adjacent to the surface and can be calculated from the electrophoretic mobility of the membranes. The zeta-potential is assumed to be proportional to surface potential but is usually lower than the latter.[2,3,5]

Summary of the Expected Characteristics of the Phenomena Governed by the Membrane Surface Charges

Any processes affected by the change of surface potential should show the following characteristics according to the Gouy–Chapman theory.[2–8]

1. Higher valence cations/anions are more effective in inducing the effect when membrane is negative/positive. The difference of the effectiveness enlarges as the net surface charge increases.

2. The salt effect is determined mainly by the valence of cations and is essentially independent of the anion species with negative membrane surface. The situation is reversed in the case of a positive membrane.

3. No dependence on the chemical nature of the ions except their valence is expected.

4. The direction of the salt effect does not change with the change of salt concentration, since the increase of salt concentration decreases the magnitude of the surface potential but not its sign.

5. The direction of the salt effect changes with the change of the sign of net surface charge density, which can be induced by the change of bulk pH or by the binding of hydrophobic ions.

6. The salt effect of the higher valence cations/anions with the negative/positive membrane often deviates from that expected from Eq. (2), due to the nonidealistic behavior.[7,8]

7. Even at very low electrolyte concentrations, the concentration of cations/anions on the negative/positive membrane surface is usually very high due to the high surface potential. Thus, the washing of membranes in low ionic medium does not mean depletion of ions from the membranes. Washing in high electrolyte medium rather facilitates the exchange of surface ions; concentration of preexisting cation on the membrane surface can be decreased only in the presence of other cations as shown in Fig. 2B (dashed lines).

8. If the process is found to be regulated by surface potential, the value of surface potential or charge density can be easily obtained by Eqs. (2)–(7). One simple way is to get a set of concentrations of mono- and divalent ion salts, at which concentrations these salts exhibit the same effect on the process. Then the q and Ψ_0 values can be obtained by solving Eq. (3) or (4). The other way is to get the best fit parameters by assuming some specific relations between the value of surface potential and the process, as indicated in the next section.

It should be kept in mind that the net surface charge density is also a function of surface potential, since the degree of protonation of surface groups changes with the change of proton concentration on the surface, which is related to that in the outer medium by the surface potential.

Methods for the Measurement of Surface Potential

A direct electrical measurement of surface potential is possible at the surface of lipid monolayers developed in the air–water interface.[7,8] Measurements of membrane conductance also give information for the surface potential in the case of planar bilayer lipid membranes.[7,8] There are a number of electrokinetic phenomena which can give information for the surface potential of large planes and colloids.[2,7] However, the surface potential of small vesicles of biological energy transducing membrane cannot be correctly measured by these types of measurements. A reliable, relatively direct, method of estimating surface potential is to measure electrophoretic mobility of vesicles using the microelectrophoresis apparatus.[3] The mobility gives the value of zeta-potential by the Helmholtz–Smoluchowski equation. The zeta-potential, which is the electrostatic potential at the hydrodynamic plane of shear within an aqueous phase at a few angstroms from the surface, is smaller than the surface potential [as judged from Eq. (8)], although it is proportional to the surface potential. Some ambiguity remains in the estimation of surface potential from the zeta-potential, since the proportionality factor depends on the density of surface charge and on the ionic conditions.

Apart from these electrical methods of measuring surface potential, this chapter covers the indirect but reliable methods of measuring surface potential, which can easily be applicable to the energy transducing membrane systems, and briefly summarizes the roles and importance of surface potential in the energy transduction. One simple way of estimating surface potential is to measure ion concentration on the membrane surface and calculate the potential value using Eq. (1). The difference of electrical potential at the surface is compensated by the change of surface concentration of each ion under the equilibrated conditions. Surface concentrations of ions (including protons, charged dyes, and electrons) can be estimated from the change of reaction rate at the membrane surface, change of absorbance or fluorescence, shift of apparent pK value of the surface dissociable groups, change of redox state of membrane electron transfer components, etc. Information on the intramembrane electric field, which can be obtained by the measurement of the band shift of intrinsic carotenoids, can also be used to estimate surface potential in photosynthetic membranes. Various intrinsic and extrinsic parameters, which can be used for the measurement of surface potential, are discussed in the following sections. The table summarizes the surface potential values estimated by these indirect methods.

Surface Ion Concentration and the Salt Effect on the Surface Reactions

Electron transfer reaction between a component, A_1, on the membrane and a redox reactant, A_2, in the outer medium can be represented as follows:

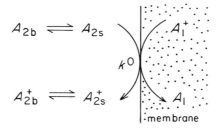

Subscripts b and s denote molecules in the bulk phase and at the surface, respectively. As far as the initial rate is concerned, the rate of the reverse reaction at the surface can be ignored. So the rate of the reaction is expected to be proportional to the activity of A_2 at the membrane surface if the exchange of A_2, or its reaction product, between the surface and bulk phase is not rate determining, i.e., if the rate-limiting step is the reaction at the membrane surface. This can be readily tested by checking the effect of stirring the medium or by checking the linearity of the dependence of the rate on the A_{2b} concentration. The surface activity of A_2 molecule can be obtained from the Boltzmann distribution [Eq. (7)].

$$A_{2s} = A_{2b} \exp(-Z_2 F \Psi_0 / RT) \tag{9}$$

We calculate the apparent rate constant, k, from the bulk concentration of A_2. Then the relation between the apparent rate constant k and the actual rate constant k^0, defined with respect to the surface activity of A_2,

VALUES OF NET SURFACE CHARGE DENSITY OF VARIOUS MEMBRANES AT
NEUTRAL pH (6.5–8.5)

Membranes	q (μC/cm^2)	Methods
Inner membrane of chloroplast (thylakoid membrane) (spinach and pea)		
Outer surface	$-1.1 \sim -1.4$	Reactions of Q,[4] cyt f, cyt b_{559}[10], and P700[9]
	-0.95	Change of pH optimum of Fd-NADP reductase[12]
Inner surface	$-0.8 \sim -1.1$	Reaction of P700 in thylakoid fragments[9–11,14–18]
Total membrane	-3.0	Merocyanine dye partition[37]
	-2.5	Fluorescence of 9-aminoacridine[5,40] and intrinsic chlorophyll[5,41]
	-1.6	Ion-exchange ability[27]
Chromatophore membrane of photosynthetic bacteria		
Outer (periplasmic) surface	-3.0	Carotenoid absorbance change[33] ($R.$ $sphaeroides$)
	-2.2	Reaction of cyt c_2[22] ($R.$ $sphaeroides$)
Inner (cytoplasmic) surface	-4.6	Carotenoid absorbance change[35] ($R.$ $sphaeroides$)
Intramembrane components	-1.6	E_m of cyt c_2, bacteriochlorophyll dimer[22] ($R.$ $sphaeroides$)
	-1.2	E_m of cyt c_{555}[31] ($C.$ $vinosum$)
Purple membrane of $H.$ $halobium$		
Surface of bacteriorhodopsin layer	-13.9	Absorption change of bacteriorhodopsin chromophore[66]
Mitochondrial inner membrane		
Outer surface	-0.38	Phosphorescence of Tb^{3+} [46] (rat liver)
Outer surface	-3.3	Fluorescence of 9-aminoacridine[24] (Jerusalem artichoke)
Smooth microsome membrane (rat liver)		
Outer surface	-0.62	Reaction of arylsulfatase[25]
Total membrane	-1.73	Ion-exchange ability[27]
Plasmalemma membrane (wheat root)		
Outer surface	-1.6	Fluorescence of 9-aminoacridine[65]
Inner surface	-3.4	Fluorescence of 9-aminoacridine[65]

becomes as follows,

$$k = k^0\gamma_b \exp(-Z_2 F\Psi_0/RT) \tag{10}$$

or

$$\ln k/\gamma_b = \ln k^0 - (Z_2 F/RT)\Psi_0 \tag{11}$$

where γ_b is the activity coefficient of A_2 in the bulk phase which can be calculated by the extended Debye–Hückel expression.

Substitution of Eq. (10) by Eq. (4) gives a linear approximation equation.

$$\ln k/\gamma_b = \ln k^0 - Z_2/Z(2\pi/RT\varepsilon)^{1/2} C_b^{-1/2}q \tag{12}$$

Numerical substitutions give (q expressed in $\mu C/cm^2$)

$$\log k/\gamma_b = \log k^0 - 0.078(Z_2/Z)C_b^{-1/2}q \qquad \text{(at 25°)} \quad (13)$$

In practical cases in biological membranes Eq. (13) can only be used with monovalent symmetrical ion salts. Equation (13) indicates that the plot of log k/γ_b value against $C_b^{-1/2}$ gives a straight line, which gives k^0 as the extrapolated intercept value, and that the slope of the plot depends on Z_2 and the net surface charge density of the membrane.

With net negative membrane surface ($\Psi_0 < 0$), surface activities of high-valence cations are much higher than those of monovalent ones [Eq. (7)]. Thus, the valence of the cation mainly determines the effectiveness of the salt in decreasing the extent of negative surface potential (and hence in changing the k value). In the case of a positively charged membrane surface, the converse holds. The practical application of this type of analysis has been initially done in chloroplast membranes by Itoh[4] and has been shown to be applicable to other membranes (see table).

The reaction may also be analyzed as a bimolecular one between a membrane and a small reactant molecule by the Brønsted theory.[9] The membrane system can be treated as a large spherical molecule (with a diameter of several hundred angstroms), whose activity coefficient can be calculated according to the extended Debye–Hückel expression as in the case of small protein molecules. Then, after suitable approximations, an equation with different salt-independent terms, and the same salt concentration-dependent term as Eq. (12) can be obtained in the case of symmetrical ion salts.[9] However, the latter treatment is rather artificial and does not seem to give useful information for the reaction of components on the membrane.

[9] S. Itoh, *Biochim. Biophys. Acta* **548**, 579 (1979).

Examples of the Effect of Salt Concentration on the Surface
   Redox Reaction

Figure 3 shows an example of the salt effects: a salt-induced increase
of the apparent rate constant of the reaction between the membrane-
bound electron carrier, P700 (photosystem I reaction center chlorophyll
situated on the inner surface of the thylakoid membrane of chloroplast)
and ferrocyanide ($Z_i = -4$).[9] The increase of the k value was determined
mainly by the concentration and valence of cations added and not by
those of anions at pH 7.8. Divalent cation salts are more effective than
monovalent ones. Thus the characteristics of the salt effects can be understood by considering the change of activity of ferrocyanide on the negative membrane surface. Salt additions will decrease the value of negative
surface potential and will increase the surface activity of ferrocyanide
according to Eq. (7). Decreases of the reaction rate at the high ionic
concentrations are due to the decrease of γ_b value. Figure 4A shows the
plot of log k/γ_b value versus inverse square root of KCl concentration. As
expected from Eq. (13), the plot gave a straight line with a slope indicating
a net surface charge density value of the surface ($-0.8\ \mu C/cm^2$) in the

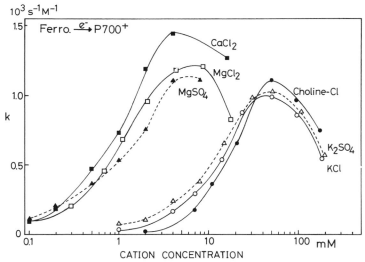

FIG. 3. Effects of various salts on the apparent rate constant (k) of the reduction of the
photosystem I reaction center chlorophyll, P700, bound to fragmented chloroplast membranes by ferrocyanide. Reaction mixture contained 0.2 mM K$_4$Fe(CN)$_6$, 5 mM tricine
buffer, pH 7.8, 19 μM DCMU, 8 μM methyl viologen, sonicated chloroplasts equivalent to
68 μg chlorophyll/ml, and varied concentration of salts as indicated in the figure. Data were
obtained from Itoh.[9]

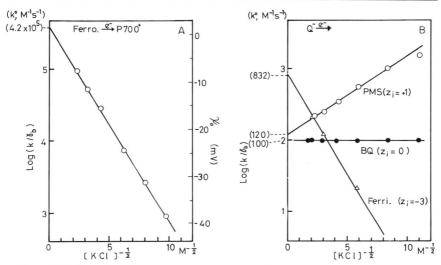

FIG. 4. (A) Plot of log k/γ_b vs (KCl concentration)$^{-1/2}$ in the reaction of P700 bound to the fragmented thylakoid membranes and ferrocyanide. Contributions of other monovalent cations are included in KCl concentration. Other conditions are the same as in Fig. 2 (modified from Fig. 4 in Itoh[9]). (B) Plot of log k/γ_b versus (KCl)$^{-1/2}$ in the reaction between the membrane-bound photosystem II primary acceptor, Q, and ferricyanide (Ferri), phenazine methosulfate (PMS), and benzoquinone (BQ). Reaction mixture contained 10 μM DCMU, 0.4 M sucrose, 5 mM Na–tricine buffer, pH 7.6, and chloroplast membranes equivalent to 3 μg chlorophyll/ml; 1 mM $K_3Fe(CN)_6$, 80 μM phenazine methosulfate, or 0.4 mM benzoquinone was added to the mixture (after Itoh *et al.*[10]).

vicinity of P700 molecules. From the intercept value at the infinite ionic concentration the true rate constant k^0 defined with respect to the surface activity of ferrocyanide can be obtained. In this type of plot, on the other hand, the value of log k/γ_b gives the value of surface potential at each salt concentration [Eq. (11)] as indicated on the right-hand ordinate.

Another example of the reaction at the membrane surface is shown in Fig. 4B: a similar plot of log k/γ_b value in the reactions of Q (a special quinone acting as the primary electron acceptor in the reducing side of photosystem II, situated close to the outer surface of the thylakoid membrane) with ferricyanide ($Z_i = -3$), phenazine methosulfate (PMS, $Z_i = +1$ in its oxidized state), and benzoquinone ($Z_i = 0$).[4] Straight lines with relative slope ratios of $-3 : +1 : 0$ were obtained for these reagents, respectively, with the same net surface charge density value of -1.4 μC/cm^2. With the increase of KCl concentration the order of apparent rate constants changed from PMS > benzoquinone > ferricyanide to ferricyanide > PMS > benzoquinone. It is clear that the k^0 value is essential when the reaction mechanism at the reaction site on the surface is discussed and that the apparent reactivity of the charged reactant is highly sensitive to the surface potential.

This type of analysis gives information about the net surface charge density in the local domain in the vicinity of the membrane component and can probe the surface with respect to the charge distribution in addition to the actual reaction rate constant on the surface. The size of the domain probed may be with a radius comparable to the Debye–Hückel length, and the length will vary depending on the ionic condition. Various studies have been done in chloroplast and in chromatophore membranes. As far as the reactions were performed on the outer membrane surface of chloroplast thylakoid (Q,[4] P700,[9] fast- and slow-type reactions of cytochrome b_{559},[10] cytochrome f,[10] ferredoxin-NADP reductase,[11,12] and plastoquinone pools measured by Hill reaction[13]) surface charge density values of about $-1.1 \sim -1.4$ $\mu C/cm^2$ were obtained in the neutral pH region. On the other hand, the reactions performed mainly at the inner surface of the thylakoid (reaction of P700 with endogenous electron donor plastocyanin[14-17] and redox reagents[18]) indicated lower surface charge density values of -0.8 to -1.1 $\mu C/cm^2$. The results suggest that the inner surface of the thylakoid membrane has a smaller amount of negative charges.[18]

These analyses of the salt effect according to the Gouy–Chapman theory are also applicable to the reaction of small proteins at the membrane surface. Reaction of a water-soluble protein, plastocyanin, with P700 is an example of the surface potential-dependent regulation under physiological conditions. The reaction has been extensively studied by number of research groups in intact[14] and in fragmented membrane systems[15-20] since the reaction rate in the intact thylakoids seems to be regulated by the pH change of the inner thylakoid space during illumination.[19] It can be concluded that the main factor which regulates the reaction rate is the surface potential-induced change in the number of plastocyanin

[10] S. Itoh, N. Tamura, K. Hashimoto, and M. Nishimura, in "The Oxygen Evolving System of Photosynthesis," (Y. Inoue, A. R. Crofts, Govindjee, N. Murata, G. Renger, and K. Satoh, eds.), p. 421. Academic Press Japan, Tokyo, 1983.

[11] N. Tamura, S. Itoh, and M. Nishimura, *Plant Cell Physiol.* **24**, 215 (1983).

[12] N. Tamura, S. Itoh, and M. Nishimura, *Plant Cell Physiol.* **25**, 589 (1984).

[13] S. Itoh, *Plant Cell Physiol.* **23**, 595 (1978).

[14] K. Matsuura and S. Itoh, *Plant Cell Physiol.* **26**, 1057 (1985).

[15] N. Tamura, Y. Yamamoto, and M. Nishimura, *Biochim. Biophys. Acta* **592**, 536 (1980).

[16] N. Tamura, S. Itoh, Y. Yamamoto, and M. Nishimura, *Plant Cell Physiol.* **22**, 603 (1981).

[17] L. F. Olsen and R. P. Cox, *Biochim. Biophys. Acta* **679**, 436 (1982).

[18] S. Itoh, *Biochim. Biophys. Acta* **548**, 596 (1979).

[19] W. Haehnel, V. Hesse, and A. Propper, *FEBS Lett.* **111**, 79 (1980).

[20] T. Takabe, H. Ishikawa, S. Niwa, and S. Itoh, *J. Biochem.* **94**, 1901 (1983).

molecules bound to the active site of P700 on the inner thylakoid surface.[14–18]

In the chromatophore membranes of photosynthetic bacteria, reactions of membrane-bound cytochrome c_{555} in *Chromatium vinosum*,[21] reaction center bacteriochlorphyll dimer and cytochrome c_2 bound to the membrane in *Rhodopseudomonas sphaeroides* with added redox reagent[22] have been interpreted by considering surface potential. Interestingly, the reaction between cytochrome c_2, which is water soluble and detachable from the membrane, and the bacteriochlorophyll dimer was rather insensitive to the change of surface potential in fresh chromatophores, in contrast to the case of similar reaction between plastocyanin and P700.[14] It suggests the formation of a supramolecular complex between the cytochrome and the reaction center on the membrane surface.[14]

In mitochondrial membranes, the role of surface potential on the salt-dependent change of the NADH oxidation rate has been clearly shown by the extensive work of Møller *et al.* in one of the NAD(P)H reaction sites.[23,24] The activity of arylsulfatase in smooth microsome membranes was also shown to be regulated by surface potential.[25]

pH Dependence of the Salt Effects: Change of Surface Charge Density

Salt additions decrease the magnitude of negative or positive Ψ_0 value by simply screening the surface charges without changes in the surface charge density if there is no specific binding of the ions to the membrane. On the other hand, change of pH of the bulk reaction medium or specific adsorption of hydrophobic ions alters the surface charge density and, therefore, the Ψ_0 value. Decrease of pH usually makes q and Ψ_0 values more positive by inducing the protonation of surface dissociable groups. Thus, in the neutral pH region, a decrease of medium pH induces effects similar to those of salt addition by inducing a shift of surface potential to a more positive value.[4,9,12] However, the critical difference between the salt and pH effect is that the decrease of pH from neutral to acidic range across the isoelectric pH of the membrane surface (usually in a pH range between 4 and 5) changes the sign of net charge density from negative to positive. This induces the inversion of the direction of salt effect.

Figure 5 shows the pH dependence of the reaction rate of water-soluble horse heart cytochrome c (isoelectric pH is about 10) and P700 on

[21] K. Hashimoto, S. Itoh, K. Takamiya, and M. Nishimura, *J. Biochem.* **91,** 1111 (1982).
[22] K. Matsuura, K. Takamiya, S. Itoh, and M. Nishimura, *J. Biochem.* **87,** 1431 (1980).
[23] S. P. Johnston, I. M. Møller, and J. M. Palmer, *FEBS Lett.* **108,** 28 (1979).
[24] I. M. Møller, W. S. Chow, J. M. Palmer, and J. Barber, *Biochem. J.* **193,** 37 (1981).
[25] K. Masamoto, *J. Biochem.* **95,** 715 (1984).

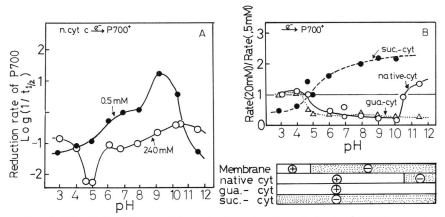

FIG. 5. (A) pH dependence of the reduction rate of membrane-bound P700 by native horse heart cytochrome c at low (0.5 mM, closed circles) and high (240 mM, open circles) NaCl concentrations. Logarithms of the reciprocals of the half-reduction times (in sec) are plotted against pH. Reaction mixture contained TSF-I preparation of spinach chloroplast membrane fragments equivalent to 10 μM chlorophyll, 10 μM methyl viologen, and varied concentrations of NaCl and cytochrome c at pH 7. (B) pH dependence of the effects of salt on the P700 reduction rate by native and modified cytochromes c. The "salt effect" is expressed by the ratio of the P700 reduction rate at 20 mM NaCl to that at 0.5 mM NaCl. Estimated signs of net charges of the membrane fragments (TSF-I particles), native, suc-cinylated, and guanidylated cytochrome c are shown at the bottom of the figure. (○) Native cytochrome c; (△) guanidylated cytochrome c; (●) succinylated cytochrome c (after Tamura et al.[11]).

the surface of the P700-enriched membrane fragments, measured at two different salt concentrations: a typical example of the pH and salt effect.[11] The reaction rate changed more significantly depending on bulk pH at the lower ionic concentration than at the higher ionic concentration. In the pH range between 5 and 10, at which the membrane is net negative and the cytochrome is positive, the rate was decreased by the salt addition, while in the pH range below 5 (membrane becomes positive) or above 10 (cytochrome c becomes net negative), the salt addition increased the rate. It was also confirmed that when the membrane was net negative (above pH 5) the salt effects were determined mainly by the concentration and valency of cations. The pH-dependent change of the direction of the salt effect can be readily understood by considering the electrostatic interaction between the membrane surface and the cytochrome. The salt effect is summarized in Fig. 5B. It is clear that the salt effect changes its direction each time when the bulk pH changes across the isoelectric pH of the membrane or cytochrome as expected.

In the same figure, the salt effects on the guanidylated (isoelectric point shifted to above pH 11) and succinylated (below pH 2) cytochromes are also shown. Also in these cases, pH dependences of the directions of the salt effects were determined by the net charge densities of the reactants. However, the pH dependences of the rate constants were very different among these three types of cytochromes at low ionic conditions. They became almost the same at the higher ionic conditions (not shown). This indicates that pH dependence of the interaction between the local small domains on the cytochrome and on the P700 protein, which are more important in considering the reaction mechanism, can be studied correctly only at high ionic conditions (with lower contribution of net charges). It seems to be concluded that the pH dependence of the interaction between the active sites on the large reactant molecules can be studied at high salt conditions. At low salt conditions electrostatic interaction between charges situated far from the active site will modify the pH dependency.

Proton Concentration on the Membrane Surface

Concentration of protons on the membrane surface is also different from that in the bulk outer medium. This induces the apparent changes in the values of pK_a of surface dissociable groups when the surface potential value is changed by the salt additions, even when the bulk pH is maintained constant. The relation between the effective values of bulk and surface pH (pH_b and pH_s, respectively) can be given as follows from Eq. (7) at 25°:

$$pH_s = pH_b + \Psi_0/59 \tag{14}$$

when Ψ_0 is expressed in mV. The dissociation constant of surface negative groups ($AH \rightleftharpoons A^- + H^+$) defined on the basis of the bulk pH (K_b) is related to that defined on pH_s (K_s, which equals K_b at zero surface potential):

$$K_b = K_s \exp(\Psi_0/59) \tag{15}$$

Thus the change of ionic composition in the medium leads to the change of dissociation of ionic groups at the membrane surface. Therefore, pH indicator dyes bound to the membrane surface give information for the pH_s and respond to the Ψ_0 change.

Evidence for the salt-induced pH_b (K_b) change was demonstrated in membranes of chloroplasts,[26] microsomes,[25] and in liposomes.[7] When the

[26] K. Masamoto, S. Itoh, and M. Nishimura, *Biochim. Biophys. Acta* **591,** 142 (1980).

salt was added to the unbuffered membrane solution, release of protons from the membrane surface into the bulk medium (ion-exchange ability of the membrane) was detected by a glass electrode as a consequence of the change of dissociation of the surface groups responding to the pH_s changes. Schematic representation of the relation between the pH_s and pH_b estimated for the thylakoid membranes is shown in Fig. 6. Number and pK values of surface dissociable groups are estimated from the pH and salt concentration dependences of the ion-exchange capacity and electrophoretic mobility of the membrane.[26] It can be seen that pH_s is significantly different from pH_b at low ionic conditions and that the change of pH_b induces only a small pH_s change at low salt concentrations. It indicates that the pH difference between the inner and outer surfaces of the membranes is usually lower than that between the bulk aqueous phases on both sides of the membranes. The chemical term (proton concentration) of the free energy change is converted to the electrical term (surface potential change) at the membrane surface under low ionic conditions. Similar estimations were also done in microsome membranes.[25]

Upon the application of pH difference by a pH jump, a technique often used for the study of ATP synthesis, a part of the change of proton chemical potential in the bulk phase is converted to the change of electri-

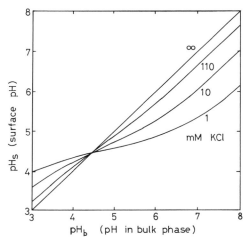

FIG. 6. Calculated relationship between effective pH at the membrane surface pH_s, and in the bulk aqueous phase pH_b at various KCl concentrations in chloroplast membranes. The relation was calculated by Eqs. (2), (4), (11), and (12) from the estimated number of carboxyl groups on the surface (1.7×10^{-3} carboxyl group/Å^2 with pK 4.6), and measured isoelectric pH (4.4) of the membranes. The latter values are estimated from the ion-exchange capacity and electrophoretic mobility of the membranes (after Masamoto *et al.*[26]).

cal potential at the surface, depending on the ionic conditions. Thus, the pH jump induces the jump of electrical potential, at the same time, on the membrane-bound components or enzymes. The magnitude of pH jump itself is expected to be smaller on the membrane component. This point is important when the forces acting on the membrane molecules are considered in the energy conversion.[4-10,28] A salt-induced shift of the pH optimum of the activity of membrane-bound enzymes can also be interpreted by a change of pH_s.[12,25]

Redox Potential (Activity of Electron) on the Membrane and the Surface Potential

When an electrostatic potential difference ($\Delta\varphi$) exists between the bulk solution (in which the redox potential is measured by electrodes) and the site of an electron carrier on the membrane, the midpoint potential E_m of the redox couple on the membrane takes a different value from the standard potential, E_0. The change in chemical potential of electrons (redox potential) should compensate the difference in the electrical potential for the equilibration of electrochemical potential of electrons as shown by Hinkel and Mitchell[29] for the case of membrane potential change, or as discussed by Walz.[30] The following relationship should exist between the values of E_m and E_0 and the potential difference, $\Delta\varphi$.

$$E_m = E_0 + \Delta\varphi \tag{16}$$

$\Delta\varphi$ can also be the surface potential. The value of E_m of a nonprotonating redox component on the negative surface at the equilibrium can be given from Eqs. (2) and (3).

$$E_m = E_0 + \Psi_0 = E_0 + 2RT/ZF \sinh^{-1}(q/2AC_b^{1/2}) \tag{17}$$

On the other hand, for a component which binds proton(s) upon reduction, the relation below will hold when it is in equilibrium with protons in the bulk solution.

$$\begin{aligned} E_m &= E_0 + \Psi_0 - (2.3RT/F)pH_s \\ &= E_0 + (2.3RT/F)pH_b \end{aligned} \tag{18}$$

Equations (17) and (18) indicate that the change of electrolyte composition in the bulk medium will affect the redox state of the nonprotonating

[27] K. Masamoto, *J. Biochem.* **92**, 365 (1982).
[28] B. Rumberg and H. Mühle, *Bioelectrochem. Bioenerg.* **3**, 393 (1976).
[29] P. Hinkel and P. Mitchell, *Bioenergetics* **1**, 45 (1970).
[30] D. Walz, *Biochim. Biophys. Acta* **505**, 279 (1979).

membrane component even at the same bulk redox potential if there exists an appropriate redox mediator which equilibrates electrons. This process is a simple interchange between electrical and chemical terms of electrochemical potential of electrons, and does not induce any free energy change.

An experimental approach to demonstrate this relation has initially been done for cytochrome c_{555} in chromatophore membranes of *C. vinosum*,[31] for cytochrome c_2 (when it is bound to the membrane), and for the reaction center bacteriochlorophyll dimer in chromatophores of *R. sphaeroides*.[22] E_m values of these components showed no pH dependence at high ionic concentrations, while under low ionic conditions they became more negative with the increase of pH reflecting the negative shift of surface potential. Salt addition induced opposite effects above and below the isoelectric point of the membrane (pH 5.2) as expected.[31] E_m value of cytochrome c_{555} shifted from 310 to 380 mV at pH 8.5 and 400 to 380 mV at pH 4, respectively, by the addition of 0.3 M KCl to 1 mM sodium phosphate buffer. The salt effect was strictly dependent on the cation composition of the salt at pH 7.8, at which the membrane is net negative, as expected from the effects of the cations on the surface potential. The salt sensitivity of the E_m value of this membrane-bound cytochrome was initially observed by Case and Parson.[32] They have discussed the effect of salt on the activity coefficient of this cytochrome according to the Debye–Hückel theory, assuming the protonation of the cytochrome accompanying its reduction. Later reexamination of the effect by Itoh[31] has shown that the effects can be better and quantitatively understood by considering surface potential change of the entire membrane.

Change of Surface Potential and Intramembrane Electrical Field

Changing the surface potential does not induce a change of intramembrane electrical field unless it induces the intramembrane movement of charges. However, when the membrane has appropriate ionic conductance, the intramembrane electrical field can be changed by changing surface potential. Interconversion of the electrical and chemical terms of the electrochemical potential of membrane permeable ions is expected to occur also inside the membrane, since the intramembrane concentration of the ions reequilibrates with the newly developed surface concentrations. A typical case of this situation can be observed in the presence of ionophores, which increase membrane conductance and clamp membrane

[31] S. Itoh, *Biochim. Biophys. Acta* **591,** 346 (1980).
[32] G. D. Case and W. W. Parson, *Biochim. Biophys. Acta* **292,** 677 (1973).

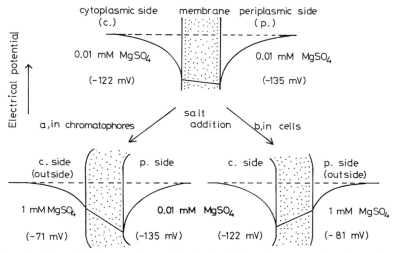

FIG. 7. Schematic diagram for electrical potential profiles across the chromatophore membrane of *Rhodopseudomonas sphaeroides* before and after salt addition to each side of the membrane in the presence of CCCP. Reaction mixture contained 1 mM NaCl and 2 mM tricine/NaOH buffer (pH 7.8). The values of surface potential indicated in parentheses were calculated using Eq. (3) with net surface charge density values of -3 and -4.6 μC/cm², estimated from the response of cartenoid, of the cytoplasmic and periplasmic surfaces of the membrane. Permeability of membrane to the salt added (1 mM MgSO$_4$), and the presence of fixed charges in periplasmic space or inside of chromatophores are ignored (after Matsuura et al.[35]).

potential (the potential difference between the bulk aqueous phases on both sides of the membrane) to the level determined by the difference of bulk concentrations of the permeable ion (Nernst potential). Figure 7 shows the schematic representation of the effect of surface potential change on the intramembrane electrical field in the presence of the proton conductor carbonylcyanide *m*-chlorophenylhydrazone (CCCP) which clamps membrane potential to the level determined by ΔpH. Under this condition, a change of surface potential on one side of the membrane will lead to the change of intramembrane electrical field. It means that the intramembrane electrical field can be altered by changing the surface potential without changing the membrane potential. However, it does not create any free energy. The change of intramembrane electrical field simply reflects the interconversion of the chemical potential term of proton (or CCCP) to the electrical term on the surface and inside of the membrane in order to maintain the equilibrium. This type of intramembrane electrical field will only be sensed by molecules in the membrane.

An experimental attempt to demonstrate this possibility in the energy-transducing membrane was first done by Matsuura et al.[33] in chromatophore membrane of *Rhodopseudomonas sphaeroides* by the exploitation of the carotenoid absorption band shift, which is the intrinsic sensor of the intramembrane electrical field in the photosynthetic membrane.[34] They showed that the addition of membrane-impermeable ions to the suspension of chromatophore vesicles in the presence of CCCP induced the carotenoid band shift with the change of surface potential (Fig. 7).[33] The response of carotenoid was opposite when the surface potential on the other side of the membrane was changed by the use of cell or spheroplast membranes,[35] which have the opposite membrane siddedness to that in chromatophores (Fig. 7, case b). From these studies, surface charge density values of chromatophore membranes on both the inner and outer surfaces have been estimated as shown in Fig. 7.[33,35] It was also shown that an attention must be paid to the change of surface potential in the calibration of membrane potential with the carotenoid band shift.[33,35,36]

Probes of Surface Potential and the Detection of Surface Potential Change During Energization

As has been presented in the preceding sections, surface potential values can be estimated by measurements of the salt effects on the various intrinsic probes. Various organic or inorganic molecules with negative or positive charges have been used as extrinsic probes of the surface potential. These include various merocyanine dyes,[37–39] 9-amino-acridine,[5,40–43] acriflavin,[37] safranin O,[37] 8-anilinonaphthalenesulfonate

[33] K. Matsuura, K. Masamoto, S. Itoh, and M. Nishimura, *Biochim. Biophys. Acta* **547**, 91 (1979).

[34] J. B. Jackson and A. R. Crofts, *FEBS Lett.* **4**, 185 (1969).

[35] K. Matsuura, K. Masamoto, S. Itoh, and M. Nishimura, *Biochim. Biophys. Acta* **592**, 121 (1980).

[36] M. Symons, A. Nuyten, and C. Sybesma, *FEBS Lett.* **107**, 10 (1979).

[37] K. Masamoto, K. Matsuura, S. Itoh, and M. Nishimura, *J. Biochem.* **89**, 397 (1981).

[38] K. Masamoto, K. Matsuura, S. Itoh, and M. Nishimura, *Biochim. Biophys. Acta* **638**, 108 (1981).

[39] S. Itoh, *in* "Advances in Photosynthesis Research" (C. Sybesma, ed.), Vol. 2, p. 355, Martinus Nijhoff/Dr. W. Junk, The Hague, 1984.

[40] G. F. W. Searle and J. Barber, *Biochim. Biophys. Acta* **502**, 309 (1978).

[41] W. S. Chow and J. Barber, *Biochim. Biophys. Acta* **591**, 82 (1980).

[42] Y. Yamamoto, R. C. Ford, W. S. Chow, and J. Barber, *Photobiochem. Photobiophys.* **1**, 271 (1980).

[43] Y. Yamamoto and B. Ke, *Biochim. Biophys. Acta* **636**, 175 (1981).

(ANS),[37,44] auramine O,[37] oxonol,[45] cyanines,[45] ethidium,[45] Tb^{3+},[46] EPR probes,[47,48] etc. Almost all charged molecules can be used as surface potential probes when some change of measurable parameters is present to indicate the amounts bound to the membranes. The response to the change of environmental factors other than surface potential must be negligible to get correct surface potential. This condition is generally difficult to attain, since organic dye molecules usually creep into the membrane to a greater or lesser extent and respond to the difference of free energy between the surface and interior of the membrane as well as to that between the bulk phase and surface. Nonspecific binding of the probes[37] and nonidealistic behavior of the ions should always be considered.[4–8]

Reliable information on the change of surface potential during energization of photosynthetic membranes is obtained by the use of merocyanines,[38,39] which have a localized negative charge on one end and show very low membrane permeability, with a due caution about the nonspecific binding. Use of the dyes indicated that the light-induced energization of chromatophore membranes induced only a small change of the surface potential on the outer surface of chromatophores (2–20 mV positive shift at 1–200 mM NaCl).[38,39] The dyes were assumed to respond to the local intramembrane electrical field change in the edge part of the membrane[38,39]: a portion of the membrane potential change. This conclusion seems to be in line with that obtained in mitochondrial membranes estimated from the response of 8-anilinonaphthalenesulfonate (ANS).[44] It suggests, on the other hand, that the large change of surface potential estimated by EPR probe in chloroplasts[47] might have detected some other process than the surface potential change of the entire membrane.

One major difficulty in the study of surface potential change is to define the surface plane between the aqueous phase and the actual membrane which has a complex three-dimensional structure. Each probe is detecting an electrochemical potential change at a different site inside the membrane boundary region. A probe may detect change of local charge distribution due to the conformation change of the specific binding site, a change of surface potential in the local limited area, or a change of surface potential delocalized over the entire membrane. In order to distinguish

[44] D. E. Robertson and H. Rottenberg, *J. Biol. Chem.* **258,** 11039 (1983).
[45] K. Masamoto, K. Matsuura, and S. Itoh, unpublished data.
[46] K. Hashimoto and H. Rottenberg, *Biochemistry* **22,** 5738 (1983).
[47] A. T. Quintahilha and L. Packer, *FEBS Lett.* **78,** 161 (1977).
[48] S. Tokutomi, T. Iwasa, T. Yoshizawa, and S.-I. Ohnishi, *FEBS Lett.* **114,** 145 (1980).

the changes of localized and delocalized surface potential it seems to be indispensable to check the probe responses to the salt-induced change of delocalized surface potential and to compare the responses of various probes.

Surface Potential and Electrochemical Potential of Proton across Membrane during Energization

Proton pumping during energization induces a change of intravesicular pH of energy-transducing membranes. The pH change induces the change of net surface charge density on the inner surface of the vesicles. It results in a change of surface potential and a change of intramembrane electrical field, especially under low ionic conditions. Figure 8 shows the ΔpH and electrical potential profiles across the chromatophore membrane of *R. sphaeroides* during energization by light.[49] The change of net surface charge density of the inner surface was estimated from the ΔpH value (measured by the 9-aminoacridine fluorescence-quenching method[50]) and the pH dependence of net surface charge density (measured in spheroplasts with right-side-out membrane sidedness[35]). When ΔpH was small during energization in the absence of valinomycin, the change of intramembrane electrical potential ($\Delta\varphi_{s-s}$, measured by the band shift of intrinsic carotenoid) almost equaled the change of membrane potential (bulk-to-bulk potential difference, $\Delta\varphi_{b-b}$). However, with a large ΔpH created in the presence of valinomycin, the change of intramembrane electrical field was mainly due to the change of Ψ_0 of inner surface and was estimated to be larger than the real $\Delta\varphi_{b-b}$ value.[49] The result indicates that the protein or lipid constituents of the membrane experience the electrical potential difference and ΔpH which are different from those between the bulk aqueous phases with a high activity of proton pumping.

It has not been seriously considered until the prediction of Rumberg and Muhle[28] that the change of intramembrane electrical potential is different from that between the bulk aqueous phases. Matsuura *et al.*[33,35] gave experimental evidence for this possibility by showing that the intramembrane electrical field can be changed independently from the membrane potential (between the bulk aqueous phases) as discussed in the preceding section. This phenomenon is very important when the accurate protonmotive force must be measured in the discussion of energy coupling, since some electrical potential probes or ΔpH probes (such as fluorescence probes) measure electrical potential or pH values on the mem-

[49] S. Itoh, *Plant Cell Physiol.* **23**, 595 (1982).
[50] R. Casadio, A. Baccarini-Melandri, and B. A. Melandri, *Eur. J. Biochem.* **47**, 121 (1974).

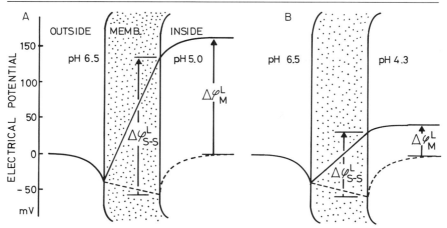

FIG. 8. Schematic profiles of the electrical potential across the chromatophore membrane in the dark (broken lines) and during illumination (solid lines) in the presence of 0.1 M KCl. (A) Without valinomycin; (B) with 1 μM valinomycin. The values for $\Delta\varphi^L_{S-S}$ and $\Delta\varphi^L_M$ were calculated from the $\Delta\varphi^L_{S-S}$ value measured by carotenoid band shift and ΔpH values estimated from the measurement of 9-aminoacridine fluorescence quenching (after Itoh[49]).

brane surface and some (such as measurements of distribution of inorganic ions) measure those in the bulk aqueous medium. In the calculation of $\Delta\bar{\mu}_{H^+}$, the combination of surface-to-surface ΔpH/$\Delta\varphi$ value and the bulk-to-bulk $\Delta\varphi$/ΔpH value will lead to an erroneous estimation. Depending on the calibration method, the obtained values of $\Delta\varphi$, ΔpH, and $\Delta\bar{\mu}_{H^+}$ can be over- or underestimated. In addition to this problem in the techniques for $\Delta\bar{\mu}_{H^+}$ measurement, the surface potential-induced interchange of electrical and chemical terms of electrochemical potential of proton or other ionic species on both surfaces of membrane will affect the rate of electron or ion transfer at the surface or across the membrane. This may also become very important under some circumstances, especially under artificial low ionic conditions.

Interaction of Small Protein Molecules and Membrane Surface

As has been discussed in the preceding section, interactions of small peripheral protein molecules and the membrane can be treated by considering the surface potential as in the case of reaction between small ions and the membrane component. It has been shown that salt effects on the reaction of plastocyanin with P700 on various types of thylakoid membranes,[14–20] have been successfully interpreted by considering the concentration change of this protein at the reaction site on the membrane. Behavior of water-soluble cytochromes[11] can also be interpreted as in the case of plastocyanin. In the case of interaction between cytochrome c_2 and the reaction center bacteriochlorophyll on the chromatophore membrane of *R. sphaeroides*, the situation was a little different suggesting the existence of net-charge-independent binding of the cytochrome to the reaction site.[14] However, even in this case once the cytochrome was detached from the reaction center, reaction of the cytochrome with the reaction center became more sensitive to salts.[14]

Changing the electrostatic interaction of peripheral proteins with the membrane surface sometimes induces a wide range of effects. One well-known example is the case of the CF_1 molecule (one moiety of the ATP synthetase) attached to the thylakoid membrane surface. This net negative protein can be detached from its counterpart (CF_0) on the membrane under low salt conditions,[51,52] and can be reassociated in the presence of K^+ or Mg^{2+}.[53,54] These processes affect the coupling of electron transport and phosphorylation, a case of the ionic control of electron transport reaction in the nonphysiological medium.[54]

Another recent hot field of the salt effects on the peripheral proteins can be found in the oxygen-evolving system of chloroplast, in which selective removal of the three peripheral proteins by the use of different types of high salt conditions (1 M NaCl,[55,56] 2 M CaCl$_2$,[57] 0.8 M Tris–Cl,[58] etc.) has now greatly facilitated the investigation of the water oxidation mechanism on the molecular level. The binding of Cl^- which is indispensable for O_2 evolution, to the active site on the membrane surface, seems to be regulated by these proteins.[59]

It seems that some of the salt effects on peripheral proteins can be interpreted by assuming the long range interaction between the net charges on the protein and the membrane according to the Gouy–Chapman theory under low ionic conditions. On the other hand, investigations of the salt effects at high ionic concentrations, at which the Coulombic interactions between charges are restricted to a shorter distance than the diameter of protein or membrane, give information about the interaction of charges in the local domains on the interacting surfaces of the protein and the membrane.

[51] A. T. Jagendorf and M. Smith, *Plant Physiol.* **37**, 135 (1962).
[52] M. Avron, *Biochim. Biophys. Acta* **77**, 699 (1963).
[53] A. T. Jagendorf, *in* "Bioenergetics of Photosynthesis" (Govindjee, ed.), p. 413. Academic Press, New York, 1975.
[54] A. Telfer, J. Baber, and A. T. Jagendorf, *Biochim. Biophys. Acta* **591**, 331 (1980).
[55] H. E. Åkerland and C. Jansson, *FEBS Lett.* **124**, 229 (1981).
[56] T. Kuwabara and N. Murata, *Plant Cell Physiol.* **23**, 533 (1982).
[57] T. Ono and Y. Inoue, *FEBS Lett.* **168**, 281 (1984).
[58] Y. Yamamoto, S. Shimada, and M. Nishimura, *FEBS Lett.* **151**, 49 (1983).

Other Phenomena Related to the Surface Charge Effect

Another surface charge-related phenomenon which can be interpreted by the application of the Gouy–Chapman theory and indirectly affects the redox reactions on the membrane is the regulation of interactions between the membranes. van der Waals force attracts membranes to each other, while the repulsive Coulombic force, which is related to the surface potential, prevents aggregation. Thus, the screening of the surface charges by salts leads to the adhesion of membrane as has been studied in colloid chemistry.[2,7] It shows the valence-specific characteristics of cations as seen with various phenomena with net negative membranes. This effect has been studied in detail in the study of stacking mechanism of thylakoid membrane of chloroplast, as reviewed by Barber.[5] It has been shown that the repulsive force can be quantitatively estimated by the consideration of the density of charges in the diffuse layer between two closely located membrane surfaces.[60,61] Not only the cation valence specificity of the salt effect but also the antagonism between the effects of mono- and divalent cation salts (effects of monovalent salt change their direction depending on the level of background divalent cations) have been interpreted by the estimation of space charge density (amount of screening charges) derived from the integration of the ion concentrations in the aqueous space between the membrane surfaces, using the surface potential value.[60,61] Stacking of the thylakoid membranes induces the rearrangements of the light-harvesting antenna and the reaction center complexes, and introduces the change in distribution of light energy between photosystems I and II.

In other types of biological membranes, the regulation of membrane stacking has not been extensively studied as in thylakoid membranes. However, even without stacking, lateral rearrangements of protein complexes on the surface will be regulated by surface ion concentrations and by surface pH values. Short range interaction of charges can be studied more adequately under higher ionic conditions. Actual pK values of the dissociable groups situated in the vicinity of active site can be more adequately estimated from the pH dependence of the reaction at high ionic concentrations as studied in the case of pH dependence of the reaction between ferredoxin and NADP-reductase on the membrane surface[12] or that of cytochrome (Fig. 5). Other physicochemical characteristics of the membrane such as membrane thickness, surface area, and phase transition temperature of the membrane are also affected by the surface potential.[7]

[59] S. Itoh and S. Uwano, *Plant Cell Physiol.* **27,** in press (1986).
[60] J. Barber, *FEBS Lett.* **118,** 1 (1980).
[61] H. Y. Nakatani, J. Barber, and M. J. Minski, *Biochim. Biophys. Acta* **545,** 24 (1979).

Concluding Remarks

As has been reviewed briefly, immobilized charges on the membrane surface induce different ionic concentrations, electrical potential, proton concentration, and redox potential at the membrane surface, and influence energy coupling processes at the surface or inside of the membrane. It is clear that these points should always be kept in mind when we design the *in vitro* experimental conditions. On the other hand, we can set up our special experimental conditions, which can never be attained under natural physiological conditions, by using the feature of surface charge effects. Although most of the examples in this article concern photosynthetic membranes, on which extensive work has been carried out, it is emphasized that the surface charge-related phenomena shown in preceding sections are common and general in all biological energy-transducing membranes.

The table summarizes the values of net surface charge density estimated in various membrane systems. In chloroplast membranes, analysis of the effects of monovalent salts on the reaction rates of various membrane component gave larger values of net surface charge density for the outer surface ($-1.1 \sim -1.4 \ \mu C/cm^2$) than the inner surface ($-0.8 \sim -1.1 \ \mu C/cm^2$), suggesting different distribution of charges on these surfaces. On the other hand, larger values of net surface charge density were estimated for the total membrane (average of inner and outer surfaces) from the comparison of mono- and divalent cation effects on the fluorescence yield of chlorophyll or 9-aminoacridine. The values obtained by the reaction method seem to be somewhat underestimated since they do not fit the case well with divalent cation salts as mentioned in the preceding section, although the effects of monovalent salts can be most correctly predicted with these values. On the other hand, the values obtained by fluorescence probes may be a slight overestimation due to the nonspecific binding of divalent cations to the surface. Binding of fluorescence probes also contributes to the surface charge density.

Our experience tells us that the larger surface charge density value is obtained when we use the difference of the effectivenesses of mono- and divalent salts in the estimation of net surface charge density. The value which is the best fit for the cases with mono- and divalent cation salts, in turn, cannot correctly explain the case with trivalent cations. The surface domain measured by the reaction-rate method, which monitors the local area in the vicinity of the membrane components, may be different from the entire surface. Thus, the comparison of the values of net surface charge density seems to be meaningful only among values obtained by similar estimation methods. The condition of the membrane may be also important since the depletion of peripheral proteins seems to change the net surface charge density of thylakoid inner surface.[62] In chromatophore membrane of photosynthetic bacteria, *Rhodopseudomonas sphaeroides,* different surface charge density values of -3.0 and -4.6 $\mu C/cm^2$ were estimated for the outer and inner surfaces from the comparison of the effects of mono- and divalent cation salts on the intramembrane electrical field, sensed by the carotenoid band shift.[33,35] Also in this membrane lower values were obtained from the analysis of kinetic parameters.[22] Net surface charge density values have also been estimated in membranes of rat liver mitochondria,[46] Jerusalem artichoke mitochondria,[24] and rat liver smooth microsomes.[25,27]

The values of charge density estimated for various biological membranes (see the table) are not very large ($-0.4 \sim -4.6$ $\mu C/cm^2$, which correspond to $0.25 \sim 2.9$ electronic charges per 1000 square angstroms) except for the case of *Halobacterium halobium* which grows at very high salt concentrations. The values of surface potential under physiological conditions, if it may be represented by 0.1 M monovalent salt, are estimated to be -27 mV with -2 $\mu C/cm^2$. Increase of the surface charge density to -4 $\mu C/cm^2$ gives -48 mV under the same conditions. It indicates that, even under physiological conditions, a few to several 10 mV of negative surface potential is expected to exist. Movement of divalent cations across the membrane, or change of the intravesicular pH during energization, is expected to change the surface potential value a few 10 mV. It is not very large but is not negligible either, especially when the correct values of electrochemical free energy difference involved in the functioning of the membrane molecules are required. However, in the presence of an adequate level of salt, electrochemical potential of proton or other ion seems to be almost in equilibrium with that in the bulk aqueous phase far from the surfaces, even under the energized conditions.[10,38,39] The electrochemical potential difference in the aqueous phase between the outer bulk phase and the membrane surface, as proposed in the "electrodic view" by Kell,[63] does not seem to occur under normal conditions. The interchange of the electrical and chemical terms of the electrochemical potential at the surface or inside of the membrane, as shown in the studies of the surface potential, is important in the understanding of the molecular mechanism of the energy coupling, since each redox or enzymatic component on the membrane is functioning in a microenvironment with different local electrical and chemical potentials. Development of methods for the modulation of the local electrical or

[62] C. T. Yerkes and G. T. Babcock, *Biochim. Biophys. Acta* **590,** 360 (1980).
[63] D. B. Kell, *Biochim. Biophys. Acta* **549,** 55 (1979).

chemical potential at the desired point inside the membrane, such as those recently carried out with hydrophobic ions which modulate the local electrical field in the hydrophobic part of the membrane,[10,64] may serve for the study of local coupling mechanisms.

Acknowledgments

The authors thank Drs. K. Masamoto, K. Matsuura, and Y Fujita for their kind discussions and help in preparation of the manuscript. The work is supported by a grant from the Ministry of Education, Science and Culture of Japan.

[64] S. Itoh, *Biochim. Biophys. Acta* **766,** 464 (1984).
[65] I. M. Møller, T. Lundborg, and A. Bérczi, *FEBS Lett.* **167,** 282 (1984).
[66] Y. Kimura, A. Ikegami, and W. Stoeckenius, *Photochem. Photobiol.* **40,** 641 (1984).

[7] Use of N,N'-Dicyclohexylcarbodiimide to Study Membrane-Bound Enzymes

By Maceij J. Nałęcz, Robert P. Casey,[1] and Angelo Azzi

Introduction

The use of N,N'-dicyclohexylcarbodiimide (DCCD) to study membrane-bound enzymes started with the observation by Beechey and co-workers[1a] that DCCD strongly inhibits the proton-translocating ATPase in mitochondria. It is now clear that this hydrophobic carbodiimide covalently binds to the hydrophobic part of the enzyme and blocks the transmembrane proton movements catalyzed by the ATPase. In recent years many other membrane-bound enzymes have been also shown to covalently bind and to be inhibited by DCCD and, in several cases, such observations led to the important conclusions about their reaction mechanisms (for review see Azzi *et al.*[2]). The summary of methods applied as well as comments concerning the use of DCCD will be given here.

[1] Deceased.
[1a] R. B. Beechey, A. M. Robertson, C. T. Holloway, and J. G. Knight, *Biochemistry* **6,** 3867 (1967).
[2] A. Azzi, R. B. Casey, and M. J. Nałęcz, *Biochim. Biophys. Acta* **768,** 209 (1984).

Chemical Aspects

A very comprehensive review on the chemistry of carbodiimides[3] together with more specialized ones on the use of these substances in biochemistry, by Hoare and Koshland,[4] Carraway and Koshland,[5] and Khorana,[6] have been published. Here we mention only a few basic characteristics of DCCD and its reactions (see also[2,7]).

DCCD at room temperature is solid, but may be purified by vacuum distillation at relatively low temperature. The compound is stable over long periods and does not tend to polymerize. Pure DCCD is characterized by a single infrared absorption band with λ_{max} at 2130 cm^{-1} and a single signal in ^{14}N-NMR.[3] Its melting point is at 35–36° and this property may be easily used for a rapid purity control of the compound. DCCD is not soluble in water, but is easily soluble in methanol, ethanol, or dimethylformamide.

Carbodiimides are highly reactive toward many organic functional groups[3] (see Fig. 1). A certain specificity, however, may be induced by the conditions of the reaction.[3–5] In aqueous solutions at acidic and neutral pH and at mild temperatures the predominant groups reacting with carbodiimides are carboxyls and sulfhydryls[5] [Fig. 1, reactions (3) and (4)]. For proteins it was found that under those conditions when —COOH and —SH groups are modified, also tyrosines react with carbodiimides[8] [Fig. 1, reaction (2)]. However, although the rates of reaction with sulfhydryl and carboxyl compounds are approximately equal, tyrosines react more slowly.[5] The reactivity of carbodiimides toward serine residues in proteins has also been reported.[9]

The reaction of DCCD with the carboxyl group alone may also have a number of different consequences [Fig. 1, reaction (4)]. An initial binding of DCCD to a single carboxyl group in the protein primarily formes dicyclohexyl-O-isourea which is highly unstable and vulnerable to a nucleophilic attack. In the case where the nuclephile is water [reaction (4), route A], the carboxyl will be regenerated with the conversion of the carbodiimide molecule to dicyclohexylurea and the protein will thus not be modified. It is also possible that an attack from a nucleophile different from water and located near the activated carboxyl occurs [reaction (4), route

[3] F. Kurzer and K. Duraghi-Zadeh, *Chem. Rev.* **67,** 107 (1967).
[4] D. G. Hoare and D. E. Koshland, *J. Biol. Chem.* **242,** 2447 (1967).
[5] K. L. Carraway and D. E. Koshland, this series, Vol. 25, p. 616.
[6] H. G. Khorana, *in* "Cellular Responses to Molecular Modulators," Miami Winter Symposium, XIIIth, Feodor Lynen Lecture, Vol. 1, p. 1, 1981.
[7] M. Solioz, *TIBS* **9,** 309 (1984).
[8] K. L. Carraway and D. E. Koshland, *Biochim. Biophys. Acta* **160,** 272 (1968).
[9] T. E. Banks, B. K. Blossey, and J. A. Shafer, *J. Biol. Chem.* **244,** 6323 (1969).

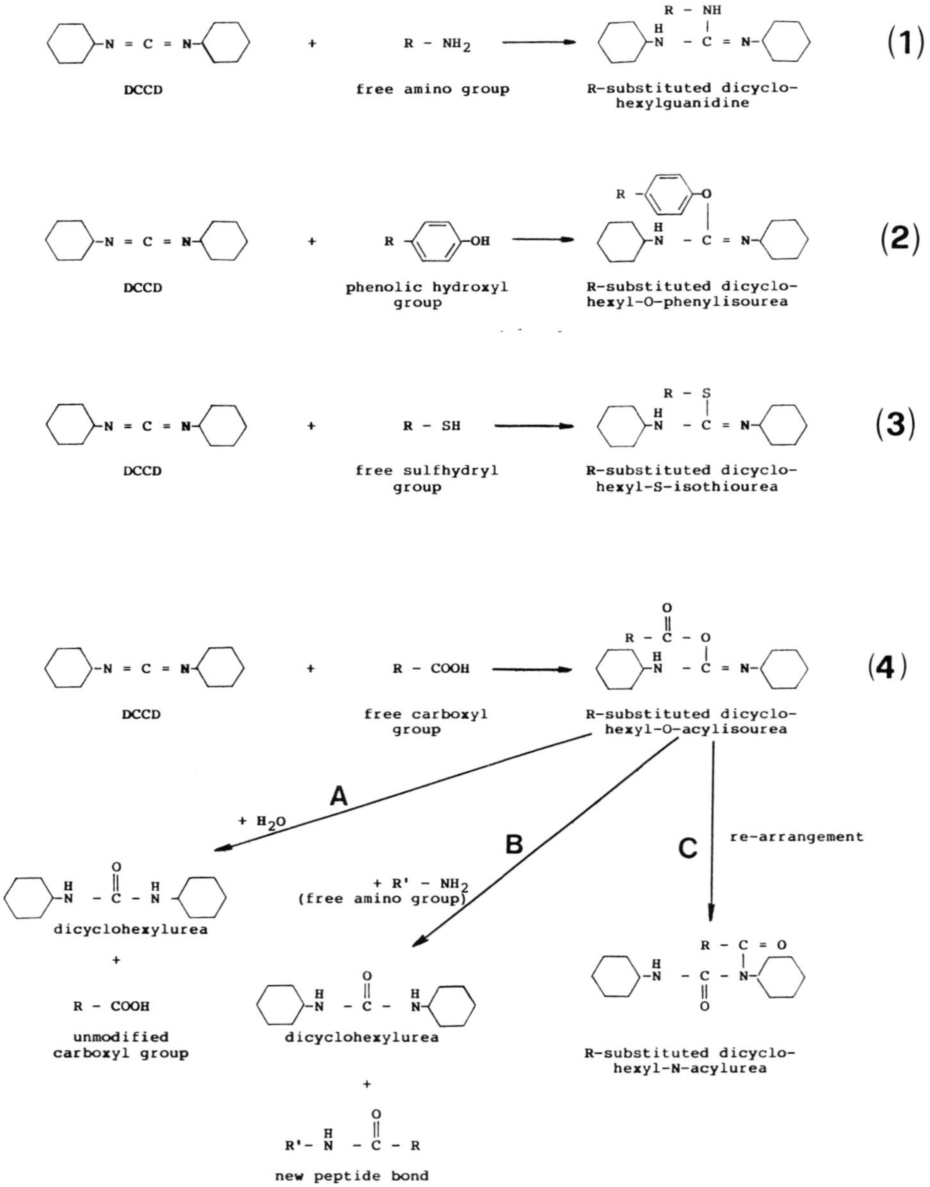

FIG. 1. Schematic representation of possible reactions between DCCD and functional groups of proteins. (1) Covalent binding to an amino group[3,9]; (2) covalent binding to a phenolic hydroxyl group[3,9]; (3) covalent binding to a sulfhydryl group[3]; (4) possible reactions following an initial binding of DCCD to a carboxyl group in a protein.[2] For details see text.

B]. If the attacking nucleophile is an ε-amino group of the neighboring amino acid the reaction results in the formation of a new peptide bond within the protein, with the release of free dicyclohexylurea. Such a cross-linking may occur intra- or intermolecularly, the latter phenomenon being possible for polypeptides located in close proximity. A third possible process [reaction (4), route C] is the shift of the activated acyl group to one of the N atoms (the so-called N-acyl shift) resulting in the stable binding of dicyclohexylurea to the protein. This is the only reaction leading to the incorporation of radioactivity in the case of [^{14}C]DCCD. Since this process requires the absence of nucleophiles in the vicinity of the activated group it follows that practically it can occur only in a hydrophobic environment.

Generally, unspecific reactions of carbodiimides may be restricted not only by the experimental conditions such as solvent, temperature, time, pH, and presence of catalysts,[3] but also by the nature of the carbodiimide. DCCD, for example, being hydrophobic, preferentially penetrates into lipids and hydrophobic domains in proteins. Its reactivity is thus restricted to certain sites located in the hydrophobic environment. Water-soluble carbodiimides, on the other hand, react preferentially at sites exposed to the water phase or at least located in the hydropholic domains of proteins. Usually DCCD does not label the same sites as water-soluble carbodiimides and thus the latter compounds are often used for negative controls in DCCD labeling experiments.

In order to obtain more information about the environment of the DCCD binding site a number of derivatives of this substance have been employed, e.g., N-(2,2,6,6-tetramethylpiperidyl-1-oxyl)-N'-cyclohexylcarbodiimide (NCCD), a spin-label derivative. This has been used to provide information about localization, mobility, and polarity in the DCCD-binding regions of the mitochondrial ATPase and cytochrome c oxidase.[2] A derivative of DCCD which becomes fluorescent only after interaction with the protein has also been reported.[10]

General Procedures

Incubation with DCCD

DCCD is a commercial product obtained from Fluka, Sigma, or other chemical companies. Since commercially available preparations of DCCD differ significantly in purity, an estimation of the melting point or thin-layer chromatography (TLC) should be applied to every new batch of the

---

[10] C. C. Chadwick and E. W. Thomas, *Biochim. Biophys. Acta* **730**, 201 (1983).

reagent. The latter procedure is usually performed on aluminum oxide G plates (neutral, type E, Merck) using benzene as eluent.[11] Pure DCCD should show a single spot after color development with I_2 vapors, with an R_f of 0.71. If needed, DCCD may be additionally purified by a distillation under reduced pressure.

[^{14}C]DCCD is a commercial product from Amersham (Buckinghamshire, England) or CEA (Gif sur Yvette, France).

DCCD is most often used by diluting in the appropriate buffer small aliquots of stock solutions prepared of 100–500 mmol/liter in ethanol. Methanol or dimethylformamide can also be employed. A fresh solution of carbodiimide should be prepared any time that precise titration data are required, since, once in solution, DCCD tends to decompose to dicyclohexylurea (appearing as a white precipitate at the bottom of the tube) during storage even at low temperatures.

It is important to prepare and store DCCD solutions in glass containers since the carbodiimide is easily adsorbed at plastic surfaces.

The experimental conditions to be employed in order to modify a protein with DCCD vary from system to system (see below). The time of incubation is of the order of 12 to 18 hr at around 0°, 2 to 3 hr at 12–13° and less than an hour at room temperature and above. The solvent itself may also influence many membrane and protein systems, therefore it appears essential to perform control experiments in the presence of the same amount of solvent as used for the addition of DCCD.

If it is expected that DCCD modifies some functional groups located in or near the active site of the enzyme, protection by the substrate or by the inhibitor during incubation with DCCD may give useful information.

In general, more reagent should be used with membranous systems due to the presence of lipid in which DCCD easily partitions, and in comparative studies the same amount of lipid should be present. It is, for example, incorrect to correlate the binding of [^{14}C]DCCD studied in an isolated enzyme with the inhibition of its activity measured in reconstituted proteoliposomes.

Stopping of the Reaction

Removal of the carbodiimide can be realized by extraction with an organic solvent,[12] a procedure which, however, denatures several proteins. A milder procedure is to remove DCCD by using rapid filtration of

---

[11] F. S. Stekhoven, R. F. Waitkus, and H. Th. B. van Moerkerk, *Biochemistry* **11**, 1144 (1972).

[12] R. P. Casey, M. Thelen, and A. Azzi, *J. Biol. Chem.* **255**, 3994 (1980).

the reaction mixture through a small Sephadex G-50 column,[13] which, however, requires controls to assure that all nonreacted DCCD has been removed. Addition to the DCCD-treated sample of bovine serum albumin (about 10 mg/ml) followed by ultracentrifugation was shown to be also an efficient way of removing the carbodiimide.[14] This procedure, which requires a relatively long execution time for small samples, may be accelerated by the use of an airfuge.

Another way of stopping the reaction is by diluting the sample with a concentrated solution of a carbodiimide-reacting reagent such as ammonium acetate or glycine methyl ester in the presence of acetic acid.[4,5] The resulting O-acetylisourea, however, may cause acetylatation of protein side chains.[5]

If the reaction with DCCD is performed at room temperature, it can be stopped by diluting the sample with buffer without DCCD and placing it in ice.[15] The possibility that the reaction proceeds further to a significant rate during activity measurements, centrifugations or SDS–gel electrophoresis has to be checked in every single case.

If the treatment with DCCD was performed with a system containing a large amount of lipids (e.g., proteoliposomes), that has to be removed prior to further studies. The following techniques are recommended: (1) extraction with organic solvent[16]; (2) solubilization with detergent followed by sedimentation of the pure protein through a layer of 10% sucrose by ultracentrifugation[12,15]; (3) solubilization with detergent and precipitation of the protein with trichloroacetic acid followed by centrifugation.[16] A combination of these three procedures may also be used.[12,16]

Side Reactions

Since carbodiimides react with a large number of functional groups,[2,3] it is impossible to establish a priori with which group the reaction will take place in a given system.[17] When carboxyl groups are to be activated, the prior use of reversible reagents to protect sulfhydryl groups is advisable.[5]

The reaction of DCCD with phospholipid molecules has also been described[16] and as a consequence of this the presence of phospholipids and their amount relative to protein cannot be neglected when planning protein modification experiments using DCCD.

[13] H. S. Penefsky, *J. Biol. Chem.* **242,** 5789 (1967).
[14] K. Sigrist-Nelson and A. Azzi, *J. Biol. Chem.* **255,** 10638 (1980).
[15] M. J. Nałęcz, R. P. Casey, and A. Azzi, *Biochim. Biophys. Acta* **724,** 75 (1983).
[16] D. S. Beattie, L. Clejan, and C. G. Bosch, *J. Biol. Chem.* **259,** 10526 (1984).
[17] A. Azzi and M. J. Nałęcz, *Trends Biochem. Sci.* **9,** 513 (1984).

Specific Procedures

Use of DCCD as an Inhibitor of Oxidative Phosphorylation in Mitochondria

DCCD, when incubated with intact mitochondria under conditions described below, inhibits coupled respiration at sites I, II, and III of the respiratory chain. Uncoupling agents relieve this inhibition. Under similar conditions DCCD inhibits also the adenosinetriphosphatase and P_i-ATP exchange reactions in submitochondrial particles. The mode of action of DCCD is therefore similar to that of oligomycin. However, the inhibition by DCCD is not reversible since the binding of carbodiimide is covalent.[1,18]

Membrane Preparations and Reagents. Beef heart, rat heart, or rat liver mitochondria as well as submitochondrial particles are prepared by standard methods.[19] The particles are suspended in a buffer of 250 mM sucrose and 10 mM Tris–chloride (pH 7.4).

Measurement of Respiratory Rates. This is done with a Clark-type oxygen electrode, e.g., as described previously.[1]

Measurement of ATPase Activity in Submitochondrial Particles. Two methods may be used. In the first, the consumption of ATP is monitored by measuring the P_i content in the reaction mixture after stopping the reaction by the addition of trichloroacetic acid followed by centrifugation.[1] In the second, the ATPase activity is measured spectrophotometrically according to the method of Pullman *et al.*[20]

Measurement of DCCD-Induced Inhibitory Effects. DCCD, when added at the concentration of about 40 nmol of DCCD per mg of protein, produces an immediate inhibition of ADP-stimulated (state 3) mitochondrial respiration as well as the inhibition of ATPase activity in submitochondrial particles. For such a measurement DCCD is added directly to the appropriate reaction mixture containing already mitochondria or submitochondrial particles. The reaction is thus started by addition of substrate.

It is also possible to lower substantially the concentration of DCCD needed for inhibition by preincubating mitochondria and submitochondrial particles with the carbodiimide:

1. For mitochondria, the preincubation is performed by suspending the organelles at the concentration of about 75 mg of protein/ml in 250 mM sucrose and treating the aliquots with either ethanol (control) or 100 mM

[18] K. J. Cattell, C. R. Lindop, I. G. Knight, and R. B. Beechey, *Biochem. J.* **125,** 169 (1971).
[19] This series, Vol. 10, respective chapters.
[20] M. E. Pullman, H. S. Penefsky, A. Datta, and E. Racker, *J. Biol. Chem.* **235,** 3322 (1960).

solution of DCCD in ethanol to give the final concentration of 2 nmol of DCCD per mg of protein. After mixing, the mitochondrial suspensions are stored at 0° for about 18 hr. DCCD-treated organelles are unable to develop state 3 respiration, though their state 4 as well as uncoupler-stimulated respiratory rates are as those of the control.

2. For submitochondrial particles, the preincubation is performed by suspending the membranes at the concentration of about 35 mg of protein/ml in 250 mM sucrose and treating the aliquots, as in the case of mitochondria, with ethanol (control) or with a solution of DCCD in ethanol to give a final concentration of 2 nmol of DCCD per mg of protein. The membrane suspension is then stored at 0° for about 20 hr. The treatment with DCCD results in full inhibition of the ATPase activity.

Labeling of Mitochondrial Membranes with [^{14}C]DCCD. Mitochondrial membranes (submitochondrial particles) are suspended at a protein concentration of about 40 mg of protein/ml in a medium containing (final concentrations) 250 mM sucrose, 1 mM ATP, 1 mM succinate, 5 mM MgCl$_2$, and 10 mM Tris–chloride buffer (pH 7.4) and incubated for 16–20 hr at 0° with the required amount of [^{14}C]DCCD, usually at a final concentration of 2 nmol/mg of protein. The membranes are then sedimented by centrifuging at 105,000 g for 30 min and are washed twice with 250 mM sucrose–10 mM Tris–chloride buffer (pH 7.4). First washing may be performed with the buffer containing additionally bovine serum albumin (BSA, 10 mg/ml) to remove noncovalently bound carbodiimide, though no changes in the final labeling pattern were observed when washing with or without BSA was used. After final resuspension the labeled membranes may be stored at −20°.

Identification of the DCCD-Binding Polypeptide(s). Samples of labeled membranes are dissolved at a protein concentration of ~1 mg of protein/ml by incubating at 40° for 14 hr in a solution containing (final concentrations) 3% (w/v) of sodium dodecyl sulfate (SDS), 5 mM EDTA, 5 mM dithiothreitol, and 100 mM sodium phosphate buffer (pH 7.0). Suspensions of membranes containing 60 mg of protein/ml are diluted directly into this solution.

The SDS–gel electrophoresis of the solubilized membrane material may be performed with different acrylamide gel systems. After slicing of the gels into 1-mm discs, each disc is dissolved in 1 ml of 28% (w/v) hydrogen peroxide, the solutions are heated at 80° for 14 hr, and then their radioactivity is measured. About 90% of the radioactivity is associated with a single protein band of M_r about 8 kDa. DCCD-binding protein can be extracted from the mitochondrial membranes with 25 volumes of chloroform–methanol (2:1, v/v), as for the extraction of lipids, and as a consequence of this it was named "proteolipid."[18]

Other Applications and Comments. DCCD, when used at very low concentrations, reacts preferentially at a single site, i.e., the proteolipid subunit of the mitochondrial ATPase. At higher concentrations, however, several other modifications of mitochondrial functions can occur, such as inhibition of electron flow at the level of both succinate- and ubiquinol-cytochrome *c* reductases in beef heart mitochondria,[21] inhibition of the formation of succinate-driven transmembrane pH gradient in submitochondrial particles,[22] inhibition of F_1-ATPase (see below), and inhibition of the mitochondrial nicotinamide nucleotide transhydrogenase (for review see Azzi *et al.*[2]). Stimulation by DCCD of succinate-ubiquinone reductase activity[23] and of the electron flow from succinate to oxygen[24] in rat liver mitochondria was also reported. In addition, Beattie and Villalobo[25] observed an increase in H^+ permeability of rat liver mitochondria treated with DCCD, suggesting that DCCD may act like an uncoupling agent. It seems possible that DCCD in high concentrations may induce multiple modifications of mitochondrial membranes.[2]

Reaction of DCCD with the F_1F_0-Type H^+ Translocating ATPases

The effects of DCCD observed with whole mitochondria led to the studies on the isolated F_1F_0-ATPase. Already at an early stage the inhibition of both the ATPase and the ATP synthase activity of the enzyme was attributed to the binding of DCCD to the hydrophobic polypeptide (proteolipid) representing a low M_r component of the membrane sector of the enzyme, F_0.[18] A similar binding and inhibition was observed in the H^+ translocating ATPases from *Neurospora crassa* and yeast mitochondria, from chloroplasts and from plasma membranes of *E. coli* and of the thermophilic bacterium PS3 (for review see Azzi *et al.*[2]). Later it was discovered that not only the hydrophobic F_0, but also the water-soluble part of the enzyme, F_1, may be modified with DCCD, though under different conditions and at much higher carbodiimide concentrations.[26]

Since details on the labeling of the F_0 sector of the ATPase, on the extraction of the proteolipid, its identification as the H^+ translocating channel and further characterization of its reactions with DCCD are de-

---

[21] M. Degli Esposti, G. Parenti-Castelli, and G. Lenaz, *It. J. Biochem.* **30,** 453 (1981).

[22] G. Lenaz, M. Degli Esposti, and G. Parenti-Castelli, *Biochem. Biophys. Res. Commun.* **105,** 589 (1982).

[23] M. Degli Esposti, E. M. M. Meier, J. Timoneda, and G. Lenaz, *Biochim. Biophys. Acta* **725,** 349 (1983).

[24] B. D. Price and M. D. Brand, *Biochem. J.* **206,** 419 (1982).

[25] D. S. Beattie and A. Villalobo, *J. Biol. Chem.* **257,** 14745 (1982).

[26] R. Pougeois, M. Satre, and P. V. Vignais, *Biochemistry* **18,** 1408 (1979).

scribed elsewhere[27]; here only the modification of the F_1 part of the enzyme will be considered.

Mitochondrial Water-Soluble Complex (F_1). The critical factor which controls the binding of DCCD to the F_1 sector of the ATPase is the pH of the medium. F_1 binds DCCD preferentially at acidic pH, contrary to F_0 which reacts with this carbodiimide in a much broader pH range. A half-maximal effect of DCCD on F_1-ATPase is observed around pH 7.5 and no modification occurs above pH 8.5. The rate of inhibition induced by DCCD is decreased in the presence of ATP and ADP and increased in the presence of $MgCl_2$.[26]

Preparation of the Mitochondrial F_1-ATPase. Beef heart mitochondrial F_1-ATPase is prepared and stored as an ammonium sulfate suspension according to a published method.[28] Prior to incubation with DCCD the enzyme is subjected to elution-centrifugation,[13] in order to remove ammonium sulfate. The activity of ATPase is measured with one of the usual methods, as described above for submitochondrial particles.

Incubation with DCCD. F_1-ATPase (0.7 mg/ml) is preincubated at 24° in tubes containing 0.1 ml of a medium composed of 50 mM Tris, 50 mM MOPS [3-(N-morpholino)propanesulfonic acid], 2 mM EDTA, and 4 mM ATP (final pH 7.0). An ethanol solution of DCCD is added to give a final concentration of 200 nmol/mg protein (the lowest amount of carbodiimide necessary to obtain maximal inhibition[26]). A parallel incubation is carried out with ethanol, in the absence of DCCD (control). Aliquots of the incubation mixtures are assayed at various times for the ATPase activity. Under these conditions the enzyme is fully inhibited by DCCD after about 60 min of incubation. The half-time of inactivation (about 30 min) is shorter in the presence of either 10 mM ATP or 10 mM ADP (25 min) and prolonged in the presence of 10 mM $MgCl_2$ (45 min). The kinetic analysis of the inhibitory effect of DCCD shows that the carbodiimide acts as a noncompetitive inhibitor of the enzyme, without affecting the K_m value for ATP.

Binding of [^{14}C]DCCD. Incubation of F_1-ATPase with [^{14}C]DCCD is performed as before, but using a double amount of the material (0.2 ml). The reaction with carbodiimide is stopped by passing the mixture through a small column of Sephadex G-50 (fine).[13] The radioactivity in the eluate (bound [^{14}C]DCCD) is corrected for the background by omission of the enzyme in the reaction mixture. Aliquots of the eluate prepared at various times of the incubation are assayed for ATPase activity and [^{14}C]DCCD

[27] A. Azzi, K. Sigrist-Nelson, and N. Nelson, this series, in press.
[28] A. F. Knowles and H. S. Penefsky, *J. Biol. Chem.* **242,** 2447 (1972).

incorporation. The linear relationship between binding of DCCD and AT-Pase inhibition is observed, suggesting that both phenomena are coupled.

Identification of the Binding Subunit. F_1-ATPase (1 mg of protein) is incubated with [^{14}C]DCCD as above, except that the pH is 7.5. The incubation is prolonged to about 2 hr, when the enzyme is found inactivated to about 85%. After precipitation by ammonium sulfate (50% final concentration) the pellet is solubilzied in 40 mM Tris–chloride, 2 mM EDTA, 4 mM ATP (pH 7.5) and dialyzed for 1 hr at 20° against the same buffer. A sample (70 μg protein) is analyzed by SDS–polyacrylamide gel electrophoresis. After staining and destaining, the gel is scanned and the distribution of radioactivity is determined.[26] The profile of radioactivity shows one peak associated with the polypeptide of an apparent M_r of 50,000, corresponding to the β-subunit of F_1-ATPase.

Remarks. F_1-ATPase, though representing a water-soluble enzyme, is still specifically interacting and being inhibited by DCCD, but at much higher concentrations than required for inhibition of the whole membrane-bound F_0F_1 complex. The inhibition of F_1-ATPase was shown to correlate with the binding of carbodiimide to the β-subunit of the enzyme and it was suggested, on the basis of kinetic analysis of inactivation, that the interaction with DCCD involves a single modification of an essential amino acid residue.[26] Indeed, it was found later that DCCD binds to a specific glutamic acid residue in the β-subunit.[29]

Reaction of DCCD with the Ca^{2+}-ATPase from Sarcoplasmic Reticulum

Another membrane-bound enzyme known to be inhibited by DCCD is the Ca^{2+}-ATPase from sarcoplasmic reticulum. The inhibition of the enzyme occurs only in the presence of calcium chelating agents and is prevented by calcium, suggesting that calcium ions bind to the residue which is modified by carbodiimide.[30]

Enzyme Preparations and Their Activity. Sarcoplasmic reticulum is prepared according to MacLennan.[31] Ca^{2+} uptake by isolated sarcoplasmic reticulum vesicles is assayed as described,[30] by using $^{45}CaCl_2$ and measuring the radioactivity associated with membrane vesicles quickly passed through a minicolumn of Sephadex G-50 (fine) by centrifugation.[13]

Ca^{2+}-ATPase is purified from sarcoplasmic reticulum as described in Zimniak and Racker.[32] The ATPase activity of the membrane-bound as

---

[29] F. S. Esch, P. Bohlen, A. S. Otsuka, M. Yoshida, and W. S. Allison, *J. Biol. Chem.* **256,** 9084 (1981).

[30] U. Pick and E. Racker, *Biochemistry* **18,** 108 (1979).

[31] D. H. MacLennan, *J. Biol. Chem.* **245,** 4508 (1970).

[32] P. Zimniak and E. Racker, *J. Biol. Chem.* **253,** 4631 (1978).

well as of the isolated enzyme is assayed by measuring of $^{32}P_i$ released from [γ-^{32}P]ATP during incubation for 5 min at 37°.[30]

Ca^{2+} Binding. Ca^{2+} binding to the purified ATPase (in the absence of ATP) is assayed by measuring the absorption changes of arsenazo III upon addition of increasing concentrations of CaCl$_2$ with a dual-wave-length spectrophotometer.[30] The absorption changes of arsenazo III recorded in the absence of protein are used as standards.

Incubation with DCCD. The inhibition of Ca^{2+}-ATPase by DCCD is pH dependent, with maximal effects at pH 6.0.[30] This pH is therefore chosen for routine incubations with DCCD. Of the enzyme (sarcoplasmic reticulum or purified Ca^{2+}-ATPase) 0.3–0.4 mg is incubated at room temperature in 5 ml of the medium containing 100 mM KCl, 30 mM Tris/(N-morpholino)ethanesulfonic acid (Mes) of pH 6.0, 1 mM ethylene glycol bis(aminoethyl ether)-N,N,N',N'-tetraacetic acid (EGTA), or 1 mM CaCl$_2$ and DCCD to give a final concentration of about 250 nmol/mg of protein. After 1 hr the reaction mixture is cooled to 4° and centrifuged at 105,000 g for 45 min. The sedimented enzyme is resuspended in 0.2 ml of 200 mM sucrose, 50 mM Tris–chloride (pH 8.0) and used for further studies. A parallel sample of the enzyme is incubated as above, in the absence of DCCD, as a control for thermal inactivation of the Ca^{2+}-ATPase.

The treatment with DCCD results in substantial inhibition (60–85%) of the ATPase activity measured either with the sarcoplasmic reticulum membranes or with the isolated enzyme. The inhibitory effects are even more pronounced in the presence of the inonophore A-23187, but almost completely absent in the presence of 1 mM CaCl$_2$. Similarly inhibited is also the uptake of ^{45}Ca^{2+} by sarcoplasmic reticulum vesicles.[30] Binding of Ca^{2+} is also diminished to approximately one-half, after treatment of the isolated enzyme with DCCD.

Binding of [^{14}C]DCCD to the Isolated Enzyme. The Ca^{2+}-ATPase (0.66 mg protein) is incubated as described above, with [^{14}C]DCCD. At different times the aliquots of the reaction mixture are withdrawn and added to precooled tubes containing either trichloroacetic acid or CaCl$_2$ to give a final concentration of 5% and 3 mM, respectively. The former is used to determine DCCD binding; the latter is washed by centrifugation and assayed for ATPase activity. For DCCD-binding studies the enzyme precipitate is collected by centrifugation (145,000 for 30 min) and washed sequentially with 5 ml of (1) 10% trichloroacetic acid plus 2% sodium deoxycholate, (2) 40% ethanol, and (3) diethyl ether to remove noncovalently bound DCCD. Finally, the pellet is dried, redissolved in a detergent-containing buffer, and protein-bound radioactivity is counted.

Binding of [^{14}C]DCCD correlates kinetically with the inhibition of the ATPase activity, suggesting that the two phenomena are coupled. The presence of CaCl$_2$ during incubation of the Ca^{2+}-ATPase with [^{14}C]DCCD substantially lowers the amount of radioactivity incorporated into the protein. Complete inhibition of the isolated ATPase is accompanied by binding 4–5 nmol of [^{14}C]DCCD per mg protein in the absence of Ca^{2+}, compared with 2 nmol bound per mg protein in the presence of Ca^{2+}. In the latter case no ATPase inhibition is observed.[30]

Identification of [^{14}C]DCCD-Binding Region. Ca^{2+}-ATPase from sarcoplasmic reticulum consists of a single subunit and therefore the identification of the region involved in DCCD binding may be obtained only by protein fragmentation. The incubation with [^{14}C]DCCD is performed as described above. The reaction mixture containing about 0.4 mg of protein is cooled in the presence of 3 mM CaCl$_2$, sedimented by centrifugation, and resuspended in 1 ml of a medium containing 1 M sucrose, 100 mM KCl, and 20 mM Tris-maleate (pH 7.0). Trypsin (20 μg) is added, followed by the addition of trypsin inhibitor (50 μg) after 30 min of incubation at 30°. The further procedure is as described by MacLennan.[31] The precipitated protein is centrifuged, washed as for the binding studies, and finally subjected to SDS–polyacrylamide gel electrophoresis. After staining and destaining of the gel, the distribution of radioactivity is measured.

Trypsin digestion of the Ca^{2+}-ATPase results in the formation of four major protein bands, the smallest of which, a fragment of M_r 20,000, contains covalently bound DCCD.

Remarks. Since the sarcoplasmic reticulum Ca^{2+}-ATPase has an M_r of 100,000, it may be calculated that 1 mol of DCCD binds to 4 mol of the enzyme. On this basis it was proposed that the functional enzyme unit is a tetramer.[30] In addition, the protecting effect of Ca^{2+} on the inhibition and binding of DCCD suggested that the carbodiimide reacts specifically with the Ca^{2+}-binding site, which in subsequent studies[33] was identified as a carboxylic group.

Reaction of DCCD with Cytochrome c Oxidase

The reaction of DCCD with the cytochrome c oxidase is well established and characterized. Studies on the interaction of this reagent with the purified detergent–oxidase complex have been reported, showing multiple binding sites for the reagent.[12,34] The enzyme reconstituted in vesicles is labeled only at one subunit with parallel inhibition of the proton

[33] A. J. Murphy, *J. Biol. Chem.* **256**, 12046 (1981).
[34] L. J. Prochaska, R. Bisson, R. A. Capaldi, G. C. M. Steffens, and G. Buse, *Biochim. Biophys. Acta* **637**, 360 (1981).

translocation properties of the oxidase. The modification of the reconstituted enzyme and its assay will be considered here.

Modification of Cytochrome c Oxidase with DCCD. Enzymic Preparations and Reagents. Cytochrome *c* oxidase is prepared from bovine heart mitochondria by the method of Yu *et al.*[35] and stored at −80°.

L-Phosphatidylcholine from soybean phospholipids (asolectin, type II-S, Sigma) is further purified by acetone–ether treatment[36] before use.

Potassium cholate is prepared from recrystallized cholic acid (Fluka AG, Switzerland).

Preparation of Cytochrome c Oxidase Vesicles. Cytochrome *c* oxidase is added to a sonicated suspension of asolectin (40 mg/ml) in 24.5 mM potassium cholate, 0.1 M HEPES (pH 7.4) to give a final heme *a* concentration of 7.5 μM and the mixture dialyzed as described in Casey *et al.*[37] The respiratory control ratio (the resting electron transfer rate versus the maximal rate after the addition of a protonophore) should range between 4 and 10.

Inhibition of the H⁺ Pump. DCCD is added to cytochrome *c* oxidase vesicles (0.25 to 0.5 nmol of oxidase) suspended in 75 mM choline chloride, 25 mM KCl and, for the spectrophotometric measurements of the proton pump, 50 μM Phenol Red. Up to 100–200 mol DCCD per mol enzyme the inhibition of proton translocation is observed associated with the labeling of a single subunit. Above 1–2% ethanol nonspecific modifications are seen.

During the incubation at 12–15° for 3 hr before the measurement the vesicles maintain a low proton permeability, although the activity of the control sample is somewhat decreased. If the incubation is carried out at 4° for 20–24 hr, the membrane proton permeability is increased but the enzymatic activity of the control is fully retained.

[¹⁴C]DCCD Labeling. To an 0.5 ml sample of the stock vesicle suspension containing 5 nmol of heme *a* is added 210 nmol of unlabeled DCCD and 20 nmol of ¹⁴C-labeled DCCD (final specific activity of 4 Ci/mol) and the sample is incubated as for the activity measurements. Then 0.05 ml of 20% potassium cholate is added and the sample is applied to approximately 3.2 ml of 10% sucrose in a centrifuge tube and centrifuged at 4° and approximately 200,000 *g* for 15–22 hr. The oxidase should form a fairly tight pellet. The supernatant is discarded.

Functional and Structural Analysis of the DCCD-Modified Oxidase. Materials and Apparatus. Medium for proton translocation measure-

[35] C. Yu, L. Yu, and T. E. King, *J. Biol. Chem.* **250**, 1383 (1975).
[36] Y. Kagawa and E. Racker, *J. Biol. Chem.* **246**, 5477 (1971).
[37] R. P. Casey, J. B. Chappell, and A. Azzi, *Biochem. J.* **182**, 149 (1979).

ments: 75 mM choline chloride, 25 mM KCl, 50 μM Phenol Red (for spectrophotometric measurements) or 75 mM choline chloride, 25 mM KCl (for pH electrode measurements).

Medium for electron transfer measurements: 30 mM sodium ascorbate, 40 μM cytochrome c, 90 μM TMPD, 10 mM HEPES, pH 7.4 (with or without Phenol Red to correspond to the proton translocation medium).

Valinomycin: ethanol solution (0.2 mM).

Carbonyl cyanide m-chlorophenylhydrazone (CCCP): ethanol solution (1 mM).

Reduced cytochrome c: prepared from horse heart ferrocytochrome c (Sigma, type VI) by dithionite reduction and Sephadex G-25 chromatography, concentration 0.5–1 mM.

Oxalic acid: 0.5 mM, freshly prepared.

A dual-wavelength spectrophotometer fitted with a thermostat and a magnetic stirring device *or* a sensitive, rapidly responding pH electrode and pH meter.

A thermostatted Clark-type O_2 electrode.

Measurement of Proton Pumping Activity. The sample, following incubation, is supplemented with 0.5 nmol of valinomycin and subsequently pH changes are monitored using a pH electrode or spectrophotometrically, as the change in the absorbance of Phenol Red at 556.5 minus 504.5 nm.[37] In both cases, the sample pH is set at 7.4 and the change in the external proton activity resulting from the oxidation of ferrocytochrome c (1–2 nmol) is measured. In both control and DCCD-treated samples an acidification is observed, resulting from proton pumping by the oxidase, though this is much smaller in the latter case. The pH changes are calibrated by additions of oxalic acid. CCCP (5 nmol) is then added, and the subsequent oxidation of ferrocytochrome c now leads, in both types of sample, to an uptake of 1 proton per electron transferred (see Fig. 2).

Measurement of Electron Transfer Activity. The cuvette of the O_2 electrode is filled with the measuring medium and closed. Of the vesicle suspension (approximately 0.05 nmol of heme a) 0.01 ml is added and the rate of O_2 consumption monitored at 25°. The measurement is repeated in the presence of 0.5 nmol of valinomycin and 5 nmol of CCCP. Under both coupled and uncoupled conditions, O_2 consumption is slower in the DCCD-treated sample than in the control.

Determination of DCCD-Binding Subunit(s). The pellet of [^{14}C]DCCD-labeled cytochrome c oxidase is suspended in 50 to 100 μl of 25 mM Tris acetate, 2 mM EDTA, 3% SDS, and then electrophoresed according to Swank and Munkres[38] on a slab gel containing 10% poly-

[38] R. T. Swank and K. D. Munkres, *Anal. Biochem.* **39**, 462 (1971).

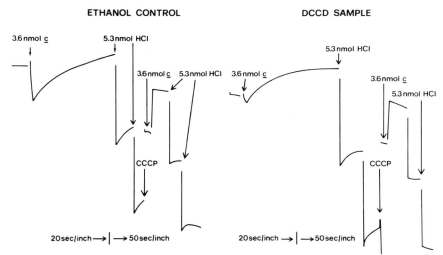

FIG. 2. Inhibition by DCCD of ferrocytochrome c-induced H^+ extrusion from cytochrome c oxidase vesicles is not an uncoupling phenomenon. Aliquots (0.1 ml) of cytochrome c oxidase vesicles (0.3 nmol cytochrome c oxidase) were added to 1.3 ml of 75 mM choline chloride, 25 mM KCl, 50 μM Phenol Red (pH 7.4), followed by 11 μl of 20 mM DCCD in ethanol (DCCD sample) or 11 μl of ethanol (control sample). The suspensions were incubated at 4° for 22 hr, and then the ferrocytochrome c-induced H^+ translocation was measured as described in the text. The vertical discontinuities in the absorbance deflections represent corrections for artifacts due to addition of cytochrome c or HCl.

acrylamide. After staining and destaining according to Weber and Osborn,[39] the gel is soaked for 2 days in 7% acetic acid, 10% ethanol. The gel is then prepared for autoradiography and exposed according to the method of Bonner and Laskey.[40] After development of the exposed film, the amounts of incorporated radioactivity are assessed by densitometry. Another way is to cut the gel into slices and measure radioactivity of every single slice as described in the legend for Fig. 3.

Remarks. Figures 2 and 3 show the results of representative determinations of the effects of DCCD on reconstituted cytochrome c oxidase. There is a potent DCCD-induced inhibition of proton translocation (Fig. 2) accompanied by a considerably smaller reduction in the rate of electron transfer. It is clear from the meager effect of DCCD on the rate of backflux of the extruded protons (Fig. 2) that the inhibition of proton pumping is not an artifact reflecting increased proton permeability. In fact, as shown in Fig. 3, the inhibition is accompanied by the binding of DCCD with a high degree of specificity to subunit III. There is also some labeling

[39] K. Weber and M. Osborn, *J. Biol. Chem.* **244**, 4406 (1969).
[40] W. M. Bonner and R. A. Laskey, *Eur. J. Biochem.* **46**, (1974).

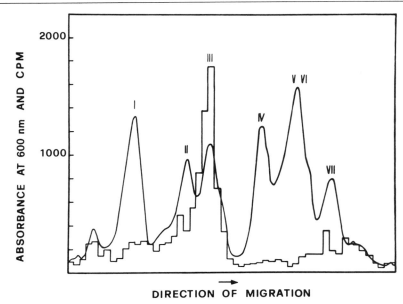

FIG. 3. Subunit analysis of DCCD-labeled cytochrome c oxidase reconstituted into proteoliposomes. To 0.5 ml of cytochrome c oxidase vesicles in 30 mM KCl, 79 mM sucrose, 1 mM HEPES (pH 7.4), containing 4.27 nmol enzyme was added 8.5 μl of [^{14}C]DCCD (10 mM, 50 Ci/mol). The mixtures were incubated at 4° for 15 hr, and then the lipid and unbound DCCD were removed by centrifugation as described in the text. The resulting pellets were dissolved in 30 μl of 25 mM Tris–acetate, 2 mM EDTA, 3% SDS, and then underwent electrophoresis. The gels, after being stained and destained (see text), were scanned for protein distribution at 600 nm using a Gelman ACD-15 gel scanner (smooth line). Following dialysis for 60 hr versus 7% acetic acid, 10% ethanol, the gels were sliced and the slices were heated for 5 hr at 80° with 0.2 ml of 30% H_2O_2, 1% concentrated ammonia solution. The samples were assayed for radioactivity after the addition of 7 ml of the liquid scintillation mixture Lipoluma, Lumasolve (Lumac System AG, Basel), water (50:5:1, v/v). The other line represents the distribution of radioactivity in the gel.

of subunits I and II but this accounts for only 13% of the total radioactivity incorporated.

It has been shown[12] that the above-described binding of DCCD to subunit III correlates kinetically with the onset of inhibition of proton pumping. Further studies on the DCCD-binding site within subunit III have shown it to be restricted to a single glutamic acid residue at position 90.[34]

Reaction of DCCD with Mitochondrial Cytochrome bc_1 Complex

Reactions of DCCD with the mitochondrial cytochrome bc_1 complex were extensively studied during the last few years. While different

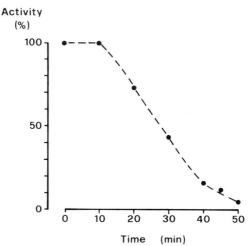

FIG. 4. Inhibition by DCCD of electron transfer activity in the isolated bc_1 complex. For experimental details see text.

DCCD-induced functional alterations of the bc_1 region of mitochondria and submitochondrial particles were observed[2] (see also section concerning DCCD effects on oxidative phosphorylation), the interactions with the isolated and reconstituted bc_1 complex are better characterized and thus will be considered here.

Modification of the Isolated bc_1 Complex. The ubiquinol:cytochrome c reductase activity of the isolated mitochondrial cytochrome bc_1 complex is strongly inhibited by DCCD[41] (see Fig. 4). This inhibitory effect is associated with the binding of the carbodiimide at several sites of the enzyme and with cross-linking between two or three subunits.[2,15,23,42]

Enzymic Preparation. Cytochrome bc_1 complex from bovine heart mitochondria is isolated according to the method of Rieske[43] and stored at $-80°$. The usual preparation contains about 6.8–7.0 nmol heme b and about 3.6–3.8 nmol heme c_1/mg of protein.

Measurement of Electron Transfer Activity. The electron transfer activity is determined by monitoring the reduction of cytochrome c at 550–540 nm using a dual-wavelength spectrophotometer fitted with a magnetic stirring device. The experimental sample contains 50 mM potassium phosphate buffer (pH 7.4), 5 μM cytochrome c, 0.4 μM cytochrome bc_1

[41] M. J. Nałęcz, R. P. Casey, and A. Azzi, *Biochimie* **65**, 513 (1983).
[42] M. Lorusso, D. Gatti, D. Boffoli, E. Bellomo, and S. Papa, *Eur. J. Biochem.* **137**, 413 (1983).
[43] J. S. Rieske, this series, Vol. 10, p. 239.

complex, and 0.25 mg/ml dispersed asolectin as recommended by Rieske.[43] The reaction is started by addition of quinol, at a final concentration of about 20 μM. Several different quinols may be used as electron donors. Ubiquinol 10 gives a relatively large antimycin-insensitive reaction. Much lower autoxidation rates are shown by quinols having a shorter side chain such as Q_1 and Q_2. Also duroquinol is relatively stable, especially when kept in acidified solutions. Widely used are synthetic analogs of quinols such as 2,3-dimethoxy-5-methyl-6-decyl-1,4-benzoquinone (DBH, to be synthesized according to the published method[44]) or 3-methyl-3-undecyl-1,4-naphthoquinone (Aldrich, Library of Rare Chemicals), characterized by the high stability of the reduced form and low rates of nonenzymatic reaction with cytochrome c.

Incubation with DCCD. The sample of the bc_1 complex is diluted to about 10 μM heme c_1 with 50 mM potassium phosphate buffer (pH 7.4). DCCD is added to give a ratio of about 100 nmol/nmol heme b, the sample is vigorously mixed and placed in a water bath at 35°. Small samples of the mixture are taken after the desired time for measurements of enzymatic activity. In the case of samples taken for SDS–polyacrylamide gel electrophoresis, the reaction with DCCD is stopped by diluting the mixture 1 : 1 (v/v) with the sample buffer prepared according to Laemmli[45] and placing the diluted mixture in ice. Control samples are incubated with an equal amount of ethanol. SDS–gel electrophoresis is performed using slab or cylindrical gels.[45]

Figure 4 presents the inhibitory effect of DCCD on the activity of the bc_1 complex. Under conditions described above the antimycin-sensitive quinol : cytochrome c reductase activity of the enzyme is almost completely inhibited by DCCD after 50 min of incubation (half-maximal inhibition time is about 30 min). All data shown on Fig. 4 represent the percentage of inhibition of initial rates of cytochrome c reduction measured with DCCD-treated enzyme (the activity at "zero incubation time" is taken as 100%). The values are corrected for a slight thermal inactivation of the control enzyme during its incubation with ethanol.

Duplicate samples of the cytochrome bc_1 complex are subjected to SDS–gel electrophoresis following incubation with either DCCD or ethanol. Substantial modification of the electrophoretic migration pattern of the enzyme treated with DCCD is observed. A new band appears between bands II and III whereas bands V (iron–sulfur protein), VII, and VIII are largely diminished.[15,23,42] The kinetics of this DCCD-induced cross-linking

[44] W. P. Wan, R. H. Williams, K. Folkers, K. H. Leung, and E. Racker, *Biochem. Biophys. Res. Commun.* **63,** 11 (1975).

[45] U. K. Laemmli, *Nature (London)* **227,** 680 (1970).

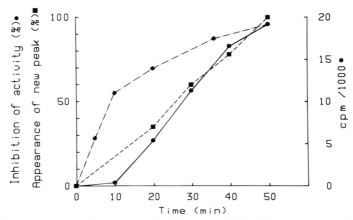

FIG. 5. Comparison of time courses of inhibition by DCCD of cytochrome c reduction by the isolated bc_1 complex, of appearance of cross-linked material, and of binding of [^{14}C]DCCD to the isolated enzyme. DCCD was incubated with the enzyme as described in the text. The initial rates of cytochrome c reduction were corrected for antimycin-insensitive activity and for the thermal inactivation of the control sample. The decrease in the reaction rate during incubation with DCCD is shown as percentage of the rate measured at zero incubation time (solid line, closed circles). Samples of the incubated enzyme were taken at the times indicated and either subjected to electrophoresis or used for estimation of binding of [^{14}C]DCCD to the enzyme as described. The band corresponding to the cross-linked material was integrated and the integrals are shown as a percentage of the value after 50 min of incubation (solid line, closed squares). Radioactivity of covalently bound [^{14}C]DCCD is presented in cpm following corrections for protein content of the samples and the radioactivity of the zero incubation time (dashed line, stars).

correlates with the time course of the inhibition of the quinol-cytochrome c reductase activity[15,41] (see Fig. 5).

Binding of [^{14}C]DCCD. [^{14}C]DCCD of specific activity about 2.0 Ci/ mol is incubated with the isolated bc_1 complex as described above. After the desired time the reaction with the carbodiimide is stopped by addition of 5 ml of an ice-cold chloroform/methanol mixture (1:4, v/v). The sample is centrifuged at 4000 rpm for 5 min on a bench centrifuge. The supernatant is discarded and a further 5 ml of the mixture is added. Following mixing, the sample is centrifuged again and the pellet is dispersed in 0.4 ml of 25 mM Tris–acetate (pH 8.2), 3% SDS, and 2 mM EDTA under gentle sonication. Aliquots of this dispersion are taken for measurements of radioactivity following the addition of 5 ml of a universal liquid scintillation coctail INSTA-GEL (Packard Inst. Co. Inc., Downers Growe, IL). Parallel samples are assayed for protein content. The onset of binding of DCCD is more rapid than that of the DCCD-induced inhibition, this difference being especially noticeable during the first minutes of incubation (see Fig. 5).

Identification of the [^{14}C]DCCD-Binding Subunit(s). This is performed after subjecting the [^{14}C]DCCD-treated enzyme to SDS–polyacrylamide gel electrophoresis followed by determination of the distribution of radioactivity by autoradiography.[15] The dried gel is scanned for Coomassie blue at 570–530 nm and the autoradiogram is scanned at 590–480 nm using a special attachment to the dual-wevelength spectrophotometer.[46] Binding of [^{14}C]DCCD to the bc_1 complex under the experimental conditions described above is found highly unspecific, with all enzyme subunits labeled approximately to the same extent.

Remarks. The DCCD-induced modifications of the isolated bc_1 complex appear more complicated than those observed for all other systems described previously. It has to be noticed that binding of [^{14}C]DCCD to the isolated complex differs upon the experimental conditions and the source of enzyme.[2] For example, a short incubation of the bovine bc_1 complex with relatively low concentrations of DCCD (10–30 min, 25°, 20–30 nmol DCCD per nmol heme b) leads to the preferential binding of the carbodiimide to subunit VIII,[23,42] whereas different conditions of incubation applied to the enzyme purified from yeast (3 hr, 10–12°, 100–200 nmol DCCD per heme b) result in the preferential binding of carbodiimide to cytochrome b.[47] Moreover, no cross-linking was observed with the enzyme isolated from yeast mitochondria, consistent with a much weaker inhibition by DCCD of the electron flow measured in this system.[16] DCCD was also reported to lower antimycin binding to the yeast bc_1 complex, thus inducing a substantial increase of the antimycin-insensitive rate of cytochrome c reduction.[48]

Modification of the bc_1 Complex Reconstituted into Proteoliposomes. The use of the reconstituted system allows studies on DCCD-induced modifications of the other important function of the enzyme, namely the H$^+$-translocating activity of the bc_1 complex.

Reconstitution Procedure. A suspension of 105 mg of soybean phospholipids/ml in 24.5 mM potassium cholate, 100 mM HEPES (pH 7.2) is sonicated with cooling and under nitrogen flux at 6 μm peak to peak in an MSE sonicator for 1 min/ml of suspension. Cytochrome bc_1 complex is added to give a final concentration of 7.5 μM heme b and the suspension is dialyzed for 4 hr against 100 volumes of 100 mM HEPES (pH 7.2), then for a further 4 hr against 200 volumes of 10 mM HEPES, 27 mM KCl, and

[46] C. Broger, P. Allemann, and A. Azzi, *J. Appl. Biochem.* **1,** 455 (1979).
[47] D. S. Beattie and L. Clejan, *FEBS Lett.* **149,** 245 (1982).
[48] L. Clejan and D. S. Beattie, *J. Biol. Chem.* **258,** 14271 (1983).

73 mM sucrose (pH 7.2) and finally for 12 hr against 200 volumes of 1 mM HEPES (pH 7.2), 30 mM KCl, and 79 mM sucrose. All solutions are adjusted to the indicated pH with KOH. The procedure is carried out at about 4°. Proteoliposomes prepared by this procedure show a ratio of the rate of electron flow in the resting state and in the presence of a H$^+$-conducting agent (respiratory control ratio) of about 5.

Measurements of H$^+$ Translocation and Electron Flow. Reconstituted bc_1 proteoliposomes are suspended to give a final concentration of 0.2 μM heme b in a buffer containing 75 mM choline chloride, 25 mM KCl, 5 μM cytochrome c, and 0.13 μM valinomycin. The final pH of the reaction mixture is adjusted to about 7.0 with HCl and/or KOH and the reaction is started by the addition of quinol. The electron transfer activity (reduction of cytochrome c) is monitored spectrophotometrically (see above), simultaneously with the measurement of H$^+$ translocation. The latter is performed by inserting a 2-mm-thick, rapidly responding pH electrode into the sample cuvette. Each measurement of pH is calibrated after the reaction by subsequent additions of freshly prepared 1 mM oxalic acid. As a control for nonenzymatic reactions parallel measurements are performed in the presence of 5 μM antimycin. Carbonyl cyanide m-chlorophenylhydrazone (CCCP), used as a protonophore, is present in the concentration of 2.5 μM. The value of H$^+/e^-$ ratio calculated from the above measurements should be about 2 in the absence and about 1 in the presence of CCCP.

Incubation with DCCD. DCCD is added directly to the vesicle suspension from 0.5 M solution in ethanol to give a ratio of 400 mol/mol heme b. The suspension is mixed and placed in a water bath at 35°. Control samples are incubated with ethanol alone. After the desired time, 50-μl aliquots of both mixtures are taken for measurements of H$^+$-translocation and cytochrome c reduction activities as described above. In the case of samples taken for SDS–gel electrophoresis, 0.5-ml aliquots of the reaction mixture are placed in ice, solubilized with 2.5% (final concentration) of sodium cholate (pH 7.8) and layered onto 3.2 ml of 10% sucrose. They are then centrifuged for 20 hr at 4° and 250,000 g. The resulting pellets are dissolved in 100 μl of the sample buffer prepared according to Laemmli[45] and subjected to SDS–gel electrophoresis.

Incubation of bc_1 proteoliposomes with DCCD results in an inhibition of both H$^+$ translocation and electron transport activities of the enzyme. DCCD induces, in the reconstituted system, a cross-linking phenomenon similar to that observed with the isolated enzyme[15] (see above). Moreover, cross-linking occurs with a time course similar to that of inhibition.

Binding of [^{14}C]DCCD to the Reconstituted Enzyme. If the incubation of the bc_1 proteoliposomes is performed in the presence of [^{14}C]DCCD, the samples taken for SDS–gel electrophoresis may also be used for identification of the DCCD-binding subunit(s). The resulting polyacrylamide gel, after being stained with Coomassie blue, is dried, subjected to autoradiography, and then analyzed as described above for the isolated enzyme. The amount of radioactivity incorporated into the protein can be estimated by the procedure described for the isolated enzyme. DCCD binds preferentially to cytochrome b in the reconstituted enzyme,[15] though kinetics of binding have been described to be much more rapid than those of the inhibition.[15]

Remarks. Incubation conditions with relatively high concentration of DCCD lead to parallel inhibition of H^+ translocation and electron transport activities which has been correlated with the cross-linking between subunits, but not with binding of DCCD to the complex.[15]

Different incubation conditions with DCCD have resulted in different observations.[2] Low concentrations of carbodiimide and very short incubation times produce stimulation of electron flow and inhibition of H^+ translocation with the consequent substantial drop at H^+/e^- ratio.[23,42]

Another important difference concerns the bc_1 complex from yeast mitochondria. In general, this enzyme appears less sensitive to structural modifications by DCCD, since no cross-linking occurs in this system even at high concentrations of carbodiimide.[16] Binding of DCCD seems to involve only cytochrome b and phospholipids[16] and leads to structural modification of the enzyme manifested by diminished binding of antimycin to the DCCD-treated complex.[48]

The multiplicity of structural alterations caused by DCCD in the bc_1 complex suggests that extreme care is required in correlating them with functional effects of this substance.

Final Comments

It has been shown that DCCD can be utilized with success to study a number of membrane-bound enzymes. It is striking that, besides its broad reactivity, DCCD in some enzymes under precise conditions reacts with only one or very few sites. The observation that DCCD can be used as a cross-linker may also provide valuable information as to neighboring groups or subunits in a complex protein. The use of binding and inhibition by DCCD, which have been successfully demonstrated to occur in some cation-transporting proteins, has no intrinsic predictive value and cannot, therefore, be used as a diagnostic tool for similar systems.[2,17]

[8] Voltammetric Measurements of Quinones

By ROGER C. PRINCE, PAUL LLOYD-WILLIAMS, J. MALCOLM BRUCE,
and P. LESLIE DUTTON

Quinones are cyclic conjugated carbonyl compounds which may be formally derived from benzene, fused polycyclic aromatics, or certain heterocycles. The parent members of the series are 1,4-benzoquinone (or *p*-benzoquinone) and 1,2-benzoquinone (or *o*-benzoquinone), obtainable by oxidative removal of the two hydroxy hydrogens from the corresponding hydroquinones, respectively 1,4-dihydroxybenzene (hydroquinone) and 1,2-dihydroxybenzene (catechol). 1,3-Benzoquinone (*m*-benzoquinone) does not exist, although its "hydroquinone," 1,3-dihydroxybenzene (resorcinol), is well known.

A quinone (Q) yields a resonance stabilized, *aromatic,* semiquinone ($Q^{\bar{}}$) when reduced by a single electron.

1,4-Naphthoquinone and 1,4- and 9,10-anthraquinones can be regarded as benzologs of 1,4-benzoquinone, and there is an analogous relationship with the series beginning with 1,2-benzoquinone. Substituents attached to the quinonoid nucleus control its chemical and electrochemical behavior, which for a large number of reactions, are inextricably linked.

Quinones play vital roles in many biological systems. The majority of those which occur naturally belong to the *para* series, although *o*-quinones are well represented.[1] Oxidized catecholamines may be considered as substituted nascent 1,2-benzoquinones.[2]

Quinones are involved in membrane electron transfer systems,[3] some soluble dehydrogenases,[4] blood coagulation,[5] defensive chemistry,[6,7] and pigmentation.[7] Some are important antineoplastic agents.[8-10] In many of

[1] R. H. Thomson, "Naturally Occurring Quinones." Academic Press, New York, 1971.
[2] P. M. Plotsky, this series, Vol. 103, p. 469.
[3] B. L. Trumpower, "Function of Quinones in Energy Conserving Systems." Academic Press, New York, 1982.
[4] J. A. Duine, J. Frank, and P. E. J. Verweil, *Eur. J. Biochem.* **118,** 395 (1981).
[5] J. W. Suttie, *Proc. Soc. Exp. Biol. Med.* **39,** 2730 (1980).
[6] T. Eisner, T. H. Jones, D. J. Aneshansley, W. R. Tschinkel, R. E. Silberglied, and J. Meinwald, *J. Insect Physiol.* **23,** 1383 (1977).
[7] G. Britton, "The Biochemistry of Natural Pigments." Cambridge Univ. Press, London and New York, 1983.
[8] W. A. Remers, "The Chemistry of Antitumour Antibiotics." Academic Press, New York, 1979.

METHODS IN ENZYMOLOGY, VOL. 125

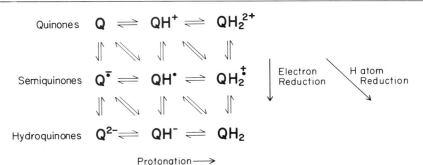

FIG. 1. The nine possible redox states of a quinone. Horizontal arrows involve protons, vertical ones involve electrons, and diagonal lines involve H-atom transfer.

these roles the quinone undergoes consecutive reduction–oxidation (redox) processes which, as illustrated in Fig. 1, can involve a variety of pathways and redox couples.

Quinones are generally assumed to undergo redox reactions from the unprotonated quinone (Q) to the hydroquinone (QH_2), but this need not always be the case. For example the primary quinone of the bacterial reaction center, Q_A, functions between only the $Q_A/Q_{\bar{A}}$ states, although at equilibrium the latter can protonate to $Q_{\bar{A}}H$.[11] The secondary quinone, Q_B, seems to have a stable protonated semiquinone, at least at equilibrium, such that it exhibits two one electron, one proton reactions, $Q_B/Q_{\dot{B}}H$ and $Q_{\dot{B}}H/Q_BH_2$ at physiological pH.[12] In contrast, the quinones of succinate dehydrogenase, an interacting pair known as Q_S, and a quinone of the cytochrome bc_1 complex, known as Q_c, have stable anionic semiquinones, and function between the $Q/Q^{\bar{}}$ and $Q^{\bar{}}/QH_2$ forms.[13,14] Meanwhile the quinone that seems to function as the reductant of chloroplast photosystem II and the oxidant of the water-splitting enzyme (which can be detected by EPR as Signal II) seems to behave as $QH_2^{\dot{+}}/QH_2$.[15] On the other hand, another quinone of the bc_1 complex, known as Q_Z, and the

[9] F. Arcamone, "Doxorubicin Anticancer Antibiotics." Academic Press, New York, 1981.

[10] H. S. El Khadem, "Anthracycline Antibiotics." Academic Press, New York, 1982.

[11] R. C. Prince and P. L. Dutton, *in* "The Photosynthetic Bacteria" (R. K. Clayton and W. R. Sistrom, eds.), p. 439. Plenum, New York, 1978.

[12] A. W. Rutherford and M. C. W. Evans, *FEBS Lett.* **110**, 257 (1979).

[13] J. C. Salerno and T. Ohnishi, *Biochem. J.* **192**, 769 (1980).

[14] D. E. Robertson, R. C. Prince, J. R. Bowyer, K. Matsuura, P. L. Dutton, and T. Ohnishi, *J. Biol. Chem.* **259**, 1758 (1984).

[15] P. J. O'Malley, G. T. Babcock, and R. C. Prince, *Biochim. Biophys. Acta* **766**, 283 (1984).

quinones of the "quinone-pool" seem to have no stable semiquinone, and function between the Q and QH_2 forms.[16,17] These different redox couples have very different oxidation–reduction midpoint potentials (E_m). For example the quinone functioning as the reductant of Photosystem II probably has an effective E_m of $\sim +1$ V,[18] while that functioning as the Q_A of Photosystem II probably has an E_m of -130 mV.[11,19] Yet these are apparently both molecules of plastoquinone-9, so it is obvious that the binding site of the quinone severely modulates the redox chemistry.

This chapter describes electrochemical methods of obtaining thermodynamic data from quinones. In contrast to potentiometric titrations, where chemical oxidant or reductant is added and the redox potential is measured with null-current electrodes,[20] electrochemical techniques provide or withdraw electrons at the electrodes. By definition then, these are surface, and not bulk phase, techniques. While electrochemical methods can be used, with mediators, to study protein- or membrane-bound systems,[21,22] perhaps their most powerful applications lie in the analysis of homogeneous solutions of compounds that interact directly with the electrodes. Electrochemical techniques are thus ideally suited for analysis of quinones in homogeneous solution. For example, they can be used to examine the effects of different substituents on the quinone nucleus, to compare a series of quinones under identical conditions, and to examine the effects of the solute dielectric and proticity. In some cases they can be used to identify unknown quinones.

Two related techniques have proven most useful for routine electrochemical measurements, cyclic voltammetry and dc polarography.[23,24] Conceptually, both techniques scan the potential at the working electrode and monitor the current required. Very little current flows to scan the potential over a wide range of potential in a solution of electrolyte. However, if an oxidized redox-active compound is present in the solution, it will be reduced at the working electrode as the potential becomes appropriate, and current will flow to provide the reductant. Similarly if a re-

[16] R. C. Prince, C. L. Bashford, K. Takamiya, W. H. van den Berg, and P. L. Dutton, *J. Biol. Chem.* **253**, 4137 (1978).
[17] K. Takamiya and P. L. Dutton, *Biochim. Biophys. Acta* **546**, 1 (1977).
[18] A. Boussac and A. L. Etienne, *Biochim. Biophys. Acta* **766**, 576 (1984).
[19] D. B. Knaff, *FEBS Lett.* **60**, 331 (1975).
[20] P. L. Dutton, this series, Vol. 54, p. 411.
[21] G. S. Wilson, this series, Vol. 54, p. 396.
[22] D. F. Wilson and D. Nelson, *Biochim. Biophys. Acta* **680**, 233 (1982).
[23] A. M. Bond, "Modern Polarographic Methods in Analytical Chemistry." Dekker, New York, 1980.
[24] P. T. Kissinger and W. R. Heineman, "Laboratory Techniques in Electroanalytical Chemistry." Dekker, New York, 1984.

duced species is present, it will be oxidized at potentials above its E_m. The experimental apparatus has three electrodes: a working electrode, a counter electrode, and a reference electrode. Current flows between the working and counter electrodes. The reference electrode is usually a saturated calomel electrode (although other systems may be used), and the counter electrode is usually a piece of platinum wire. The most common working electrodes are platinum, gold, mercury, glassy carbon, or tin oxide, but other materials may also be used. Optically transparent electrodes allow spectroscopy of the electrochemical intermediates.

Many polarographic analyzers are on the market, varying in price from a few hundred to many thousand dollars, with a concomitant range of sophistication. Relatively simple devices seem adequate for routine measurements of the kind discussed here.

Cyclic Voltammetry

As the name implies, this technique scans the potential repetitively, so that quinones at the electrode surface are repeatedly reduced and reoxidized. The technique works well in both protic and aprotic solvents, and a typical result, in an aprotic solvent, is shown in Fig. 2. Note that this is a surface technique at a single electrode, and that in aprotic solvents the redox reactions are constrained to the $Q/Q^{\overline{\cdot}}$ and $Q^{\overline{\cdot}}/Q^{2-}$ couples. Inspection of Fig. 2 reveals several phenomenologically distinct regions. No current flows at the positive potentials of region I, because all the quinone is in the oxidized form. The slight difference in current in the two directions of the scan indicates the current flow to polarize the electrode assembly. In region II, Q is reduced to $Q^{\overline{\cdot}}$, and reoxidized on the return scan. The reductive and oxidative waves peak at different potentials; for a truly electrochemically reversible system, where the rate of reduction is the same as the rate of oxidation, and the diffusion rates of the oxidized and reduced species are the same, the difference in peak positions will be approximately $59/n$ mV at room temperature, where n is the number of electrons transferred.[23,24]

Once Q is reduced to $Q^{\overline{\cdot}}$, ideally no further current should flow until the potential is low enough to reduce $Q^{\overline{\cdot}}$ to Q^{2-}. As can be seen in Fig. 2, region III, there is in fact some current flow in this region, due to the diffusion of $Q^{\overline{\cdot}}$ away from the electrode, and its replacement by Q from the bulk phase which is promptly reduced. Since the bulk phase is Q throughout the experiment, this effect is seen in the oxidative scan as well. In region IV the potential is scanned through the second reduction of the quinone, and again the reductive and oxidative waves are offset, by approximately $59/n$ mV.[23,24] Finally in region V all the quinone at the

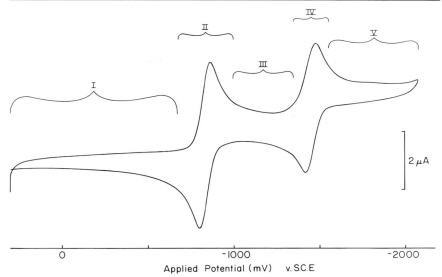

FIG. 2. Cyclic voltammetry of 9,10-anthraquinone. Approximately 3 mM quinone in DMF containing 50 mM tetrabutylammonium tetrafluoroborate. Scan rate 10 mV/sec. The $E_{1/2}$ of the Q/Q$^{\bar{\cdot}}$ and Q$^{\bar{\cdot}}$/Q^{2-} couples are -831 and -1443 mV (with respect to a saturated calomel electrode), respectively. Under identical conditions, ferrocene had an $E_{1/2}$ of $+524$ mV.

electrode surface is in the Q^{2-} state, and further current flows merely to reduce new Q molecules that diffuse to the electrode.

The half-wave potentials for the two couples Q/Q$^{\bar{\cdot}}$ and Q$^{\bar{\cdot}}$/Q^{2-} are the average of the reductive and oxidative wave of each reaction. In principle this is true only if the reactions are electrochemically reversible, as discussed above, but in practice little error is introduced if the reactions are "quasi-reversible" with a separation of less than ~100 mV.

Not all quinones behave as ideally as shown in Fig. 2. For example, those carrying carboxy groups yield polarograms that suggest that secondary reactions occur in the reduced species, perhaps a cleavage process.[25] 2-Hydroxy-1,4-benzo- and 2-hydroxy-1,4-naphthoquinones show regular reductive waves, but the reoxidation of Q$^{\bar{\cdot}}$ occurs at a much higher potential than expected. α-Hydroxy-9,10-anthraquinones behave normally, but β-hydroxy-9,10-anthraquinones show very broad reoxidative waves for Q$^{\bar{\cdot}}$.[26] Some substituents, such as nitro groups, undergo their own additional redox reactions.

[25] R. M. Cervino, W. E. Triaca, and A. J. Arvia, *J. Electroanal. Chem.* **172**, 255 (1984).
[26] A. Ashnagar, J. M. Bruce, P. L. Dutton, and R. C. Prince, *Biochim. Biophys. Acta* **801**, 351 (1984).

Cyclic voltammetry can also be run in aqueous solution, but quinones often display anomalous behavior. At neutral pH the overall redox reaction seen by voltammetry is usually the Q/QH$_2$ couple, because the E_m of the first electron is lower than the second, and the overall E_m is the average of the two one-electron couples. Unfortunately the involvement of protons means that the reactions are rarely electrochemically reversible, because the electrons and protons have different kinetics.[27] Even relatively water-soluble quinones adsorb to the electrode, and may not see the bulk pH.[28]

dc Polarography

In its simplest form, dc polarography is half of cyclic voltammetry, and the technique can be applied at a solid electrode. Scan rates should be comparable, but should also be varied to test for electrochemical reversibility. The peak position for a reduction should be approximately $59/2n$ mV lower than $E_{1/2}$, and independent of sweep rate.[23,24] Variations of a few tens of millivolts indicate "quasi-reversibility," and scans should then be run as slowly as possible. Obviously there is a limit on how slow the scan can be performed and still yield an interpretable current flow,[24] and some compounds cannot be reliably measured with linear sweep voltammetry.

The technique becomes most useful, however, when used with a dropping mercury electrode. Here the electrode, a hanging drop of metallic mercury, is replaced perhaps every second. Thus while this is still a surface technique, each new drop of mercury is formed in the original solution. As can be seen in Fig. 3, the resulting experimental data look very different from those obtained with cyclic voltammetry. In dc polarography at the dropping mercury electrode the current is measured continuously, and oscillates as the mercury drop grows, and is then knocked off. There are two contributions to this current; capacitance current to charge the surface of the growing electrode, and Faradaic current to reduce electroactive solutes. Both vary as the mercury drop grows, yielding the polarogram shown in Fig. 3A. Note that the current flowing reaches a limiting value, which is a function of the drop size and the concentration of electroactive reactant. This is shown more clearly in Fig. 3B, where the current was measured just before the drop was displaced, so that the capacitance current was eliminated; this is known as sampled dc polarography. For an electrochemically reversible system, defined as

[27] P. R. Rich, *Faraday Discuss. Chem. Soc.* **74**, 349 (1984).
[28] S. I. Bailey, *Chem. Aust.* **50**, 202 (1983).

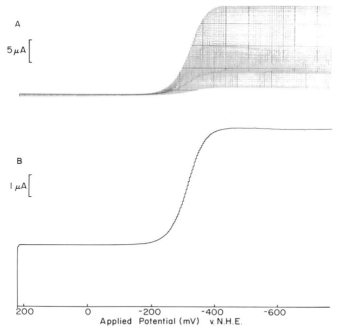

FIG. 3. Polarography of sodium 9,10-anthraquinone-2-sulfonate. Quinone (2.4 mM) in 50 mM 3-cyclohexylamino-2-hydroxypropane sulfonate (CAPSO), 100 mM KCl, pH 10. Scan rate 2 mV/sec. The $E_{1/2}$ is −319 mV. (A) dc polarography; (B) sampled dc.

above, the applied potential *(E)* is related to the polarographic current *(i)* by

$$E = E_{1/2} + RT/nF \ln(i_T - i)/i$$

where $E_{1/2}$ is the half-wave potential, and i_T is the limiting, or maximum, current flow.

Practical Considerations

Working Electrode. We have found little difference in the electrochemistry of quinones at gold, platinum, and glassy carbon electrodes in aprotic solvents, and routinely use the latter because it is easily polished. Mercury for dropping electrodes should be of very high purity, at least triply distilled.

Reference Electrode. We routinely use a saturated calomel electrode, although other couples, such as silver/silver chloride are also widely used.

In aqueous solutions, potentials are usually referred to the hydrogen electrode (e.g., by adding 247 mV to values obtained using a saturated calomel reference system), but this is not possible in aprotic solvents, so in these solvents potentials are usually referred to the reference electrode system without any secondary calculations. A major problem in assessing literature data is the junction potential between the aqueous reference electrode and the bulk solution. This can present a difficulty even for aqueous solution, but is most obvious in aprotic solvents.[29] Fortunately the International Union of Pure and Applied Chemistry has recently promulgated recommendations on how to deal with the problem in aprotic solvents.[30] The recommendation is to run standard compounds, with solvent independent redox reactions, and report these along with other data. The two chosen are

Ferrocene/Ferricinium ion

Bis(biphenyl)chromium(I)/Bis(biphenyl)chromium(0)

While only the former is readily commercially available, quoting its potential under identical conditions to those used for measuring quinones will allow ready comparison between laboratories.

Solvents. Many solvents can and have been used. Prime considerations are electrochemical inertness in the region of interest, and the solubility of quinone and electrolyte. A corollary of this latter point is that solvents of very low dielectric constant often provide too high a resistance to current flow, the so-called iR drop. Solvents such as dimethylformamide (DMF), acetonitrile (AN), and dimethyl sulfoxide (DMSO) are close to being universal solvents for quinones. The highest possible purity is essential; materials sold as HPLC grade are usually adequate, but they must be dried before use. Drying over molecular sieves (3A or 4A) is most convenient, but is unlikely to lower the water concentration below millimolar[30] (commercial DMF may be 100 mM in water). As can be seen in Fig. 4, low concentrations of water have little effect on the $E_{1/2}$ of the Q/Q$^{\bar{}}$ couple, but radically affect that of Q$^{\bar{}}$/Q^{2-}. It is worth remembering that water is itself very concentrated; even 1 μl of water in 5 ml of solvent is 11 mM, which should be compared to a typical quinone concentration of 1–5 mM. Drier solvents can be redistilled from calcium hydride[31]; reduced pressure is advisable for DMF and DMSO.

[29] J. Q. Chambers, *in* "The Chemistry of Quinonoid Compounds" (S. Patai, ed.), Part 2, p. 737. Wiley, New York, 1974.

[30] G. Gritzner, and J. Kuta, *Pure Appl. Chem.* **56**, 461 (1984).

[31] A. J. Fry and W. E. Britton, cited in ref. 24, p. 367.

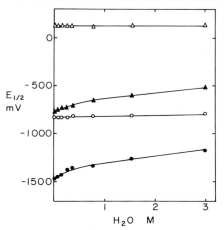

FIG. 4. The effect of water on the half-wave potentials of tetrachloro-1,4-benzoquinone and 9,10-anthraquinone. Triangles, tetrachlorobenzoquinone; circles, anthraquinone; open symbols, $Q/Q^{\overline{\cdot}}$; solid symbols, $Q^{\overline{\cdot}}/Q^{2-}$.

Solvents should be made anaerobic, which can be conveniently done by bubbling with dry argon. Since this will also evaporate some of the solvent, the experimental apparatus should be installed in a hood.

Electrolyte. The concentration of electrolyte should be 50- to 100-fold that of the quinone under study to prevent electrophoretic movement of charged quinone species.[23,24] In aqueous solution there is an almost infinite array of potential electrolytes, but few will disolve in aprotic media. Tetrabutylammonium tetrafluoroborate (TBATFB) is a good electrolyte for many aprotic solvents, and is commercially available. It should be recrystallized from 4:1 by volume water–methanol before use, and dried, preferably *in vacuo,* finally over P_2O_5. Tetrabutylammonium perchlorate has also been widely used, but like all perchlorates it can be a safety hazard, and there seems no good reason not to abandon its use.[31]

Typical concentrations of electrolyte are 50–200 mM.

Electrochemistry. Obviously the electrochemical properties of the solvent, electrodes and electrolyte should be determined before the addition of the quinone under study. Mercury electrodes cannot be used much above 0 mV in aqueous media because the electrode itself is oxidized. Typical concentrations of quinone are in the millimolar range, and typical aprotic cyclic voltammagrams might encompass +1 to −2 V. It is wise to look at both the first cycle and subsequent ones to see if the quinone adsorbed to the electrode is changing its properties during the experiment. In some cases the electrochemistry of $Q/Q^{\mathbf{=}}$ is altered if the semi-quinone is further reduced to Q^{2-},[26] and the electrochemistry of substituents is not always chemically reversible. Typical scan rates are in the 1 mV–1 V per second range.

Treating the Data

Data obtained in aqueous solution may be treated in an analogous way to that obtained by redox potentiometry[20]; the $E_{1/2}$ can be obtained as a function of pH, and the data interpreted to reveal pK values on oxidized or reduced forms.[20] Data obtained with the dropping mercury electrode can also readily be inspected to reveal n values.[20] Unfortunately the various problems associated with the aqueous electrochemistry of quinones, such as their relative insolubility, and the fact that their electrochemistry is rarely "electrochemically reversible" (see above), may render the data uninterpretable, and definitive conclusions should be drawn with care (see Rich[27]).

Measurements made in aprotic solvents are more simply interpreted because by definition no protons are involved in the redox reactions, and the redox reaction is more likely to be "electrochemically reversible." Once a significant number of quinones have been measured, the Hammett equation[32] allows semiempirical predictions of the properties of unmeasured compounds.

$$\Delta E_{1/2} = \rho \Sigma \sigma$$

where ρ is the reaction constant and σ is the substituent constant. As has been shown by Zuman,[33] the appropriate substituent constants for quinones are those for *para* substituents in benzene derivatives.[34] Calculations are simplest in the case of 1,4-benzoquinones, where all substituent positions in the parent are equivalent. For example, from measurements on over 100 benzoquinones in DMF, we have measured ρ to be +521 mV for the $Q/Q^{\mathbf{=}}$ couple,[35] but even here predictions must be made with caution, for the substituent constants were originally obtained "in isolation," where, for example, free rotation of the substituent is allowed. Thus methyl derivatives behave as expected, but methoxy substituents display anomalous effects when they are perturbed by bulky neighbors.[35]

[32] C. D. Johnson, "The Hammett Equation." Cambridge Univ. Press, London and New York, 1973.

[33] P. Zuman, "The Elucidation of Organic Electrode Processes." Academic Press, New York, 1969.

[34] C. Hansch and A. J. Leo, "Substituent Constants for Correlation Analysis in Chemistry and Biology." Wiley, New York, 1979.

[35] R. C. Prince, J. M. Bruce, and P. L. Dutton, *FEBS Lett.* **160,** 273 (1983).

9,10-Anthraquinones have two sets of substituent positions; those alpha to the quinone function (positions 1, 4, 5, and 8), and those beta (positions 2, 3, 6, and 7). Rather different reaction constants would be expected for these two groups, and in some cases, such as with hydroxy substituents, there is a dramatic effect of position due to the geometry of possible hydrogen bonds.[26] 1,4-Naphthaquinones have three sets of substituent sites; *ortho* (positions 2 and 3), alpha (positions 5 and 8) and beta (positions 6 and 7). These would all be expected to exhibit different reaction constants.

Acknowledgments

This work was supported by grants from USPHS (GM 27309), NSF (PCM 82-09292), and DOE (DE-AC02-80-ER 10590).

[9] Determination of Partition Coefficients of Quinones by High-Performance Liquid Chromatography

By BARRY S. BRAUN, ULI BENBOW, PAUL LLOYD-WILLIAMS, J. MALCOLM BRUCE, and P. LESLIE DUTTON

Introduction

The hydrophobicity of a molecule is well known to control its distribution between a membrane lipid or protein phase, and an aqueous phase. The hydrophobic characters, taken as a general property or as local characteristics of a molecule, can have profound significance regarding its biological function and activity.[1]

The traditional shake-flask method for the determination of hydrophobicity involves shaking the solute in a mixture of two immiscible liquids which have been thoroughly presaturated with each other, and measuring its concentration in each phase; the ratio gives the partition coefficient, P. A large number of P values derived using the shake method have been published.[2] However, in addition to the issues of the meaning of these

[1] C. Tanford, "The Hydrophobic Effect: Formation of Micelles and Biological Membranes." Wiley, New York, 1980.
[2] C. Hansch and A. J. Leo, "Substituent Constants for Correlation Analysis in Chemistry and Biology." Wiley, New York, 1979.

METHODS IN ENZYMOLOGY, VOL. 125

values,[3] there are some technical drawbacks: for example, establishment of equilibrium takes time, significant amounts of pure material are needed, accurate measurement of concentration in each phase is often difficult, and for P values greater than 10^5 the methods have proven to be a failure. Nevertheless, a large data base has accumulated, and together with a careful series of measurements for compounds, has yielded fundamental values for common substituents and groups from which the overall P of a molecule can be calculated.[2,4,5] These methods have some accuracy in predicting changes in P for small, well-defined changes in molecular structures. However, like many of these empirically derived parameters, they cannot reliably account for the proximity (steric) effects of multiple substituent groups on complex molecules or for the position dependent H-bonding and dipolar interactions introduced by many substituents.

Reversed-phase high-performance liquid chromatography (RP-HPLC) offers an alternative way to measure partition coefficients. On the practical side RP-HPLC is fast, uses extremely small amounts of material that need not necessarily be pure, and does not involve concentration measurement.[6-9] This is especially desirable for the many biologically important compounds which are not suitable for partition coefficient determination by the shake-flask method; they are often available only in very small quantities, and tend to be ill-defined with respect to their spectrometric properties or are otherwise difficult to label for appropriate concentration measurement.

Partition coefficients of many common compounds measured by the shake-flask method have been compared with their retention times on a RP-HPLC column with promising results.[6-10] In this report we demonstrate the practicalities of the RP-HPLC application to determine partition coefficients. Quinones have been chosen to illustrate the method for a number of reasons. A sufficient number of them have had their P values determined by the shake-flask method to provide a suitable basis for comparison and correlation of their retention times with actual partition coefficients between defined solvents. Further, quinonoid compounds were chosen because they appear in many key positions in living systems,

[3] R. Wolfenden, *Science* **222**, 1087 (1983).

[4] T. Fujita, J. Iwasa, and C. Hansch, *J. Am. Chem. Soc.* **86**, 5175 (1964).

[5] R. F. Rekker, "The Hydrophobic Fragmental Constant." Elsevier, Amsterdam, 1977.

[6] A. Nahum and C. Horvath, *J. Chromatogr.* **192**, 315 (1980).

[7] K. Miyake and H. Terada, *J. Chromatogr.* **240**, (1982).

[8] M. S. Mirrless, S. J. Moulton, C. T. Murphy, and P. J. Taylor, *J. Med. Chem.* **19**, 615 (1970).

[9] R. Kaliszan, *J. Chromatogr.* **220**, 71 (1981).

[10] K. Karch, I. Sebestian, I. Halasz, and H. Englehardt, *J. Chromatogr.* **122**, 171 (1976).

as discussed by Prince *et al.* in this volume [8], and display a very wide range of hydrophobic character.

Experimental Methods

Basic HPLC Equipment

The minimum equipment requirements are as follows: two pumps capable of delivering a precise ratio of a two solvent mixture through the chromatographic column at a constant flow rate, a solute injection system, a reversed-phase HPLC column consisting of alkyl residues covalently bonded to a silica support, an ultraviolet absorbance detector, and a chart recorder. The system we have used, all from Waters Scientific Company, is comprised of two dual head solvent pumps (model 6000A), an injector (model U6K), a spectrophotometric detector set to measure absorbance at 254 nm (model 441), a radial compression module (RCM-100) containing a Waters 10 cm long × 0.8 cm (internal diameter) C_{18} reversed-phase Radial-PAK column.

Mobile Phase

The mobile phase is a mixture of water and a miscible organic modifier. So far we have used acetonitrile or methanol as the modifier; both have a low absorbance at 254 nm, the wavelength used to assay the quinone in the eluent. The composition of the solvent mixture in the mobile phase governs the retention time of a compound on the column. Thus, a suitable mixture can be chosen so that a particular compound emerges from the column at an optimal time after injection. In general, higher proportions of organic modifier in the mobile phase will decrease the retention time.

The retention time of a compound on an HPLC column is usually expressed as a "capacity factor" (k') which is free of units and permits ready comparison from one chromatographic system to another; it is defined as follows:

$$k' = (t_R - t_0)/t_0$$

where t_R is the retention time of the compound on the column and t_0 is the time spent in the apparatus by a compound that is unretained on the column.

Sample Preparation and Chromatographic Conditions

All test compounds were dissolved in 100% ethanol at a concentration of 0.01 mg/ml. A compound commonly used to establish the t_0 is potassium iodide which is present as an internal standard (1 mM) in the solutions of each test compound. A 5-μl quantity of solution is typically injected onto the column. We have found that 3.0 ml/min is a convenient flow rate for the mobile phase. In our work we have routinely maintained the temperature of the column module in a simple temperature-controlled environment.

Peaks eluted at a k' of less than 0.5 are often difficult to resolve from t_0 with sufficient accuracy. At the other extreme, peaks which have a k' of more than 10–15 have a tendency to broaden, which decreases resolution, and wastes expensive solvent. Values of k' can usually be kept within these boundaries by varying the conditions of measurement. Relatively water-soluble compounds are best eluted with lower levels of organic component in the mobile phase to increase retention and avoid confusion with the unretained standard. More hydrophobic compounds require a higher organic percentage in order to elute them in a reasonable amount of time. Retention time is inversely proportional to flow rate, but there is a danger of upsetting the bonded phase-solute-mobile phase equilibrium if the flow rate is too fast. Another parameter that effects the t_R is temperature. Increasing column temperature will result in an overall decrease in retention times up to a limit of about 80° with most columns.[11] The performance of many columns is reported to be affected by small temperature changes making strict temperature control mandatory, at least for absolute time determinations.[11] However, since we found no relative changes in the order of elution of our compounds at temperatures ranging from 26–40° we conclude that for comparative analysis of the kind used in our work it is only necessary to maintain a constant temperature during measurements for a series of quinone systems which include suitable internal references. The retention time is also related to the nature of the hydrocarbon bonded to the silica support such that k' increases with a change from columns based on phenyl to octyl to octadecyl derivatives. So far we have only explored the C_{18} columns in any detail.

Correlation of Retention Time With a Scale of Partition Coefficients Determined by the Shake-Flask Method

There are a number of quinones for which partition coefficients have been determined by the shake-flask method. We have taken these from the large compilation of data of Hansch and Leo.[2] Most determinations with quinones have been done with the cyclohexane–water system but a sufficient number have been done with the octanol–water system to permit comparison.

[11] R. W. Yost, L. S. Ettre, and R. D. Conlon, "Practical Liquid Chromatography." Perkin Elmer, Norwalk, Connecticut, 1980.

At this stage in the development of the method we have restricted our attention to the simpler quinones for which the most reproducible shake-flask derived data exist. Fourteen unsubstituted and methyl-substituted quinones were chosen for which partition coefficients in the cyclohexane–water system are established. These quinones have P values which span more than four orders of magnitude, and hence provide a suitable base for correlation with retention time on the RP-HPLC column. Seven of the compounds in this set have also been determined for the octanol–water system.

Figure 1 shows the correlation of partition coefficient, P, with k'. The data are presented logarithmically as a matter of convenience due to the

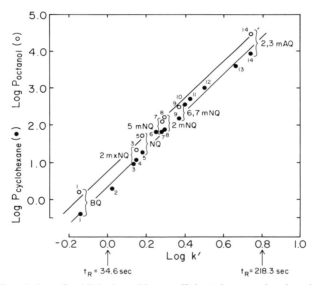

FIG. 1. Correlation of published partition coefficients in octanol and cyclohexane with column retention time. The compounds used are indicated on the figure as numbers. Compounds for which partition coefficient data are available for both solvent systems have their abbreviated names on the figure. The key is as follows: (1) 1,4-benzoquinone (BQ), (2) 2-methyl-1,4-benzoquinone (2mBQ), (3) 2-methoxy-1,4-naphthoquinone (2mxNQ), (4) 2,5-dimethyl-1,4-benzoquinone (2,5mBQ), (5) 1,4-naphthoquinone (NQ), (6) 6-methyl-1,4-naphthoquinone (6mNQ), (7) 5-methyl-1,4-naphthoquinone (5mNQ), (8) 2-methyl-1,4-naphthoquinone (2mNQ), (9) 6,7-dimethyl-1,4-naphthoquinone (6,7mNQ), (10) 2,3,5,6-tetramethyl-1,4-benzoquinone, duroquinone (DQ), (11) 2,3-dimethyl-1,4-naphthoquinone (2,3mNQ), (12) 9,10-anthraquinone (AQ), (13) 1-methyl-9,10-anthraquinone (1mAQ), (14) 2,3-dimethyl-9,10-anthraquinone (2,3mAQ). Chromatographic conditions in all figures unless otherwise noted are as follows: mobile phase = 70% acetonitrile/30% H_2O by volume; temperature = 26°; flow rate = 3.0 ml/min; column = Waters C_{18} Radial-PAK 10 × 0.8 cm i.d.

wide range of values presented. Two points are clear: first, the relationship between log k' and log P values is linear over the four orders of magnitude. Second, for the two organic phases used in shake-flask measurements the lines are close to parallel, displaying the anticipated similar log P/log k' relationship. The displacement of the lines simply represents the differing partition behavior of the quinone compounds between water and cyclohexane or octanol. Figure 2 shows that the linearity is maintained for different acetonitrile–water mobile phase mixtures ranging from 50 to 100% acetonitrile. This provides confidence that different solvent systems can be used to adjust t_R into the optimal analytical range when attention is directed to a particular region of log P values. It is worth noting, however, that as the organic percentage is increased to measure values of more hydrophobic compounds, the slope of the line defining the relationship steepens and the resolving power of the method diminishes.

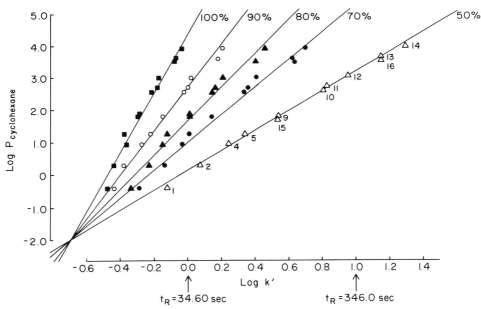

FIG. 2. Correlation of retention times versus log P for different acetonitrile–water mixtures. The numbers refer to the compounds listed in Fig. 1; additional compounds are as follows: (15) 2,3,5-trimethyl-1,4-benzoquinone (2,3,5mBQ), (16) 2,3,6,7-tetramethyl-1,4-naphthoquinone (2,3,6,7mNQ). The percentages above the lines correspond to the fraction of acetonitrile in the mobile phase. An idea of the actual retention time (t_R) on the column corresponding to a particular log k' is indicated on the abscissa.

Routine Calibration of the System

The system described forms the basis for calibration of a column to be used to determine partition coefficients of other quinones. All values presented in this report are measured in the cyclohexane–water solvent system.

We have found it advisable to calibrate the system using the quinone standards each time the mobile phase or column is changed. At least five of these standards should be run daily even when chromatographic conditions are not changed as a check for system problems like column aging, flow disruption, etc. Once the retention characteristics of the standards become familiar, they can be injected in aliquots containing four or five compounds at a time, as long as all of the peaks within an aliquot are well resolved. It is often helpful if one of the standards (we used 9,10-anthraquinone) is injected together with a test compound as a check for aberrant behavior in the system.

Some Examples

Alkyl Side Chain. Incremental increase in chain length of a hydrocarbon substituent has been previously reported to yield a regular increase in hydrophobicity.[1,3] The nature of the parent molecule (i.e., aliphatic or aromatic) as well as the position of substitution on it exert a strong influence on the magnitude of hydrophobic change for the initial alkyl substituent, but subsequent alkyl additions are characterized by uniform increases in log P. Shake-flask determined values for the serial addition of methylene groups to a variety of aromatic compounds show an increase in the range of 0.55–0.60 log P units per group, i.e., a molecule becomes 3.6–4.0 times more hydrophobic for each methylene group addition. Our HPLC-determined increases in partition coefficient attributable to the first four methylene group additions (CH_3 to C_4H_9), derived from three different homologous series of p-quinones with alkyl side chains of different length were 0.62 ± 0.02, 0.61 ± 0.02, 0.62 ± 0.02, and 0.65 ± 0.03 log P units. However, the fifth methylene group increased log P by 0.70 ± 0.04 and subsequent additions elicited even higher values so that at C_9H_{19} the per methylene increment was 0.78 ± 0.03; the basis for this trend needs further investigation.

Isoprene Side Chain of Ubiquinones and Menaquinones. Regular increase in partition coefficient as a hydrocarbon chain is lengthened incrementally is also observed via RP-HPLC in extremely hydrophobic larger molecules (see Fig. 3). Ubiquinones, which share a 2,3-dimethoxy-5-methyl-1,4-benzoquinone head group, and menaquinones, which share a 2-methyl-1,4-naphthoquinone head group, can be obtained as a homologous series with from 0 to 12 isoprene (C_5H_8) units at position 6 for the benzoquinones and position 3 for the naphthoquinones. The addition of the first isoprene unit to either molecule generates an increase in hydrophobic character of 2.26 ± 0.15 log P units. The second and third isoprene substituents have a slightly greater effect on the partition coefficient ($+2.49 \pm 0.11$ and $+2.55 \pm 0.09$, respectively). However, in similar fashion to the alkyl chain additions, subsequent isoprene substitutions induce a larger increase in partition coefficient than expected (2.90 ± 0.10 for the fourth isoprene, 3.00 ± 0.08 and 3.01 ± 0.07 for the fifth and sixth ones, respectively). Empirical calculations suggest an isoprene value between 1.8 and 2.4.

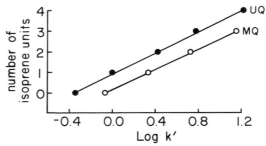

FIG. 3. Changes in log k' as prenyl (C_5H_8) units are added to ubiquinone (UQ) and menaquinone (MQ) homologs. The mobile phase used consisted of 90% acetonitrile/10% H_2O by volume. All other conditions are as described in Fig. 1.

Methyl Group Addition to Quinone Ring Systems. In assessing the magnitude of hydrophobic change exerted by methyl groups, electronic and steric effects are significant and no single value for a methyl substituent can be assigned without consideration of its local molecular environment. In the simplest case of singular methyl substitution, calculated values, shake-flask data, and our RP-HPLC results all generate a Δlog P in the range of $+0.55$ to $+0.60$ for phenol, benzene, and quinone parent molecules, independent of position. However, subsequent methyl additions can increase the partition coefficient by more than 0.80 or less than 0.40 log P units per substitution, dependent on position. As mentioned by Tanford in reference to normal alkanes, the effect of adding a methyl group is dependent on its proximity to one which is already present.[3] Methyl groups which are adjacent tend to increase log P by a greater amount than the sum of the individual groups. For example, the difference betwen 2-methyl-1,4-naphthoquinone and 1,4-naphthoquinone is 0.61 log P units by HPLC (0.62 by shake-flask), but the difference between 2,3-dimethyl-1,4-naphthoquinone and 2-methyl-1,4-naphthoquinone is 0.78 log P units by HPLC (0.82 by shake-flask). Furthermore, methyl groups substituted on the other naphthoquinone ring behave quite differently. The difference induced by the initial methyl residue in the case of 6-methyl-1,4-naphthoquinone and 1,4-naphthoquinone (0.59 log P units by HPLC, 0.56 by shake-flask) is greater than that exerted by the second addition between 6,7-dimethyl-1,4-naphthoquinone and 6-methyl-1,4-naphthoquinone (0.36 log P units by HPLC, 0.35 by shake-flask). While interesting in their own right, for our present purposes these data convincingly demonstrate parallel behavior between the shake-flask and the RP-HPLC methods.

Methoxy Group Substitution

The introduction of a methoxy group is also position dependent in its effects on partition cofficient. As shown in Fig. 4d, methoxy substituents in an α position decrease log P (Δlog P = -0.80 between 1mxAQ and AQ; -0.53 between NQ and 5mxNQ) (see legend to Fig. 1 for explanation of the abbreviations used). A second methoxy addition in any other α position has the same effect (Δlog P = -1.66 between 9,10AQ and 1,4 or 1,5mxAQ; -1.12 between NQ and 5,8mxNQ). However, substitution in a β position, shown in Fig. 4e, increases log P (Δlog P = $+0.30$ between AQ and 2mxAQ). Again, this effect is additive (Δlog P = $+0.55$ between AQ and 2,7mxAQ; $+0.59$ between AQ and 2,6mxAQ). In the case of a substitution ortho to the carbonyl on the p-quinone ring, there is a small decrease in hydrophobicity for the initial methoxy group (Δlog P = -0.09 between BQ and 2mxBQ; -0.07 between NQ and 2mxNQ), and a similar decrement is encountered on addition of a second methoxy group in a nonadjacent position (Δlog P = -0.17 between BQ and 2,5mxBQ; -0.13 between BQ and 2,6mxBQ). However, a methoxy substitution adjacent to the initial substituent induces a net *increase* in log P (Δlog P = $+0.13$ between BQ and 2,3mxBQ; $+0.09$ between NQ and 2,3mxNQ). None of these results is qualitatively surprising; but they would have been difficult to appraise quantitatively. Thus, an α-methoxy system may be more highly solvated by water because hydrogen bonding can occur simultaneously with the two *peri*-oxygen functions to give a six-membered ring, a situation which is not possible with the β-methoxy compounds. In the case of methoxy groups substituted in adjacent positions, it is known[12] that hindrance from the bulk of the methoxy groups can interfere with resonance-induced charge separation causing the molecule to become less polar than expected; such effects are more commonly seen in the $E_{1/2}$ values induced by each methoxy substituent, which shift in concert with the partition coefficient data.

[12] R. C. Prince, P. L. Dutton, and J. M. Bruce, *FEBS Lett.* **160**, 273 (1983).

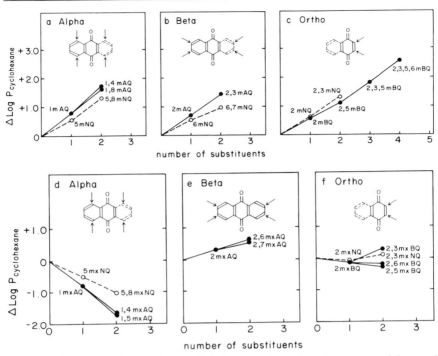

FIG. 4. The change in partition coefficient upon substitution of benzo-, naphtho-, and anthraquinones in the α, β, and ortho positions. Parts a, b, and c show the effect of methyl substitution; parts d, e, and f show the effect of methoxy substitution. These values are obtained from the retention times of a calibrated HPLC column as described in Figs. 1 and 2.

Concluding Remarks

The measurement of partition coefficients by RP-HPLC is a viable alternative to shake-flask or empirical methods. Once the calibration lines for a given class of compounds have been characterized, the retention time of previously undetermined compounds and single or multiple substituent groups can be assigned a partition coefficient based on a defined solvent system. The method is rapid and simple, and within certain boundaries seems reliable.

The procedure outlined has been confined to quinones. The calibration curves for the quinones were found to be coincident with those of other compounds. It seems clear that the carbonyl group of the quinone exerts no significant effect that would cause deviation in the calibration curve from that of the simplest hydrocarbon molecules. However, it is well known that substituents, particularly hydrogen bonding groups such as hydroxyl, cause the parent compound to adopt an independent set of calibration curves. Thus, when a group is added to the quinone, tests must be run to establish whether the construction of a new calibration curve incorporating the group is warranted.

Finally, it is worth pointing out an observation regarding the partition coefficients in general. The shake-flask method provides partition coefficients for systems, compounds, and solutes that unavoidably include water. There is reason to believe that the presence or absence of water has an extremely powerful effect on partitioning behavior, the "water-logged molecule" effect described by Wolfenden.[3] Our attempts to use 100% acetonitrile as a mobile phase to derive log P values for the extremely hydrophobic larger ubiquinones[5-10] and menaquinones are at present unsuccessful. The lack of an aqueous component appears to change the chromatographic environment drastically, and values obtained from this system are as yet unreliable. Anhydrous HPLC may give additional insight into the special interactions between various parts of a solute molecule and solvent water.

Acknowledgments

We are grateful to Marilyn R. Gunner and Dr. Dennis R. Gere for helpful discussions and to Ms. Michele A. Vitale for preparation of the manuscript. This work was supported by Grant GM 27309-06 from the National Institutes of Health, by Grant DE-AC02-80-ER 10590 from the United States Department of Energy, and by Grant PCM 82-09292 from the National Science Foundation.

[10] Function of Protonmotive Force in Translocation of Protein across Membranes

By LINDA L. RANDALL

Introduction

During growth of a cell, whether it be a prokaryote or eukaryote, there are proteins, synthesized in the cytosol, that must be transported to final destinations in membranes or in aqueous compartments bounded by membranes. This complex phenomenon of localization includes, in addition to synthesis of the polypeptide, targeting of the protein to the proper compartment and translocation of that protein into or through the membrane once it has arrived at the proper site.

Tne signal hypothesis, originally proposed to account for secretion in eukaryotes, also applies to assembly of the plasma membrane and lysosomal proteins. A mechanism is postulated that tightly couples synthesis of a polypeptide to its insertion into or through the membrane.[1] In this model, targeting of appropriate nascent polypeptides to the membrane of the rough endoplasmic reticulum is accomplished via interaction of an amino-terminal signal sequence with a proteinaceous complex, the signal recognition particle. Synthesis of the polypeptide is blocked in the cytosol by this interaction and proceeds only when the complex interacts with a receptor, the docking protein, on the membrane of the rough endoplasmic reticulum. Insertion of the growing polypeptide into or through the membrane is mechanistically linked to elongation of the chain. No additional energy is needed in this vectorial translation model.

Although export of protein in bacteria shares properties with eukaryotic secretion, i.e., the amino terminal signal or leader sequences resemble eukaryotic signals and bacterial proteins can be secreted in eukaryotic systems (see Randall and Hardy[2] for a review), in bacteria the translocation step is independent of elongation of the protein. In this respect bacterial export resembles the posttranslational uptake of precursors by chloroplasts and mitochondria.[3] The transfer of protein across membranes in these three systems requires energy in addition to the energy of protein synthesis. The nature of that energy requirement is the subject of this review.

Uptake of Polypeptides by Mitochondria

The earliest observation of an energy requirement for the uptake of protein by mitochondria was made by Neupert and co-workers[4,5] in their studies of *Neurospora*. They reported that the uncoupler, carbonyl cyanide *m*-chlorophenylhydrazone (CCCP), inhibited the posttranslational transport of the ADP/ATP carrier into mitochondria *in vivo*.[4] The uncoupler was also used *in vitro* to block import of the ADP/ATP carrier[5,6] into

---

[1] P. Walter, R. Gilmore, and G. Blobel, *Cell* **38,** 5 (1984).

[2] L. L. Randall and S. J. S. Hardy, *in* "Modern Cell Biology" (B. Satir, ed.), Vol. 3, p. 1. Liss, New York, 1984.

[3] M. Teintze and W. Neupert, *in* "Cell Membranes, Methods and Reviews" (E. Elson, W. Frazier, and L. Glaser, eds.), p. 39. Plenum, New York, 1983.

[4] G. Hallermayer and W. Neupert, *in* "Genetics and Biogenesis of Chloroplasts and Mitochondria" (T. Bucher *et al.,* eds.), p. 807. Elsevier, Amsterdam, 1976.

[5] M. A. Harmey, G. Hallermayer, and W. Neupert, *in* "Genetics and Biogenesis of Chloroplasts and Mitochondria" (T. Bucher *et al.,* eds.), p. 813. Elsevier, Amsterdam, 1976.

[6] R. Zimmerman and W. Neupert, *Eur. J. Biochem.* **109,** 217 (1980).

isolated mitochondria. These early experiments did not allow conclusions to be made about the direct source of energy since collapse of the electrochemical membrane potential results in depletion of intramitochondrial ATP as the proton translocating ATPase hydrolyzes ATP in a futile attempt to restore the protonmotive force.

Nelson and Schatz[7] mistakenly suggested that ATP was the direct source of energy for protein translocation. Their conclusion was based on the observation *in vivo* that protein is imported into mitochondria of *rho⁻* mutants of *Saccharomyces cerevisiae*. These mutants, which lack both a functional respiratory chain and the proton-translocating ATPase, should not have been able to generate protonmotive force. Thus it seemed unlikely that protonmotive force was the source of energy. The mitochondria of *rho⁻* mutants are able to transport ATP into the matrix via the ADP/ATP exchange system. When transport of ATP was blocked, import and processing of cytochrome c_1, and the α, β, and γ subunits of the F_1-ATPase was inhibited. Thus it was concluded that the ATP was the direct source of energy. Subsequent work (discussed below) done *in vitro* by the same laboratory[8,9] as well as others[10] showed that the requirement is not for ATP but for protonmotive force. The ability of *rho⁻* mutants to transfer protein into mitochondria might be explained if protonmotive force were generated by a pathway other than respiration or hydrolysis of ATP. The transport of ATP into the matrix via the ADP/ATP exchanger is electrogenic and might generate sufficient membrane potential to allow import of polypeptides.

Definitive demonstration that protonmotive force is the required form of energy entailed investigations of import under conditions in which protonmotive force was dissipated while ATP was present and conversely under conditions in which the level of ATP was depleted while protonmotive force was maintained. Gasser *et al.,*[9] using isolated yeast mitochondria *in vitro,* inhibited import of the β subunit of the F_1-ATPase by collapsing protonmotive force (with either valinomycin plus K^+ or CCCP) in the presence of oligomycin to inhibit hydrolysis of matrix ATP by the F_1-ATPase. Similar experiments were carried out *in vitro* using isolated *Neurospora* mitochondria.[10] Transfer of the ADP/ATP carrier and the subunit 9 of the ATPase was inhibited in the presence of oligomycin by either uncouplers (CCCP or dinitrophenol) or valinomycin plus K^+. In these experiments protonmotive force was dissipated while the matrix levels of

[7] N. Nelson and G. Schatz, *Proc. Natl. Acad. Sci. U.S.A.* **76,** 4365 (1979).

[8] S. M. Gasser, A. Ohashi, G. Daum, P. Böhni, J. Gibson, G. A. Reid, T. Yonetani, and G. Schatz, *Proc. Natl. Acad. Sci. U.S.A.* **79,** 267 (1982).

[9] S. M. Gasser, G. Daum, and G. Schatz, *J. Biol. Chem.* **257,** 13034 (1982).

[10] M. Schleyer, B. Schmidt, and W. Neupert, *Eur. J. Biochem.* **125,** 109 (1982).

ATP were high. In other experiments the electrochemical membrane potential was unperturbed while the intramitochondrial ATP pool was depleted by addition of both oligomycin to inhibit the ATPase and carboxyatractyloside to block the ADP/ATP exchanger. Under these conditions import was not inhibited. The same pattern of inhibition was seen in similar experiments using *Neurospora* in which import of cytochrome c_1, and subunits I, V, and VII of the cytochrome bc_1 complex were examined.[11] Thus it is clear that protonmotive force and not ATP is the primary source of energy utilized during import of these proteins.

Uncouplers have been used to inhibit import of numerous mitochondrial proteins.[12-15] Even though the secondary effects of ATP depletion were not directly excluded in all cases, it seems safe to assume that in general the requirement for energy is satisfied by the protonmotive force.

Proteins are imported into all the compartments of mitochondria: the matrix, the inner membrane, the outer membrane, and the intermembrane space. A requirement for energy is correlated not to the final destination, but rather to the pathway taken to reach that destination. If during localization a protein crosses the inner membrane, energy is required. This is true whether or not the protein is proteolytically processed. Thus the insertion of the ADP/ATP carrier into the inner membrane requires energy even though it is not processed. Cytochrome c and porin, proteins that do not cross the inner membrane, do not require energy for their localization.[12,16] Contrary to what one might imagine, import of some proteins into the intermembrane space is dependent on energy. Cytochrome b_2 reaches the intermembrane space via a pathway that involves two proteolytic cleavages. The first cleavage is mediated by the same soluble protease that is responsible for maturation of proteins localized to the matrix. Thus at least the amino-terminal portion of the precursor crosses the inner membrane in a process that requires protonmotive force. The intermediate form that is generated is membrane associated. A second protease, presumably on the outer face of the inner membrane, cleaves the intermediate to release the mature protein into the intermembrane space.[8,17]

It seems likely that cytochrome c peroxidase is imported to the intermembrane space by the same two-step mechanism. Import of the precur-

[11] M. Teintze, M. Slaughter, H. Weiss, and W. Neupert, *J. Biol. Chem.* **257**, 10364 (1982).
[12] R. Zimmerman, B. Hennig, and W. Neupert, *Eur. J. Biochem.* **116**, 455 (1981).
[13] M. Mori, T. Morita, S. Miura, and M. Tatibana, *J. Biol. Chem.* **256**, 8263 (1981).
[14] T. Morita, S. Miura, M. Mori, and M. Tatibana, *Eur. J. Biochem.* **122**, 501 (1982).
[15] A. Autor, *J. Biol. Chem.* **257**, 2713 (1982).
[16] H. Freitag, M. Janes, and W. Neupert, *Eur. J. Biochem.* **126**, 197 (1982).
[17] G. Daum, S. M. Gasser, and G. Schatz, *J. Biol. Chem.* **257**, 13075 (1982).

sor was blocked *in vivo* by CCCP; however, an intermediate form was not observed.[18] Possibly, the kinetics of the first and second steps do not allow detection of the intermediate.

The processing of cytochrome c_1 requires energy and occurs in two steps.[11,19] This protein is an inner membrane protein that has a large domain protruding into the intermembrane space. It has been proposed[17] that the mechanism of localization of cytochrome c_1 is related to that employed by proteins free in the intermembrane space and differs only in that after the two cleavages at the amino-terminus, the protein remains anchored to the inner membrane through a hydrophobic carboxy-terminus.

Precursors bind to mitochondria[6,10,12,20] under conditions in which there is no protonmotive force but they are not imported or processed. The precursor of the ADP/ATP carrier could be isolated bound to mitochondria in the absence of membrane potential. When protonmotive force was reestablished the bound carrier was imported.[20] Thus the energy-requiring step is translocation.

Export of Protein in Bacteria

Wickner and his colleagues[21] first demonstrated a requirement for protonmotive force during export of protein by bacteria in their studies of the insertion of the coat protein of phage M13 into the cytoplasmic membrane of *E. coli*. Treatment of cells with arsenate to deplete the cellular pools of high energy phosphate had no effect on conversion of the precursor form of the coat protein to the mature transmembrane form. When the electrochemical potential gradient of protons was dissipated by uncouplers, the precursor was found bound to the membrane on the cytoplasmic surface.[22] Thus, integration into the transmembrane conformation requires protonmotive force.

Uncouplers have been used *in vivo* to prevent processing of precursors of numerous exported proteins. A continually expanding list includes the periplasmic proteins maltose-binding protein,[23,24] arabinose-binding

[18] G. A. Reid, T. Yonetani, and G. Schatz, *J. Biol. Chem.* **257,** 13068 (1982).

[19] A. Ohashi, J. Gibson, I. Gregor, and G. Schatz, *J. Biol. Chem.* **257,** 13042 (1982).

[20] C. Zwizinski, M. Schleyer, and W. Neupert, *J. Biol. Chem.* **258,** 4071 (1983).

[21] T. Date, C. Zwizinski, S. Ludmerer, and W. Wickner, *Proc. Natl. Acad. Sci. U.S.A.* **77,** 827 (1980).

[22] T. Date, J. M. Goodman, and W. T. Wickner, *Proc. Natl. Acad. Sci. U.S.A.* **77,** 4669 (1980).

[23] H. G. Enequist, T. R. Hirst, S. Harayama, S. J. S. Hardy, and L. L. Randall, *Eur. J. Biochem.* **116,** 227 (1981).

[24] J. M. Pages and C. Lazdunski, *Eur. J. Biochem.* **124,** 561 (1982).

protein,[23] heat-labile enterotoxin,[25] leucine-specific binding protein,[26] TEM β-lactamase,[24,26] alkaline phosphatase,[27] and the outer membrane proteins OmpF,[23,24] OmpA,[23,24,28] and LamB.[23,24]

The only cytoplasmic membrane protein other than the M13 coat protein that has been studied with respect to energetics of insertion is leader peptidase, the enzyme that proteolytically processes exported proteins. It is synthesized without a leader sequence but requires protonmotive force for assembly as a transmembrane protein.[29]

Studies of bacterial protein export have demonstrated directly that uncouplers have their effect via dissipation of protonmotive force and not through secondary effects on ATP levels.[21,23,30] However, this establishment of a requirement for protonmotive force *in vivo* does not necessarily exclude an additional requirement for ATP. Work with cell-free systems indicates that ATP may also be required (P. C. Tai, personal communication).

What Is the Role of Protonmotive Force?

What role protonmotive force plays in translocation of proteins through membranes is an intriguing and unsolved question. It is not even known whether protonmotive force has a direct role or whether its influence on translocation is indirect through an effect on the properties of the membrane. Schatz and Butow[31] have suggested that the effect might be indirect since protonmotive force is established in *E. coli* and in mitochondria with the same polarity, positive outside, and yet the movement of polypeptides is directed outward in *E. coli* and inward in mitochondria. Perhaps the stability of the bilayer is altered to facilitate spontaneous insertion of the polypeptides, or if insertion is mediated by proteins, the conformation of a proteinaceous translocator might be altered by the state of the membrane. Although the mechanism is undefined, it is clear that the energy state of the cell does affect properties of the membrane. There are reports correlating deenergization with changes in the binding of hy-

---

[25] E. T. Palva, T. R. Hirst, S. J. S. Hardy, J. Holmgren, and L. Randall, *J. Bacteriol.* **146**, 325 (1981).

[26] C. J. Daniels, D. G. Bole, S. C. Quay, and D. L. Oxender, *Proc. Natl. Acad. Sci. U.S.A.* **78**, 5396 (1981).

[27] J. M. Pages and C. Lazdunski, *FEBS Lett.* **149**, 51 (1982).

[28] R. Zimmermann and W. Wickner, *J. Biol. Chem.* **258**, 3920 (1983).

[29] P. B. Wolfe and W. Wickner, *Cell* **36**, 1067 (1984).

[30] E. P. Bakker and L. L. Randall, *EMBO J.* **3**, 895 (1984).

[31] G. Schatz and R. A. Butow, *Cell* **32**, 316 (1983).

drophobic probes to the cell surface[32] and with altered susceptibility of the membrane to disruption by detergents.[33,34]

Polypeptides might be imported by mitochondria and exported by *E. coli* at special sites of contact between the inner and outer membranes. If this were the case, protonmotive force might be involved in the maintenance of close apposition of the two membranes as was proposed by Konisky[35] to explain why killing of *E. coli* by colicin Ia requires an energized membrane. Such a model might apply to mitochondria and gram-negative bacteria; however it fails to explain why secretion in gram-positive bacteria, which have only one membrane, requires protonmotive force (E. M. Murén and L. L. Randall, unpublished; Tweten and Iandolo[36]).

At this time, while it is impossible to eliminate an indirect role for protonmotive force, direct roles seem to be equally likely. Oxender and his colleagues have shown that CCCP inhibits processing and export of the periplasmic leucine-specific binding protein.[26] Although their experiments do not distinguish between a specific requirement for the electrical membrane potential ($\Delta\Psi$) and a requirement fulfilled by total protonmotive force (including ΔpH), they favor a model in which membrane potential is itself the motive force of export. They propose that the amino-terminal amino acids of the precursor polypeptide form a hairpin structure made of two antiparallel α-helices. This hairpin structure has a net dipole which would be aligned in a transmembrane orientation by the electric field associated with the protonmotive force. In addition, they suggest that a negatively charged domain of the polypeptide might move across the membrane by electrophoresis in response to the membrane potential.[26,37,38] Such a model excludes any role for the chemical potential component of protonmotive force, ΔpH.

A study of export of β-lactamase in *E. coli* directly addressed the question of specificity of the energy requirement.[30] It was concluded that the requirement is not specific for the membrane potential, but that the pH gradient can serve in place of the membrane potential. Thus, the system responds to total protonmotive force, and not only to $\Delta\Psi$.

[32] D. Nieva-Gomez and R. B. Gennis, *Proc. Natl. Acad. Sci. U.S.A.* **74,** 1811 (1977).

[33] J. K. Lanyi, *Biochemistry* **12,** 1433 (1973).

[34] E. Komor, H. Weber, and W. Tanner, *Proc. Natl. Acad. Sci. U.S.A.* **76,** 1814 (1979).

[35] J. Konisky, *in* "Bacterial Outer Membranes, Biogenesis and Function" (M. Inouye, ed.), p. 319. Wiley, New York, 1979.

[36] R. K. Tweten and J. J. Iandolo, *J. Bacteriol.* **153,** 297 (1983).

[37] R. C. Landick, C. J. Daniels, and D. L. Oxender, this series, Vol. 97, p. 146.

[38] B. R. Copeland, R. Landick, P. M. Nazos, and D. L. Oxender, *J. Cell. Biochem.* **24,** 345 (1984).

No proteins other than β-lactamase have been studied to determine whether their export responds specifically to $\Delta\Psi$ or to total protonmotive force. It is not sufficient to demonstrate that export is blocked by treatment of cells with valinomycin and K^+, which specifically dissipates $\Delta\Psi$. Even though ΔpH would not be affected, there might not be sufficient ΔpH present to satisfy the requirement for energy. In addition, demonstration that export proceeds normally in the absence of a pH gradient across the membrane does not mean the ΔpH would not substitute for $\Delta\Psi$ if it were present. It remains an open question whether export of all bacterial proteins is responsive to total protonmotive force. The question has not been addressed in mitochondrial systems.

If ΔpH can fulfill the energy requirement, the simplest model for the bacterial systems would be a mechanism that directly couples the translocation of the polypeptide through the membrane to the movement of protons in the opposite direction (or to the movement of hydroxyl ions in the same direction). In mitochondria such a mechanism would require symport of the polypeptides with protons or antiport with hydroxyl ions. Models of this type would involve proteinaceous translocators. In the bacterial system candidates for such a translocator would be products of the genes known to be necessary for export such as prlA (also called secY).[39]

The recent development of an energy-dependent cell-free system for translocation of bacterial proteins into vesicles[40,41] should allow the identification of the components necessary for translocation and provide information that will aid in the elucidation of the role of protonmotive force in the process of protein export.

Is Protonmotive Force Required in All Protein-Translocating Systems?

Proteins that are synthesized in the cytosol of plant cells are taken up by chloroplasts via a posttranslational mechanism that appears to be closely related to the mechanism of import in mitochondria.[42] Definitive experiments concerning the energetics of the chloroplast system are difficult to perform since there are transmembrane electrochemical potentials

[39] V. A. Bankaitis, J. P. Ryan, B. A. Rasmussen, and P. J. Bassford, Jr., in "Membrane Protein Biosynthesis and Turnover" (P. A. Knauf and J. Cook, eds.), p. 105. Academic Press, New York, 1985.

[40] D. B. Rhoads, P. C. Tai, and B. D. Davis, J. Bacteriol. 159, 63 (1984).

[41] L. Chen, D. Rhoads, and P. C. Tai, J. Bacteriol., 161, 973 (1985).

[42] A. Grossman, S. Bartlett, and N.-H. Chua, Nature (London) 285, 625 (1980).

across the envelope as well as across the internal energy-transducing membranes, the thylakoids.[43] However, the study that was carried out[42] led to the conclusion that ATP, not protonmotive force, was the form of energy utilized. It was shown that light stimulated the uptake into intact chloroplasts of proteins synthesized *in vitro*. Addition of uncouplers to the system inhibited transport of proteins located both in the stroma and in the thylakoid membranes. Addition of ATP to the uptake mixture reversed the inhibition. Since uncouplers should dissipate the electrochemical potential across the thylakoid membrane and the envelope membranes, it seems that ATP supplies the required energy.

The energy-dependent transfer of polypeptides in bacteria and mitochondria occurs across energy transducing membranes. Chloroplasts differ in that the initial transfer is through the envelope membrane whereas the energy-transducing membranes are internal. Thus the chloroplast might not have access to sufficient protonmotive force to drive uptake so that an alternative energy source, ATP, is used. However, this argument cannot account for the observation that integration into the thylakoids was dependent on ATP.[42]

Transfer of polypeptides into the lumen of the rough endoplasmic reticulum during secretion in eukaryotes is tightly coupled to elongation of the polypeptide and thus differs from the systems discussed above. If this coupling is mechanistic, the energy of elongation of the polypeptide might be used to transfer the growing nascent chain through the membrane. The current models for secretion assume no additional requirement for energy. However, since transfer is observed only under conditions compatible with protein synthesis, it is impossible to eliminate a requirement for ATP in the translocation step. It is also difficult using a cell-free system to eliminate definitively a role for protonmotive force *in vivo*. *In vitro,* in the absence of energy, translocation might proceed slowly and protonmotive force might be necessary in order for transfer to occur *in vivo* at a biologically significant rate. In this regard it is interesting to note that the presence of a proton-translocating ATPase in the rough endoplasmic reticulum has been reported.[44] Thus *in vivo* there may be an electrochemical potential gradient of protons across the membrane of the rough endoplasmic reticulum. The energetics of the secretory system will be particularly difficult to elucidate, not only because translocation requires ongoing protein synthesis but also because isolated microsomes may be altered in permeability relative to the *in vivo* situation.

[43] U. Heber and H. W. Heldt, *Annu. Rev. Plant Physiol.* **32,** 139 (1981).
[44] R. Rees-Jones and Q. Al-Awqati, *Biochemistry* **23,** 2236 (1984).

Concluding Remarks

The work summarized here represents only the beginning. The straightforward experiments that have been done allow us to state conclusively that protonmotive force is involved in translocation of protein through membranes in bacteria and in mitochondria. The challenge that lies ahead is to determine the precise function of the protonmotive force.

[11] Use of Secretion Cloning Vectors for Guiding the Localization of Proteins in *Escherichia coli*

By CHARLES A. LUNN, MASAYASU TAKAHARA, and MASAYORI INOUYE

The gram-negative bacterial cell is divided into four major subcellular compartments: the cytoplasm, the cytoplasmic (inner) membrane, the periplasmic space, and the outer membrane. Each of these compartments contains a unique set of enzymes and structural proteins which carry out the processes specific for that compartment. All of these proteins are synthesized on ribosomes contained within the cytoplasm. This requires that the cell possess mechanisms to translocate hydrophilic periplasmic proteins, as well as hydrophobic outer membrane proteins, across a hydrophobic inner membrane, and to segregate these proteins to their correct site of action.

Structural Information Required for Localization

Understanding the process by which cells transport proteins from their site of synthesis in the cytoplasm to their ultimate site of action *in vivo* has become the interest of many laboratories. Most proteins translocated across a bacterial membrane are synthesized as a protein precursor, containing an approximately 20 amino acid extension at the amino terminus called the signal peptide.[1] This peptide, involved in the initial interaction between the membrane and the translocated protein, must have certain structural characteristics to efficiently function as a signal peptide.[2] The amino-terminus of the signal peptide must contain several positively

[1] T. Silhavy, S. A. Benson, and S. D. Emr, *Microbiol. Rev.* **47**, 313 (1983).
[2] G. P. Vlasuk, S. Inouye, H. Ito, K. Itakura, and M. Inouye, *J. Biol. Chem.* **258**, 7142 (1983).

charged amino acids. This is followed by a 14 to 20 amino acid hydrophobic domain, with centrally located proline or glycine residues. Together,
these residues are proposed to form a hydrophobic loop through the membrane.[3] Serine and/or threonine are usually located on the carboxy-terminus of this proposed loop structure. Finally, correct processing by the
signal peptidase requires that the carboxy-terminal residue be either
alanine or glycine.

The presence of a signal peptide is necessary to direct a protein into
the inner membrane, and to translocate the amino-terminus of the protein
from the interior to the exterior surface of the inner membrane. The
ultimate localization of the protein within the inner membrane, periplasm,
or outer membrane requires structural information contained within the
protein itself.[3] Thus, cytoplasmic β-galactosidase cannot be converted
into an outer membrane protein simply by fusing this hydrophilic protein
with the signal sequence of *lamB*, an outer membrane protein.[4] Similarly,
the signal sequence from a periplasmic protein cannot direct a hydrophobic protein into the periplasm.[5] This suggests that directing a protein to a
desired compartment may require fusing the protein with critical portions
of a representative protein from that compartment.

We have constructed a battery of cloning vectors designed to affect
the localization of cloned gene products. This was accomplished by constructing hybrid proteins containing a signal sequence (for periplasmic
localization), or a signal sequence plus part of the major lipoprotein (for
outer membrane localization) to the cloned gene. We will describe the
construction and characteristics of these secretion cloning vehicles, and
will discuss how to use these secretion vectors for the expression of
foreign genes in *E. coli.*

pIN III: Multipurpose Expression Vehicle in *E. coli*

We have used the pIN III high expression vector system[6] as a basis for
constructing several types of secretion cloning vectors. This expression
system possesses several characteristics of an ideal cloning vehicle. The
vector is based on pBR322, a plasmid produced in roughly 30 copies per
bacterial cell.[7] The pIN III vector utilizes a high efficiency *lpp* promoter,

[3] M. Inouye and S. Halegoua, *CRC Crit. Rev. Biochem.* **7,** 339 (1980).
[4] F. Moreno, A. V. Fowler, M. Hall, T. J. Silhavy, I. Zabin, and M. Schwartz, *Nature (London)* **286,** 356 (1980).
[5] J. Tommassen, H. van Tol, and B. Lugtenberg, *EMBO J.* **2,** 1275 (1983).
[6] Y. Masui, T. Mizuno, and M. Inouye, *Bio/Technology* **2,** 81 (1984).
[7] D. Stueber and H. Bujard, *EMBO J.* **1,** 1399 (1982).

one of the strongest promoters in *E. coli*,[3] to efficiently initiate transcription of the cloned gene. These two factors ensure production of large amounts of protein from the cloned gene. Transcription via this *lpp* promoter is controlled by the *lac* UV5 promoter-operator, inserted downstream from the *lpp* promoter. Therefore, the expression of the cloned gene is repressed by the *lac* repressor (*lacI* gene product) in the absence of a *lac* inducer. This allows controlled induction of the cloned gene under desired growth conditions (e.g., when amplification of the cloned gene products is least detrimental to the host cell metabolism). The *lacI* gene is inserted into the plasmid, so that expression of the cloned gene is regulated by a *lac* repressor produced by the vector itself. Details of the construction of the pIN III vector system are presented elsewhere.[6]

All of the secretion cloning vectors based on the pIN III system (pIN III-A, pIN III-B, pIN III-C, and pIN III-*ompA*) are designed to contain identical multirestriction sites. This permits exchange of cloned DNA fragments between vectors, from a unique upstream *Xba*I site through the multirestriction sites. This *Xba*I site also provides a convenient site for sequencing the junction between the cloning vehicle and the cloned gene.

Secretion Cloning Vectors for Periplasmic Localization

pIN III-ompA

To direct hydrophilic proteins into the periplasmic space, we constructed a set of vectors which fuse the *ompA* signal sequence directly to the cloned gene.[8,9] A summary of the production of this vehicle is presented in Fig. 1. A 27-bp fragment of pIN III-A3,[6] between *Xba*I and *Eco*RI restriction sites, was removed and replaced with the *ompA* signal peptide sequence. This signal peptide directs the translocation of OmpA protein across the cytoplasmic membrane.[10] The signal sequence is known to be cleaved off from the *ompA* precursor protein by signal peptidase I,[10] a nonspecific signal peptidase which is considered to process all secreted proteins, except for the lipoproteins.[11–14] The *ompA* signal pep-

[8] Y. Masui, J. Coleman, and M. Inouye, *in* "Experimental Manipulation of Gene Expression" (M. Inouye, ed.), p. 15. Academic Press, New York, 1983.

[9] J. Ghrayeb, H. Kimura, H. Takahara, H. Hsiung, Y. Masui, and M. Inouye, *EMBO J.* **3**, 2437 (1984).

[10] R. Zimmermann and W. Wickner, *J. Biol. Chem.* **258**, 3920 (1983).

[11] H. Yamagata, C. Ippolito, M. Inukai, and M. Inouye, *J. Bacteriol.* **152**, 1163 (1982).

[12] H. Yamagata, K. Daishima, and S. Mizushima, *FEBS Lett.* **158**, 301 (1983).

[13] M. Tokunaga, J. M. Loranger, P. B. Wolfe, and H. C. Wu, *J. Biol. Chem.* **257**, 9922 (1982).

[14] W. Wickner, *Trends Biol. Sci.* **8**, 90 (1983).

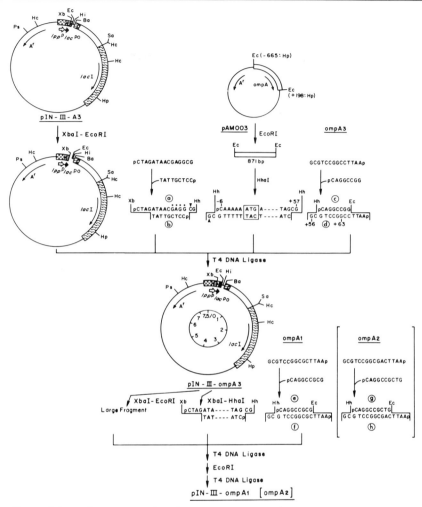

FIG. 1. Schematic representation of the construction of secretion vectors pIN III-*ompA*1 (*ompA*2 and *ompA*3). Detailed description of the production of these vectors is presented elsewhere.[8] Dots indicate the Shine–Dalgarno sequence. Ba, *Bam*HI; Ec, *Eco*RI; Hc, *Hin*cII; Hh, *Hha*I; Hi, *Hin*dIII; Hp, *Hpa*I; Ps, *Pst*I; Sa, *Sal*I; and Xb, *Xba*I; lpp[p], *lpp* promoter, and lac[po], *lac* promoter–operator.

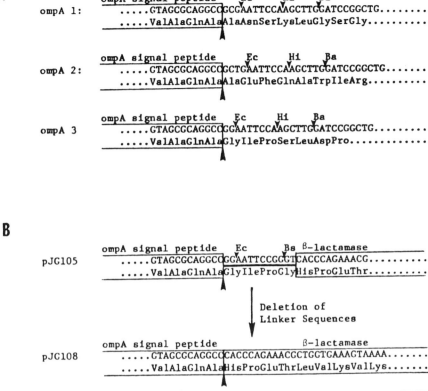

FIG. 2. Partial DNA sequences of linker polynucleotides. (A) pIN III-*ompA*1, pIN III-*ompA*2, pIN III-*ompA*3. (B) pJG105 and pJG108. Abbreviations are the same as in Fig. 1. Arrows indicate the cleavatge site of the *ompA* signal peptide. The linker sequence of pJG105 is underlined.

tide sequence was prepared from a 65-bp *Hha*I fragment from the cloned *ompA* gene (P. J. Green and M. Inouye, unpublished results) plus two synthetic oligonucleotides coding for the Shine–Dalgarno ribosome binding site,[15] and the *ompA* signal peptide cleavage site. This vector (pIN III-*ompA*3) has three unique restriction sites, *Eco*RI, *Hin*dIII, and *Bam*HI, immediately after the sequence of the *ompA* signal peptide. In order to express any foreign DNA fragments with any reading frame, two other vectors (pIN III-*ompA*1 and pIN III-*ompA*2) have been constructed.

---

[15] J. Shine and L. Dalgarno, *Nature (London)* **254,** 34 (1975).

These two vectors are identical to pIN III-*ompA*3, but have the multiple cloning sites in two different reading frames (see Fig. 1). The resulting junctional sequences for each vector, containing the unique *Eco*RI, *Hin*dIII, and *Bam*HI cloning sites, is shown in Fig. 2A. Because of these junctional sequences, each cloned gene product contains several extra amino acids at its amino-terminal end. However, the sequences coding for these extra amino acids can be easily removed by oligonucleotide-directed site-specific mutagenesis (Fig. 2B; and see Morinaga *et al.*[16]).

Secretion Cloning Vector for Outer Membrane Localization

pIN III-C

Localization of a protein within the outer membrane requires more structural information than that contained within the signal sequence.[3] Our studies of the major lipoprotein suggested that the signal sequence, plus a short peptide (8 to 9 amino acids) from the amino-terminus of the lipoprotein are sufficient to translocate β-lactamase to the outer membrane.[17] Based on this result, we constructed the pIN III-C cloning vectors. These vectors contain three unique restriction sites (*Eco*RI, *Hin*dIII, and *Bam*HI) for cloning a foreign DNA fragment located immediately after the *lpp* signal sequence, plus eight amino acids from the *lpp* structural gene. Construction of the pIN III-C vectors was described elsewhere.[18] As with the pIN III-*ompA* vectors described above, the unique *Eco*RI, *Hin*dIII, and *Bam*HI sites for cloning are constructed in three different reading frames (pIN III-C1, pIN III-C2, and pIN III-C3). The sequences of these vectors at the multirestriction sites are shown in Fig. 3.

Use of the Secretion Cloning Vectors

Cloning β-Lactamase into pIN III-ompA

The structural gene for β-lactamase (minus its own signal sequence) was inserted into the *Eco*RI site of pIN III-*ompA*3 to yield pJG105.[9] This cloning strategy predicts the presence of 12 extra base pairs between the *ompA* signal peptide sequence and the structural gene for β-lactamase

[16] Y. Morinaga, T. Franceschini, S. Inouye, and M. Inouye, *Bio/Technology* **2**, 636 (1984).
[17] J. Ghrayeb and M. Inouye, *J. Biol. Chem.* **259**, 463 (1984).
[18] M. Inouye, K. Nakamura, S. Inouye, and Y. Masui, *in* "The Future in Nucleic Acid Research" (I. Watanabe, ed.), p. 419. Academic Press Japan, Tokyo, 1983.

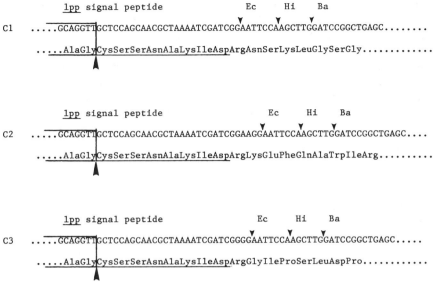

Fig. 3. Partial DNA sequences of linker polynucleotides of pIN III-C1, pIN III-C2, and pIN III-C3. Abbreviations are the same as in Fig. 1. Arrows indicate the cleavage site of the *lpp* signal peptide. The sequences underlined represent the amino-terminal region of lipoprotein.

(see Fig. 2B). Thus, four extra amino acids (Gly-Ile-Pro-Gly) were presumed to be attached to the amino terminus of the amplified gene product. Figure 4 shows that pJG105 produced a single product, with an apparent molecular weight slightly higher than larger authentic β-lactamase, after induction with 2 mM isopropyl-β-D-galactopyranoside (lane 3). No product was observed in the absence of inducer (lane 1). The protein yield reached approximately 20% of total cellular protein after induction for 3 hr. The new product exhibited β-lactamase activity, and was immunoprecipitated with anti-β-lactamase serum.

We next probed the amino acid sequence of the amino terminus of the pJG105 gene product. Following induction, the cloned gene product was labeled with [³H]proline, purified, then subjected to sequential Edman degradation. Significant radioactivity was observed after cycle 3 and cycle 6. This indicates that the *ompA* signal peptide was cleaved at its normal cleavage site, yielding β-lactamase gene product with four extra amino acids (Gly-Ile-Pro-Gly), as predicted from the construction (Fig. 2). The conclusion was further supported by Edman degradation of the pJG105 gene product labeled with [³H]glycine (radioactivity appeared at

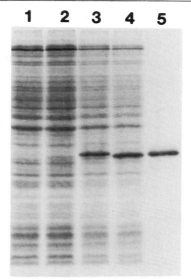

FIG. 4. SDS–polyacrylamide gel electrophoresis of total cell fractions of *E. coli* JA221 *lpp⁻*/F' *lac*I^q/pJG105 and pJG108. Cells were grown at 37° in M9 medium to a Klett reading of 30. IPTG (2 m*M*) was then added to half of the culture. After 3 hr, the cells were collected by centrifugation, washed twice with 10 m*M* Tris–HCl, pH 7.0, lysed by sonication, and aliquots submitted to SDS–polyacrylamide gel electrophoresis. Lanes 1 and 3: 0.1 ml culture of *E. coli* JA221 *lpp⁻*/F' *lac*I^q/pJG105 grown with (lane 1) and without (lane 3) IPTG. Lanes 2 and 4: 0.1 ml culture of *E. coli* JA221 *lpp⁻*/F' *lac*I^q/pJG108 grown with (lane 2) and without (lane 4) IPTG. Lane 5: purified β-lactamase.

cycle 1 and cycle 4). These results indicate that no cleavage occurred at histidine, the normal amino-terminus of β-lactamase.

The four extra amino acids at the amino-terminus of the pJG105 gene product using oligonucleotide-directed site-specific mutagenesis. For this purpose, a 24-mer oligonucleotide (GTAGCGCAGGCCCACCCA-GAAACG) was synthesized, and used to remove the linker sequence as described.[16] The DNA sequence of one resulting transformant, pJG108, is shown in Fig. 2B. As with pJG105, the induction of pJG108 yielded a single product (Fig. 4, lanes 2 and 4). The product migrated to the same position as authentic β-lactamase (lane 5). This protein was also produced to 20% of the total cellular protein. Sequential Edman degradation of the gene product labeled with [³H]lysine yielded significant radioactivity at cycles 7 and 9. This result demonstrates that the *ompA* signal peptide was correctly removed at a cleavage site immediately preceeding the amino-terminal histidine (Fig. 2B).

In *E. coli,* the β-lactamase encoded by pBR322 is localized within the periplasmic space.[19] Because the cloned β-lactamase gene produced a product identical to the natural gene product, we expected the pJG108 gene product would be contained within the periplasmic space. However, neither the pJG108 nor the pJG105 gene products were released from IPTG-induced cells following osmotic shock or digestion of the outer cell envelope with lysozyme-EDTA. Both proteins were recovered in a low speed pellet fraction following sonication. The pellet contained mainly outer membrane proteins. Pure (greater than 96% pure) β-lactamase was extracted from the low speed pellet fraction with 0.3% sodium lauryl sarcosinate. We concluded that overproduction of β-lactamase leads to its aggregation, presumably within the periplasmic space. Protein aggregates associated with the peptidoglycan layer of the outer membrane were also observed upon overproduction of a hybrid lipoprotein.[20] In our case, aggregation was not the result of intermolecular disulfide bonds, as no high-molecular-weight forms of the pJG105 gene product were observed upon SDS–polyacrylamide gel electrophoresis of samples prepared without reducing agent. Because correct disulfide bonds are formed as β-lactamase is translocated into the periplasmic space,[21] this result supports periplasmic localization of the cloned protein.

Cloning Staphylococcal Nuclease A into pIN III-ompA

The structural gene for staphylococcal nuclease A was inserted into the *Bam*HI site of pIN III-*ompA*3.[22] The gene was included within a 518-bp *Sau*3AI fragment containing the nuclease structural gene, a 6-residue amino-terminal extension, and a 53-bp 3'-nontranslated region. This produced a plasmid, pIN III-*ompA*3-#98, with 33 extra base pairs between the *ompA* signal peptide sequence and the structural gene for nuclease A. The extra bases derive from the linker sequence (15 bp) and from the amino terminal extension of nuclease A (18 bp) present within the cloned *Sau*3AI fragment. The protein produced by this plasmid was characterized. Figure 5 shows that pIN III-*ompA*3-#98 produced a single product comprising 3% of total cellular protein, after induction for 70 min with 2 mM IPTG. Little of this protein was present without induction. The product exhibited nuclease activity, and was precipitated with antiserum raised against nuclease A. Edman degradation demonstrated that the

[19] D. Koshland and D. Botstein, *Cell* **20,** 749 (1980).
[20] F. Yu, H. Furukawa, K. Nakamura, and S. Mizushima, *J. Biol. Chem.* **259,** 6013 (1984).
[21] S. Pollitt and H. Zalkin, *J. Bacteriol.* **153,** 27 (1983).
[22] M. Takahara, D. W. Hibler, P. J. Barr, J. A. Gerlt, and M. Inouye, *J. Biol. Chem.* **260,** 2670 (1985).

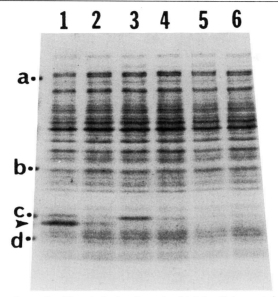

FIG. 5. SDS–polyacrylamide gel electrophoresis of total cellular proteins. *E. coli* JA221 *lpp⁻* harboring pONF1 (lanes 1 and 2), pIN III-*ompA*-#98 (lanes 3 and 4), or pIN III-*ompA*3 (lanes 5 and 6) was grown at 37° in L-broth containing 50 μg/ml ampicillin. At approximately 1 × 10⁸ cells/ml, IPTG was added to a final concentration of 2 m*M*, and the cultures were further incubated for 110 min. The cells were solubilized, and subjected to SDS–polyacrylamide gel electrophoresis. The gel was then stained with Coomassie brilliant blue. Lanes 1, 3, and 5 show total cellular proteins produced in the presence of IPTG; lanes 2, 4, and 6 show proteins produced in the absence of IPTG. The letters represent migration positions of molecular weight standards: a, phosphorylase *b* (92,500); b, carbonate dehydratase (31,000); c, soybean trypsin inhibitor (21,500); d, egg white lysozyme (14,400). An arrow indicates the migration position of commercially obtained staphylococcal nuclease A.

ompA signal peptide was removed at its normal cleavage site, yielding a protein composed of nuclease A with an 11 amino acid extension. This protein was localized within the periplasmic space.

The 33-bp linker sequence between the *ompA* signal peptide sequence and the nuclease A structural gene was again removed by oligonucleotide-directed site-specific mutagenesis. The plasmid resulting from this manipulation (pONF1) produced a one peptide comprising 10% of total cellular protein (mature nuclease A) and one peptide, comprising 2% of total cellular protein, with a larger molecular weight (unprocessed nuclease A). The processed nuclease was localized within the periplasmic space. The reason for the 4-fold increase in protein production after removal of the linker sequence is not known.

Cloning Staphylococcal Nuclease A into pIN III-C

The same *Sau*3AI fragment used above was also inserted into the *Bam*HI site of pIN III-C3 (M. Takahara and M. Inouye, unpublished results). The localization of the gene product from this plasmid (pIN III-C3 nuclease A) was then compared to that of the pONF1 plasmid. Cells harboring these two plasmids were induced with 2 mM IPTG for 3 hr, then collected by centrifugation, lysed by sonication, and fractionated into a soluble and membrane fraction. Each fraction was then analyzed by SDS–polyacrylamide gel electrophoresis. Figure 6 shows that the pIN III-C3-nuclease A gene product is found exclusively in the membrane fraction, migrating with an apparent molecular weight of 18,800. The pONF1 gene

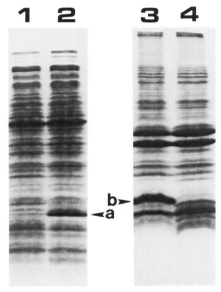

FIG. 6. Localization of staphylococcal nuclease A gene product after cloning into pIN III-C or pIN III-*ompA*. Staphylococcal nuclease A was cloned into pIN III-C (yielding pIN III-C3-nuclease A) or pIN III-*ompA* (yielding pONF1). These plasmids were used to transform *E. coli* JA221 *lpp*⁻/F' *lac*I^q. Selected transformants were then grown at 37° in M9 medium. At a Klett of 50, IPTG (2 mM) was added, and cells incubated for an additional 3 hr. Cells were then collected by centrifugation, separated into a soluble and a membrane fraction, and aliquots analyzed by electrophoresis on an SDS–polyacrylamide gel using a Tris buffering system. Lanes 1 and 2: 0.15 ml culture of soluble fraction from cells harboring pIN III-C3-nuclease A (lane 1) or pONF1 (lane 2). Lanes 3 and 4: 0.45 ml culture of membrane fraction from cells harboring pIN III-C3-nuclease A (lane 3) or pONF1 (lane 4). Arrows indicate positions of mature staphylococcal nuclease A (a) and the membrane associated form of nuclease A (b).

product, on the other hand, resides exclusively in the soluble (including periplasmic) fraction at an apparent molecular weight of 17,000. The increased size of nuclease A cloned in pIN III-C3 reflects addition to the nuclease gene of a 24-bp portion of the *lpp* structural gene. This fragment, along with the *lpp* signal sequence, was shown to be sufficient to direct a β-lactamase hybrid protein into the outer membrane, after modification of the terminal cysteine with a glyceride group, and processing via the globomycin-sensitive signal peptidase II.[17] Thus, the β-lactamase protein with nine amino acids from the lipoprotein is processed as is the major lipoprotein. Further characterization of the pIN III-C3-nuclease gene product showed that the protein was was modified by a glyceride group, and its processing was sensitive to globomycin. The hybrid protein was localized in both the outer membrane fraction, and the inner cytoplasmic membrane fraction.

Comments

We described three cloning systems which direct the cloned gene product out of the cytoplasm. These vectors provide several advantages not afforded by standard vector systems for cloning foreign genes into *E. coli*. First, the transport of cloned proteins out of the cytoplasm allows cloning of genes whose products are lethal to the host cell (such as nucleases and proteases). Second, gene products sensitive to the high proteolytic activity present in the bacterial cytoplasm can be translocated to the periplasmic space. Third, production of proteins in the cytoplasm requires an initiator methionine at the polypeptides amino-terminus. In cases (such as staphylococcal nuclease A) in which the terminal amino acid is not methionine, the processing that occurs during translocation provides a method of resurrecting a native gene product from a transcriptionally active precursor. Finally, transport out of the cytoplasm affords a severalfold purification of the cloned gene product, based on the relative protein concentration in cytoplasmic fractions versus membrane or periplasmic fractions. For all of these reasons, secretion cloning vectors provide powerful tools for the production of biologically interesting proteins.

Acknowledgments

Our thanks to Paul E. March for his critical reading of this manuscript.

[12] Use of *lac* Gene Fusions to Study Transport Proteins

By HOWARD A. SHUMAN

Membrane proteins which mediate the active transport of solutes are often difficult to detect, isolate, and characterize biochemically. Many transport proteins do not have ligand-binding characteristics which can be used as a "handle" to study these molecules. We have developed a general method for labeling proteins genetically with a carboxy-terminal fragment of active β-galactosidase. Gene fusions are constructed between the gene that codes for the transport protein and the structural gene (*lacZ*) which codes for β-galactosidase. The gene fusion results in the production of a hybrid protein which may or may not retain the biological activity of the transport protein. This method allows one to study the synthesis and regulation of the transport protein *in vivo* as well as facilitating its biochemical detection in subcellular extracts with antibodies generated against the purified transport–β-galactosidase hybrid protein. This article describes how *lac* gene fusions have been used to study the MalF and MalK proteins, two cytoplasmic membrane components of the maltose transport system of *Escherichia coli* K-12.

Methods

Construction of lac Gene Fusions. In 1976 Malcolm Casadaban described a technique for fusing the 5' end of any *E. coli* gene to the *lacZ* gene.[1] Since then many other techniques have been developed to fuse genes to *lacZ*. Some of these take advantage of the ease and flexibility with which one can manipulate the *lac* genes *in vivo*[2,3] and others entail the use of recombinant DNA technology.[4] One recent variation takes advantage of both *in vitro* technology and *in vivo* manipulations to conveniently generate a collection of gene fusions on a high-copy plasmid vector.[5]

The properties of β-galactosidase which make it ideal as an *in vivo* labeling agent are (1) the ability to assay its enzymatic activity conven-

[1] M. J. Casadaban, *J. Mol. Biol.* **104,** 541 (1976).
[2] M. J. Casadaban and J. Chou, *Proc. Natl. Acad. Sci. U.S.A.* **81,** 535 (1980).
[3] E. Bremer, T. J. Silhavy, J. M. Wiseman, and G. M. Weinstock, *J. Bacteriol.* **158,** 1084 (1984).
[4] M. J. Casadaban, J. Chou, and S. N. Cohen, *J. Bacteriol.* **143,** 971 (1980).
[5] M. L. Berman and D. E. Jackson, *J. Bacteriol.* **159,** 750 (1984).

iently in both living cells and cell free extracts by colorimetric methods,[6] (2) the large molecular weight of the enzyme (116,000) allows the purification of the hybrid proteins by molecular seive chromatography or preparative gel electrophoresis,[7] and (3) the substrate specificity of the enzyme allows the rapid high-yield purification of some β-galactosidase hybrids by affinity chromatography.[8]

Any of the techniques mentioned above can be used to generate *lacZ* gene fusions which retain β-galactosidase activity. In these cases, the amino-terminal 20–30 residues of the enzyme are replaced by an amino-terminal fragment derived from the protein of interest. An unusual property of β-galactosidase is that almost any polypeptide sequence can substitute for the amino-terminal 30 amino acids without a drastic reduction in the catalytic efficiency of the enzyme.[9] Exceptions to this rule occur when a long amino-terminal sequence is derived from some secreted proteins such as maltose-binding protein or the LamB protein. In such cases, the localization of the hybrid protein may interfere with efficient tetramerization and thus lower the specific activity of the enzyme.[10,11]

In order to illustrate how *lacZ* fusions have been used to identify and characterize transport proteins, the purification of a MalF-LacZ hybrid protein and its use are described below.

Purification of MalF-LacZ Hybrid 11-1. One of the first *mal–lacZ* gene fusions to be described was ϕ *(malF–lacZ)* hyb 11-1 (see Fig. 1).[12] The β-galactosidase activity in strains containing this fusion is maltose inducible and located in the cytoplasmic membrane of the cell.[12] The apparent molecular weight of the hybrid β-galactosidase is 150,000, based on its electrophoretic mobility on polyacrylamide gels containing SDS. Since the molecular weight of wild-type β-galactosidase is 116,000, a substantial amount of the MalF protein is present at the amino-terminal portion of the 11-1 MalF–LacZ hybrid protein.[13] This conclusion was substantiated when the nucleotide sequences of the *malF* gene and the *malF–lacZ* 11-1 gene fusion were determined. Of the 514 amino acid

[6] J. H. Miller, "Experiments in Molecular Genetics." Cold Spring Harbor Laboratory, Cold Spring Harbor, New York, 1972.
[7] A. V. Fowler and I. Zabin, *J. Biol. Chem.* **253**, 5521 (1978).
[8] A. Ullmann, *Gene* **29**, 27 (1984).
[9] A. V. Fowler and I. Zabin, *J. Biol. Chem.* **258**, 14354 (1978).
[10] P. J. Bassford, Jr., T. J. Silhavy, and J. R. Beckwith, *J. Bacteriol.* **139**, 14 (1979).
[11] M. N. Hall, M. Schwartz, and T. J. Silhavy, *J. Mol. Biol.* **156**, 93 (1982).
[12] T. J. Silhavy, M. J. Casadaban, H. A. Shuman, and J. R. Beckwith, *Proc. Natl. Acad. Sci. U.S.A.* **73**, 3423 (1976).
[13] H. A. Shuman, T. J. Silhavy, and J. R. Beckwith, *J. Biol. Chem.* **255**, 168 (1980).

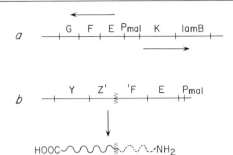

FIG. 1. The *malB* region of the *Escherichia coli* chromosome. (a) The wild-type region. This region comprises five structural genes arranged in two operons which are transcribed divergently from a central control region (Pmal). All five gene products participate in the active transport of maltose and longer maltodextrins. The *malE* gene encodes a water-soluble, periplasmic, maltose-binding protein. The *lamB* gene encodes an outer membrane protein that facilitates the movement of maltose and maltodextrins across the outer membrane. This protein is also the receptor for bacteriophage lambda. The *malF*, *malG*, and *malK* genes encode cytoplasmic membrane proteins which catalyze the active transport of maltose and maltodextrins across the inner membrane. (b) The *malF–lacZ* gene fusion. Expression of the *malF–lacZ* hybrid gene and the *lacY* gene is under control of Pmal. The *malF–lacZ* hybrid gene codes for a protein with an amino-terminal region coded by the 5' portion of the *malF* gene and a carboxy-terminal region coded by the 3' portion of the *lacZ* gene. The wavy line indicates the fusion joint.

residues in the entire MalF protein, 287 of them are present in the 11-1 MalF–LacZ hybrid.[14]

In order to purify this protein, advantage was taken of the membrane location of the hybrid protein and its large molecular weight. The 11-1 MalF–LacZ hybrid was purified from a strain (HS6503) that was diploid for the gene fusion and produced about 1000 units of β-galactosidase activity.[15] Cells were grown in M63 medium containing 2% maltose as the source of carbon. The culture (100 liters) was harvested at the end of exponential growth and the bacterial pellet was frozen at $-20°$. The cells (200 g) were thawed in 20 mM Tris–HCl, pH 7.4, containing 0.1 mM MgSO$_4$ and broken by two passages through a Mantin-Gaulin Mill at maximum pressure. After removal of unbroken cells by centrifugation at 10,000 g for 20 min, the membranes were harvested by centrifugation at 150,000 g for 2 hr at 4° in the Beckman 45Ti rotor. The pellet material was washed twice by resuspension in buffer and recentrifugation. The initial supernatant fraction contained a substantial amount of cytoplasmic mem-

[14] S. Froshauer, *J. Biol. Chem.* **259,** 10896 (1984).
[15] H. A. Shuman and J. R. Beckwith, *J. Bacteriol.* **137,** 365 (1979).

brane and malF–LacZ hybrid protein. These were recovered by precipitation with 50% (w/v) ammonium sulfate. Soluble proteins redissolved from the precipitated material, but the membrane fragments were recovered in a high-speed pellet. Aliquots of the membrane fractions were solubilized by treating the membranes at 100° for 10 min with 1% SDS (w/v) and 5 mM 2-mercaptoethanol at a protein concentration of 5 mg/ml. After cooling, any insoluble material was removed by centrifugation at 100,000 for 1 hr at 20°. The supernatant was then concentrated with PEG 6000 and dialyzed against Tris buffer containing SDS and 2-mercaptoethanol. In order to separate the MalF–LacZ hybrid from the other membrane polypeptides, the mixture was chromatographed on a column of 6% beaded agarose (Bio-Rad A5m) equilibrated in Tris–SDS–2-mercaptoethanol. Fractions containing the MalF–LacZ hybrid were identified by electrophoresis of a small aliquot of each fraction on SDS–polyacrylamide gels (see Fig. 2). Two chromatographic runs were necessary to obtain large amounts of homogeneous material. The overall yield was 5 μg/ml of total membrane protein.

Production of Anti-MalF Antibodies and Identification of the MalF Protein. The purified MalF–LacZ hybrid protein was then used to immunize rabbits. The resulting antisera were adsorbed with a crude extract of a *lacZ, ∆malF* strain to remove antibodies that recognize β-galactosidase and *E. coli* proteins other than MalF. After adsorbtion, the antibodies were used to identify the MalF protein in radioactive extracts from *mal$^+$* strains which do not contain the *malF–lacZ* gene fusion. The antibodies precipitated a maltose-inducible protein with an apparent molecular weight of 40,000. The protein was not found in strains which carry insertion mutations or chain-termination mutations in *malF* or polar mutation in DNA sequences upstream from *malF*. In addition, the radioactive 40,000 molecular weight protein cross-reacted with purified MalF–LacZ hybrid protein but not with wild-type β-galactosidase. This protein is found in the cytoplasmic membrane of the cell; it can be solubilized from the membrane by Triton X-100 in the absence of EDTA.

The 40,000 apparent molecular weight of MalF protein from SDS–polyacrylamide gel electrophoresis differs from the molecular weight predicted from the nucleotide sequence (56,947).[14] The anomalous observed molecular weight may be due to some secondary structure that is retained in SDS. Recently it has been found that denaturation of MalF protein with 70% formic acid prior to electrophoresis results in a change in the apparent molecular weight of the protein to 60,000 (P. Oh and H. Shuman, unpublished data). The antibodies have also been used to quantitate the MalF protein. Maltose-grown bacteria contain approximately 1000 copies of MalF per cell. This is about 50-fold less than the amount of periplasmic

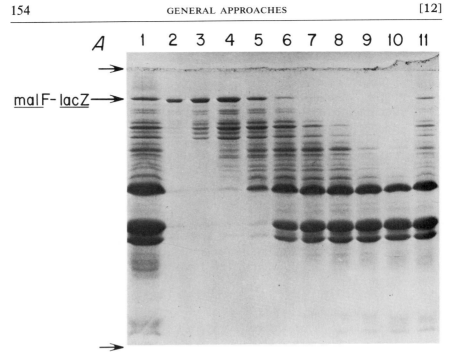

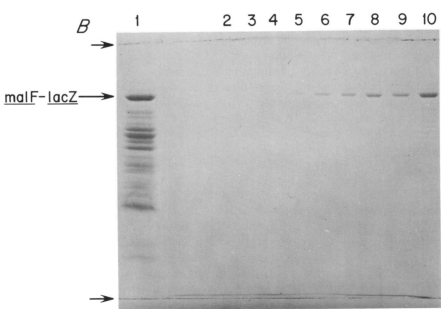

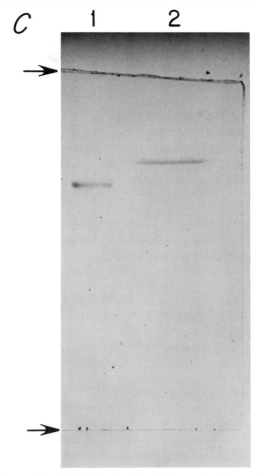

FIG. 2. Purification of the 11-1 MalF–LacZ hybrid protein. Electrophoresis of proteins and staining of gels were carried out as described.[13] (A) Separation of large and small polypeptides from membranes of HS6503. Lanes 1 and 11 contain an aliquot of the sample that was applied to a column of beaded agarose (see text). Lanes 2 to 10 contain 50 μl of successive fractions eluting from the column starting with the fraction after the void volume. (B) Gel filtration of the large polypeptides obtained from HS6503 membranes. Lane 1 contains an aliquot of the material that was obtained as a result of pooling the fractions depicted in lanes 2 through 5 of A. Lanes 2 to 8 contain successive fractions from the column. (C) Comparison of purified wild-type β-galactosidase (lane 1) and purified MalF–LacZ hybrid (lane 2). Short arrows indicate the top of the gel and the dye front.

maltose-binding protein. At this level it is not surprising that the MalF protein is not detectable by SDS–gel electrophoresis of crude membrane preparations.

In order to find out if the low level of MalF protein is the result of degradation, pulse-chase experiments were performed. It was found that the MalF protein labeled during a 1 min pulse was stable over an entire doubling time. Therefore the low level of MalF protein appears to be the result of decreased gene expression rather than degradation.

Identification of the MalK Protein. A similar approach was taken to identify the product of the *malK* gene. The MalK–LacZ hybrid protein that was used as an antigen contained the entire *malK* gene encoded sequence and retained MalK activity.[16,17] The anti-MalK antibodies precipitated a protein of 40,000 molecular weight maltose-inducible protein from the cytoplasmic membrane of wild-type cells. When the location of the MalK protein was examined in *malG* mutants, it was found that the MalK protein was located in the cytoplasm in a soluble form.[18] One likely explanation for this result is that the MalK protein is located in the membrane via a physical interaction with the MalG protein which is also a cytoplasmic membrane protein (H. Shuman, unpublished observation).

Other Uses of lac Gene Fusions in the Study of Transport Proteins. Most transport proteins exist in extracytoplasmic locations. *lac* gene fusions have been extremely helpful in defining the genetic information within the structural genes for secreted and membrane proteins, which determines their cellular location.[19–21]

In addition, *lac* gene fusions have been used to study the regulation of the structural genes for the outer membrane porins, OmpF and OmpC. The expression of these genes is regulated at the level of transcription in a complex manner which depends on the osmolarity of the external medium. Fusions of the *ompF* and *ompC* genes to the *lacZ* gene have been useful not only in quantifying these effects, but also in selecting for mutant strains in which the porin regulatory proteins, OmpR and EnvZ, can be studied.[22,23]

[16] S. D. Emr and T. J. Silhavy, *J. Mol. Biol.* **141,** 63 (1980).

[17] E. Gilson, H. Nikaido, and M. Hofnung, *Nucleic Acids Res.* **1,** 7449 (1982).

[18] H. A. Shuman and T. J. Silhavy, *J. Biol. Chem.* **256,** 560 (1981).

[19] P. J. Bassford, Jr., J. R. Beckwith, M. Berman, E. Brickman, M. J. Casadaban, L. Guarente, I. St. Girons, A. Sarthy, M. Schwartz, H. A. Shuman, and T. J. Silhavy, *in* "The Operon" (J. H. Miller and W. Reznikoff, eds.), p. 245. Cold Spring Harbor Laboratory, Cold Spring Harbor, New York, 1978.

[20] J. R. Beckwith and T. J. Silhavy, this series, Vol. 97, p. 3.

[21] S. Michaelis and J. R. Beckwith, *Annu. Rev. Microbiol.* **36,** 435 (1982).

[22] M. N. Hall and T. J. Silhavy, *J. Bacteriol.* **140,** 342 (1979).

[23] M. N. Hall and T. J. Silhavy, *J. Mol. Biol.* **146,** 23 (1981).

[13] Use of Transposons to Isolate and Characterize Mutants Lacking Membrane Proteins, Illustrated by the Sugar Transport Systems of *Escherichia coli*

By Maurice C. Jones-Mortimer and Peter J. F. Henderson

This chapter describes strategies for obtaining mutations in genes of selected membrane transport proteins. The emphasis is on strategy, not tactics, because poor experimental design is not compensated by good execution, and because full experimental details are given in a series of manuals published by the Cold Spring Harbor Laboratory[1-3] and elsewhere. For details of mechanisms of genetic events mediated by transposons references 4–7 should be consulted—the field is too large for an adequate summary to be included here.

Emphasis is placed on mutations that completely prevent production of a specified protein, or that profoundly alter its relative molecular weight (M_r), so that the absence of the protein can be recognized after separation of the membrane proteins by gel electrophoresis. Sensitive methods for recognition are detailed in this volume [31]. Our examples in both chapters are restricted to sugar transport proteins of *Escherichia coli*. However, in principle the methodologies are applicable to any membrane protein (or even cytoplasmic protein) in any organism in which a combination of genetical and biochemical approaches is feasible.

The Choice of a Suitable Strain

Not every laboratory strain of *Escherichia coli* K12 is equally suitable for every kind of experiment. In the past the proverbial Irishman's response to the traveler seeking directions—I wouldn't start from here if I were you—has often proved to be sound advice. The only justification for

[1] J. H. Miller, "Experiments in Molecular Genetics," p. 230. Cold Spring Harbor Laboratory, Cold Spring Harbor, New York, 1972.

[2] R. W. Davis, D. Botstein, and J. R. Roth, "Advanced Bacterial Genetics." Cold Spring Harbor Laboratory, Cold Spring Harbor, New York, 1980.

[3] T. J. Silhavy, M. L. Berman, and L. W. Enquist, "Experiments with Gene Fusions." Cold Spring Harbor Laboratory, Cold Spring Harbor, New York, 1984.

[4] N. Kleckner, J. Roth, and D. Botstein, *J. Mol. Biol.* **116**, 125 (1977).

[5] A. I. Bukhari, J. A. Shapiro, and S. L. Adhya, "DNA Insertion Elements, Plasmids and Episomes." Cold Spring Harbor Laboratory, Cold Spring Harbor, New York, 1977.

[6] D. E. Berg and C. M. Berg, *Biotechnology* **1**, 417 (1983).

[7] J. A. Shapiro, "Mobile Genetic Elements." Academic Press, New York, 1983.

METHODS IN ENZYMOLOGY, VOL. 125

using *E. coli* (or the closely related organism *Salmonella typhimurium*), rather than a gram-positive organism, for the investigation of membrane proteins is the advanced state of knowledge of the genetics of the former organisms. The discovery and development of transposons and of the mutator phage Mu have done much to make the old advice appear outdated. Nevertheless some starting places remain better than others. As a result the most useful comparisons to make are often not those between a mutant and its immediate parent, but those between derivatives of each that differ in the corresponding way. These derivatives must be as closely related as possible. *E. coli* K12 has been cultivated in the laboratory since 1922 and since then different lines have diverged, often very markedly.

We list below a number of the properties of a desirable strain. Unfortunately features that are desirable for biochemical experiments are not always compatible with the requirements of genetical experiments. We hope that the subsequent sections of this chapter may help the reader to assess the relative importance of these properties for the experiments that he proposes to perform.

1. For biochemical experiments a high level of gene expression is desirable. However, some strains showing enhanced expression do so as the result of duplications of the relevant structural genes. They are unsuitable for isolating mutants, and are inherently unstable.

2. For biochemical experiments constitutive expression of the transport system may be a convenience or an economy. Some genetical experiments require the constitutivity, others the inducibility, of the transport system. A means of interconverting inducible and constitutive strains is therefore desirable.

3. The strain should grow readily both in liquid and on solid minimal medium (we have encountered strains that fail to do one or the other). The former is essential for biochemical experiments, the latter for genetic experiments.

4. The strain should be free of prophages (caution: some of these are cryptic), transposons, and plasmids (including the sex factor, F, in its integrated, Hfr, state).

5. The strain should not carry an amber suppressor mutation.

6. The strain must be sensitive to the phages that will be used, P1, T4, λ and $\phi80$, and should possess any host factors necessary for the propagation of these phages or for transposon insertion.

7. The strain should not be excessively mucoid, nor mutate at a high frequency to be mucoid.

8. The lactose *(lac)* operon should be deleted. Several types of genetic experiment, besides those outlined below, are simplified by the use of such deletions.

9. It should be easy to distinguish the strains being used from contaminants. The more (irrelevant) genetic markers a strain has the easier it is to recognize by growth tests. Conversely the medium required for growth is more complex.

Our experience is that nonmucoid derivatives of MC4100[8] have been suitable for all the types of experiment[9] we have done, while other strains have proved to be better for particular purposes.

In all experiments involving sugars it is important to remember that even the best commercial preparations are not absolutely pure. Disaccharides are often readily hydrolyzed in solution. Sugars are often difficult to recrystallize without substantial losses. For the genetic experiments it is usually satisfactory to use commercial preparations at the lowest concentration that will yield good growth of colonies on plates (5–10 mM of a monosaccharide). For the growth of organisms in liquid culture the impurities may be preferentially utilized.

Positive Selection Techniques for Sugar-Transport Mutants

Mutants may be obtained from a mixed population of wild-type and mutant cells by either negative or positive selection techniques. Negative selection techniques, of which the best known is the penicillin-enrichment method,[1,10,11] depend on lethal synthesis in the wild-type (but not in the mutant) organism, leading to the more rapid death of the former. Positive selection methods are those in which the multiplication of mutant cells in the population is permitted, while that of wild-type cells is inhibited. The latter approach is always technically simpler, since it requires only that the mixed culture be incubated (usually on solid growth medium, but sometimes in liquid) for the desired mutants in the population to outgrow the commoner wild-type organisms. Whichever approach is adopted we cannot emphasize too strongly the necessity, when a mutagen is employed, for subsequent genotypic segregation[12] and phenotype expression[12] before imposing the selection. Genotypic segregation is required since any mutation that gives rise to a defective structural gene for a transport system is probably recessive and all organisms contain (at least during part of the cell cycle) more than one copy of any one gene as a

[8] M. J. Casadaban, *J. Mol. Biol.* **104,** 541 (1976).

[9] *fda* (JM2087), *fda*[ts] (JM2314), *ppc* (JM2390=NJ27), and *eda* (JM2125) are available from the Coli Genetic Stock Center, Department of Human Genetics, Yale University.

[10] B. D. Davis, *J. Am. Chem. Soc.* **70,** 4267 (1948).

[11] J. Lederberg and N. Zinder, *J. Am. Chem. Soc.* **70,** 4267 (1948).

[12] F. J. Ryan and L. K. Wainwright, *J. Gen. Microbiol.* **11,** 364 (1954).

result of DNA replication.[12] Phenotypic expression, the diluting-out by cell division of the products expressed from a gene before the mutational event occurred, is not in general a problem with inducible transport systems. Problems may, however, arise in selecting mutants in strains that express constitutively the transport system under consideration. This situation occurs in practice more frequently than might be envisaged, for two reasons. First, the use of a constitutive strain simplifies the distinction between lesions in structural genes for individual enzymes of transport or metabolism and lesions in positive regulatory genes. Second, positive selection techniques employing nonmetabolizable analogs often work only in constitutive strains.[13]

Accumulation of Nonmetabolizable Sugar Phosphates

Intracellular accumulation of nonmetabolizable sugar phosphates and related compounds inhibits the growth of any organism.[14] The intracellular accumulation of free sugars, except in the case of L-arabinose[15] in *E. coli,* is insufficiently deleterious to the organism to allow the convenient selection of mutants. Accumulation of sugar phosphates is in general bacteriostatic, though UDPgalactose[16] and *N*-acetylglucosamine 6-phosphate[17] are bactericidal. The site of action of the inhibitors is not known, though it may be the other transport system required for the continued growth of the organism,[18] the pathway of phospholipid biosynthesis,[19] or another enzyme allosterically regulated by a sugar phosphate.

Nonmetabolizable Sugar Analogs. Sugars may be chemically modified in such a way that they are unable to be catabolized. 2-Deoxyglucose, for instance, can be phosphorylated by a phosphoenolpyruvate sugar phosphotransferase system[20] but the product, 2-deoxyglucose 6-phosphate, cannot be a substrate for glucose-6-phosphate isomerase (phosphoglucose isomerase, EC 5.3.1.9). Inhibitory intracellular concentrations of 2-deoxyglucose 6-phosphate may therefore accumulate, preventing growth on whatever carbon source may be present. Failure to

---

[13] M. C. Jones-Mortimer and H. L. Kornberg, *J. Gen. Microbiol.* **96,** 383 (1976).

[14] T. Ferenci and H. L. Kornberg, *Biochem. J.* **132,** 341 (1973).

[15] J. Gross and E. Englesberg, *Virology* **9,** 314 (1959).

[16] M. B. Yarmolinski, H. Wiesmayer, H. M. Kalckar, and E. Jordan, *Proc. Natl. Acad. Sci. U.S.A.* **45,** 1786 (1959).

[17] R. J. White, *Biochem. J.* **106,** 847 (1968).

[18] H. L. Kornberg, *in* "Rate Control of Biological Processes," Symposium of the Society for Experimental Biology (D. D. Davies, ed.), Vol. 27, p. 174. Cambridge Univ. Press, London and New York, 1973.

[19] C.-H. L. Su, J. P. Merlie, and H. Goldfine, *J. Bacteriol.* **122,** 565 (1975).

[20] S. J. Curtis and W. Epstein, *J. Bacteriol.* **122,** 1189 (1975).

translocate or phosphorylate the analog may result either from the loss of the sugar-specific, membrane-bound, components of the phosphotransferase system (in this case the "mannose-specific"[21] Enzyme II[20]) or from the loss of the common components of the system (Enzyme I, EC 2.7.3.9 or HPr, no EC number). The isolation of Enzyme I- or HPr-negative mutants in such a selection can be avoided by the use of a carbon source, such as sorbitol, that requires these functions for catabolism. [Enzyme I mutants[22] and HPr mutants[23] are most conveniently selected on sorbitol tetrazolium indicator plates containing fosfomycin (~0.5 mM, but the concentration required is strain dependent). Red colonies are defective in Enzyme I or HPr; the (larger) white colonies are usually defective in the transport system for sn-glycerol 1-phosphate.[24]]

Other nonmetabolizable sugar analogs have been used to select mutants defective in other transport systems (see the table). With a very few exceptions[24,25] the analogs are inhibitory only if the relevant transport system is expressed constitutively. It may be that those parts of the molecule recognized by a transport system differ from those recognized by the regulatory mechanism for the synthesis of that transport system. Though such dual determination of specificity would be of value to the organism as a result of its better selectivity it poses severe problems to the designer of toxic analogs.

The main limitation to the use of toxic sugar analogs is that they are not often commercially available. When they are available, selection for resistant mutants is the method of choice, provided that the mutants obtained can be satisfactorily characterized in a constitutive strain.[26] In cases where the position of the regulatory gene on the chromosome is well separated from the structural gene(s) for the transport system the interconversion of inducible and constitutive stocks poses no problem. If, however, the two genes are close together it may be difficult to obtain a recombination event between the two loci.

Mutations Conferring Sensitivity to Specific Sugars. The limitations of the preceding approach can be circumvented if the natural substrate of a transport system can be converted by the organism itself into a phosphorylated derivative that cannot be further metabolized. This situation is obtained in mutants blocked in sugar catabolism at a stage subsequent to phosphorylation. Such mutants will be considered in two groups: those

---

[21] Glucosamine is a better diagnostic carbon source with many strains.
[22] C. Cordaro, *Annu. Rev. Genet.* **10,** 341 (1976).
[23] M. C. Jones-Mortimer, unpublished results.
[24] F. H. Kahan, J. Kahan, P. Cassidy, and H. Kropp, *Ann. N.Y. Acad. Sci.* **235,** 364 (1974).
[25] R. J. Miles and S. J. Pirt, *J. Gen. Microbiol.* **76,** 305 (1973).
[26] Constitutive expression complicates the characterization of fusion mutants: see below.

SPECIFIC POSITIVE SELECTION TECHNIQUES FOR MUTANTS DEFECTIVE IN CARBON SOURCE
TRANSPORT AND UTILIZATION[a,b]

Carbon source	Toxic analog of carbon source	Genetic lesion causing sensitivity to carbon source	Sensitivity of fda mutant to carbon source
Acetate	2-Fluoroacetate (1)	—	R
N-Acetylglucosamine	N-Iodoacetylglucosamine (2)	$nagA$ (24)	S
	Streptozotocin (3)		
L-Arabinose	—	$araD$ (25, 26)	S
2-Deoxyribose and derivatives	—	$deoC$ (27)	R
Fructose	Xylitol, L-sorbose (4)	fpk (28)	S
L-Fucose	—	$fucA$ (29)	R
Galactonate	—	$dgoA$ (30)	R
Galacturonate	—	eda (31)	R
Galactose	2-Deoxygalactose (5, 6)	$galE, galU$ (32)	S
Galactitol	2-Deoxygalactitol (7)	$gatD$ (3), kba (33)	R
	1-Deoxy-D-galactitol (7)		
	Arabitol (4)		
Gluconate	—	eda (34)	S
Glucosamine	2-Deoxyglucose (8, 9)	$nagB$ (9)	S
	Bacteriophage λ (10)		
Glucose	3-Deoxy-3-fluoroglucose (11)	—	S
	L-Sorbose (12)		
	2-Deoxyglucose (13)		
	5-Deoxy-5-thioglucose (14)		
Glucose 6-phosphate	2-Deoxyglucose 6-phosphate (15)	—	S
Glucuronate	—	eda (31)	R
Glycerol	—	$glpD$ (35)	R
sn-Glycerol 1-phosphate	Fosfomycin (16)	$glpD$ (35)	R
Lactose	o-Nitrophenyl thiogalactoside (17)	$galE$ (17)	S
	Phenylethyl galactoside (17)		
Maltose	Bacteriophage λ (18)	—	S
Mannitol	2-Deoxyglucitol (19)	$mtlD$ (36)	S
Mannose	2-Deoxyglucose (9, 20)	$manA$ (3)	S
	Bacteriophage λ (10)		
Melibiose	—	$galE$ (17)	S
L-Rhamnose	—	$rhaD$ (37)	R
D-Serine	Cycloserine (21)	$dsdA$ (38)	R
Sorbitol	2-Deoxyglucitol (3)	$srlD$ (3)	S
	2-Aminoglucitol (22)		
	Arabitol (4)		
Succinate	3-Fluoromalate (23)	—	R

[a] This table is undoubtedly incomplete and is thus intended only as an introduction to the literature. Furthermore, the indication of a toxic analog or of a genetic lesion causing sensitivity is not intended to imply that either method has actually been used to isolate mutants. All compounds have the D (or *meso*) configuration unless otherwise stated.

[b] References: (1) T. D. K. Brown, M. C. Jones-Mortimer, and H. L. Kornberg, *J. Gen. Microbiol.* **102,** 327 (1977). (2) R. J. White and P. W. Kent, *Biochem. J.* **118,** 81 (1970). (3) J. Lengeler, *Mol.*

with sugar-specific lesions and those with lesions in the central glycolytic pathway (see below).

In the table are also listed the lesions that cause a sugar-sensitive phenotype in *E. coli*. Some of these lesions, especially those involved in the catabolism of sugars that enter by a phosphotransferase mechanism, have been extensively used for the isolation of transport-deficient mutants. In principle the same approach could be used with the sugars that enter by other transport mechanisms, though in practice this appears rarely to have been done. (Bachmann,[27] however, describes the apparently unintentionally complex *gal-6* lesion, where strains have become defective in *both* the normal galactose transport systems, encoded by *mgl* and *galP,* as a result of growing galactose-sensitive strains on galactose-containing indicator medium.)

The reason for this omission is unclear. Possibly each of these sugars is transported by more than one transport system. If so the probability of a cell losing both transport systems simultaneously by independent mutations is much lower than the probability of any single mutation that causes protection. Such single mutations in positive regulatory genes, e.g., *araC* or *xylR,* required for the expression of both transport systems for a single sugar, or in a gene such as *galK* required for the conversion of intracellu-

---

[27] B. J. Bachmann, *Bacteriol. Rev.* **36,** 525 (1972).

Gen. Genet. **179,** 49 (1980). (4) A. M. Reiner, *J. Bacteriol.* **132,** 166 (1977). (5) P. J. F. Henderson and R. A. Giddens, *Biochem. J.* **168,** 15 (1977). (6) See Ref. 63. (7) C. E. Delidakis, M. C. Jones-Mortimer, and H. L. Kornberg, *J. Gen. Microbiol.* **128,** 601 (1982). (8) Text Ref. 20. (9) M. C. Jones-Mortimer and H. L. Kornberg, *J. Gen. Microbiol.* **96,** 383 (1976). (10) J. Elliot and W. Arber, *Mol. Gen. Genet.* **161,** 1 (1978). (11) Text Ref. 25. (12) A. C. Slater, M. Jones-Mortimer, and H. L. Kornberg, *Biochim. Biophys. Acta* **646,** 365 (1981). (13) H. L. Kornberg and J. Smith, *FEBS Lett.* **20,** 270 (1972). (14) H. L. Kornberg and P. D. Watts, *FEBS Lett.* **89,** 329 (1978). (15) R. C. Essenberg and H. L. Kornberg, *J. Gen. Microbiol.* **99,** 157 (1977). (16) Text Ref. 24. (17) See Ref. 1. (18) M. Schwartz, *Ann. Inst. Pasteur* **113,** 685 (1967). (19) J. Lengeler, *J. Bacteriol.* **124,** 26 (1975). (20) Text Ref. 20. (21) R. J. Wargel, C. A. Shadur, and F. C. Neuhaus, *J. Bacteriol.* **105,** 1028, (1971). (22) A. Boronat, M. C. Jones-Mortimer, and H. L. Kornberg, *J. Gen. Microbiol.* **128,** 605 (1982). (23) W. W. Kay and H. L. Kornberg, *FEBS Lett.* **3,** 93 (1969). (24) R. J. White, *Biochem. J.* **106,** 847 (1968). (25) J. Gross and E. Englesberg, *Virology* **9,** 314 (1959). (26) E. Englesberg, J. Irr, J. Power, and N. Lee, *J. Bacteriol.* **90,** 946 (1965). (27) S. I. Ahmad and R. H. Pritchard, *Mol. Gen. Genet.* **104,** 351 (1969). (28) T. Ferenci and H. L. Kornberg, *Proc. R. Soc. London Ser. B* **187,** 105 (1974). (29) C. S. Skjold and D. H. Ezekiel, *J. Bacteriol.* **152,** 120 (1982). (30) R. A. Cooper, *Arch. Microbiol.* **118,** 119 (1978). (31) F. Stober, A. Lagarde, G. Nemoz, G. Novel, M. Novel, R. Portalier, J. Pouyssegur, and J. Robert-Baudouy, *Biochimie* **56,** 199 (1974). (32) T. A. Sundararajan, A. M. C. Rapin, and H. M. Kalckar, *Proc. Natl. Acad. Sci. U.S.A.* **48,** 2187 (1962). (33) J. Lengeler, *Mol. Gen. Genet.* **152,** 83 (1977). (34) See Ref. 41. (35) N. R. Cozzarelli, J. P. Koch, S. Hayashi, and E. C. C. Lin, *J. Bacteriol.* **90,** 1325 (1965). (36) E. Solomon and E. C. C. Lin, *J. Bacteriol.* **111,** 566 (1972). (37) J. Power, *Genetics* **55,** 557 (1967). (38) S. D. Cosloy and E. McFall, *J. Bacteriol.* **114,** 685 (1973).

lar sugar to sugar phosphate, are equally effective in protecting against inhibition.

The principal problem in applying this technique is that the mutation conferring resistance sometimes maps very close to the mutation conferring sensitivity so that the two lesions may prove difficult to separate by genetic techniques.

Mutations Conferring Sensitivity to Many Sugars, e.g., fda. Organisms lacking enzymes of the glycolytic pathway, glucose-6-phosphate isomerase (phosphoglucoisomerase EC 5.3.1.9, *pgi*), phosphofructo-kinase (EC 2.7.1.11, *fpk*), or fructose-1,6-bisphosphate aldolase (EC 4.1.2.13, *fda*) remain able to grow on gluconeogenic carbon sources. However, the growth of *pgi*[28] or *pfk*[23] strains is inhibited by sugars that feed into the glycolytic pathway above the metabolic block especially if the strain is also deficient in glucose-6-phosphate dehydrogenase (EC 1.1.1.49, *zwf*) activity. This requirement for two metabolic blocks makes the selection system somewhat cumbersome to use. Aldolase-negatve[29] (*fda*) mutants are more convenient.

Aldolase-negative mutants are sensitive to inhibition by all sugars that feed into glycolysis at or above the level of fructose-1,6-bisphosphate[30] but they grow on any carbon source that feeds into metabolism below the block.[29] Thus mutants resistant to inhibition by sugars may readily be selected on a variety of carbon sources. Unfortunately the rationale for this selection remains obscure. We do not know the mechanism by which the accumulated sugar phosphates inhibit growth (see above). Nor is it clear why *fda* mutants are able to grow on gluconeogenic carbon sources, though the following hypotheses may be proposed.

1. A second fructose-1,6-bisphosphate aldolase has been demonstrated[31] which, since it is activated by citrate and phosphoenolpyruvate, presumably acts principally in the gluconeogenic direction *in vivo*. Its existence has been shown in *E. coli* strain Crookes but not in *E. coli* K12.[31]

2. There are conditions under which a thermodynamically reversible enzymatic reaction may be kinetically irreversible.[32] The available aldolase-negative mutants might fulfill them. However, extracts of the mutated strains have never been shown to possess activity in either direction *in vitro*.

3. The lesions may be "leaky"—there may be some residual enzymatic activity. The carbon flux required through the aldolase reaction

[28] D. G. Frankel and S. R. Levisohn, *J. Bacteriol.* **93,** 1571 (1967).
[29] A. Böck and F. C. Neidhardt, *J. Bacteriol.* **92,** 464 (1966).
[30] M. C. Jones-Mortimer and H. L. Kornberg, *Proc. R. Soc. London Ser. B* **193,** 313 (1976).
[31] S. A. Baldwin and R. N. Perham, *Biochem. J.* **169,** 643 (1978).

during glycolysis is about 10 times that required for growth on a gluconeogenic carbon source[33] when the direction of the reaction is reversed. Thus a leaky mutant might have sufficient aldolase for growth on a gluconeogenic carbon source but not on a glycolytic one.

4. There are (as indicated in the table) several other aldolases in *E. coli* that normally act on other substrates. One, or a combination, of these might be sufficiently active *in vivo* (and have an appropriate lack of specificity) to fulfill the requirements of gluconeogenic growth.

This lack of understanding of the basis of the aldolase selection technique has not limited its utility. Its advantages are as follows.

1. It eliminates the need for any chemical synthesis of sugar analogs.

2. It does not require the constitutive expression of the target transport system.

3. The *fda* lesion is sufficiently far away from any of the genes for sugar transport systems, and prevents growth on such a number of different carbon sources that the wild-type allele may be substituted for the mutant allele without inconvenience.

4. There exists not only an absolute[34,35] mutant allele of *fda* [which is useful in conjunction with the thermoinducible mutator phage Mud-(Ap^R*lac*)I—see below] but also a thermosensitive[29] mutant allele. The properties of the thermosensitive allele make it particularly useful for selecting mutants blocked in sugar utilization. At 30° strains carrying the thermosensitive allele will grow on any sugar at a rate approaching that of an isogenic wild-type strain. At 42° the same sugar will inhibit growth on a gluconeogenic carbon source, thus permitting the selection of resistant mutants.[30] Such mutants may then be screened at 30° for their ability to utilize the sugar.

5. The gluconeogenic carbon source used in the selection may be chosen to achieve any desired degree of *catabolite repression*[36] in the culture. (Catabolite repression determines differentially the maximum level of induction of an inducible system.)

These properties of strains with the temperature-sensitive *fda* lesion may be illustrated by the investigation of the genetics of the proton-linked

[32] J. B. S. Haldane, "Enzymes." Longmans, Green, London, 1930.

[33] R. B. Roberts, D. B. Cowie, P. H. Abelson, E. T. Bolton, and R. J. Britten, "Studies of Biosynthesis in *Escherichia coli*." Carnegie Institution, Washington, D.C., 1955.

[34] R. A. Cooper, *FEBS Symp.* **19,** 99 (1970).

[35] E. O. Davis, M. C. Jones-Mortimer, and P. J. F. Henderson, *J. Biol. Chem.* **259,** 1520 (1984).

[36] B. Magasanik, *in* "The Lactose Operon" (J. R. Beckwith and D. Zipser, eds.), p. 189. Cold Spring Harbor Laboratory, Cold Spring Harbor, New York, 1970.

transport of arabinose.[37] Arabinose inhibited the growth of fda^{ts} mutants on glycerol at 42°. Of the mutants selected as resistant to arabinose inhibition at 42° an unexpectedly high proportion, about one-quarter of the independent isolates, was still able to grow with arabinose as sole carbon source at 30°. Since fda^+ derivatives of the mutants also grew on arabinose at 42° the original lesions causing arabinose resistance could not have had a thermosensitive lesion in the conversion of intracellular arabinose to (toxic) sugar phosphates. Genetic mapping of the lesions indicated that they lay near (or perhaps in) a gene araE known to be involved in arabinose transport.[38,39] Subsequent biochemical investigation confirmed the loss of the arabinose/H^+ symport transport system.[37] Thus it is possible using this technique to isolate mutants that lack only one of the two transport systems for a single sugar. Analogous results have been obtained with xylose-resistant mutants.[35,40] Again, unfortunately, it is not entirely clear why the selection method should yield mutants affected in pentose utilization. However the properties of the araE mutants are entirely consistent with those predicted for a lesion that abolishes the function of only one of two transport systems for a single sugar. Perhaps the glycerol used as the carbon source for growth exerts sufficient catabolite repression on the alternative transport systems for arabinose (and xylose) to prevent their being expressed under the conditions of selection of mutants.

The same approach, but starting with a mutant blocked in the aldolase of the Entner-Doudoroff pathway (phospho-2-dehydro-3-deoxygluconate aldolase, 2-deoxy-3-oxogluconate-6-phosphate aldolase, EC 4.1.2.14, eda) has been used to isolate mutants defective in gluconate transport,[41] though this is again complicated by the existence of two transport systems.[42]

Inhibition of the Glyoxalate Cycle in ppc Mutants

The techniques indicated in the previous section suffice for the isolation of mutants in the transport of all the naturally occurring disaccharides, aldoses, ketoses, and alditols (and the uronic acids) utilized by E. coli as carbon sources. Mutants defective in transport and metabolism of

[37] A. J. S. Macpherson, M. C. Jones-Mortimer, and P. J. F. Henderson, *Biochem. J.* **196**, 269 (1981).

[38] D. Isaacson and E. Englesberg, *Bacteriol. Proc.* 113 (1964).

[39] C. E. Brown and R. W. Hogg, *J. Bacteriol.* **111**, 606 (1972).

[40] E. O. Davis, unpublished results.

[41] P. Faik and H. L. Kornberg, *FEBS Lett.* **32**, 260 (1973).

[42] B. Bächi and H. L. Kornberg, *J. Gen. Microbiol.* **90**, 321 (1975).

those deoxy sugars that serve as carbon sources cannot be isolated using the fructose-1,6-biosphosphate aldolase method since this enzyme is not required for their catabolism. The following positive selection technique[43] may be employed to obtain mutants in the transport or utilization of these sugars, and any other carbon source that is catabolzied to phosphoenolpyruvate (PEP). The selection is thus expected to work for any sugar.

Wild-type *E. coli* when grown on a carbon source that is a PEP precursor use the anaplerotic enzyme phosphoenolpyruvate carboxylase (EC 4.1.1.31), specified by the gene *ppc*, to replenish the intermediates of the tricarboxylic acid cycle used as biosynthetic precursors.[44] Mutants lacking this enzyme do not grow on sugars as sole carbon source unless a source of C_4 acids is provided.[44] In practice aspartate is used, since *E. coli* K12 does not grow with aspartate as sole carbon source at an appreciable rate. A *ppc* mutant will grow on acetate, since it then uses the glyoxalate cycle[44,45] as its anaplerotic route. However, high intracellular concentrations of pyruvate and/or phosphoenolpyruvate inhibit the function of the glyoxalate cycle.[45] Hence the presence in the growth medium of compounds that can be catabolized to pyruvate inhibits growth on acetate. Selection for growth on acetate in the presence of a sugar therefore leads to the isolation of mutants blocked in the transport of that sugar[43,44,46] (or in any subsequent stage of its metabolism prior to phosphorylation).

This selection system has been less widely used in *E. coli* K12 than the *fda* selection system primarily because acetate allows only very slow growth of the strain. Since acetate is the only carbon source that can be used it is not possible to control the catabolite repression status of the culture. This can lead to difficulties when there is more than one transport system for the sugar.

Isolation of Mutants by Targeted Mutagenesis

Once the map location of a mutation is known further mutant alleles can readily be generated by targeted mutagenesis. In general when positive mutant selection techniques are available they may be more convenient than targeted mutagenesis, but in some circumstances it is advantageous to combine the two techniques.

Targeted mutagenesis involves the mutagenesis at random of the

[43] H. L. Kornberg and J. Smith, *Nature (London)* **224,** 1261 (1969).
[44] J. M. Ashworth and H. L. Kornberg, *Proc. R. Soc. London Ser. B* **165,** 179 (1966).
[45] H. L. Kornberg, *Biochem. J.* **99,** 1 (1966).
[46] A. M. Roberton, P. A. Sullivan, M. C. Jones-Mortimer, and H. L. Kornberg, *J. Gen. Microbiol.* **117,** 377 (1980).

genome, followed by the introduction into a nonmutagenized organism of DNA from a specific region of the genome of the mutagenized organism. Transfer of DNA is usually achieved by generalized transduction. Recombinants are then screened for having the desired phenotype.

Three variations of the method are available. Hong and Ames[47] in the original version (using phage P22 in *Salmonella typhimurium*) grew the phage on a wild-type organism, treated the transducing lysate with a mutagen and then used it to infect a recipient culture, selecting for the inheritance of a donor gene close to the region that they wanted to mutate. Alternatively[48] the strain to be used as the donor may be treated with a mutagen prior to propagation of the transducing phage on it. A protocol for this method is given by Silhavy *et al.*, pp. 59–62.[3]

A third variation of the approach is sometimes of value when mutants mapping at different sites have phenotypes that are not readily distinguishable. For example one might wish to isolate a series of *galP* mutants in a strain that was *mgl lacY*. All positive selections for *galP* mutants also yield *galK* mutants, but the lesions are well separated on the chromosome. Thus by (1) isolation of a pool of mutants of both types (by a positive selection in liquid culture), (2) propagating a transducing phage on this pool, (3) transducing the *galP* lesion into an appropriate strain selecting for a marker linked to *galP* but not *galK*, and finally (4) screening the recombinants for failure to grow on galactose, it should be possible to isolate *galP* mutants without simultaneously obtaining *galK* mutants.

Advantages and Disadvantages of Different Mutagenic Techniques

Having decided upon a selection scheme two important considerations remain. First, one requires a convenient way of demonstrating that the mutation is in the gene of interest. This can be a much more difficult problem for integral membrane proteins with transport functions than for soluble enzymes. Second, the mutation should alter the gene product in a recognizable way, preferably by a substantial change in M_r or complete deletion which can be detected by comparison with wild-type proteins separated by gel electrophoresis. The first criterion is probably best fulfilled by the use of phage Mu-mediated gene fusions with β-galactosidase, but the second can also be satisfied by selecting amber,[49,50] deletion, or transposon-insertion[4] mutants. These strategies are now described and their relative advantages discussed.

[47] J. S. Hong and B. N. Ames, *Proc. Natl. Acad. Sci. U.S.A.* **68,** 3168 (1971).
[48] E. Hawrot and E. P. Kennedy, *Mol. Gen. Genet.* **148,** 271 (1976).
[49] A. S. Sarabhai, A. O. W. Stretton, S. Brenner, and A. Bolle, *Nature (London)* **201,** 13 (1964).
[50] A. O. W. Stretton and S. Brenner, *J. Mol. Biol.* **12,** 456 (1965).

Spontaneous Mutants

The spontaneous mutation rate in *E. coli* is about 10^{-9} per base pair per generation.[51] Thus if the apparent mutation rate for the loss of a transport system is significantly lower than 10^{-6} per generation (say about 10^{-8} per generation) the selection method is for some reason or other inappropriate. By this we mean that, even though the selection yields mutants of the desired phenotype the genetic nature of the lesion may be complex and therefore difficult to ascertain.[52] For example, the phenotype of the lesion might result from two independent mutational events, or from a relatively rare type of mutation such as the *superrepressor* mutations of the lactose operon.[53] Under these circumstances it is probably always desirable to try variations of the selection technique that may lead to an apparent increase in the mutation rate: a different carbon source or a different selective agent.

Though many types of spontaneous mutational event produce a change in the M_r of the gene product, the easiest to characterize are the chain-terminating *amber* mutations, though under favorable circumstances *deletion* mutations may sometimes be easily recognized. However, it is important to remember that such lesions are frequently polar,[54] that is the site affected by the lesion is not necessarily within the structural gene for the missing protein but may be in a promoter-proximal gene in the same operon.

Amber Mutants. Amber mutants may be recognized by the introduction of an amber suppressor gene.[50] Such genes code for mutant transfer RNAs the anticodons of which insert an amino acid into the polypeptide chain at the point where otherwise chain elongation would have terminated.[55,56] Detection of every amber mutant is unlikely by this technique; insertion of an inappropriate amino acid may not lead to the recovery of protein's catalytic function and the efficiency with which polypeptide chain elongation is restored is variable. However, under favorable circumstances it may be as high as 60%.[57] There are two convenient techniques for introducing amber suppressors into mutant stocks.

Specialized transducing phages are available that carry suppressor

---

[51] The apparent mutation rate is not a direct function of the proportion of mutant cells in a culture, since it depends on the size of the culture. The most convenient estimate of the mutation rate is obtained by Method 2 of Luria and Delbrück[52] who give a graphic solution of their equation.

[52] S. E. Luria and M. Delbrück, *Genetics* **28**, 492 (1943).

[53] C. D. Willson, D. Perrin, M. Cohn, F. Jacob, and J. Monod, *J. Mol. Biol.* **8**, 582 (1964).

[54] C. Yanofsky and J. Ito, *J. Mol. Biol.* **21**, 313 (1966).

[55] S. Benzer and S. P. Champe, *Proc. Natl. Acad. Sci. U.S.A.* **48**, 1114 (1962).

[56] A. Garen and O. Siddiqui, *Proc. Natl. Acad. Sci. U.S.A.* **48**, 1121 (1962).

[57] A. Garen, *Science* **160**, 149 (1968).

transfer RNA genes incorporated into the phage genome. The first such phage to be isolated,[58] ϕ80p*suIII,* is unable to lysogenize its host; the suppressor gene can, however, be integrated into the host genome by homologous recombination.[58] More recently lysogenizing transducing phages have been constructed by *in vitro* techniques and may be used to introduce the amber suppressor.[59] Such phages have two advantages; first, strains carrying them may be selected by virtue of their lysogeny and second, the prophage (and hence the amber suppressor) may subsequently be eliminated from the stock by heteroimmune superinfection curing.[60] Alternatively a strain carrying a thermosensitive amber suppressor may be used for the isolation of mutants. In such a background an amber mutation produces the mutant phenotype at the higher, restrictive, temperature but not at lower, permissive temperatures. Unfortunately the usefulness of this latter approach is limited. The strains used in Oeschger's system[61] do not, in our experience, grow well on minimal medium, and the lesions required cannot readily be transferred to other strains. In the alternative system, using ϕ80p*suIII*A2P,[62] suppression is not very efficient.

The characterization of amber mutations, at least in genes for membrane proteins of *E. coli,* is probably now obsolete. However, if we were now to undertake it, screening with a lysogenising specialized transducing phage such as $\lambda\Delta$[srI 1-2]Σ[su$^+$F]singlet *att*λ *imm21c$^+$ nin* would be our method of choice.

Deletion Mutants. Deletion mutants, from which a finite length of DNA has been lost, occur spontaneously at a reasonably high rate, about 10% of all mutants in any one gene.[63] Although certain chemical mutagenic techniques[63] increase the absolute rate at which deletions occur they do not increase the rate relative to point mutations of the same gene. So such mutagenesis is not helpful provided that a positive selection technique is available. The problem is to recognize a mutation as being due to a deletion.

One possible criterion for a deletion is the simultaneous loss of three topologically linked but functionally unrelated genes. Neither the loss of only two functions, nor the loss of metabolically related functions is suffi-

[58] R. L. Russell, J. N. Abelson, A. Landy, M. L. Gefter, S. Brenner, and J. D. Smith, *J. Mol. Biol.* **47,** 1 (1970).

[59] K. Borck, J. D. Beggs, W. J. Brammar, A. S. Hopkins, and N. E. Murray, *Mol. Gen. Genet.* **146,** 199 (1976).

[60] A. D. Kaiser and T. Masuda, *Virology* **40,** 522 (1970).

[61] M. P. Oeschger and G. T. Wiprud, *Mol. Gen. Genet.* **178,** 293 (1980).

[62] J. D. Smith, L. Barnett, S. Brenner, and R. L. Russell, *J. Mol. Biol.* **54,** 1 (1970).

[63] M. D. Alper and B. N. Ames, *J. Bacteriol.* **121,** 259 (1975).

cient criterion for a deletion, though in practice such mutants are suitable for biochemical experiments. They may, however, be unsuitable for concomitant genetic experiments. Two unrelated functions may be simultaneously lost by the inversion of a stretch of DNA. If the inversion is large it may complicate the mapping of the lesion. The simultaneous loss of two related functions is likely to be due to a polar mutation in an operon and may therefore result from a single point mutation. A further complication may occur in determining whether two phenotypic differences are or are not genetically related. This is particularly relevant to mutations of membrane proteins since such proteins may be required for phage infection processes.

The classical criterion for detecting deletions is that they fail to recombine (to yield wild-type progeny) with two or more point mutants that do so recombine. It is not always possible to apply this criterion to genes for membrane proteins because, though deletions themselves are of necessity genetically stable, the phenotype of a deletion mutant may not be stable.[13] It may be possible to overcome the negative phenotype of a mutation in a transport system by expressing a different transport system for the same substrate.[13] For example, a *galP mgl* strain cannot transport, and therefore cannot grow on, galactose. The strain will readily mutate to grow on galactose. These secondary mutants are not revertants (appropriate experiments will demonstrate that the mutant *galP* and *mgl* lesions are still present in the strains), but usually *lacI* mutants that express the lactose operon constitutively. The galactose-positive derivative mutants use the lactose transport system, which was not previously being expressed, to transport galactose. The isolation of "by-pass" mutants of this type is often a convenient way of obtaining constitutive mutations. Thus the unequivocal demonstration that a particular mutation in a transport system is a deletion may be difficult.

The isolation of spontaneous deletion mutants will in our opinion remain a useful technique particularly since it may facilitate the mapping of genetic lesions. The practical problem is that the location of the gene may need to be known before an appropriate scheme for selecting and screening deletion mutants can be devised.

Experimental Suggestions for Isolating Spontaneous Mutants. In any population of bacteria large enough to contain two or more spontaneously generated mutant organisms *either* the two mutant organisms arose by different mutational events *or* by the multiplication of the progeny of a single mutant that arose earlier in the history of the population. Thus the screening of more than one mutant from any one population by any method that is significantly more time-consuming than that described below for deletion mutants is counterproductive. One may be duplicating one's results (which should be valuable) but one cannot know that one is doing so, rather than performing the same experiment on an independent mutant. To avoid this waste of effort only one mutant from any one population should be investigated.

To isolate deletion mutants (1) set up as many (5–10) overnight cultures of the starting strain as is convenient, each from a different single colony; (2) spread a sufficient volume (about 0.1 ml) of each on a plate of the selection medium to obtain 100–500 colonies per plate; (3) when the colonies are about 1 mm in diameter (i.e., before they are fully grown) replica plate them onto the screening medium and onto a new plate of the selection medium; and (4) after appropriate incubation compare the two new plates.

If, at stage 2, the distinction between the mutant colonies and the background growth is unsatisfactory, it may often be improved simply by replica plating onto a new plate of the selection medium. If possible, in stage 3, design the screening medium so that the selection pressure imposed at stage 2 is continued. For example, selection of an *araE lysA* deletion would employ a medium containing lysine, arabinose, and glycerol for the selection and one without lysine but containing both arabinose and glycerol as the screening medium. When we have tried this method for obtaining deletions our experience has been that, if the mutation can occur at all, about 1 colony in 500 has the desired phenotype. A sufficient number of colonies therefore require to be screened.

To isolate other spontaneous mutants (including amber mutants) it is not, in practice, necessary to grow a large number of liquid cultures from individual colonies to ensure that one obtains independent mutants. It is sufficient to spread about 10^4 bacteria on a nonselective (nutrient agar) plate and when colonies have grown up on this medium to replica plate them onto the selective medium. Mutational events occur as the colonies grow on the nonselective medium, so each colony on the selective medium (since the colonies are spatially separated) must have arisen by an independent mutational event. The original inoculum of 10^4 cells is unlikely to contain a mutant. If however it does so, the colony derived from it on the replica plate will be visible about 24 hr before the colonies resulting from subsequent mutations. This approach has obvious advantages when the selection medium is expensive. Our experience indicates that it also has advantages for isolating slow-growing mutants from the population when faster growing mutants are also present.

For the past 15 years we have used filter paper (Whatman No. 1) rather than velvets for replica plating. In England, at least, the filter paper arrives sterile from the manufacturer. It is, however, important both to ensure that the plates are adequately dry and to use several layers of paper to absorb moisture to prevent smudging. (The lower layers of paper can be reused after drying.) Old colonies do not give good results.

The Use of Transposons

Transposons are genetic elements that can become inserted at different sites, more or less randomly (depending upon the particular transposon used), into the genome of a host organism.[4–7] Most of the useful transposons carry genes specifying antibiotic resistance, which simplifies the selection of strains that harbor them. The choice of transposon for a particular purpose depends upon two factors: the ease of recognizing the phenotype conferred by the genes carried by the transposon and the frequency with which the transposon can integrate into a new chromosomal site. Transposons are used for several different purposes, either as mutagens to inactivate chosen genes (insertion into which normally causes a change in the M_r of the polypeptide synthesized) or as linked markers for mapping lesions and for transferring preexisting lesions to other strains.[4] For the former purpose a relatively high frequency of transposition is desirable, for the latter a low. We do not, however, intend to imply that either kind of transposon cannot be used for either purpose. The insertion into a gene of interest of a transposon such as Tn10 that does not readily move provides a particularly valuable kind of mutant. However, the low transposition frequency makes it correspondingly difficult to isolate the mutant in the first place. Experimental design must also take into account the physical size of the transposon. This varies from <1 kb for the chloramphenicol-resistance transposon Tn9 to nearly 40 kb for bacteriophage Mu and its derivatives.

Transposons as Mutagens. Two transposons, the kanamycin(neomycin)-resistance transposon Tn5 and the bacteriophage Mu integrate into chromosomal DNA sufficiently readily and sufficiently randomly for them to be useful mutagens. In practice modern derivatives of Mu have so many advantages, especially when dealing with inducible membrane proteins, that we have never had occasion to use Tn5. A protocol for the use of Tn5 as a mutagen is given by Shaw and Berg.[64]

The temperate bacteriophage Mu transposes in the course of its lytic cycle DNA replication.[65] Lysogens of Mu are functionally equivalent to strains harboring other transposons, and will therefore be considered here. The use of wild-type Mu as a mutagen is described by Tabor *et al.*[66]; the thermoinducible derivatives Mu*cts*62 may be similarly employed. The

[64] K. J. Shaw and C. M. Berg, *Genetics* **92,** 741 (1979).
[65] A. Toussaint, M. Faelen, and A. I. Bukhari *in* Bukhari *et al.,*[5] p. 275.
[66] H. Tabor, E. W. Hafner, and C. W. Tabor, this series, Vol. 94, p. 91.

thermoinducible derivatives Mu*cts*Ap[67] and Mu*cts*Km[66] that also confer resistance to antibiotics have further advantages, but they (like Tn5) are less convenient than those described next unless the assay for the product of the mutated gene is exceptionally easy.

For the investigation of membrane proteins the transposons that we would at present recommend are the bacteriophage Mu derivatives MudAp*lac*I and λp*lac*Mu1. Since these phages are described in detail elsewhere[3,68,69] we shall only consider here the reasons why they are superior to other transposons.

1. They express β-galactosidase activity under the control of the promoter of the mutated gene provided the expression of the mutated gene is regulated. This greatly simplifies the demonstration that the mutation affects a particular gene.

2. MudAp*lac*I insertions confer ampicillin resistance. This may be used for selecting lysogens, for mapping (the ampicillin-resistance phenotype may be scored in genetic backgrounds that do not allow the scoring of the actual mutant phenotype) and for transferring the lesion to other strains.

3. The thermosensitive immunity repressor of MudAp*lac*I causes lysogens to die at 42° thus providing a positive selection for isolating deletion mutants.

4. λp*lac*Mu1, unlike MudAp*lac*I, also requires the protein synthesis initiation signals of the mutated gene for the expression of β-galactosidase. Furthermore there are no translation termination signals upstream from the structural information for β-galactosidase. Therefore the mutated β-galactosidase monomer is a hybrid polypeptide, the N-terminus of which is the N-terminus of the product of the mutated gene while the C-terminus is catalytically active β-galactosidase. If the N-terminal portion encoded by the mutated gene is sufficiently hydrophobic the β-galactosidase activity becomes associated with the membrane fraction of the cell. This is a useful way of identifying genes for inner membrane proteins.[35,70,71] When such fusions are made with outer membrane or periplasmic proteins they are frequently deleterious to the cell since they block the process of protein export.[71] We have not observed this inhibition of growth in fusions with any of the proton–symport transport systems that we have investigated.

[67] D. Leach and N. Symonds, *Mol. Gen. Genet.* **172,** 179 (1979).
[68] This volume [11].
[69] This volume [12].
[70] H. A. Shuman, T. J. Silhavy, and J. R. Beckwith, *J. Biol. Chem.* **255,** 168 (1980).
[71] T. J. Silhavy, H. A. Shuman, J. R. Beckwith, and M. Schwartz, *Proc. Natl. Acad. Sci. U.S.A.* **74,** 5411 (1977).

5. λ*plac*Mu allows the hybrid gene to be readily obtained in a specialized transducing phage. This should prove particularly useful in the investigation of any transport system the induction of which requires its activity (see below).

There exists also a transposon, MudAp*lac*II,[72] intermediate in properties between MudAp*lac*I and λ*plac*Mu1. Like the latter it yields hybrid β-galactosidase molecules, but the hybrid gene cannot be cloned directly. We suspect that its future use may be limited though a derivative of it,[73] MudII(*lacZU131*,Ap) is still useful if a protein that is believed to undergo N-terminal modification is under investigation. MudII(*lacZU131*,Ap) has an amber mutant early in *lacZ*, so that the inhibitory hybrid β-galactosidase is produced only in a strain with an active amber suppressor.

Factors Affecting the Choice of Transposons as a Linked Marker

Besides their use as mutagens to inactivate a particular gene, transposons that are integrated close to that gene without affecting its function are useful in three kinds of experiments—mapping, strain construction, and strain verification. Before describing these, we must discuss three factors which affect the design of experiments. They are the distance between the lesion and the linked transposon, the frequency at which the transposon integrates at new sites, and whether the linked transposon is inserted into coding or noncoding DNA.

To be useful the linked transposon should show an effective cotransduction frequency of at least 10% with the lesion. This corresponds[74-76] to a separation of about 1 min when P1 is the transducing phage or about 2 min with T4GT7.

When a strain carrying a transposon is used as the donor in a transduction (selecting for antibiotic resistance), conditions are favorable for the transposition of the transposon to a new site. For this kind of experiment it is desirable to use a transposon that migrates at a low rate compared to the rate of homologous recombination events that are required to integrate the transposon at its original site. For this purpose the tetracycline-resistance transposon Tn10 is the most suitable. Our experience with Tn10 is that >95% of the tetracycline-resistant progeny of the transduction arise by recombination rather than by translocation. With Tn5 how-

[72] M. J. Casadaban and J. Chou, *Proc. Natl. Acad. Sci. U.S.A.* **81,** 535 (1984).

[73] E. T. Palva and T. J. Silhavy, *Mol. Gen. Genet.* **194,** 388 (1984).

[74] T. T. Wu, *Genetics* **54,** 405 (1966).

[75] B. J. Bachmann, K. B. Low, and A. L. Taylor, *Bacteriol. Rev.* **40,** 116 (1976).

[76] G. G. Wilson, K. K. Y. Young, G. J. Edlin, and W. Konigsberg, *Nature (London)* **280,** 80 (1979).

ever the ratio of recombinants to new insertions of the transposon may be as low as 5–10% under unfavorable circumstances.

The migration rate of a transposon thus affects the apparent frequency of cotransduction with a linked marker. If the transposon was originally integrated into a known gene it may be possible to distinguish recombination events from new transpositions and thus obtain the actual cotransduction frequency. Sometimes, for instance when the original transposon insertion causes auxotrophy, it may be preferable to carry out the experiment in two stages—first introducing the auxotrophy by transduction and selection for antibiotic resistance, and then removing the transposon by selection for prototrophy.

Mapping. It is beyond the scope of this chapter to discuss all the possible strategies for using transposons in gene mapping. What is practicable depends on the phenotype and the stability of the neighboring markers. Obviously the most useful transposon insertions are those where the transposon has integrated into a gene of known function so that the phenotypes permit selection in both directions (for antibiotic resistance or for regain of gene function). Since transposons are of finite size they decrease the amount of chromosomal DNA that can be packaged in the same phage particle for transduction. This is not a problem with Tn5 or Tn10, but Mu and its derivatives are so large that they take up about half the DNA-carrying capacity of a P1 transducing particle. (Sometimes this is an advantage.) Insertion mutants reduce DNA homology at their ends so it may be difficult to obtain a recombination event close to a transposon.

Transposons can catalyze deletions adjacent to their site of insertion.[4] This may be used to isolate deletions extending into a gene of interest.[4] Tn10 is a useful transposon for this purpose since one can obtain tetracycline-sensitive derivatives by selecting for resistance to fusaric acid.[77,78] Caution is required in the interpretation of the experimental results since inversions are also known to occur, and nearby point mutants may be caused.[79,80]

Strain Construction. It is frequently necessary to introduce a particular mutation into a particular genetic background. The properties of a strain that make it suitable for one's experiments may result from the cumulative effect of several unliked genes. Therefore it is usually desirable to introduce the new mutation by transduction, rather than by conju-

[77] B. R. Bochner, H.-C. Huang, G. L. Schieven, and B. N. Ames, *J. Bacteriol.* **143,** 926 (1980).
[78] S. R. Maloy and W. D. Nunn, *J. Bacteriol.* **145,** 1110 (1981).
[79] D. Botstein and N. Kleckner *in* Bukhari *et al.,*[5] p. 185.
[80] H. W. Duckworth, E. Hatchwell, and M. C. Jones-Mortimer, *Soc. Gen. Microbiol. Q.* **8,** 244 (1981).

gation, since this minimizes the possible changes to the strain. The procedure is greatly simplified by the use of transposons,[4] the experimental approach to be employed depending mainly on whether subsequent experiments will be affected by the presence of the transposon.

The approaches may be illustrated by the techniques for introducing the *fda* lesion into a new genetic background, using the linked markers *galP*::Tn10 and *serA* (growth requirement for serine). By the first approach one would construct the doubly mutant strain *fda galP*::Tn10, use bacteriophage P1 grown on this to transduce the desired strain to tetracycline resistance, and screen the progeny for being *fda*. The other approach is first to construct the doubly mutant strain *serA galP*::Tn10, use P1 grown on this to transduce the desired strain to tetracycline resistance, and screen the progeny for being *serA*. Such a *serA galP*::Tn10 recombinant is then used as the recipient in a transduction with an *fda* donor. The serine-positive progeny are screened for having become *fda* and *galP*+ (tetracycline sensitive). Transductants cannot be satisfactorily selected for the inheritance of *galP*+ even if they are also *mgl,* for reasons explained above. Obviously, if they could, the use of the *serA* mutation would be unnecessary. The latter approach, since the final strain does not contain a transposon, is to be preferred.

Strain Verification. When a mutant has been isolated after the insertion of a transposon one must ascertain that the lesion being characterized is definitely the result of the insertion. When Mud(Ap*lac*)I or λp*lac*Mu1 has been used as the transposon this can be extremely simple—insertion of these into *galP*,[81] *xylE*,[35] or *araE*[82] yields β-galactosidase inducible by galactose, xylose, and arabinose, respectively.

However, strain verification is not always so straightforward. Evidence of β-galactosidase induction is of course never available if Tn5 (or Tn10) is used as the mutagen, nor is it always available when Mud-(Ap*lac*)I or λp*lac*Mu1 is used. For instance the selection may have been performed in a strain that originally expressed the mutated function constitutively,[83] or the mutation may itself prevent induction.[83] Furthermore the transposon may have inserted anywhere in the chromosome of a mutant of the desired phenotype which preexisted in the population (this occurs at an appreciable frequency) and with such double mutants there is no genetical evidence to indicate that the mutant should produce a protein of altered M_r. Therefore under some circumstances one is obliged to use the following genetic methods to distinguish single from double events, and these are not entirely satisfactory.

[81] P. J. F. Henderson, unpublished.
[82] M. C. J. Maiden, unpublished.
[83] A. J. S. Supramaniam, unpublished.

If the mutant under consideration is the result of a double event then the observed biochemical lesion and the transposon can be separated by recombination, but if it is the result of a single lesion the two phenotypes cannot be separated in this way. The approach to the problem depends on whether or not the genetic locus giving rise to the biochemical lesion is known (or may be assumed) from previous experiments. If it is known the new mutation will presumably have been isolated in a strain that already has an appropriate closely linked marker (if not a second strain carrying a different, suitably sited, transposon will also be required). The linked marker is transduced out (or the linked transposon in) and the recombinants scored for coinheritance of the phenotypes of the mutant. If the mutant was the result of two events the transposon is *probably* integrated at a considerable distance from the genetic locus of the biochemical lesion. This situation generally leads to the production of recombinants that have lost the biochemical lesion without losing the transposon, and is quite readily recognized, even when only a few recombinants are examined. The method obviously may fail if the loci of the lesion and the transposon are sufficiently close, but no other method is superior. It would therefore be advisable if possible to perform all subsequent biochemical experiments on more than one (independent) isolate. This kind of experiment also yields the equally important information that the mutant in question does not carry a second copy of the transposon at a different site. If λplacMu1 is being used the appropriate test is for λ immunity[3] since many of the insertions will not lead to the expression of β-galactosidase. If the genetic locus of the biochemical lesion is not known a more complicated procedure is required. One would, almost certainly, start by mapping the lesion.

Therefore our advice is that mutants should be isolated (using positive selection techniques in inducible strains) after the insertion of MudAplacI *or* λplacMu1. Linked Tn10 insertion mutants are the most convenient for mapping and strain verification.

A list of known Tn10 insertion sites has been published.[84] Otherwise Tn10 insertions near a particular gene may be isolated by targeted mutagensis. A protocol for this is given by Silhavy *et al.*[3]

Experimental Suggestions. Since references are given to protocols for all the types of experiment suggested above, we shall not repeat details here, but merely add a few further suggestions.

Bacteriophage P1 cannot be satisfactorily propagated at temperatures lower than 37°. Though Mud(Aplac)I lysogens are not completely stable at this temperature P1 can be propagated on them at 37° so that they can

---

[84] C. M. Berg and D. E. Berg, *Microbiology* **107,** (1981).

be used as transduction donors, at least when the recipient is lysogenic for Mu. Under these conditions, about 50% of the ampicillin-resistant trans-ductants arise by homologous recombination and have retained the Mud(Ap*lac*)I insertion at the original site.

Despite indications in the literature to the contrary we have found Mud(Ap*lac*)I insertions sufficiently stable at 30° for any biochemical experiments. Replacement of the Mu DNA by λ DNA[3] is not necessary. One cannot conveniently use the straightforward Mud(Ap*lac*)I insertion mutants that show inducible β-galactosidase activity to select for constitutive mutants, because the frequency of transposition to another site in the chromosome is at least as high as the mutation rate. For this purpose the lysogen must be stabilized either by conversion to the λ lysogen or by deletion of the Mu *A* and *B* genes required for transposition. [Alternatively it may be possible to use the conditionally transposition-defective derivative MudI-8(Tpn[AM]AmpLac*c*62ts).[85]] The isolation of deletions of the gene into which Mud(Ap*lac*)I is inserted is simplified if one can select simultaneously for thermoresistance (i.e., for the loss of those Mu genes that are responsible for the death of the host) and loss of ability to catabolize lactose.[35]

X-gal (5-bromo-4-chloro-3-indolyl-β-D-galactoside) is expensive and its use can often be avoided by using ONPG (*O*-nitrophenyl-β-D-galacto-side) to test cultures on plates for β-galactosidase production. We have published a method[35] for this, though the use of less chloroform (~0.5 ml per plate) is recommended. Plastic petri dishes can be used, but if so the experiment should be performed on a sheet of paper; otherwise the plates stick to the bench.

E. coli K12 strains are frequently rather mucoid at 30°. Some strains indeed are so mucoid that they slide on an agar surface. However good this may be as a party trick it is unsatisfactory in the laboratory. Mucoidy can be prevented by mutation of the gene *non*.[86] We have normally used strains with a deletion, Δ*(his gnd)*, that apparently also extends through the *non* gene. Palva and Silhavy[73] have described a *non*::Tn10 insertion mutation which presumably may be used to cure strains of mucoidy.

The identification of individual membrane transport proteins is made difficult by the following factors. Their abundance in the membrane is low compared with other proteins. Any purification procedure to reduce the proportions of such contaminating proteins may result in loss of an assay, either because of deactivation of the transport protein or because of its removal from the membrane, e.g., by solubilization with detergents. A

[85] K. T. Hughes and J. R. Roth, *J. Bacteriol.* **159,** 130 (1984).
[86] K. L. Radke and E. C. Siegel, *J. Bacteriol.* **106,** 432 (1971).

method of specifically labeling a particular transport system is rarely efficient or even available. These problems may be overcome by comparing the proteins of a wild-type strain with those in a mutant impaired only in the structural gene for the transport protein, isolated by any of the methods described here. The genetic criterion of identification is often crucial, even when a biochemical labeling method is relatively specific and ways of combining the two approaches are described in this volume [31].

Acknowledgments

We are grateful to the Science and Engineering Research Council for financial support.

[14] ECF Locus in *Escherichia coli:* Defect in Energization for ATP Synthesis and Active Transport

By JEN-SHIANG HONG

The energy coupling factor *(ecfA)* locus was discovered as the result of attempts to genetically dissect the energy-transducing machinery in *E. coli.*[1-4] Even though the gene product has yet to be isolated and the biochemical role it plays to be elucidated, it is clear that the *ecfA* gene is essential for cell growth, since all *ecfA* mutants thus far isolated are temperature-sensitive *(ts)* mutants able to grow at 25° but not at 42°. The *ecfA* locus which is closely linked with the *metC* gene maps at 64 min on the revised *E. coli* chromosome map,[2] and is distinct from the H^+-ATPase *(unc)* locus located at 83 min. EcfA mutants exhibits a high reversion frequency (10^{-5}), and about 50% of the revertants simultaneously acquire auxotrophic *metC* mutations. Whether there is a genetic relationship between the *ecfA* and *metC* gene is unknown.

The *ECFA* is intimately involved in the coupling of metabolic energy to active transport of amino acids and certain sugars. Three types of transport-defective *ecfA* mutants have been described. In the first type, the defect in transport as well as ATP synthesis is due to the inability of the cells to maintain a transmembrane potential[2] whereas in the second type the defect can be attributed to an inability to couple the electrochem-

[1] M. A. Lieberman and J.-s. Hong, *Proc. Natl. Acad. Sci. U.S.A.* **71**, 4395 (1974).
[2] M. A. Lieberman, M. Simon, and J.-s. Hong, *J. Biol. Chem.* **252**, 4056 (1977).
[3] J.-s. Hong, *J. Biol. Chem.* **252**, 8582 (1977).
[4] K.-I. Tomochika and J.-s. Hong, *J. Bacteriol.* **133**, 1008 (1978).

ical proton gradient to the transport process.[3] The third type, which is the majority, is altered in membrane proton permeability; mutants of this type excrete ATP and other nucleotides and form filament-like cells upon growth at the nonpermissive temperature (42°).[4] The methods for the isolation and characterization of *ecfA* mutants are described here.

Isolation of ecf Mutants

In the original isolation, slow 42° growing neomycin-resistant cells were screened for the desired *ecf* phenotype.[1] Since the genetic location of the *ecfA* locus is now known, isolation of *ecfA* mutants is considerably simplified. By taking advantage of the fact that the *ecfA* gene is about 90% cotransducible with the *metC* gene by P1 phage-mediated transduction, one can readily isolate *ecfA* mutants by means of the localized mutagenesis method[5] using *metC* strains as recipients for transduction.

Procedure. A *metC* strain such as JSH210 *(thi, metC)* is grown overnight on nutrient broth at 25°, and CaCl₂ and MgCl₂ are added to the culture to final concentration of 5 and 10 m*M*, respectively. P1 phage lysate prepared from a *MetC⁺* strain (JSH1) and mutagenized with hydroxylamine as described by Cunningham-Rundles and Maas[6] is then added. After 15 min at 37°, cells are pelleted and resuspended in the same volume of minimal salts medium. Aliquots (0.2 ml) are then spread onto minimal glucose plates lacking methionine and the plates are incubated at 25°. Only *Met⁺* colonies will grow on these plates, and after 36 to 48 hr small colonies appear. The plates are then transferred to a 42° incubator for 5 hr, after which they are transferred back to 25° to allow further growth for 24 hr. The resulting small colonies (which are presumably those unable to grow at 42°) are then picked with wooden applicators and spotted onto a pair of succinate plates. One is incubated at 42° and the other at 25°, both for 2 days. About 5% of the small colonies picked are found unable to grow at 42° but able to grow at 25°. These mutants are then streaked on nutrient broth plates and incubated at 42°. Those unable to grow or growing extremely slowly under these conditions are presumably *ecfA* mutants. However, definitive confirmation requires transport assay and reversion test which are described below.

Three types of *ecfA* mutants have been described; type I is exemplified by the mutants MAL300[2] and MAL321,[7] type II by JSH270,[3] and type

---

[5] J.-s. Hong and B. N. Ames, *Proc. Natl. Acad. Sci. U.S.A.* **68,** 3158 (1971).

[6] S. Cunningham-Rundles and W. K. Maas, *J. Bacteriol.* **124,** 791 (1975).

[7] J.-s. Hong, D. L. Haggerty, and M. A. Lieberman, *Antimicrob. Agents Chemother.* **11,** 881 (1977).

III by JSH267.[4] Some mutants revert at a such high rate (much higher than 10^{-5}) that they remain uncharacterized.

Isolation of Revertants

Revertants of *ecf* mutants able to grow on succinate, fumarate, malate, or lactate as a sole carbon source and regaining the ability to transport amino acids and sugars at 42° and which appear spontaneously at a rather high rate (10^{-5}) are isolated as follows. To isolate Class I revertants (*ecfA*$^+$, *MetC*$^+$), aliquots (0.2 ml) of overnight cultures grown in nutrient broth or minimal glucose medium at 25° are spread on succinate plates lacking methionine and incubated for 2 days at 42°. *Suc*$^+$ revertants that appears are purified on succinate plate by streaking using a loop. To isolate Class II revertants which are *Suc*$^+$ but *metC,* methionine (0.4 mM) must be present in the medium. Both Class I and Class II revertants are isolated under these conditions.

Transport Properties of ecfA Mutants

Transport Assay. Cultures are grown at 25° with vigorous shaking by diluting overnight cutlures 100-fold with 100 ml of minimal glucose medium in a 500-ml Erlenmeyer flask and are harvested in mid-log phase ($OD_{660} = 0.3–0.7$). The cells are collected by centrifugation, washed once with and resuspended in carbon-free salts medium (N^+C^-)[8] containing 100 μg/ml of chloramphenicol to an $OD_{660} = 3.0$. Transport is assayed in the absence of added carbon source, and performed as follows. Cells (50 μl) in test tubes (12 × 75 mm) are preincubated at 25 or 42° for 2 min and then 1 μl of a radioactive substrate (>100 mCi/mmol) is added to about 2–20 μM. At various times thereafter a tube is removed from the water bath, 2 ml of N^+C^- is added, and the cells are collected and washed once with 2 ml of the same medium on a Millipore filter (0.45-μm pore size). The filter is then dried and the radioactivity determined by liquid scintillation counting.

Properties. The transport defect exhibited by *ecfA* mutants is *specific,* affecting only those transport systems that depend on the protonmotive force for activity, and can not be attributed to a nonspecific, generalized membrane defect since α-methylglucoside transport in the *ecfA* mutants which is transported via the phosphoenolpyruvate phosphotransferase system by a mechanism that has no direct relationship to the protonmo-

---

[8] H. J. Vogel and D. M. Bonner, *J. Biol. Chem.* **218,** 97 (1956).

tive force-driven transport is unaffected in the *ecfA* mutants at 42°. The transport defect is also *pleiotropic* affecting all protonmotive force dependent transport systems examined without an affect on the functioning of transport carriers.

In the absence of an added energy source, the transport defect exhibited by the type I and type II *ecfA* mutants is easily observed after only 2 min incubation at the nonpermissive temperature (42°).[1,2] The half-time of inactivation of transport at 42° is about 1 min. Transport by the mutants is normal at 25°; however, the accumulated solute at 25° is rapidly effluxed upon a temperature shift at 42°, indicating that the mutants lose the ability to retain solutes against a concentration gradient at 42°. Heat inactivation of transport is irreversible; prolonged cooling in ice (up to 24 hr) after a 10 min incubation at 42° could not restore transport activity when assayed at 25°.

Type I *ecfA* mutations probably affect the process of coupling energy to active transport, since the transport defect in the mutant MAL300 is not due to an alteration of membrane permeability, inactive carrier molecules, a failure in respiratory chain, or a deficiency in H^+-ATPase activity.[4] Support for this contention came from the study of the effect of exogenous energy sources upon transport in the mutant. Although addition of 20 mM glucose to the cells stimulates transport at 42°, the initial rate of stimulation as well as the steady-state accumulation achieved is markedly reduced compared with that observed with wild type in the absence of exogenous energy. Moreover, if glucose is added at low concentrations (below 2 mM), transport stimulation is transient; within the first 30 sec of glucose addition, the cells are capable of accumulating solutes, yet within the next 2–3 min the accumulated substrate has completely effluxed from the cells. By examining the initial rates of transport as a function of time after glucose addition, it can be demonstrated that during the efflux period the cells can no longer transport solute. Thus, when efflux occurs the cells have lost both the ability to actively transport and to retain accumulated solutes. If a second dose of glucose is added to the cells after the efflux period, a restimulation occurs. This eliminates the possibility that a long-lived glucose metabolite may be interferring with energy-coupling process in the mutants.

Reduced Ability in Maintaining Transmembrane Potential

By measuring the membrane potential using the method described below it becomes apparent that in the absence of an exogenous energy source, the *ecfA* mutants have a very low transmembrane potential. Addition of glucose to the mutants cells allows generation of a membrane potential, and hence stimulation of amino acid transport. Since the extent of stimulation of amino acid transport parallels the extent of $TPMP^+$ accumulated for all glucose concentrations examined, it is apparent that the defect in the type I mutants lies in the generation and maintenance of the membrane potential. By comparison with starved wild-type cells (which contains an *uncA* mutation in order that energization via the ATPase pathway can not occur) we were able to show that the ability to maintain a membrane potential, in the presence of exogenous energy, is defective in the mutant. Furthermore, the mutant was also shown to be inefficient in the coupling of energy to the generation of a membrane potential.[2]

Membrane Potential Measurement. Transmembrane potential can be conveniently determined by measuring the uptake of radiolabeled lipophilic cations such as triphenylmethylphosphonium ($TPMP^+$) and tetraphenylphosphonium (TPP^+) (available from New England Nuclear).[9] Because *E. coli* cells are impermeable to these cations due to the presence of the peptidoglycan layer in the cell envelope they must be first permeabilized by treatment with Tris–EDTA. The method described is a modification of the procedure of Schuldiner and Kaback.[10] Glucose-grown cells (160 ml) are harvested in mid-exponential phase, washed twice with N^+C^-, and resuspended in 0.1 M Tris–HCl, pH 7.8, at a cell density of 2.8 to 4.2×10^9 cells/ml. The cells are immediately transferred to an Erlenmeyer flask such that the flask volume is at least 10 times that of the sample volume and the potassium salt of EDTA, adjusted to pH 7.0, was added to final concentration of 10 mM. The cells (about 15 ml) are rapidly swirled by hand in a 30° water bath for 30 sec, then diluted at least fourfold with N^+C^-. The cells are then pelleted, washed once with N^+C^-, and resuspended in the same medium containing 100 μg of chloramphenicol/ml. [^{3}H]TPMP or [^{3}H]TPP uptake is assayed as described. Cells can be stored in ice for up to 3.5 hr with no loss in their initial level of $TPMP^+$ or TPP^+ uptake.

Defective ATP Synthesis

To observe ATP synthesis, it is necessary to deplete intracellular ATP as much as possible without permanently impairing the cell's ability to effect such synthesis. The starvation method developed by Berger[11] is excellent for this purpose. Using starved cells prepared by this procedure

---

[9] S. Ramos, S. Schuldiner, and H. R. Kaback, this series, Vol. 55, p. 680.
[10] S. Schuldiner and H. R. Kaback, *Biochemistry* **14,** 5451 (1975).
[11] E. A. Berger, *Proc. Natl. Acad. Sci. U.S.A.* **70,** 1514 (1973).

as described below, similar levels of ATP synthesis can be observed in both the wild-type and the *ecfA* mutant MAL300 at 26° using the artificial electron donor ascorbate-phenazine methosulfate as the source of energy. Preheating wild-type cells at 42° does not reduce the net amount of ATP formed as compared to nonheated cells. Preheating the mutants cells, however, reduces the net amount of ATP formed, as compared to non-heated cells by 100%.[2] The defect in ATP synthesis is most probably due to the inability of the mutant to efficiently maintain the membrane potential.

Starvation of Cells. Cultures are grown on minimal glucose medium at 25°, harvested in mid-log phase, washed twice with Medium B (consisting of 1 mM MgCl$_2$, 0.1 M potassium phosphate, pH 7.0, and 100 μg of chloramphenicol/ml), and resuspended in medium B containing 2,4-dinitrophenol (twice recrystallized from water) at a cell density of 7×10^9 cells/ml. The cells are then starved by incubating in a 26° water bath for either 2.0 min (for mutants) or 4.5 hr (for wild type). The starved cells are then washed three times with Medium B and resuspended in the same at 5×10^8 cells/ml. The cells are stored on ice. One-half of the cells suspension is heated at 42° for 5 min and cooled on ice for at least 10 min prior to beginning the experiments, all of which are conducted at 26°. ATP in the starved cells is determined by the luciferin–luciferase method as described.[12]

Defect in Type II Mutant

In the type I *ecfA* mutants the transport defect can be accounted for by the inability of the mutants to maintain the membrane potential. However, in the type II mutant JSH270 a membrane potential of about −86 mV (interior negative) is established and maintained under conditions in which active transport is defective (pH 7.0 at 39°). In the wild type a potential of such magnitude under these conditions is sufficient for active transport. The observed transport defect with the mutant is indeed puzzling and is interpreted to suggest that the electrochemical gradient, although necessary as a driving force for transport, may not be sufficient to drive the active transport in the mutant. In light of this finding it was postulated that the ECF protein might be a common component to all transport systems driven by the protonmotive force with a role in sensing and responding to the protonmotive force as a proton symporter in the framework of the chemiosmotic hypothesis. It was also postulated that alternatively the ECF factor might act as an energy-transducing factor

[12] M. A. Lieberman and J.-s. Hong, *Arch. Biochem. Biophys.* **172**, 312 (1976).

transforming the protonmotive force into conformational energy which is then directly utilized to effect coupling. It is now clear that both models are no longer viable in light of the successful reconstitution of purified lactose carrier into liposomes.[13] Thus, the underlying biochemical defect observed with the ecfA mutant JSH270 remains puzzling and unclear.

Defect in Type III Mutants

The mutational lesion in the type III ecfA mutants is more extensive and severe than in the type I and type II. While the first two types have a defect in the ability either to maintain the membrane potential or to couple the membrane potential to active transport, all in the absence of an observable increased membrane permeability to protons or nucleotides, the mutations in the type III mutants affect the membrane permeability to protons and nucleotides and, presumably, to other small molecules as well, but they do not affect permeability to proteins.[4]

And unlike the first two types whose ability to transport solutes is rapidly inactivated by heat (2 min at 42°), type III mutants require cell growth at the nonpermissive conditions (42°) for the membrane lesion to be expressed, suggesting that defective membrane can be synthesized only at the nonpermissive conditions.

The permeability alterations observed with the type III ecfA mutants are strikingly similar to those observed with the T7 phage-infected male E. coli cells. Condit[14] and Britoon and Haselkorn[15] have observed that male E. coli cells lose the ability to transport amino acids and excrete nucleotides shortly after T7 infection. This permeability change requires at least two genes, one on the episome and one on T7. The molecular mechanism for this permeability change is presently unknown, but it may be related to inactivation of a membrane protein functionally similar to the ECF protein.

Acknowledgment

Work performed in this laboratory was supported by Grant GM 29843 from the National Institute of General Medical Sciences.

[13] M. J. Newman, D. Foster, T. H. Wilson, and H. R. Kaback, *J. Biol. Chem.* **256,** 11804 (1981).
[14] R. C. Condit, *J. Mol. Biol.* **98,** 45 (1975).
[15] J. R. Britton and R. Haselkorn, *Proc. Natl. Acad. Sci. U.S.A.* **72,** 2222 (1975).

[15] Mutations in the *eup* Locus of *Escherichia coli,* Energy Uncoupled Phenotype

By CHARLES A. PLATE

Three genetic loci have been identified in *Escherichia coli* in which mutations result in an energy uncoupled phenotype, i.e., lack of growth on nonfermentable carbon sources, reduced growth yields on limiting glucose, and normal electron transport. The best characterized of the three, the *unc* locus near minute 84 on the *E. coli* linkage map,[1] encodes the subunits of the BF_0F_1-ATPase.[2] Certain Unc mutants are abnormally permeable to protons resulting in a reduced protonmotive force (PMF), which in turn leads to defective proton/solute cotransport and a low level resistance to aminoglycoside antibiotics.[3] The second locus, the *ecf*A locus near minute 65, is the subject of another chapter in this volume[4] and will not be dealt with further here.

The third locus, independently identified by three laboratories, maps near minute 88 and has been variously designated *ecf*B,[5] *ssd*,[6] and *eup*[7,8] (the latter designation to be used hereafter).[9] Mutations within the *eup* locus result in strains that grow on glucose, do not grow on succinate, have reduced growth yields on limiting glucose, have normal electron transport, have normal BF_0F_1-ATPase activity, and are defective in proton/solute cotransport. In addition, Eup mutants have decreased sensitivity to colicins K and A and they exhibit low level resistance to aminoglycoside antibiotics. Although Eup mutants phenotypically resemble the proton permeable Unc mutants, they differ in one significant respect. Eup mutants are not abnormally permeable to protons and they are capable of generating and maintaining a PMF of normal magnitude.[8,10,11] The pheno-

---

[1] B. J. Bachmann, *Microbiol. Rev.* **47**, 180 (1983).

[2] J. A. Downie, F. Gibson, and G. B. Cox, *Annu. Rev. Biochem.* **48**, 103 (1979).

[3] B. P. Rosen, *J. Bacteriol.* **116**, 1124 (1973).

[4] J. S. Hong, this volume [14].

[5] S. H. Thorbjarnardottir, R. A. Magnusdottir, G. Eggertsson, S. A. Kagan, and O. S. Andresson, *Mol. Gen. Genet.* **161**, 89 (1978).

[6] E. B. Newman, N. Malik, and C. Walker, *J. Bacteriol.* **150**, 710 (1982).

[7] C. A. Plate, *J. Bacteriol.* **125**, 467 (1976).

[8] C. A. Plate and J. L. Suit, *J. Biol. Chem.* **256**, 12974 (1981).

[9] It should be noted that while it is likely that *ecfB, ssd,* and *eup* are alleles, complementation studies to definitively establish this have not yet been done.

[10] E. R. Kashket, *J. Bacteriol.* **146**, 377 (1981).

[11] G. D. Hitchens, D. B. Kell, and J. G. Morris, *J. Gen. Microbiol.* **128**, 2207 (1982).

typic similarities between Eup and Unc mutants suggests that a product of the *eup* locus is important to the process of energy coupling, but the precise nature of its involvement has yet to be determined. The purpose of this article is to describe the procedures for obtaining and identifying *eup* mutations in *E. coli*.

Selection Procedure

E. coli Strains and Growth Media. The media used in these studies are LB broth (containing per liter: tryptone, 10 g; yeast extract, 5 g; NaCl, 5 g; adjusted to pH 7.0 with 1.0 *N* NaOH) and Ozeki minimal base (containing per liter: K_2HPO_4, 10.5 g; KH_2PO_4, 4.5 g; $(NH_4)_2SO_4$, 1.0 g; $MgSO_4$, 0.05 g; sodium citrate, 0.47 g) supplemented with a carbon source (0.4%), required amino acids (50 μg/ml), and thiamine (0.5 μg/ml). Solid Ozeki minimal and LB media contain 1.5% agar and LB soft agar contains 0.65% agar.

As previously stated, one characteristic of Eup mutants is their inability to grow on succinate as sole carbon source. We and others have found that this particular Eup trait is subject to extragenic suppression.[6,8] The following Eup mutant isolation protocol assumes that the *E. coli* strain to be employed does not harbor this ill-defined suppressor and is sensitive to colicins A and K and amino glycoside antibiotics such as neomycin. We have used two *E. coli* K12 strains that satisfy these conditions: strain A279a (HfrH3000) and strain M72 [*lacZ*(Am) *trp*(Am) *thi*]. Both of these strains are available from Dr. S. E. Luria (Department of Biology, Massachusetts Institute of Technology, Cambridge, MA 02139). The required *E. coli* strains colicinogenic for colicins A, K, E1, E2, and E3 can be obtained from this same source or from the *E. coli* Genetic Stock Center (Yale University School of Medicine, New Haven, CT 06510).

Mutagenesis and Screening for Eup Mutants. Ethylmethane sulfonate mutagenesis is carried out essentially by the procedure of Miller.[12] *E. coli* cells are grown in minimal medium with glucose as carbon source to a density of 2×10^8 cells/ml. The cells are pelleted, washed, and resuspended in one-half the original volume of the minimal medium base without carbon source. To 2.0 ml of this cell suspension add 0.03 ml of ethylmethane sulfonate, mix vigorously to dissolve, and incubate the cells at 37° with shaking for 2 hr (this results in approximately a 90% reduction in viable count). The cells are then pelleted and resuspended in 3.0 ml of unsupplemented minimal medium base.

---

[12] J. H. Miller, *in* "Experiments in Molecular Genetics," p. 138. Cold Spring Harbor Laboratory, Cold Spring Harbor, New York, 1972.

Fifteen tubes, each containing 2.0 ml of LB broth supplemented with glucose (0.4%), are inoculated with 0.2 ml of the mutagenized cell suspension and incubated overnight at 37° with shaking. Dilute an aliquot of each culture 100-fold and spread 0.1 ml of each dilution onto a minimal glucose agar plate containing neomycin sulfate (20 μg/ml). The plates are incubated at 37° and, after 3 days, there should be approximately 100–200 colonies/plate.

The neomycin-resistant colonies are replica plated onto minimal succinate, minimal glucose, minimal glucose-neomycin (10 μg/ml), and LB agar plates supplemented with glucose (0.4%) and $KClO_3$ (0.2%).[13–15] All plates are incubated at 37° for 2 days. Putative Eup mutants will grow on the glucose and glucose-neomycin plates, will grow aerobically in the presence of chlorate, but will not grow on the succinate plates.

Colonies exhibiting a succinate-nonutilizing, neomycin-resistant, chlorate-resistant phenotype are then checked for colicin A and/or K sensitivity. Eup mutants exhibit a markedly reduced sensitivity to these colicins but exhibit normal sensitivity to colicins E1, E2, and E3.[7] Colonies to be checked are inoculated into LB broth and grown overnight at 37°. LB agar plates are stabbed with cells colicinogenic for colicin A or K and these also are incubated overnight at 37°. The colicinogenic cell buttons are removed with filter paper and the plates sterilized by exposure to chloroform vapors for 30 min. Aliquots of the LB cultures to be tested are inoculated into LB soft agar (maintained at 45°), spotted over the colicinogenic stabs, and the plates incubated overnight at 37°. Colicin sensitivity is indicated by a clear zone in the soft agar overlay, and Eup mutants will give a much smaller zone of growth inhibition than the wild-type starting strain. Colonies exhibiting a reduced colicin K or A sensitivity can then be checked in the same manner for colicin E1, E2, and E3 sensitivity, using *E. coli* strains colicinogenic for these colicins.

Mapping of eup Mutations. The *eup* locus maps near minute 88 on the *E. coli* linkage map and cotransduces with *met*B at a frequency of approximately 20%.[5,8] To determine if the mutation resulting in a Eup⁻ phenotype has occurred within the *eup* locus, bacteriophage P1 lysates are prepared on the putative Eup mutants and used to transduce an *E. coli met*B

---

[13] Selection for neomycin resistance can yield electron transport defective mutants, some of which are chlorate sensitive under aerobic conditions.[14,15] Nitrate reductase synthesis, which is repressed under aerobic conditions in wild-type *E. coli,* is derepressed in these mutants. The chlorate sensitivity is due to nitrate reductase reducing chlorate to chlorite which is toxic to *E. coli.*

[14] R. D. Simoni and M. K. Shallenberger, *Proc. Natl. Acad. Sci. U.S.A.* **69**, 2663 (1972).

[15] G. Giordano, L. Grillet, R. Rosset, J. H. Dou, E. Azoulay, and B. A. Haddock, *Biochem. J.* **176**, 553 (1978).

strain to MetB.$^+$. The MetB$^+$ transductants are picked onto minimal glucose, minimal succinate, and LB-neomycin (10 μg/ml) plates to score for the Eup$^-$ phenotype.

Measurements of Transport and the PMF. Eup mutants are defective in the PMF-coupled transport of lactose, proline, and alanine but are normal for the ATP-dependent transport of glutamine and arginine.[7,8] Eup mutants are also normal in their ability to generate and maintain a pH gradient (inside alkaline) and a membrane potential (inside negative), the components of the PMF.[8,10,11] Procedures for making transport measurements and measuring the components of the PMF are given in detail elsewhere in this series.[16,17]

Storage of Eup Mutants. Eup mutants readily revert to Eup$^+$ and are difficult to maintain. The most successful method that we have found for preserving Eup stocks is to make liquid cultures 20% in glycerol and store aliquots frozen at $-80°$. Eup stocks cannot be maintained as stabs.

Eup Null Phenotype. Results obtained with an *E. coli* strain deleted of the *eup* locus have shown that the *eup* null phenotype is *quasi* Eup$^+$.[18] The *eup* deletion strain grows on nonfermentable carbon sources, although not as well as its Eup$^+$ counterpart, has normal proton/solute cotransport activities, and is hypersensitive to the aminoglycoside amikacin. Apart from the constraints that this finding places on possible modes of *eup* function, it is significant in that it limits the types of mutations that might be expected to give rise to the Eup$^-$ phenotype. Thus the Eup$^-$ phenotype is not likely to result from insertion or early nonsense mutations that occur within the *eup* locus.

[16] G. F. Ames, this series, Vol. 32, p. 843.
[17] S. Ramos, S. Schuldiner, and H. R. Kaback, this series, Vol. 55, p. 680.
[18] C. A. Plate, unpublished observation (1984).

[16] Methods for Mutagenesis of the Bacterioopsin Gene

By Marie A. Gilles-Gonzalez, Neil R. Hackett, Simon J. Jones, H. Gobind Khorana, Dae-Sil Lee, Kin-Ming Lo, and John M. McCoy

Proton translocation is important in a number of biological systems and bacteriorhodopsin, an integral membrane protein, offers a simple model to study its mechanism.[1] However, investigations of the structure

[1] W. Stoeckenius and R. A. Bogomolni, *Annu. Rev. Biochem.* **52,** 587 (1982).

to function relationship for such proteins have made little progress so far. Classically structure–function studies have employed chemical modification techniques. The methods used often lack selectivity and the modifying groups may cause nonspecific perturbations. A more precise and versatile approach is that of making predetermined mutations in the gene by recombinant DNA methods, expression of the mutated genes, and functional assay of the altered proteins.

To apply this approach to the study of bacteriorhodopsin, we have cloned the gene and have developed suitable vectors to express the protein in *E. coli*.[2,3] Furthermore, we have demonstrated that the native bacteriorhodopsin, after complete denaturation, readily refolds in defined detergent/lipid mixtures to give the fully active native conformation.[4,5] Similarly, bacterioopsin synthesized in *E. coli* in the inactive form can be reconstituted with retinal to regenerate the bacteriorhodopsin-like chromophore. Therefore, this general approach allows preparation of mutant bacterioopsins to test the structural model for bacteriorhodopsin (Fig. 1) and to investigate the mechanism of the proton pump.

These are some of the questions that we would like to answer:

1. What amino acids in the folded protein cause the large red shift that is observed in the association of retinal with the opsin?[6,7] In addition, is there an electrostatic interaction between the protonated Schiff base and a carboxylate group, for example, Asp-212? Is an interaction between the polyene chain and an external negative charge important in the spectral shift?

2. What is the mechanism of proton translocation?[8,9] Is it mediated by a proton conductance channel in which functional groups of suitably placed amino acids take part?

3. What can we learn about the structure of bacteriorhodopsin as a whole? Namely, the size of the individual helices, the size and role of the

[2] R. J. Dunn, J. M. McCoy, M. Simsek, A. Majumdar, S. H. Chang, U. L. RajBhandary, and H. G. Khorana, *Proc. Natl. Acad. Sci. U.S.A.* **78**, 6744 (1981).
[3] R. J. Dunn, N. R. Hackett, K.-S. Huang, S. S. Jones, D.-S. Lee, M.-J. Liao, K.-M. Lo, J. M. McCoy, S. Noguchi, R. Radhakrishnan, U. L. RajBhandary, and H. G. Khorana, *Cold Spring Harbor Symp. Quant. Biol.* **48**, 853 (1984).
[4] K.-S. Huang, H. Bayley, M.-J. Liao, E. London, and H. G. Khorana, *J. Biol. Chem.* **256**, 3802 (1981).
[5] E. London and H. G. Khorana, *J. Biol. Chem.* **257**, 7003 (1982).
[6] K. Nakanishi, V. Balogh-Nair, M. Arnaboldi, K. Tsujimoto, and B. Honig, *J. Am. Chem. Soc.* **102**, 7945 (1980).
[7] B. Honig, T. Ebrey, R. M. Callander, V. Dinur, and M. Ottolenghi, *Proc. Natl. Acad. Sci. U.S.A.* **76**, 2503 (1979).
[8] J. F. Nagle and H. J. Morowitz, *Proc. Natl. Acad. Sci. U.S.A.* **75**, 298 (1978).
[9] T. Konishi and L. Packer, *FEBS Lett.* **92**, 1 (1978).

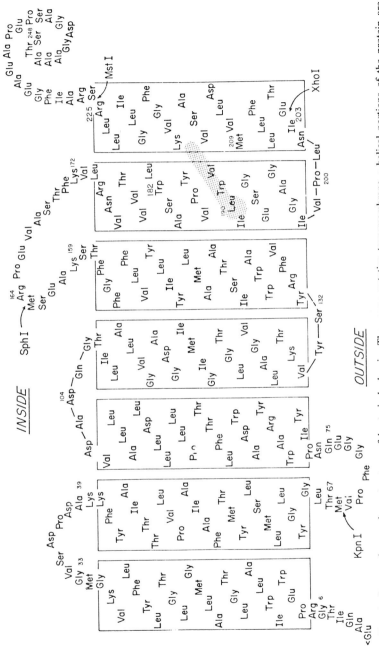

Fig. 1. Proposed secondary structure of bacteriorhodopsin. The seven putative transmembrane α-helical portions of the protein are enclosed in the vertical boxes. The shaded area suggests the position of retinal which is covalently linked to lysine-216 through a Schiff base. The positions in the protein corresponding to unique restriction sites in the gene are indicated. From K.-S. Huang et al.[11]

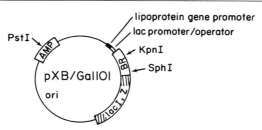

FIG. 2. Structure of the bacterioopsin/β-galactosidase fusion plasmid pXB/Gal101. The fusion plasmid pXB/Gal101 contains the origin (ori) and β-lactamase (AMP) gene of pBR322. A *lacI/lacZ* fusion is placed downstream from the *bop* gene which is transcribed from a tandem *lpp/lac* promoter/operator. The unique restriction sites used in this work are indicated. From J. M. McCoy and H. G. Khorana.[13]

loops connecting the helices, and an assessment of the hydrophobic and electrostatic interactions that contribute to the stability of the structure.[10,11]

Insights into these questions should be of general value in understanding integral membrane proteins.

Introduction of Amber Mutations into the Bacterioopsin Gene by Generalized Mutagenesis

Background and Rationale

Since most of the bacteriorhodopsin structure is of undetermined function, procedures involving generalized mutagenesis are potentially very useful in the initial stages. Unfortunately *in vivo* selection of point mutants of bacteriorhodopsin in *Halobacterium halobium* is hindered by a high frequency of mutations caused by insertion elements.[12] In addition, there is as yet no genetic transformation system available in *H. halobium*. Although bacterioopsin has been expressed in *E. coli*,[3] there is no selectable phenotype that is conferred on *E. coli* as a result of the expression of this gene. To provide an easily selectable phenotype, a fusion between the genes for bacterioopsin and β-galactosidase was constructed on a multicopy plasmid, pXB/Gal 101 (Fig. 2).[13] The fusion gene, containing the

[10] D. M. Engelman, R. Henderson, A. D. McLachlen, and B. A. Wallace, *Proc. Natl. Acad. Sci. U.S.A.* **77,** 2023 (1980).

[11] K.-S. Huang, R. Radhakrishnan, H. Bayley, and H. G. Khorana, *J. Biol. Chem.* **257,** 13616 (1982).

[12] S. DasSarma, U. L. RajBhandary, and H. G. Khorana, *Proc. Natl. Acad. Sci. U.S.A.* **80,** 2201 (1983).

[13] J. M. McCoy and H. G. Khorana, *J. Biol. Chem.* **258,** 8456 (1983).

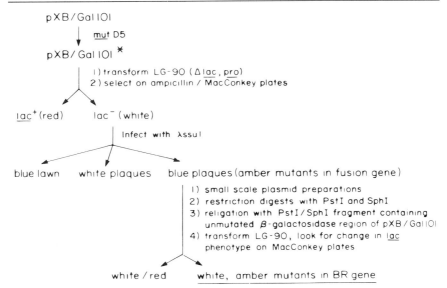

FIG. 3. Scheme outlining the procedure used for the creation and identification of amber mutations in the *bop* gene. After J. M. McCoy and H. G. Khorana.[13]

bacterioopsin gene *(bop)* fused upstream from the β-galactosidase gene, was under the control of tandem lipoprotein and *lac* gene promoters.[14] When expressed in *E. coli* the fusion protein retained β-galactosidase activity and conferred a Lac⁺ phenotype on *lac⁻* strains.

Amber mutations in the bacterioopsin portion of the fusion gene were produced by the scheme shown in Fig. 3.[13,15] The fusion plasmid was mutagenized by passage through an *E. coli* mutator strain and was then transformed into a *lac⁻* strain, selecting for ampicillin resistance and screening for Lac⁻ transformants. Amber mutants were identified by the phenotypic reversion of these transformants to Lac⁺ by lysogenization with a phage carrying an amber suppressor. Amber mutations occurring within the *bop* gene were localized by replacing the β-galactosidase region of each mutant plasmid with a β-galactosidase region which was known to be unmutated. Precise localization of the mutations was achieved first by sizing the prematurely terminated peptides produced by the mutant plasmids in an *in vitro* coupled transcription–translation system, and second, by DNA sequence analysis.

[14] K. Nakamura, Y. Masui, and M. Inouye, *J. Mol. Appl. Genet.* **1,** 289 (1982).
[15] J. H. Miller, C. Coulondre, M. Hofer, U. Schmeissner, H. Sommer, and A. Schmitz, *J. Mol. Biol.* **131,** 191 (1979).

Procedure

Mutagenesis of the Bacterioopsin/β-Galactosidase Fusion Plasmid. A scheme outlining the mutagenesis procedure is given in Fig. 3. The *E. coli* mutator strain (W3350, str^r, azi^r, $mutD5$, $galU95$) was assayed for mutagenic activity by checking for the appearance of resistant colonies on LB plates supplemented with 20 μg/ml of nalidixic acid. A fresh overnight inoculum (0.5 ml) of W3350 was added to 50 ml of LB and grown at 37° to an A_{650} of 0.5. Cells were collected by centrifugation at 3000 g for 10 min at 4° and were gently resuspended in 20 ml of ice-cold 30 mM CaCl$_2$. After being kept on ice for 1 hr, the cells were again sedimented and resuspended in 0.5 ml of ice-cold 30 mM CaCl$_2$. The bacterioopsin/β-galactosidase fusion plasmid, pXB/Gal 101 (0.5 μg dissolved in 10 μl of water), was then combined with 100 μl of the calcium chloride-treated cells and kept on ice for 10 min. The suspension was then heated at 37° for 90 sec, 1 ml of LB was added before further incubation at 37° for 1 hr. Small aliquots were removed at this time to check for both mutagenic activity and transformation efficiency. A 0.5-ml aliquot was also removed and used to inoculate 50 ml of LB supplemented with 35 μg/ml of ampicillin. This culture was incubated at 37° for 14 hr before two 10-ml aliquots were taken to inoculate two flasks, each containing 1 liter of LB supplemented with 35 μg/ml of ampicillin. These cultures were incubated at 37° to an A_{650} of 0.6 before the addition of chloramphenicol to a final concentration of 200 μg/ml. Plasmid was prepared from these cells as described.[16] The yield of pXB/Gal 101 was 600 μg/liter of cells. pXB/Gal 101 prepared from W3550 cells was used to transform[17] *E. coli* strain LG-90.[18] Transformed cells were grown at 37° on lactose MacConkey plates supplemented with 35 μg/ml of ampicillin. Ampicillin-resistant colonies with a Lac⁻ phenotype (white colonies) were picked and used for further selection procedures.

Selection for Amber Mutations in pXB/Gal 101. White (Lac⁻) LG-90 colonies which had been transformed with pXB/Gal 101 and which were resistant to ampicillin were picked and grown at 37° for 14 hr in 2-ml cultures of LB supplemented with 35 μg/ml of ampicillin. Portions (50 μl) of each of these cultures were combined with a suitable dilution of bacteriophage λssul and kept at 25° for 15 min before the addition of 1 ml of molten LB soft top agar (LB + 0.6% agar). 5-Bromo-4-chloro-3-indolyl-β-D-galactoside (20 μl of a 2% solution in dimethylformamide) and 10 μl of a 0.1 M solution of isopropyl-β-D-thiogalactopyranoside were each com-

[16] H. C. Birnboim and J. Doly *Nucleic Acids Res.* **7,** 1513 (1979).
[17] M. Dagert and S. D. Ehrlich, *Gene* **6,** 23 (1979).
[18] L. Guarente, G. Lauer, T. M. Roberts, and M. Ptashne, *Cell* **20,** 543 (1980).

bined with the agar mix before pouring out over small 3-cm-diameter LB plates. The plates were incubated for 24-48 hr at 30°, after which time plaques could easily be seen. The appearance of blue plaques indicated the presence of an amber mutation in the bacterioopsin/β-galactosidase fusion gene.

Localization of Amber Mutations to the Bacterioopsin Region of the Fusion Gene. Small-scale plasmid preparations[19] were performed on all colonies which yielded blue plaques in the λssul screen. A solution of 1 μg of each of the mutant pXB/Gal 101 plasmids was prepared in 20 μl of 20 mM Tris–HCl (pH 7.5), 50 mM NaCl, 6 mM MgCl$_2$, 6 mM 2-mercapto-ethanol containing 100 μg/ml of bovine serum albumin, 5 units of endonuclease *Pst*I, and 3.5 units of endonuclease *Sph*I. Digestion by both enzymes was complete after 12 hr at 37°. Both restriction endonucleases were heat inactivated before the DNA was recovered by ethanol precipitation. A similar double digestion was performed on a larger scale using unmutated pXB/Gal 101 as substrate. The large 6.1-kilobase fragment from this digestion was purified on a 1% low melting point agarose gel, recovered, and 0.25-μg portions combined with each of the *Pst*I/*Sph*I-cleaved mutant plasmids. The mixtures were each in 100 μl of 20 mM Tris–HCl (pH 7.5), 10 mM MgCl$_2$, 10 mM dithiothreitol containing 600 $\mu$$M$ ATP and 0.25 units of T$_4$ DNA ligase. The reactions were run at 15° for 2 hr before termination by ethanol precipitation. Each of the ligation mixtures was then used to transform[17] 100 μl of competent LG-90 cells. Transformants were again selected on lactose MacConkey plates supplemented with 35 μg/ml of ampicillin. If the colonies which grew were all white (Lac$^-$), then the amber mutation carried by the mutant plasmid was in the bacterioopsin region of the fusion gene. If a mixture of red (Lac$^+$) and white colonies appeared, the amber mutation was in the β-galactosidase region of the fusion gene.

Identification of the Sites of Amber Mutations within the Bacteriorhodopsin Gene by Coupled Transcription–Translation of Mutant Plasmids and by DNA Sequence Analysis. For approximate localization of the amber mutations, an *E. coli in vitro* coupled transcription–translation system was used. *E. coli* extracts were prepared from strain MRE 600 as described by Zubay.[20] The reactions were performed as described by Zubay,[20] but incorporating modifications suggested by Collins.[21] The mutant plasmids were used to direct protein synthesis in the *in vitro* system in the presence of [^{35}S]methionine. Ten micrograms of each plasmid was used for each reaction. ^{35}S-labeled peptidic products were sized on 15%

[19] R. D. Klein, E. Selsing, and R. D. Wells, *Plasmid* **3**, 88 (1980).
[20] G. Zubay, *Annu. Rev. Genet.* **7**, 267 (1973).
[21] J. Collins, *Gene* **6**, 29 (1979).

MUTANTS OF THE BACTERIOOPSIN GENE

Procedure	Mutant	Comments
Genetic selection of	W10 Amber	
amber mutants	W12 Amber	
	W80 Amber	
	W86 Amber	
	W137 Amber	
Deletion mutagenesis	(G65–N76)	
at *Kpn*I site	(G65–Q75)	N76 changed to D
	(G65–Q75)	
	(M60–R82)	
	(T67–Q74)	
Deletion mutagenesis	(M163–K172)	S162 changed to R
at *Sph*I site	(E161–A168)	
	(F154–M163)	
Oligonucleotide-directed	I203 L	
point mutagenesis	D96 E	
Mutagenesis through extensive	P186 L	
synthesis	E194 Q	
	D212 E	
	D212 N	
	E204 D	
	S193 A	
	S214 A	
	W189 F	
	Y185 F	
	F208 Y	
	F219 L	
	K216 R	
	K216 A, A215 K	
	K216 V, V217 K	

sodium dodecyl sulfate–polyacrylamide gels, run according to the procedure of Laemmli.[22] Gels were soaked in sodium salicylate before fluorography. The mobility of prematurely terminated, ^{35}S-labeled peptides was used to estimate the positions of the amber mutations. Precise identification of the sites of the amber mutations was by DNA sequence analysis performed as described by Maxam and Gilbert.[23]

Comments

Fifteen amber mutants were found at 6 positions in the *bop* gene from an initial screen of 10,000 colonies (see the table). One of these was a

[22] U. K. Laemmli, *Nature (London)* **227**, 680 (1970).
[23] A. M. Maxam and W. Gilbert, this series, Vol. 65, p. 499.

transversion mutation at a lysine codon; the other five were all transition mutations at tryptophan codons, codons 10, 12, 80, 86, and 137 of the bacteriorhodopsin sequence. The bias for mutations at tryptophan codons could be removed by using alternative strategies. Other methods for the generation of mutations can be used, such as chemical mutagens *in vivo* or *in vitro*, or other *E. coli* mutator strains. Further, the procedure is not limited to amber mutants, similar selections can be used to find point mutations resulting in UGA or UAA chain termination codons. Furthermore, there is no restriction to the use of *lacZ* gene as the fusion partner and the method should be generally applicable for the selection of mutants within a gene which normally has no phenotype in *E. coli*.

Certain proteolytic fragments of bacteriorhodopsin have been shown to reassociate to form native structure.[24] This provides another approach to structure–function studies. The amber mutations described in this work enable the preparation of fragments of bacteriorhodopsin which could not be obtained by proteolytic means. The amber mutants obtained can also be expressed in *E. coli* suppressor strains. In this way, specific replacements to a variety of different amino acids can be made at the site of each amber codon, depending on the particular suppressor strain used.

Introduction of Small Deletions at Predetermined Sites of the Bacterioopsin Gene

Background and Rationale

Deletion mutagenesis has been widely used to define functional domains in DNA and proteins.[25,26] In bacteriorhodopsin and other polytopic membrane proteins, this approach could be used to test structural models and to investigate systematically the size of the loops that protrude into the aqueous medium and of the domains embedded in the bilayer.

The procedure employed makes small deletions at a unique restriction site in the gene of interest. First, a plasmid containing the gene is linearized by cutting at a unique restriction site that falls within that gene (Step 1, Fig. 4). A limited exonuclease III digestion is carried out at the site of the cut (Step 3, Fig. 4), followed by S1 nuclease treatment (Step 4, Fig. 4), to shorten the linear duplex. If the initial restriction cut leaves 3′ protruding ends, treatment with a 3′-exonuclease, such as the exonuclease

[24] M.-J. Liao, E. London, and H. G. Khorana, *J. Biol. Chem.* **258**, 9949 (1983).
[25] S. L. McKnight, E. R. Gravis, R. Kingsbury, and R. Axel, *Cell* **25**, 385 (1981).
[26] M. Jasin, L. Regan, and P. Schimmel, *Nature (London)* **306**, 441 (1983).

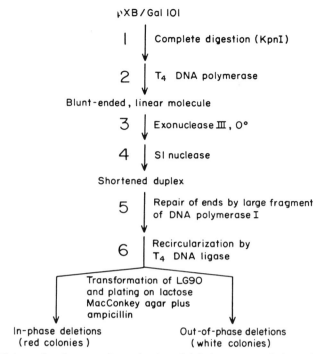

Fig. 4. Scheme for the enzymic production of deletions centered about the *Kpn*I site of the *bop* gene.

present in T$_4$ DNA polymerase (Step 2, Fig. 4) is necessary to produce ends that exonuclease III can efficiently digest. After S1 digestion, any frayed ends are filled in with the large fragment of *E. coli* DNA polymerase I (Step 5, Fig. 4), and the blunt-ended molecules are recircularized with T$_4$ DNA ligase (Step 6, Fig. 4). An appropriate *E. coli* strain is then transformed with this ligation mixture. In-phase deletions are selected by screening for the expression of a second gene, spliced in-phase, downstream from the mutagenized gene. In our case, we used the plasmid pXB/Gal 101[13] (Fig. 2), in which the bacterioopsin gene is fused to the *lacZ* gene to make an active (Lac⁺) fusion protein. Fortunately, there are a number of unique restriction sites in the part of this plasmid corresponding to the bacterioopsin gene at which deletions can be made by exonuclease III digestion. In-phase deletions were identified by selecting for Lac⁺ colonies, which were then characterized by DNA sequencing.

Procedure

Production of Linearized Blunt-Ended Plasmid DNA. pXB/Gal 101 (25 μg) was digested in 150 μl of 10 mM Tris–HCl, pH 7.4, containing 5 mM NaCl, 10 mM MgCl$_2$, 10 mM DTT, and 100 μg/ml BSA with 50 units of *Kpn*I at 37° for 2 hr (Step 1, Fig. 4). The reaction was terminated by adding EDTA (20 mM) and heating to 65° for 10 min. The DNA was recovered by phenol extraction and ethanol precipitation. It was redissolved in a mixture (30 μl) of 33 mM Tris-acetate, pH 7.9, 66 mM K-acetate, 10 mM Mg-acetate, 10 mM DTT, and 100 μg/ml BSA. T$_4$ DNA polymerase (45 units) was added and the mixture was incubated for 10 min at 37°. A mixture (0.5 μl) of dCTP, dGTP, and dTTP (6 mM each) and 10 μCi (1 μl) of [α-^{32}P]dATP (400 Ci/mmol) was added and the reaction was further incubated for 15 min. A final incubation was carried out (45 min) after the addition of 0.5 μl of 6 mM dATP (Step 2, Fig. 4). The reaction was terminated and the DNA extracted and precipitated as above.

Limited Exonuclease Digestion of the Linearized Plasmid. Linearized DNA (approximately 10 μg) was dissolved in 30 μl of 66 mM Tris–HCl containing 90 mM NaCl, 10 mM DTT, and 5 mM MgCl$_2$ and the solution was kept on ice for 5 min. An aliquot was removed and saved as a zero time point. Subsequently, 4 units of exonuclease III were added per microgram of DNA and digestion performed at 0° for 2 min (Step 3, Fig. 4). An equal volume of 20 mM EDTA was added and the DNA was extracted and precipitated. The precipitates were washed twice in 80% ethanol at room temperature. After drying, each sample was dissolved in 10 μl of 10 mM Tris–HCl, pH 8.0. Exonuclease III-treated DNA was incubated in 25 μl of 30 mM Na-acetate, pH 4.5, with 0.25 M NaCl and 1 mM ZnSO$_4$ with 0.25 units of S1 nuclease at room temperature for 40 min (Step 4, Fig. 4). The digestion was terminated and the DNA was recovered as above.

Recircularization of the Plasmid. The recovered DNA was treated with 1 unit of the large (Klenow) fragment of DNA polymerase I at 37° for 15 min (Step 5, Fig. 4) in 10 μl of 20 mM Tris–HCl, pH 7.4, containing 10 mM MgCl$_2$, 20 mM NaCl, 1 mM DTT, and 0.5 mM each dNTP. The volume was adjusted to 200 μl in the same buffer (without dNTPs) and rATP added to a final concentration of 0.2 mM. Ligation was at 14° for 16 hr with 20 units of T$_4$ DNA ligase (Step 6, Fig. 4).

Recovery and Characterization of Deletion Mutants. The ligation mixture was used directly to transform *E. coli* strain LG90[18] by the procedure of Hanahan.[27] The transformants were plated on lactose McConkey plates with 35 μg/ml ampicillin. Red colonies were chosen and grown on a small

[27] D. Hanahan, *J. Mol. Biol.* **166,** 557 (1983).

scale for plasmid DNA preparation.[16] Plasmids that were not cut by *Kpn*I were then propagated on a larger scale for DNA sequence analysis.[28]

Comments

There are a number of steps in this procedure that need to be carefully monitored for obtaining a large number of mutants. First, it is essential that the first restriction digestion be complete (Step 1, Fig. 4). We aim for a 4-fold overdigestion and monitor the progress of the reaction by running aliquots on agarose gels. At the ligation step (Step 6), the possibility of wild-type plasmid in the ligation mixture can be eliminated by redigestion with the restriction enzyme used in Step 1. In our case, the enzyme used (*Kpn*I) produced a 3' protruding end, which is a poor substrate for exonuclease III, so it was treated with T_4 DNA polymerase before proceeding (Step 2). This opportunity was taken to label the DNA by replacement synthesis, which allows the ready monitoring of the subsequent manipulations. The conditions for exonuclease III digestion (Step 3) are derived from a method devised by Donelson and Wu.[29] The 2-min time point should remove on average six bases from each end, but they suggest the digestion will be very nonuniform at 0°. Exonuclease III seems to proceed rapidly for the first minute and then stop.[29] However, a zero time point can be carried through, which will give a low frequency of very small deletions. The subsequent S1 treatment is designed to digest the other strand and was titrated by spiking the mixture with a 5' end-labeled synthetic oligonucleotide and determining the time taken for the label to be fully digested to the nucleotide level. In the plasmid pXB/Gal 101 containing the *bop/lacZ* fusion gene, it is fortunate that unique restriction sites were present within the bacterioopsin sequence. However, in cases where suitable sites are not present they can readily be produced by introducing linkers.

The procedure described above has been used to create small deletions in loop 2–3 and in loop 5–6 of BR (Fig. 1, see the table). The in-phase deletions were initially examined by restriction analysis and later by DNA sequencing. Deletions of 4 to 20 amino acids were readily obtained, although the deletion sizes tended to be clustered (see the table). This is a consequence of the fact that exonuclease III digestion is not uniform at 0°.[29] If the deletions are first sized by restriction analysis, much sequencing work can be avoided in finding the mutants that are missing the desired number of amino acids. All Lac⁺ colonies had in-frame dele-

[28] F. Sanger, S. Nicklen, and A. R. Coulson, *Proc. Natl. Acad. Sci. U.S.A.* **74**, 5463 (1977).
[29] J. E. Donelson and R. Wu, *J. Mol. Biol.* **237**, 4661 (1972).

tions, and in all cases, the deletions were well centered about the site of the original restriction cut. Therefore, this is a very effective and rapid method for making small deletions at specific sites in a protein.

Oligonucleotide-Directed Point Mutagenesis

Background and Rationale

Point mutagenesis can be achieved by a number of procedures.[30–32] The procedure of Zoller and Smith[31] is the most straightforward. This involves cloning the gene of interest into M13 and using single-stranded DNA from this clone as template for *in vitro* primer extension with the mutagenic oligonucleotide as primer. This yields double-stranded DNA which should segregate to give 50% mutant clones. In practice, the yields of the mutants vary widely and often difficulties are encountered in obtaining any mutants. For example, our first attempts at mutagenizing the *bop* gene used this approach, but we were never successful in recovering the desired mutants. Since the incorporation of 5' end-labeled primer into the correct position of covalently closed circular DNA was demonstrated, we argued that the host was preferentially repairing the strand made *in vitro* to give only the wild-type product, perhaps because this strand is under-methylated relative to the template.[33] We would point out that the 1.6 kb insert in our M13 clone contains the recognition sequence for the *dam* methylase (GATC) at 10 positions, an unusually high frequency.

Therefore, we used an alternative approach in which the template is largely double-stranded but contains a short single-stranded gap which exposes the site to be mutagenized (Fig. 5). This is made by nicking a plasmid clone of the *bop* gene by a restriction enzyme in the presence of ethidium bromide (Fig. 5, Step 1) and then using the nick to make a small gap by limited digestion with exonuclease III. This is then used as template for primer extension in the presence of the mutagenic oligonucleotide (Step 3). We argued that covalently closed circular DNA could be made which contained the mutagenic oligonucleotide but was methylated equally on both strands, and this should avoid the bias of the mismatch repair system. Mutants were then identified using a colony screening procedure and the mutagenic oligonucleotide as hybridization probe (Step 4).

[30] G. Dalbadie-McFarland, L. W. Cohen, A. D. Riggs, C. Morin, K. Itakura, and J. H. Richards, *Proc. Natl. Acad. Sci. U.S.A.* **79,** 6409 (1982).

[31] M. J. Zoller and M. Smith, this series, Vol. 100, p. 468.

[32] D. Shortle, P. Grisafi, S. J. Benkovic, and D. Botstein, *Proc. Natl. Acad. Sci. U.S.A.* **79,** 1588 (1982).

[33] W. Kramer, K. Schughart, and H.-J. Fritz, *Nucleic Acids Res.* **10,** 6475 (1982).

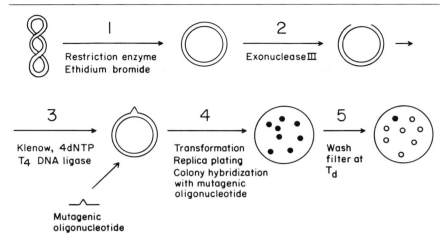

FIG. 5. Scheme outlining the procedure used for the production and identification of point mutants in the *bop* gene using gapped double-stranded DNA and a single mutagenic oligonucleotide.

Our procedure is described in detail for the production of a single nucleotide change (A to C) in the *bop* gene which changes Ile-203 to Leu. This was done to create a unique *Xho*I site which facilitates the replacement of a section of the gene by synthetic DNA.[34]

Procedure

Preparation of Double-Stranded DNA Containing Single-Stranded Gap at Specified Region. The BO expression plasmid pLBB[3] (30 μg) was nicked in 370 μl of 20 mM Tris–HCl, pH 7.5, 100 mM NaCl, 10 mM MgCl₂, 2 mM 2-mercaptoethanol, and 20 μg ethidium bromide by *Sph*I (150 units) at room temperature for 90 min. Ethidium bromide was removed by three extractions with *n*-butanol and the aqueous phase was extracted once each with phenol, chloroform, then ether, and the DNA was precipitated.

The nicked DNA (about 25 μg) was redissolved in 40 μl of Tris–HCl, pH 8.0, 5 mM MgCl₂, and 10 mM 2-mercaptoethanol and digested by exonuclease III (25 units) at room temperature for 15 min. The reaction was terminated by chloroform extraction followed by heating at 65° for 5 min. The extent of exonuclease III digestion was estimated by removing

[34] K.-M. Lo, S. J. Jones, N. R. Hackett, and H. G. Khorana, *Proc. Natl. Acad. Sci. U.S.A.* **81**, 2285 (1984).

aliquots from the reaction at intervals, stopping the reaction and precipitating the DNA. This was 3′ end labeled with the Klenow fragment of DNA polymerase I and then digested with *Eco*RI. Analysis of the products on a 4% denaturing gel showed that the average digestion under the conditions described above was 150 nucleotides.

Incorporation of the Mutagenic Oligonucleotide into Covalently Closed Circular DNA. The mutagenic oligonucleotide used was the pentadecamer, ACTTG<u>G</u>AGCTCTGCG, which has a single mismatch from the sense strand of the *bop* gene at the position underlined. It was quantitatively 5′-phosphorylated[35] on the scale of 120 pmol using a 2-fold molar excess of [γ-32P]ATP (25 mCi/mol). Aliquots (1 μl) were removed at 0, 15, 30, and 60 min and analyzed either on PEI cellulose TLC plates or on DE81 paper.[35] The PEI plates were run in 0.75 M sodium phosphate, pH 3.5 for about 30 min, air-dried, and autoradiographed. The spots may be excised and counted for a semiquantitative assay. The phosphorylated oligonucleotide (60 pmol) and the gapped DNA (4 pmol) were mixed at 4°. Primer extension was initiated without an annealing step by the addition of ATP, dATP, dCTP, dGTP, and dTTP (0.5 mM each), DTT (2 mM), the large (Klenow) fragment of *E. coli* DNA polymerase I (4 units), and T_4 DNA ligase (60 units). The reaction had a final volume of 60 μl and was incubated at 14° for 16 hr.

Transformation of E. coli and Replica Plating. The primer extension mixture from above was used directly to transform *E. coli* strain HB101 by the high efficiency procedure of Hanahan.[27] After transformation and phenotypic development, 0.1 ml of the cells was plated on top of filters (Millipore HATF, previously wetted and autoclaved between sheets of damp Whatman 3MM) on plates of LB with ampicillin.

After growth at 37° for 18 hr, replicas of the originals were taken by firmly pressing a second filter against the first between two sheets of damp Whatman 3MM. The replica was grown at 37° for 6 hr on a plate of LB with ampicillin and then transferred to another plate with the same medium plus 180 μg/ml chloramphenicol. It was incubated at 37° overnight whilst the master plate was stored at 4°.

Screening of Transformants for the Presence of the Desired Mutation. Cells were lysed and the DNA bound to the filter by washing for 10 min each with 0.5 M NaOH (twice), 1 M Tris–HCl, pH 8.0 (twice), and 1 M Tris–HCl containing 1.5 M NaCl. Washes were effected by placing the filter, colonies up, on a sheet of 3MM saturated with the appropriate solution. The filter was then allowed to dry and baked in a vacuum oven at 80° for 2 hr.

---

[35] E. L. Brown, R. Belagaje, M. J. Ryan, and H. G. Khorana, this series, Vol. 65, p. 109.

It was rewetted in 6 × SSC and prehybridized in 6 × SSC, 10 × Denhardts, 0.2% SDS, 50 μg/ml tRNA at 55° for 4 hr. Hybridization was in 6 × SSC, 10 × Denhardts using the minimum volume of solution (about 3 ml/82 mm filter) and 10^7 cpm ^{32}P of the mutagenic oligonucleotide. The oligonucleotide (10 pmol) was kinased as described above using 50 μCi [γ-^{32}P]ATP (5000 Ci/mmol) and purified on Sep-Pak. After hybridization, filters were washed in 6 × SSC for 10 min twice. The first wash was at room temperature after which an exposure to film for about 1 hr was sufficient to see an impression of all the colonies. Subsequently, it was washed at sequentially higher temperatures, and the wild-type colonies lost intensity much faster than the mutant. For oligonucleotides of this length a value T_d calculated as (4 × GC) + (2 × AT) gives the approximate temperature at which wild-type and mutant colonies will be most easily distinguished.

Putative positions were identified on the master filter by comparison of the pattern of colonies on the autoradiograph of the filter washed at room temperature with the master plate. Colonies were streaked to isolate them, picked off onto a sheet of nitrocellulose, and these were grown, replicated, and screened as described above. Positives from this second round were then used for the preparation of larger amounts of DNA for sequencing.

Comments

The procedure described is a modification of that of Dalbadie-Mc-Farland et al.[30] The progress of the enzymic manipulations of the DNA was monitored by running small aliquots on 0.8% agarose gels containing 0.6 μg/ml ethidium bromide. It should be emphasized that the reactions are not optimized for the maximum mutagenic frequency, but rather we rely on the power of the screening procedure to identify relatively rare mutagenic events.

In their original protocol Dalbadie-McFarland et al.[30] used HpaII to nick the plasmid at many positions. We wanted to nick the plasmid with SphI since it has only one recognition site in the bop gene and there is no dam site between this and the site to be mutagenized. The conditions optimum for nicking were established in small scale analytical reactions in which the ratio of DNA to ethidium bromide to enzyme was altered. Other enzymes reported to nick plasmids in the presence of ethidium bromide are EcoRI, HindIII, ClaI, BamHI,[32] and DNase I.[36] It would have been useful in the present work to use KpnI and HincII, but these enzymes were found not to nick the plasmids.

[36] L. Greenfield, L. Simpson, and D. Kaplan, *Biochim. Biophys. Acta* **407,** 365 (1975).

Under the conditions described above the exonuclease III digestion proceeded for an average of 150 nt. It is quite important to titrate this step to ensure that the site to be mutagenized is exposed to the primer. This may not be necessary when using enzymes that nick the DNA at random, such as DNase I. In the latter case it may be desirable to gel purify the nicked DNA after exonuclease III treatment to serve as template for a number of mutagenesis experiments. A number of variables in the primer extension step (Step 3) could also be considered such as an annealing step or the use of a larger excess of the mutagenic oligonucleotide.

A high efficiency transformation procedure[27] was used since it generally gave more colonies per plate. About 300 on a 100-mm plate or 1000 on a 150-mm plate are reasonable numbers to screen. If too few transformants are obtained the reaction mixture can be phenol extracted and run on a G-50 column to clean up the DNA.

The hybridization procedure is a modification of the dot blot procedure of Zoller and Smith.[31] The considerations they discuss in oligonucleotide design are less important when only a small region of the plasmid is single-stranded. Important factors, however, are the length of the oligonucleotide and the nature of the mismatch. A longer oligonucleotide may give more specific priming but will be a less sensitive probe. In addition, a purine/purine mismatch will cause maximum destabilization and show real positives better. Finally, sometimes false positives are seen. These can be identified by washing the filter well above the T_d. This will abolish all base-pairing interactions and remaining signals reflect nonspecifically bound probe.

Site-Specific Mutagenesis by Replacing Restriction Fragments with Synthetic DNA Duplexes Containing Altered Codons

Background and Rationale

Strategy for Synthesis and Construction of Expression Plasmid. In this approach restriction fragments in the region of interest are excised from the gene and replaced by totally synthetic counterparts. In the present example, a unique *Xho*I site was first created at Ile-203.[34] This allowed the excision of an *Ava*II–*Xho*I fragment (Fig. 6, amino acids Trp-182 to Leu-203). Replacement of this fragment allows studies of a portion of the gene partially encoding the presumed helix 6. This region was chosen in order to study possible interactions between the retinal chromophore and helix 6. Further, if proton translocation is mediated by proton conductance channels,[8,9] then removal of hydroxyl or charged side chains from suitable amino acids, e.g., Tyr-185 to Phe, Ser-193 to Ala, would

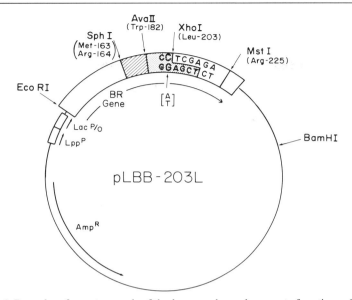

FIG. 6. Procedure for mutagenesis of the *bop* gene by replacement of portions of the gene by synthetic DNA duplexes. The expression plasmid pLBB-203L is shown which contains the *bop* gene downstream from the tandem *lac/lpp* operator/promoter. The wild-type sequence has been modified by point mutagenesis to create an *Xho*I site (CTCGAG). This allows the replacement of a *Xho*I–*Ava*II (stippled) fragment by synthetic counterparts. The gene is reconstructed using the large *Xho*I–*Sph*I fragment, a short *Sph*I–*Ava*II fragment (cross-hatched), and synthetic *Ava*II–*Xho*I fragments. In addition, the part of the gene corresponding to helix 7 of the protein can be mutagenized by replacing the *Xho*I–*Mst*I fragment by synthetic DNA. After K.-M. Lo *et al.*[34]

interrupt or block the channel with minimal perturbation to the protein. The substitution of Glu-194 by Gln is a significant change that could provide evidence for or against the external point charge model for the red shift.[6,7,11] Additionally, the removal of charges within the embedded helices would test the role of electrostatic interactions in stabilization of the membrane-embedded parts of bacteriorhodopsin.[10] Replacement of Trp-189 by Phe may also have serious consequences in structural or spectroscopic properties and the substitution of Pro-186 by Leu should test its effect on helix stability and the "proton hole" mechanism.[37]

The strategy for the excision and replacement of the *Ava*II–*Xho*I fragment involved: (1) preparation of a large *Sph*I–*Xho*I fragment from the mutated plasmid pLBB-203L, (2) preparation of a short *Sph*I–*Ava*II re-

[37] A. K. Dunker, *J. Theor. Biol.* **97,** 95 (1982).

striction fragment, (3) synthesis of mutant and native DNA duplexes corresponding to the *Ava*II–*Xho*I fragment, and (4) ligation of the three fragments to reform the *bop* gene (Fig. 6).

The plan for the synthesis of the *Ava*II–*Xho*I fragments (Duplex I) containing the native amino acid sequence and five mutant DNA duplexes (Duplex II–VI) is shown in Fig. 7. Four silent changes to A/T were introduced in the DNA duplexes to decrease the GC content of the oligonucleotides and to distinguish the modified from the native sequence. The length of the oligonucleotides synthesized was kept to only 11 to 12 bases because of the extensive substitutions planned. Codons rarely used in *E. coli* were avoided. The pairs of oligonucleotides mostly have overhangs of four nucleotides at each cohesive end,[35] self-complementarity within the internal cohesive ends was avoided to prevent self-ligation. A computer search was carried out so as to avoid self-complementarity of the oligonucleotides, and to make sure that every protruding sequence is unique and that each oligonucleotide will hybridize only to its intended partner. If the above criteria are not met, then either silent base substitutions are made or oligonucleotide components in the ligation are chosen to avoid wrong joining.

The A : T to C : G base pairs change introduced at Ile-203 (Fig. 6) enabled the *Xho*I cut to be made and to give AGCT cohesive end, but in our synthetic duplexes (*Ava*II–*Xho*I) we have kept the A : T base pair as in the native sequence (Fig. 7). The cohesive end in synthetic duplexes is designated *Xho*I′ since it will ligate with an *Xho*I cohesive end but will not regenerate the *Xho*I site. Upon reconstruction with the synthetic duplex and the *Sph*I–*Xho*I vector fragment, the Ile-203 to Leu change created by point-mutagenesis would thus be reversed. This also allows us to digest the ligation mixture with *Xho*I before transformation to reduce the background due to transformants of the parental type.

Purification of Oligonucleotides by Sep-Pak. Sep-Pak C_{18} is a cartridge containing loosely packed C_{18}, a reverse-phase gel matrix in which an octadecyl group is covalently bonded to a silica framework. In reverse-phase liquid chromatography on C_{18} columns, the sample is usually applied in an aqueous solvent. Because of the hydrophobic nature of the octadecyl group, different components in the sample are adsorbed to a varying degree according to their lipophilicity. Since the hydrophobic binding can be disrupted by an organic solvent, elution of the desired compounds can be achieved by selectively varying the concentration of a miscible organic solvent in the mobile phase, and this is usually accomplished in the form of a gradient.

Reverse phase high-performance liquid chromatography on C_{18} columns is commonly used for purification of synthetic oligonucleotides;

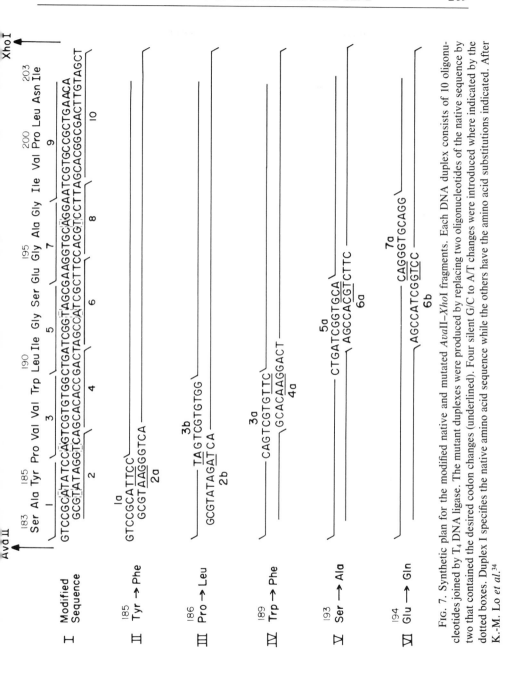

Fig. 7. Synthetic plan for the modified native and mutated AvaII–XhoI fragments. Each DNA duplex consists of 10 oligonucleotides joined by T₄ DNA ligase. The mutant duplexes were produced by replacing two oligonucleotides of the native sequence by two that contained the desired codon changes (underlined). Four silent G/C to A/T changes were introduced where indicated by the dotted boxes. Duplex I specifies the native amino acid sequence while the others have the amino acid substitutions indicated. After K.-M. Lo et al.[34]

first for the separation of the DMTr product from the 5'-hydroxyl failed sequences, and in the last step, for the final purification of the fully deprotected oligonucleotide.[38] Because the first separation exploiting the strongly lipophilic DMTr handle is a trivial one, it can be achieved preparatively on a Sep-Pak by using stepwise elution. The same principle is applied to the separation of phosphorylated oligonucleotide from ATP in a kinase reaction mixture.

Procedures

Chemical Synthesis. All oligonucleotides were synthesized in 15-ml sintered glass funnels (10–15 μm pore size) by the solid-phase "phosphite-triester" method[38] using 10 molar equiv of the phosphoramidites for each condensation. After completion of synthesis, the methoxy and the N-protecting groups were removed, and ammonia was evaporated in the presence of 1 M TEAB (0.5 ml). (Since ammonia is very volatile and the DMTr group is very labile under slightly acidic conditions, it is very important to maintain a basic pH during the evaporation.) The crude DMTr oligonucleotide(s) was finally dissolved in 10 ml of 25 mM TEAB and then purified on a μC$_{18}$ Sep-Pak cartridge.

Eluents Used in Sep-Pak Purification. Triethylammonium bicarbonate (TEAB) buffer was prepared as a 1 M solution by bubbling carbon dioxide from dry ice through glass tubing with a fritted end into a suspension of 140 ml of redistilled triethylamine in 700 ml water kept in an ice bath, until (over 4 hr) the pH dropped to below 8. It was then made up to 1 liter and stored in a dark bottle at 4°. For long-term storage of over a few months, the pH should be checked and readjusted by bubbling CO_2.

Four eluents were prepared: 25 mM TEAB as a loading buffer and for washing off orthophosphate, pyrophosphate, urea, and most salts; 5% acetonitrile in 25 mM TEAB for eluting nucleoside triphosphates; 10% acetonitrile in 25 mM TEAB for eluting failed sequences; and 30% acetonitrile in 100 mM TEAB, which was conveniently used for eluting single and double-stranded DNA and DMTr oligonucleotide.

Purification of 5'-DMTr Oligonucleotides from Failed Sequences on Sep-Pak. The Sep-Pak was connected to a 10-ml syringe with a luer tip and before use was washed successively with 10 ml of acetonitrile, 5 ml of 30% acetonitrile in 100 mM TEAB, and finally, 10 ml of 25 mM TEAB. The 10 ml of TEAB solution containing the crude DMTr oligonucleotide(s) was loaded onto the Sep-Pak (quick drops, in about 30 sec). (The total capacity of a Sep-Pak is about 100 A_{260} units. If the solution

---

[38] M. H. Caruthers, "Chemical and Enzymatic Synthesis of Gene Fragments" (H. G. Gasser and A. Lang, eds.). Verlag Chemie, Weinheim, 1982.

contains more than 50 A_{260} units, a repeat passage is necessary.) The failed sequences containing terminal 5'-hydroxyl group were eluted with 10–15 ml of 10% acetonitrile in 25 mM TEAB, and the DMTr oligonucleotide(s) was eluted with 5 ml of 30% acetonitrile in 100 mM TEAB. A simple "spot test" can be used for quick checking. The 10 and 30% acetonitrile fractions were concentrated separately and a rough estimation of their A_{260} units was made. One microliter of each (preferably this should contain about 0.5 A_{260} unit of the 10% acetonitrile fraction and only 0.05 A_{260} unit of the 30% acetonitrile fraction) was spotted onto a silica tlc plate with fluorescent indicator. Then 2 μl of a solution of 0.1 M p-toluenesulfonic acid in acetonitrile was placed on top of each spot. Upon heating, the spot corresponding to the 30% fraction appeared bright orange, indicating the presence of the DMTr cation, whereas the spot from the 10% fraction, though at a 10-fold higher concentration, remained colorless.

In a typical synthesis of a dodecamer using 3 μmol of the starting nucleoside on the solid support, there are about 200 A_{260} units in the crude DMTr oligonucleotide mixture before Sep-Pak. When this total mixture was applied to Sep-Pak, about 15 A_{260} units of UV absorbing impurities were recovered unabsorbed, about 90 A_{260} units of non-DMT-containing oligonucleotides (failed sequences) are eluted in the 10% acetonitrile/25 mM TEAB fraction, and about 90 A_{260} units of DMTr-positive material are recovered in the 30% acetonitrile/100 mM TEAB fraction.

Further Purification of Synthetic Oligonucleotides by Gel Electrophoresis. The 30% acetonitrile eluate was evaporated to dryness and 1 ml of 80% acetic acid was added. After 15 min at room temperature, the acetic acid was evaporated. Water (0.5 ml) was added back and reevaporated. This was repeated twice. (Acetic acid and water form an azeotrope with a 3:97 composition that boils at 76°, so addition of water helps to remove the last trace of acid.) The fully deprotected product was then further purified by electrophoresis on a preparative polyacrylamide gel run under denaturing conditions (Fig. 4 in ref. 34). The gel was then transferred onto a TLC plate with fluorescent indicator and visualized under UV. The desired band was excised, crushed, and extracted with 6 volumes of 1 M TEAB at 37° with shaking for at least 4 hr. The gel extract plus washings were filtered through a sintered glass funnel (medium), and the filtrate was diluted about 5 times with water. This solution was then passed through Sep-Pak. The urea and salts were removed by washing with 10 ml of 25 mM TEAB. The oligonucleotide was eluted with 3 ml of 30% acetonitrile in 100 mM TEAB. The eluate was evaporated, and 100 μl of water was added to redissolve the residue. The aqueous solution was lyophilized twice. The purified, salt-free oligonucleotide was stored in a buffer containing 10 mM Tris–HCl, pH 8, and 1 mM EDTA, and were used directly for various enzymatic reactions.

The procedure described above is equally applicable to recover double-stranded DNA from polyacrylamide gels. If Maxam and Gilbert buffer[23] is used to extract the DNA from the gel slice, the gel extract is diluted twice with water before loading. The sodium dodecyl sulfate can be first removed by washing with 5 ml of 5% acetonitrile in 25 mM TEAB.

Since oligonucleotides cannot be efficiently precipitated by ethanol, the Sep-Pak method is very useful for desalting and concentrating eluates of oligonucleotides and DNA duplexes from polyacrylamide gel slices. We have tested restriction fragments of up to 1600 base pairs (HinfI restriction digest of pBR322) and found that recovery of double-stranded DNA is often better than single-stranded. Thus, Sep-Pak is also very useful for purification of DNA from nucleoside triphosphate after tailing with terminal deoxynucleotidyl transferase.

Finally, if the DNA comigrates with xylene cyanol, the xylene cyanol can be selectively retained on the Sep-Pak. In such case, the DNA is recovered by eluting with 4 ml of 16.5% acetonitrile in 100 mM TEAB. (It takes over 18% acetonitrile to elute xylene cyanol off the Sep-Pak.)

[32]P-Labeled 5'-Phosphorylation of Oligonucleotide, Purification and Characterization. In order to ensure efficient ligation, the 5' ends of the oligonucleotides have to be quantitatively phosphorylated and labeled to the same specific activity. By having the 5' ends uniformly labeled, stoichiometric amounts (measured by cpm) of each oligonucleotide can be added to the ligation mixture; and quantitative phosphorylation ensures that there are no oligomers with free 5'-hydroxyl ends which will compete for the annealing sites but cannot participate in the joining reactions. 5'-Phosphorylation was performed as described in the preceding section[35] using 2 nmol of oligonucleotide in 100 μl with a 2-fold molar excess of [γ-^{32}P]ATP (4 mCi/mol). For purification, the kinase reaction mixture was passed through Sep-Pak, the P_i and ATP were eluted with 10 ml of 5% acetonitrile in 25 mM TEAB, and the 5'-phosphorylated oligonucleotide was then eluted with 2 ml of 30% acetonitrile in 100 mM TEAB. The purity of each ^{32}P-labeled oligonucleotide was checked by gel electrophoresis and by sequence analysis (two-dimensional fingerprinting).[39]

DNA Ligase-Catalyzed Joining of Oligonucleotides to Form DNA Duplexes. All oligonucleotides, except the two that contain the external 5' termini of the duplex, were preparatively phosphorylated at the 5' ends to a specific activity of 4 Ci/mmol, and the products were purified by using Sep-Pak. T_4 DNA ligase-catalyzed joining of oligonucleotides was per-

---

[39] M. Silberklang, A. M. Gillum, and U. L. RajBhandary, this series, Vol. 59, p. 58.

formed in two blocks: oligonucleotides 1–4 and 5–10. Before ligation, the requisite oligonucleotides, in a volume of about 20 μl of Tris–HCl buffer, pH 7.6, were heated at 70° for 2 min and then slowly cooled in the water bath to room temperature in over an hour. Ligation was performed in 25 μl containing the oligonucleotides (40 μM of each), Tris–HCl, pH 7.6 (50 mM), MgCl$_2$ (10 mM), DTT (10 mM), ATP (400 μM), and T$_4$ ligase (25 units), at 7° for 3 hr. The corresponding blocks (1–4 and 5–10) were then mixed together without purification. After the addition of 1 μl DTT (0.5 M), 1 μl ATP (5 mM), and ligase (30 units), the reaction was allowed to proceed at room temperature for 3 more hr. The products were purified on a 12% polyacrylamide gel run under denaturing conditions, recovered from gel extract, and purified by Sep-Pak. The overall yield was 35–55%.

Characterization of Synthetic DNA Duplexes. Since each oligonucleotide had been individually sequenced, it sufficed to characterize the synthetic duplexes by their electrophoretic mobility on gels and by nearest-neighbor analysis[40] to confirm the accuracy of joining.

Restriction Fragments from the BO Expression Plasmid

The SphI–AvaII Fragment (Fig. 6). The plasmid pLBB (1 mg) was double-digested with *Sph*I (300 units) and *Bam*HI (500 units) in 1 ml for 5 hr. Separation on low melting point agarose gel yielded 83 μg of the *Sph*I–*Bam*HI fragment. This fragment (20 μg) was further cut with *Ava*II (10 units) in 150 μl for 4 hr. A portion of the *Ava*II digest was first dephosphorylated with calf intestinal phosphatase and then 5'-^{32}P phosphorylated with T$_4$ polynucleotide kinase. The labeled material was mixed with the remainder and run on an 8% polyacrylamide gel under native conditions. The desired band was extracted with 1 M TEAB and the DNA precipitated with ethanol twice.

The Large SphI–XhoI Vector Fragment (Fig. 6). This was obtained by digestion of pLBB-203L (400 μg) with *Sph*I and *Xho*I (200 units each) in 500 μl for 5 hr. Isolation was by electrophoresis on low melting point agarose gel.

Reconstruction of the Plasmid and Transformation of E. coli. The large *Sph*I–*Xho*I vector fragment (0.06 pmol), the *Sph*I–*Ava*II fragment (0.3 pmol), and the synthetic duplex (0.6 pmol) with 5'-hydroxyl termini were incubated with 2 units of T$_4$ DNA ligase in 10 μl of standard ligation buffer at 15° overnight. After inactivation of the ligase, the DNA was digested with *Xho*I (1 unit) at 37° for 1 hr to reduce the background due to the parental plasmid. Aliquots of the incubation mixture were used to

---

[40] T. Sekiya, P. Besmer, T. Takeya, and H. G. Khorana, *J. Biol. Chem.* **251**, 634 (1976).

transform HB101 cells.[27] Colonies were picked and DNA was prepared by the alkaline lysis method.[16] The DNA was further purified by Elutip-d columns before restriction analysis. Colonies that yielded DNA which was cut by *Sph*I but not by *Xho*I were selected.

Subcloning of EcoRI–BamHI Fragment (Fig. 6) into M13mp8. The *Eco*RI–*Bam*HI fragment containing the *bop* gene from the selected colonies was subcloned into M13mp8. JM103 cells were made competent with calcium chloride[17] before transfection. White plaques were picked from minimal plates and single-stranded DNA was prepared.

DNA Sequencing. Phage DNA was sequenced by the dideoxy method[28] using as a primer a synthetic oligonucleotide complementary to amino acids 209–215 of the *bop* gene. All clones were sequenced through the synthetic section to the *Sph*I site to ensure the desired constructions had been obtained.

[17] Oligonucleotide-Directed Site-Specific Mutagenesis of the *lac* Permease of *Escherichia coli*

By Hemanta K. Sarkar, Paul V. Viitanen, Etana Padan, William R. Trumble, Mohindar S. Poonian, Warren McComas, and H. Ronald Kaback

In order to determine the role of amino acid residues or sequences in the function of a particular protein or enzyme, it is helpful to be able to make specific alterations in the protein in a highly precise fashion. With the recent advent of site-directed mutagenesis, this goal can be realized,[1-9] and it is the purpose of this article to describe an application of the technique to the *lac* permease of *Escherichia coli*.

[1] G. Dalbadie-McFarland, L. W. Cohen, A. D. Riggs, C. Morin, K. Itakura, and J. H. Richards, *Proc. Natl. Acad. Sci. U.S.A.* **79,** 6409 (1982).

[2] K. Norris, F. Norris, L. Christiansen, and N. Fiil, *Nucleic Acids Res.* **11,** 5103 (1983).

[3] M. Schold, A. Colombero, A. A. Reyes, and R. B. Wallace, *DNA* **3,** 469 (1984).

[4] D. Shortle, D. DiMaio, and D. Nathans, *Annu. Rev. Genet.* **15,** 265 (1981).

[5] M. Smith, *TIBS* **7,** 440 (1982).

[6] M. Smith and S. Gillam, *in* "Genetic Engineering" (J. K. Setlow and A. Hollaender, eds.), Vol. 3, p. 1. Plenum, New York, 1981.

[7] Y. Morinaga, T. Franceschini, S. Inouye, and M. Inouye, *Bio/Technology* **2,** 636 (1984).

[8] M. J. Zoller and M. Smith, this series, Vol. 100, 468.

[9] M. J. Zoller and M. Smith, *DNA* **3,** 479 (1984).

The *lac* permease is an intrinsic membrane protein, encoded by the *lacY* gene, that catalyzes symport (cotransport) of β-galactosides with protons (cf. Kaback[10,11] and Overath and Wright[12] for recent reviews). The *lacY* gene has been cloned[13] and sequenced,[14] and the permease has been purified to homogeneity in a completely functional state.[15] Furthermore, a putative secondary structure model has been proposed based on the circular dichroic spectrum of the purified protein and on the sequential hydropathic character of the amino acid sequence.[16] According to the model, the protein consists of 12 hydrophobic segments in α-helical conformation that traverse the membrane in a zigzag fashion connected by shorter, hydrophilic segments. Chemical modification studies with *N*-ethylmaleimide (NEM) have identified a sulfhydryl group (Cys-148) that is protected by substrate,[17,18] and studies with diethylpyrocarbonate (DEPC) and rose bengal suggest that histidyl residue(s) in the permease may play an important role in the symport mechanism.[19–21] Although chemical modification of specific amino acid residues in a protein can provide important information, there are obvious drawbacks to this approach. For this reason, we have utilized oligonucleotide-directed, site-specific mutagenesis to study the structure and function of the *lac* permease.

As a brief overview, the methodology utilized follows that described by Zoller and Smith[8] with certain modifications (Fig. 1). The *lacY* gene is cloned from plasmid pGM21[22] into the replicative form (RF) of the single-stranded (ss) DNA phage, M13, which is used as a template (steps 1–3). Accordingly, a synthetic mutagenic primer with a mismatched base(s) is annealed with ssDNA containing the *lacY* gene (step 4), and the primer is

[10] H. R. Kaback, *J. Membr. Biol.* **76**, 95 (1983).

[11] H. R. Kaback, *in* "Physiology of Membrane Disorders" (T. E. Andreoli, J. F. Hoffman, D. D. Fanestil, and S. G. Schultz, eds.), p. 387. Plenum, New York, 1985.

[12] P. Overath and J. K. Wright, *TIBS* **8**, 404 (1983).

[13] R. M. Teather, B. Müller-Hill, U. Abrutsch, G. Aichele, and P. Overath, *Mol. Gen. Genet.* **159**, 239 (1978).

[14] D. E. Büchel, B. Gronenborn, and B. Müller-Hill, *Nature (London)* **283**, 541 (1980).

[15] P. V. Viitanen, M. J. Newman, D. L. Foster, T. H. Wilson, and H. R. Kaback, this volume [32].

[16] D. L. Foster, M. Boublik, and H. R. Kaback, *J. Biol. Chem.* **258**, 31 (1983).

[17] C. F. Fox and E. P. Kennedy, *Proc. Natl. Acad. Sci. U.S.A.* **54**, 891 (1965).

[18] K. Beyreuther, B. Bieseler, R. Ehring, and B. Müller-Hill, *in* "Methods in Protein Sequence Analysis" (M. Elzina, ed.), p. 139. Humana Press, Clifton, New Jersey, 1981.

[19] E. Padan, L. Patel, and H. R. Kaback, *Proc. Natl. Acad. Sci. U.S.A.* **76**, 6221 (1979).

[20] M. L. Garcia, L. Patel, E. Padan, and H. R. Kaback, *Biochemistry* **21**, 5800 (1982).

[21] L. Patel, M. L. Garcia, and H. R. Kaback, *Biochemistry* **21**, 5805 (1982).

[22] R. M. Teather, J. Bramhall, I. Riede, J. K. Wright, M. Fürst, G. Aichele, U. Wilhelm, and P. Overath, *Eur. J. Biochem.* **108**, 223 (1980).

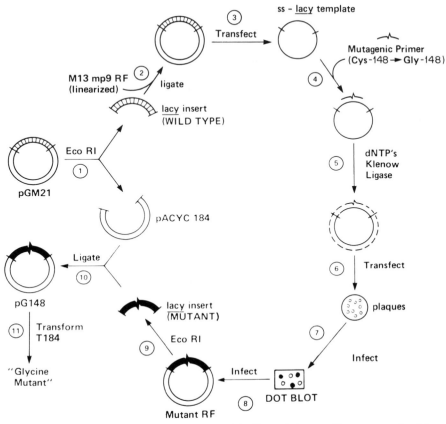

FIG. 1. A simplified scheme for oligonucleotide-directed site-specific mutagenesis of the *lac* permease of *Escherichia coli*. The scheme depicts the strategy employed to convert Cys-148 to Gly-148 in the primary sequence of *lac* permease as an example. Details of each step and additional modifications employed are described in the text.

extended and ligated *in vitro* to yield a closed heteroduplex in which one strand of the DNA contains the mutation (step 5). After transfection of an appropriate host, ss phage DNA from individual plaques is screened for mutated *lacY* DNA by dot-blot hybridization with the mutagenic primer, and positive phage ssDNA is sequenced to verify the mutation (steps 6 and 7). Finally, mutated *lacY* is recloned into the original vector which is used to transform a *lacY⁻* strain of *E. coli* (steps 8–11). The cells are then tested for lactose transport activity and expression of the *lac Y* gene product.

M13 Host Strains

JM101[23]: Δ*lacpro*, *supE*, *thi*/F' *traD*36, *proAB*, *lacI^qZ*Δ*M*15.

JM103[23]: Δ*lacpro*, *supE*, *thi*, *strA*, *sbcB*15, *endA*, *hspR*4/F' *traD*36, *proAB*, *lacI^qZ*Δ*M*15.

JM105[24]: Δ*lacpro*, *thi*, *strA*, *hsdR*4, *sbcB* 15 *endA*/F'*traD*36, *proAB*, *lacI^qZ*Δ*M*15.

A few additional remarks regarding these M13 strains are notable. *E. coli* JM101 and JM103 are r_k^+. On the other hand, both K12 and P1 restriction systems are expressed in JM103, while JM101 does not contain the P1 restriction system[24] and therefore should be used in preference to JM103. *E. coli* JM105 is r_k^- and does not contain a *supE* amber suppressor. Thus, it will not support propagation of amber phage (e.g., M13mp7, M13mp8, and M13mp9).

It is particularly important to verify host strains for genetic markers before use. For example, in addition to its inability to support propagation of double amber phage, *E. coli* JM105 should be streptomycin resistant. However, several putative JM105 strains obtained from commercial sources do not exhibit both of these properties.

Other Bacterial Strains

T206[22]: $lacI^+O^+Z^-Y^-(A^+)$, *rpsL*, *met*⁻, *thr*⁻, *recA*, *hsdM*, *hsd R*/F' *lac* $I^qO^+Z^{U118}$ (Y^+A^+). This strain harbors the plasmid pGM21 which is *lac* $\Delta(I)O^+P^+\Delta(Z)Y^+\Delta(A)$, *tet*^r.

T184[22]: This strain, which is devoid of plasmid pGM21, was regenerated from *E. coli* T206 by growth for about 50 generations in minimal medium without antibiotics.[22,25]

HB101[26]: *hsdS*20 (r_B^-, m_B^-), *recA*13, *ara*-14, *proA*2, *lacY*1, *galK*2, *rpsL*20 (Sm^r), *xyl*-5, *mtl*-1, *supE*44, λ⁻/F⁻.

CS71[27]: *gltC*, *metB*, *lacY*1/Hfr.

M13 Vectors

mp9: the ss phage DNA has 7599 nucleotides and contains two amber mutations, the positions of which are unknown.[28] Due to the presence of

[23] J. Messing, this series, Vol. 101, 23.
[24] J. Felton, *BioTechniques* **March–April,** 42 (1983).
[25] W. R. Trumble, P. V. Viitanen, H. K. Sarkar, M. S. Poonian, and H. R. Kaback, *Biochem. Biophys. Res. Commun.* **119,** 860 (1984).
[26] H. W. Boyer and D. Roulland-Dussoix, *J. Mol. Biol.* **41,** 459 (1969).
[27] D. Zilberstein, V. Agmon, S. Schuldiner, and E. Padan, *J. Biol. Chem.* **257,** 3687 (1982).
[28] One CAG codon in gene I and one in gene II have been converted into TAG (amber) codon ("M13 Cloning/Dideoxy Sequencing Manual," BRL).

the amber codons, the vector can be propagated in host strains containing amber suppressors (e.g., JM101 or JM103), but not in wild-type strains (e.g., JM105). The RF form of the phage DNA may be obtained commercially from BRL or New England Biolabs.

mp19: a wild-type vector containing 7249 bp[28a] that can be propagated in JM101, JM103, or JM105. The RF form of the phage DNA may be obtained from P-L Biochemicals.

Plasmid Vector

pACYC184[29]: carries an *Eco*RI site in the *cap*[r] gene.

Synthetic Mutagenic Oligonucleotides

For a comprehensive discussion on the design of mutagenic primers for site-directed mutagenesis, the reader is referred to Zoller and Smith[8] and two recent papers by Hirose *et al.*[30] and Kramer *et al.*[31] The deoxyoligonucleotides listed in Table I represent examples of *lacY* mutagenic primers used thus far to generate given alterations. In most instances, the primers contain one mismatch, but in one instance, two mismatches were incorporated so that two closely approximated His codons were simultaneously altered. In another case, the primer was designed to cause deletion of the last putative α-helical segment in the secondary-structure model of the permease. The deoxyoligonucleotides, which vary in length from 18 to 30 bases, were synthesized by the phosphoramidate method[32,33] on a Syntek automatic synthesizer. Each oligonucleotide was then purified either by preparative gel electrophoresis on 20% polyacrylamide under denaturing conditions or by C_{18} reverse-phase, high-performance liquid chromatography.

Synthetic Commercial Oligonucleotides

In order to increase the efficiency of closed circular heteroduplex synthesis, either one or both of the following commercially available syn-

---

[28a] C. Yanisch-Perron, J. Vieira, and J. Messing, *Gene* **33,** 103 (1985).

[29] A. C. Y. Chang and S. N. Cohen, *J. Bacteriol.* **134,** 1141 (1978).

[30] S. Hirose, K. Takeuchi, H. Hori, T. Hirose, S. Inayama, and Y. Suzuki, *Proc. Natl. Acad. Sci. U.S.A.* **81,** 1394 (1984).

[31] W. Kramer, V. Drutsa, H.-W. Jansen, B. Kramer, M. Pflugfelder, and H.-J. Fritz, *Nucleic Acids Res.* **12,** 9441 (1984).

[32] S. L. Beaucage and M. H. Caruthers, *Tetrahedron Lett.* **22,** 1859 (1981).

[33] M. D. Matteucci and M. H. Caruthers, *J. Am. Chem. Soc.* **103,** 3185 (1981).

TABLE I
SYNTHETIC MUTAGENIC PRIMERS FOR SITE-DIRECTED MUTAGENESIS OF *lac* PERMEASE

Synthetic mutagenic (mismatched base underlined)	Position and nature of change in amino acid sequence of *lac* permease
(i) 5'-CCAGCCAACACCGCCAAACAT-3'[a]	Cys-148 → Gly-148[25]
(ii) 5'-CAGCCAACAGAGCCAAACAT-3'[a]	Cys-148 → Ser-148
(iii) 5'-GGCACTCAGCGCCCAGCC-3'	Cys-154 → Ser-154
(iv) 5'-GGCACCCAGCGCCCAGCC-3'	Cys-154 → Gly-154
(v) 5'-GCGGTTCGAATAATAGCG-3'[b]	Gln-60 → Glu-60
(vi) 5'-TGATACGGTTGATGTCACGTAGC-3'[c]	His-35,39 → Arg-35,39
(vii) 5'-AACATACGCAGCGTTTTC-3'[c]	His-322 → Arg-322
(viii) 5'-GCCGAACGGTTGGCACCT-3'[c]	His-205 → Arg-205
(ix) 5'-CGCAGCAGGGAAAGCTTGCCCGCCAGTACA-3'[d]	Δ(Met-372 → Pro-405)

[a] Replaces a sulfhydryl group that is alkylated by NEM and protected by substrate.
[b] Creates a negative charge in putative transmembrane α-helix II and an additional *Taq*I site in *lacY*; also used as a sequencing primer for His-35,39 → Arg-35,39 mutant.
[c] Evaluates the role of His residues in lactose : proton symport.
[d] Deletes putative transmembrane α-helix XII and creates a unique *Hin*dIII site in *lacY*.

thetic primers complementary to the 3' and 5' ends of the *Eco*RI site of the M13 vector were utilized concomitant with the mutagenic primer: M13 15-base primer: 5'-AGTCACGACGTTGTA-3' (BRL); M13 hybridization probe primer: 5'-CACAATTCCACACAAC-3' (New England BioLabs).

Phosphorylation of Oligonucleotides

Prior to use, both the mutagenic and nonmutagenic primers were phosphorylated. Thus, 200 pmol of mutagenic primer, M13 15-base primer, or M13 hybridization probe primer was phosphorylated by incubation in 100 mM Tris–HCl (pH 8.0)/10 mM MgCl$_2$/5 mM dithiothreitol (DTT)/0.1 mM ATP containing 4.5 U of T4 polynucleotide kinase (Amersham/Searle) (30 μl, final volume) for 45 min at 37°. The reaction was stopped by incubation at 65° for 10 min and the kinased oligonucleotides were used directly for the mutagenesis reaction as described.[8]

For use as probes during dot-blot hydridization, 25 pmol of synthetic mutagenic primer was phosphorylated as described above, except that 0.1 mM ATP was replaced with 25 pmol of [γ-^{32}P]ATP (Amersham/Searle; 5000 Ci/mmol, 10 μCi/μl). Phosphorylated oligonucleotides were separated from [γ-^{32}P]ATP with Elutip-d columns (Schleicher and Schuell) using the manufacturer's protocol with about 50% yield.

Steps 1 and 2 (Fig. 1): Cloning *lacY* into M13 RF DNA

The 2.3 kbp DNA fragment containing *lacY* was isolated from the plasmid pGM21[22] by restricting the plasmid with *Eco*RI (BRL) in 100 mM Tris–HCl (pH 7.2)/5 mM MgCl$_2$/50 mM NaCl/2 mM mercaptoethanol at 37° for 1 hr. The fragments were then separated by preparative gel electrophoresis on agarose (Sea Plaque Agarose, FMC Corp.), and the 2.3 kbp piece was isolated from the gel by using an Elutip-d column following the procedure described by Schmitt and Cohen.[34] The 2.3 kbp fragment (~0.25 μg DNA) was then cloned into RF DNA from M13mp9 or M13mp19 (~0.15 μg DNA) that had been linearized previously with *Eco*RI. Ligation was accomplished by incubating the two DNA fragments with 2.5 U T4 ligase (BRL) in 50 mM Tris–HCl (pH 7.5)/10 mM MgCl$_2$/10 mM DTT/1 mM spermidine/1 mM ATP/1 mg/ml bovine serum albumin for 18–20 hr at 15°. The larger DNA fragment resulting from restriction of pGM21 with *Eco*RI (linearized pACYC184) was also purified by Elutip-d column chromatography and treated with bacterial alkaline phosphatase (BRL) according to manufacturer's protocol. The alkaline phosphatase was removed by phenol extraction; the DNA was purified by preparative agarose gel electrophoresis and Elutip-d column chromatography and put aside for subsequent use as a vector for cloning mutant *lacY* genes (cf. below).

A portion of the ligation mixture described above was used to transfect either JM101 or JM103. Phage were isolated from individual plaques,[8] spotted on nitrocellulose (BA85 or PH79; Schleicher and Schuell), and the presence of the antisense strand of *lacY* was detected by dot-blot analysis[35-37] using a 5'-^{32}P-labeled synthetic oligonucleotide complementary to the antisense strand of *lacY*. One recombinant was chosen and used as a source of ss template DNA for mutagenesis.

Step 3 (Fig. 1): Preparation of Single-Stranded Template DNA

Preparation of recombinant DNA in ss form consists essentially of two steps: (1) concentrating phage by polyethylene glycol (PEG) precipitation, and (2) extracting with phenol to remove viral coat proteins.

[34] J. J. Schmitt and B. N. Cohen, *Anal. Biochem.* **133**, 462 (1983).
[35] S. Gillam, K. Waterman, and M. Smith, *Nucleic Acids Res.* **2**, 625 (1975).
[36] R. B. Wallace, M. Schold, M. J. Johnson, P. Dembek, and K. Itakura, *Nucleic Acids Res.* **9**, 3647 (1981).
[37] G. Winter, A. R. Fersht, A. J. Wilkinson, M. Zoller, and M. Smith, *Nature (London)* **299**, 756 (1982).

Clear supernatant (~1.2 ml) containing phage from an infected cell culture was mixed with 27% PEG-8000/3.3 M NaCl (200 μl), agitated vigorously on a vortex mixer, and incubated at room temperature for 15 min. After centrifugation at room temperature in a microfuge (Beckman, Microfuge 12), the supernatant was discarded, and the pellet was resuspended in 100 μl of 10 mM Tris–HCl (pH 8.0)/0.1 mM ethylenediaminetetraacetic acid (EDTA). The sample was then extracted sequentially with equal volumes of phenol, chloroform, and then with two volumes of ether. The DNA was precipitated with 2.5 volumes of cold ethanol ($-20°$) after adding 0.1 volume of 3 M NaOAc (pH 4.8), incubated in liquid nitrogen for 30 min, and collected by centrifugation for 15 min at 4° in a microfuge (Eppendorf 5414, Brinkmann). The pellet was washed once with cold ethanol ($-20°$) avoiding resuspension, dried in a Speed-Vac (Savant), and dissolved finally in 10 mM Tris–HCl (pH 8.0)/0.1 mM EDTA.

Preliminary Test for Specific Hybridization of Mutagenic Primers

In most cases, mutagenic primers were tested as sequencing primers prior to use in heteroduplex synthesis in order to ensure proper annealing of the primer to *lacY* at the target site. Thus, the sequence of the ss *lacY* DNA template 50 to 100 bases upstream from the primer binding site was determined by dideoxy chain termination.[38,39]

Steps 4 and 5 (Fig. 1): Primer Annealing and Extension–Ligation

Primer annealing and extension–ligation were performed essentially as described,[8] except that annealing was carried out using a 2- to 3-fold molar excess of primer over the ss template rather than a 10- to 30-fold molar excess. In our hands, a 2- to 3-fold molar excess is preferable, as it is more restrictive and avoids false priming, as judged by the preliminary sequencing test for specific targeting.

In a typical reaction, ss M13 recombinant DNA bearing *lacY* (0.5 pmol) was mixed with 1.15 pmol of 5′-P-mutagenic primer and 1.15 pmol of 5′-P-M13 hybridization probe primer (New England BioLabs). The mixture was dried and resuspended in 20 μl of 20 mM Tris–HCl (pH 7.5)/ 10 mM MgCl$_2$/50 mM NaCl/1.0 mM DTT. Annealing was achieved by heating the mixture at 90° for 5 min, followed by cooling to room tempera-

---

[38] F. Sanger, S. Nicklen, and A. R. Coulson, *Proc. Natl. Acad. Sci. U.S.A.* **74**, 5463 (1977).
[39] F. Sanger and A. R. Coulson, *FEBS Lett.* **87**, 107 (1978).

ture over a period of 60 min.[40] Twenty microliters of 20 mM Tris–HCl (pH 7.5)/10 mM MgCl$_2$/10 mM DTT/1 mM dCTP/1 mM dGTP/1 mM dTTP/5 μM dATP/1 mM ATP/3.49 μM [α-^{32}P]dATP (410 Ci/mmol, 10 μCi/μl)/6 U T4 DNA ligase (BRL) was added to the annealed DNA, followed by addition of 6 U of Klenow cocktail.[41] The sample was incubated at room temperature, and after 5 min, 2 μl of 10 mM dATP was added and the mixture was incubated at 15° for 20 hr.

Enrichment of Closed Circular (cc) Heteroduplex DNA

Unincorporated [α-^{32}P]dATP was removed from the DNA synthesized *in vitro* by precipitating the DNA with 6.5% PEG-8000/0.8 M NaCl (final concentration).[8] The precipitate was resuspended in 10 mM Tris–HCl (pH 8.0)/0.1 mM EDTA, and NaOH was added to 0.2 M final concentration. The sample was incubated at room temperature for 5 min and applied to an alkaline sucrose density gradient (5 to 20% sucrose containing 1.0 M NaCl/0.2 M NaOH/2.0 mM EDTA).[8] Centrifugation was performed in an SW 56 rotor at 37,000 rpm for 2.5 hr at 4°. Aliquots ($\sim$150 μl) were collected by puncturing the tube at the bottom, and each fraction was assayed for Cherenkov radiation. Fractions containing cc heteroduplex DNA (faster migrating peak) were pooled and neutralized with a solution of 0.5 M HCl prepared in 10 mM Tris–HCl (pH 8.0). The material was then dialyzed for 48 hr at 4° against 6 liters of 2 mM Tris–HCl (pH 8.0)/0.1 mM EDTA with three changes. Extensive dialysis of the ccDNA after alkaline sucrose density gradient centrifugation is essential, as undialyzed samples exhibit extremely low efficiency of transfection. As a more rapid alternative to dialysis, ccDNA from the sucrose density gradient can be precipitated with ethanol after adding carrier tRNA and dissolved in 10 mM Tris–HCl (pH 8.0)/0.1 mM EDTA.[42]

---

[40] In the case of the deletion mutant generated with the synthetic 30-mer oligonucleotide (Table I, ix), the conditions for annealing were as follows: 1 pmol of wild-type ssDNA template (M13mp19 · *lacY*) was mixed with 10 pmol of each phosphorylated primer (mutagenic 30-mer, M13 15-base primer, and M13 hybridization probe primer), and annealing was carried out in 20 mM Tris–HCl (pH 7.5)/10 mM MgCl$_2$/50 mM NaCl/1 mM DTT by placing the mixture at 80° for 5 min; subsequently, the sample was incubated at 50° for 5 min and finally at room temperature for 5 min. Extension–ligation was carried out as described in the text.

[41] Even though the polymerase reaction started from the desired site, premature termination occurred frequently. The problem was overcome by preparing a mixture of three polymerases as follows: Klenow (BRL) (5 U/μl) : Klenow (New England Biolabs) (5 U/μl) : Klenow (Boehringer Manheim) (5 U/μl)

[42] P. J. Carter, G. Winter, A. J. Wilkinson, and A. Fersht, *Cell* **38**, 835 (1984).

PEG-precipitated DNA was also extracted with acid-phenol in order to enrich for ccDNA.[43] Thus, DNA precipitated from the extension–ligation mixture with PEG/NaCl was resuspended in 10 mM Tris–HCl (pH 8.0)/0.1 mM EDTA, and 0.05 volumes of 1.0 M NaOAc (pH 4.0) and 0.05 volumes of 1.5 M NaCl were added. The mixture was extracted at least twice with equal volumes of phenol that was preequilibrated with 50 mM NaOAc (pH 4.0), followed by two extractions with 4 volumes of ether. After adding 0.1 volume of 3.0 M NaOAc (pH 4.8) and 5 μg of carrier tRNA, the DNA was precipitated with 2.5 volumes of cold ethanol. The precipitate was washed once with cold 80% ethanol and finally dissolved in 100 μl of 10 mM Tris–HCl (pH 8.0)/0.1 mM EDTA.

Step 6 (Fig. 1): Transfection with cc Heteroduplex DNA and Segregation of Phage Containing Mutant DNA

Aliquots of ccDNA prepared either by alkaline sucrose density gradient centrifugation or by acid-phenol extraction were used to transfect CaCl$_2$-treated *E. coli* JM101 or JM103.[8,23,44]

For segregating wild-type and mutant genomes of the heteroduplex, transfected cells were grown overnight at 37° in YT medium,[45] rather than plating out immediately, and ssDNA was isolated from the culture supernatants as described above. Since the ssDNA is a mixed population of wild-type and mutant ssDNAs, transfection of JM101 or JM103 followed by plating results simultaneously in mutant segregation and plaque purification (cf. Plaque Purification).

Nicking of cc Heteroduplex DNA by *Hin*dIII/Ethidium Bromide (EtBr) and Transfection with Nicked DNA

As discussed below, host stains often have the ability to recognize subtle alterations in the secondary structure of ccDNA and are able to correct such alterations, thereby oblating the mutation. Since *Hin*dIII makes single-stranded cuts in the presence of EtBr[46] and transfection can be accomplished with ssDNA, this step was found to be useful in increasing the efficiency of mutant recovery.

---

[43] M. Zasloff, G. D. Ginder, and G. Felsenfeld, *Nucleic Acids Res.* **5,** 1139 (1978).
[44] M. Dagert and S. D. Ehrlich, *Gene* **6,** 23 (1979).
[45] YT Medium: Bactotryptone, 8 g/liter; Bacto yeast extract, 5 g/liter; NaCl, 5 g/liter.
[46] M. Österlund, H. Luthman, S. V. Nilsson, and G. Magnusson, *Gene* **20,** 121 (1982).

Closed circular heteroduplex DNA synthesized *in vitro* (20 μl) was treated with 50 μg/ml of EtBr and digested with 3 U of *Hin*dIII (BRL) in BRL core buffer[47] for 2 hr at 37° in the dark (40 μl, total volume). The reaction was stopped by extraction with an equal volume of phenol, followed by extraction with equal volumes of chloroform and ether. Finally, the DNA was precipitated with cold ethanol after adding 5 μg of carrier tRNA and 0.1 volumes of 3 *M* NaOAc (pH 4.8) and dissolved in 20 μl of 10 m*M* Tris–HCl (pH 8.0)/0.1 m*M* EDTA.

Nicked heteroduplex DNA was denatured by incubation at 90° for 5 min, followed by rapid cooling in ice water. The mixture was then used directly to transfect CaCl$_2$-treated host cells, and samples were grown on plates containing 5-bromo-4-chloro-3-indolyl-β-D-galactoside (X-Gal; BRL) as indicator.[8,23]

Step 7 (Fig. 1): Screening for Mutant Phage by Dot-Blot Hybridization

Recombinant phage were screened for mutations by dot-blot analysis.[35–37] Plaque-purified phage were grown in 1.5 ml cultures and concentrated by precipitation with 27% PEG-8000/3.3 *M* NaCl as described above (cf. Preparation of Single-Stranded Template DNA) and resuspension in 100 μl of 10 m*M* Tris–HCl (pH 8.0)/0.1 m*M* EDTA. An aliquot (2 μl) was spotted on nitrocellulose (BA85, Schleicher and Schuell) and baked *in vacuo* at 80° for 2 hr. The material was then treated with 6 × SSC[48]/10 × Denhardt's solution[49]/0.2% sodium dodecyl sulfate (SDS) at room temperature for 1 hr. Subsequently, the filter was washed with 50 ml of 6 × SSC for 1 min, hybridized with 5′-^{32}P-labeled mutagenic primer (>10^6 cpm) in 6 × SSC/10 × Denhardt for 1 hr at room temperature, washed 3 times with 50 ml of 6 × SSC at room temperature for a total period of 10 min, blotted dry, and autoradiographed for 1 hr between two sheets of Saran wrap. The filter was then washed sequentially in preheated 6 × SSC for 5 min, and the temperature of each wash was increased 5–10° in a stepwise manner. After each wash, the filter was autoradiographed for 1 hr. At room temperature, the primer anneals to both mutant and wild-type DNAs, but by washing at higher temperatures, the primer can be selectively dissociated from wild-type DNA.[35–37]

---

[47] 10 × BRL core buffer: 500 m*M* Tris–HCl (pH 8.0)/100 m*M* MgCl$_2$/500 m*M* NaCl.

[48] 20 × SSC (stock solution): 3 *M* NaCl/0.3 *M* sodium citrate/0.01 *M* EDTA. Adjust pH to 7.2 with HCl.

[49] 100 × Denhardt's solution (stock solution): 2% bovine serum albumin/2% poly(vinylpyrrolidone)/2% Ficoll.

Screening of Mutants by Restriction Enzyme Digestion

In certain instances, introduction of a point mutation can create a new or unique restriction site(s), a possibility that can be determined with readily available computer programs (e.g., Genetic Software Associates). Where indicated (see Table I), RF DNA was prepared from mutants suspected of harboring a new restriction site(s), purified by Elutip-d chromatography, treated with the appropriate restriction enzyme according to the protocol provided by the supplier and analyzed by agarose gel electrophoresis.

Additional Screening of Mutants by Indicator Plates

E. coli K12 CS71 (Z^+Y^-) was infected with the mutant phage and plated on lactose/eosin methylene blue indicator plates. Formation of red plaques indicated the presence of functional *lac* permease.

Plaque Purification

Recombinant M13 phage (wild type and mutant) were plaque purified by isolating ssDNA from the phage and transfecting $CaCl_2$-treated cells. Since ssDNA has only about one-tenth the transfection efficiency of RF DNA (only about one ssDNA molecule in 50,000 leads to transfection),[23] plaque purification with ssDNA is achieved easily with very low probability of multiple transfection. Alternatively, plaque purification may be carried out by dilution of phage and plating. In either case, purified plaques were analyzed by dot-blot analysis.

DNA Sequencing

DNA sequencing of the *lacY* gene was performed by dideoxy chain termination[38,39] using recombinant M13 ssDNA as template. A 2.5-fold molar excess of a suitable synthetic primer (usually an 18-mer complementary to a region 60 to 150 bases downstream from the mutation in *lacY*) was annealed to the ssDNA template by placing the mixture in a water-bath at 90° for 5 min and cooling slowly for one hr at room temperature. Primer elongation and chain termination were carried out following the "M13 Cloning/Sequencing Manual" (BRL) with a slight modification. When sequencing GC-rich regions (e.g., His-322 → Arg and the helix XII deletion mutants), dideoxy sequencing with the Klenow fragment was performed at a higher temperature (37°). Electrophoresis was carried out at 30–35 mA in a BRL S-1030 gel electrophoresis apparatus using 8% polyacrylamide/8 M urea gels (0.4 mm thick).

Step 8 (Fig. 1): Preparation of Recombinant M13/*lacY* RF DNA

M13/*lacY* recombinant RF DNA was prepared by following a scaled-up version of the rapid, small-scale alkaline-lysis method.[50,51] Cultures of early log phase JM101 or JM103 in YT medium (35 ml) were infected with 35 μl of phage ($\sim 10^{11}$ PFU/ml) and growth was continued at 37° with vigorous shaking for 5 hr. Cells were collected in polypropylene tubes in a Sorvall SS-34 rotor by centrifugation at 7000 rpm for 7 min at 4°, and the supernatant was saved to prepare ssDNA. The pellet was resuspended in 4 ml of ice-cold 50 mM glucose/10 mM EDTA/25 mM Tris–HCl (pH 8.0) containing 4 mg/ml lysozyme and incubated at room temperature for about 5 min. Eight milliliters of 0.2 N NaOH/1.0% was added to the tube which was then covered with parafilm. The contents were mixed by inverting the tube rapidly five or six times, followed by incubation on ice for 10 min. Six milliliters of 5 M potassium acetate (pH 4.8)[51] was added and mixed as above. After incubation on ice for 15 min, the lysis mixture was cleared by centrifugation in a Sorvall SS-34 rotor at 15,000 rpm for 20 min at 4°. The cleared supernatant was then extracted with an equal volume of phenol : chloroform (1 : 1, v/v), and the DNA was precipitated by adding 2 volumes of ethanol (room temperature) and incubating at room temperature for 20 min. Precipitated DNA was collected by centrifugation in a Sorvall SS-34 rotor at 18,000 rpm for 30 min at 23°. The pellet was washed once with 20 ml of 70% ethanol (room temperature), dried in a vacuum desiccator, and dissolved in 1.5 ml of 10 mM Tris–HCl (pH 8.0)/0.1 mM EDTA containing 40 μg/ml DNase-free pancreatic RNase (Sigma).

Purification of Recombinant M13 RF DNA

When necessary, purification of recombinant M13 RF DNA was carried out by Elutip-d column chromatography. An Elutip-d column was rinsed according to the manufacturer's instructions with 2 ml of high-salt buffer [0.02 M Tris–HCl (pH 7.4)/1 mM EDTA/1 M NaCl] and equilibrated with 7 ml of low-salt buffer [0.02 M Tris–HCl (pH 7.4)/1 mM EDTA/0.2 M NaCl]. The RF DNA from the minipreparation was mixed with 5 ml of low-salt buffer and applied to the Elutip-d column at a steady rate of approximately 2 drops/sec. The column was then rinsed with 4 ml of low-salt buffer, and bound DNA was eluted with 0.4 ml of elution buffer that was prepared by mixing 2 ml of low-salt buffer with 3.5 ml of high-salt buffer. DNA was precipitated with ethanol and dissolved in 10 mM Tris–HCl (pH 8.0)/0.1 mM EDTA.

[50] H. C. Birnboim and J. Doly, *Nucleic Acids Res.* **7**, 1513 (1979).
[51] T. Maniatis, E. F. Fritsch, and J. Sambrook (eds.), "Molecular Cloning: A Laboratory Manual." Cold Spring Harbor Laboratory, Cold Spring Harbor, New York, 1982.

Steps 9–11 (Fig. 1): Cloning the Mutated *lacY* into pACYC184

The 2.3 kbp DNA fragment containing the mutant *lacY* gene was restricted from recombinant M13 RF DNA with *Eco*RI, purified by agarose gel electrophoresis,[34] and ligated into linearized, alkaline phosphatase-treated pACYC184 using a 3-fold molar excess of *lacY* to pACYC184. A portion of the ligation mixture was used to transform *E. coli* T184 as described.[52] Transformed T184 harboring recombinant plasmids carrying a mutated *lacY* were selected based on tetracycline resistance and sensitivity to chloramphenicol.

Orientation of the Mutant *lacY* Gene in Recombinant Plasmids

Clones of T184 harboring recombinant plasmids with a mutated *lacY* were grown overnight at 37° in LB medium containing 100 μg/ml streptomycin and 20 μg/ml tetracycline (2 ml cultures). Plasmid DNA was prepared from 1.5 ml of culture by alkaline lysis.[50,51] The orientation of the mutant *lacY* gene in the recombinant plasmid was determined by *Hinc*II restriction enzyme analysis. Plasmid DNA (10 μl) was digested with 4 U of *Hinc*II (BRL) in 25 mM Tris–HCl (pH 8.0)/50 mM NaCl/10 mM MgCl$_2$/1 mM DTT/bovine serum albumin (100 μg/ml). Digested DNA was electrophoresed on 0.8% agarose (SeaKem, FMC) gels containing 0.8 μg/ml EtBr in 40 mM Tris–acetate (pH 7.8)/5 mM NaOAc/10 mM EDTA. Plasmids containing *lacY* in the correct orientation were identified from the size of the restriction fragments relative to those obtained with pGM21.[22]

Transport Assays

For qualitative assays, *E. coli* HB101 (Z^+Y^-) were transformed with a given plasmid and grown on lactose/eosin methylene blue indicator plates. For quantitative assays, transformed *E. coli* T184 were assayed for transport of radioactive lactose by rapid filtration as described.[53] The [^{14}C]lactose transport activities of given mutants are listed in Table II.

Antibody Binding Assays

In order to quantitate the amount of permease present in the membrane of transformed cells, binding studies were performed with ^{125}I-labeled monoclonal antibodies or site-directed polyclonal antibodies and

[52] S. R. Kushner, *in* "Genetic Engineering" (H. W. Boyer and S. Nicosia, eds.), p. 17. Elsevier, Amsterdam, 1978.
[53] H. R. Kaback, this series, Vol. 22, 99.

TABLE II
TRANSPORT ACTIVITIES OF *E. coli* CELLS
HARBORING PLASMIDS WITH WILD-TYPE AND
MUTATED *LacY* INSERT[a]

Lac Permease	$\Delta\bar{\mu}_{H^+}$ Driven lactose transport (initial rate, %)
Wild type	100
Cys-148 → Gly-148[25]	25
Cys-148 → Ser-148	50–110
Cys-154 → Gly-154[53a]	<1
Cys-154 → Ser-154[53a]	10
Gln-60 → Glu-60	80–100
His-35,39 → Arg-35,39[53b]	80–100
His-322 → Arg-322 ⎫ [53b]	<1
His-205 → Arg-205 ⎭	
Δ(Met-372 → Pro-405)[b]	<1

[a] All the mutants have been verified by dideoxy sequence method of Sanger.[38,39]

[b] As determined by antibody binding studies, *lac* permease in this mutant is either not inserted into the membrane or is proteolyzed after synthesis. In any event, a very small amount of protein was detected in membrane preparations.

right-side-out or inside-out cytoplasmic membrane vesicles.[54,55] Alternatively, the amount of permease present was determined by immunoblotting.[54,55]

Increasing the Efficiency of Site-Specific Mutagenesis

Theoretically, transfection with cc heteroduplex DNA containing a single base mismatch should yield a population of M13 in which 50% of the phage contain DNA with the mutation. In practice, however, this is seldom the case. Rather, the efficiency at which mutant DNA is obtained is quite variable, and, in many instances, no mutants are detected.

[53a] D. R. Menick, K. Sarkar, M. S. Poonian, and H. R. Kaback, *Biochem. Biophys. Res. Commun.* **132,** 162 (1985).

[53b] E. Padan, H. K. Sarkar, P. V. Viitanen, M. S. Poonian, and H. R. Kaback, *Proc. Natl. Acad. Sci. U.S.A.* **82,** 6765 (1985).

[54] N. Carrasco, D. Herzlinger, and H. R. Kaback, this volume [33].

[55] D. Herzlinger, N. Carrasco, and H. R. Kaback, *Biochemistry* **24,** 221 (1985).

One or more of the following reasons might contribute to a low efficiency of mutant recovery. (1) Ambiguous primer annealing under nonstringent conditions resulting in a high frequency of mutants with large deletions and insertions.[56] We have also observed deletions in high frequency with nonstringent primer-annealing conditions, but the use of a primer : template molar ratio of about 2.5 seems to obviate this problem to a large extent. (2) Recombinant ssDNA isolated from M13 (i.e., the DNA used as template) is methylated at GATC sites, while the complementary strand containing the mutation which is synthesized *in vitro* is not. This difference may bias mismatch repair toward the wild-type strand *in vivo*.[57–60] In this context, it is noteworthy that the M13 genome in the M13mp vectors contains three GATC sites.[61] Furthermore, there are at least four GATC sites in the 2.3 kbp segment containing the *lacY* gene.[13] (3) Depending on the nature of the single base-pair mismatch (i.e., pyrimidine : pyrimidine, pyrimidine : purine, or purine : purine), the local secondary structure of the DNA may be distorted at the site of the mutation, and the distortion may be recognized and corrected *in vivo*. In a recent systematic study,[62] different types of base-pair mismatches were shown to be repaired with markedly different efficiencies.

Efforts to circumvent these problems and increase the efficiency of mutant recovery rely generally on construction of heteroduplexes in which a selective advantage is conveyed to the DNA strand containing the mutation. For example, Kramer *et al.*[57] were able to isolate mutants at a high frequency by using gapped duplex DNA in which one strand (M13 ssDNA containing the template) was isolated from a *dam⁻* host bacterium and the other strand (linearized M13 ssDNA without an insert) was isolated from a *dam⁺* strain. By this means, the mutation is contained in the methylated strand of DNA, thereby conveying a selective advantage. In another instance,[31] gapped duplex DNA was constructed in which the (+) strand of M13mp9 ssDNA containing two amber mutations was hybridized with linearized wild-type M13mp9 ssDNA. In this case, the mutation is built into the DNA strand devoid of amber mutations, and a marked selective advantage is achieved when the construct is propagated in a

[56] K. A. Osinga, A. M. van der Bliek, G. van der Horst, M. J. A. Groot Koerkamp, H. F. Tabak, G. H. Veeneman, and J. H. van Boom, *Nucleic Acids Res.* **11**, 8595 (1983).

[57] W. Kramer, K. Schughart, and H.-J. Fritz, *Nucleic Acids Res.* **10**, 6475 (1982).

[58] C. Dohet, R. Wagner, and M. Radman, *Proc. Natl. Acad. Sci. U.S.A.* **82**, 503 (1985).

[59] P. J. Pukkila, J. Peterson, G. Herman, P. Modrich, and M. Meselson, *Genetics* **104**, 571 (1983).

[60] B. W. Glickman and M. Radman, *Proc. Natl. Acad. Sci. U.S.A.* **77**, 1063 (1980).

[61] P. M. G. F. van Wezenbeek, T. J. M. Hulsebos, and J. G. G. Schoenmakers, *Gene* **11**, 129 (1980).

[62] B. Kramer, W. Kramer, and H.-J. Fritz, *Cell* **38**, 879 (1984).

wild-type host. Other variations on this general scheme have also been used.[30,63,64]

In the protocol described here, mismatch repair *in vivo* is circumvented by nicking the heteroduplex DNA synthesized *in vitro* with *Hin*dIII in the presence of EtBr. Subsequently, the nicked DNA is heat denatured to form linear and closed circular ssDNA molecules, and the mixed population is used to transfect *E. coli* JM101. By this means, the host is transfected with closed circular ssDNA (both wild-type and mutant) which acts as template for the synthesis of the complementary strand. Thus, RF DNA is made *in vivo,* and mismatch repair is avoided. Using this procedure, mutants have been obtained at a frequency of 3 to 10%, which is clearly less than the theoretical value of 50%. However, it should be stressed that transfection with closed circular ssDNA is about an order of magnitude less than that observed with RF DNA and that the procedure has been successful when direct transfection with cc heteroduplex DNA was completely negative.

Finally, it is noteworthy that we have succeeded recently in recovering mutants with a frequency of about 6% by methylating cc heteroduplex DNA *in vitro* with *S*-adenosyl-L-methionine in the presence of *dam* methylase (New England BioLabs) and then transfecting JM101 with the methylated cc heteroduplex DNA.[65] Moreover, in more than one case, the frequency of mutant recovery was improved to about 30% by nicking cc heteroduplex DNA methylated *in vitro* with *Hin*dIII/EtBr followed by heat denaturation.

[63] A. Marmenout, E. Remaut, J. van Boom, and W. Fiers, *Mol. Gen. Genet.* **195,** 126 (1984).
[64] M. Kozak, *Nature (London)* **308,** 241 (1984).
[65] This approach was suggested by Dr. M. G. Marinus, University of Massachusetts, Worcester, MA.

[18] DNA Sequence and Transcription of Genes for ATP Synthase Subunits from Photosynthetic Bacteria

By VICTOR L. J. TYBULEWICZ, GUNNAR FALK, and JOHN E. WALKER

The genes for the eight constituent polypeptides of *Escherichia coli* ATP synthase are arranged in a single transcriptional unit, the *unc* operon.[1,2] The operon contains subclusters of genes corresponding to the

[1] F. Gibson, *Proc. R. Soc. London Ser. B* **215,** 1 (1982).
[2] J. E. Walker, M. Saraste, and N. J. Gay, *Biochem. Biophys. Acta* **768,** 164 (1984).

F_0 and F_1 segments of the enzyme complex. It also contains a ninth gene, *unc*I[3] of obscure biological function.[4] In order to find out if gene order is maintained and if a homolog of *unc*I is associated with other ATP synthase operons, we have investigated two photosynthetic purple non-sulfur bacteria, *Rhodopseudomonas blastica* and *Rhodospirillum rubrum*. From their genomes we have cloned and determined the DNA sequences around regions that hybridize to segments of genes for α and β subunits of the *E. coli* complex.[5,6] The DNA sequence analysis has been determined by primed synthesis in the presence of dideoxy chain terminators[7] of DNA fragments generated by sonication.[8] We have also studied the transcription of the *R. blastica* genes by the methods of primer extension and mapping with S1 nuclease.[5] The methodologies employed in the cloning, sequence analysis, and transcriptional studies are described below.

Cloning and DNA Sequencing

Media

$2 \times$ TY broth: 16 g Bacto-tryptone, 10 g Bacto yeast extract, 5 g NaCl, 1 liter H_2O adjusted to pH 7.4.

R. blastica growth medium: 0.2% (w/v yeast extract) 0.02% glucose, 4.2 mM Na_2HPO_4, 2.2 mM KH_2PO_4, 0.8 mM NaCl, 1.8 mM NH_4Cl, 0.2 mM $MgSO_4$, and 0.01 mM $CaCl_2$.

R. rubrum growth medium[9]: KH_2PO_4, 600 mg; K_2HPO_4, 900 mg; $MgSO_4 \cdot 7H_2O$, 11.8 mg; trace element solution (containing per 100 ml of deionized H_2O: H_3BO_3, 280 mg; $MnSO_4 \cdot 4H_2O$, 210 mg; $Na_2MoO_4 \cdot 2H_2O$, 75 mg; $ZnSO_4 \cdot 7H_2O$, 24 mg; $Cu(NO_3)_2 \cdot 3H_2O$, 4 mg), 1 ml; ethylenediaminetetraacetic acid, 20 mg; biotin, 15 μg; DL-malic acid, 6 g and $(NH_4)_2SO_4$, 1.25 g dissolved in 1 liter deionized water, pH 6.8.

TYE agar: 15 g Bacto agar, 10 g Bacto tryptone, 5 g Bacto yeast extract, 8 g NaCl, 1 liter H_2O.

CY broth: 10 g Bacto casamino acids, 5 g Bacto yeast extract, 3 g NaCl, 2 g KCl, 1 liter H_2O adjusted to pH 7.4.

H-bottom agar: 10 g Bacto agar, 10 g Bacto tryptone, 8 g NaCl, 1 liter H_2O.

[3] N. J. Gay and J. E. Walker, *Nucleic Acids Res.* **9**, 3919 (1981).
[4] N. J. Gay, *J. Bacteriol.* **158**, 820 (1984).
[5] V. L. J. Tybulewicz, G. Falk, and J. E. Walker, *J. Mol. Biol.* **179**, 185 (1984).
[6] G. Falk, A. Hampe, and J. E. Walker, *Biochem. J.* **228**, 391 (1985).
[7] F. Sanger, S. Nicklen, and A. R. Coulson, *Proc. Natl. Acad. Sci. U.S.A.* **74**, 5463 (1977).
[8] P. L. Deininger, *Anal. Biochem.* **129**, 216 (1983).
[9] S. K. Bose, H. Gest, and J. G. Ormerod, *J. Biol. Chem.* **236**, PC13 (1961).

H-top agar: 8 g Bacto agar, 10 g Bacto tryptone, 8 g NaCl, 1 liter H_2O.

Bacterial Strains

E. coli
    JM101[10]   Δ*lac-pro*, *supE*, *thi⁻*, F'*traD*36, *proAB⁻*, *lacI*q, *lacZ* M15
    Q358[11] r_k^-, m_k^+, su_{II}^+, 80ᴿ
    Q359[11] r_k^-, m_k^+, su_{II}^+, 80ᴿ, P2

R. *blastica*[12]: RQ14 from the collection of S. Brenner, MRC Laboratory of Molecular Biology, Cambridge, England.

R. *rubrum*: strain S1

Vectors

λ*1059*[11]:   *h*λ*sbam*1° *b*189⟨*int*29 *ninL*44 *c*I857 pACL29⟩
           Δ[*int-c*III]KH54*s*RI4° *nin*5 *chi*3

This phage is a *Bam*HI substitution vector which will accommodate DNA fragments 6–24 kb long. The central segment contains the *red* and γ genes of λ. These stop the bacteriophage from growing on strains lysogenic for bacteriophage P2 (e.g., Q359). So only recombinants which have lost the central segment, but not nonrecombinants, will grow on Q359. Genomic libraries are made by ligating partial *Sau*3A digestion products of a genome directly into *Bam*HI-cut λ1059. *Sau*3A and *Bam*HI produce the same cohesive ends. Note that this does not necessarily recreate the *Bam*HI sites that delineate the ends of the arms of the vector.

    λ*2001*[13]: This phage, a derivative of λ2053, carries a Δ[*int*-cIII]KH54*s*RI4° *nin*5 *s*RI5 *sHin*dIII 6° *chi*C right arm and has polylinker sequence TCTAGAATTCAAGCTTGGATCCTCGAGCTC-TAGA cloned into the *Xba*I sites. It is a vector for *Eco*RI, *Hin*dIII, *Bam*HI, *Xho*I, *Sac*I, and *Xba*I.

M13mp8 and mp9

These are single-stranded M13 bacteriophage vectors used extensively for DNA sequencing.[14] Both vectors contain a fragment of the *E. coli lac* operon, the *lac* promoter and the N-terminal α-peptide of β-galactosidase

---

[10] J. Messing, *Recomb. DNA Tech. Bull.* **2**, 43 (1979).
[11] J. Karn, S. Brenner, L. Barnett, and G. Cesareni, *Proc. Natl. Acad. Sci. U.S.A.* **77**, 5172 (1980).
[12] K. Eckersley and C. S. Dow, *J. Gen. Microbiol.* **119**, 465 (1980).
[13] J. Karn, S. Brenner, and L. Barnett, this series, Vol. 101, p. 3.
[14] J. Messing and J. Vieira, *Gene* **19**, 269 (1982).

(*lacZ'*). When infected into *E. coli* JM101, this α-peptide complements the other half of the β-galactosidase protein being produced by the *lacZ* M15 gene on an F′ episome in *E. coli* JM101. This can be demonstrated by plating out in the presence of the gratuitous inducer, isopropyl-β-D-thiogalactopyranoside (IPTG) and the indicator, 5-bromo-4-chloro-3-indolyl-β-D-galactopyranoside (BCIG). A functional β-galactosidase hydrolyses BCIG liberating a blue color and hence mp8 and mp9 plaques on *E. coli* JM101 are blue. Both phages have synthetic multiple cloning sites (MCS) inserted at the N-terminus of the *lacZ'* gene. These MCSs contain many different restriction sites not found elsewhere in the M13 genome. Insertion at any of these sites usually inactivates the *lacZ'* gene and thus recombinant phage give colorless plaques when plated out in the presence of IPTG and BCIG. M13mp8 and mp9 are complementary vectors in that the MCSs are the same in both but in opposite orientations. This can be exploited to control the orientation of a fragment being cloned (see Clone Turnaround).

pUC8 and pUC9

These plasmid vectors[15] are derived from pBR322 and carry ampicillin resistance as a selectable marker. They contain the same MCSs in the *lacZ'* gene as M13mp8 and M13mp9. This allows for the screening of recombinants at the MCSs by the blue/white color assay for β-galactosidase. They too are propagated in *E. coli* JM101.

Preparation of DNA from Photosynthetic Bacteria

Cells of *R. blastica* were grown aerobically and those of *R. rubrum* anaerobically at 30° in the media described above. They were harvested toward the end of logarithmic growth. Washed cells were suspended in buffer (5 g/50 ml) containing 0.15 *M* NaCl, 0.1 *M* EDTA, 0.05 *M* Tris, pH 8.0 and then DNA extracted essentially as described by Marmur.[16]

Construction of Genomic Libraries

The genomic library of *R. blastica* was constructed by Dr. J. Sulston by cloning a partial *Sau*3A digest into bacteriophage λ1059. It was kindly supplied to us for the present work in the form of purified DNA from individual λ recombinants.

R. rubrum DNA was also partially digested with *Sau*3A. DNA in the size range 15–20 kb was recovered by phenol extraction and ethanol

[15] J. Vieira and J. Messing, *Gene* **19**, 259 (1982).
[16] J. Marmur, *J. Mol. Biol.* **3**, 208 (1961).

precipitation following agarose gel electrophoresis in 0.5% LGT agarose. It was then cloned into bacteriophage λ2001. Vector DNA was first cleaved with *Bam*HI and *Eco*RI and the fragments of *R. rubrum* DNA ligated to it. Viable phage particles were recovered after *in vitro* packaging. The library contained about 50,000 PFU. The phage were amplified in *E. coli* Q359 and stocks were stored at 4° in λ dil (10 m*M* Tris–HCl, pH 7.4, 5 m*M* MgSO$_4$, 0.2 *M* NaCl, 0.1% gelatin) over chloroform.

Preparation of Bacteriophage λ DNA

λ1059 and λ2001 recombinant phage were grown in *E. coli* Q358 in CY medium supplemented with 10 m*M* MgCl$_2$, 25 m*M* Tris–HCl, pH 7.4, and 0.2% maltose.[11] A better yield of bacteriophage DNA was obtained in Q358 than in Q359, though the phages were plated on Q359 to select for recombinants. A stationary phase culture of Q358 was diluted 1 : 100 into supplemented CY medium (10 ml) and phage stock (300 μl, ~10^{10} phage/ml) was added. The cultures were shaken at 37° for 5 hr during which time complete lysis occurred. A sample (5 ml) of it was added to an exponentially growing culture of *E. coli* Q358 (500 ml, OD$_{600}$ 0.5) in supplemented CY medium. Again, complete lysis usually occurred within 5 hr. Chloroform (10 ml) was added and the culture shaken for 5 min. After centrifugation to remove cellular debris (11,000 *g*, 10 min), pancreatic ribonuclease A and pancreatic deoxyribonuclease I were added to a concentration of 10 μg/ml and the broth incubated at 37° for 60 min. To precipitate the phage, NaCl (29.2 g) and PEG 6000 (50 g) were dissolved per 600 ml culture and the suspension left at 0° for 60 min. The phage was centrifuged (11,000 *g*, 10 min) and the pellet resuspended in buffer (10 ml) containing 10 m*M* Tris–HCl, 10 m*M* EDTA, pH 7.5. The DNA was extracted three times with an equal volume of phenol and precipitated with ethanol by addition of 0.1 vol 3 *M* sodium acetate (pH 5.0) and 2.5 vol 95% ethanol, followed by freezing in a dry ice/ethanol bath and centrifugation (1000 *g*, 10 min). The pellet was washed with 95% ethanol and redissolved in TE buffer (300 μl) consisting of 10 m*M* Tris–HCl, 0.1 m*M* EDTA, pH 8.0. Typical yields were 300 μg DNA/500 ml culture.

Preparation of pUC Plasmid DNA

Plasmid DNA was prepared by the alkaline SDS method.[17] Single colonies of pUC recombinants were transferred with toothpicks into 2 × TY broth (5 ml) supplemented with ampicillin (100 μg/ml). An overnight culture was diluted 1 : 100 into 2 × TY broth (50 ml), again supplemented with ampicillin (100 μg/ml). This culture was grown at 37° to OD$_{600}$ 0.5.

[17] H. C. Birnboim and J. Doly, *Nucleic Acids Res.* **7**, 1513 (1979).

Chloramphenicol (150 μg/ml) was then added and the cells incubated for a further 18 hr.[18] The cells were harvested and resuspended in buffer (1 ml) containing 25 mM Tris–HCl, pH 8.0, 10 mM EDTA, 60 mM glucose, and 2 mg/ml lysozyme. After incubation for 30 min at 0°, a solution (2 ml) containing 0.2 M NaOH and 1% SDS was added. The solution was mixed thoroughly and left on ice for 5 min. Then 3 M sodium acetate, pH 5.0 (1.5 ml) was added and the solution incubated at 0° for 90 min. Cell debris and precipitated chromosomal DNA were removed by centrifugation (29,000 g, 10 min).

Plasmid DNA was precipitated from the supernatant with 95% ethanol (10 ml), with cooling at −20° for 20 min. The precipitate was centrifuged (12,000 g, 10 min) and the DNA pellet redissolved in reagent containing 0.1 M sodium acetate, 1 mM EDTA, 0.1% SDS, 40 mM Tris–HCl, pH 8.0 and extracted with phenol : chloroform (2 ml, 1 : 1 w/v). The aqueous phase was removed and the phenolic phase reextracted with a further portion (1 ml) of the SDS-acetate buffer. The aqueous fractions were combined and DNA was precipitated by addition of 95% ethanol (7.5 ml), with cooling to −20° for 20 min. DNA was collected by centrifugation (12,000 g, 10 min). The pellet was redissolved in H$_2$O (400 μl) and the DNA again precipitated by addition of buffer (60 μl) containing 1 M sodium acetate pH 8.0 and 95% ethanol (1 ml). Then the suspension was kept at −20° for 20 min and DNA collected by centrifugation (10,000 g, 5 min). This last step was repeated and the final pellet redissolved in H$_2$O (200 μl). The DNA was incubated at 37° for 30 min with pancreatic ribonuclease A (20 μl, 1 mg/ml) (which had been preheated to 100° for 5 min to remove deoxyribonuclease activity). Finally, the DNA was reprecipitated by addition of buffer (7.5 μl) containing 4 M sodium acetate, pH 5.0 and 95% ethanol (300 μl). After incubation at 25° for 15 min, the plasmid was collected by centrifugation (10,000 g, 5 min). The pellet was washed with 95% ethanol (500 μl), dried *in vacuo,* and redissolved in TE buffer (200 μl). Typical yields were 300 μg plasmid DNA/50 ml culture.

Digestion of DNA with Restriction Endonucleases

All digests were performed as recommended by the New England Biolabs catalogue.

Agarose Gel Electrophoresis of DNA

Agarose gel electrophoresis was carried out either in a 10 × 10 cm minigel apparatus or in a 20 × 20 cm flat bed apparatus. All gels were run in a buffer (1 × TBE) containing 90 mM Tris, 90 mM boric acid, 2.5 mM

---

[18] D. B. Clewell, *J. Bacteriol.* **110,** 667 (1972).

EDTA, pH 8.3. Analytical gels contained 1% high gelling temperature (HGT) agarose and preparative gels were made from 1% low gelling temperature (LGT) agarose. Minigels were usually run at ~5 V/cm, and the large gels at 1 V/cm. Both the gels and the running buffer contained ethidium bromide (1 μg/ml); to visualize the DNA, the gels were viewed under short wave UV light.

Transfer of DNA to Nitrocellulose and Hybridization ("Southern Blotting")

DNA digests were fractionated by electrophoresis in 1% agarose gels containing 1 × TBE. DNA fragments were transferred to nitrocellulose filters according to Southern,[19] by blotting through with 20 × SSC (1 × SSC contains 150 mM NaCl and 15 mM sodium citrate). The filters were baked at 80° *in vacuo* for 2 hr and then shaken at 65° for 30 min in a solution containing 6 × SSC, 0.02% bovine serum albumin fraction V, 0.02% Ficoll, 0.02% poly(vinylpyrrolidone), 0.1% N-laurylsarcosine, and 50 μg/ml yeast RNA. The solution was replaced with a fresh solution of the same composition containing 10% dextran sulfate. Radioactive probe was added to at least 10^6 cpm/ml and hybridization allowed to proceed at 65° for at least 18 hr in a shaking water bath.

Then the filters were washed four times for 20 min each at 65° in a solution containing 2 × SSC, 0.02% bovine serum albumin fraction V, 0.02% Ficoll, 0.02% poly(vinylpyrrolidone), 0.1% N-laurylsarcosine, 50 μg/ml yeast RNA. The filters were air-dried and autoradiographed with preflashed X-ray film and intensifying screen for 1–5 days at −70°.

In Vitro Labeling of DNA by "Nick-Translation"

Radioactive probes for hydridization to nitrocellulose filters were made from double-stranded DNA restriction fragments by the "nick-translation" procedure.[20]

Isolation of DNA Restriction Fragments after Preparative Agarose Gel Electrophoresis

The required fragments were excised from LGT agarose gels after electrophoresis. The gel slice was melted at 80° and then extracted 3 times with phenol saturated with TE buffer (1 vol). The aqueous phase was extracted twice with an equal volume of diethyl ether saturated with H_2O and the DNA precipitated by addition of 0.1 vol of 3 M sodium acetate,

[19] E. M. Southern, *J. Mol. Biol.* **98**, 503 (1975).
[20] P. W. J. Rigby, M. Dieckmann, C. Rhodes, and P. Berg, *J. Mol. Biol.* **113**, 237 (1977).

pH 5.0 and 2.5 vol of 95% ethanol. The solution was frozen in a dry-ice/ ethanol bath and then centrifuged (10 min at 10,000 *g*). The pellet was washed in 95% ethanol, dried *in vacuo* and resuspended in an appropriate volume of TE buffer (10–50 μl).

Cloning DNA Restriction Fragments

Vectors were prepared for cloning by simultaneously treating with a restriction endonuclease and calf alkaline phosphatase. This produces linear vector DNA with the 5'-phosphate groups removed which is unable to recircularize in the subsequent ligation. Typically 5 μg of supercoiled vector DNA was treated at 37° for 1 hr with 10 units of restriction endonuclease and a quantity of calf alkaline phosphatase empirically determined to be sufficient, in a buffer containing 50 mM NaCl, 10 mM Tris, 10 mM $MgCl_2$, and 1 mM dithiothreitol (DTT), pH 7.5. The DNA was extracted twice with 1 vol of phenol saturated with TE buffer, once with 1 vol of diethyl ether saturated with H_2O, and then precipitated with ethanol as described before. The DNA was redissolved in TE buffer.

Ligations of DNA fragments were done in a buffer containing 50 mM Tris–Cl, 10 mM DTT, 10 mM $MgCl_2$, and 1.0 mM ATP, pH 7.8; 5 units of T_4 DNA ligase per μg DNA was added for cohesive-ended ligation and 200 units per μg DNA for blunt-ended ligation. Typically a twofold excess of fragment over vector was used. Ligations of cohesive ends were carried out for 2 hr at 15° and those of blunt ends for 18 hr at 15°.

Transformation and Transfection of E. coli Cells

Transformation and transfection of *E. coli* cells were carried out by the method of Hanahan.[21] *E. coli* cells were grown in 2 × TY broth with vigorous shaking to OD_{600} 0.4. The cells were harvested, resuspended in 0.5 vol of transformation buffer (10 mM K-MES, 100 mM KCl, 15 mM $MnCl_2$, 10 mM $CaCl_2$, 3 mM $Co(NH_3)_6Cl_3$, pH 6.2) and placed on ice. After 15 min they were spun down and resuspended in 0.05 vol of transformation buffer. After a further 15 min on ice, dimethylformamide (DMF) was added (7 μl per 200 μl cells). The cells were left on ice for 5 min, then 2.25 M DTT was added (7 μl/200 μl cells) and after a further 10 min on ice an additional quantity of DMF (7 μl/200 μl cells) was added. The cells were then competent for transformation.

Usually the DNA for transformation was added to 200 μl competent cells. The mixture was left for 30 min on ice, then incubated at 42° for 90 sec and finally placed at room temperature.

[21] D. Hanahan, *J. Mol. Biol.* **166,** 557 (1983).

For transformation of plasmid DNA, 0.5 ml of 2 × TY broth was added and the cells were shaken at 37° for 30 min. In the case of pUC plasmids 6 μl of ampicillin (5 mg/ml), 60 μl of 5-bromo-4-chloro-3-indolyl-β-D-galactopyranoside (BCIG, 25 mg/ml) in DMF, and 60 μl of 100 mM isopropyl-β-D-thiogalactopyranoside (IPTG) were added. Samples were spread on selective plates: for pUC plasmids, TYE agar plates supplemented with ampicillin (100 μg/ml) were employed.

For M13 transfection, the competent cells + DNA were heat-shocked at 42° and then added to 3 ml of H-top agar supplemented with 50 μl of 25 mg/ml BCIG in DMF and 20 μl of 100 mM IPTG, mixed and overlayed on H-agar plates. Generally, plates were incubated for 18 hr at 37°.

Screening Genomic Libraries

To screen genomic libraries in bacteriophage λ, plaques were transferred to nitrocellulose filters. The filters were left in contact with the plaques for 30 sec, then denatured for 45 sec in a solution containing 0.5 M NaOH, 1.5 M NaCl and neutralized in a solution containing 1.5 M NaCl, 0.5 M Tris–Cl, pH 6.0 for 5 min. The filters were rinsed in 2 × SSC, allowed to dry for 15 min, and baked at 80° *in vacuo* for 2 hr. The conditions for hybridization and washing were the same as those described under Southern blotting.

DNA Sequencing and Analysis

DNA sequence was determined using the dideoxynucleotide chain termination method[7] incorporating recent modifications.[22] A rapid random sequencing strategy was employed.

Shotgun Cloning into Bacteriophage M13

To generate DNA fragments of a predetermined size range with a random distribution throughout the fragment, the sonication procedure was used, whereby DNA is sheared by ultrasound and then size fractionated on agarose gels.[8] The ends of the fragment clone preferentially and would be overrepresented in the random library. To avoid this the fragment is first self-ligated, giving either circles or linear oligomers of the fragment.

The DNA fragment to be sequenced (10 μg) was purified by electrophoresis through LGT agarose. The DNA was resuspended in 20 μl of TE buffer in a 1.5 ml Sarstedt tube. After addition of buffer (3 μl) containing

[22] M. D. Biggin, T. J. Gibson, and G. F. Hong, *Proc. Natl. Acad. Sci. U.S.A.* **80**, 3963 (1983).

500 mM Tris–HCl, pH 7.5, 100 mM MgCl$_2$, 100 mM DTT, 100 mM ATP, and 20 units T$_4$ DNA ligase, the solution was incubated for 2 hr at 15°. The DNA was then sonicated in a cup-horn sonicator (Heat Systems-Ultrasonics, Inc., Model W-375) at maximum power for four bursts of 40 sec. In between each burst the tube was centrifuged to bring the solution back down to the bottom of the tube.

Sonication produces ragged ends and so it is necessary to repair the ends of the sheared fragments using T$_4$ DNA polymerase before ligation. To the sonicated DNA were added 2 μl of 0.25 mM dNTPs (i.e., containing all 4 deoxynucleoside triphosphates), 3 μl of buffer containing 100 mM Tris–HCl, pH 8.0, 50 mM MgCl$_2$, and 2 μl of T$_4$ DNA polymerase (10 units/μl). The reaction mixture was kept for 4 hr at 15°.

The sequences of the clones can be read out to 300, and sometimes to 350–400, nucleotides. Therefore the sizes of the fragments to be inserted were kept larger than this to avoid reading noncontiguous sequences from multiply ligated fragments, by size fractionation of the end-repaired sonicated DNA fragments before ligation into the sequencing vector. This is done by electrophoresis through a 1.5% HGT agarose minigel with a *Sau*3A digest of pBR322 as a size marker. The DNA was run into the gel at 30 mA for 30 min and a trough (1 mm wide) cut in the path of the DNA at a position corresponding to fragments ~350 nucleotides in length. The trough was refilled with TBE buffer and electrophoresis continued at 20 mA. Thereafter, the buffer, containing eluted DNA, was removed from the trough and replaced with fresh buffer every 60 sec. The eluted DNA usually covered the size range 350–700 nucleotides. The DNA was extracted once with phenol and then precipitated with ethanol by addition of 0.1 vol of 3 M sodium acetate pH 5.0 and 2.5 vol of 95% ethanol. The DNA was pelleted, washed with 95% ethanol (500 μl), dried, and redissolved in TE buffer (20 μl).

Then the DNA was ligated into M13mp8 that had previously been cut with *Sma*I and treated with phosphatase. A range of amounts of DNA was used in order to find an optimum ratio of DNA : vector. Typically, a ligation contained the following:

    Sonicated DNA 0.1, 1 or 2 μl

    M13mp8 (20 ng/μl) (cut with *Sma*I, treated with phosphatase), 1 μl

    500 mM Tris–HCl, pH 7.5, containing 100 mM MgCl$_2$, 100 mM DTT, 1 μl

    100 mM ATP, 1 μl

    T$_4$ DNA ligase (100 units/μl), 1 μl

    H$_2$O to 10 μl

The mixture was incubated for 15 hr at 15° and then transfected into competent cells of *E. coli* JM101. Along with these ligations, controls were also set up. A ligation with no added sonicated DNA served as a negative control and one with 50 ng of bacteriophage λ DNA digested with *Alu*I served as a positive control for the ligation. Typically, ligations of the sonicated DNA gave 50–250 white plaques/μl sonicated DNA.

Preparation of Template DNA

White plaques were transferred with a toothpick into 2 × TY broth (1.5 ml) containing a 1 : 100 dilution of a stationary phase culture of *E. coli* JM101. The culture tubes were shaken vigorously at 37° for 4.25 hr. The cells were centrifuged (10,000 *g*, 5 min) and the supernatant carefully poured off into a 1.5-ml Sarstedt tube containing a solution of 20% polyethylene glycol 6000 (PEG 6000) and 2.5 *M* NaCl (200 μl). The solutions were mixed well, allowed to stand at room temperature for 10 min, and again centrifuged (10,000 *g*, 5 min). The supernatant was sucked off with an aspirator and the tubes recentrifuged in order to bring down any remaining PEG solution from the walls of the tube. The last drop of liquid was removed by suction and the viral pellet redissolved in TE buffer (100 μl). The DNA was extracted once with phenol saturated with TE buffer (50 μl) and then precipitated with ethanol as described above. The DNA pellet was finally redissolved in TE buffer (30 μl).

Sequencing Reactions and Electrophoresis

Hybridization of a synthetic 17-mer oligonucleotide[23] complementary to a region of DNA adjacent to the cloning site, primes the template DNA for DNA synthesis by the Klenow fragment of DNA polymerase. To anneal the primer to the template, 5 μl of template DNA, 2 μl of buffer containing 50 m*M* Tris–HCl, pH 8.0, 25 m*M* MgCl$_2$, 2 μl of the primer (0.15 pmol/μl), and 2 μl of H$_2$O were mixed together and incubated at 60° for 60 min.

All sequencing reactions were carried out in uncapped 1.5 ml Sarstedt tubes in plastic 10-hole centrifuge racks. These racks fit vertically into an IEC Centra 3 centrifuge. This enables the rack to be centrifuged briefly to mix the sequencing solutions, and then to be placed in a boiling water bath to denature and to concentrate the samples prior to loading onto the gel. All samples were dispensed onto the side of the tube using a Hamilton repetitive dispenser which delivers 2 μl, and hence all volumes were adjusted to 2 μl for ease of handling.

[23] M. L. Duckworth, M. J. Gait, P. Goelet, G. F. Hong, M. Singh, and R. C. Titmas, *Nucleic Acids Res.* **9,** 1691 (1981).

NTP Mixes. The following solutions were used for the sequencing reactions.

NTP mixes	Volume dispensed (μl)			
	T	C	G	A
0.5 m*M* dTTP	25	500	500	500
0.5 m*M* dCTP	500	25	500	500
0.5 m*M* dGTP	500	500	25	500
10 m*M* ddTTP	50	—	—	—
10 m*M* ddCTP	—	8	—	—
10 m*M* ddGTP	—	—	16	—
10 m*M* ddATP	—	—	—	1
TE buffer	1000	1000	1000	1000

Enzyme-Radioactive Label Cocktail. This solution is kept ice-cold and made up just prior to use. The amounts are given for sequencing one clone only. Typically as many as 42 clones were sequenced simultaneously and hence this cocktail would be scaled up to make enough for ~44 clones.

Reagent	Volume (μl)
5 units/μl Klenow polymerase	0.1
8 μCi/μl [α-^{35}S]dATP (410 Ci/mmol)	0.75
0.1 *M* DTT	1.0
10 m*M* Tris, pH 8.0	5.85
	8.0

"Chase" Solution

0.25 m*M* dTTP
0.25 m*M* dCTP
0.25 m*M* dGTP
0.25 m*M* dATP
All in TE buffer

Formamide Dye Mix

98% deionized formamide
2% 0.5 *M* EDTA containing 0.1% (w/v) xylene cyanol FF and 0.1% (w/v) bromphenol blue

Sequencing was carried out by distributing the annealed template : primer mix (2 μl) into 4 tubes labelled T, C, G, and A. This was followed by the appropriate NTP mix solution (2 μl) into each tube, e.g., 2 μl of T mix into the T tube, etc. To start the reactions, enzyme-label cocktail (2 μl) was added and the tubes centrifuged. After 20 min reaction at room temperature, chase solution (2 μl) was added to each tube and the reactants mixed by centrifugation. After 15 min at room temperature, formamide dye mix (2 μl) was added and the tubes placed in a boiling water bath for 4 min to denature the DNA.

The sequencing reactions were analyzed by electrophoresis through 6% polyacrylamide gels containing 7 M urea.[24] The gels were 50 cm long and 0.4 mm thick. They were poured with a buffer gradient[22] with 0.5 × TBE at the top of the gel and 5 × TBE at the bottom. Use of 50 cm buffer gradient gels along with [α-³⁵S]dATP as the radioactive label allows sequences to be read routinely up to 300 nucleotides and often as far as 350 or 400 nucleotides from a single gel.

The sequencing reaction mixtures were denatured by boiling in formamide as described above and loaded onto these buffer gradient gels. Electrophoresis was performed at 37 W for 3.5 hr. Then the front plate of the gel was removed and the gel was soaked for 10 min in a solution containing 10% acetic acid and 10% methanol. This allowed most of the urea in the gel to leach out. The gel was then transferred to Whatman 3MM paper, dried by heating *in vacuo* and autoradiographed with Fuji RX film for 18 hr at −70°. Note that the film was not preflashed and no fluorescent screen was used, since both these procedures lower the quality of the final autoradiograph.

Compressions

The biggest problem in the interpretation of DNA sequencing gels is the problem of band compression in G : C-rich areas. These are areas on the gel where the band spacing is compressed so that bands are running together or may even turn out to be running in inverted order. Compressions occur when the sequence at the end of a newly synthesized chain can form a hairpin loop usually requiring the presence of at least 3–4 consecutive G : C residues with an exact complementary sequence present within about 3 to 15 nucleotides. Under normal fractionation conditions these hairpin loops appear to be stable when they are at the 3' ends of the synthesized chains and so these chains migrate anomalously with decreased apparent sizes.

Unless there are multiple, overlapping potential base-paired structures then the sequence of the compressed region can be deduced from the

---

[24] F. Sanger and A. R. Coulson, *FEBS Lett.* **87,** 107 (1978).

sequence of the complementary strand. Although a compression will be seen on both strands, due to the presence of the complementary sequences, it will occur on opposite sides of the center of the base pairing on the opposite strands, thus enabling the sequence of one side of the complementary sequences to be deduced from one strand and the other side from the other strand. The gel distortion due to compressions may sometimes be very slight, e.g., 2 bands may coalesce into 1 and may pass unnoticed if only sequenced on one strand. For this reason it is essential that all parts of the sequence are determined at least once on each strand.

Severe compressions were resolved by substituting inosine triphosphate (dITP) for dGTP in the sequence reactions thus replacing the G : C base pairs with the weaker I : C base pair with only one hydrogen bond. This usually abolishes the compression entirely, but the sequences tend to have a higher background and can generally only be read for the first 150 nucleotides or so. Thus, to resolve a compression with dITP, it must occur near the start of the insert.

The NTP mixes are the same as before except that 2.0 mM dITP replaced 0.5 mM dGTP and the ddGTP in the G mix is reduced from 16 to 2 μl. All other solutions remained unchanged.

Clone Turnaround

The random approach to sequencing a fragment rapidly produces a contiguous sequence with the majority of the sequence (>90%) determined on both strands. However, often there remain a few areas of sequence determined only on a single strand. At this point the random approach becomes inefficient and so a directed method is employed. This involves reversal of the orientation of a specific cloned, sonicated fragment such that now the complementary strand can be sequenced.

The method is facilitated by the complementary vectors M13mp8 and M13mp9; the sonicated fragments having been cloned into the *Sma*I site of M13mp8 were cut out with *Bam*HI and *Eco*RI and then cloned into M13mp9.

To make a double-stranded template, 5 μl of single-stranded template DNA was added to sequencing primer (4.5 μl, 0.2 pmol/μl), buffer (1.5 μl) containing 50 mM Tris–HCl, pH 7.5 and 26 mM MgCl$_2$ and annealed at 60° for 60 min. Then 4 μl of 0.5 mM dNTP (i.e., all 4), 0.5 μl of Klenow polymerase (5 units/μl), and 13 μl of H$_2$O were added and the solution incubated at 37° for 30 min. It was extracted twice with phenol saturated with TE buffer and precipitated with ethanol as usual. The pellet was redissolved in buffer (10 μl) containing 10 mM NaCl, 5 mM Tris–HCl, pH 7.5, 1 mM MgCl$_2$, 0.1 mM DTT. Then 15 units each of *Eco*RI and *Bam*HI was added. The digest was incubated at 37° for 60 min and then loaded onto a 0.8% LGT agarose minigel containing 1 μg/ml ethidium bromide. Electrophoresis was carried out at 30 mA for 20 min and the gel was viewed under UV light. A rapidly migrating band, corresponding in size to ~500 nucleotides, was seen and excised from the gel. The DNA was extracted from the gel slice as described previously. It was then ligated into M13mp9 which had been cut previously with both *Bam*HI and *Eco*RI and treated with phosphatase. The ligation mixtures were transfected into *E. coli* JM101 and the resulting white plaques sequenced in the normal manner.

Compilation and Analysis of Data

The random sequencing strategy relies on computing facilities to compile and analyze the data. The following programs were employed.

GELIN[25] allows gels to be read in from a sonic digitizer. A pen generates an ultrasonic signal when touched to a band on the gel. Then two microphones triangulate the position of the pen. The information is processed by GELIN into a sequence.

SCREENV[26] compares gel readings to known vector sequences such as M13 or pBR322.

SCREENR[27] screens gel readings for any clones that contain the self-ligated ends by searching for the restriction enzyme sequence used to generate the DNA fragment being sequenced.

DBAUTO[26] builds up a database of overlapping gel readings and where appropriate inserts padding characters to align mismatched sequences.

DBUTIL[26] allows the database to be edited and permits gel readings that cannot be handled by DBAUTO to be entered manually. It also calculates a consensus of the sequence.

ANALYSEQ[28] is a versatile sequence analysis program. It will translate nucleic acid sequences, search for promoters, ribosome binding sites, repeats, tRNAs, etc. One of its most useful functions is to predict the location of protein coding regions within a nucleic acid sequence by a variety of statistical methods. One procedure uses the codon preference method[29] which assumes that codon usage is nonrandom and that the codon usage of neighboring genes is similar. The codon usage of a known gene is used as a standard and compared with that of the nucleic acid

[25] R. Staden, *Nucleic Acids Res.* **12**, 499 (1984).
[26] R. Staden, *Nucleic Acids Res.* **10**, 4731 (1982).
[27] R. Staden, unpublished work.
[28] R. Staden, *Nucleic Acids Res.* **12**, 521 (1984).
[29] R. Staden, and A. D. McLachlan, *Nucleic Acids Res.* **10**, 141 (1982).

sequence being investigated. A high degree of similarity is interpreted as a high probability that the nucleic acid sequence codes for a protein.

HYDROPLOT[27] uses the algorithm of Kyte and Doolittle[30] to calculate the relative hydrophobicity of short spans of amino acids along the length of a protein sequence. The program is useful for predicting which segments of a protein might be embedded in a membrane.

DIAGON[31] compares two nucleic acid or protein sequences. The program looks for segments of perfect identity or for regions where, using a scoring matrix, a minimum value is exceeded. The results of the comparisons are presented as a matrix displayed on a graphics terminal. The program uses the algorithm of McLachlan,[32] incorporating the amino acid substitution scoring matrix of Dayhoff.[33] It can detect homologies in distantly related sequences which have undergone many conservative amino acid substitutions. Such homologies would not be detected by searches for identities.

PROTSCAN[34] carries out a DIAGON-type comparison of a protein sequence of interest against a protein sequence data bank using the substitution scoring matrix of Dayhoff.

WILBURLIP[35] runs a comparison of a protein sequence under investigation against a protein sequence data bank, but scores only for identities. Thus, it is less powerful than PROTSCAN, since it overlooks conservative substitutions.

NEWAT[36] is a protein sequence data bank containing 658 entries. It contains a representative selection of the Dayhoff protein sequence databank, e.g., only one cytochrome *c* sequence, only one IgG heavy chain, etc. All searches for homologies using PROTSCAN and WILBURLIP were done against this databank.

Transcriptional Mapping Techniques

Preparation of RNA from R. blastica

Cultures of *R. blastica* (25 ml) were grown aerobically at 30° in the dark in broth containing 0.2% (w/v) yeast extract, 0.02% glucose, 4.2 mM

[30] J. Kyte and R. F. Doolittle, *J. Mol. Biol.* **157,** 105 (1982).
[31] R. Staden, *Nucleic Acids Res.* **10,** 2951 (1982).
[32] A. D. McLachlan, *J. Mol. Biol.* **61,** 409 (1971).
[33] M. O. Dayhoff, "Atlas of Protein Sequence and Structure." National Biomedical Research Foundation, Silver Spring, Maryland, 1969.
[34] A. D. McLachlan, unpublished work.
[35] W. J. Wilbur and D. J. Lipmann, *Proc. Natl. Acad. Sci. U.S.A.* **80,** 726 (1983).
[36] R. F. Doolittle, *Science* **214,** 149 (1981).

Na_2HPO_4, 2.2 mM KH_2PO_4, 0.8 mM NaCl, 1.8 mM NH_4Cl, 0.2 mM $MgSO_4$, and 0.01 mM $CaCl_2$. Cells were harvested in mid-log phase at OD_{600} 0.3. The bacteria were resuspended in lysis buffer (2 ml) containing 0.5% SDS, 20 mM sodium acetate, and 1 mM EDTA, pH 5.5, and then extracted twice at 60° with phenol (5 ml, equilibrated with lysis buffer). RNA was precipitated by addition of 3 M sodium acetate, pH 5.0 (0.1 vol) and 95% ethanol (2.5 vol), followed by freezing in a dry-ice/ethanol bath and centrifugation (10,000 g, 10 min). The pellet was washed with 95% ethanol (500 μl), dried *in vacuo*, and redissolved in H_2O (500 μl).

"Prime-Cut" Probes

The random DNA sequencing strategy produces a comprehensive library of M13 clones covering the entire DNA sequence being studied, in both orientations of the DNA. These can be used to generate single-stranded radioactively labeled DNA probes from any region of the DNA by the "prime-cut" method.

Single-stranded M13 template DNA (5 μl), the 17-mer sequencing primer (5 μl, 0.2 pmol/μl), and buffer (1 μl) containing 100 mM Tris–HCl, pH 7.5 and 50 mM MgCl were mixed and incubated at 60° for 1 hr to anneal the primer to the template. Then, 4 μl of solution containing 2.5 mM dTTP, dCTP, and dGTP, 4 μl of 10 μCi/μl [α-^{32}P]dATP (800 Ci/mmol), and 1 μl of Klenow polymerase (5 units/μl) were added and synthesis carried out for 10 min at room temperature. The reaction was chased with 4 μl of 2.5 mM dATP at room temperature for a further 15 min. The DNA was digested with a restriction endonuclease, the choice of enzyme being governed by the precise probe required. Usually a site within the insert was chosen. The radioactive DNA was then desalted by centrifugation (1600 g, 1 min) through a 1 ml column of Sephadex G-50 (coarse). The eluent was mixed with 1 vol of formamide dye mix and heated in a boiling water bath for 4 min. The DNA was subjected to electrophoresis (37 W, 1.5 hr) through a thin polyacrylamide gel (0.4 mm thickness, containing 6% acrylamide, 7 M urea, 1 × TBE). The gel was autoradiographed for 5 min and the radioactive band cut out, macerated, and the DNA eluted in H_2O (500 μl) at 50° for 60 min after which acrylamide was removed by centrifugation. A 500 nucleotide-long "prime-cut" probe typically contained 1–5 × 10^7 cpm/5 μl M13 template in the original synthesis.

S1 Nuclease Mapping

Single-stranded "prime-cut" probe (~10^5 cpm) and bacterial RNA solution (5 μl) were coprecipitated with ethanol and resuspended in buffer (15 μl) containing 80% formamide, 0.5 M NaCl, 40 mM PIPES, pH 6.4,

1 mM EDTA.[37] The RNA was denatured at 80° for 10 min and then annealed to the probe for 14 hr at 52–57° (annealing temperature was empirically varied according to length and G/C content of the probe). The RNA:DNA duplexes were digested at 37° for 30 min with S1 nuclease (400 units) in buffer (200 μl) containing ice-cold 30 mM sodium acetate, pH 4.5, 0.25 M NaCl, 1 mM ZnSO$_4$, 5% glycerol. The DNA was extracted with phenol saturated with TE buffer, precipitated with ethanol, washed with 95% ethanol, dried, and redissolved in 4 μl of formamide dye mix. The sample was heated in a boiling water bath for 4 min and analyzed by electrophoresis (37 W, 90 min) through a thin gel (6% acrylamide, 7 M urea) containing 1 × TBE buffer. The gel was placed in 10% acetic acid for 10 min, dried, and autoradiographed. An *Msp*I digest of pBR322 DNA, end-labeled by incubating the digest with [α^{32}P]dCTP, unlabeled dGTP and Klenow polymerase was run alongside as a size marker.

Primer Extension Analysis

As described for S1 nuclease mapping, 10^5 cpm of "prime-cut" probe and bacterial RNA solution (5 μl) were coprecipitated with ethanol and redissolved in a solution (15 μl) containing 80% formamide, 0.5 M NaCl, 40 mM PIPES, and 1 mM EDTA, pH 6.4. The solution was kept at 80° for 10 min and then at 55° for 14 hr. The annealed mixture was again precipitated with ethanol, dried, and redissolved in buffer (10 μl) containing 0.1 M Tris–HCl, pH 8.3, 0.15 M KCl, 10 mM MgCl$_2$, 10 mM DTT, and 1 mM of each of the four nonradioactive deoxynucleoside triphosphates. The probe was extended with 5 units of reverse transcriptase for 30 min at 42°. The reactions products were again precipitated with ethanol, washed with 95% ethanol, dried, and analyzed by electrophoresis on denaturing gels containing 6% acrylamide, 1 × TBE, 7 M urea. The sequencing reactions of the M13 clone from which the "prime-cut" probe was derived were used as markers.

Results

A 983 bp *Eco*RI–*Pvu*II fragment from the *unc*D gene (β-ATPase) of *E. coli* was used as a hybridization probe.[38] The probe extended from amino acid 17 to amino acid 344 of the β-subunit. This encompasses a region that is highly conserved in sequence in the *E. coli* and bovine[39] subunits.

[37] A. J. Berk and P. A. Sharp, *Proc. Natl. Acad. Sci. U.S.A.* **75,** 1274 (1978).
[38] M. Saraste, N. J. Gay, A. Eberle, M. J. Runswick, and J. E. Walker, *Nucleic Acids Res.* **9,** 5287–5296 (1981).
[39] M. J. Runswick and J. E. Walker, *J. Biol. Chem.* **258,** 3081 (1983).

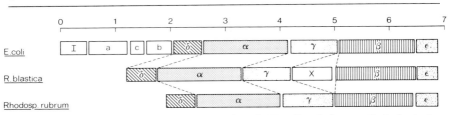

FIG. 1. Organization of genes for ATP synthase in *E. coli* and photosynthetic bacteria. The symbol "X" represents a gene of unknown function in the midst of the F_1 cluster of *R. blastica*. The scale is in kilobases. The genes are shown 5'–3', left to right.

Digests of *R. blastica* DNA were probed with this restriction fragment and a *Bam*HI fragment (3.3 kb) and two *Pst*I fragments (1.4 and 0.7 kb) hybridized under the conditions employed. Then the probe was used to screen a library of the *R. blastica* genome in lambda by hybridization to plaques transferred to nitrocellulose. However, this experiment failed because of hybridization between the probe and *E. coli* genomic DNA liberated in each plaque by the lambda infection. Even though each *E. coli* cell liberates on lysis about 200 copies of the infecting lambda genome and only one copy of the bacterial genome is present, hybridization is far stronger to the bacterial genome because of its perfect match with the probe. Therefore background hybridization drowns out the signal from positively hybridizing recombinant phage.

This problem was overcome by probing restriction digests of DNA isolated from individual lambda recombinants. Given a random distribution of lambda recombinants over the bacterial genome, screening 1000 recombinants gives a 99% probability of finding the fragment of interest. By screening 250 recombinants two positive clones were isolated.

On the other hand the plaque hybridization technique was used successfully to identify phage harboring genes for the α and β subunits of *R. rubrum* ATPase. About 4000 recombinants were screened using prime-cut probes derived from the genes for the α and β subunits of *R. blastica* ATPase.[6] Seven positive plaques were obtained of which three hybridized to both probes.

The DNA sequences of the hybridizing regions in *R. blastica* and *R. rubrum* have been presented elsewhere.[5,6] These experiments have shown that the cloned loci contain genes for F_1-ATPase subunits only: genes for the F_0 part of the ATP synthase complex apparently lie elsewhere in the chromosome of these bacteria. The F_1 genes cluster and it has been shown in the case of *R. blastica*[5] and also in *R. rubrum*[40] that they are all co-

[40] G. Falk and J. E. Walker, *Biochem. J.* **229**, 663 (1985).

transcribed from a single promoter; β and ε are also transcribed from a second internal promoter. The order of the genes in these clusters is summarized in Fig. 1: it is the same as in the F_1 cluster in the *E. coli unc* operon, except that in *R. blastica* an additional gene of unknown function is interposed between those for the γ and β subunits.

Addendum. Recently we have found that in the cyanobacterium, *Synechococcus* 6301, ATP-synthase genes are arranged in two clusters in the order $a : c : b : b' : \delta : \alpha : \gamma$ and $\beta : \varepsilon$; b' is a duplicated and diverged form of b (A. Cozens and J. E. Walker, unpublished results).

Section II

Bacterial Transport

[19] Membrane Transport in Rickettsiae

By HERBERT H. WINKLER

Members of the genus *Rickettsia* are obligate intracellular bacterial parasites and include some significant human pathogens. They resemble gram-negative bacteria in that they have both inner and outer membranes.[1] Rickettsiae can only be cultivated within eukaryotic cells where their normal ecological niche is the cytoplasm and occasionally the nucleoplasm. Unlike other obligate and facultative intracellular bacteria, rickettsiae grow directly in the host cell's cytoplasm unbounded by either a phagosomal or phagolysosomal membrane. The rickettsial envelope has cytoplasm on both sides: eukaryotic cytoplasm on the outside and rickettsial cytoplasm on the inside. Thus, rickettsiae have had the unusual opportunity to evolve transport systems for those metabolites that are present in host-cell cytoplasm but would be absent from the extracellular milieu of free living bacteria. This unique environment is the source of both the problems and opportunities in studying transport by these organisms.

Overall, the methodology of studying transport in rickettsiae consists of (1) the preparation of a defined suspension of purified rickettsiae and (2) the assay of transport in the purified rickettsiae by methods that are similar to those used with *Escherichia coli* or mitochondria. Special regard should be given to miniaturization to save the precious preparation and to the possibility that the transport activity measured in the rickettsial preparation is artifactually due to contaminating eukaryotic organelles.

Taxonomy

There are three groups within the genus *Rickettsia*. The typhus group includes *R. prowazekii* and *R. typhi,* the etiological agents of epidemic typhus and endemic typhus, respectively. The scrub typhus group has *R. tsutsugamushi* as its only member. The spotted fever group includes *R. rickettsii* and *R. conorii,* the etiological agents of Rocky Mountain spotted fever and Mediterranean spotted fever, respectively. *Rochalimaea quintana, Ehrlichia sennetsu,* and *Coxiella burnetii* formerly were classified as *Rickettsia* but have been removed from this genus.

Almost all transport studies in rickettsiae to date have employed the typhus group, especially *R. prowazekii*. This article will, therefore, em-

---

[1] D. K. Smith and H. H. Winkler, *J. Bacteriol.* **137,** 963 (1979).

phasize *R. prowazekii*. Once the special problems of yield and purification for the other species have been satisfied, the transport methods used with *R. prowazekii* can be readily adapted to studies with any rickettsial species.

Biohazard

Rickettsiae are Class 3 pathogens and must be handled with respect for their ability to cause infections that are potentially lethal. Virulent rickettsiae are usually handled with physical containment at the P3 level. The most important route of laboratory infection is through breaks in the skin, especially those inflicted by hypodermic needles, although infection by the respiratory route is possible. Proper education of laboratory personnel is the most effective safeguard, and it is important to include in this program auxiliary personnel such as glassware washers. Members of the genus *Rickettsia* are labile when purified; hence, laboratories will not experience an increase in the level of contamination with time (this is not true of *Coxiella*). Tetracycline is the antibiotic of choice for treating infections; however, this antibiotic is not useful prophylatically. Vaccination with the available vaccines is not recommended since it could lead to a false sense of security and complicate serological testing.

The Madrid E strain of *R. prowazekii* is avirulent,[2] but biochemically similar to virulent strains, and is a suitable organism for most transport studies. The avirulent strain can be used routinely and then limited studies can confirm that a particular transport system is also present in the virulent strain.

Purification of Rickettsiae

Cultivation of rickettsiae in the cells of the yolk sac of the embryonated chicken egg is the most reasonable method for obtaining the quantities necessary for transport and enzymological studies. It is essential to use eggs from flocks fed an antibiotic-free diet. Rickettsiae are often sensitive to the level of antibiotics in the eggs from standard hatcheries and hence, cannot be grown reproducibly in such eggs. The yolk sacs of normal 6-day embryos are inoculated with 0.2 ml of an appropriate dilution of a stock suspension of rickettsiae such that after eight days further incubation at 35° about half of the embryos will have died from the rickettsial infection. The yolk sacs of the embryos that are alive 8 days after inoculation are aseptically removed, and a small portion of yolk sac tissue

---

[2] F. Perez Gallardo and J. P. Fox, *Am. J. Hyg.* **48**, 6 (1948).

(stained by the Gimenez method[3]) is examined microscopically to confirm the presence of rickettsiae. The cells of the infected yolk sacs are disrupted by blending the yolk sacs in SPG buffer (0.218 M sucrose, 3.76 mM KH_2PO_4, 7.1 mM K_2HPO_4, and 5 mM glutamic acid at pH 7.0) at a ratio of 1 g yolk sac to 4 ml SPG.[4] Purification of the rickettsiae is accomplished by differential centrifugation,[5,6] selective removal of yolk sac contaminants by Celite and bovine serum albumin,[5,6] selective precipitation of mitochondrial contaminants by the addition of $MgCl_2$ (10 mM),[7] and passage of the rickettsial suspension through AP20 glass prefilters (Millipore Corp).[8] Final purification may be enhanced by centrifuging the rickettsiae through a continuous gradient of Renografin (obtained from C. R. Squibb & Sons, Inc. Princeton, N.J. 08540) 25 to 45% (v/v) in SPG, in an SW27 rotor at 25,000 g for 1 hr.[8,9] The rickettsiae, which band at 1.2 g/cm^3, are removed, diluted 1/10 in SPG, and washed free of the Renografin before use. Only freshly prepared rather than frozen rickettsiae should be used for transport studies. For membrane transport studies the most significant contaminants are mitochondria since they may also have the transport system under study. The extent of mitochondrial contamination should be monitored. This can be accomplished spectrophotometrically by the cytochrome c oxidase assay of Minnaert.[10] Each step in the purification procedure is a trade-off of purity, viability, and yield. Satisfactory preparations, essentially mitochondria-free, often can be prepared without the Renografin step. Uninfected yolk sacs subjected to the same purification procedure should be assayed to ensure that significant transport activity is absent in these sham preparations. Approximately 26 mg of purified rickettsiae (Lowry protein) can be obtained from 35 harvested yolk sacs by a skilled technician in 4 hr.

Viability

The harvest of the yolk sacs from the living embryos at the time when half of the inoculated embryos have died of the infection provides the

[3] D. F. Gimenez, *Stain. Technol.* **39,** 135 (1964).

[4] M. R. Bovarnick, J. C. Miller, and J. C. Snyder, *J. Bacteriol.* **59,** 509 (1950).

[5] M. R. Bovarnick and J. C. Snyder, *J. Exp. Med.* **89,** 561 (1949).

[6] C. L. Wisseman, Jr., E. B. Jackson, F. E. Hahn, A. C. Ley, and J. E. Smadel, *J. Immunol.* **67,** 123 (1951).

[7] H. H. Winkler, *J. Biol. Chem.* **251,** 389 (1976).

[8] E. Weiss, J. C. Coolbaugh, and J. C. Williams, *Appl. Microbiol.* **30,** 456 (1975).

[9] B. A. Hanson, C. L. Wisseman, Jr., A. Waddell, and D. J. Silverman, *Infect. Immun.* **34,** 596 (1981).

[10] K. Minnaert, *Biochim. Biophys. Acta* **50,** 23 (1961).

maximal yield of rickettsiae in an active growth phase. The number of viable rickettsiae in the purified preparation can be determined by plaque formation[11] but this is labrious, provides only retrospective information, and is not routinely done. The hemolytic activity is an index of the metabolic activity of typhus rickettsiae and can be quickly estimated by adding rickettsiae to sheep erythrocytes, incubating this mixture for 30 min at 34°, centrifuging and determining colorimetrically the hemoglobin released into the supernatant fluid.[12] This procedure has been modified so that every active rickettsia lyses one and only one erythrocyte to quantitatively determine the number of metabolically active typhus rickettsiae.[13] Although much faster than plaquing the rickettsiae, this last procedure is still too time consuming to be employed on a routine basis. The investigator should be aware that perhaps 50% of the purified rickettsiae may not be viable or metabolically active. The total rickettsial bodies (viable or not) can be determined by microscopic particle counts after nebulization onto glass slides[14,15] or filtration onto black Nuclepore filters.[16] Purified rickettsiae are labile and should be suspended in a buffer such as SPG, which is more typical of buffers used with mitochondria than bacteria. In addition, to maintain viability the rickettsial suspension should be held on ice and at a rickettsial protein concentration higher than 100 μg/ml.

Cell Water

The rickettsial cell water must be determined in order to calculate the intracellular concentration of a transported substrate. This can be done by determining the [^{14}C]sucrose space and the ^{3}H$_2$O space of rickettsiae after separating the rickettsiae from the bulk of the extracellular fluid by centrifugation in a capillary tube[17] or by centrifugation through nonaqueous layers.[18] The rickettsial inner membrane is impermeable to sucrose so that the water space minus the sucrose space equals the intracellular water. Typically, 55 μl of rickettsiae (4 mg/ml in SPG) which have been incubated for 20 min with radioactive sucrose and water are added to a 250-μl microfuge tube containing (bottom to top) 30 μl of 14% HClO$_4$, 100 μl dibutyl phthalate, and 30 μl Dow 560 silicone fluid. The tube is centri-

[11] D. A. Wike, G. Tallent, M. G. Peacock, and R. A. Ormsbee, *Infect. Immun.* **5,** 715 (1972).
[12] L. E. Ramm and H. H. Winkler, *Infect. Immun.* **7,** 550 (1973).
[13] T. S. Walker and H. H. Winkler, *J. Clin. Microbiol.* **9,** 645 (1979).
[14] R. Silberman and P. Fiset, *J. Bacteriol.* **95,** 259 (1968).
[15] D. J. Silverman, P. Fiset, and C. L. Wisseman, Jr., *J. Clin. Microbiol.* **9,** 437 (1979).
[16] H. H. Winkler and E. T. Miller, *Infect. Immun.* **45,** 577 (1984).
[17] H. H. Winkler, *Appl. Environ. Microbiol.* **31,** 146 (1976).

fuged for 10 min at 4° in a microfuge, then frozen at −80°. The HClO₄ layer is then cut off and counted by liquid scintillation spectrometry. The space of each isotope is calculated from the dpm/μl of that isotope in the suspension. We have found R. *prowazekii* to have an intracellular space of ~1.6 μl/mg protein.

Metabolism

In the study of the transport of a particular substrate, the investigator wants to eliminate metabolism of the substrate so that the intracellular material that had been transported is chemically identical to that added to the medium. The limited metabolic capacity of the purified rickettsiae helps in this regard. Neither periplasmic enzymes nor glycolytic activities have been found in rickettsiae, and special supplements must be added to obtain protein synthesis.[19] Purified rickettsiae readily metabolize glutamate through the Krebs cycle to yield CO_2 from all carbon atoms and to generate ATP via oxidative phosphorylation.[20]

Space Measurements

In cases of carrier-mediated facilitated diffusion the inadvertant efflux of transported intracellular substrate during the wash step of a filtration assay (see below) may prevent its detection and erroneously indicate that the substrate is not transported. By modifying the technique described above for cell water, one may determine the distribution of a radioactive substrate without washing the cells.[18] Briefly, if the transport substrate is tritium-labeled, then the system employed is ³H-labeled substrate and [¹⁴C]sucrose in the first set of tubes and ³H₂O and [¹⁴C]sucrose in the second set. If the transport substrate is ¹⁴C labeled, then the first set will have ¹⁴C-labeled substrate and ³H₂O. After incubation for 30 to 60 min the cells are centrifuged through the nonaqueous layers and the substrate, water, and sucrose spaces determined (that is, the volume in the HClO₄ layer that would be required to contain the amount of the compound present if the concentration of that chemical were the same as that in the extracellular medium). In the simplest case where the substrate is neither bound nor metabolized, if the substrate space is greater than the water

[18] W. H. Atkinson and H. H. Winkler, *in* "Rickettsia and Rickettsial Diseases" (W. Burgdorfer and R. L. Anacker, eds.), p. 411. Academic Press, New York, 1981.
[19] M. R. Bovarnick and L. Schneider, *J. Biol. Chem.* **235,** 1727 (1960).
[20] M. R. Bovarnick, *J. Biol. Chem.* **220,** 353 (1956).

space then it is actively transported. If the substrate space is equal to the intracellular water, then it is a permeant and has equilibrated during the time of incubation. If the substrate space is equal to the sucrose space, then the substrate has not entered the intracellular water but has equilibrated in the periplasmic space.

Microporous Filtration

Filtration of a suspension of rickettsiae and a radioactive substrate through an 0.45-μm filter, such as a Millipore HAWP025 in a XX1002500 holder, is the most convenient and widely used technique to study rickettsial membrane transport. To determine influx, the rickettsiae (500 μg protein per ml in SPG buffer) are incubated with radioactive substrate for an appropriate time and temperature to obtain initial rates. To stop the reaction, a 100 μl portion of the suspension is placed on the center of a filter (which has already been prewetted with SPG and is under vacuum) and a wash of 5 ml of SPG at the same temperature as the incubation is added as quickly as possible. The filter is immediately removed from its holder while still under vacuum and placed in a liquid scintillation vial. It is important that the suspension be added to the center of the prewetted filter since poor washing will occur if the suspension comes in contact with the edge of the holder funnel. In contrast, the wash should be added over the entire filter surface. The wash buffer should not be at a temperature greatly different from the incubation temperature and it must not contain nonradioactive substrate. If the substrate is ^{14}C labeled, the filter is dried before adding 5 ml of a nonaqueous based counting fluid so that the radioactivity remains on the filter. If the substrate is ^{3}H labeled, self-adsorption can be a problem; therefore, the filter is dissolved in Filter Count (United Technologies—Packard). Disposable minivials are most convenient for experiments with ^{3}H-labeled substrates, but reusable glass 20 ml vials are more economical for experiments with ^{14}C-labeled substrates since the vials are easily washed because the radioactivity remains associated with the filter rather than the glass vial.

To measure efflux, the rickettsial reaction mixture is the same as that used for influx studies. However, the incubation time is lengthened so that a steady state is approached in which influx equals efflux and hence, there is no longer any net uptake. At this time, efflux is initiated by diluting the suspension at least 250-fold with SPG buffer. After the efflux has proceeded for the desired time a volume equal to 0.1 ml multiplied by the dilution factor is filtered. The caveats concerning the placement of the sample on the filter and the need for a wash do not apply because of the large dilution.

Bioenergetics

One of the most interesting aspects of membrane transport in rickettsiae is that *R. prowazekii* has an obligatory exchange system for ATP/ADP.[7] This system, unlike a similar system found in mitochondria (including yolk sac mitochondria), is not inhibited by atractyloside.[7] By virtue of the ATP/ADP transport system, rickettsiae can be poisoned by KCN or starved for their energy source (glutamate) so that active transport is abolished and then reenergized by the addition of ATP.[21]

The membrane electrical potential across the cell of *R. prowazekii* can be measured from the distribution of the lipid-soluble cation tetraphenylphosphonium. A miniature apparatus[22] is used for flow dialysis measurements,[23] which are performed basically as described for *E. coli* membrane vesicles.[24]

[21] D. K. Smith and H. H. Winkler, *J. Bacteriol.* **129,** 1349 (1977).
[22] R. J. Zahorchak and H. H. Winkler, *J. Bacteriol.* **153,** 665 (1983).
[23] S. P. Colowick and F. C. Womack, *J. Biol. Chem.* **244,** 774 (1969).
[24] S. Ramos, S. Schuldiner, and H. R. Kaback, *Proc. Natl. Acad. Sci. U.S.A.* **73,** 1892 (1976).

[20] Transport in Mycoplasmas

By SHLOMO ROTTEM and VINCENT P. CIRILLO

Introduction

The mycoplasmas are the smallest and simplest self-replicating prokaryotes. These organisms have the smallest genome of any living cells and, unlike other prokaryotic microorganisms, have no cell wall or intracellular membrane structures.[1] The absence of a cell wall in mycoplasmas is a characteristic of outstanding importance to which the mycoplasmas owe many of their peculiarities; for example, their morphological instability, osmotic sensitivity, resistance to antibiotics that interfere with cell wall biosynthesis, susceptibility to detergents, and organic solvents. In the recent classification of the prokaryotes that was based on the cell envelope, the mycoplasmas were given the status of a major division named Mollicutes ("soft skin") along with three other divisions: Gracillicutes, the gram-negative bacteria, Firmacutes, the gram-positive bacteria, and Mendocutes, bacteria with incomplete cell walls. The number of

[1] S. Razin, *Microbiol. Rev.* **42,** 414 (1978).

newly recognized species of Mollicutes is continuously increasing due to the improvement in cultivation and identification techniques. Currently, there are over 75 established species classified in several families, and genera. (Classic mycoplasmas include members of the genera *Mycoplasma* and *Acholeplasma*.)

The fact that the mycoplasma cells contain only one membrane type, the plasma membrane, constitutes one of their most useful properties for membrane and transport studies. Another advantage stems from the possibility to alter the membrane lipid composition of mycoplasmas in a controlled manner. This results from the partial or total inability of the mycoplasmas to synthesize fatty acids or cholesterol making them dependent on the supply of these lipid molecules in the growth medium. The ability to introduce controlled alterations in the fatty acid composition and cholesterol content of mycoplasma membranes has been intensively utilized over the past decade to study the relationship between the molecular organization and physical state of membrane lipids and various transport mechanisms.[2-4]

Mycoplasmas exhibit the same variety of transport processes shown by other bacteria, namely simple diffusion, active transport and group translocation (see the table refs. 4–17). The active transport processes are driven by a protonmotive force generated solely by the membrane-bound ATPase that differs from the BF_0-BF_1-ATPase of bacteria.[3,4] The group translocation of sugars is mediated by a phosphoenolpyruvate-dependent phosphotransferase system (PTS) very similar to the IIA/IIB type of *E. coli*. The PTS has only been found in sugar fermenting species of the genus *Mycoplasma,* it is absent in *Acholeplasma* species and in the non-fermenting mycoplasmas.[4]

[2] S. Rottem, *Biochim. Biophys. Acta* **604,** 65 (1980).

[3] R. N. McElhaney, *Biochim. Biophys. Acta* **779,** 1 (1984).

[4] V. P. Cirillo, *in* "The Mycoplasmas" (M. F. Barile and S. Razin, eds.), Vol. 1, p. 323. Academic Press, New York, 1979.

[5] R. N. A. H. Lewis and R. N. McElhaney, *Biochim. Biophys. Acta* **735,** 113 (1983).

[6] J. W. Chen, Q. Sun, and F. Hwang, *Biochim. Biophys. Acta* **777,** 151 (1984).

[7] G. Leblanc and C. Le Grimellec, *Biochim. Biophys. Acta* **554,** 168 (1979).

[8] C. Le Grimellec and G. Leblanc, *Biochim. Biophys. Acta* **599,** 639 (1980).

[9] M. Benyoucef, J. L. Rigaud, and G. Leblanc, *Biochem. J.* **208,** 529 (1982).

[10] M. Benyoucef, J. L. Rigaud, and G. Leblanc, *Biochem. J.* **208,** 539 (1982).

[11] S. Rottem, C. Linker, and T. H. Wilson, *J. Bacteriol.* **145,** 1299 (1981).

[12] T. H. Wilson, C. Linker, and S. Rottem, *Isr. J. Med. Sci.* **20,** 800 (1984).

[13] M. A. Tarshis, A. G. Bekkouzjin, and V. G. Ladygina, *Arch. Microbiol.* **109,** 295 (1976).

[14] M. A. Tarshis, A. G. Bekkouzjin, V. G. Ladygina, and L. F. Pauchenko, *J. Bacteriol.* **125,** 1 (1976).

[15] A. H. Jaffor Ullah, and V. P. Cirillo, *J. Bacteriol.* **127,** 1298 (1976).

[16] A. H. Jaffor Ullah, and V. P. Cirillo, *J. Bacteriol.* **131,** 988 (1977).

[17] U. Mugharbil and V. P. Cirillo, *J. Bacteriol.* **133,** 203 (1978).

TRANSPORT PROCESSES IN MYCOPLASMAS

Solute	Type	Organisms	Reference
1 Ions (K⁺, Na⁺, H⁺)	Active transport	*Acholeplasma laidlawii*	4–6
		Mycoplasma mycoides subsp. *capri*	4, 7–10
		M. gallisepticum	11, 12
2 Amino acids (histidine, methionine)	Active transport	*M. hominis*	4
		M. fermentans	4
3 Polyhydric alcohols	Unmediated diffusion	*A. laidlawii*	4
4 Sugars (glucose, fructose, mannose)	Active transport	*A. laidlawii*	4, 13, 14
	Group translocation	*M. gallisepticum*	4
		M. capricolum	4, 15–17
		M. mycoides subsp. *capri*	4
		M. mycoides subsp. *mycoides*	4

Growth Media for Mycoplasmas

In keeping with their small genome size, it is to be expected that most mycoplasmas will have limited biosynthetic capabilities and accordingly will require a wide array of precursor molecules for macromolecule biosynthesis. Vitamins, nucleic acid precursors, amino acids, and lipids may have to be supplied. Classic mycoplasmas will grow reasonably well on a modified Edward medium.[18,19] The essential constituents for this medium are beef heart infusion, peptone, yeast extract, and serum. In some cases, the serum can be replaced by albumin, fatty acids, and cholesterol. Completely defined media have been described for only two species: *M. mycoides* subsp. *mycoides* and *A. laidlawii*.[20]

Modified Edward Medium[19]

Basal Medium

Beef heart infusion, 13 g
Bacto-peptone, 5 g
NaCl, 2.5 g

---

[18] D. G. ff. Edward, *J. Gen. Microbiol.* **1**, 238 (1947).

[19] S. Razin and S. Rottem, *in* "Biochemical Analysis of Membranes" (A. H. Maddy, ed.), p. 3. Chapman & Hall, London, 1976.

[20] A. W. Rodwell, *in* "Methods in Mycoplasmology" (S. Razin and J. G. Tully, eds.), Vol. 1, p. 163. Academic Press, New York, 1983.

Deionized water, 680 ml
Adjust pH to 8.0 and sterilize.
Add asseptically:
Horse serum, 40–200 ml
Dextrose (50%, w/v), 10 ml
Bacto-yeast extract (7%, w/v) or fresh yeast extract (25%, w/v), 100 ml
K_2HPO_4 (25%, w/v), 10 ml
Benzylpenicillin (400,000 units/ml), 1 ml
Adjust volume to 1 liter with sterile deionized water and pH to the optimal pH.

To cultivate the more fastidious mycoplasmas, the Bacto-yeast extract is replaced by fresh yeast extract prepared from dry yeast and commercially available from several sources. The addition of 20 mM L-arginine to the medium greatly improves the growth of the nonfermentative mycoplasmas that possess the arginine deiminase pathway. Growth of *Ureaplasma* is most markedly improved by the addition of 0.5% urea to the medium. *Spiroplasma* will best grow in MD1 or SP-4 media.[21]

The concentration of horse serum in the media may vary from 4 to 20% depending on the species and strain. To modify membrane lipid composition, the horse serum is replaced by 0.5% (w/v) fatty acid poor bovine serum albumin. The medium for the growth of the cholesterol nonrequiring *Acholeplasma* species is supplemented with the desired mixture of fatty acids in ethanol (10 mg/ml) to a final concentration of 0.08 mM. For the growth of the cholesterol requiring *Mycoplasma* species, the medium is supplemented with both the fatty acid mixture and cholesterol (0.06 mM).

In many cases, the labeling of a cellular component during growth is desired. As mycoplasmas depend on an external supply of nutrients, the specific labeling of cell components can be easily achieved. Usually labels must be added at high levels to maintain a high enough specific activity in the rich mycoplasma media. However, because of the low turnover of mycoplasma membrane lipids and the relatively low levels of free fatty acids in the growth medium, labeling of cellular lipids using radioactive fatty acids is most common. For labeling, either saturated or unsaturated fatty acid can be used depending on the fatty acid requirement of the organism. Most species are efficiently labeled with palmitic acid. Adding 0.01–0.05 μCi/ml of radioactive palmitate will result in labeling to a level of 20,000–100,000 dpm/mg cell protein.

[21] R. F. Whitcomb, *in* "Methods in Mycoplasmology" (S. Razin and J. G. Tully, eds.), Vol. 1, p. 147. Academic Press, New York, 1983.

Growth Conditions

Although most mycoplasma species are facultative anaerobes, they grow better aerobically. The optimal pH for mycoplasma growth varies between 6.0 and 8.5. The fermentative *Acholeplasma* species usually produce higher yields when the pH of the medium is adjusted to 8.0–8.5, while the arginine-hydrolyzing *Mycoplasma* species and the urea-splitting *Ureaplasma* strains produce higher yields when the initial pH of the medium is adjusted to 6.5. Growth is carried out statically at 37° and can be followed by pH changes in the medium and by turbidity measurements at 640 nm. With the fast-growing mycoplasmas, yields of up to 100–150 mg cell protein per liter of medium may be expected within 18–24 hr of growth after inoculation with 0.01–0.1% (v/v) of an overnight culture.

Harvest and Washing

The absence of a protective cell wall renders the mycoplasmas much more fragile than ordinary bacteria. Thus, care must be taken not to damage mycoplasmas during harvesting and washing. Some mycoplasmas may lyse even in the growth medium, others may lyse if washed in unsuitable media or harsh centrifugation. Most mycoplasmas tend to be leaky and lose cofactors upon extensive washings, even in isosmotic solutions. Cells are usually harvested in the cold by centrifugation at 9,000 *g* for 15 min. The cell pellet is washed only once and resuspended to 2% of the volume of the growth medium in cold 0.25 *M* NaCl (or 0.4 *M* sucrose) solution containing 0.01 *M* MgCl$_2$ and 0.025 *M* HEPES buffer adjusted to pH 7.5. Since some lysis is unavoidable, the washed cells may be sticky and aggregated due to the release of DNA. The addition of deoxyribonuclease (to a final concentration of 25 μg/ml) to the washed cell suspension will reduce this problem.

Transport

All transport experiments have been carried out so far with intact cells. Attempts to prepare sealed membrane vesicles have so far failed. Transport can be carried out with the washed cell suspensions (1 mg cell protein/ml) by the addition of the solute to be tested under the appropriate conditions of temperature, pH, and energy supply. The energy requirement of the fermentative mycoplasmas can be satisfied by glucose (1–10 m*M*) and of the nonfermentative species by L-arginine (20 m*M*). Cells may be separated from the uptake media by either filtration or centrifugation.[11] Samples of 0.2 ml are removed at intervals, filtered, and washed on Millipore membrane filters. Due to their small size and plasticity, variable number of cells of most mycoplasmas will pass through 450- or even 220-nm membranes; only 100-nm filters will totally retain the cells.[22] However, as filtration through 100 nm is too slow, 220- or 450-nm filters are used. The cell loss is minimized by using a low negative pressure of 60 nm of mercury, but, in any case, the amount of cell loss must be determined. This can best be achieved by using radioactively labeled cells. To separate cells by centrifugation, 1 ml samples are pipetted onto the surface of 0.5 ml silicone oil [3:1 (v/v) ratio of No. 550 and No. 510 silicone fluids, Dow Corning Co., Midland, MI] in 1.5-ml microfuge tubes and centrifuged in an Eppendorf centrifuge (12,800 g) for 1–2 min. Under these conditions, 99% of the cells pass through the silicone oil forming a pellet at the bottom of the tube with the aqueous phase remaining above the oil. The aqueous phase and oil are then carefully removed by suction, and the tip of the tube containing the cells is cut off and analyzed.

The transport of polyhydric alcohols through the cell membrane of all mycoplasmas so far tested occurs by an unmediated diffusion. Lacking a cell wall, the permeability can be measured by determining the initial swelling rates of the cells.[23] The cell swelling is followed by measuring the absorbance of the cell suspension at 500 nm or by measuring the intracellular water at various time intervals by the microfuge method described below. The unmediated diffusion was intensively investigated with *A. laidlawii* which behave as nearly ideal osmometers.

Measurement of Cell Water

Intracellular cell water of mycoplasmas can be measured in the uptake media using ^{3}H-labeled water.[11] Cells (about 1 mg cell protein/ml) are incubated with ^3H_2O (4 μCi/ml) and [^{14}C]polyethylene glycol (PEG, 0.4 μCi/ml, 100 μg/ml) for various time intervals. The cells are separated from the aqueous medium by the mcirofuge method described above, and ^{3}H and ^{14}C radioactivity is determined. The ^3H_2O is used to measure total water space in the pellet. As PEG does not penetrate the cells, the [^{14}C]PEG measures the extracellular space in the pellet. With several mycoplasmas that bind PEG, [^{14}C]inulin should be used. The water space minus the PEG space represents the intracellular water space.

[22] J. G. Tully, *in* "Methods in Mycoplasmology" (S. Razin and J. G. Tully, eds.), Vol. 1, p. 179. Academic Press, New York, 1983.
[23] R. N. McElhaney, J. De Gier, and E. C. M. Van der Neut-kok, *Biochim. Biophys. Acta* **298,** 500 (1970).

[21] Transport through the Outer Membrane of Bacteria

By Hiroshi Nikaido

Bacterial outer membrane is located outside the cytoplasmic, or the plasma, membrane.[1] It is necessary, therefore, for any nutrients or antibiotic molecules to permeate across the outer membrane before they reach the cytoplasm or even the cytoplasmic membrane. Transport across the outer membrane has been studied extensively and numerous reviews on this topic are available.[2-4]

Mechanistically, transport through the outer membrane can take place by any of three possible pathways: (1) diffusion through the lipid domains of the outer membrane, (2) diffusion through the water-filled channels of porin, a characteristic component of the outer membrane capable of producing largely nonspecific, transmembrane channels, and (3) transport catalyzed by substrate-specific transport systems.

It should be emphasized that processes (1) and (2) are *passive* processes, without any coupling to the expenditure of energy. Most of the specific processes [classified as (3), above] may also be passive processes, although there has been one report[5] claiming that the transport of vitamin B_{12} across the outer membrane occurred against very high concentration gradients. An important characteristic of the passive transport process is that the rate is proportional to the difference in substrate concentrations outside (C_o) and periplasmic side (C_p), that is,

$$V = PA(C_o - C_p) \tag{1}$$

as defined by Fick's first law of diffusion, where V, P, and A denote the net rate of diffusion, permeability coefficient of the membrane toward the particular substrate, and the area of the membrane, respectively. Once the substrate crosses the outer membrane, it is often removed rapidly from the periplasm (i.e., the space between the outer and inner, cytoplasmic membrane) either by periplasmic hydrolytic enzymes or by active transport systems located in the cytoplasmic membrane. Thus, for many systems, C_p is much smaller than C_o, and we can say, as a first approxima-

[1] A. M. Glauert and M. J. Thornley, *Annu. Rev. Microbiol.* **23**, 159 (1969).
[2] H. Nikaido and T. Nakae, *Adv. Microb. Physiol.* **20**, 163 (1979).
[3] H. Nikaido, in "Bacterial Outer Membranes" (M. Inouye, ed.), p. 361. Wiley, New York, 1979.
[4] H. Nikaido and M. Vaara, *Microbiol. Rev.* **49**, 1 (1985).
[5] P. R. Reynolds, G. P. Mottur, and C. Bradbeer, *J. Biol. Chem.* **255**, 4313 (1980).

tion, that V is roughly proportional to C_o. This consideration explains why the bacterial cell contains so many porin molecules, for example, 10^5 molecules/cell for *E. coli*. At high substrate concentrations, for example 10^{-2} M glucose one usually adds initially to minimal media, C_o is so high that V can be very high even without the requirement for high values of permeability. Very high values of P are required, however in order to maintain sufficient levels of V when C_o decreases, for example 10,000-fold, to micromolar levels; in order to achieve these high values of P, very high numbers of porin are required.

Similarly, some of the proteins involved in substrate-specific transport processes, such as the LamB protein (or maltoporin) for maltose transport[6] and proteins involved in the transport of iron-chelator complexes,[7] are produced in very large copy numbers under the physiological conditions requiring the maximal functioning of these transport pathways. The Btu protein, involved in the transport of vitamin B_{12}, may be an exception in this regard; the presence of this protein only in low copy numbers may be related to the exceptionally high affinity of this protein to its substrate.[8]

Because the measurement of permeability using reconstituted systems has been already described in this series,[9] this chapter will be limited to discussions of systems using intact outer membrane in whole cells. It is important to realize that we are measuring essentially the permeability through the water-filled porin channels when small, hydrophilic solutes are used, whereas with bulky, very hydrophobic compounds the penetration through the lipid bilayer plays a predominant role because the bilayer permeability increases with the hydrophobicity of the solute, and because the diffusion through the porin channel becomes slower with the more hydrophobic compounds.[4]

Measurement of Outer Membrane Permeability by Isotope Equilibration[10]

Because of the juxtaposition of the two membrane layers on their surface, a major problem in the study of membrane permeability in gram-negative cells becomes that of distinguishing which membrane layer acts as the limiting step in the permeation process. One solution to this difficulty is the use of small radiolabeled, hydrophilic compounds, which are not transported by the specific transport systems of the cytoplasmic mem-

[6] M. Schwartz, this series, Vol. 97, p. 100.
[7] J. B. Neilands, *Annu. Rev. Microbiol.* **36**, 285 (1982).
[8] C. D. Holroyd and C. Bradbeer, *in* "Microbiology—1984" (L. Leive and D. Schlessinger, eds.), p. 21. American Society for Microbiology, Washington D.C., 1984.
[9] H. Nikaido, this series, Vol. 97, p. 85 (1983).

brane. Because the phospholipid bilayer region of the cytoplasmic membrane is essentially impermeable to hydrophilic solutes, one can determine whether the substrate has crossed the outer membrane barrier into the periplasm, or not, by measuring the amount of radioactivity found in the cell pellet. One can correct for the radioactivity due to the material trapped in intercellular space, by determining the size of this space with the use of macromolecules, such as dextran, that cannot penetrate through the outer membrane. In order to increase the precision of the assay, one can also expand the periplasmic volume by a mild plasmolysis. The following procedure takes advantage of the fact that most strains of *Escherichia coli* and *Salmonella typhimurium* cannot transport sucrose and the oligosaccharides of the raffinose-stachyose series across the ctyoplasmic membrane.

Materials

Methoxy[^{14}C]dextran. Commercial preparations often contain low-molecular-weight fractions that penetrate through the outer membrane porin channel. For this reason, it should be purified either by gel filtration on a column of BioGel P-30[10] or by thorough dialysis in membranes with the molecular weight cut-off of 10,000 or higher.

[^{3}H]Sucrose and [^{3}H]raffinose. Commercially available.

^{3}H-labeled higher oligosaccharides. Nonradioactive stachyose (a tetrasaccharide) is commercially available. Verbascose (a pentasaccharide) can be prepared from lentils.[10] These sugars can be labeled with ^{3}H by first oxidizing the 6-CH$_2$OH group of the nonreducing galactose residue to an aldehyde with galactose oxidase, and then reducing it with sodium boro[^{3}H]hydride.[10] A number of radioactive contaminants are created in the process, apparently because commercial preparations of galactose oxidase contain many oligosaccharides. Thus it is essential to purify the radioactive product by gel filtration followed by paper chromatography (see ref. 10).

Procedure

S. typhimurium or *E. coli* cells are grown in L broth (10 g Difco Tryptone, 10 g Difco yeast extract, 5 g NaCl/liter) with aeration by shaking. When the cell density reaches about 1 mg/ml as judged by the turbidity, the cells are harvested by centrifugation at 6000 *g* for 5 min. This as well as all following procedures are performed at room temperature. The cells are washed once with 25 m*M* sodium phosphate buffer, pH 7.0,

---

[10] G. M. Decad and H. Nikaido, *J. Bacteriol.* **128,** 325 (1976).

resuspended in the same buffer at the concentration of about 150 mg wet weight/ml, and are used immediately. To this suspension, small volumes of [^{14}C]dextran (0.15 μCi) and [^{3}H]oligosaccharide (0.4 μCi) are added, as well as 0.33 volume of 2 M sucrose. After mixing the suspension (final volume: 0.5 ml) thoroughly in an Eppendorf plastic centrifuge tube, the tube is centrifuged at 1790 g for 10 min, the clear part of the supernatant fluid is removed to another tube, and the remaining supernatant fluid is removed by wiping carefully with tissue paper. The centrifuge tube containing the cell pellet is weighed, and the volume of the pellet is calculated approximately by assuming that the density of the pellet is 1.1. Water or 25 mM sodium phosphate buffer is then added to make the final volume 1.00 ml, the suspension is mixed with a vortex mixer, and the tube is centrifuged again.

A portion of the first supernatant (100 μl) plus 400 μl of water, as well as 400 μl of the second supernatant plus 100 μl of water are transferred to separate vials, and the ^{14}C and ^{3}H radioactivity in each vial is determined with a liquid scintillation spectrometer. Because the method relies on small differences between large numbers, the gain and the windows of the spectrometer should be carefully adjusted to produce minimal levels of spillover and near-optimal counting efficiency. It is also necessary to use, as controls, the supernatants of the reaction mixture containing all ingredients except one of the isotopes, rather than [^{3}H]toluene or [^{14}C]toluene, because even when the counting mixture appears quite transparent there are apparently some heterogeneity in the solution, which produces significant effects on the quenching behavior.[10]

Calculation. The space (in microliters) permeable to any solute can be calculated by the following formula:

Permeable space = (total radioactivity in the second supernatant)/(radioactivity in 1 μl of the first supernatant)

= 250(radioactivity in 400 μl of second supernatant)/(radioactivity in 100 μl of the first supernatant)

This calculation for ^{14}C radioactivity will yield the size of intercellular space; that for ^{3}H will give a value which is the sum of intercellular space plus periplasmic space, if the ^{3}H-labeled compound penetrates across the outer membrane.

The results of these experiments showed that sucrose (M_r 342) and raffinose (M_r 504) penetrated fully into the periplasmic space, but there was only a partial or insignificant penetration by stachyose (M_r 666) or verbascose (M_r 828),[10] when *S. typhimurium* or *E. coli* cells were used. (However with *Pseudomonas aeruginosa,* polysaccharides of a few thousand daltons were shown to permeate through the outer membrane.[10,11] This dependence on solute size, as well as a number of other pieces of evidence,[2] showed that these small, hydrophilic solutes are penetrating almost entirely through the water-filled channels of porin.

Comments

This method has been useful in the initial description of the molecular-sieving properties of bacterial outer membrane. However, it suffers from a very poor time resolution, and attempts to use this method to measure the *kinetics* of diffusion across the outer membrane were unsuccessful in our hands. Since then, modification of the method using centrifugation in silicone oil[12] or measurement of efflux rates from preequilibrated cells[13] have been proposed, and in each case the authors believed that they succeeded in measuring the kinetics of permeation. However, the conclusions obtained were in contradiction with the results obtained with reconstitution methods giving a far better time resolution. This discrepancy can now be understood. Because of the very large number of porin channels present on cell surface, the equilibration between the medium and periplasm is very rapid. We can calculate,[14] from the known permeability coefficients of the *E. coli* outer membrane,[15] that the half-equilibration time across the outer membrane is less than 1 sec even for such large, and thus slowly penetrating solutes as disaccharides. Thus what is observed on the time scale of 5–60 min[12,13] is very unlikely to have anything to do with the permeation of solutes through the outer membrane; possibly it reflects the flux into or out of the cytoplasm of injured cells.

Measurement of Outer Membrane Permeability toward Very Hydrophobic Compounds

We have described, in the preceding section, the use of small, hydrophilic compounds that do not penetrate through the inner membrane, for the measurement of outer membrane permeability. Another way to avoid the complication arising from the existence of the inner membrane is to use compounds that penetrate through the inner membrane extremely

[11] R. E. W. Hancock, G. M. Decad, and H. Nikaido, *Biochim. Biophys. Acta* **554,** 323 (1979).

[12] K. B. Heller and T. H. Wilson, *FEBS Lett.* **129,** 253 (1981).

[13] C. A. Caulcott, M. R. W. Brown, and I. Gonda, *FEMS Microbiol. Lett.* **21,** 119 (1984).

[14] H. Nikaido, E. Y. Rosenberg, and J. Foulds, *J. Bacteriol.* **153,** 232 (1983).

[15] H. Nikaido and E. Y. Rosenberg, *J. Gen. Physiol.* **77,** 121 (1981).

rapidly, so that the penetration through the outer membrane will become the rate-limiting process. Strongly hydrophobic compounds fit this requirement, because (1) such compounds penetrate through the phospholipid bilayer of the inner membrane very rapidly,[16] (2) they diffuse through the porin channels with negligible rates,[14] and (3) the asymmetric lipid bilayer of the outer membrane has a lower permeability toward these compounds,[2–4,17] and therefore is likely to become the rate-limiting step in the overall influx process.

One of the earlier efforts to use this system was that of Gustafsson *et al.*[18] who used the uptake of cystal violet, a hydrophobic dye, by *E. coli* cells as the indicator of cell wall permeability. Crystal violet, however, is a cationic molecule that is adsorbed strongly to the negatively charged groups on the cell surface as well as to the polyanionic molecules in the cytoplasm, and this introduces practical as well as theoretical difficulties in the experiments. We have avoided this difficulty by using an anionic substance, nafcillin[17]; we also use very high concentration of this compound so that we get little disturbance from the sequestration of the compound within the membrane or from the effect of membrane potential.

Materials

Nafcillin, a semisynthetic, highly hydrophobic penicillin, is available from Wyeth Pharmaceuticals.

Procedure[17]

E. coli or *S. typhimurium* cells are grown to mid-exponential phase in L broth (see above), harvested rapidly by centrifugation, are washed once with 0.1 *M* NaCl, and resuspended in 0.05 *M* Na-phosphate buffer, pH 7.0. All operations are carried out at room temperature. It is essential to carry out these procedures rapidly. If it takes more than 30 min to prepare the cell suspension, the outer membrane tends to become "leaky"[17]; this phenomenon is presumably related to the drastic increase in outer membrane permeability to hydrophobic compounds, observed in energy-starved cells.[19–21] Suspension containing 1 g (wet weight) of cells in 2.1 ml was mixed with 0.3 ml of 50 mg/ml solution of sodium nafcillin, and at

[16] W. D. Stein, "The Movement of Molecules across Cell Membranes." Academic Press, New York, 1967.

[17] H. Nikaido, *Biochim. Biophys. Acta* **433**, 118 (1976).

[18] P. Gustafsson, K. Nordstrom, and S. Normark, *J. Bacteriol.* **116**, 893 (1973).

[19] S. L. Helgerson and W. A. Cramer, *Biochemistry* **16**, 4109 (1977).

[20] E. S. Tecoma and D. Wu, *J. Bacteriol.* **142**, 931 (1980).

[21] M. K. Wolf and J. Konisky, *FEMS Microbiol. Lett.* **21**, 59 (1984).

intervals 0.3 ml portions were transferred to heavy-walled glass centrifuge tubes immersed in an ice bath, to stop the diffusion process. The tubes are centrifuged in the cold, and the concentration of nafcillin in the extracellular fluid (C_{ext}) was determined from the optical density of the diluted supernatants at 254 nm. The concentration of nafcillin when it has been distributed at equal concentrations inside and outside the cells (C^∞) was predicted from that of the mixture containing 0.3 ml of nafcillin and 2.1 ml of buffer. Theory predicts[17] that the plot of log($C_{ext} - C^\infty$) versus time will be linear, and the (negative) slope of the line will be proportional to the first-order rate constant governing the process of influx. This is indeed what was found.[17]

Comments

With the wild-type cells of *S. typhimurium*, the rate of influx of nafcillin was negligible because the lipopolysaccharide–phospholipid bilayer of the outer membrane has very low permeability for hydrophobic compounds. However, consideration of the physical chemistry of the diffusion process[16] predicts that increasing hydrophobicity of the solute would increase permeation rates. Thus it is probable that even the wild-type outer membrane will allow the penetration, at significant rates, of compounds of very high hydrophobicity. Evidence favoring this idea include the growth of *P. aeruginosa* on hydrocarbons such as dodecane and hexadecane[22] as well as the rapid penetration of hydrophobic tetracycline derivatives such as minocycline into *E. coli* cells.[23]

Nafcillin rapidly penetrates through the outer membrane of mutants producing lipopolysaccharides of very defective structure (so-called "deep rough" mutants).[17] This is presumably due to the weakened lateral interaction between lipopolysaccharide molecules, or the presence of phospholipid bilayer patches, or both.[4] The permeation process is strongly dependent on temperature, suggesting that it takes place through the lipid interior of the outer membrane, rather than through the pores.[17]

Measurement by Using Periplasmic Hydrolytic Enzymes as the "Sink"

The magnitude of outer membrane permeability is usually so large, as described above, and the periplasmic volume so small, that equilibration across the outer membrane takes place too rapidly for small, hydrophilic compounds for measurement of their permeation rates by conventional

[22] R. Y. Stanier, N. J. Palleroni, and M. Doudoroff, *J. Gen. Microbiol.* **43**, 159 (1966).
[23] L. M. McMurry, J. C. Cullinane, and S. B. Levy, *Antimicrob. Agents Chemother.* **22**, 791 (1982).

methods. In situations like this, the standard biochemical approach is to add a "sink" to the periplasmic side, which has an effect formally equivalent to that of enlarging the periplasmic space almost to infinity, so that diffusion of the solute molecule through the outer membrane will continue at a constant rate (i.e., "steady state") rather than coming to a halt within a second or so due to the approach to the equilibrium.

This approach was pioneered by W. Zimmermann and A. Rosselet,[24] who determined the permeability of *E. coli* outer membrane to β-lactam antibiotics by using periplasmic β-lactamase as the "sink." This still remains the most rigorously correct approach for the determination of outer membrane permeability. Below we give a procedure slightly modified and extended by us.[14]

Materials

Bacterial Strains. Although *E. coli* K12 produces a β-lactamase coded by a chromosomal gene *(ampC)*,[25] its activity is too weak (without gene amplification) to be useful as the sink. We therefore introduce an R-factor, for example R_{471a}, which produces very high levels of periplasmic β-lactamase,[26] by standard conjugational transfer procedure.[27] R-factors coding for the production of high levels of β-lactamase are available for most species of gram-negative bacteria. Although at one time some R-factors were believed to alter the permeability of the outer membrane drastically,[28] it is believed currently that this does not happen with most R-factors.[29] Nevertheless, if it is desired to exclude the effect of the R-factors completely, one may use chromosomally coded β-lactamases. However, many of these enzymes must be induced by growth in the presence of β-lactams, and one has to be careful not to damage the cell wall complex and produce the leakage of the enzyme into the medium during the induction process.

β-Lactams. It is advantageous to use those β-lactams that are hydrolyzed rapidly by the periplasmic β-lactamase, and also penetrate the outer membrane (porin chennel) rapidly. For the TEM type β-lactamase, coded for by such R-factors as R_1, R_{471a}, and RP1, cephaloridine, a zwitterionic compound, is an excellent substrate, and is also commercially available.

[24] W. Zimmermann and A. Rosselet, *Antimicrob. Agents Chemother.* **12**, 368 (1977).
[25] E. B. Lindstrom, H. G. Boman, and B. B. Steele, *J. Bacteriol.* **101**, 218 (1970).
[26] R. W. Hedges, N. Datta, P. Kontomichalou, and J. T. Smith, *J. Bacteriol.* **117**, 56 (1974).
[27] J. H. Miller, "Experiments in Molecular Genetics," p. 82. Cold Spring Harbor Laboratory, Cold Spring Harbor, New York, 1972.
[28] N. A. C. Curtis and M. H. Richmond, *Antimicrob. Agents Chemother.* **6**, 666 (1974).
[29] I. Crowlesmith and T. G. B. Howe, *Antimicrob. Agents Chemother.* **18**, 667 (1980).

Among monoanionic compounds, cephacetrile (available from CIBA-GEIGY, Basel, Switzerland) has the highest rate of penetration through *E. coli* porin channels.[14] The solutions of β-lactams are best prepared fresh.

Procedure[14]

The strains containing the R-factors are grown overnight as a small (1–5 ml) preculture in L Broth (see above). It is advisable to incorporate ampicillin (20–50 μg/ml) in this medium so that the culture will not be overgrown by those cells that have lost the R-factor plasmids. The preculture is diluted 10- to 5-fold with fresh, prewarmed L broth which contains 5 mM MgSO$_4$ but does not contain any β-lactam. The culture is incubated at 37° with aeration, and the cells are harvested at late exponential phase by centrifugation for 5 min at 3000 g. This and all subsequent operations are carried out at room temperature, because sudden chilling of bacterial cells often produces leakage of periplasmic components. The cells are washed immediately with 10 mM sodium phosphate buffer, pH 7.0, containing 5 mM MgCl$_2$, and are resuspended in the same buffer at the concentration of 1.5 mg (dry weight)/ml.

Although the hydrolysis of β-lactams can be followed by microiodometry (see ref. 24), it is time consuming and requires some practice. We find it far easier to use a spectrophotometric assay. Because one records continuously the optical density of the reaction mixture, the precision and the reliability of the data are increased very significantly. The major difficulty in the use of spectrophotometric assay is the large amount of light scattering by bacterial cells, which limits the amount of cells that can be used. We circumvent this difficulty by using cuvettes of 1 mm light path, which allows the use of 10-fold higher concentrations of both the cell suspension and the substrate, a condition producing an approximately 10-fold higher sensitivity for the assay (for details, see ref. 14). Cell suspension (0.1 ml containing 0.15 mg (dry weight) cells) is mixed with 0.4 ml of 10 mM sodium phosphate buffer, pH 7.0–5 mM MgCl$_2$ containing 0.5 μmol of cephaloridine, and a portion of the mixture was quickly transferred to a cuvette of 1 mm light path and the optical density is recorded for 5 min at 260 nm at 25°.

A portion of the cell suspension is sonicated until translucent using a probe-type sonicator, and the assay of cephaloridine hydrolysis is carried out exactly as above, but using the sonicated cell suspension instead of the suspension of intact cells. It is also necessary, at the beginning of a series of experiments, to determine hydrolysis rates of various concentrations of cephaloridine by a sonic extract, and calculate the value of K_m.

Calculation

1. We first calculate the cephaloridine concentration in the periplasm (C_p) in intact cells incubated in the presence of this compound, as follows.

    a. We calculate the V_{max} of the enzyme contained in 0.15 mg (dry weight) of cells. With the R_{471a} enzyme the K_m for cephaloridine is 930 μM.[14] Let us assume that the rate of change of OD_{260} with the sonicated cells (0.15 mg) at the substrate concentration *(C)* of 1000 μM was 0.1 min^{-1}; this means that the rate of hydrolysis *(V)* was $(0.1/0.8) \times 0.5/0.15 = 0.413$ μmol/min/mg or 413 nmol/min/mg, because complete hydrolysis of 1 mM cephaloridine would decrease the OD_{260} by 0.8 in a cuvette of 1 mm thickness.[30] (The factor 0.5 is necessary because the total volume of the mixture was 0.5 ml.)

By substituting these values of K_m, *C,* and *V* into the Michaelis–Menten equation,

$$V = V_{max}C/(K_m + C) \tag{2}$$

one can calculate V_{max}, which in this case is 0.8 μmol/min/mg or 800 nmol/min/mg.

    b. Let us assume that the OD_{260} decreased at the rate of 0.088 min^{-1} when intact cells, rather than the sonic extract was used. This corresponds to the hydrolysis rate (V_{cells}) of $(0.088/0.8) \times 0.5/0.15 = 0.367$ μmol or 367 nmol/min/mg. Because we know the V_{max} (800 nmol/min/mg) and K_m (930 μM) of the periplasmic β-lactamase, we can substitute these values into the Michaelis–Menten equation (2) and obtain the substrate concentration in the periplasm, C_p, as 787 μM.

2. As pointed out by Zimmermann and Rosselet,[24] at the steady state the net rate of influx of cephaloridine *(V)* is equal to the rate of its hydrolysis in the periplasmic space (V_{cells}). Fick's first law of diffusion states that the net influx rate *(V)* is proportional to the permeability coefficient of the membrane *(P)*, area of the membrane *(A)*, and the concnetration difference of the solute between the outside medium and the periplasm, as seen in Eq. (1), if we denote the external concentration of cephaloridine as C_o. Substituting, into this equation, the values of V (V_{cells}, i.e., 367 nmol/min/mg), A (132 cm^2/mg),[31] C_o (1000 nmol/cm^3), and C_p (787 nmol/cm^3),

[30] P. V. Demarco and R. Nagarajan, *in* "Cephalosporins and Penicillins: Chemistry and Biology" (E. H. Flynn, ed.), p. 311. Academic Press, New York, 1972.
[31] J. Smit, Y. Kamio, and H. Nikaido, *J. Bacteriol.* **124**, 942 (1975).

$$P = 367 \text{ nmol/min/mg}/(132 \text{ cm}^2/\text{mg} \times (1000 - 787) \text{ nmol/cm}^3)$$
$$= 0.0127 \text{ cm/min}$$
$$= 2.1 \times 10^{-4} \text{ cm/sec}$$

Comments. There are several prerequisites for the successful execution of this method.

1. If the enzyme leaks out into the medium, or if outer membrane of some cells becomes damaged in such a way as to compromise its barrier properties, the enzymes in the medium or in such damaged cells will have a full access to the substrate. Under these conditions, hydrolysis by the enzyme leaked out into the medium or in damaged cells will become a particularly significant part of the total hydrolysis rate by cell suspensions.

If the bacterial strain has an outer membrane with low permeabilty (e.g., *Pseudomonas aeruginosa*[32,33]) or if the substrate used has difficulty in permeating through the porin channel because of its bulky size, high degree of hydrophobicity, or the presence of multiple negative charges (see ref. 34), the influx and consequently the hydrolysis by the periplasmic enzyme occur only slowly.

For this reason, we try to minimize the leakage of the enzyme by growing the cells, and carrying out the assay, in the presence of Mg^{2+} (see refs. 35, 36). Furthermore, we centrifuge a portion of the bacterial suspension used for the assay, measure the rate of hydrolysis by the supernatant, and correct V_{cells} for this rate, due to the enzyme leaked out into the medium. With *E. coli* and with such rapidly penetrating drugs as cephaloridine, this correction usually amounts to only 5% or less of the V_{cells}. However, it can become very large in less favorable situations, as described above. Furthermore, we cannot correct for hydrolysis by cells with damaged outer membranes, and the precision of the results obtained under these conditions becomes rather poor.

2. One major assumption of the method is that the K_m and V_{max} of the enzyme determined with the sonic extracts of the cells are valid for the enzyme in the periplasm. We should note, however, that the ionic composition of the periplasm may be quite different from that of the external medium, owing to the Donnan potential that exists across the outer mem-

[32] B. L. Angus, A. M. Carey, D. A. Caron, A. M. B. Kropinski, and R. E. W. Hancock, *Antimicrob. Agents Chemother.* **21**, 299 (1982).
[33] F. Yoshimura and H. Nikaido, *J. Bacteriol.* **152**, 636 (1982).
[34] H. Nikaido and E. Y. Rosenberg, *J. Bacteriol.* **153**, 241 (1983).
[35] H. Nikaido, P. Bavoil, and Y. Hirota, *J. Bacteriol.* **132**, 1045 (1977).
[36] H. Stan-Lotter, M. Gupta, and K. E. Sanderson, *Can. J. Microbiol.* **25**, 475 (1979).

brane.[37] In dilute buffers, concentrations of small cations may be an order of magnitude higher in the periplasm, and its pH may be lower by as much as 1 pH unit.[37] The activity of the TEM-type β-lactamase, coded for by many R-factors, fortunately appears to be little affected by changes in pH (in the range 5–7) or salt concentration.[38]

3. This method is not limited to β-lactams. Wherever active hydrolytic enzymes are found in the periplasmic space, and the substrate is not rapidly transported into the cytoplasm, the same approach can be used. For example, periplasmic phosphatase was used for measuring the rates of diffusion of phosphate esters across the outer membrane of *Pseudomonas aeruginosa*.[33]

4. A problem with this method is that one can use only those β-lactam that are rapidly hydrolyzed by the periplasmic β-lactamase. An ingenious method to overcome this difficulty has been proposed.[39] Here one measures the rate of hydrolysis of a rapidly hydrolyzed cephalosporin in the presence of a slowly hydrolyzed cephalosporin, which often acts as a competitive inhibitor. From these inhibited rates, one can calculate the periplasmic concentration of the inhibitor cephalosporin, and its permeability can be deduced if one knows its hydrolysis rates by the sonic extracts of cells. In practice, however, it appears rather difficult to execute this method successfully, probably because it requires that many parameters be determined with the degree of precision which are difficult to achieve.[40]

5. Several other methods have been proposed for the purpose of determining the permeabilty of outer membrane to antibiotics. These include the simple comparison of β-lactam hydrolysis rates by intact cells vs extracts,[41] comparison with "hyperpermeable mutants,"[42] or EDTA-treated cells,[43] and measurement of β-lactam binding to penicillin-binding proteins in intact cells.[44,45] Although these methods sometimes give excellent qualitative information, they cannot, in general, be used as rigorous, quantitative methods because of the reasons described elsewhere.[40]

[37] J. B. Stock, B. Rauch, and S. Roseman, *J. Biol. Chem.* **252,** 1850 (1977).
[38] J. Hellman and H. Nikaido, unpublished.
[39] H. Kojo, Y. Shigi, and M. Nishida, *J. Antibiot.* **33,** 310 (1980).
[40] H. Nikaido, *Pharmacol. Ther.* **27,** 197 (1985).
[41] M. H. Richmond and R. B. Sykes, *Adv. Microb. Physiol.* **9,** 31 (1973).
[42] M. H. Richmond, D. C. Clark, and S. Wotton, *Antimicrob. Agents Chemother.* **10,** 215 (1976).
[43] R. A. Scudamore, T. J. Beveridge, and M. Goldner, *Antimicrob. Agents Chemother.* **15,** 182 (1979).
[44] W. Zimmermann, *Antimicrob. Agents Chemother.* **18,** 94 (1980).
[45] A. Rodriguez-Tébar, F. Rojo, J. C. Montilla, and D. Vazquez, *FEMS Microbiol. Lett.* **14,** 295 (1982).

Estimation of the Lower Limit of Outer Membrane Permeabilty by
Coupling with Active Transport Systems of the Inner Membrane

We have described how the permeability of the outer membrane to
small, hydrophilic solutes can be determined with precision by coupling
the influx of the solutes across the outer membrane with their removal via
hydrolysis by periplasmic enzymes. This method, however, can be applied only to a few good substrates of periplasmic enzymes. Alternatively,
it should be possible to determine the outer membrane permeability by
using, as the "sink," the removal of substrates from the periplasm by
active transport systems located in the inner, cytoplasmic membrane:
such a method will allow the determination of outer membrane permeabilty to a number of compounds of physiological interest.

The major drawback of such an approach is that it is very difficult to
determine the true K_m and V_{max} of the active transport process with precision. The values of these parameters, especially those of K_m, determined
with intact cells, are unlikely to represent the true parameters of the
active transport systems per se, as they may be limited by the diffusion
through the outer membrane. In principle it should be possible to determine these parameters by using spheroplasts, but deformation and
stretching of the cytoplasmic membrane accompanying spheroplast formation may affect the active transport process profoundly. Because of
these reasons, the approach often becomes limited to the estimation of the
lower limit of the outer membrane permeability by *assuming* that the
active transport machinery has very high affinity or low K_m. This assumption indeed appears to be justified in *E. coli,* because even the values of
"overall" transport K_m, potentially limited by the outer membrane, usually are lower than a few μM.[45]

As an example, let us consider the uptake of glucose by *E. coli.* The
V_{max} for this process is reported to be 6 nmol/mg/sec.[46] In whole cells, the
apparent K_m for this process was 3 μM according to one study.[15] Thus at
this external concentration the rate of transport of glucose is one-half of
V_{max}, or 3 nmol/mg/sec. Using this value for V, 132 cm^2/mg for A (see
above), and 3 nmol/cm^3 for C_o in Eq. (2), one gets 3 (nmol/mg/sec) = $P \cdot$
132 (cm^2/mg)(6 $-$ C_p) (nmol/mg/sec). Thus P = 3/132/(6 $-$ C_p) > 3/132/6
(cm/sec).

One obviously cannot calculate the precise value of P, but we can get
its minimal estimate by neglecting C_p, which is equal to the true K_m of the
active transport system located in the cytoplasmic membrane. This minimal estimate is 3.8 × 10^{-3} cm/sec. The actual value of permeability coeffi-

[46] A. L. Koch, *Adv. Microb. Physiol.* **6,** 147 (1971).

cient can be much higher: it will be 7.6×10^{-3} cm/sec, if the true K_m of the active transport system is 3 μM, for example.

As a special case, one can obtain a much less ambiguous estimate of P when one is certain that C_p is negligible in comparison with C_o. With wild-type cells, such a situation is unlikely to be found because the active transport systems of the inner membrane must have evolved in association with the transport systems of the outer membrane, so that the former usually has poor affinity toward substrates whenever the latter is rather inefficient toward these substrates (see the case for lactose[3,46]). However, in mutants producing greatly decreased amounts of porins, the apparent K_m values for transport of various substances increase tremendously,[47] for example in one mutant the overall K_m for glucose transport increases to 500 μM. In comparison with this value of C_o, the value of C_p, i.e., the true K_m of the inner membrane transport system, is certainly less than 6 μM (i.e., the overall K_m in the wild type) and is almost negligible. Thus, P(mutant, glucose) = 3/132/500 = 4.5×10^{-5} cm/sec. If the mutant produces only 1% of the amount of the porin found in the wild type, P(wild type, glucose) = $4.5 \times 10^{-5}/0.01 = 4.5 \times 10^{-3}$ cm/sec.

Comments

As described above, this method gives values of reasonable accuracy only in porin-deficient mutants, in which it is difficult, in turn, to quantitate the amounts of porin with precision. Yet it is often the only method available for estimating outer membrane permeability for physiologically important nutrients. It should also be emphasized that the permeability coefficients calculated for glucose (in the range of 2–5 $\times 10^{-3}$ cm/sec)[15,47] are in a good agreement with the permeabilty coefficient calculated from the number of porin channels and their cross-section.[15] This suggests that in *E. coli* most, or at least a major fraction, of the channels are "open," despite a claim to the contrary,[48] and in a striking contrast to *P. aeruginosa* in which most of the channels appear to be closed.[32,33]

[47] P. Bavoil, H. Nikaido, and K. von Meyenburg, *Mol. Gen. Genet.* **158**, 23 (1977).
[48] H. Schindler and J. P. Rosenbusch, *Proc. Natl. Acad. Sci. U.S.A.* **75**, 3751 (1978).

[22] Binding Protein-Dependent Active Transport in *Escherichia coli* and *Salmonella typhimurium*

By CLEMENT E. FURLONG

Types of Transport Systems

The mechanisms of the systems that actively transport nutrients across the bacterial membrane are not yet fully understood. However, in recent years, a great deal has been learned about the proteins that make up specific systems. One way to categorize the different transport systems is on the basis of mechanism and the protein components of a given system. Figure 1 presents a schematic representation of 3 different types of active transport systems. The phosphotransferase systems require cytoplasmic and membrane-bound proteins, whereas the systems that are active in membrane vesicles require only membrane-bound proteins.[1-4] The osmotic shock-sensitive systems require at least periplasmic and membrane bound proteins.[5-13] The periplasmic protein components of the osmotic shock-sensitive systems appear to serve as the substrate recognition components and have, for lack of a better term, been referred to as binding proteins. A number of the membrane-bound protein components of several of the shock-sensitive systems have been identified genetically and/or biochemically.[5-13] It appears that between two and four membrane-associated proteins are required for interaction with a given binding protein, depending on the specific system. It should be noted that a given nutrient can be, and most often is, transported by more than one system

[1] M. H. Saier, Jr., "Mechanisms and Regulation of Carbohydrate Transport." Academic Press, New York, 1985.

[2] S. S. Dills, A. Apperson, M. R. Schmidt, and M. H. Saier, Jr., *Microbiol. Rev.* **44,** 385 (1980).

[3] J. B. Hayes, *in* "Bacterial Transport" (B. P. Rosen, ed.), p. 43. Dekker, New York, 1980.

[4] H. R. Kaback, *Science* **186,** 882 (1978).

[5] G. F.-L. Ames, *Microbiology* 13 (1984).

[6] C. F. Higgins, *Microbiology* 17 (1984).

[7] P. M. Nazos, T. Z. Su, R. Landick, and D. L. Oxender, *Microbiology* 24 (1984).

[8] P. Duplay, H. Bedouelle, A. Charbit, J. M. Clement, D. E. Gilson, W. Saurin, and M. Hofnung, *Microbiology* 29 (1984).

[9] W. W. Kay, J. M. Somers, G. D. Sweet, and K. A. Widenhorn, *Microbiology* 34 (1984).

[10] R. W. Hogg, *Microbiology* 38 (1984).

[11] B. Rotman and R. Guzman, *Microbiology* 57 (1984).

[12] J. E. Lopilato, J. L. Garwin, S. D. Emr, T. J. Silhavy, and J. R. Beckwith, *J. Bacteriol.* **158,** 665 (1984).

[13] A. Iida, S. Harayama, T. Iino, and G. L. Hazelbauer, *J. Bacteriol.* **158,** 674 (1984).

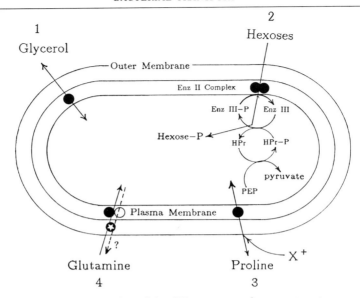

FIG. 1. Schematic representation of the different types of transport systems present in the gram-negative bacteria. System 1, typified by the glycerol facilitator, is a facilitated diffusion system.[1] System 2 is a phosphotransferase (PTS) uptake system and is typified by the transport systems for heoxses and hexitols. The salient point about this type of transport system is that it requires both cytoplasmic and membrane-associated components. The nutrients accumulated through this system are said to be taken up by a mechanism of group translocation, where the incoming substrate is phosphorylated through the interaction of small phosphorylated proteins (HPr or Enz III) or with the membrane-associated proteins. Phosphoenolpyruvate serves as the source of high energy phosphate. The reader is referred to recent reviews for a detailed description of these interesting systems.[2-4] System three is typified by the proline and lactose transport systems. This type of system appears to depend on a single membrane-associated protein for function and derives its energy from chemiosmotic gradients.[5-7] Proton cotransport is most often observed with these systems, although Na+ cotransport is observed with some systems. In contrast to the other types of energy coupled systems, the membrane-associated transport systems are active in osmotically shocked cells and membrane vesicles. Several excellent reviews include descriptions of the

[1] E. C. C. Lin, *Annu. Rev. Microbiol.* **30**, 535 (1978).
[2] M. H. Saier, Jr., "Mechanisms and Regulation of Carbohydrate Transport." Academic Press, New York, 1985.
[3] S. S. Dills, A. Apperson, M. R. Schmidt, and M. H. Saier, Jr., *Microbiol. Rev.* **44**, 385 (1980).
[4] J. B. Hayes, *in* "Bacterial Transport" (B. P. Rosen, ed.), p. 43. Dekker, New York, 1980.
[5] R. D. Simoni and P. W. Postma, *Annu. Rev. Biochem.* **44**, 523 (1975).
[6] B. P. Rosen, ed., "Bacterial Transport." Dekker, New York, 1978.
[7] B. P. Rosen, *in* "*Escherichia coli* and *Salmonella typhimurium*, Cellular and Molecular Biology" (J. L. Ingraham, K. B. Low, B. Magasanik, F. C. Neidhardt, M. Schaechter, and H. E. Umbarger, eds.). American Society of Microbiology, Washington, D.C., 1986.

or type of system.[14] This overview will deal primarily with the osmotic shock-sensitive or binding protein-dependent nutrient transport systems of *Escherichia coli* and *Salmonella typhimuium.*

Protein Components of the Binding Protein-Dependent Nutrient Transport Systems

Organization of the Protein Components

In Fig. 1, the binding protein-dependent transport systems are depicted as having one periplasmic protein component and two or more

[14] C. Furlong and G. D. Schellenberg, *in* "Microorganisms and Nitrogen Sources" (J. W. Payne, ed.). Wiley, New York, 1980.

---

membrane-bound type systems.[6,8-10] The fourth type of system is represented by binding protein dependent transport systems.[6,8,9,11,12] This type of system requires a soluble periplasmic substrate recognition component and membrane-associated protein components. The periplasmic protein is initially synthesized with an amino-terminal signal sequence that "guides" the protein to the periplasmic space where it is processed to a very stable mature substrate recognition component (usually referred to as a binding protein). Recent reviews and papers describe the present state of knowledge of binding protein secretion.[13-16] In addition to the periplasmic binding protein, two or more membrane-associated proteins are required for the function of this fourth type of transport system. The binding proteins may be released from cells by a gentle osmotic shock procedure[17] (or by a recently developed simple chloroform extraction procedure[18]). Thus, these systems are often referred to as osmotic shock-sensitive systems. The binding protein-dependent transport systems appear to utilize some form of high energy phosphate that may be ATP or a closely associated high energy metabolite.[19,20] The binding protein-dependent transport systems of *E. coli* and *S. typhimurium* are the major focus of this overview. As noted in the text, a given nutrient may be, and most often is transported through one or more of these different types of systems.

[8] D. Wilson, *Annu. Rev. Biochem.* **47,** 933 (1978).
[9] T. J. Silhavy, T. Ferenci, and W. Boos, *in* "Bacterial Transport" (B. P. Rosen, ed.). Dekker, New York, 1978.
[10] H. R. Kaback, *Science* **186,** 882 (1978).
[11] C. E. Furlong and G. D. Schllenberg, *in* "Microorganisms and Nitrogen Sources" (J. W. Payne, ed.), p. 89. Wiley, New York, 1980.
[12] D. L. Oxender, *Annu. Rev. Biochem.* **41,** 777 (1972).
[13] L. L. Randall and S. J. S. Hardy, *Microbiol. Rev.* **48,** 290 (1984).
[14] L. L. Randall and J. J. S. Hardy, *Modern Cell Biol.* **3,** 1 (1984).
[15] M. Inoye and S. Halegoua, *Crit. Rev. Biochem.* **7,** 339 (1980).
[16] M. Müller and G. Blobel, *Proc. Natl. Acad. Sci. U.S.A.* **81,** 7421 (1984).
[17] H. C. Neu and L. A. Heppel, *J. Biol. Chem.* **240,** 3685 (1965).
[18] G. F.-L. Ames, C. Prody, and S. Kustu, *J. Bacteriol.* **160,** 1181 (1984).
[19] E. A. Berger and L. A. Heppel, *J. Biol. Chem.* **249,** 7747 (1974).
[20] A. G. Hunt and J.-s. Hong, *J. Biol. Chem.* **256,** 11988 (1981).

membrane-bound protein components. This model is a consensus of the existing data. The evidence for the existence and function of the membrane-associated proteins and periplasmic protein(s) in transport varies for each system. In general, biochemical and genetic experiments have provided evidence for the involvement of the binding proteins in the uptake of their respective ligands. The role of the specific membrane-associated proteins has been much more difficult to establish, in part because the proteins appear to be associated with the plasma membrane, and in part because the membrane proteins are present in much lower quantities than their periplasmic counterparts.[5,7,8] Genetic experiments and molecular genetic experiments have provided most of the evidence for the involvement of the membrane-associated proteins in the function of specific binding protein-dependent transport systems.

Isolation of Specific Periplasmic Binding Proteins

The concept that a protein associated with a high affinity transport system should bind its substrate with high affinity led to the isolation of a sulfate-binding protein from *Salmonella typhimurium* by Pardee and co-workers.[15] They concluded that the sulfate-binding protein appeared to be localized in the periplasmic space since the binding protein was released when the cells were converted to spheroplasts or subjected to an osmotic shock procedure developed by Heppel and co-workers.[16] The osmotic shock procedure involves the plasmolysis of bacterial cells in the presence of EDTA and sucrose, harvesting the plasmolyzed cells, then resuspending the plasomlyzed cells in deionized water or a dilute solution of MgCl$_2$. The osmotic shock procedure selectively releases proteins located within the periplasmic space. The isolated sulfate-binding protein had the same substrate specificity and affinity as the sulfate permease.[15] Further, growth of cells under conditions that altered the level of the sulfate transport system also altered the level of the binding protein. In addition, some of the sulfate transport mutants had low levels of the binding protein, whereas others did not. Since three cistrons appeared to be involved in the sulfate transport system is was reasonable that not all transport mutants should be affected in the binding protein gene.

Following the isolation of the sulfate-binding protein, Oxender and co-workers isolated a binding protein that bound leucine, isoleucine, valine, threonine, or serine (LIV-BP).[17] This protein also had biochemical prop-

[15] A. B. Pardee, L. S. Prestidge, M. B. Whipple, and J. Dreyfuss, *J. Biol. Chem.* **241,** 3962 (1966).

[16] H. C. Neu and L. A. Heppel, *J. Biol. Chem.* **240,** 3685 (1965).

[17] W. R. Penrose, G. E. Nicholalds, J. R. Piperno, and D. L. Oxender, *J. Biol. Chem.* **243,** 5921 (1968).

erties that corresponded with the whole cell transport system. In addition, the presence of leucine in the medium repressed both the level of the transport system and the binding protein.

Anraku, working in Heppel's laboratory, independently isolated the LIV-BP and a galctose/glucose-binding protein (Galactose-BP).[18] Both of these proteins were released by the osmotic shock procedure.

Ames and co-workers used both biochemical and genetic approaches to demonstrate that a histidine-binding protein was an essential component of the high affinity histidine transport system of *Salmonella typhimurium*.[19] In addition, they provided evidence for the existence of an additional protein component required for binding protein-dependent transport. Their early studies also pointed out the importance of the multiplicity of transport systems for a given nutrient.[20]

Subsequently, many other binding proteins have been isolated from the *E. coli* and *S. typhimurium*. The table lists the binding proteins so far isolated and their properties. Most binding proteins have a number of properties in common. They are usually isolated as single subunit proteins with molecular weights between 23,000 and 52,000. The binding proteins whose crystal structures have been elucidated all seem to have similar two domain structures with the ligand binding site between the two domains.[21] Most are very resistant to heat or proteases and many have unusually broad pH optima of binding. Most bind their substrates with high affinity over a broad range of ionic strengths with high affinity. K_D values are usually in the micromolar range or below and correspond well to the observed K_m values for uptake.

Reconstitution Studies

Since the binding proteins could be removed from the periplasmic space by a gentle osmotic shock procedure with a concomitant reduction in transport through a specific binding protein-dependent system, it seemed reasonable to restore transport in the osmotically shocked cells simply by adding back osmotic shock fluid or highly purified binding protein. Early reports indicated success with this approach in reconstituting leucine,[22] galactose,[23] phosphate,[24] arginine,[25] and glutamate[26] trans-

[18] Y. Anraku, *J. Biol. Chem.* **243**, 3116 (1968).
[19] G. F.-L. Ames and J. E. Lever, *J. Biol. Chem.* **247**, 4309 (1972).
[20] G. F.-L. Ames, *ICN–UCLA Symp. Mol. Biol.* (1972).
[21] F. A. Quiocho and N. K. Vyas, *Nature (London)* **310**, 381 (1984).
[22] Y. Anraku, *J. Biol. Chem.* **243**, 3128 (1968).
[23] Y. Anraku, *J. Biol. Chem.* **242**, 793 (1967).
[24] N. Medveczky and H. Rosenberg, *Biochim. Biophys. Acta* **211**, 158 (1972).
[25] O. H. Wilson and J. T. Holden, *J. Biol. Chem.* **244**, 2743 (1969).
[26] H. Barash and Y. Halpern, *Biochem. Biophys. Res. Commun.* **45**, 681 (1971).

Binding protein	Purification and organism	Molecular weight	pI	K_D (μM)
Amino acids				
Arginine	*E. coli*[1-4]	28–33K	5.1[1,3]	Arg 0.03–0.1[1,3]
Ornithine				
Lysine	*E. coli*[5]	26–30K	5.1	Arg 0.15
Arginine				Lys 3
Ornithine	*S. typhimurium*[6]	26K[7]		Orn 5
Cystine	*E. coli*[9,10]	27–28K[9]		Cys 0.01[9]
Glutamine	*E. coli*[11-13]	23–29K[11,12]	8.6[11,12]	0.15–0.3[11,12]
	S. typhimurium[6]	23K[6]		
Glutamate	*E. coli*[16-18]	29–32K[16,17]	9.1–9.7[16,17]	0.8–6[16,17]
Aspartate	*S. typhimurium*[6,19]	30K[6,19]		Asp 1
Histidine	*E. coli*[20]	25–31K[20,21]		0.8
	S. typhimurium[23,24]	25K[7,23,24]	5.5[23]	0.15–1.5[23,24]
Leucine	*E. coli*[27-29]	36.7K[30]	4.8[29]	0.2–2[27-29]
Isoleucine				
Valine				
Threonine	*S. typhimurium*[33]	35–39K[33]	4.94[33]	Leu 0.43[33]
				Ile 0.15
				Val 0.89
Leucine-specific	*E. coli*[29,35]	37K[31,35]		0.7[35]
	S. typhimurium[33]	34–38K[33]	4.74[33]	0.54
Peptides				
Oligopeptide	*E. coli*	52K[38a]		
	S. typhimurium	52K[38a]		
Sugars				
Arabinose	*E. coli*[39-41]	33K[42]		0.2–2[39,40]
Galactose	*E. coli*[28,41]	32K[49]		1[28]
(Glucose)				
	S. typhimurium[56]	33K[56]		0.38[56]
Maltose	*E. coli*[57,58]	40.7K[59]		1[60]
Ribose	*E. coli*[65,66]	29.5[65]	6.6	0.13[65]
	S. typhimurium[69]	29K	7.3	0.33
Xylose	*E. coli*[70,71]	37K[70]	7.4[70]	0.6[70]
Anions				
Citrate	*S. typhimurium*[72]	28K	6.1	1–2.6
Phosphate	*E. coli*[75]	34K[76,77]		0.8[75]
sn-Glycerol-3-phosphate	*E. coli*[79]	45K	7	0.2
Sulfate	*S. typhimurium*[82]	34.7[83]		0.02[82]
Vitamins				
B$_{12}$	*E. coli* (partial)[85]	22K[85]		0.005[86]
Thiamine	*E. coli* (partial)[87,88]			0.03–0.1[87,88]

K_m transport (μM)	Map position of structural gene(s) (min)	Amino acid sequence	Gene sequence	Signal sequence	Membrane associated proteins[b]
Arg 0.15[3] Orn 4.0 Arg 0.5	60[2]				
	46[7,8]	Inferred[7]	—[7]	Inferred[7]	Same as histidine
0.1–0.3[9] 0.5–0.8[11,12]	17.7[14]	Partial[15]			
0.5[18]					
	50[22]				
0.04–0.4[25]	46[7,8]	—[26]	—[7]	Inferred[7]	hisQ 24.5K[27] hisM 26K hisP 28.7K
0.1–0.5[27]	74.5[31]	—[30]	—[32]	—[32]	livH 30K[31] livM 27K livG 23K
0.3–1[34]					
0.3[36]	74.5[31]		—[32]	—[32,37]	Same as LIV-BP system[31]
	27[38b]				(oppB)[38b]
	34[38b]				(oppC) (oppD) (in both organisms)
3–8[40,43]	45[22,44,45]	—[46]	—[47]	Inferred[47]	
0.5[50]	45[22,51]	—[53]		Inferred[53]	mglA 52K[54,55] mglC 38K
1[60]	91[22,61]	—[59]	—[59]	—[59,62,63]	malG 24K?[64] malF 57K malK 41K lamB 47K molA 13K?
0.3[65]	84[22]	—[67]	—[67]	Inferred[67]	rbsA 50K[68] rbsC 27K
5[70]	82[8]				
3	59[73]				40K?[74]
0.2[78]	74[22,79]		—[76,77]	—[76]	
2[80]	75.3[81]				
36[84]	(See text)				
0.01[85] 0.83[89]					

(*continued*)

References to TABLE

a References:
1. B. P. Rosen, *J. Biol. Chem.* **248,** 1211 (1973).
2. R. T. F. Celis, *J. Biol. Chem.* **256,** 773 (1981).
3. T. F. R. Celis, *J. Bacteriol.* **130,** 1244 (1977).
4. O. H. Wilson and J. T. Holden, *J. Biol. Chem.* **244,** 2743.
5. B. P. Rosen, *J. Biol. Chem.* **246,** 3653 (1971).
6. S. G. Kustu, N. C. McFarland, S. P. Hui, B. Esmon, and G. F. Ames, *J. Bacteriol.* **138,** 218 (1979).
7. C. F. Higgins and G. F.-L. Ames, *Proc. Natl. Acad. Sci. U.S.A.* **78,** 6038 (1981).
8. K. Sanderson and J. R. Roth, *in* "Genetic Maps 1984" (S. J. O'Brien, ed.), Vol. 3. Cold Spring Harbor Laboratory, Cold Spring Harbor, New York, 1984.
9. A. Burger and L. A. Heppel, *J. Biol. Chem.* **247,** 7684 (1972).
10. R. G. Oshima, R. C. Willis, C. E. Furlong, and J. A. Schneider, *J. Biol. Chem.* **249,** 6033 (1974).
11. J. H. Weiner, C. E. Furlong, and L. A. Heppel, *Arch. Biochem. Biophys.* **124,** 715 (1971).
12. J. H. Weiner and L. A. Heppel, *J. Biol. Chem.* **246,** 6933 (1971).
13. R. C. Willis and J. E. Seegmiller, *Anal. Biochem.* **72,** 66 (1976).
14. P. S. Masters and J.-S. Hong, *J. Bacteriol.* **147,** 805 (1981).
15. B. Marty, M. Bruschi, M. Ragot, and C. Gaudin, *C.R. Acad. Sci. Paris* **292,** III-987 (1981).
16. R. C. Willis and C. E. Furlong, *J. Biol. Chem.* **250,** 2574 (1975).
17. H. Barash and Y. S. Halpern, *Biochim. Biophys. Acta* **386,** 168 (1975).
18. G. D. Schellenberg and C. E. Furlong, *J. Biol. Chem.* **252,** 9055 (1977).
19. R. R. Aksamit, B. J. Howlett, and D. E. Koshland, Jr., *J. Bacteriol.* **123,** 1000 (1975).
20. L. A. Beck and C. E. Furlong, unpublished results.
21. F. A. Ardeshir and G. F. Ames, *J. Supramol. Struct.* **13,** 117 (1980).
22. B. Bachman, *in* "Genetic Maps 1984" (S. J. O'Brien, ed.), p. 145. Cold Spring Harbor Laboratory, Cold Spring Harbor, New York, 1984.
23. J. A. Lever, *J. Biol. Chem.* **247,** 4317 (1972).
24. B. P. Rosen and F. Vasington, *J. Biol. Chem.* **246,** 5351 (1971).
25. K. Krajewska-Grynkiewicz, W. Walczak, and T. Klopotowski, *J. Bacteriol.* **105,** 28 (1971).
26. R. W. Hogg, *J. Biol. Chem.* **256,** 1935 (1981).
27. W. R. Penrose, G. E. Nichoalds, J. R. Piperno, and D. L. Oxender, *J. Biol. Chem.* **243,** 5921 (1968).
28. Y. Anraku, *J. Biol. Chem.* **243,** 3116 (1968).
29. C. E. Furlong and L. A. Heppel, this series, Vol. 17B, p. 639.
30. Y. A. Ovchinnikov, N. A. Aldanova, V. A. Grinkevich, N. M. Arzamazova, and I. N. Moroz, *FEBS Lett.* **78,** 313 (1977).
31. P. M. Nazos, T. Z. Su, R. Landick, and D. L. Oxender, *Microbiology* 24 (1984).
32. D. L. Oxender and R. Landick, *J. Biol. Chem.* **260,** 8257 (1985).
33. K. Ohnishi and K. Kiritani, *J. Biochem. (Tokyo)* **94,** 433 (1983).
34. K. Ohnishi, K. Murata, and K. Kiratani, *Jpn. J. Genet.* **55,** 349 (1980).
35. C. E. Furlong and J. H. Weiner, *Biochem. Biophys. Res. Commun.* **38,** 1076 (1970).
36. C. E. Furlong, unpublished data.
37. D. L. Oxender, J. J. Anderson, C. J. Daniels, R. Landick, R. P. Gunzalus, G. Zurawski, and C. Yanofsky, *Proc. Natl. Acad. Sci. U.S.A.* **77,** 2005 (1980).
38a. C. F. Higgins and M. M. Hardie, *J. Bacteriol.* **155,** 1434 (1983).
38b. M. M. Gibson, M. Price, and C. F. Higgins, *J. Bacteriol.* **160,** 122 (1984).
39. R. W. Hogg and E. Englesberg, *J. Bacteriol.* **100,** 429 (1969).
40. R. Schleif, *J. Mol. Biol.* **46,** 185 (1969).
41. R. W. Hogg, this series, Vol. 90 [76].

References to TABLE (*continued*)

[42] F. A. Quiocho and N. K. Vyas, *Nature (London)* **310,** 381 (1984).
[43] C. E. Brown and R. W. Hogg, *J. Bacteriol.* **111,** 606 (1972).
[44] R. W. Hogg, *J. Supramol. Struct.* **6,** 411 (1977).
[45] A. Clark and R. W. Hogg, *J. Bacteriol.* **147,** 920 (1981).
[46] R. W. Hogg and M. A. Hermodson, *J. Biol. Chem.* **252,** 4135 (1977).
[47] R. W. Hogg, *Microbiology* 38 (1984).
[48] F. A. Quiocho, W. E. Meador, and J. W. Pflugrath, *J. Mol. Biol.* **133,** 181 (1979).
[49] W. Boos, *Curr. Top. Membr. Transp.* **5,** 51 (1974).
[50] B. Rotman and J. Radojkovic, *J. Biol. Chem.* **239,** 3153 (1964).
[51] A. K. Ganesan and B. Rotman, *J. Mol. Biol.* **16,** 42 (1965).
[52] W. C. Mahoney, R. W. Hogg, and M. A. Hermodson, *J. Biol. Chem.* **256,** 4350 (1981).
[53] J. B. Scripture and R. W. Hogg, *J. Biol. Chem.* **258,** 10853 (1983).
[54] B. Rotman and R. Guzman, *J. Biol. Chem.* **257,** 9030 (1982).
[55] S. Harayama, J. Bollinger, R. Iino, and G. L. Hazelbauer, *J. Bacteriol.* **153,** 408 (1983).
[56] R. S. Zukin, P. G. Strange, L. R. Heavey, and D. E. Koshland, Jr., *Biochemistry* **16,** 381 (1977).
[57] O. Kellerman and S. Szmelcman, *Eur. J. Biochem.* **47,** 139 (1974).
[58] O. Kellerman and T. Ferenci, this series, Vol. 90 [75].
[59] P. Duplay, H. Bedouelle, A Fowler, I. Zabin, W. Saurin, and M. Hofnung, *J. Biol. Chem.* **259,** 10606 (1984).
[60] S. Szmelcman, M. Schwartz, T. J. Silhavy, and W. Boos, *Eur. J. Biochem.* **65,** 13 (1976).
[61] M. Hofnung, M. Schwartz, and D. Hatfield, *J. Bacteriol.* **117,** 40 (1974).
[62] H. Bedouelle, P. J. Bassford, Jr., A. V. Fowler, I. Zabin, J. Beckwith, and M. Hofnung, *Nature (London)* **285,** 78 (1980).
[63] K. Ito, *J. Biol. Chem.* **257,** 9895 (1982).
[64] P. Duplay, H. Bedouelle, A. Charbit, J. M. Clement, D. E. Gilson, W. Saurin, and M. Hofnung, *Microbiology* 29 (1984).
[65] R. C. Willis and C. E. Furlong, *J. Biol. Chem.* **249,** 6926 (1973).
[66] C. E. Furlong, this series, Vol. 90 [77].
[67] J. M. Groarke, W. C. Mahoney, J. N. Hope, C. E. Furlong, F. T. Robb, H. Zalkin, and M. A. Hermodson, *J. Biol. Chem.* **258,** 12952 (1983).
[68] A. Iida, S. Harayama, T. Iino, and G. L. Hazelbauer, *J. Bacteriol.* **158,** 674 (1984).
[69] R. R. Aksamit and D. E. Koshland, *Biochem. Biophys. Res. Commun.* **48,** 1348 (1972).
[70] C. Ahlem, W. Huisman, G. Neslund, and A. S. Dahms, *J. Biol. Chem.* **257,** 2926 (1982).
[71] A. S. Dahms, W. Huisman, G. Neslund, and C. Ahlem, this series, Vol. 90, p. 473.
[72] G. D. Sweet, J. M. Somers, and W. W. Kay, *Can. J. Biochem.* **57,** 710 (1979).
[73] J. M. Sommers, G. D. Sweet, and W. W. Kay, *Mol. Gen. Genet.* **181,** 338 (1981).
[74] W. W. Kay, J. M. Somers, G. D. Sweet, and K. A. Widenhorn, *Microbiology* 34 (1984).
[75] N. Medveczky and H. Rosenberg, *Biochim. Biophys. Acta* **211,** 158 (1970).
[76] B. P. Surin, D. A. Jans, A. L. Fimmel, D. C. Shaw, G. B. Cox, and H. Rosenberg, *J. Bacteriol.* **157,** 772 (1984).
[77] K. Magota, N. Otsuje, T. Miki, T. Horiuchi, S. Tsunasawa, J. Kondo, F. Sakiyama, M. Amemura, T. Morita, H. Shinagawa, and A. Nakata, *J. Bacteriol.* **157,** 909 (1984).
[78] H. Rosenberg, R. G. Gerdes, and K. Chegwidden, *J. Bacteriol.* **131,** 505 (1977).
[79] M. Argast and W. Boos, *J. Biol. Chem.* **254,** 10931 (1979).
[80] H. Schweizer, M. Argast, and W. Boos, *J. Bacteriol.* **150,** 1154 (1982).
[81] H. Schweizer, T. Grussenmeyer, and W. Boos, *J. Bacteriol.* **150,** 1164 (1982).
[82] A. B. Pardee, *J. Biol. Chem.* **241,** 5886 (1966).
[83] H. Isihara and R. W. Hogg, *J. Biol. Chem.* **255,** 4614 (1980).

(continued)

port. However, a number of laboratories had difficulties in reproducing these experiments.[27-29] Three major problems contributed to these difficulties. First, osmotic shock should remove the entire small molecule pool from the cells.[30] Since the binding protein-dependent transport systems are energy dependent, anything that supplied energy to the shocked cells would appear to "reconstitute" transport.[14] Second, the multiplicity of transport systems with different mechanisms for the same nutrient was not sufficiently understood at that time.[14] Third, the binding protein: ligand complex bound to nitrocellulose filters with sufficient affinity that the system could be used as a binding assay.[31] These three problems contributed significantly to making the reconstitution studies difficult to interpret.

Recently, progress has been made in reconstituting binding protein dependent transport in several different systems.[11,29,32-35]

Regulation of the Binding Protein-Dependent Transport Systems

Most, if not all of the binding protein-dependent transport systems (as well as other transport systems) are subject to one or more forms of regulation—specific induction, repression or general carbon, nitrogen, sulfur, or phosphate control, or combinations of any of these. The term regulation was originally proposed to describe systems that were under

[27] D. L. Oxender, *Annu. Rev. Biochem.* **41**, 777 (1972).
[28] L. A. Heppel, B. P. Rosen, I. Friedberg, E. A. Berger, and J. H. Weiner, *in* "The Molecular Basis of Biological Transport" (J. F. Woessner, Jr. and F. Huijing, eds.), p. 133. Academic Press, New York, 1972.
[29] R. G. Gerdes, K. P. Strickland, and H. Rosenberg, *J. Bacteriol.* **131**, 512 (1977).
[30] R. J. Britten and F. T. McClure, *Bacteriol. Rev.* **26**, 292 (1962).
[31] C. E. Furlong and J. H. Weiner, *Biochem. Biophys. Res. Commun.* **38**, 1076 (1970).
[32] P. S. Masters and J.-s. Hong, *Biochemistry* **20**, 4900 (1981).
[33] A. G. Hunt and J.-s. Hong, *Biochemistry* **22**, 851 (1983).
[34] J. M. Brass and M. D. Manson, *J. Bacteriol.* **157**, 881 (1984).
[35] J. T. Robb and C. E. Furlong, *J. Supramol. Struct.* **13**, 183 (1979).

References to TABLE (*continued*)

[84] J. Dreyfuss, *J. Biol. Chem.* **239**, 2292 (1964).
[85] R. T. Taylor, S. A. Norrell, and M. L. Hanna, *Arch. Biochem. Biophys.* **148**, 366 (1972).
[86] C. Bradbeer, J. S. Kenley, D. R. Di Masi, and M. Leighton, *J. Biol. Chem.* **253**, 1347 (1978).
[87] T. Nishimune and R. Hayashi, *Biochim. Biophys. Acta* **244**, 573 (1971).
[88] A. Iwashima, A. Matsuura, and Y. Nose, *J. Bacteriol.* **108**, 1419 (1971).
[89] T. Kawasaki, I. Miyata, K. Esaki, and Y. Nose, *Arch. Biochem. Biophys.* **131**, 223 (1969).
[b] Genetic and/or biochemical identification.

the control of a single repressor,[36] but has gained acceptance as a term to describe collections of operons that share even positive control elements.[37] The expression of many of the binding protein-dependent transport systems appear to be under the control of one or more regulons.

The Binding Protein-Dependent Transport Systems of *Escherichia coli* and *Salmonella typhimurium*

The table lists the binding proteins from *E. coli* and *S. typhimurium* that have been characterized so far. In addition to the binding proteins and their properties, references are provided for the membrane-associated proteins that have been identified biochemically or genetically. Where known, the chromosomal location of the genes encoding the binding proteins and associated membrane proteins are noted.

[36] W. K. Maas and A. J. Clark, *J. Mol. Biol.* **8**, 365 (1964).
[37] B. L. Wanner, *J. Mol. Biol.* **166**, 283 (1983).

[23] Calcium-Induced Permeabilization of the Outer Membrane: A Method for Reconstitution of Periplasmic Binding Protein-Dependent Transport Systems in *Escherichia coli* and *Salmonella typhimurium*

By JOHANN M. BRASS

The cell envelope of gram-negative bacteria is composed of three layers or compartments: (1) the outer membrane, (2) the inner or cytoplasmic membrane, and (3) the periplasm between these two membranes, containing soluble proteins and the peptidoglycan layer. The outer membrane protects the cell against detergents and degrading enzymes which are present in the colon, the normal habitat of enteric bacteria. The outer leaflet of this membrane contains the amphipathic lipopolysaccharide (LPS) consisting of a hydrophobic lipid A moiety and a hydrophilic polysaccharide. Phospholipids are found only in the inner leaflet of this membrane. The core region of the polysaccharide carries 4–5 negatively charged residues that allow tight association of LPS after cross-bridging by divalent cations.[1,2] The fact that the LPS molecules are held together

[1] L. Leive, *Proc. Natl. Acad. Sci. U.S.A.* **53**, 745 (1965).
[2] M. Schindler and M. J. Osborn, *Biochemistry* **18**, 4425 (1979).

by ionic interactions as well as by hydrophobic interactions is responsible for the permeability barrier function of the outer membrane against hydrophobic and hydrophilic solutes. Passive entry of small substrate molecules and minerals through the outer membrane into the periplasm occurs through unspecific and substrate specific pore proteins (porins).[3,4]

The membrane-surrounded compartment of the periplasm harbors some enzymes and a variety of soluble-binding proteins involved in active transport of nutrients like sugars and amino acids or ions.[5,6] Some binding proteins, such as those for maltose, ribose, and galactose, function in addition as chemoreceptors for these nutrients in bacterial chemotaxis.[7] One example of a binding protein-dependent transport system, the maltose transport system of *E. coli* is described in Fig. 1. Other binding protein-dependent transport systems seem to be composed in a similar way, operating with a soluble periplasmic binding protein with high substrate specificity and 2–3 inner membrane transport proteins, which together accomplish accumulation of substrate in the cytoplasm.

For isolation of periplasmic binding proteins, efforts were made to find conditions for selective permeabilization of the outer membrane. The permeability barrier of this membrane for macromolecules can be overcome by exposure of cells to EDTA, which complexes divalent cations bound to LPS.[1] EDTA treatment combined with an osmotic shock at low temperature leads to release of the soluble periplasmic proteins in *E. coli* and other gram-negative bacteria.[8]

The structure of the outer membrane can also be destabilized by treatment of gram-negative cells at low temperature with high concentrations of divalent cations. Ca^{2+}-induced LPS aggregates can be seen in the outer membrane of *E. coli* by electron microscopy.[9] Cells treated with millimolar concentrations of Ca^{2+} are competent to take up DNA.[10,11] We found recently that proteins can also easily be introduced into the periplasm of *E. coli* and *S. typhimurium* during a Ca^{2+} treatment at low temperature.[12-14]

[3] T. Nakae, *Biochem. Biophys. Res. Commun.* **64,** 1124 (1975).

[4] H. Nikaido, this volume [21].

[5] D. Oxender and S. Quay, *Methods Membr. Biochem.* **6,** 183 (1976).

[6] R. Hengge and W. Boos, *Biochim. Biophys. Acta* **737,** 443 (1983).

[7] G. L. Hazelbauer and S. Harayama, *Int. Rev. Cytol.* **81,** 33 (1983).

[8] H. C. Neu and L. A. Heppel, *J. Biol. Chem.* **240,** 3685 (1965).

[9] L. Van Alphen, A. Verkleij, J. Leunissen-Bijvelt, and B. Lugtenberg, *J. Bacteriol.* **134,** 1089 (1978).

[10] M. Mandel and A. Higa, *J. Mol. Biol.* **53,** 159 (1970).

[11] A. Taketo, *J. Biochem.* **75,** 895 (1974).

[12] J. M. Brass, W. Boos, and R. Hengge, *J. Bacteriol.* **146,** 10 (1981).

[13] J. M. Brass, U. Ehmann, and B. Bukau, *J. Bacteriol.* **155,** 97 (1983).

[14] B. Bukau, J. M. Brass, and W. Boos, *J. Bacteriol.* **163,** 61 (1985).

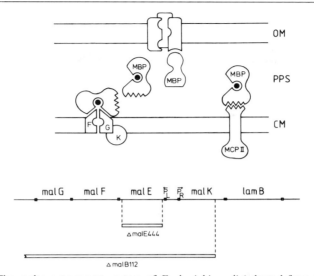

FIG. 1. The maltose transport system of *Escherichia coli* (adapted from Hengge and Boos[6] and Brass and Manson[30]). The localization and the interaction of *malB* gene products which are responsible for uptake of maltodextrins as well as the genetic organization of the *malB* region is shown. The *malB* region consists of two divergent operons transcribed from promoters P_L and P_R. Transcription is dependent on the presence of the *malT* gene product and is induced by binding of maltose to this activator protein. The *malE* gene codes for the maltose-binding protein (MBP) which is localized in the periplasmic space (PPS), *lamB* codes for the outer membrane maltoporin (or λ receptor), and *malF*, *malG*, and *malK* code for the cytoplasmic membrane components of the transport system. The *malE444* deletion is nonpolar and the *malB112* deletion eliminates expression of all *malB* gene products. Malto-porin specifically facilitates permeation of maltose and maltodextrins through the outer membrane (OM), and its maltodextrin binding sites are indicated. MBP undergoes a confor-mational change upon binding substrate. The substrate-bound form is believed to interact with the complex formed by the *malF*, *G*, and *K* gene products. Active transport is achieved by translocation of substrate to a hypothetical substrate-binding site of this complex in synchrony with conformational changes of all the components. Energization requires a metabolite related to ATP. Maltose-loaded MBP can also interact with methyl-accepting chemotaxis protein II (MCPII), product of the *tar* gene, to initiate the maltose chemotactic response. (For references see the above cited review of R. Hengge and W. Boos.[6])

Reconstitution of Binding Protein-Dependent Transport in Spheroplasts and Ca²⁺-Treated Cells

Much of the evidence for the involvement of periplasmic binding pro-teins as essential substrate recognition components of specific transport systems (and as chemoreceptors in chemotaxis) has been circumstantial. Most of our knowledge is based on genetic data. The direct biochemical proof for involvement of binding proteins requires a demonstration of restoration of transport by addition of a purified binding protein to cells lacking this component.

In early studies, binding protein was removed from the periplasm by the osmotic shock procedure.[8] The stimulation of transport which was seen after addition of binding protein to shocked cells was not due to transport reconstitution but rather to artifacts arising from residual and resynthesized binding protein.[15,16] Successful reconstitution of phosphate, glutamate, and ribose transport in E. coli was reported by addition of the respective binding proteins to spheroplasts.[16–18] Reconstitution using spheroplasts derived form a binding protein-deficient mutant, however, was reported only in the case of the ribose transport system.[19] Several disadvantages are inherent to the spheroplast procedure. The preparation of spheroplasts is laborious, the quality of preparations is variable, and transport assays have to be performed in the presence of high concentrations of binding protein and in the presence of high concentrations of sucrose to prevent lysis of the spheroplasts.

An important development in reconstitution procedures was achieved when inner membrane vesicles were used instead of spheroplasts. This interesting technique which allows studies about the energization of binding protein-mediated transport systems[20,21] is outlined in a previous section of this volume.

In the present work we describe a new method for reconstitution of binding protein-mediated transport in whole cells which is based on the reversible Ca^{2+}-induced permeabilization of the outer membrane. The introduction of exogenous binding protein into the periplasm of binding protein-deficient mutants is accomplished by incubating cells and binding protein in the presence of 300 mM Ca^{2+} at 0°.

Several advantages of this approach are evident.

1. Reconstitution experiments with whole cells are highly reproducible, much simpler than those with spheroplasts and vesicles, and reflect a more natural situation.

2. Due to resealing of the outer membrane after transfer of the cells into Ca^{2+}-free medium, extracellular binding protein can be washed off

[15] A. S. Rae, K. P. Strickland, N. Medveczky, and H. Rosenberg, Biochim. Biophys. Acta 433, 555 (1976).
[16] R. G. Gerdes, K. P. Strickland, and H. Rosenberg, J. Bacteriol. 131, 512 (1977).
[17] H. Barash and Y. S. Halpern, Biochem. Biophys. Res. Commun. 45, 681 (1971).
[18] D. R. Galloway, and C. E. Furlong, Arch. Biochem. Biophys. 197, 158 (1979).
[19] F. T. Robb and C. E. Furlong, J. Supramol. Struct. 13, 183 (1980).
[20] A. G. Hunt and J. S. Hong, J. Biol. Chem. 256, 11988 (1981).
[21] B. Rotman and R. Guzman, Microbiology 17A, 57 (1984).

prior to the transport assay; the cells retain for hours their reconstituted transport ability.

3. The efficiency of the technique as judged by the K_m and V_{max} values of reconstituted maltose transport, e.g., is high: K_m is identical with the wild-type value, V_{max} is about 30% of the wild-type value.

4. The technique also allows reconstitution of binding protein-mediated chemotaxis.

Standard Procedure for Reconstitution of Binding Protein-Mediated Transport in Ca^{2+}-Treated Cells of E. coli and S. typhimurium

Reconstitution of binding protein-mediated transport in whole cells of E. coli and S. typhimurium can be measured in strains lacking a specific binding protein but expressing the inner membrane components of the respective transport system. The expression of the mglA, mglC, and mglE gene products of the binding protein-dependent galactose transport system[21,22] of E. coli can be induced in strains carrying nonpolar mglB mutations, missing the galactose-binding protein (GBP), by growth in 10^{-3} M fucose. The expression of the hisM, hisP, and hisQ gene products of the histidine transport system[23] of S. typhimurium can be increased in strains carrying nonpolar hisJ mutations, missing the histidine-binding protein, by nitrogen starvation[24] or by use of dhuA regulator gene mutations.[25] Constitutive expression of the malF, malG, malK, and lamB gene products of the binding protein-dependent maltose transport system of E. coli (see Fig. 1) can be accomplished in strains carrying the nonpolar malE444 deletion, missing the maltose-binding protein (MBP), by use of $malT^c$ activator gene mutations.[26,27] The following optimized reconstitution procedure is a modification of that described by us recently.[13]

Materials

Minimal medium A (MMA): for a 10-fold stock solution add 105 g K_2HPO_4, 45 g KH_2PO_4, 4 g Na-citrate · $2H_2O$, 1 g $MgSO_4$ · $7H_2O$, and 10 g

[22] N. Müller, H.-G. Heine, and W. Boos, J. Bacteriol. **163,** 37 (1985).
[23] G. F.-L. Ames, Arch. Biochem. Biophys. **104,** 1 (1964).
[24] S. G. Kustu, N. C. McFarland, S. P. Hui, B. Esmon, and G. F.-L. Ames, J. Bacteriol. **138,** 218 (1979).
[25] G. F.-L. Ames, K. D. Noel, H. Taber, E. Negri Spudich, K. Nikaido, J. Afong, and F. Ardeshir, J. Bacteriol. **129,** 1289 (1977).
[26] M. Debarbouille, H. A. Shuman, T. J. Silhavy, and M. Schwartz, J. Mol. Biol. **124,** 359 (1978).
[27] H. A. Shuman, J. Biol. Chem. **257,** 5455 (1982).

$(NH_4)_2SO_4$ to 1 liter of distilled water; after autoclaving, dilute 1/10 with distilled sterile water and add a sterile solution of glycerol up to a final concentration of 0.4%.

Tris buffer: 100 mM Tris–HCl, pH 7.5.

Tris–Ca^{2+} buffer: 100 mM Tris–HCl, pH 7.5, containing 300 mM CaCl$_2$.

Phosphate buffer: 100 mM potassium phosphate, pH 7.5.

Shock fluids: shock fluids containing between 25 and 50% MBP of total protein are prepared by the osmotic shock procedure[8] from a maltose-induced Mal$^+$ $E.$ $coli$ strain. Shock fluid containing up to 90% galactose-binding protein (GBP) is prepared from strains carrying a pACYC184 derivative, harboring the cloned mgl region.[28] Shock fluid containing the histidine-binding protein from $S.$ $typhimurium$ is conveniently prepared from a $dhuA$ strain with constitutive expression of the $hisJ$ gene.[25] Lyophilized shock fluids are dialyzed once against 10 mM Tris–HCl, pH 7.5, containing 0.02% sodium azide and twice against 1 mM Tris–HCl, pH 7.5.

MBP: MBP is purified to homogeneity by amylose affinity chromatography[29] and dialyzed three times as described above for shock fluids. After dialysis, all protein preparations, shock fluids, and purified binding protein are centrifuged for 20 min at 40,000 g before lyophilization to remove particulate matter.

The biological activity of freeze dried shock fluid and MBP as measured in reconstitution of transport declines slowly ($t_{1/2}$ of about 2 month).

Procedure

Cells are grown ın minimal medium A (MMA) containing 0.4% glycerol, appropriate supplements, and inducers at 37°. Cells are harvested in exponential phase at optical densities between 0.5 and 1 at 578 nm. A total of 1×10^9 cells are pelleted in an Eppendorf centrifuge. The supernatant is decanted and the remaining drops of the medium are removed after a second short run in the rotor. The cells are now subjected to two alternative pretreatments (see comments).

$Tris–Ca^{2+}$ $Pretreatment.$ The cells are resuspended and washed once with 1 ml of ice-cold 100 mM Tris–HCl buffer, pH 7.5, containing 300 mM CaCl$_2$.

$Tris–PO_4^{2-}$ $Pretreatment.$ The cells are resuspended and washed at room temperature first in 1 ml 100 mM Tris–HCl buffer, pH 7.5, and second in 1 ml 100 mM potassium phosphate buffer, pH 7.5.

$Introduction$ of $Binding$ $Protein.$ Cells subjected to the Tris–Ca^{2+} pre-

[28] N. Müller, H. G. Heine, and W. Boos, $Mol.$ $Gen.$ $Genet.$ **185,** 473 (1982).
[29] T. Ferenci and U. Klotz, $FEBS$ $Lett.$ **94,** 213 (1978).

treatment or the Tris–PO$_4^{2-}$ pretreatment are centrifuged, the supernatant is completely removed, and the cells are resuspended in 50 μl of ice cold 100 mM Tris–HCl buffer, pH 7.5, containing 300 mM CaCl$_2$ and purified binding protein (final concentration between 20 and 50 mg/ml) or crude shock fluid (final concentration between 40 and 100 mg/ml). The reconstitution mixture is shaken at 0° for 30 min. After washing the cells in 1 ml 0.9% NaCl the cells are resuspended in 1 ml MMA and kept at room temperature before transport assays are performed. Under these conditions the reconstituted competence for transport (or chemotaxis) is stable for more than 2 hr.[13,30]

Comments

The efficiency of reconstitution of MMA grown cells subjected to the Tris–PO$_4^{2-}$ pretreatment is somewhat higher than that of cells subjected to the Tris–Ca^{2+} pretreatment.[13,14]

Cells grown in other media (Vogel–Bonner minimal medium, double yeast extract tryptone (2 × YT) or Luria broth (LB) are also competent for uptake of binding proteins. The reconstitution efficiency of cells grown in rich media is high, provided that the Tris–PO$_4^{2-}$ pretreatment is applied.

Transport Assays

For determination of the initial rate of maltose uptake in *E. coli*, 1 × 10^9 cells are resuspended in 1 ml MMA at room temperature. At time zero [^{14}C]maltose (specific activity 6 mCi/mmol; final concentration 5 μM) is added. Samples (150 μl) are withdrawn at 10, 25, 40, 60, and 80 sec, and the cells are filtered onto membrane filters (0.45 μm pore size; Millipore Corp.) and washed three times with 5 ml portions of MMA at room temperature. The radioactivity of the dried filters is counted in a toluene-based scintillation fluid.

For determination of the initial rate of galactose uptake in *E. coli*, 1 × 10^8 cells are resuspended in 1 ml MMA at room temperature. At time zero [^{14}C]galactose (specific activity 50 mCi/mmol; final concentration 0.4 μM) is added. Processing of the samples is the same as described for maltose transport. At this low substrate concentration, binding protein-mediated transport is half saturated, and galactose uptake by additional transport systems is negligible.

For determination of the initial rate of histidine uptake in *S. typhimurium,* cells are assayed for incorporation of labeled histidine into protein.[23]

---

[30] J. M. Brass and M. D. Manson, *J. Bacteriol.* **157,** 881 (1984).

Conditions Affecting Entry and Function of Exogenous Maltose-Binding Protein in Ca^{2+}-Treated *malE* Mutants of *E. coli*

Permeabilization of the outer membrane as measured by reconstitution of maltose transport in $\Delta malE444$ $malT^c$ mutants of *E. coli* after addition of exogenous maltose-binding protein is dependent on the presence of high concentrations of divalent cations.[13] Monovalent cations are completely inefficient. Besides Ca^{2+}, other divalent cations such as Ba^{2+}, Sr^{2+}, Mg^{2+}, and Ni^{2+} are also able to permeabilize with the efficiency decreasing in that order.[14]

The calcium concentration in the reconstitution mixture is a critical factor. Reconstituted transport activity increases linearly with the Ca^{2+} concentration up to 400 mM when cells are pretreated with MBP at 0°. The viability of the cells remains high (80%) at all Ca^{2+} concentrations tested.[13] The inner membrane which is free of LPS and not bound to the rigid peptidoglycan matrix is by far less sensitive to Ca^{2+}-induced permeabilization than the outer membrane. Recently, it has become clear that the plasmolysis of the cells caused by these high Ca^{2+} concentrations is a prerequisite for optimal reconstitution. Reconstitution is equivalent in cells incubated with 50 mM Ca^{2+} plus 400 mM sucrose and in cells treated with 300 mM Ca^{2+}.[14] The plasmolysis effect may allow uptake of large amounts of MBP into the enlarged periplasm of plasmolyzed cells. Plasmolysis in the presence of MBP and varying concentrations of EDTA (1 × 10^{-4} to 1 × 10^{-3} M) did not result in reconstitution.

Ca^{2+} by itself is inefficient in promoting reconstitution. The combined action of the Ca^{2+} and phosphate ions seems to be necessary.[14] Cells cannot be reconstituted when they are washed free from the phosphate-containing growth medium in Tris buffer or in NaCl prior to the addition of Ca^{2+}. Compared to cells washed in Tris alone, cells washed in Tris plus increasing phosphate concentrations (pH 7.5) show a dramatic increase in transport activity. The phosphate optimum in this pretreatment is around 100 mM. Thus, it is perhaps more accurate to speak of calcium phosphate-induced, rather than Ca^{2+}-induced reconstitution. A similar combined action of Ca^{2+} and phosphate is seen in studies about fusion of phospholipid vesicles and erythrocyte membranes.[31,32]

Temperature strongly influences reconstitution which becomes dramatically better when the temperature is lowered from 20 to 0°. The pH optimum for reconstitution of maltose transport is about pH 7.5 and is very sharp; reconstitution is quite poor below pH 7.0 and above pH 8.0.[13]

[31] R. Frayley, J. Wilschut, N. Duzgunes, C. Smith, and D. Papahadjopoulos, *Biochemistry* **19**, 6021 (1980).

[32] D. Hoekstra, J. Wilschut, and G. Scherphof, *Biochim. Biophys. Acta* **732**, 327 (1983).

MBP enters the periplasm rapidly; high rates of reconstituted transport are already seen in cells incubated with MBP for less than 1 min. Cells incubated for longer periods (up to 30 min) show only slightly higher transport activity. The reconstituted transport activity remains stable even after repeated washing of the cells over the next 2 hr. The rapid entry of MBP and its very slow exit after shift of the cells into NaCl or minimal medium (little loss of transport activity in 2 hr) indicate either that the outer membrane reseals immediately after removal of Ca^{2+} or that there are high-affinity binding sites for MBP in the periplasm.[13]

About 50–70% of the cells from an exponentially growing culture are competent for Ca^{2+}-induced uptake of protein. This can be concluded from microscopic observation of cells incubated with rhodamine isothiocyanate-labeled, fluorescent-binding protein.[33] Part of the competent cells (10–20%) showed strongly increased fluorescence at the pole caps of the cells.

The competence of cells for reconstitution decreases at the end of the exponential growth phase.[13] This decrease is not seen in *lpo* mutants, which lack lipoprotein and therefore the normal covalent linkage of the outer membrane with the rigid murein layer.[34]

Most of the parameters affecting reconstitution of transport in *E. coli*, such as growth phase of the cells used, pH, temperature, ion specificity, and stimulation by plasmolysis, were found to be similar to those affecting transformation of *E. coli*.[11,13,14,35] One difference was that the short heat pulse necessary for transformation (2 min, 42°), which is required to get DNA through the outer and inner membrane does not stimulate reconstitution of transport.[14]

Characteristics of Reconstituted Maltose Transport in Δ*malE malT*^{c} Mutants of *E. coli*

Addition of exogenous MBP allows reconstitution of maltose transport in *malT*^{c} strains carrying the nonpolar deletion Δ*malE444*, but not in the *malT*^{c} Δ*malB112* control strain carrying a deletion covering four genes necessary for maltose uptake (Figs. 1 and 2). Addition of sodium azide (20 mM) to the uptake mixture completely inhibits maltose uptake in reconstituted cells, showing that reconstituted transport is due to active transport. Purified MBP (20 mg/ml) and crude shock fluid (40 mg/ml) have about the same activity in reconstitution (Fig. 2).

[33] C. F. Higgins, J. M. Brass, M. Foley, P. Rugman, and P. Garland, *J. Bacteriol.*, in press (1986).

[34] Y. Hirota, H. Suzuki, Y. Nishimura, and S. Yasuda. *Proc. Natl. Acad. Sci. U.S.A.* **74,** 1417 (1977).

[35] H. E. N. Bergmans, J. M. van Die, and W. P. M. Hoekstra, *J. Bacteriol.* **146,** 564 (1981).

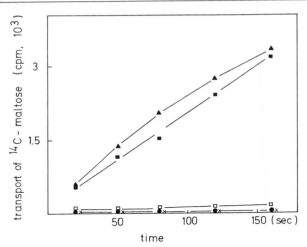

FIG. 2. Reconstitution of active maltose transport in whole cells of strain HS3018 ($\Delta malE444$ $malT^c$-1) with purified MBP and crude shock fluid. Cells (1×10^9) were subjected to the standard reconstitution procedure (Tris–Ca^{2+} pretreatment). After the cells were washed free of Ca^{2+} and MBP, maltose uptake was measured. (●) HS3018 without protein in the reconstitution mixture; (▲) HS3018 plus 40 mg of crude shock fluid/ml containing 20 mg of MBP/ml; (■) HS3018 plus 20 mg of MBP/ml; (□) HS3018 plus 20 mg of MBP/ml in the reconstitution mixture and 20 mM NaN$_3$ in the uptake medium; (×) control strain pop1740 ($\Delta malB112$) plus 20 mg of MBP/ml. The results are given as the amount of maltose taken up by 7.5×10^7 cells.

The K_m for maltose transport in reconstituted cells is 2 μM, a value which is identical to that of wild-type cells (Fig. 3).[13,36] This shows that the described reconstitution procedure is a gentle method allowing the normal functioning of all components involved in maltose transport.

The half-optimal concentration of purified MBP in the reconstitution mixture is about 0.5 mM (Fig. 4), a value which comes close to the estimated MBP concentration in the periplasm of maltose-induced wild-type cells.[37] The extrapolated maximal rate of transport at saturating MBP concentrations is around 30% that of the rate of Ca^{2+}-treated $malE^+$ $malT^c$ cells.[13] The threefold reduced V_{max} value is due to the fact that only about 50% of the cells from an exponentially growing culture are competent for Ca^{2+}-induced uptake of protein (see above). $\Delta malE$ $lamB$ $malT^c$ mutants lacking MPB and maltoporin are also competent for reconstitu-

[36] S. Szmelzman, M. Schwartz, T. J. Silhavy, and W. Boos, *Eur. J. Biochem.* **65,** 13 (1976).
[37] I. Dietzel, V. Kolb, and W. Boos, *Arch. Microbiol.* **118,** 207 (1979).

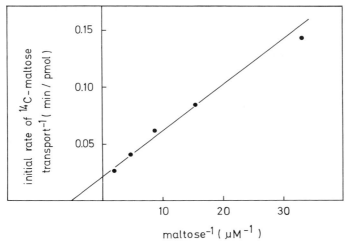

FIG. 3. Determination of the K_m of maltose transport in reconstituted cells. Samples of 1×10^9 cells from strain HS3018 ($\Delta malE444\ malT^c$-1) were subjected to the standard reconstitution procedure (Tris–Ca²⁺ pretreatment) applying shock fluid containing 10 mg of MBP/ml. The K_m of maltose transport was determined by measuring the initial rates of maltose uptake at maltose concentrations between 3×10^{-7} and $5.15 \times 10^{-6}\ M$.

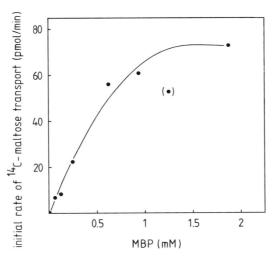

FIG. 4. Determination of the concentration optimum for purified MBP in reconstitution. Samples of 1×10^9 cells from strain HS3018 ($\Delta malE444\ malT^c$-1) were subjected to the standard reconstitution procedure (Tris–Ca²⁺ pretreatment) except that different concentrations of purified MBP (from 0 to 75 mg/ml) were added to the cells in the reconstitution mixture. The efficiency of reconstitution was determined by measuring the initial rates of maltose uptake.

tion and have a similar MBP concentration optimum for reconstitution.[38] The MBP concentration optimum for reconstitution of maltose chemotaxis in $\Delta malE$ mutants is also in the same mM range (half-optimal at 0.25 mM MBP[30]). This value comes close to the half-optimal periplasmic MBP concentration for chemotaxis (0.2 mM) determined by an independent approach using $malE$ signal sequence mutants with reduced periplasmic MBP concentrations.[39]

General Applicability of the Ca^{2+}-Dependent Reconstitution Procedure for Whole Cells

The Ca^{2+}-dependent reconstitution procedure is generally applicable. Besides maltose transport in $malE$ mutants of $E.$ $coli$, the binding protein-dependent galactose transport of $E.$ $coli$ can be reconstituted by addition of shock fluid containing high amounts (90%) of galactose-binding protein (GBP). The method is fast and easy enough to be used as screening procedure to distinguish mutations affecting only periplasmic components of the transport system from those affecting also additional inner membrane transport proteins. Strains lacking only GBP (LA 6021 and LA 6022) can be easily discriminated from those affected also in the $mglA$ and $mglC$ gene products (LA 6028) by their competence or incompetence for reconstitution of galactose transport (see the table).[13]

The procedure is also applicable to $S.$ $typhimurium$. Histidine transport in $S.$ $typhimurium$ strain TR 2918 ($hisJ$) lacking the histidine-binding protein can be restored by addition of shock fluid from wild-type strains. A deletion mutant covering the $hisP$ gene in addition to $hisJ$ (strain TA 1835) cannot be reconstituted (see the table).[13]

The described procedure for import of macromolecules into the periplasm is also applicable to a variety of other gram-negative bacteria. Import of fluorescently labeled proteins (GBP, MBP, and nonperiplasmic proteins like bovine serum albumin or catalase) was observed with $S.$ $typhimurium,$ $Enterobacter$ $aerogenes,$ and $Pseudomonas$ $aeruginosa$ (J. M. Brass, unpublished results, 1985).

Summary and Conclusion

Reconstitution of transport and chemotaxis in $E.$ $coli$ and $S.$ $typhimu$-$rium$ by introduction of exogenous binding protein into the periplasm of

---

[38] W. Beck and J. M. Brass, unpublished results (1983); transport assays were performed in these experiments at 5×10^{-4} M [^{14}C]maltose; at this high concentration entry of maltose through the OmpF and OmpC porins is sufficient to saturate the maltose transport system.

[39] M. D. Manson, W. Boos, Ph. J. Bassford, and B. A. Rasmussen, $J.$ $Biol.$ $Chem.$ **260,** 9729 (1985).

GENERAL APPLICABILITY OF THE Ca^{2+}-DEPENDENT RECONSTITUTION PROCEDURE WITH WHOLE CELLS

Strain	Relevant genotype	Treatment[a]	Concentration (mg/ml) of shock fluid (strain)	Initial rate of transport[a] (pmol/min per 7.5×10^7 cells) of		
				Maltose ($6 \times 10^{-6}\,M$)	Galactose ($5 \times 10^{-7}\,M$)	Histidine ($3 \times 10^{-8}\,M$)
E. coli						
JB3018-2	*malE⁺*	MMA	—[b]	597		
		Ca^{2+}	—	423		
HS3018	*ΔmalE*	Ca^{2+}	—	0.22		
		Ca^{2+}	20 (pop1080)	94.8		
		Ca^{2+}	20 (LT2)	79.8		
LA5539	*mgl⁺*	MMA	—		81.1	
LA6021	*mgl*	Ca^{2+}	—		6.7	
		Ca^{2+}	—		0.16	
		Ca^{2+}	40 [LA5709(pHG4)]		17.5	
LA6022	*mgl*	Ca^{2+}	—		0.00	
		Ca^{2+}	40 [LA5709(pHG4)]		20.4	
LA6028	*mgl*	Ca^{2+}	—		0.00	
		Ca^{2+}	40 [LA5709(pHG4)]		0.02	
S. typhimurium						
TA271	*hisJ⁺ dhuA*	Vogel–Bonner				0.80
		Ca^{2+}				0.56
TA1835	*hisJ hisP*	Ca^{2+}				0.014
		Ca^{2+}	40 (TA271)			0.00
TA2918	*hisJ*	Ca^{2+}	—			0.29
		Ca^{2+}	40 (TA271)			0.80

[a] Reconstitution and transport assays were done as described by Brass *et al.*[13]

[b] —, None.

Ca^{2+}-treated cells of binding protein-deficient mutants is an easy and generally applicable method which provides biochemical proof for the essential role of a specific binding protein in these processes. The described procedure opens the periplasm for a series of new studies since it allows the introduction of a variety of different macromolecules, including periplasmic binding proteins, which were chemically modified *in vitro*. We are now in the position to study *in situ* the rate of diffusion of binding proteins by laser bleaching analysis after introduction of binding protein labeled with a fluorescent probe.[33] These experiments enable us to investigate in detail the various aspects of the structure and functions of the periplasmic compartment of gram-negative bacteria.

Acknowledgments

I am grateful to M. Villarejo, W. Boos, and E. Bremer for critically reading the manuscript. The work was supported by grants from SFB 156 of the Deutsche Forschungsgemeinschaft and the Fonds der Chemischen Industrie.

[24] Reconstitution of Periplasmic Binding Protein-Dependent Glutamine Transport in Vesicles

By Arthur G. Hunt and Jen-shiang Hong

Several bacterial active transport systems require, in addition to one or more membrane-bound components, periplasmic binding proteins.[1,2] Genetic studies, experiments with intact cells, and investigations using spheroplasts have all established the unique role that these proteins play in the transport process. We have developed an *in vitro* system, utilizing isolated membrane vesicles, for studying binding protein-dependent active transport systems, in particular the glutamine transport system,[3,4] from *Escherichia coli*. We describe here methods for preparing isolated membrane vesicles capable of binding protein-dependent glutamine transport, and for assaying glutamine transport in such vesicles.

[1] D. B. Wilson and J. B. Smith, *in* "Bacterial Transport" (B. P. Rosen, ed.), p. 495. Dekker, New York, 1978.

[2] C. Furlong, this volume [22].

[3] A. G. Hunt and J.-s. Hong, *J. Biol. Chem.* **256,** 11988 (1981).

[4] A. G. Hunt and J.-s. Hong, *Biochemistry* **22,** 844 (1983).

Bacterial Strains and Media

Two different kinds of mutants are essential for the study of binding protein-dependent transport in isolated membrane vesicles. First, a strain that overproduces the various components of the system under study is needed, both to provide a ready source of the substrate binding protein, and to serve as a parent for the second type of strain. Second, a derivative of the overproducing strain that has a point mutation in the binding protein gene (and therefore retains functional membrane-bound components) is needed to serve as a source of membrane vesicles. This mutation is necessary to completely eliminate any effects that trapped binding protein may contribute to substrate uptake and retention by vesicles (discussed below). To study glutamine transport in vesicles, the following strains are used. Glutamine-binding protein is purified from a strain that overproduces the various components of the glutamine transport system (JSH505: *thi, metC, gluP$_o$*, also glutamine requiring). Membrane vesicles are prepared from a derivative of *E. coli* that contains a point mutation in the glutamine binding protein (PSM 116: *thi, metC, gluP$_o$, glnP*).[5,6]

These strains are routinely grown at 37° in minimal medium E[7] (per liter: 0.2 g $MgSO_4 \cdot 7H_2O$, 2 g citric acid–H_2O, 10 g K_2HPO_4, and 3.5 g $NaHNH_4PO_4$) supplemented with glycerol (0.5%), methionine (0.4 mM), vitamin B_1 (0.04 mM), and for JSH505, glutamine (2 mM).

Purification of the Glutamine-Binding Protein

Like other periplasmic proteins, the glutamine-binding protein is most easily purified after its release from cells by a cold osmotic shock treatment.[8] The rapid purification procedure described here involves a minimum number of steps: cold osmotic shock, concentration of the shock fluid, and DEAE-cellulose chromatography.

Growth, Harvest, and Cold Osmotic Shock of Cells. Fifteen liters of JSH505, grown at 37° in a fermenter, are harvested by a centrifugation. A culture of this size yields roughly 80 g of cells (wet weight). These cells are resuspended in 500 ml of 33 mM Tris–HCl, pH 7.3 at room temperature. Then 500 ml of 33 mM Tris–HCl, pH 7.3, containing 40% sucrose and 2 mM EDTA, is added to the stirring cell suspension. The cells are then harvested by centrifugation. The cell pellet is then transferred, with a spatula, to a Waring blendor and the cells resuspended at low speed in

[5] P. S. Masters and J.-s. Hong, *J. Bacteriol.* **147**, 805 (1981).
[6] P. S. Masters and J.-s. Hong, *Biochemistry* **20**, 4900 (1981).
[7] H. J. Vogel and D. M. Bonner, *J. Biol. Chem.* **218**, 97 (1956).
[8] H. C. Neu and L. A. Heppel, *J. Biol. Chem.* **140**, 3685 (1965).

1 liter of ice-cold distilled water. After 30 sec, $MgSO_4$ is added to 1 mM and the blending continued for an additional 30 sec. The cells are then removed by centrifugation and discarded.

Acid Fractionation and Concentration of Shock Fluid. The pH of the resulting shock fluid is adjusted to 4.5 with 7.5% acetic acid, and the fluid stirred in the cold for 30 min. The precipitated proteins are removed by centrifugation and discarded. The shock fluid is brought to pH 8.0 with 5% NH_4OH and concentrated by ultrafiltration (Amicon) to roughly 100 ml. Ammonium sulfate is added to 100% saturation and after at least 2 hr at 0° protein collected by centrifugation. The precipitate is taken up in 10 ml of 10 mM Tris–HCl, pH 7.6 and dialyzed against two changes of 2 liters of the same buffer.

DEAE-Cellulose Chromatography. The ammonium sulfate-precipitated protein is applied to a 70 ml DEAE-cellulose column that has been equilibrated with 10 mM Tris–HCl, pH 7.6. Glutamine-binding protein is not retained by DEAE-cellulose, but is present in the flow-through. The flow-through is collected in 5 ml fractions, and protein-containing fractions (as determined by absorbance at 280 nm) are pooled.

Properties of Purified Glutamine-Binding Protein. At this state, the glutamine-binding protein is greater than 95% pure, as judged by polyacrylamide gel electrophoresis and is suitable for the study of glutamine transport *in vitro*. Fifteen liters of cells routinely yields 100 mg of protein. Binding protein so purified retains full transport-restoring activity after 2 days of storage at 29°, or 2 years of storage in liquid nitrogen.

Assay of Glutamine-Binding Protein

Glutamine binding to its binding protein is routinely assayed with a filter binding assay,[6] taking advantage of the high affinity of the glutamine-binding proteins for nitrocellulose.

Briefly, 50 μl of binding protein sample, 50 μl of [³H]glutamine (20 μM, 916 mCi/mmol), and 50 μl of buffer are mixed and incubated at 26° for 2 min (we routinely use a buffer consisting of per liter NH_4Cl (1.0 g), K_2SO_4 (1.0 g), K_2HPO_4 (13.5 g), KH_2PO_4 (4.7 g), and $MgSO_4 \cdot 7H_2O$ (0.1 g) for these assays; however, any buffer in the pH range of 6–8 can be used when assaying the binding of glutamine to its binding protein). Of this mixture 100 μl is then applied to a prewetted nitrocellulose filter. Suction is applied, then removed. One milliliter of buffer is added and suction reapplied. The filter is then dried and the radioactivity retained is determined by liquid scintillation counting. After correcting for background radioactivity (obtained by performing an assay without added binding protein) the quantity of glutamine-binding protein is expressed in units of nmol glutamine retained on the filter.

Preparation of Membrane Vesicles

Membrane vesicles from *E. coli* are prepared, with some modifications, according to the basic procedures described by Kaback.[9]

Growth and Harvest Cells. One liter of PSM116, grown to late logarithmic phase at 37° in a 2-liter flask with vigorous shaking, is harvested by centrifugation at 12,000 *g* for 10 min, washed twice with 50 ml of 10 m*M* Tris–HCl, pH 8.0, and resuspended in 80 ml/g (wet weight) of cells of 30 m*M* Tris–HCl, pH 8.0 (a 1 liter culture so grown routinely yields about 3 g of cells).

Preparation of Spheroplasts. Spheroplasts are prepared by the lysozyme-EDTA method described by Kaback.[9] Potassium EDTA, pH 7.0, and lysozyme are added to 10 m*M* and 50 μg/ml, respectively. This suspension is stirred at room temperature for 30 min. The resulting spheroplasts are harvested by centrifugation at 16,000 *g* for 15 min and resuspended with 1 to 10 ml of 0.1 *M* potassium phosphate, pH 7.0, containing 10 m*M* MgSO$_4$ and 20% sucrose. DNase and RNase (2 mg of each) may be added to aid the resuspension.

Lysis of Spheroplasts to Form Membrane Vesicles. Vesicles are formed by adding 0.25 to 2 ml of the spheroplast suspension to 50 to 1000 ml of a rapidly stirring solution of 50 m*M* potassium phosphate, pH 7.0. The volume of the lysis buffer as well as the concentration of the spheroplast suspension is determined by the desired "cleanliness" of the vesicles (see below). Also, compounds and enzymes to be incorporated into vesicles, such as NAD or pyruvate kinase, are included in the lysis buffer at the desired concentrations.

After 15 min of stirring at room temperature, potassium EDTA, pH 7.0, is added to 10 m*M*. After another 15 min, MgSO$_4$ is added to 15 m*M*. After a further 15 min, the vesicles are collected by centrifugation at 27,000 *g* for 10 min and resuspended in 10 ml of 0.1 *M* potassium phosphate, pH 7.0, and 10 m*M* EDTA. After a low-speed spin (100 *g*, 10 min) to remove unlysed cells and debris, the vesicles are collected by centrifugation, washed 2 times with this same buffer, and finally resuspended in 1 ml of this buffer.

Properties of Vesicle Preparations. One liter of cells generally yields 5 ml of vesicles at 5 mg/ml of protein. Vesicles so prepared can be stored in liquid nitrogen for at least 12 months without substantial loss of glutamine transport activity, except when pyruvate- or D-lactate-driven transport is measured. In these instances, vesicles are much less stable, losing about 50% of their activity in 1 week. This instability presumably reflects the cold sensitivity of acetate kinase, an enzyme required for pyruvate or D-

---

[9] H. R. Kaback, this series, Vol. 22, p. 99.

lactate stimulated transport.[3] In our hands, vesicles prepared using the above procedures are very consistent with respect to glutamine transport activity. Routinely, "dirty" vesicles[3] (see below) prepared by dilution of concentrated suspension of spheroplasts into 50–100 ml of lysis buffer accumulate between 0.75 and 2.0 nmol glutamine per mg of vesicle protein, depending on the exogenous energy source used and the presence or absence of incorporated NAD. These vesicles can also use the artificial electron donor ascorbate-phenazine methosulfate (ASC-PMS) to drive glutamine transport.[10] "Cleaner" vesicles, prepared by dilution of dilute suspension of spheroplasts into a large lysis buffer, cannot use D-lactate, succinate, or pyruvate to drive glutamine transport even in the presence of incorporated NAD,[10] but can still use ASC-PMS to energize glutamine transport. Extremely clean vesicles can also be prepared by dilution of very dilute spheroplast suspension into a very large lysis buffer. Vesicles so prepared no longer can use ASC-PMS to drive transport.[10]

Glutamine Transport Assays

Glutamine transport by isolated membrane vesicles is assayed by incubating vesicles, binding protein, and glutamine for given lengths of time, then rapidly quenching suspensions and collecting vesicles on cellulose acetate filters. Equal volumes of vesicle suspension (3–6 mg/ml) and purified glutamine-binding protein (0.5–2 mg/ml), each prepared as described above, are mixed, $MgSO_4$ added to 10 mM, KH_2PO_4 added so that the final pH is 6.1, and the mixture divided into 25-μl aliquots. Exogenous energy source is then added as 1 μl of a 50× stock solution, taking care to maintain the pH at 6.1 and the suspensions incubated at 37° for 10 min. Radioactive glutamine (L-[3,4-^{3}H]glutamine, 99.8 mCi/mmol) is added to 36.4 μM, and the incubations continued at 37°. Samples are quenched at appropriate times with 2 ml of 0.1 M LiCl and immediately filtered through a prewetted cellulose acetate membrane filter (0.5 μm pore size; nitrocellulose filters cannot be used since glutamine-binding protein is retained by nitrocellulose; this binding leads to a substantial retention of glutamine to filters through binding to its binding protein), and washed with an additional 1 ml of 0.1 M LiCl. The filters are then dried and counted in a suitable scintillation solvent.

Properties of Glutamine Transport in Vesicles

Glutamine transport in isolated membrane vesicles is saturable with respect to glutamine with a K_m (3×10^{-7} M) similar to that seen in whole

---

[10] J.-s. Hong, unpublished results.

cells.[3,11] Inhibitors of glutamine metabolism such as azaserine have no effect on glutamine transport in vesicles,[3] indicating that subsequent conversion of glutamine to other compounds does not play a role in this transport.

Glutamine transport is absolutely dependent on added binding protein, with maximal stimulation seen at roughly 2 mg/ml added binding protein.[3] Inhibitors of the binding of glutamine to its binding protein, such as γ-glutamyl hydrazide, inhibit glutamine transport in vesicles. Other components of the glutamine transport system are required for transport in vesicles since vesicles prepared from a strain of *E. coli* carrying a deletion that spans the *glnP* locus (PSM223) show no such activity.[3]

Glutamine transport activity in membrane vesicles displays a sharp pH optimum of 6.0.[4] This behavior is not seen in the binding properties of the glutamine-binding protein. Thus pH optimum therefore reflects some other step in the transport process, possibly the interaction of the glutamine-binding protein with the membrane-bound components of the glutamine transport system.[12] Glutamine transport is also inhibited by high concentrations (0.5 M) of KCl and by complete replacement of K^+ with Na^+ in the preparation of membrane vesicles and subsequent assays. Once again, since these properties are in distinct contrast to the binding properties of the glutamine-binding protein, they must reflect other steps of the transport process in vesicles.

Glutamine transport in vesicles is energy dependent, being inhibited by the energy poisons CCCP, CN^-, azide, and arsenate.[4] An exogenous source of energy is also required. However, the method of preparation of vesicles has a large bearing on the ability of various compounds to drive transport vesicles. Vesicles prepared by dilution of concentrated suspension of spheroplasts into small volumes of hypotonic lysis buffer are able to use D-lactate, succinate, pyruvate, and ASC-PMS to drive glutamine transport, whereas vesicles prepared by dilution of dilute spheroplast suspension into relatively large volumes of buffer cannot use D-lactate, succinate, or pyruvate to energize glutamine transport. The ability of the clean vesicles to use ASC-PMS to drive glutamine transport depends upon the degree of cleanliness of vesicle preparations. These differences reflect the metabolic capabilities of various vesicles preparations.[10] Apparently, "dirty" vesicle preparations are able to generate the energy donor for glutamine transport from a variety of compounds.

The range of compounds that can be used to drive glutamine transport in vesicles can be affected by incorporating various cofactors, enzymes, or even transport systems into vesicles. For example, vesicles containing

[11] J. H. Weiner and L. A. Heppel, *J. Biol. Chem.* **246**, 6933 (1971).
[12] A. G. Hunt and J.-s. Hong, *Biochemistry* **22**, 851 (1983).

0.1 mM NAD are capable of utilizing pyruvate to stimulate transport, whereas vesicles lacking NAD do not possess this capability.[3] Vesicles capable of transporting PEP, prepared from a strain of *E. coli* carrying the PEP transport system from *Salmonella typhimurium* on a recombinant plasmid, are capable of using exogenous PEP to drive glutamine transport.[4]

The importance of eliminating binding protein activity from vesicle preparations by mutation is demonstrated by comparing the glutamine transport properties of vesicles prepared from a strain of *E. coli* that overproduces the various components of the glutamine transport system, including the glutamine-binding protein (PSM2). In such vesicles, glutamine uptake does not require added binding protein, is marginally stimulated by exogenous energy sources, is not inhibited by arsenate, and is not affected by glutamate, azaserine, or γ-glutamyl hydrazide.[4] Furthermore, the K_m for this uptake is greater than 2×10^{-6} M,[4] nearly an order of magnitude greater than that seen in PSM116 vesicles or in whole cells.[3,11] Since glutamine-binding protein can be detected in vesicles prepared from PSM2,[4] the most likely explanation for these observations is that substantial binding protein is retained by vesicles during their preparation, either because of insufficient dilution or some sort of association of binding protein with the vesicles. This retained binding protein, probably localized inside of vesicles, is able to sequester glutamine and give rise to the properties described above.

Studying Binding Protein-Dependent Transport in Isolated Membrane Vesicles

We have used the system described here to study several aspects of glutamine transport in *E. coli*. By modifying the glutamine-binding protein with diethyl pyrocarbonate (DEPC) and *N*-bromosuccinimide and assaying for glutamine binding and transport-restoring activity in vesicles, we have shown that different domains of the glutamine-binding protein are involved in glutamine binding and in its interaction with the membrane-bound components of the glutamine transport system.[12] Studies using recombinant DNA technology should define these domains in greater detail.

By adapting the procedures described here to small cultures, we have screened several glutamine transport mutants for transport activity in vesicles. We have seen that glutamine transport mutants fall into one of three classes: those defective in binding protein activity, but retaining transport activity in vesicles (such as PSM116); most predominantly, those retaining binding protein activity but yielding vesicles incapable of transport; and those lacking active binding protein and yielding inactive vesicles.

The availability of a vesicle system in which to study binding protein-dependent transport promises to speed investigations into several key aspects of these systems. A clear, unambiguous definition of the energetics of binding protein-dependent transport is now possible, since the ablity of various compounds to drive transport can now be directly tested in an *in vitro* system. Furthermore, the purification, in active form, of the membrane-bound components of binding protein-dependent transport systems, using techniques developed for the *lac* transport system,[13,14] is now feasible.

Acknowledgment

Work performed in this laboratory was supported by Grant GM 29843 from the National Institute of General Medical Sciences.

[13] D. L. Foster, M. L. Garcia, M. J. Newman, L. Patel, and H. R. Kaback, *Biochemistry* **21,** 5634 (1982).
[14] H. K. Sarkar, P. V. Viitanen, E. Padan, W. R. Trumble, M. S. Poonian, W. McComas, and H. R. Kaback, this volume [17].

[25] Isolation and Crystallization of Bacterial Porin

By R. Michael Garavito and Jürg P. Rosenbusch

Introduction

Several advances have been made in the preparation of three-dimensional crystals of membrane proteins.[1-5] The protocols used for such crystallizations are essentially conventional, using methods such as vapor diffusion or microdialysis with high salt or high-molecular-weight polyethylene glycols (PEG) as precipitants.[6,7] A significant difference between

[1] H. Michel and D. Oesterhelt, *Proc. Natl. Acad. Sci. U.S.A.* **77,** 1283 (1980).
[2] R. M. Garavito and J. P. Rosenbusch, *J. Cell Biol.* **86,** 327 (1980).
[3] H. Michel, *J. Mol. Biol.* **158,** 567 (1982).
[4] H. Michel and D. Oesterhelt, this series, Vol. 88, p. 111.
[5] R. M. Garavito, J. A. Jenkins, J. N. Jansonius, R. Karlsson, and J. P. Rosenbusch, *J. Mol. Biol.* **164,** 313 (1983).
[6] A. McPherson, "Preparation and Analysis of Protein Crystals," p. 82. Wiley, New York, 1982.
[7] G. L. Gilliland and D. R. Davies, this series, Vol. 104, p. 370.

soluble and integral membrane proteins is that integral proteins exist in an intrinsically anisotropic environment within the membrane and must therefore first be solubilized into an isotropic state by detergents which themselves form isotropic micelles. Such amphiphiles displace the original bilayer-forming lipids and yield completely solubilized components. Although membrane proteins bind detergents to their hydrophobic surfaces,[8] they cannot necessarily be assumed to be monodisperse, as is usually the case with homogeneous soluble proteins. Purified protein–detergent complexes are monodisperse only inasmuch as the detergent used has well-characterized colloidal properties which are adequate for the handling of membrane proteins.[9]

Three classes of membrane proteins have so far allowed advanced crystallographic analysis: bacteriorhodopsin,[4,10] reaction centers from photosynthetic bacteria,[3,11–13] and porin from *Escherichia coli* outer membranes.[5] In our laboratories, crystallization of porin was investigated because it is a functionally well-characterized protein,[14] and yields structural information from two-dimensional crystals using electron microscopy and image processing.[15] Porin is a trimer (M_r 110,000) consisting of three identical polypeptides[16,17] whose sequence has been established in several strains by amino acid[18] and nucleotide sequencing.[19] Porin is unusual in that it exhibits extensive β-pleated sheet structure[20] and is very polar for a transmembrane protein.[21] Most pertinent in the present context is the unusual stability of porin.[16] Its trimers are resistant to proteases, detergents [including sodium dodecyl sulfate (SDS) and cetyltrimethylammonium halides], and organic solvents, and at pH values over a range from 2 to 12 and temperatures $\geq 70°$. Porin is easy to handle and allows testing a wide variety of crystallization conditions. Using oc-

[8] A. Helenius, D. R. McCaslin, E. Fries, and C. Tanford, this series, Vol. 56, p. 734.

[9] M. Zulauf and J. P. Rosenbusch, *J. Phys. Chem.* **87,** 856 (1983).

[10] H. Michel, *EMBO J.* **1,** 1267 (1982).

[11] H. Michel, *Trends Biochem. Sci.* **8,** 56 (1983).

[12] J. Deisenhofer, O. Epp, K. Miki, R. Huber, and H. Michel, *J. Mol. Biol.* **180,** 385 (1984).

[13] J. P. Allen and G. Feher, *Proc. Natl. Acad. Sci. U.S.A.* **81,** 4795 (1984).

[14] H. Schindler and J. P. Rosenbusch, *Proc. Natl. Acad. Sci. U.S.A.* **78,** 2302 (1981).

[15] D. L. Dorset, A. Engel, A. Massalski, and J. P. Rosenbusch, *Biophys. J.* **45,** 128 (1984).

[16] J. P. Rosenbusch, *J. Biol. Chem.* **249,** 8019 (1974).

[17] J. P. Rosenbusch, R. M. Garavito, D. L. Dorset, and A. Engel, *in* "Protides of the Biological Fluids" (H. Peeters, ed.), p. 171. Pergamon, Oxford, 1984.

[18] R. Chen, C. Kraemer, W. Schmidmayr, K. Chen-Schmeisser, and U. Henning, *Biochem. J.* **203,** 33 (1982).

[19] K. Inokuchi, N. Muton, S. Matsuyama, and S. Mizushima, *Nucleic Acids Res.* **10,** 6957 (1982).

[20] B. Kleffel, R. M. Garavito, W. Baumeister, and J. P. Rosenbusch, *EMBO J.* **4,** 1589 (1985).

[21] C. Paul and J. P. Rosenbusch, *EMBO J.* **4,** 1593 (1985).

tyl-POE, a nonionic octyloligooxyethylene detergent, porin is solubilized to trimers.[22] The transition from the anisotropic to an isotropic state does not irreversibly affect structure and function of porin, since it can be reconstituted in its active form into lipid bilayers for functional[14] as well as for structural studies.[15]

Preparation of Porin Suitable for Crystallization

Isolation of porin consists of two steps. The first, *solubilization,* is unique to membrane proteins, and causes, by the presence of excess detergents, dissociation of proteins and lipids into isotropic components complexed with detergent. The second step is *purification* of porin. If the detergent used is freely miscible with aqueous solutions, if it is nonionic, has a high critical micellar concentration (CMC), and forms small, well-defined micelles, both steps can be carried out efficiently. The alkyl-oligooxyethylene, octyl-POE (Table I), was synthesized to satisfy these requirements[16] and has proven useful for the investigation of porin, as well as for several other membrane proteins of different origin.[23–27] In the case of porin, it allows differential (selective) solubilization: after preextraction, subsequent extractions yield fractions highly enriched in porin (see below). The critical temperature of the demixing of the detergent (T_d; also known as cloud point) is high (58°), and thus allows biochemical studies within a range of 0–40°. Its physicochemical properties are essentially those of the homogeneous octyloxyethylene (Table I).[28,30]

Solubilization

Crude cell envelope preparations, including both plasma (inner) membrane and outer membrane components, are used for fractional extraction

[22] J. P. Rosenbusch, A. C. Steven, M. Alkan, and M. Regenass, *in* "Electron Microscopy at Molecular Dimensions" (W. Baumeister and W. Vogell, eds.), p. 1. Springer-Verlag, Berlin and New York, 1980.

[23] H. Arad, J. P. Rosenbusch, and A. Levitzki, *Proc. Natl. Acad. Sci. U.S.A.* **81,** 6579 (1984).

[24] B. Erni, H. Trachsel, P. W. Postma, and J. P. Rosenbusch, *J. Biol. Chem.* **257,** 13726 (1982).

[25] P. Ott, A. Lustig, U. Brodbeck, and J. P. Rosenbusch, *FEBS Lett.* **138,** 187 (1982).

[26] B. Ludwig, M. Grabo, I. Gregor, A. Lustig, M. Regenass, and J. P. Rosenbusch, *J. Biol. Chem.* **257,** 5576 (1982).

[27] S. Schenkman, A. Tsugita, M. Schwartz, and J. P. Rosenbusch, *J. Biol. Chem.* **259,** 7570 (1984).

[28] M. Grabo, Ph.D. thesis, University of Basel, 1982.

[29] K. W. Herrmann, J. G. Brushmoller, and W. L. Courchene, *J. Phys. Chem.* **70,** 2909 (1966).

[30] D. L. Dorset and J. P. Rosenbusch, *Chem. Phys. Lipids* **29,** 299 (1981).

TABLE I

Detergent		CMC[a]		
Short designation	Name (and abbreviation)	(mM)	Source	Application
C_xE_y[b]	Alkyloxyethylene			
C_8E_4	Octyltetraoxyethylene	7.2	Bachem[c]	Analytical ultracentrifugation, light scattering and neutron spin echo resonance[9]
C_8E_5	Octylpentaoxyethylene	4.3	Bachem	
$C_8E_{\bar{5}}$	Octyl(polydisperse)-oligooxyethylene; octyl-POE	6.6	d	Solubilization, purification, crystallization
C_8E_{6-11}	Octylpleiooxyethylene[e]	n.d.		Crystallization
C_8E_8	Octyloctaoxyethylene	n.d.	f	Crystallization
$C_{12}E_8$	Dodecyloctaoxyethylene		Nikko Chemicals, Tokyo; Calbiochem	Crystallization
C_xZ[g]				
$C_8(HE)SO$	Octyl(hydroxyethane)sulfoxide	29.9	Bachem[c]	Crystallization
C_8-glc	1-n-Octyl-β-D-glucopyranoside (octyl-glucoside, also β-OG or OG)[h]	20	Bachem[c]; Calbiochem, Riedel-de Haën, Sigma	Solubilization, purification, crystallization
$C_{12}(DM)NO$	Dodecyl(dimethyl)amine oxide (DDAO, LDAO)	2.4	Oxyl[i]	Solubilization, purification, crystallization

[a] CMC, critical concentration of micellization. The values indicated were determined by the surface tension method and are from Grabo.[28] n.d. in this column indicates not determined. The CMC of $C_{12}(DM)NO$ is from Herrmann et al.[29]

[b] The alkyl moiety is given as C, with x (subscript) the number of carbons in the chain. E indicates the polar oxyethylene head group, with y (subscript) the number of units (if homodisperse) or the peak of a distribution (indicated by a dash over the number).

[c] Bachem, Hauptstr. 144, CH-4416 Bubendorf, Switzerland. Related detergents (C_7E_5, C_6E_4, etc.) have not proven particularly advantageous for porin crystallization so far.

[d] Octyl-POE is the main fraction of molecular distillation after oxyethylation of n-octanol by ethylene oxide.[17] It presents a distribution of octyloxyethylenes with a sharp peak at 5 units, which falls off rapidly on either side to 3 and 11 units. Samples of this detergent are available from J. Rosenbusch for specified experiments (Biozentrum, Klingelbergstr. 70, CH-4056 Basel, Switzerland).

[e] This mixture is the product of a further fraction of molecular distillation using the residues after octyl-POE distillation. The head groups contain between 6 and 11 oxyethylene units. (-pleio) stands for intermediate (oxyethylene chain length).

with octyl-POE (Table II, step a). As shown in Fig. 1, preextractions with 0.5% detergent solubilize substantial quantities of unrelated proteins. Five consecutive extractions with 3% octyl-POE are highly enriched in porin, but minor protein components are still present (Fig. 1). Gel electrophoresis of the solubilized pellet of the last extraction in SDS sample buffer shows that essentially quantitative solubilization of porin has been achieved. The routine selective extraction procedure presented in Table II was established empirically with a batch size of 150 g *E. coli* cells. It has proven necessary to establish different conditions for different batch sizes. Once established, volumes as well as detergent concentrations of extraction solutions have been rigorously kept constant to achieve good reproducibility. Variation of extraction conditions, which may well be required for other membrane proteins, could be the addition of EDTA, or variation of ionic strength, pH, or temperature. In the procedure used for porin, most phospholipids are extracted during preextraction (Fig. 1B), while lipopolysaccharides are extracted mostly in the porin-containing extract.[17] Glycolipids and phospholipids were followed by growing small cell batches (1 liter cultures) with $^{32}P_i$, and resolved from one another by thin-layer chromatography.[22]

Purification

The main steps applied are DEAE column chromatography, chromatofocusing, and gel filtration, with ethanol precipitations and ultrafiltrations as means to concentrate the solutions (Table II, step c). Although at 50–66% ethanol most of the detergent is removed (monitoring with [^{14}C]octyl-POE), the aggregated porin can be resolubilized without detectable loss of activity. The initial precipitation was chosen at a 1 : 1 (v/v) ratio of ethanol, because porin is precipitated essentially quantitatively while EDTA remains in the supernatant (monitored by [^{14}C]EDTA). At a 1 : 2 volume ratio (66% ethanol), the initially high concentration of EDTA would precipitate also. Its removal is essential for the following ion-exchange chromatography. If sizable concentrations of EDTA remain in the porin solution, part of the porin is eluted in the wash, presumably due to exceeding the anion-exchange capacity of the resin. In the last step of

---

[f] The kind gifts from Dr. Tiddy, Unilever Research Laboratory, are gratefully acknowledged.

[g] Z stands for any polar nonionic head group (except E).

[h] The abbreviation used here is octylglucoside. Although the α-anomer was used in earlier work,[2] it proved, due to its critical temperature of crystallization ($T_c = 42°$; cf. Dorset and Rosenbusch[30]) of little practical value. The β-anomer does not exhibit this transition. The CMC of octylglucoside (20 mM) corresponds to 0.7% (w/v).

[i] Oxyl, Peter Henlein Str. 11, D-8903 Bobingen, F.R.G.

TABLE II

PURIFICATION OF PORIN FROM *E. coli* STRAIN B[E]3000 *ompA*[−] (BZB 6/3)[a]

Step a. Envelope preparation

Cells were grown in a 100-liter fermentor in M-9 medium, containing 0.5% casamino acids and 20 μM ferricitrate, and harvested in the late exponential growth phase. This yielded 150 g cell wet weight which were stored at −70°. Batches of 150 g were thawed in 600 ml 1 mM EDTA, pH 7.5 (overnight at 4°, or with strong shaking at 37°). Cells were collected in 1-liter bottles at 4200 rpm for 30 min in a DPR 6000 centrifuge (IEC). The pellet was resuspended by means of a Sorvall mixer in breaking buffer[b] containing 1.5 mg PMSF-treated DNase (Serva, 2000 U/mg). Bacteria were broken in a French Pressure Cell (90 atm). Breakage was checked by light microscopy and found to be >95%. Unbroken cells were removed by centrifugation (10 min at 10,000 rpm) in a Sorvall SS34 rotor. Crude membranes were collected by centrifuging the supernatant at 18,000 rpm for 90 min (SS34). For combined DNase/RNase treatment, cells were suspended in 150 ml K-phosphate buffer,[c] containing 1.5 mg each of PMSF-treated DNase and RNase. Incubation was for 30 min at 37°. Envelopes were then collected by centrifugation (90 min, 18,000 rpm, SS34).

Step b. Fractional (selective) detergent extraction of porin

1. Preextractions, each with 150 ml extraction buffer[d] containing 0.5% (v/v) octyl-POE, were performed repeatedly (8×; see Fig. 1) either at 37° for 1 hr, or overnight at 4°. Most protein bands were removed without solubilizing porin.

2. Five extractions (150 ml each, with 3% octyl-POE, also for 1 hr at 37° or overnight in the cold room) followed until all porin was solubilized (Fig. 1). After each extraction, centrifugation was performed for 90 min at 18,000 rpm in a Sorvall SS34. Pellets were resuspended by agitation in a Sorvall mixer for 30 sec at top speed. All supernatants were examined separately by SDS–PAGE (Fig. 1). Supernatants 1–5 from 3% extractions in Fig. 1 were then pooled (750 ml) and centrifuged. EDTA (0.5 M, pH 7.6) was added to a final concentration of 0.1 M. To precipitate porin, ethanol (94%, precooled at −70°) was added to the solution (at room temperature) to a final concentration of 50% (v/v). The precipitate was collected by centrifugation (GS3 rotor in a Sorvall centrifuge) for 30 min at 9000 rpm. Resuspension of the pellet was in 750 ml column buffer,[e] without EDTA, followed by the addition of two volumes of ethanol to a final concentration of 66% (v/v). After centrifugation of this second precipitation, the pellet was resuspended in 350 ml column buffer. If solubilization was not immediate, neat octyl-POE was added to a final concentration of 4% and the preparation was bath sonicated and dialyzed. After solubilization, preparations were routinely subjected to dialysis against a 10-fold volume of column buffer containing 0.2% octyl-POE for 15 hr. Before column chromatography, samples were filtered through a 100 ml Sybron filter (0.45 μm pore size, Nalgene, Rochester, NY).

Step c. Purification

1. DEAE column chromatography at pH 7.6. Resin (100 ml) (DEAE Merck) was pretreated with 10-fold concentrated column buffer until the pH was stable at 7.6. It was then equilibrated with 2 column volumes of nonconcentrated buffer and filled into a column of 18 cm length. Pooled extracts (350 ml) were applied and subsequently washed with the same buffer (~200 ml) until A_{280} returned to baseline. Elution was with 250 ml

TABLE II (*continued*)

column buffer including 10 mM EDTA. Remaining protein was eluted with (200 ml) column buffer containing 30 mM EDTA and 0.1 M NaCl. Peak fractions of each elution were concentrated separately in an ultrafiltration cell (volume 50 ml; PM 30 Amicon filters) to 5 ml. Buffer was added again to the ultrafiltration cell to 50 ml and concentration resumed to 5 ml. Final addition to 50 ml and concentration (third cycle) to 20 ml provided best results with regard to removal of lipopolysaccharides. The high salt pool was dialyzed (as the last step in b above) and rechromatographed on an identical DEAE column. The 10 mM EDTA pool, concentrated as described, was directly used for the following step.

2. Chromatofocusing on a PBE94 column (Pharmacia). A 30 ml column (length 20 cm) was used which had been preequilibrated with a 10-fold concentrated PBE buffer[f] until the pH was stable at 5.5. Subsequently, 50 ml PBE buffer was applied to equilibrate the column before sample application. Only then was the protein solution (20 ml) applied and washed with 50 ml PBE buffer. Elution was with a pH gradient using 200 ml of PB74 buffer.[g] Porin elutes at pH 4.65. The porin solution was then concentrated again by ultrafiltration in an Amicon cell with a PM30 membrane. To the concentrated solution (5 ml), 1% octylglucoside was added with an additional 0.8 mg octylglucoside per mg protein. Subsequent dialysis against gel filtration buffer[h] was for 24 hr.

3. Gel filtration chromatography. A column (volume 175 ml, 100 cm long) containing Ultrogel AcA34 was washed with 2 column volumes of gel filtration buffer.[h] The protein solution was chromatographed in two batches of 3 ml each. Eluted peak fractions were pooled, concentrated, and dialyzed for ≥24 hr against gel filtration buffer. If desired, the 20 mM HEPES[h] were replaced by 20 mM Na-phosphate, pH 7.2. Dialysis may also be used for detergent exchange. If quantitative removal is required, gel filtration is more efficient.

Step d. Yield

Final yield was about 650 mg. Maximization of recovery at the expense of purity was not desirable. Purity was tested with biochemical methods[16] and by crystallization.[5] Nearly 20 g of porin has so far been purified with this method.

---

[a] Available from J. Rosenbusch (Biozentrum, Klingelbergstr. 70, CH-4056 Basel, Switzerland).

[b] Breaking buffer: 50 mM Na-phosphate, 0.1 M NaCl, 3 mM NaN$_3$ (0.02%), 5% sucrose, pH 7.6.

[c] K-phosphate buffer: 0.1 M K-phosphate, 20 mM MgSO$_4$, 3 mM NaN$_3$, pH 6.6.

[d] Extraction buffer: 20 mM Na-phosphate, 3 mM NaN$_3$, 3 mM dithiothreitol, ocytl-POE as indicated, pH 7.6.

[e] Column buffer: 5 mM Na-phosphate, 3 mM NaN$_3$, 1% octyl-POE, pH 7.6.

[f] PBE buffer: 25 mM histidine–HCl, 5 mM EDTA, 3 mM NaN$_3$, 1% octyl-POE, pH 5.5.

[g] PB74 buffer: 12.5% Polybuffer 74, 5 mM EDTA, 3 mM NaN$_3$, 1% octyl-POE, pH 3.4. The resin was regenerated with 50 ml of a buffer containing 0.08 N HCl, 1 M NaCl, 3 mM NaN$_3$, and 1% octyl-POE, and subsequently equilibrated with 10-fold concentrated PBE buffer until pH was stable (5.5) and further washed with 2 column volumes of PBE buffer.

[h] Gel filtration buffer: 20 mM HEPES (N-2-hydroxyethylpiperazine-N'-2-ethanesulfonic acid), 0.1 M NaCl, 3 mM NaN$_3$, 1% octylglucoside, pH 7.2.

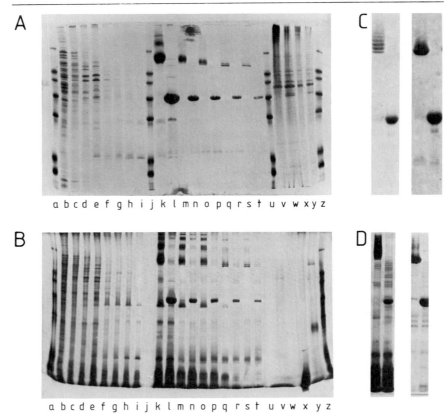

FIG. 1. Selective extraction and subsequent purification of porin from *E. coli* envelopes with octyl-POE. Polyacrylamide slab gels were stained with Coomassie blue (A and C) or with silver (B and D). Lanes b–i in A and B show 8 consecutive preextractions with 0.5% octyl-POE in extraction buffer (Table II). The samples shown were not heat treated. Exposure to 100° did not reveal any extracted porin (not shown). While solubilization of contaminant proteins peters out after the fourth extraction, silver staining indicates that the continuing extraction of phospholipids (distinguished from lipopolysaccharides by determination of glucosamine content, see Rosenbusch[16]) still proceeds (quantitation of components by silver staining is not possible). Lanes k–t present extractions of highly enriched porin with 3% octyl-POE. Silver staining indicates massive extraction of remaining phospholipids and glycolipids (lipopolysaccharides). Lanes k, m, o, q, and s were applied to the gel without treatment. Under these conditions, porin migrates as trimers.[17] Samples in lanes l, n, p, r, and t were heated to 100° which causes porin to migrate as monomers.[16] Lanes a, j, u, and z in A represent standards [phosphorylase *a* (95 kDa), catalase (57.5), fumarase (48.5), glyceraldehyde-3-phosphate dehydrogenase (36.5), bovine carbonic anhydrase B (29), myoglobin (17), and cytochrome *c* (11.7)]. Lanes v and w in A represent the last pellet, redissolved in a volume of SDS sample application buffer equivalent in volume to that used for extractions (Table II). x and y were diluted threefold with sample buffer. v, x, unboiled. w, y,

purification, gel filtration on Ultragel AcA34, residual contaminants of low molecular weight are removed and octyl-POE is replaced quantitatively by octylglucoside. Although octyl-POE (or octylpleiooxyethylene) is used for crystallization (see below), quantitative replacement of octyl-POE by octylglucoside in this step allows full control of detergent conditions.

Characterization

Biochemical *homogeneity* is primarily ascertained by SDS–polyacrylamide gel electrophoresis (Fig. 1). At all steps of the procedure, aliquots are taken and electrophoresed on two gel slabs for staining with Coomassie blue and silver stain to visualize proteins as well as lipopolysaccharides. Onto each gel, aliquots were applied in duplicate, without or with heat treatment at 100° (2 min). In the untreated form, porin migrates as a trimer,[17,31] while its mobility (after boiling) reflects the true molecular mass.[16] Heat modification thus renders porin identification and the determination of its purity simple and straightforward. The two most obstinate impurities of porin proved to be lipopolysaccharides and lipoprotein. The preparations of porin purified for initial crystallization[2] were often contaminated with lipopolysaccharides[14] which escaped detection because they are not stained with Coomassie blue. As seen in Fig. 1B, they can be readily visualized by silver staining and eventually eliminated (Fig. 1D). Lipoprotein apparently interacts very tightly with porin.[33] It is not stained

[31] R. M. Garavito, J. A. Jenkins, J.-M. Neuhaus, A. P. Pugsley, and J. P. Rosenbusch, *Ann. Microbiol. (Institut Pasteur)* **133A**, 37 (1982).
[32] M. Schindler and J. P. Rosenbusch, *FEBS Lett.* **173**, 85 (1984).

---

heated to 100° for 2 min. [The splotches in lanes m (bottom) and n (top) are artifacts.] In B, lanes a and z contain the same standard proteins listed above. Lanes j and u were left empty and lane y contains a sample which is irrelevant in this context. Lanes v and w are untreated and heat-treated samples of ultrafiltration eluent after extensive concentration (Table II). They contain lipopolysaccharides which, if massively overloaded, take on the appearance shown in lane x (purified lipopolysaccharide). C shows porin essentially pure (≥95%) with respect to other proteins. Pairs of samples show untreated (left) and heat-treated (right) preparations. The preparation on the left in C is massively contaminated with lipopolysaccharide, as seen by the ladder in the untreated sample, while the probe on the right is essentially lipopolysaccharide free. Silver staining (D) reveals massive contamination with lipopolysaccharides (which in the concentration shown inhibits crystallization). On the right, no lipopolysaccharides are detected. This sample which, after Coomassie blue staining, appears homogeneous illustrates that the notion of purity changes with increasing sensitivity of detection methods. The sample shown behaves normally during crystallization. Polyacrylamide slab gels contained 12% acrylamide and 0.32% bisacrylamide. C is slightly different in scale.

by Coomassie blue either, but can be detected with silver staining (Fig. 1B). While the former can be removed by recycling on DEAE column chromatography, chromatofocusing is more effective in removing lipoprotein quantitatively. As final criterion for purity, amino-terminal sequence analysis was performed.[34]

Monodispersity of protein–detergent complexes was ascertained by sedimentation equilibrium and sedimentation velocity in the analytical ultracentrifuge. This yields a protein mass of 110 kDa and a single sharp peak in sedimentation velocity experiments. The sedimentation coefficient of the protein–detergent complex is $s_{20,w} = 4.7$ S after extrapolation to atmospheric pressure.[17] Both results, performed over wide protein concentration ranges, demonstrate that porin–detergent complexes in solution are monodisperse at 20°.

Porin Crystallization

The first step in preparing a crystallization experiment for membrane proteins is the judicious choice of a detergent system. If a protein is well characterized and the property of the solubilizing detergent well understood,[9,17] conventional procedures[6,7] allow an appropriate starting point for crystallization. In most of our experiments, the best porin crystals grew when octylglucoside was used as the primary detergent. The most successful precipitant was the polymer polyethylene glycol (PEG). Ammonium sulfate can be used to obtain crystalline material, but no *single* crystals of porin have been observed (R. M. Garavito, unpublished). The addition of a precipitant to a detergent solution alters its micellar behavior[35,36] and can induce the appearance of new phase boundaries where the solution undergoes a separation into a detergent-rich and a detergent-poor phase. For the PEG/octylglucoside system, an upper consolute boundary appears (Fig. 2a). A similar behavior also occurs in dodecyldimethylamine oxide/PEG solutions (R. M. Garavito and M. Zulauf, unpublished). Either increasing PEG concentrations or increasing PEG molecular weight (at constant weight percentage) raises this phase boundary. High ionic strength also can affect critical temperatures of demixing (T_d; consolution boundary) drastically, yielding phase separation at room tem-

[33] J. Wirz, *Experimentia* **39**, 677 (1983).

[34] A. Tsugita and J. P. Rosenbusch, in preparation (1985).

[35] M. Zulauf, *in* ''Physics of Amphiphiles: Micelles, Vesicles and Microemulsions'' (V. Degiorgio and M. Corti, eds.), p. 663. Elsevier, Amsterdam, 1985.

[36] R. M. Garavito and J. A. Jenkins, *in* ''Structure and Function of Membrane Proteins'' (E. Quagliariello and F. Palmieri, eds.), p. 205. Elsevier, Amsterdam 1983.

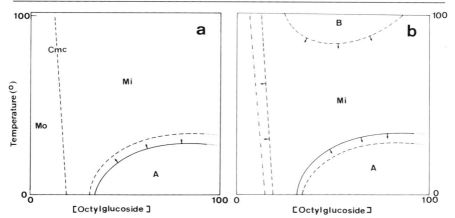

FIG. 2. Phase diagrams of octylglucoside in the presence of polyethylene glycol (PEG). Phases are shown as functions of temperature and detergent concentration. A pure octylglucoside solution (a) shows a region of detergent phase separation (A) at low temperature upon the addition of PEG. An increase in concentration or molecular weight of PEG causes an expansion of this phase (small arrows). When an octyloligooxyethylene detergent is added in increasing amounts to a PEG/octylglucoside solution (b), the CMC and the boundaries of the A phase are depressed and a new region of phase separation B (top) appears at higher temperature. Mo and Mi indicate monomer and micellar detergent phases below and above the CMC, respectively. a was adapted from Zulauf.[35]

perature or below.[35] As compared to changes in PEG concentration, the effects of moderate variations in ionic strength on octylglucoside solutions are rather small. The addition of another detergent component such as octyloligooxyethylene depresses the boundary (Fig. 2b) in a manner related to the head group size.

As the phase boundary is approached by lowering the temperature or increasing the PEG concentration, micelle aggregation seems to occur. When micelle aggregates become too large to remain in solution, they separate to form a second detergent-rich liquid.[35] This phenomenon is qualitatively analogous to PEG-induced protein aggregation and precipitation. The addition of porin to the detergent solution does not seem to affect phase boundaries significantly in the concentration ranges used. Micellar behavior has not been related directly to porin crystallization, but the behavior of crystal growth and stability of porin is predictable on the basis of the micellar behavior of the detergent system. This seems to imply that porin crystallization is a cooperative process of protein–protein and detergent–detergent interactions[35,36] and could reflect the property of membrane proteins which bind detergents over a significant portion of their surfaces.[28]

Vapor Diffusion

Initial crystallization of porin used the method of vapor diffusion to screen for suitable conditions as well as to produce large single crystals for X-ray diffraction. Table III gives the general conditions for crystallization. The hanging drop technique of vapor diffusion[6] affords a very quick way to assay new conditions. Since the presence of detergent in the protein sample drastically reduces the surface tension in the hanging drop, its initial volume is limited to 5–8 μl. The growth of large single crystals under these conditions is correspondingly restricted by the drop size. Sitting drop vapor diffusion[6] does not suffer from this limitation, since initial sample volumes may be varied from 5 to 200 μl. In this case, the rate of water vapor exchange, and therefore the rate of crystallization, can be carefully controlled by adjusting the surface area-to-volume ratio. Large crystals of porin suitable for X-ray diffraction have been obtained by this method (Fig. 3, Table IV).

The primary variable responsible for porin crystallization seems to be PEG. We have looked at the reproducibility and quality of crystal growth

TABLE III

INITIAL CONDITIONS FOR PORIN
CRYSTALLIZATION BY VAPOR DIFFUSION AT
21–23° AND pH 6.5–7.0

Component	Concentration
Protein containing compartment	
PEG 2000 (Merck)	7.5% (w/v)
NaCl	20–250 mM
Sodium phosphate	20–50 mM
Octylglucoside (detergent)	0.8% (w/v)
Additive detergent[a]	0–0.08% (w/v)
NaN$_3$	0.01% (w/v)
Porin	6–9 mg/ml
Reservoir[b]	
PEG 2000	15%
NaCl	40–500 mM
Sodium phosphate	40–100 mM
NaN$_3$	0.01%

[a] Usually C$_8$E$_{6-11}$.
[b] Solute concentrations are twice the initial concentration in the protein containing compartment.

FIG. 3. Examples of porin crystals grown from detergent solutions containing PEG. Vapor diffusion was used to obtain the various crystal forms shown. Composition of the crystallization solutions was optimized according to the criteria described in the text. a shows the tetragonal crystal form; b in a monoclinic habit; and c hexagonal crystals grown in octylglucoside solutions. Triclinic crystals (d) were grown in octyl(hydroxyethane) sulfoxide. Bars represent 0.2 mm in all photographs.

when polyethylene glycols of different molecular weights were used.[37] PEG 4000 and 6000 (analytical grade; Merck) induced extensive crystallization, but at a rate which favored crystal defects and nucleation of multiple crystals. Using PEG 2000, the rate of crystallization was reduced and the yield of single crystals increased. Due to the smaller polymer size, detergent phase separation appears reduced or suppressed with PEG 2000. The second detergent component (or additives; see Tables III and IV) had an effect on the crystallization as well. In pure octylglucoside,

[37] R. M. Garavito and J. A. Jenkins, "Progress in Biophysics and Molecular Biology," in press, 1985.

TABLE IV
CRYSTALS OBTAINED AND THEIR SPECIFIC GROWTH CONDITIONS

Crystal form	pH[a]	Initial conditions		
		Phosphate (mM)	NaCl (mM)	Detergent[b]
$P4_2$ (tetragonal)	6.5	50	250	0.08% C_8E_{6-11}
$P6_322$ (hexagonal)	5–9	20	20	0
$P2_1$ (monoclinic)	4–7	20	100–150	0
$C2$ (monoclinic)	4–7	20	100–150	0
$P1$ (triclinic)	~9	50[c]	250	1.0% $C_8(HE)SO$

[a] Optimal region or midpoint is given.

[b] The detergents listed here are minor components added to the octylglucoside except for the triclinic form in which $C_8(HE)SO$ was the sole detergent.

[c] Tris is used instead of phosphate as the buffer system.

crystallization was quite rapid, while increasing amounts of additive detergents reduced the rate until a ~1:1 detergent ratio was reached. Changing the head group in the additive had no observable effect on vapor diffusion experiments.

Ionic strength plays an important role in determining the crystal *form*. Altering NaCl concentration from 40 to 500 mM, one obtains four different space groups of porin crystals (Fig. 3 and Table IV). The system can be illustrated by a PEG:NaCl phase diagram for crystallization (Fig. 4). The phase diagram remains qualitatively unaltered regardless of PEG molecular weight (2000–6000), whether phosphate or HEPES buffer (see footnote h in Table II) are present, and whether or not detergent additive is included. Figure 4 also displays the change of the system as vapor diffusion or microdialysis experiments proceed. It should be noted that in the early experiments,[2] the pathway followed ($I_v \rightarrow F_v$) skirts the monoclinic:tetragonal phase boundary and finishes in the tetragonal crystal region beyond the detergent phase separation boundary. Hence, tetragonal crystals were obtained consistently in a two liquid phase system.

The crystal phase diagram can be significantly altered if the primary detergent or pH is changed. The triclinic crystal form of porin (Fig. 3d and Table IV) occurs at pH ~9 or in octyl(hydroxyethane) sulfoxide. At pH <6 in octylglucoside, the hexagonal and monoclinic forms are favored over the tetragonal habit (Z. Markovic and R. M. Garavito, unpublished). Crystallization experiments on other membrane proteins have used small amphiphilic molecules like butyloxyethanol or 1,2,3-heptanetriol to promote crystal growth.[3,11,36,37] For porin, they appear to have little useful effect on crystal formation. However, the addition of 5–10% (v/v) of

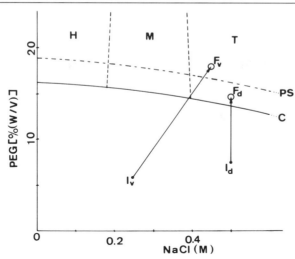

FIG. 4. Phase diagram for porin crystallization. The boundaries shown are approximate since they vary with changes in detergent concentration and temperature. C and PS refer to crystallization, and phase separation boundaries of the detergent, respectively. The appearance of three types of porin crystal forms (hexagonal, H; monoclinic, M; tetragonal, T) depends primarily on the NaCl concentration. The arrow connecting I_v to F_v depicts the pathway from the initial to the final state for crystallization by vapor diffusion (for conditions, see Table III). The corresponding pathway for microdialysis (Table V) is indicated by the arrow between I_d and F_d.

organic solvents such as dioxane or 1,6-hexanediol biases crystallization toward a monoclinic form (Z. Markovic and R. M. Garavito, unpublished).

Microdialysis

The vapor diffusion method, though successful in producing single crystals for X-ray analysis, suffers from poor reproducibility and often yields crystals that have defects or exhibit distorted morphologies. The method has the further drawback that all components of the system change their concentration as the experiment progresses. This is particularly pertinent to membrane protein crystallization, where detergent concentrations need to exceed the CMC to assure solubility and crystallization of the protein. Thus, initial concentrations of octylglucoside must exceed 20 mM[28] which, toward the end of the experiment, may result in uncontrolled micellar effects due to the excess detergent. Microdialysis offers more control over the various components in the system and has been used successfully on soluble proteins.[6,7]

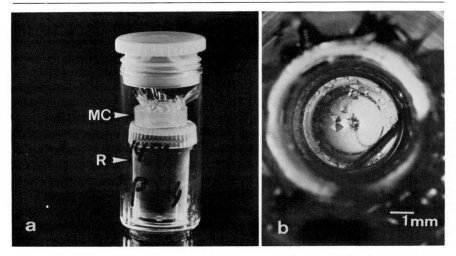

Fig. 5. The microdialysis cell used for porin crystallization. In a, MC is the microdialysis cell. It consists of a 7×7 mm piece of glass tubing (inner diameter, 3–5 mm) with a dialysis membrane held over one end and fixed by a collar made of poly(vinyl chloride) tubing. R (reservoir) is a small plastic container holding 2.3 ml of reservoir buffer. The volume capacity for protein solution is 20–150 μl in this setup. B shows that the wide tube bore facilitates observation and handling of the crystals.

Figure 5 shows our microdialysis system, a modified version of the Zeppezauer method.[38] Crystallization is induced by increasing the PEG concentration in the reservoir. As the dialysis membrane is only semipermeable to PEG 2000, and impermeable to porin, the volume of the protein sample decreases, thus causing a corresponding increase in the concentration of both components. The optimized conditions for porin crystallization with microdialysis are given in Table V. The yield of high quality single crystals is very good and reproducible with this technique. Furthermore, the availability of 2–3 ml of reservoir solution in equilibrium with the protein compartment allows more freedom in crystal manipulation. This feature is critical for crystals of membrane proteins grown from detergent solutions, as it is usually not possible to predict the equilibrium concentration of detergents in vapor diffusion systems. Thus, when buffer was needed for crystal handling or mounting, or for heavy atom derivative experiments (Table VI), all attempts to generate such buffers for vapor diffusion-grown crystals invariably failed, causing irreversible crystal disorder. This behavior might be a general property of membrane protein

[38] M. Zeppezauer, H. Eklund, and E. S. Zeppezauer, Arch. Biochem. Biophys. **126,** 564 (1968).

TABLE V
CONDITIONS FOR PORIN CRYSTALLIZATION
(SPACE GROUP $P4_2$) BY MICRODIALYSIS AT 21–23°

The following solution is dialyzed against protein-
free buffer overnight:
  7.5% PEG 2000
  0.5 M NaCl
  0.1 M Na-phosphate
  0.9% octylglucoside
  0.09% C_8E_{6-11}
  0.01%NaN$_3$
  8–10 mg/ml porin
  pH 6.5[a]
The solution is placed against a protein-free reservoir with all components at the same concentration except for the presence of 13.5% PEG
After 24 hr, the PEG concentration was raised further to 13.7–14.25%

[a] Crystals can be obtained over a range of pH
6–7.

crystals, as they contain large amounts of detergent.[5] Optimization of
several variables in crystal growth from detergent solution has been performed using the microdialysis technique as discussed presently.

Detergent Concentration. The slowest rate of crystal growth (1–2

TABLE VI
OTHER BUFFER SYSTEMS SUITABLE FOR THE $P4_2$ CRYSTAL FORM[a]

Component	Stabilization buffer	Phosphate free	High pH
PEG	15%	15%	15%
NaCl	0.5 M	0.9 M	0.9 M
Na-phosphate	0.1 M	—	—
Buffer	—	20 mM MES[b]	50 mM triethanolamine
Octylglucoside	1%	1%	1%
C_8E_{6-11}	0.1%	0.1%	0.1%
NaN$_3$[c]	0.01%	0.01%	0.01%
pH	6.5	6.1	8–8.5

[a] Microdialysis was performed at 21–23° as in Table V. Subsequent exchange into the
new buffer system was performed in 4 steps over 36 hr.
[b] MES, 2-(N-morpholino)ethanesulfonic acid.
[c] Corresponds to 1.5 mM, may be omitted if desired.

weeks) was achieved when crystallization was performed at an octylglucoside concentration of 0.8–0.9% (w/v), which is just above the CMC of the detergent (0.7%; see Table I). Under these conditions, most of the detergent is part of protein–detergent mixed micelles. Decreasing the detergent concentration would disrupt the micellar system, while increasing it would cause pure detergent micelle concentration to rise. In both circumstances, extensive nucleation and rapid growth rates occur, along with an increase in crystal defects.

Detergent Composition. As in vapor diffusion experiments, using octylglucoside as sole detergent led to rapid and extensive crystallization, with crystals generally neither large nor well ordered. Adding the detergent additive, octyloligooxyethylene in a 1 : 10 weight percent (see Table V) substantially reduces the rate of crystallization at a given PEG concentration. Our currently favored additive is C_8E_{6-11}, though octyloctaoxyethylene (C_8E_8, see Table I) has similar effects. The resulting crystals consequently grow quite large and are of high quality. Changing head group size (C_8E_8, C_8E_5, C_8E_4) also markedly changed the PEG concentration at which crystallization occurred, even though the additive is a minor component. As a first approximation, for every oxyethylene unit less in the head group, the PEG concentration must be increased by ~1% (w/v).

Variation of PEG Concentration and Temperature. With many soluble proteins, the PEG concentration range in which single crystals grow is often quite broad.[6] In contradistinction, the PEG concentration range for suitable porin crystal growth extends only over about 1% (w/v) of PEG from no crystal growth to crystal showers. At our optimal conditions for crystallization (Table V), this means 13.5–14.5% PEG 2000. Careful preparation and storage of all buffers is therefore required for reproducibility. An additional complexity arises from the significant effects of temperature on crystallization. The conditions given in Table V yield crystals only at the temperature indicated (21–23°). At 25°, no crystals grow and existing crystals begin to dissolve. This temperature effect can be compensated by changing the PEG concentration. Thus, more PEG is needed at higher temperatures and vice versa. Crystallization at 4° occurs in 10% PEG 2000 and 10 mg/ml porin. Large crystals have been obtained when equilibration of microdialysis experiments was allowed at 30°, and the temperature then lowered gradually to 21° over several days.

Comments and Conclusions

The unusual aspects of porin crystallization are consistent with the micellar interactions involved in crystal formation.[9,36] The resulting porin crystals are relatively sensitive to the environment. Care must be taken in handling and mounting crystals for X-ray analysis to avoid significant or abrupt changes in the solvent/detergent environment as well as uncontrolled variations of temperature. After growth, crystals can be stabilized by adding small amounts of reservoir buffer to the microdialysis cell (thus reducing the protein concentration) and subsequently increasing the PEG concentration by ~1% (w/v). Under these conditions, crystals can be handled with relative ease, and the buffer systems can be altered to allow chemical modification or heavy atom derivative preparation (Table VI) under optimal conditions.

Three recent reviews and commentaries on the role of detergents in membrane protein crystallizations all emphasize the effect of the micellar structure and behavior of the protein–detergent complex in both nucleation and growth of crystals.[11,36,37] In the earlier work,[2,5] several assumptions have been made relative to the crystallization of porin, and conclusions drawn which in hindsight have proven not to be completely true or of limited validity. We reported that porin crystallization occurs primarily in a two-phase detergent–liquid system.[2] While this occurs under the conditions of the original experiments (see Fig. 4, $I_v \longrightarrow F_v$), redesigning the system demonstrated that crystals could be obtained easily in a single and apparently isotropic phase.[36] The micellar processes involved in the critical phenomena such as phase separations[9,35] presumably also occur under the new crystallization conditions, though they can be better controlled, and may even contribute toward proper crystal nucleation.[35,36]

Porin has proven rather useful for the optimization of conditions to crystallize membrane proteins since it is very stable over a wide range of conditions.[32] Indeed, crystallization experiments (at 20°) were successful also using several other detergents with a moderate to high CMC, e.g., dodecyldimethylamine oxide (LDAO), octyl(hydroxyethane) sulfoxide [C_8(HE)SO], C_8E_4, C_8E_5, and $C_{12}E_8$ (R. M. Garavito, unpublished). Initial problems with reproducibility of porin crystallization[2] revealed that a high degree of purity of this protein was required, and that the removal of tightly bound lipoprotein and glycolipid[14] was critical for success. It should be noted that in other membrane proteins, removal of bound lipid moieties may cause inactivation. In such instances, a compromise between detergent properties and protein activity will have to be found if meaningful crystal structures are to be attained. So far, where appropriate conditions could be found, success has been forthcoming with other membrane proteins. Crystals of other outer membrane proteins from *E. coli* have been obtained,[39] as well as from bacteriorhodopsin[4,10] and with reaction centers of photosynthetic bacteria[3,11–13] using related techniques. Al-

[39] R. M. Garavito, U. Hinz, and J. M. Neuhaus, *J. Biol. Chem.* **259**, 4254 (1984).

though current experience is limited, it seems fair to conclude that with a better understanding of protein–amphiphile interactions, of micellar physical chemistry, and with improvement in crystallization techniques, the outlook for understanding membrane protein structure appears bright.

Acknowledgments

The expertise and active participation in preparing and optimizing octyloxyethylene detergents by Drs. P. Liechti, J. Merz, and V. Arnold (Ciba-Geigy Corporation, Basel) and by E. Boss and Dr. J. Gosteli (Bachem, Bubendorf) are most gratefully acknowledged. Dr. H. Fierz (Ciba-Geigy) kindly took responsibility for analysis and characterization of the resulting detergents whose development was possible only thanks to the effective support of Drs. H. Wegmüller and J. Nüesch (Ciba-Geigy). Ciba-Geigy Corporation generously funded these detergent syntheses. Without the energetic collaboration of J. Beltzer, U. Brütsch, M. Meins, M. Regenass, and C. Widmer-Otz, the elaboration of the procedures would not have been possible. We thank Drs. J. Jenkins, J. N. Jansonius, H. Michel, and M. Zulauf for fruitful discussions. C. Barber (EMBL) was instrumental in preparing this manuscript. Grants from the Swiss National Science Foundation (3.152.77 and 3.656.80 to J.P.R.; 3.201.82 and 3.655.84 to R.M.G.) are gratefully acknowledged. M. Meins was supported by the Emil Barell Foundation.

[26] Ion Extrusion Systems in *Escherichia coli*

By BARRY P. ROSEN

Introduction

Bacterial cells, like eukaryotic cells, have a variety of ion transport systems which catalyze the movement of cations and anions into and out of the cells.[1] This chapter will consider the systems responsible for export of ions from the bacterium *Escherichia coli*. This bacterium uses both primary and secondary extrusion systems. Primary systems couple chemical energy directly to the performance of electrochemical work, or, in the reverse mode, couple osmotic energy to chemical bond formation. The most widely recognized of the *E. coli* primary extrusion pumps are the H^+-translocating ATPase (F_0F_1) and the H^+-translocating respiratory chain. Both primary pumps establish electrochemical proton gradients, acid and positive outside of the cell. When the two run simultaneously, the force generated by respiration exceeds the amount necessary to drive

---

[1] B. P. Rosen and E. R. Kashket, *in* "Bacterial Transport" (B. P. Rosen, ed.), p. 550. Dekker, New York, 1978.

METHODS IN ENZYMOLOGY, VOL. 125

protons inward through the F_0F_1, resulting in ATP synthesis. The coupling of these two pumps is the process of oxidative phosphorylation.

The force generated by proton pumping can also be used for osmotic work. Secondary transport systems use the energy of the proton gradient to move other solutes. In the case of secondary extrusion systems, this is accomplished by exchange of protons for another ion, thus coupling downhill proton movement to uphill transport of a heterologous ion. These exchangers are called *antiporters*. Four *E. coli* antiport systems have been identified.[2,3] A calcium/proton antiporter (CHA) catalyzes electrophoretic exchange of one calcium for at least three protons.[4] A calcium phosphate symporter/proton antiporter (CPS) cotransports calcium and phosphate in a 1 : 1 ratio.[3] A sodium/proton antiporter (NHA) exchanges more than one proton per sodium ion.[5] A potassium/proton antiporter (KHA) catalyzes electroneutral exchange of the two cations.[6]

The KHA system has been implicated in pH regulation.[6,7] The intracellular pH is maintained at approximately pH 7.5 over a wide extracellular pH range.[8] When the extracellular pH is alkaline relative to the cytosol, the rate of appearance of protons intracellularly must be greater than the rate of proton pumping. A number of mechanisms could contribute, including metabolic acid production and proton reentry through specific proton channels. Since intracellular pK^+ is approximately 1, this huge intracellular potassium pool can serve as a pH buffer by K^+/H^+ exchange catalyzed by the KHA system. Since the KHA system is electroneutral, K^+/H^+ exchange does not dissipate the membrane potential. In support of this concept, a mutant specifically lacking KHA activity is unable to grow at alkaline pH.[7]

The sodium/proton antiporter can create a sodium gradient through exchange with protons. The sodium gradient provides a driving force for several sodium–solute cotransport systems, including Na^+-melibiose[9] and Na^+-glutamate[10] symporters. Since Na^+/H^+ exchange also brings protons into the cytosol, its participation in pH regulation has been pro-

[2] R. N. Brey, J. C. Beck, and B. P. Rosen, *Biochem. Biophys. Res. Commun.* **83,** 1588 (1978).
[3] R. N. Brey and B. P. Rosen, *J. Biol. Chem.* **254,** 1957 (1979).
[4] S. V. Ambudkar, G. W. Zlotnick, and B. P. Rosen, *J. Biol. Chem.* **259,** 6142 (1984).
[5] J. C. Beck and B. P. Rosen, *Arch. Biochem. Biophys.* **194,** 208 (1978).
[6] R. N. Brey, B. P. Rosen, and E. N. Sorensen, *J. Biol. Chem.* **255,** 39 (1980).
[7] R. H. Plack, Jr. and B. P. Rosen, *J. Biol. Chem.* **255,** 3824 (1980).
[8] J. L. Slonczewski, B. P. Rosen, J. R. Alger, and R. M. Macnab, *Proc. Natl. Acad. Sci. U.S.A.* **78,** 6271 (1981).
[9] T. Tsuchiya, J. Raven, and T. H. Wilson, *Biochem. Biophys. Res. Commun.* **76,** 26 (1971).
[10] S. M. Hasan and T. Tsuchiya, *Biochem. Biophys. Res. Commun.* **78,** 122 (1977).

posed.[11] A pleiotropic mutant defective in sodium cotransport and exchange systems was also shown to be unable to grow at alkaline pH.[12] However, we have found that everted membrane vesicles prepared from this mutant strain exhibit normal Na^+/H^+ exchange activity (B. Rosen, unpublished results). Additional considerations include the relatively small Na^+ pool compared to the K^+ pool and the fact that the presence or absence of sodium makes no difference in growth at alkaline pH. Also, the NHA system is electrophoretic, which means that Na^+/H^+ exchange uses both the membrane potential and pH gradient, tending to lower the overall protonmotive force. Thus, while the *phs* mutation appears to affect a number of processes including intracellular pH homeostasis, the physiological importance of the NHA system as a regulator of cytosolic pH is not clear.

Other than to lower intracellular calcium, no function for the two calcium systems. Indeed, calcium extrusion systems have been found in all bacteria examined, and no general role for calcium in intracellular regulation or metabolism has been found in bacteria.[13]

In addition to the chromosomally encoded transport systems discussed above, several plasmid-encoded extrusion systems have been identified. Resistance plasmids or R-factors produce resistance to a wide variety of antibiotics and heavy metals. The spread of promiscuous extrachromosomal elements among different bacterial species in hospital settings is a serious clinical problem. One mechanism of resistance is exclusion or extrusion of the antibiotic or heavy metal from the cell. Resistances to tetracycline,[14] cadmium,[15] mercury,[16] arsenate,[17] and arsenite[18] have each been shown to involve synthesis of transport proteins coded for by plasmid DNA. The tetracycline and cadmium transport systems are secondary proton antiporters.[14,15] In contrast the arsenate and arsenite systems are primary ATP-driven anion extrusion pumps.[18,19] The genes for the arsenate and arsenite transport systems have been cloned

[11] D. Zilberstein, E. Padan, and S. Schuldiner, *FEBS Lett.* **116**, 177 (1980).

[12] D. Zilberstein, V. Agmon, S. Schuldiner, and E. Padan, *J. Biol. Chem.* **257**, 3687 (1982).

[13] B. P. Rosen, *in* "Membrane Transport of Calcium" (E. Carafoli, ed.), p. 187. Academic Press, New York, 1982.

[14] L. McMurry, R. E. Petrucci, Jr., and S. B. Levy, *Proc. Natl. Acad. Sci. U.S.A.* **77**, 3974 (1980).

[15] Z. Tynecka, Z. Gos, and J. Zajac, *J. Bacteriol.* **147**, 313 (1981).

[16] A. O. Summers and S. Silver, *Annu. Rev. Microbiol.* **32**, 637 (1978).

[17] S. Silver, K. Budd, K. Leahy, W. Shaw, D. Hammond, R. Novick, G. Willsky, M. Malamy, and H. Rosenberg, *J. Bacteriol.* **146**, 983 (1981).

[18] M. G. Borbolla and B. P. Rosen, *Biochem. Biophys. Res. Commun.* **124**, 760 (1984).

[19] H. L. T. Mobley and B. P. Rosen, *Proc. Natl. Acad. Sci. U.S.A.* **79**, 6119 (1982).

within a 4.3 kb pair fragment of DNA which has been inserted into small multicopy plasmids.[20] Both systems are encoded by a single operon, with the promoter-proximal 2/3 of the DNA involved in arsenite transport and the promoter-distal 1/3 in arsenate transport. A 64,000 Da protein has been identified as a component of the arsenite pump, and a 16,000 Da protein identified as a component of the arsenate pump.[20]

Determination of Energy Coupling *in Vivo*

To distinguish between primary and secondary coupling of energy to transport in intact cells, the sources of metabolic energy must be identified and isolated. *E. coli* is a facultative anaerobe. When grown aerobically, cells generate ATP through both glycolysis and oxidative phosphorylation. Cells can also generate a protonmotive force by proton pumping using either the redox energy of respiration or ATP hydrolysis via the F_0F_1. Since the F_0F_1 is reversible, it can equilibrate the protonmotive force and ATP. During growth the levels of protonmotive force and intracellular ATP are in an interdependent steady state. The equilibration of these pools by the F_0F_1 allows either energy source to buffer the other. Thus, under normal conditions any transport system appears to use ATP or a protonmotive force interchangeably.

To determine whether a particular transport system is coupled to chemical or electrochemical energy, the two sources of energy must be isolated from each other. Berger showed that it is possible to create intracellular conditions where either chemical energy alone or electrochemical energy alone is available for transport.[21] First, cells are depleted of endogenous energy reserves. The nature of the reserves is unknown, although *E.coli* is able to store glycogen. This is accomplished by forcing the cells to work at a futile cycle. Cells are incubated aerobically at 37° in growth medium lacking a carbon source but containing the protonophore 2,4-dinitrophenol. The cells undergo a futile cycle of respiratory-driven proton pumping and ionophore-mediated proton reuptake, until the endogenous stores of energy are depleted. Second, the equilibrium between the proton motive force and ATP is broken using a mutant (*unc*) defective in the F_0F_1. Starved *unc* cells are metabolically inactive until supplied with an energy source. Glucose will generate both ATP through glycolysis and a protonmotive force through respiration of NADH, succinate, and other products of glucose metabolism. If the electron transport chain is

[20] H. L. T. Mobley, C.-M. Chen, and B. P. Rosen, *Mol. Gen. Genet.* **191**, 421 (1983).
[21] E. A. Berger, *Proc. Natl. Acad. Sci. U.S.A.* **70**, 1514 (1973).

poisoned, for example, with cyanide, ATP is still produced through glycolysis but a protonmotive force is no longer generated. The protonmotive force can also be dissipated with a protonophoric uncoupler. (Note that other forms of chemical energy are also produced by glucose metabolism, e.g., phosphoenolpyruvate. This physiological dissection of energy sources does not distinguish between the various chemical bonds. However, the pool of phosphate bond energy exists mostly in the form of ATP, making it a reasonable candidate for energy donor. Rigorous identification of source of chemical energy to transport requires *in vitro* study.)

Conversely, if respiratory substrates are provided, a protonmotive force is established without ATP synthesis. Care must be taken to use the proper respiratory substrate. We have found that lactate oxidation produces ATP through substrate level phosphorylations. No ATP is generated during oxidation of reduced phenazine methosulfate or succinate. To use succinate efficiently, the cells must be induced for high levels of succinooxidase activity by inclusion of succinate in the growth medium. Since *unc* strains are unable to grow on succinate as sole carbon source, we usually grow cells in a minimal medium containing 0.5% glycerol as the major carbon source. Glucose is avoided because of catabolite repression of succinooxidase. Cells grown in several different basal salts media have yielded equivalent results. Enriched media have been used successfully, although cells grown in such media can exhibit cyanide-insensitive respiration.

Extrusion systems can be measured only in cells depleted of endogenous energy reserves, otherwise it is not possible to load the cells with substrate. The cells are passively equilibrated with the radioactive ion of interest. Transport against a concentration gradient (uphill transport) is initiated by addition of an energy source to the cells. Transport down a concentration gradient (downhill transport) is measured by dilution of the cells 100-fold into ion-free medium containing an energy source. By proper choice of energy source and inhibitors it is possible to distinguish whether extrusion either against or down a concentration gradient requires chemical energy, electrochemical energy, or both. Using this method we have shown that calcium and sodium extrusion requires only a protonmotive force and not phosphate bond energy.[22,23] On the other hand, extrusion of arsenate or arsenite from *E. coli* bearing an arsenate resistance plasmid was independent of the protonmotive force.[18,19] Chemical energy was required, and a temporal relationship between ATP synthesis and arsenate extrusion was observed.[19]

[22] H. Tsujibo and B. P. Rosen, *J. Bacteriol.* **154,** 854 (1983).
[23] M. G. Borbolla and B. P. Rosen, *Arch. Biochem. Biophys.* **229,** 98 (1984).

Method for Measurement of Extrusion of $^{22}Na^+$ from Intact Cells of E. coli

Growth of Cells. E. coli AN120 (*unc*A401, *arg*E3, *thi*-1, *rps*L)[24] is grown to stationary phase in 50 ml of basal salts medium[25] supplemented with 0.5% glycerol, 0.1 mg/ml arginine, and 2.5 μg/ml of thiamine. If desired, succinate is added to 5 mM to induce succinate dehydrogenase.

Depletion of Endogenous Energy Reserves. The cells are centrifuged and suspended in 10 mM Tris buffer, pH 7.5, containing 140 mM NaCl and 1 mM MgCl$_2$ (Buffer A) plus 5 mM Na$^+$ 2,4-dinitrophenol, pH 7.5. After 2-hr of shaking at 37° the cells are centrifuged, washed three times with Buffer A, and suspended with 0.2 ml of buffer, all at room temperature. Chilling of the cells should be avoided. At this point the cells exhibit no endogenous respiration, as measured with a Clarke-type electrode, and have no measurable ATP, as measured in a luciferin–luciferase assay. The cells respire normally if glucose is added, and ATP reaches normal levels within 2 min after addition of glucose.[19]

Transport Assay. One microcurie of carrier-free $^{22}Na^+$ is added, and the suspension is incubated at room temperature for 1 hr. This is sufficient time for passive equilibration of the $^{22}Na^+$ with the unlabeled intracellular sodium. Downhill efflux is initiated by dilution of 50 μl of cells into 5 ml of Buffer A in which NaCl is replaced with KCl. At intervals 0.5 or 1 ml portions are withdrawn and filtered through 0.45-μm pore size nitrocellulose filters. The filters are washed once with 5 ml of the same buffer, dried, and counted either in a liquid scintillation or gamma counter. Energy sources included in the dilution buffer can be 10 mM glucose, 20 mM potassium succinate, or 0.2 mM phenazine methosulfate (maintained in a reduced state with 20 mM potassium ascorbate). KCN (10 mM) is used as an inhibitor of the electron transport chain. Cyanide solutions should be neutralized, refrigerated, but not frozen during storage, and made fresh weekly. Because the lipid-rich outer membrane of gram-negative bacteria acts as a sink for hydrophobic compounds, high concentrations of ionophores and uncouplers must be used. The protonophore carbonyl cyanide *p*-trifluoromethoxyphenylhydrazone (FCCP) is used as an uncoupler at 10 μM. *N,N'*-Dicyclohexylcarbodiimide (DCCD), an inhibitor of the F$_0$F$_1$, is used at 0.2 mM. FCCP and DCCD are stored as ethanolic solutions at −20°.

We have found that an energy source is required for both uphill and downhill extrusion of sodium,[23] suggesting that the protonmotive force does more than provide the energy to establish a concentration gradient.

[24] J. A. Butlin, G. B. Cox, and F. Gibson, *Biochem. J.* **124,** 75 (1971).
[25] S. Tanaka, S. A. Lerner, and E. C. C. Lin, *J. Bacteriol.* **93,** 642 (1967).

A protonmotive force-dependent conformational change is one possibility. Voltage-sensitive gating is observed with other ion transport systems. Potassium is required in the dilution buffer for maximal extrusion. Other monovalent cations, including sodium, are not nearly as effective. Uptake of a counterion is probably required, and potassium can serve this role because there are specific transport systems for rapid K^+ entry. The fact that sodium cannot serve as its own counterion suggests that the NHA system does not catalyze sodium/sodium exchange.

In Vitro Measurements of Antiporter Activity

In vivo studies provide inferences on the nature of energy coupling, but the mechanism can only be determined using *in vitro* systems. We have developed everted membrane vesicle systems for the study of antiporters.[26] Intact cells form a protonmotive force acid and positive external and extrude the substrates of antiporters. The vesicles, on the other hand, are quantitatively inside-out or everted. Protons are pumped inward during respiration or F_0F_1-catalyzed ATP hydrolysis, polarizing the vesicles acid and positive interior.

Method for Preparation of Everted Membrane Vesicles

Cells (2 liters) are grown in enriched or minimal medium to stationary phase at 37° and harvested by centrifugation at 5000 g for 10 min. All centrifugation steps are performed at 4°. The cells are suspended in a minimum volume of 10 mM Tris–HCl, pH 7.4 containing 0.14 M choline chloride, 0.5 mM dithiothreitol, and 0.25 M sucrose (Buffer B), transferred to a weighed centrifuge tube and pelleted. The tube is wiped dry and weighed. The cells are suspended in 5–10 volumes of Buffer B per gram of pellet weight and lysed by a single passage through a French pressure cell at 4000 psi. MgCl$_2$ and DNase are added to 5 mM and 10 μg/ml, respectively, and the mixture incubated for 10 min at room temperature. Unbroken cells are removed by centrifugation once or twice at 5000 g for 10 min. To increase the yield of vesicles the unbroken cells can be resuspended in the original volume of Buffer B and passed a second time through the French pressure cell. The membrane vesicles are pelleted by centrifugation at 100,000 g for 1 hr and washed twice with Buffer B. The pellet is suspended in 1 ml of Buffer B per gram of original wet cell weight and stored in small amounts at −80°.

Cations extruded by cells are taken up by vesicles. Cation/proton antiporter activity can be assayed in two ways. First is direct measure-

26 B. P. Rosen and T. Tsuchiya, this series, Vol. 56, p. 233.

ment of accumulation of isotopically labeled ion. Second is through examination of the effect of cations on the steady-state proton gradient: exchange of cation for proton perturbs the proton gradient.

Accumulation of $^{45}Ca^{2+}$, $^{32}P_i$, $^{22}Na^+$, and $^{204}Tl^+$ into everted vesicles has been measured.[2,5,6,26] In our earlier method glycerol was used as a cryoprotectant in place of sucrose.[26] These vesicles are rather leaky to ions, and accumulation of isotopically labeled ion was difficult to visualize without trapping the ion inside of the vesicle. Accumulation of either $^{45}Ca^{2+}$ and $^{32}P_i$ could be observed only when both ions were present simultaneously, and apparently intravesicular formation of calcium phosphate occurred. Similarly, accumulation of $^{204}Tl^+$ via the KHA system was observed only when Cl^- was present in the buffer. Because the membrane potential is positive interior, Cl^- is accumulated. As the concentrations of Tl^+ and Cl^- exceed the solubility product of $TlCl$, the salt would precipitate inside of the vesicle. In contrast, accumulation of $^{86}Rb^+$, a more soluble substrate of the KHA system, was not observed.

Vesicles prepared in sucrose-containing buffer appear to be less leaky by several criteria. First, their ability to form a pH gradient is greater, as measured by respiration-driven quenching of acridine orange fluorescence (discussed below). Since the respiratory rate is the same in vesicles made either in glycerol or sucrose buffer, this implies that the back leak of protons is slower. Second, uptake of $^{45}Ca^{2+}$ is observed in the absence of phosphate, and the accumulated calcium is in a free state inside of the vesicles, as shown by the fact that it can exchange with Mg^{2+} catalyzed by A23187.[3]

Assay of $^{45}Ca^{2+}$ Uptake into Everted Membrane Vesicles

Two systems, CHA and CPS, exist for the transport of calcium into vesicles. Since the CPS system requires simultaneous presence of both calcium and phosphate, the CHA system can be measured without interference from the CPS system by omission of phosphate out of the assay buffer. The CPS system can be measured without contribution by the CHA system by pretreatment of the vesicles with trypsin. Vesicles (0.2 mg protein) are incubated in a buffer consisting of 10 mM Tris–HCl, pH 8, containing 0.14 M choline chloride and 5 mM $MgCl_2$ (Buffer C) with trypsin (0.2 mg) for 5 min at room temperature. Trypsin digestion is terminated by addition of excess trypsin inhibitor.

Calcium uptake via the CHA system is assayed in 1 ml of Buffer C containing 0.2 mg of vesicles at 23°. Energy sources include 20 mM Tris DL-lactate, 5 mM NADH, or 5 mM ATP, each adjusted to pH 8. The energy source is added to the reaction mixture just prior to initiation of the assay. Usually a regenerating system for NADH or ATP is included. The NADH regenerating system consists of 0.16 M ethanol and 1 mg of alcohol dehydrogenase. The ATP regenerating system consists of 10 mM phosphocreatine and 1 unit of creatine kinase. Transport activity is initiated by the addition of ^{45}CaCl$_2$ to 0.5 mM. Samples of 0.18 ml are withdrawn at intervals and filtered through prewetted nitrocellulose filters (0.45 μm pore size). (To ensure thorough wetting and removal of contaminants, the filters are boiled for 5 min in Buffer C, with one change of buffer.) The filters are washed once with 5 ml of Buffer C, dried on a hot plate, and counted in a liquid scintillation counter. Calcium uptake through the CPS system is assayed by with a slightly different buffer system: choline is replaced with potassium, MgCl$_2$ is eliminated, and 5 mM potassium phosphate, pH 8.0, is added.

A second assay was devised for the measurement of antiporter activity.[2,3] Since the vesicles form a ΔpH which is acid interior during respiration or ATP-driven proton pumping, lipophilic weak bases accumulate. At equilibrium the distribution of weak base is proportion to the ΔpH. Acridine orange is a lipophilic weak base whose fluorescence is quenched during accumulation. Thus, changes in ΔpH can be estimated from the energy-dependent change in the energy-dependent fluorescence quenching. The relationship can be quantified by preparing vesicles at pH 6 and diluting into a series of buffers at higher pH (B. P. Rosen, unpublished results); however, the assay is most often used qualitatively. In the steady state the ΔpH is a function of the rate of proton pumping and back leak of protons. Antiporters, which exchange cations and protons, increase the leak rate and therefore perturb the steady state ΔpH. This is reflected in a reversal of the quenching of acridine orange fluorescence. The rate of fluorescence enhancement can be related to antiporter activity.

Fluorescent Assay of Antiporter Activity

The reaction mixture consists of Buffer C and 2 $\mu$$M$ acridine orange in a volume of 2 ml. Each assay contains 0.02 to 0.04 mg of protein for sucrose vesicles or 0.15 to 0.2 mg of protein for glycerol vesicles. The cuvette is stirred, and additions are made through a port by injection of small volumes with Hamilton syringes. Quenching is initiated by addition of 10 mM DL-lactate, 2.5 mM Tris–ATP, or 5 mM NADH, each adjusted to pH 8.0 with Tris. Once a steady state ΔpH is attained, antiporter activity can be estimated from the rate of fluorescence enhancement following addition of a salt, for example, 10 mM KCl, 10 mM NaCl, or 0.1 mM CaCl$_2$. Fluorescence is measured with excitation and emission pairs of either 493 and 530 nm or 430 and 570 nm.

[27] Intracellular pH Regulation in Bacterial Cells

By ETANA PADAN and SHIMON SCHULDINER

Role of pH Homeostasis in Relation to Cell Physiology

Regulation of cytoplasmic pH is a necessity for every living cell. The proton more than any other ion is involved in almost every physicochemical as well as biochemical reaction. A narrow range of optimum pH values around neutrality limits the activity and/or stability of most proteins. Bacteria and other microorganisms face challenges that are not experienced by most cells in higher animals. Thus, the proton concentration of their extracellular environments range between pH 1 at the acidic sulfur springs to pH 11 at some soda lakes. Moreover, owing to biological activity marked fluctuations of pH values (up to three pH units) occur in many aquatic systems within a diurnal cycle (for references see Padan et al.[1]). The extracellular protons are in direct contact with the bacterial outer membrane as well as the periplasmic space and cytoplasmic membrane since the cell wall excludes only molecules of molecular weight above 1000. Also, unlike eukaryotes, there is no compartmentalization inside the cell and the protons produced and/or consumed in metabolic reactions may affect directly the cytoplasm.

The primary proton pumps, which are linked to electron transport and photochemical reactions, are located in the cytoplasmic membrane and these pumps maintain a proton electrochemical gradient ($\Delta \bar{\mu}_{H^+}$) across the membrane. Therefore, an efficient homeostatic mechanism has to also compensate for any changes in the chemical component of $\Delta \bar{\mu}_{H^+}$ with corresponding changes in $\Delta \psi$. As in all cells the $\Delta \bar{\mu}_{H^+}$ serves as the main driving force for ATP synthesis in a chemiosmotic coupling mechanism.[2,3] In the prokaryotic cell the reliance on proton gradients for energy extends to other endergonic reactions, including active transport and even locomotion.[3]

It is evident that the knowledge of pH_i (intracellular pH) and its regulation bears on many aspects of prokaryotic cell physiology including the evaluation of ΔpH which is required for the calculation of the $\Delta \bar{\mu}_{H^+}$. With this recognition in mind, development and application of methods for

[1] E. Padan, D. Zilberstein, and S. Schuldiner, *Biochim. Biophys. Acta* **650**, 151 (1981).
[2] P. Mitchell, *in* "Chemiosmotic Coupling and Energy Transduction." Glynn Research, Badmin, U.K., 1968.
[3] F. M. Harold, *Curr. Top. Bioenerg.* **16**, 486 (1982).

determination of pH_i in bacteria have been undertaken. As yet all these methods are indirect since bacteria like organelles are too small for insertion of available pH electrodes. Most of these methods are thoroughly discussed in excellent and exhaustive reviews.[4-6] We will deal only with the basic principles and some of the specific probes used in bacteria.

Cross-Membrane Equilibrium Distribution of Weak Acids and Bases Allows the Calculation of pH_i

Principle of Measurement and Choice of the Probe

The distribution method is thoroughly discussed in references 4–6. This method is based on the assumption that the uncharged form (AH) of many weak acids and bases is almost freely permeable through the membrane, whereas the respective ionized forms (A^-) are almost impermeable. If the inner compartment is basic inside as compared to the outer compartment, the undissociated form of the acid will penetrate the membrane and dissociate inside according to its pK and inner pH values. This process will continue until equilibrium is reached.

At equilibrium there is no gradient of the unionized form (AH) and it can be easily shown[4,5] that the distribution ratio of the acid will be given by the following equations:

$$\frac{A_i^T}{A_o^T} = \left(\frac{1}{K_a} + \frac{1}{H_i^+}\right)\bigg/\left(\frac{1}{K_a} + \frac{1}{H_o^+}\right) \tag{1}$$

where A^T is the total acid concentration ($A^T = A^- + AH$); K_a is its dissociation constant and the subscripts i and o indicate the inside and outside compartments.

Solving the above for the internal pH:

$$pH_i = pK + \log(A_i^T/A_o^T)(10^{pH_o-pK} + 1) - 1 \tag{2}$$

We have plotted the distribution ratio of a weak acid at various pH values as a function of the difference between its pK and pH_o (extracellular pH) (Fig. 1). We should first discuss the low limits of Eq. (1): when $pH_o > pK_a + 1$ then

$$A_i^T/A_o^T \cong H_o^+/H_i^+ \tag{3}$$

when $pH_o < pK_a - 1$ then

[4] H. Rottenberg, this series, Vol. 55, p. 547.
[5] A. Ross and W. F. Boron, *Physiol. Rev.* **61**, 296 (1981).
[6] G. F. Azzone, D. Pietrobon, and M. Zoretti, *Curr. Top. Bionenerg.* **13**, 1 (1984).

$$A_i^T/A_o^T \cong 1 \qquad (4)$$

When similar considerations are used for bases (B) it can be shown that

$$B_i^T/B_o^T = (K_a + H_i^+)/(K_a + H_o^+) \qquad (5)$$

When the inner compartment is acid inside, as compared to the outer compartment, a base is preferred for convenience. It is apparent that determination of the equilibrium concentration ratio of a weak acid or base allows the calculation of pH_i.

In choosing the probe and concentrations for a certain system, several criteria must be taken into account. (1) The pK should allow 90 to 99% dissociation of the probe in the extracellular and intracellular millieu. When the dissociation is lower, the acid ratio becomes less sensitive to the pH due to differential increase in the permeable fraction (AH) that distributes according to the chemical potential (Fig. 1). More extensive dissociation, however, may decrease the concentration of the permeant species so much that equilibration may be extremely slow; (2) the probe should exhibit negligible nonspecific binding to the cells; (3) it should not affect cell metabolism nor be itself metabolized or actively transported.

In a control experiment conducted with varying concentrations a probe fulfilling all these criteria should exhibit a constant acid distribution ratio over a wide range of concentrations. In such experiments, functions

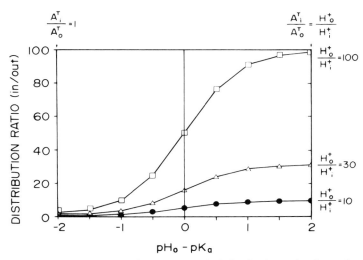

FIG. 1. Distribution of a weak acid. The theoretical distribution ratio of a weak acid with a given pK is plotted as a function of ($pH_o - pK$) according to Eq. (2). The distribution ratios are calculated for three different ΔpH values.

such as respiration rate and cell viability should be tested and found identical to that of a system lacking the probe. It is advisable to extract the intracellular probe and verify its authenticity by comparing its physicochemical properties to a standard. Chromatography can be used for this purpose. It is essential to verify that protonophores (or the combination of a protonophore and another ionophore) totally dissipate the concentration gradient of the acid. Finally, if possible two probes can be shown to yield identical results. Table I lists some of the most widely used ΔpH probes, their pK, and concentration range of operation.

Determination of Intracellular Water Volume

The cytoplasmic membrane of the gram-negative bacteria is surrounded by the periplasmic space and the cell wall. Since the pH$_i$ probes are of molecular weights lower than 1000, it is assumed that the extracel-

TABLE I
DISTRIBUTION PROBES COMMONLY USED TO MEASURE pH$_i$

Probe	pK_a	Concentration range (μM)	References[a]
Salicylic acid	3	100	18
Acetylsalicylic acid	3.5	10–60	43, 44
Benzoic acid	4.2	2–30	22, 30
Acetic acid[b]	4.75	2–200	10, 19
Butyric acid[b]	4.81	2–40	10, 19
Propionic acid[b]	4.87	2–40	10, 19
DMO	6.3	2–500	23, 24
Methylamine, ethylamine, ethanolamine	10.5–10.7	1–25	22, 41, 45
Acridine derivatives (9-aminoacridine, acridine orange, atebrin)[c]	7.9–10.5	0.5–20	41, 46

[a] This is not given as an exhaustive review but rather to guide the reader to some specific examples.
[b] Used mostly in membrane vesicle preparations in which metabolism does not interfere with the measurements.
[c] Used mostly in membrane vesicle preparations. There is one report of their use in whole cells.[14] In some systems, when total fluorescence is measured, they seem to overestimate ΔpH.[41] At low ionic strength they probably bind to surface charge on the membrane.[42]

lular and periplasmic concentrations of the probe are identical. Given that the extracellular concentration of the probe is known, calculation of cross-membrane distribution necessitates measurements of the amount of the probe and the water volume internal to the cytoplasmic membrane.

Convenient techniques to measure the micro internal water volume of organelles and vesicles have been developed and thoroughly discussed.[4–6] In general, they are based on the use of a combination of labeled permeable and impermeable solutes. Tritiated water is most commonly used as the permeant species to label the total water and a membrane impermeable solute to label the extra membrane space. The internal volume is calculated from the solute excluded space. Dextran and inuline are often used as membrane impermeable solutes. They have also been used with intact bacteria (for a summary of the probes used see Azzone *et al.*[6]). The high-molecular-weight probes are excluded by the cell wall and therefore will monitor the periplasmic space as part of the cytoplasmic space leading to overestimation of the latter. It should be noted, however, that due to the logarithmic dependence of ΔpH on the concentration ratio the value of pH_i will be relatively insensitive to small inaccuracies in internal volume.

Assuming that sucrose is impermeable through the cytoplasmic membrane the volume of the periplasmic space in *E. coli* and *S. typhimurium* was estimated[7] to comprise 20–40% of the total cell volume. Subsequently, however, sucrose was shown to be actively transported by the *lac* carrier in *E. coli*.[8] Hence, new probes for accurate measurement of the periplasmic space in bacteria are needed.

Recently, an innovative approach for direct measurement of micro volumes has been advanced and thoroughly discussed.[9] This method is based on spin-labeled metabolic inert small molecules that rapidly equilibrate across the membrane. Such a molecule is Tempone (2,2,6,6-tetramethyl-4-oxopiperidinoxy free radical). The procedure for volume measurements consists of obtaining a Tempon ESR spectrum in the presence of the microspecimen and a second spectrum after differential quenching of the external probe with impermeable spin quenchers. From the nonquenched probe concentration the internal volume is directly calculated. These probes are very promising for internal volume calculation. In particular, they allow monitoring fast changes in volume that may accompany fast ion movements.

[7] J. B. Stock, B. Rauch, and S. Roseman, *J. Biol. Chem.* **252,** 7850 (1977).

[8] K. B. Heller and T. H. Wilson, *J. Bacteriol.* **140,** 395 (1979).

[9] R. J. Mehlhorn, P. Gandau, and L. Packer, this series, Vol. 88, p. 751.

Determination of Probe Distribution

Given that the probe concentration in the external medium and the internal volume of the bacteria are known it is necessary to measure the internal amount of the probe for the calculation of the probe distribution across the membrane.

Since cell concentration at the logarithmic stage of growth amounts to no more than 100–200 μg cell protein/ml and the cell volume is approximately 5 μl/mg cell protein, the total amount trapped inside the cells is small as compared to the amount that remains outside. Hence most of the techniques for the determination of the taken up probe were developed for resting and concentrated cells, i.e., 1–2 mg cell protein/ml. These techniques have been discussed[4] and include two approaches. The first involves rapid separation of the cells from the medium by simple centrifugation or centrifugation through silicone oil. The amount taken up by the microspecimen is calculated from the amount present in the pellet after correction for the quantity in the pellet trapped external medium. In the method involving separation through silicone oil this correction is very small. During the centrifugation step nonphysiological conditions such as lack of O_2 or light limitation may prevail. These often lead to underestimation of the probe distribution ratio in respiring and photosynthesizing cells, respectively.[4,10]

Filtration of the cells through Millipore filters have also been used for cell separation.[11] However, since these filters are clogged by low bacterial mass of above 100 μg cell protein the sensitivity of this method is limited. The intracellular probe retained by the filter is not much more than the background in the trapped external medium.

The other approach for determination of the internal amount of the probe avoids manipulation of the cells during the experiment. The amount of the probe taken up is calculated from the change in the concentration of the extracellular probe during the experiment. This can be achieved if the intracellular form of the probe can be differentiated from the external form. Fluorescent[12,13] and ESR tagged molecules[9] have been found to have different characteristics in the two forms. The fluorescence of various acridine derivatives is quenched in the intracellular millieu[14,15] while ESR signals broaden.[9] It should be noted, however, that quantitation of

[10] S. Ramos, S. Schuldiner, and H. R. Kaback, *Proc. Natl. Acad. Sci. U.S.A.* **73**, 1892 (1976).
[11] D. G. Nicholls, *Eur. J. Biochem.* **50**, 305 (1974).
[12] D. W. Deamer, R. C. Prince, and A. R. Crofts, *Biochim. Biophys. Acta* **274**, 323 (1972).
[13] S. Schuldiner, H. Rottenberg, and M. Avron, *Eur. J. Biochem.* **25**, 64 (1972).
[14] E. D. Puchkov, I. S. Bulatov, and V. P. Zinchenko, *FEMS Lett.* **20**, 41 (1983).
[15] E. Blumwald, J. M. Wolosin, and L. Packer, in press (1984).

these techniques is very hard mainly due to light scattering and background of different ESR signals and interactions of the cells, respectively.

Utilizing an ion-specific electrode pH has been monitored in chromatophores of photosynthetic bacteria.[16] Even though ammonia cannot be used in intact cells, this approach could be further exploited. Recently electrodes for salicylate have been built.[17,18]

Another way to monitor changes in the extracellular fluid without cell separation is the flow dialysis method which has been thoroughly discussed.[10,19] In this system the cells are kept in one compartment which is separated from a second one by dialysis membrane. In the latter there are no cells yet the medium is equilibrated with the reaction medium through the membrane. It allows, therefore, continuous monitoring of changes in the extracellular fluid.

Since the latter techniques do not require manipulation of the cells through the experiments the values obtained are usually higher than those obtained after cell separation. Nevertheless, due to limited sensitivity these techniques are useful only for resting and concentrated cells 1–2 mg cell protein/ml.

Recently, a useful way has been found to monitor probe distribution in growing cells and to avoid most of the drawbacks mentioned. This is filtration utilizing glass filter papers (Whatman, GFC). These filters allow rapid flowthrough of 10 ml of logarithmic cells and therefore can retain on the filters as much as 1–2 mg cell protein.[20–22] The large amount of cell protein retained on the filters leads to a high amount of internal probe as compared to the extracellular probe trapped by the filters. Therefore, washing of the filtrated cells can be avoided. Cells treated with uncoupler can serve as the blank control. In a control experiment, pH_i values obtained by this technique with resting cells were found identical to values obtained by either flow dialysis[20,23] or centrifugation.[24]

NMR Spectroscopy, Colorimetry, and Fluorimetry

Although we will not discuss in detail other techniques to measure pH_i, some of them deserve at least a short description. The application of

[16] Z. Gromet-Elhanan, *TIBS* **2**, 274 (1977).
[17] W. M. Haynes and J. H. Wagenknecht, *Anal. Lett.* **4**, 491 (1971).
[18] R. W. Hendler, O. H. Setty, R. I. Shrager, D. C. Songco, and W. S. Friauf, *Rev. Sci. Instrum.* **54**, 1749 (1983).
[19] S. Ramos, S. Schuldiner, and H. R. Kaback, this series, Vol. 55, p. 680.
[20] D. Zilberstein, V. Agmon, S. Schuldiner, and E. Padan, *J. Bacteriol.* **158**, 246 (1984).
[21] D. Zilberstein, V. Agmon, S. Schuldiner, and E. Padan, *J. Biol. Chem.* **257**, 3687 (1982).
[22] E. Kashket, *Biochemistry* **21**, 5534 (1982).
[23] D. Zilberstein, S. Schuldiner, and E. Padan, *Biochemistry* **18**, 669 (1979).
[24] E. Padan, D. Zilberstein, and H. Rottenberg, *Eur. J. Biochem.* **63**, 533 (1976).

TABLE II
SELECTED MEASUREMENTS OF pH_i IN VARIOUS *E. coli* STRAINS

Conditions	Techniques	Probe	pH_o range	pH_i	References
Growing cells	Filtration	DMO, benzoic acid, methyl-amine	6.0–8.6	7.7–8.0	20–22
	β-Galactosidase activity in a pH conditional mutant	—	5.9–7.8	7.8–8.0	48
Resting aerated cells supplied with exogenous energy sources	Centrifugation Flow dialysis NMR spectroscopy	DMO, benzoic acid, methyl-amine	5.9–8.5	7.7–8.0	22–24
		^{31}P and methyl phosphonate	6–8.7	7.5–7.8	29, 31
	Proton movements in the external medium		6.6	7.8	47

^{31}P NMR spectroscopy in intact cells has been the subject of several reviews.[25-27] Its usefulness for estimation of pH_i is due to the pH sensitivity of the chemical shifts of phosphorus-containing compounds. The most commonly used is inorganic phosphate, which is generally present in a high enough concentration and has been successfully employed also in bacteria.[28-30] However, its chemical shift is practically pH insensitive above pH 7.8 (pK_a 6.8). Slonczewski *et al.* used a synthetic probe, methyl phosphonate, which is taken up by the cells and has a pK_a of 7.6.[31] The results of their studies are in excellent agreement with those obtained with the ion distribution techniques (see Table II).

[25] C. T. Burt, S. M. Cohen, and M. Barany, *Annu. Rev. Biophys. Bioeng.* **8,** 1 (1979).
[26] G. K. Radda and P. J. Seeley, *Annu. Rev. Physiol.* **41,** 749 (1979).
[27] R. G. Shulman, T. R. Brown, K. Ugurbil, S. Ogawa, S. M. Cohen, and J. A. den Hollander, *Science* **205,** 160 (1979).
[28] G. Navon, S. Ogawa, R. G. Shulman, and T. Yamane, *Proc. Natl. Acad. Sci. U.S.A.* **74,** 888 (1977).
[29] S. Ogawa, R. G. Shulman, P. Glynn, T. Yamane, and G. Navon, *Biochim. Biophys. Acta* **502,** 45 (1978).
[30] K. Nicolay, R. Kaptein, K. J. Hellingwerf, and W. N. Konings, *Eur. J. Biochem.* **116,** 191 (1981).

The advantages of ^{31}P NMR spectroscopy are quite apparent. In addition to being noninvasive and nondestructive its sensitivity and time resolution under optimal conditions are superior to other techniques. The main drawback is that since the absolute amount of P_i must be sufficient for detection, the cell concentration is well above the physiological one. Another obvious difficulty is the complex instrumentation required.

Fluorescent and spectroscopic probes have become extremely popular in cell biology in recent years. In addition to the distribution probes mentioned above two new approaches have had a strong influence in the above developments: (1) targeting of probes as introduced by Ohkuma and Poole in their studies of intralysosomal pH[32] and (2) generation of probes *in situ,* an approach first described by Thomas *et al.*[33] Obviously, only the latter technique has some potential use in bacteria. However, the pK_a values of most available probes are somewhat lower than desirable. Schechter and collaborators have estimated pH_i in several *E. coli* strains based on the pH-dependent rate of hydrolysis of fluorescein diacetate.[34]

Steady-State Conditions and pH_o Transients

Resting Cells

The phenomenology of pH_i regulation has been characterized by perturbing pH_i. A very simple and successful way was to expose resting cells to rapid changes in pH_o in either the acid or alkaline direction. Utilizing the probe distribution techniques this approach was first applied to resting cells of *E. coli*[24] and *Streptococcus faecalis.*[35] The resulting perturbation of pH_i and the recovery to normal steady value can then be studied.

With respect to *E. coli,* cells grown on minimal medium A with glycerol as carbon source are harvested, washed, and resuspended (1–2 mg cell protein/ml) in 100 mM potassium phosphate buffer containing 1 mM $MgSO_4$ (pH 7), and kept at 4°. The reaction mixtures for pH_i determination are prepared at different pH values and the reaction is commenced by introduction of the cells into the reaction to yield a cell concentration of around 1–2 mg protein/ml (for further details see Padan *et al.*[24]).

[31] J. L. Slonczweski, B. P. Rosen, J. R. Alger, and R. M. Macnab, *Proc. Natl. Acad. Sci. U.S.A.* **78,** 6271 (1981).
[32] S. Ohkuma and B. Poole, *Proc. Natl. Acad. Sci. U.S.A.* **75,** 3327 (1978).
[33] J. A. Thomas, R. N. Buchsbaum, A. Zimnick, and E. Racker, *Biochemistry* **18,** 2210 (1979).
[34] E. Shechter, L. Letellier, and E. R. Simons, *FEBS Lett.* **139,** 121 (1982).
[35] F. M. Harold, E. Pavlasova, and J. R. Baarda, *Biochim. Biophys. Acta* **196,** 235 (1970).

Similar experiments conducted with very many bacteria demonstrated that pH homeostasis is very efficient and is established very fast. Steady state of pH_i is attained within 1–2 min and is poised at around pH_i of 7.6 in the neutrophiles (pH_o 6–9), at pH 6.5 in the acidophiles (pH_o 2–6), and in the alkalophiles at pH_i of 9.5 (pH_o 9–11).[1]

In Table II we have tabulated some representative measurements of pH_i in *E. coli* at various conditions. We do not include measurements in EDTA-treated cells in which pH_i is somewhat lower[24] or experiments in which the energy supply is limiting. Interestingly, pH_i regulation seems to be less strict in anaerobically growing *E. coli* cultures and in resting cells derived from them.

It should be emphasized that in the experiments conducted with resting cells, the physiological state of the cells should be tested. A viable count can be done at the end of the experiment and compared to the initial viability. These data are particularly important when employing various ions and buffer composition at extreme pH. Unfortunately, most of the published works do not include such information. In any event, it should be always kept in mind that the mechanism found in resting cells need not necessarily represent the situation in growing cells.

Growing Cells

An experimental system allowing study of pH homeostasis in growing cells has only recently been advanced.[20–22] Such a system with *E. coli* is described. Cells are inoculated (0.02 mg cell protein/ml) into growth medium A^{36} in which the $MgSO_4$ concentration is reduced to 0.001% (to avoid precipitation of Mg salts at the alkaline pH). Cells are grown in the 1 liter flask of Bio Flo Model C30 chemostat (New Brunswick Scientific) as batch cultures at 37°. The pH of the medium is controlled by means of a Modcon (Kiryat Motzkin, Israel) pH titrator. When required, KOH or HCl is added at a rate of 2.25 or 3.6 mEq/min, respectively. For pH transients this flow is also suggested since faster titration may harm the cells.[20] At various times 10 ml of the cell suspension (0.1–0.17 mg of cell protein/ml) is rapidly transferred into a prewarmed (37°) 100 ml flask containing 9 μM [^{14}C]methylamine (10 μCi/mol) or 0.32 mM [^{14}C]DMO (5,5-dimethylazolidine-2,4-dione, 120 mCi/mol). The suspension is incubated for 1 min with continuous shaking at 37° and filtered through a glass fiber filter (GF/C, Whatman, 25 mm diameter). The filters are transferred into toluene-Triton scintillation liquid and assayed for radioactivity. The amount of radioactivity retained on the filters in the absence of ΔpH, namely at pH_o 7.6 or at pH 7 in the presence of carbonyl cyanide

[36] G. D. Davis and E. S. Mingioli, *J. Bacteriol.* **60**, 17 (1950).

p-trifluoromethoxyphenylhydrazone is identical and serves as a blank. This background correction can be as high as 50% at low ΔpH values. The steady-state level of uptake of the pH probe at all pH$_o$ values between 6 and 9 is attained within 75 sec.

Growing cells exhibit the same pattern of interchangeability between steady-state values of $\Delta\psi$ and ΔpH so the $\Delta\bar{\mu}_{H^+}$ remains constant.

Detailed analysis of the transients from pH$_o$ of 7.2 to more alkaline pH$_o$ in growing cells allows the detection of the sequence of events leading to pH homeostasis. The shift in pH$_o$ artificially imposes a ΔpH directed outward (acid inside). This ΔpH collapses to 0 within 1 min. Immediately upon the shift $\Delta\psi$ rises and reaches maximal value when ΔpH becomes zero. The rise in $\Delta\psi$ is more than can be accounted for by the decrease in ΔpH bringing about a transient increase in $\Delta\bar{\mu}_{H^+}$. Since the rate of respiration does not significantly change after the shift it is possible that the shift elicits a reduction in some dissipative electrogenic leaks of H^+ or other ions across the membrane yielding this increase in $\Delta\bar{\mu}_{H^+}$. Alternatively, the transient hyperpolarization could also be due to increased electrogenic flows in the proper directions. None of the described alternatives has experimental support. It is interesting, however, in this context that Slonczewski et al.[31] found in their studies (with resting cells) that after the transition to the alkaline pH the level of nucleotides markedly decreases. Although it remains to be shown whether the same occurs in growing cells it is tempting to speculate on some role of the H^+-ATPase in the transient hyperpolarization.

Within a few minutes $\Delta\psi$ decreases to its new steady-state value. During this step acidification occurs and the steady state pH$_i$ is established. Since at this step $\Delta\psi$ serves as the driving force it is suggested that the mechanism responsible for acidification is driven by the $\Delta\psi$.

Mechanism of pH Homeostasis

Primary Proton Pumps

With both neutrophiles and alkalophiles the propensity of pH$_i$ homeostasis can be demonstrated both in intact cells[1,37] as well as in isolated membrane vesicles.[10,38] It may be concluded that the mechanism of pH regulation is built into the membrane.

Utilizing specific inhibitors of respiration, H^+/ATPase as well as photosynthetic electron transport, it is possible to show both in resting and

[37] T. A. Krulwich and A. A. Guffanti, Adv. Microbiol. Physiol. **24**, 173 (1983).
[38] K. G. Mandel, A. A. Guffanti, and T. A. Krulwich, J. Biol. Chem. **225**, 7391 (1980).

growing cells that the primary proton pumps are involved in the maintenance of pH homeostasis. When these pumps that are involved do not operate, the protons equilibrate across the membrane. This excludes the acidophiles in which the contribution of passive components such as Donnan potential in the maintenance of pH_i is not negligible.[37,39,40]

The involvement of the primary proton pumps implies that $\Delta\psi$ must be very closely related to pH regulation since it is also maintained by these pumps. Measurement of $\Delta\psi$ (techniques thoroughly discussed in Rottenberg[4]) after shift of resting cells of *E. coli* to various pH_o showed that at new steady state $\Delta\psi$ changes in compensatory fashion with the ΔpH. It increases from 80 mV at pH_o 6 to its highest value of 160 mV at pH_o 7.8 while ΔpH decreases from about 2 to 0 over the same pH_o range. Beyond pH_o 7.8 the $\Delta\psi$ is constant whereas ΔpH reverses orientation and becomes almost 1 pH unit acid inside at pH_o 8.8.

Utilizing this technique it has been found that between pH 6 and 8.8 at steady state growing *E. coli* cells maintain the same pH_i as resting cells, i.e., pH_i 7.6–7.8 (see Table II).[41–50] The pattern of presteady state of events occurring after shifts in pH_o has also been detected. It was found to depend both on the span of pH_o change as well as the rate at which the change occurs. The transient state after stepwise transfer obtained by titration [with KOH or HCl at 2.25 to 3.6 mEq/min (3 min)] of growing cells from pH 7.2 to 8.3 to 8.6 and then to 8.8 or from 7.2 to 6.4 initiated by reduction of ΔpH across the membrane to zero and growth ceased. Subsequently, within 6 min at the most the ΔpH was built up to a magnitude that yielded the steady-state pH_i. Thereafter, growth resumed at the initial rate. However, if the shift was made abruptly from pH 7.2 to 8.6 the lag was longer and the buildup of the ΔpH was slower. The shift between pH 7 to 8.8 appears to be the limit of the pH homeostasis capacity of the *E. coli* strain we use (CS 71) since the wild type grows normally when allowed to adapt by step transfers over this range and failed to restore both

[39] A. Matin, B. Wilson, E. Zychlinsky, and M. Matin, *J. Bacteriol.* **150,** 582 (1982).

[40] T. Hackstadt, *J. Bacteriol.* **154,** 591 (1983).

[41] W. W. Reenstra, L. Patel, H. Rottenberg, and H. R. Kaback, *Biochemistry* **19,** 1 (1980).

[42] J. Barber, *Biochim. Biophys. Acta* **594,** 253 (1980).

[43] T. A. Krulwich, L. F. Davidson, S. J. Filep, R. S. Zuckerman, and A. A. Guffanti, *J. Biol. Chem.* **253,** 4599 (1978).

[44] H. Kobayashi, N. Murakami, and T. Unemoto, *J. Biol. Chem.* **257,** 13246 (1982).

[45] H. Tokuda and T. Unemoto, *J. Biol. Chem.* **257,** 10007 (1982).

[46] S. Schuldiner and H. Fishkes, *Biochemistry* **17,** 706 (1978).

[47] S. Collins and W. A. Hamilton, *J. Bacteriol.* **126,** 1224 (1976).

[48] M. Colb and L. Shapiro, *Proc. Natl. Acad. Sci. U.S.A.* **74,** 5637 (1977).

[49] E. R. Kashket, *FEBS Lett.* **154,** 343 (1983).

[50] K. J. Hellingwerf, J. G. M. Bolscher, and W. N. Konings, *Eur. J. Biochem.* **113,** 369 (1981).

normal internal pH and optimal growth if the transition was made in one step (within 30 sec titration by 13.5 to 21.6 mEq/min). Rapid titration was deleterious to the cells even if the span was smaller.

Hence, at least two electrogenic processes are elicited by the pH_o shift to alkaline pH: one is involved in the initial increase in $\Delta\psi$; the second consumes $\Delta\psi$ and is responsible for the acidification of the cytoplasm at alkaline pH.

Ion Transport

In order to simplify the treatment of the problem of pH_i homeostasis the existing models artifically and arbitrarily divide the range of homeostasis in two: below pH_o 7.8 and above pH_o 7.8.

Below pH_o 7.8, the chemical gradient of protons is always in the same direction of the pumping. Changes in the chemical gradient are compensated by corresponding changes in the electrical gradient, a mechanism which ensures a constant $\Delta\bar{\mu}_{H^+}$. The models that have been put forward to explain the interconversion of the two components of $\Delta\bar{\mu}_{H^+}$ suggest the existence of pH_o-dependent changes in the rate of electrogenic transport of a counterion, most likely K^+.[51-56]

At the pH_o range above 7.8 the main problem of pH_i regulation is the generation of a pH gradient, interior acid, i.e., of a polarity opposed to the polarity of pumping. The two solutions that have been proposed are based on the functioning of cation–proton antiporters: Na^+[1,24,46] and K^+,[1,57] respectively.

K^+ *Transport.* One approach to elucidate the participation of a specific ion in pH_i homeostasis is to perturb pH_i and follow the behavior of the cells with respect to pH_i in the absence of the ion as compared to its presence. This approach has been undertaken to study the involvement of K^+ in pH_i regulation.[51-56] In one such typical experiment *E. coli* cells were depleted of K^+ by overnight growth in glucose minimal medium[54] which was 100 μM with respect to K^+. Subsequently they were diluted with 20 μM K^+ medium. This treatment decreased the internal K^+ from 282 ng ion K^+/mg dry weight to 133 ng ion K^+/mg dry weight and also derepressed the Kdp system.

[51] F. M. Harold, E. Ravlasova, and J. R. Baarada, *Biochim. Biophys. Acta* **196**, 235 (1970).
[52] F. M. Harold and D. Papineau, *J. Membr. Biol.* **8**, 45 (1972).
[53] E. R. Kashket and S. L. Barker, *J. Bacteriol.* **130**, 1017 (1977).
[54] R. G. Kroll and I. R. Booth, *Biochem. J.* **198**, 691 (1981).
[55] E. D. Bakker and W. E. Mangerich, *J. Bacteriol.* **147**, 820 (1981).
[56] H. Tokuda, T. Nakamura, and T. Unemoto, *Biochemistry* **20**, 4198 (1981).
[57] R. N. Brey, J. C. Beck, and B. P. Rosen, *Biochem. Biophys. Res. Commun.* **83**, 1588 (1978).

Cells were harvested, washed, and resuspended in 3 mM Mes, 6.9 mM Tris, and 150 mM choline chloride at pH 6.3, containing 50 μg/ml chloramphenicol and stored at 4°. The reactions were conducted at 24° in this medium at 2.3 mg cell dry weight/ml. To study effect of K$^+$ it was added at various concentrations (1 μM–1 mM) and ΔpH probe distribution was determined by centrifugation. The rate of K$^+$ uptake was monitored in the experimental system by observing the disappearance of the ion from the medium by flame photometry. The steady state for K$^+$ uptake was obtained after 5 min and at this time probe distribution was monitored. The more K$^+$ uptake obtained the more ΔpH observed and concomitant depolarization of $\Delta\psi$ occurred.

Similar experiments conducted in the presence of glucose with cells grown in 20 mM K$^+$ also exhibited K$_o^+$ dependency. However, no net K$^+$ accumulation could be detected. A K$^+$ cycle has been suggested for this case.[54]

The effect of K$^+$ is independent of the mechanism of K$^+$ transport. Identical results were obtained with the wild type and two mutants each lacking either the Kdp or the Trk system, two of the K$^+$ transport systems in *E. coli*.[54,55] In a strain lacking both systems K$^+$ ions did not have any effect on $\Delta\psi$ or ΔpH.

In contrast to the results described above, cells depleted of K$^+$ generate a larger ΔpH (acid inside) at the alkaline pH range.[55,58] This has been explained as the result of the functioning of the K$^+$/H$^+$ antiporter.[58,59] In several studies the ability of the K$^+$/H$^+$ antiporter to generate a pH gradient across the cell membrane has been demonstrated.[57,58,60] In membrane vesicles from alkalophiles, K$^+$ collapses a ΔpH alkaline inside.[38]

In conclusion, the experimental evidence presented support the contention that electrogenic K$^+$ transport systems play a role in pH regulation in the acid pH$_o$ range. At the alkaline pH$_o$ range, the K$^+$/H$^+$ antiporter seems to play a role as well. However, the details of the mechanisms proposed are still quite obscure and we lack evidence to support this model at a more physiological level. For example, since the K$^+$/H$^+$ antiporter is electroneutral[57,60] at the alkaline pH$_o$ range, it can be driven only by the ΔpK. Thus, one would expect to see effects of high external K$^+$ on the homeostatic process and on the span of growth at the alkaline pH$_o$. These have not been reported. Moreover, mutants impaired in the various K$^+$ transport systems should show also some kind of failure in pH$_i$ homeostasis, and therefore impaired growth at the proper pH$_o$. This has not

[58] T. Nakamura, H. Tokuda, and T. Unemoto, *Biochim. Biophys. Acta* **776**, 330 (1984).
[59] I. R. Booth and R. G. Kroll, *Biochem. Soc. Trans.* **11**, 70 (1983).
[60] R. N. Brey, B. P. Rosen, and E. N. Sorensen, *J. Biol. Chem.* **255**, 39 (1980).

been observed. The question also remains open as to whether pH_i regulation is merely the balance of steady states or whether there is some kind of "active" regulation.[59]

Na+ Transport. A second approach to elucidate the participation of a specific ion in pH_i homeostasis is to isolate mutants in the ion transport(s) system(s) and to test the behavior of these mutants with regard to the homeostatic process. This approach has been exploited in *E. coli*[1,20,21] and in *Bacillus alcalophilus*[38,61] and has helped to strongly substantiate the participation of the Na^+/H^+ antiporter in pH_i regulation. In both cases the mutants have an impaired Na^+ efflux capacity from intact cells, they generate a normal $\Delta\psi$, fail to acidify their internal pH, and have lost their ability to grow at alkaline pH_o. The *E. coli* mutant displays normal growth and pH_i homeostasis at the acid pH_o range. The *B. alcalophilus* mutant gained the capacity to grow at lower pH values (5.5–9). Membrane vesicles prepared from the latter lacked the Na^+/H^+ antiporter activity but still display normal K^+/H^+ antiporter activity. Both mutants, however, show a pleiotropic inactivation of other Na^+-linked transport systems, owing to the loss of a postulated common Na^+ coupling subunit or to some other defect.[62,63] This pleiotropicity has raised criticisms of the model proposed, since it is not clearly shown that the effect of the mutation on the antiporter is indeed the direct cause of the defect on pH_i regulation. It is noteworthy in this context that the two mutants isolated, both in *E. coli* and in *B. alcalophilus,* show the same pleiotropicity.

Another criticism to the participation of the Na^+/H^+ antiporter in pH_i regulation is the failure to demonstrate dependence for either growth or pH_i homeostasis on external Na^+.[59] This is most probably due to the relatively low K_m for Na^+ of the antiporter and the Na^+ contamination of the growth medium.[1,38]

In the case of the alkalophiles most species do show Na^+ dependence for growth and for cytoplasm acidification and the exceptions appear to be those species whose Na^+-dependent processes have unusual high affinity for Na.[64,65] When cells of *Bacillus firmus* RAB (growing at pH_o 10.5) are resuspended in Na^+-free medium pH_i rises to 10.5 and viability is lost.[66] In

[61] T. A. Krulwich, K. G. Mandel, R. F. Bornstein, and A. A. Guffanti, *Biochem. Biophys. Res. Commun.* **91,** 58 (1979).

[62] A. A. Guffanti, D. E. Cohn, H. R. Kaback, and T. A. Krulwich, *Proc. Natl. Acad. Sci. U.S.A.* **78,** 1481 (1981).

[63] D. Zilberstein, I. J. Ophir, E. Padan, and S. Schuldiner, *J. Biol. Chem.* **257,** 3692 (1982).

[64] T. A. Krulwich, A. A. Guffanti, R. F. Bornstein, and J. Hoffstein, *J. Biol. Chem.* **257,** 1885 (1982).

[65] T. A. Krulwich, *Biochim. Biophys. Acta* **726,** 245 (1983).

[66] M. Kitada, A. A. Guffanti, and T. A. Krulwich, *J. Bacteriol.* **152,** 1096 (1982).

the absence of added Na^+, at pH 9.0, respiring cells generate a pH alkaline inside. Upon addition of Na^+ the polarity of ΔpH is reverted.[66]

In conclusion, the evidence supporting a role for the K^+ transport systems, the K^+/H^+ antiporter and the Na^+/H^+ antiporter is mounting. The details of the overall mechanisms and the possible interactions between them are yet to be elucidated. Thus, it is not clear whether the two antiporters function in a concerted way or whether they are alternative mechanisms functioning at different physiological conditions. Are there other systems (other ions or leaks of Na^+ and K^+) part of the regulatory systems as well? After failure of the homeostatic response, is there need for a signal to activate the regulatory mechanism (pH$_o$, pH$_i$, $\Delta\psi$) or is the new steady state reached by a mere rearrangement of the new rates of transport of the various ions? It is clear that nothing is known about the structure and regulation of the synthesis of the antiporters. Finally, the coupling between pH$_i$ and cell division is very striking: relatively small changes in pH$_i$ bring about an immediate cessation of cell division. From the results published it would seem that there is a specific and sensitive sensor of the hydrogen ion concentration that somehow controls cell division. Elucidation of the nature of this control mechanism may shed light on other processes in which cytoplasmic pH has been postulated as a signal.

[28] Regulation of Internal pH in Acidophilic and Alkalophilic Bacteria

By T. A. KRULWICH and A. A. GUFFANTI

As noted in the preceding chapter of this volume[1] and in other recent reviews,[2,3] many if not all bacteria can maintain an astonishingly constant cytoplasmic pH (pH$_{in}$) when incubated or grown at very different external pH values and even after being subjected to rather dramatic shifts in external pH. This ability to regulate pH$_{in}$ is a crucial physiological function that is made all the more important—even when the vicissitudes in external pH are not that great—by the likelihood that at least some criti-

---

[1] E. Padan and S. Schuldiner, this volume [27].
[2] E. Padan, D. Zilberstein, and S. Schuldiner, *Biochim. Biophys. Acta* **650,** 151 (1981).
[3] I. Booth and R. G. Kroll, *Biochem. Soc. Trans.* **11,** 70 (1983).

cal cell process(es) is tightly controlled by pH_{in}.[4] In neutralophilic bacteria a large array of primary ion pumps and secondary antiport-, symport-, and uniport-mediated ion movements have been proposed to play a role in pH homeostasis.[1,3,5–9] Presumably, different mechanisms are required to raise pH_{in} from those that serve to lower pH_{in}. Moreover, a single organism may possess dominant and auxiliary mechanisms whose relative importance could depend upon the specific pH range and/or special circumstances such as light. There are a few mutants whose phenotype supports a role for a particular pump or porter in pH homeostasis in a neutralophile; ultimately, clarification of cytoplasmic pH regulation may depend upon the availability of larger numbers of well-characterized mutants. At present, the precise roles of potassium and sodium in facilitating acidification of the interior during alkali stress of a neutralophile remain somewhat controversial. Similarly, in neutralophiles, the relative roles of proton-pumping ATPases, respiration, and other possible means of raising pH_{in} during acid stress are still very incompletely determined. Examination of acidophilic and alkalophilic bacteria offers an opportunity to study the problems of pH homeostasis *in extremis*. Bacteria which grow well at extremes of pH may have evolved special mechanisms for coping with their most obvious biological problem; such mechanisms may or may not be qualitatively similar to those employed in less demanding growth conditions. Even if they are not, studies in the extreme ranges should indicate directions for experimentation in the more subtle region of neutrality. Moreover, extreme acidophiles and alkalophiles are of intrinsic interest because of their broader bioenergetic problems and characteristics, the properties of their surfaces, and those processes that are exposed to the external *milieu*, their industrial potential, and their relationships to the ecosystem.[10–13]

Our own studies have focused upon species of *Bacillus* that grow obligately at very low (pH 2–3.5) or very high (pH 10–11) pH values. In the acid range, we have studied the thermoacidophile *Bacillus acido-*

[4] D. Zilberstein, V. Agmon, S. Schuldiner, and E. Padan, *J. Bacteriol.* **158**, 246 (1984).
[5] T. A. Krulwich, *Biochim. Biophys. Acta* **726**, 245 (1983).
[6] H. Kobayashi and T. Unemoto, *J. Bacteriol.* **143**, 1187 (1980).
[7] J. Lanyi, *Biochim. Biophys. Acta* **559**, 377 (1984).
[8] S. Kan-Dron, R. Shnaiderman, and Y. Avi-Dor, *Arch. Biochem. Biophys.* **229**, 640 (1984).
[9] H. Tokuda and T. Unemoto, *J. Biol. Chem.* **257**, 10007 (1982).
[10] T. A. Krulwich and A. A. Guffanti, *Adv. Microb. Physiol.* **24**, 173 (1983).
[11] K. Horikoshi and T. Akiba, "Alkalophilic Microorganism." Springer-Verlag, Berlin and New York, 1982.
[12] J. G. Cobley and J. C. Cox, *Microbiol. Rev.* **47**, 579 (1983).
[13] T. A. Langworthy, *in* "Microbial Life in Extreme Environment" (D. J. Kushner, ed.). Academic Press, New York, 1978.

caldarius, first isolated and described by Darland and Brock.[14] The obligate alkalophiles have included *Bacillus alcalophilus*[15] and *Bacillus firmus* RAB.[16] Such obligately alkalophilic organisms can be isolated with surprising ease from various soil samples, perhaps reflecting the high alkaline clay particle content of many soils. In this review of methodology, we will outline the protocols used to characterize features and mechanisms of pH homeostasis in the obligately acidophilic and alkalophilic bacilli; information derived from studies by other investigators of gram-negative acidophiles and of alkaline-tolerant organisms (those with a range of growth pH values that is lower than that of the obligate alkalophiles and includes near-neutral values) will be introduced only peripherally to indicate where important differences in experimental detail may obtain. Also, we will confine this review to those methods that have employed whole cells. It should be noted, however, that studies of pH regulation in bacteria have been conducted both *in vivo* and *in vitro*. The *in vitro* studies, using isolated membrane vesicles of the type first prepared by Kaback,[17] have made some major contributions to this area. In both the acidophile *B. acidocaldarius* and in alkalophilic *B. alcalophilus* and *B. firmus* RAB, work with right-side-out membrane vesicles has provided unequivocal evidence of outward, respiration-dependent proton translocation,[18,19] and has been part of the battery of approaches to the characterization of antiport activities relevant to pH regulation.[18–20]

Acidophilic Bacteria

It has been clearly established that any one of several acidophilic bacteria will maintain an almost constant cytoplasmic pH, at some value between 5.8 and 7.1, over a range of external pH values including pH 2–3.[12,21] There is general agreement that respiration-dependent proton translocation is required for the generation of the enormous ΔpH.[12] It has also been generally found that acidophiles, in the very acid range of pH, pos-

[14] G. Darland and T. Brock, *J. Gen. Microbiol.* **67**, 9 (1971).
[15] R. E. Buchanan and N. E. Gibbons (eds.), "Bergeys' Manual of Determinative Bacteriology." Williams & Wilkins, Baltimore.
[16] A. A. Guffanti, R. Blanco, R. A. Benenson, and T. A. Krulwich, *J. Gen. Microbiol.* **119**, 79 (1980).
[17] H. R. Kaback, this series, Vol. 22, p. 99.
[18] T. A. Krulwich, A. A. Guffanti, R. F. Bornstein, and J. Hoffstein, *J. Biol. Chem.* **257**, 1885 (1982).
[19] A. A. Guffanti, M. Mann, T. L. Sherman, and T. A. Krulwich, *J. Bacteriol.* **159**, 448 (1984).
[20] K. G. Mandel, A. A. Guffanti, and T. A. Krulwich, *J. Biol. Chem.* **255**, 7391 (1980).
[21] M. Michels and E. P. Bakker, *J. Bacteriol.* **161**, 231 (1985).

sess a "reversed $\Delta\psi$," a transmembrane electrical potential that is positive in.[10,12] Most likely, the reversed $\Delta\psi$ has a role in the maintenance and, perhaps, the generation of the ΔpH.[10,12,22] The nature and origin of the reversed $\Delta\psi$ are as yet unknown, and pH shift experiments, that might clarify aspects of ΔpH and $\Delta\psi$ generation, have yet to be reported for acidophiles. Thus, we will present protocols for measurement of the ΔpH, from which calculations of pH_{in} are made, and look forward to the application of these methods to physiological experiments in which cells equilibrated at neutral pH are shifted to highly acidic media. Because of suggestions that cytoplasmic buffering capacity and a special impermeability of the acidophile membrane to cations may be involved in pH homeostasis in at least some gram-negative acidophiles,[12,22–25] we will also consider the methodological approaches to determinations of these parameters.

Determinations of ΔpH

The principles involved in measurements of ΔpH have been elucidated and described in detail by many others, including a review by Rottenberg[26] in an earlier volume; the equations required for calculation can be found there. In acidophilic as in conventional bacteria, the distribution of weak acids is used to assay the pH gradient across the membrane; since the pH_{out} is known, the pH_{in} can then be readily ascertained.[26] The special considerations for work with acidophiles include the choice of buffer, the choice of weak acid, and the controls used to assess binding of the weak acid (as opposed to its localization, free, within the cytoplasmic component). In any measurements of the ΔpH, it should be recognized that the amount of the accumulated probe that is truly free in the cytoplasm is not easily evaluated.

We have used citrate buffer, 25 to 50 mM, to maintain the pH_{out} at low pH values in experiments with *B. acidocaldarius*.[27] Matin and his colleagues[23] have used 100 mM β-alanine buffer in some of their determinations of ΔpH in the gram-negative acidophile *Thiobacillus acidophilus*, but it has recently been suggested that use of β-alanine buffer may not be optimal if protonophore effects are to be examined as part of the experimental series.[21] The external pH of buffered cell suspensions should be

[22] W. J. Ingeldew, *Biochim. Biophys. Acta* **683**, 89 (1982).
[23] A. Matin, B. Wilson, E. Zychlinsky, and M. Matin, *J. Bacteriol.* **150**, 582 (1982).
[24] E. Zychlinsky and A. Matin, *J. Bacteriol.* **153**, 371 (1983).
[25] E. Zychlinsky and A. Matin, *J. Bacteriol.* **156**, 1352 (1983).
[26] H. Rottenberg, this series, Vol. 55, p. 547.
[27] T. A. Krulwich, L. F. Davidson, S. J. Filip, Jr., R. S. Zuckerman, and A. A. Guffanti, *J. Biol. Chem.* **253**, 4599 (1978).

monitored to assess dirft and account for it in the calculations. With *B. acidocaldarius,* suspensions are prepared from logarithmically growing cells (100–200 Klett units with a red filter) that are harvested by centrifugation and washed at least once with the buffer of choice. For determinations using a flow dialysis assay, cells are resuspended to a final concentration of 5 mg cell protein/ml; when the determination of ΔpH is to be made using a filtration assay, the cells are resuspended to a final concentration of approximately 0.5 mg cell protein/ml. It is necessary to know the internal cell volume in order to calculate the distribution of weak acid probe. We have used tritiated water to measure the total water space and either radiolabeled inulin or dextran to measure the external water space as described by Rottenberg[26]; the difference between the two values is the internal water volume. In flow dialysis assays, it is important for this volume to represent a significant part of the experimental chamber volume.

The weak acid probes of ΔpH that are most commonly used for conventional bacteria, e.g., 5,5-dimethyl-2,4-oxazolidinedione (DMO), are inappropriate for use at very acid external pH values.[21,28] Since the pK_a of DMO is 6.3, an accumulation ratio, inside to out, of only 3 to 5 would be expected at typical values of pH_{out} 2 and pH_{in} 6.5. Such a small ratio is difficult to measure accurately. Therefore, either benzoic acid (pK_a 4.17) or acetylsalicylic acid (pK_a 3.48) are the probes of choice for work with acidophiles. The latter weak acid is usually commercially available with a radioactive label in the acetyl group; it may be important to show that extramembranal cleavage of the probe does not occur. As in all determinations of ΔpH from the distribution of weak acids, it is also essential that the acid be nonmetabolizable and that the entry of the acid be entirely by passive diffusion. The weak acids should be used at concentrations less than 10 μM to avoid dissipation of the gradient that is being measured.

The accumulation of the weak acid can be measured by filtration through GFC filters (Whatman) as described by Zilberstein *et al.*[4] With the aerobic thermoacidophile, *B. acidocaldarius,* the cell suspensions are incubated at the growth temperature with rapid mixing to ensure aeration. Either [14C]acetylsalicylic acid (30 mCi/mmol) or [14C]benzoic acid (20 mCi/mmol) are added at concentrations of 1 to 10 μM. The ΔpH determined should be independent of the concentration of probe and of cell protein used within the experimental range. Samples, usually 1 ml from a total incubation mixture of 5 to 10 ml, are removed at various times after the addition of the labeled acid. Samples must be taken until a steady-state level of accumulation is reached, generally 5 to 10 min. The samples

[28] J. C. Cox, D. G. Nicholls, and W. J. Ingledew, *Biochem. J.* **178**, 195 (1979).

should filter rapidly (around 5 sec). We have had best results with these filters in experiments in which the filtered samples were not washed, thereby avoiding efflux of the accumulated acid. Controls for nonspecific binding are critical. Organic solvents, especially 5% (v/v) n-butanol,[29,30] can be used to permeabilize the cells and give a background for nonspecific binding. It should also be possible to use cells treated with ionophores such as gramicidin as a control, as long as the treatment reliably abolishes the ΔpH. We have used 2,4-dinitrophenol treatment of *B. acidocaldarius,* and found the results to be the same with protonophore (50 μM, 2,4-dinitrophenol) alone as with protonophore together with either NaSCN (5 mM) or valinomycin (2 μM).[27] In a different protocol, Michels and Bakker[21] found that 2,4-dinitrophenol decreases the ΔpH by greater than 80% in *B. acidocaldarius,* leaving a small residual pH gradient. If uncouplers are to be used as a binding control in experiments with acidophiles, however, it is important to consider that protonophores may be less effective at very low pH values than in the neutral range,[12] requiring higher than usual concentrations and substantial incubation times for effective use with at least some acidophiles. Some gram-negative and archaebacterial acidophiles, especially, may retain a substantial pH even after treatment with protonophores[12] unless a high protonophore concentration is used and the specific buffering conditions allow counterion movement to offset the effect of proton influx on the $\Delta\psi$. For example, abolition of at least 85% of the ΔpH of *Thermoplasma acidophilum* has been observed by Michels and Bakker[21] using high concentrations of 2,4-dinitrophenol and citrate buffer, but lower protonophore concentrations or use of β-alanine buffer resulted in a much more substantial residual ΔpH. Their experimental findings may provide the basis for explaining a number of earlier observations of a much larger residual ΔpH in some gram-negative acidophiles upon protonophore treatment or extensive starvation in β-alanine buffer or water.[23,24] However, such observations are still suggestive of the ability of at least certain gram-negative acidophiles to maintain an appreciable ΔpH by passive means under some conditions. It will be of interest to try to examine the relative importance of passive mechanisms to both viability and pH homeostasis under ionic conditions that approximate those of growth media. Determinations of viability are critical, since a passive mechanism for maintaining a constant pH_{in} which does not facilitate the preservation of viability could hardly represent a true adaptation of the organism, and is more likely to be a laboratory phenomenon.

[29] E. R. Kashket, A. G. Blanchard, and W. C. Metzger, *J. Bacteriol.* **143,** 128 (1980).
[30] E. P. Bakker, *Biochim. Biophys. Acta* **681,** 474 (1982).

Sample filters from the filtration assays of ΔpH should be dried, and the radioactivity counted. We use scintillation spectrometry, and include a control for quenching by the buffer in each experiment, i.e., an aliquot of the radioactive probe suspended in the buffer used is placed on a dry GFC filter and the filter is counted. The cpm recorded for a known amount of acid probe in the presence of the appropriate buffer is then used to calculate the concentration of accumulated acid.

For assays of ΔpH employing a flow dialysis method, the procedures developed by Ramos et al.[31] for use with membrane vesicles are applied. As already noted, we have used protonophore-treated cells of B. acidocaldarius as a control; killed cells should give comparable results. Vigorous aeration of the thick suspensions of aerobic cells is important, and can be achieved by blowing a stream of water-saturated oxygen over a rapidly mixing suspension. Nonetheless, it has been suggested that this procedure results in consistent but somewhat artificially low values for pH_{in} because of suboptimal aeration relative to that achieved in the filtration procedure.[21] A falsely low value of pH_{in} would lead to an underestimate of the capacity to exclude pumped protons and might also result in falsely low calculated values of the total magnitude of the protonmotive force (unless the experimental conditions reduced the magnitude of the reversed $\Delta\psi$ to the same extent).

Determinations of Cytoplasmic Buffering Capacity and Passive Proton Permeability

The observations suggesting that some acidophiles can maintain a substantial ΔpH by passive mechanisms led to consideration of the possibility that these species possess a particularly cation-impermeable membrane and/or are able to resist changes in cytoplasmic pH in part because of a high cytoplasmic buffering capacity.[10,25] Mitchell and colleagues[32-34] have described, in detail, the conceptual and mathematical considerations in determinations of cytoplasmic buffering capacity; as part of those studies, they reported experimental results using methods in which bacterial cells are permeabilized to ascertain the total buffering capacity (B_t) and methods in which the capacity is determined from the pattern of pH change after addition of a small acid pulse to a suspension of intact cells in KCl. Briefly, the internal buffering capacity (B_i) is indirectly determined

[31] S. Ramos, S. Schuldiner, and H. R. Kaback, this series, Vol. 55, p. 680.
[32] P. Mitchell and J. Moyle, Biochem. J. **104**, 588 (1967).
[33] P. Mitchell and J. Moyle, Biochem. J. **105**, 1147 (1967).
[34] P. Scholes and P. Mitchell, Bioenergetics **1**, 61 (1970).

from the total buffering capacity (B_t) and the buffering capacity of the outside surface(s) of the cell (B_o).

In a recent study,[35] we have employed both experimental approaches to measurements in bacilli that grow in various ranges of pH and in *Escherichia coli*. In methods using permeabilized cells, the B_o is first determined on an untreated cell suspension of bacteria at about 5 mg cell protein/ml of water or 200 mM KCl. B_o is calculated from the initial, rapid change in the pH of the medium upon addition of successive, small aliquots of HCl (usually 10 to 20 μl of 0.05 M HCl). For determinations of B_t, the cell suspensions are then treated with 5–10% Triton X-100,[25,35] 5% *n*-butanol,[3,35] or with a combination of agents that abolishes the ability of the cells to maintain a protonmotive force.[36] The changes in pH caused by aliquots of acid are then recorded for the treated suspension. In our hands, the presence or absence or carbonate dehydratase in these suspensions made no difference. However, it is important that additions of KOH (titrating up) give the same results as HCl (titrating down) over the same range and that several cell concentrations are examined. It is also desirable to employ several different methods for permeabilizing the cells, since at least some bacteria seem to open up more fully, for example, with *n*-butanol than with Triton (see Table I). For comparisons between different bacteria, care should be taken to calculate from measurements taken at the same initial and final pH values since B_o and B_i are highly pH dependent. In connection with the data shown in Table I, it is notable that alkalophiles, at alkaline pH, exhibit among the highest cytoplasmic buffering capacities found in our survey, values that are comparable to the relatively high B_i of the acidophile in our study at acid pH. Nonetheless, as described in the next section, a failure of the active mechanisms of pH homeostasis results in an immediate rise in the pH_{in} of *B. firmus* RAB so that it equals the pH_{out}, usually 10.5. Thus while a high B_i (and the substantial B_o found in all the species we have examined) might play a role in buffering minor perturbations in pH_{in} or in protecting some specific organelle or molecule, it is unlikely to be of importance in resisting large permanent shifts in pH.

The second approach to measurements of cytoplasmic buffering capacity employs intact cells suspended in 200 mM KCl, and involves following the decay of the pH perturbation caused by a small acid pulse. Scholes and Mitchell[34] and Maloney[37] have described this method and the

[35] T. A. Krulwich, R. A. Agus, M. Schneier, and A. A. Guffanti, *J. Bacteriol.* **162**, 768 (1985).
[36] S. H. Collins and W. A. Hamilton, *J. Bacteriol.* **126**, 1224 (1976).
[37] P. C. Maloney, *J. Bacteriol.* **140**, 197 (1979).

TABLE I

REPRESENTATIVE VALUES REPORTED FOR CYTOPLASMIC BUFFERING CAPACITY (B_i) OF MICROORGANISMS

Organism	Initial pH	Method employed	Cytoplasmic buffering capacity (B_i)		
			nmol H^+/unit/mg protein	nmol H^+/unit/mg dry wt	mM H^+/unit
Bacillus stearothermophilus[a]	9.5	Triton- or butanol-R_x	90		
	7		50		
	6		100		
	5		200		
Bacillus acidocaldarius[a]	7	Triton- or butanol-R_x	120		
	6		180		
	5		275		
Bacillus alcalophilus[a]	9.5	Triton- or butanol-R_x	500		
Bacillus firmus RAB[a]	9.5	Triton- or butanol-R_x	531		
Micrococcus denitrificans[b]	8	Butanol-R_x averaged with acid pulse		55	
	7			85	
	6			115	
Staphylococcus aureus[c]	8	Ionophore-R_x		76	
	7			90	
	6			130	
Streptococcus lactis[d]	8	Acid pulse		38	
	7			72	
	6			106	
Neurospora crassa[e]	8	pH electrode and weak acid/base			50
	7				45
	6.5				85
Escherichia coli	6[f]	Triton-R_x	33		
	6[a]	Triton-R_x	35		
	6[a]	Butanol-R_x	115		
	5[a]	Butanol-R_x	369		
Thiobacillus acidophilus[f]	6	Triton-R_x	97		
	5		260		

[a] Krulwich et al.[35]
[b] Scholes and Mitchell.[34]
[c] Collins and Hamilton.[36]
[d] Maloney.[37]
[e] Sanders and C. L. Slayman, J. Gen. Physiol. **80**, 377 (1982).
[f] Zychlinsky and Matin.[24]

relevant calculations in detail. B_o is calculated from a value for the initial pH deflection that is obtained by extrapolation back from the decay curve. B_t is determined from the final equilibration pH. Maloney[37] has indicated the importance of controls to show that equilibration has indeed occurred. In work with aerobes, it may be important to add an inhibitor of the respiratory chain in addition to the KSCN, valinomycin, and carbonate dehydratase included by Maloney. At least some bacilli that we have examined fail to achieve a flat baseline value even after prolonged incubation of the cell suspension under N_2 in the presence of such an inhibitor. If the cells are more rigorously starved, to eliminate the residual proton pumping, care must be taken to determine that the membrane is uncompromised. These considerations notwithstanding, there is an advantage to the acid pulse method. That is, a value for the passive membrane conductance to protons ($C_m^{H^+}$) can be obtained. $C_m^{H^+}$ is calculated using a $t_{1/2}$ for the time required for equilibrium to be reached, as described by Mitchell and Moyle.[32]

In initial determinations in our laboratory, the $C_m^{H^+}$ of *B. acidocaldarius* has been found to be in the same range as values reported for other bacteria. These values are preliminary, however, and no attempts have yet been made to systematically examine the passive permeability to other cations. Such determinations will be of interest and, as noted earlier, should include measurements made under various conditions of external pH and composition of the suspending medium.

Alkalophilic Bacteria

Obligately alkalophilic bacteria maintain a cytoplasmic pH of no higher than 9.6 when suspended in Na^+-containing alkaline buffers at pH values as high as 11 (Table II).[16,38] Under growth conditions, or if a substrate such as L-malate is added to the buffer, an even lower pH_{in} of about 8.5 is maintained.[39] Cells of *B. firmus* RAB that have been equilibrated at pH 8.5 also exhibit impressive homeostasis after a rapid shift in pH_{out} to 10.5 as long as Na^+ is present in the medium.[40] Evidence from *in vitro* studies[18,20] and *in vivo* experiments with both wild type and mutant strains[39–41] supports the conclusion that (1) Na^+ is required for pH homeostasis in obligate alkalophiles; (2) Na^+_{in} is exchanged electrogenically for H^+_{out} (H^+ translocated in > Na^+ translocated out) by a Na^+/H^+ antiporter

[38] A. A. Guffanti, P. Susman, R. Blanco, and T. A. Krulwich, *J. Biol. Chem.* **253**, 708 (1978).
[39] M. Kitada, A. A. Guffanti, and T. A. Krulwich, *J. Bacteriol.* **152**, 1096 (1982).
[40] T. A. Krulwich, J. G. Federbush, and A. A. Guffanti, *J. Biol. Chem.* **260**, 4055 (1985).
[41] M. L. Garcia, A. A. Guffanti, and T. A. Krulwich, *J. Bacteriol.* **156**, 1151 (1983).

TABLE II

CYTOPLASMIC pH OF BUFFERED CELL SUSPENSIONS OF
ALKALOPHILIC BACTERIA

	pH_{in}	
pH_{out}	*Bacillus alcalophilus*[a]	*Bacillus firmus* RAB[b]
9.0	9.0	9.0
10.0	9.5	9.3
10.5	9.5	9.3
11.0	9.6	9.5

[a] Guffanti *et al.*[38]
[b] Guffanti *et al.*[16]

that is active only above a cytoplasmic pH of about 8; and (3) Na^+-coupled solute uptake systems play a physiologically significant role in pH homeostasis by providing an entry route for Na^+, albeit as yet unclear whether other entry routes exist. Although some obligate alkalophiles do not exhibit marked growth dependence upon added Na^+ in the medium, they probably utilize the same kind of Na^+ cycle (perhaps augmented by incompletely defined auxiliary mechanisms) except that the affinity for Na^+ is so high that it is difficult to achieve an operational "Na^+-free" condition.[18,40,41] Work by other investigators with several other alkalophilic and alkaline-tolerant organisms indicates that a Na^+/H^+ antiporter may have a widespread role, and usually a dominant one, in pH homeostasis in the highly alkaline range.[42-47] We will outline the special considerations in measuring ΔpH in alkalophiles and will describe several protocols whereby the Na^+ requirement for pH homeostasis can be demonstrated and characterized.

Determinations of ΔpH

The pH_{in} of alkalophiles is calculated from measurements of the ΔpH using the same principles and methods as described above except that the probe, buffers, and controls require accommodation to the alkaline pH

[42] D. McLaggan, M. J. Selwyn, and A. P. Dawson, *FEBS Lett.* **165**, 254 (1984).
[43] A. G. Miller, D. H. Turpin, and D. T. Canvin, *J. Bacteriol.* **159**, 100 (1984).
[44] B. V. Chernyak, P. A. Dibrov, A. N. Glagolev, M. Yu Sherman, and V. P. Skulachev, *FEBS Lett.* **164**, 38 (1983).
[45] T. Nakamura, H. Tokuda, and T. Unemoto, *Biochim. Biophys. Acta* **776**, 330 (1984).
[46] M. Kitada and K. Horikoshi, *J. Biochem.* **87**, 1279 (1980).
[47] A. Ando, I. Kusawa, and S. Fukui, *J. Gen. Microbiol.* **128**, 1057 (1982).

range. We have employed carbonate buffers at pH 9.0 and above; it can be shown, using less optimal Tris buffers, that carbonate per se is not required for homeostasis or Na^+/H^+ antiport. For assays of a $pH_{acid\,in}$, weak bases rather than weak acids are employed. [^{14}C]Methylamine, with a pK_a of 10.6, is generally used. The probe concentration is kept at 25 μM or below (52 mCi/mmol) in order to avoid dissipating the ΔpH. Since some bacteria may have transport systems for methylamine/ammonium, it is important to verify for each organism that the weak base enters by diffusion and that its accumulation ratio reflects the ΔpH. Towards that end, methylamine uptake should be examined over a wide range of concentrations of the base to assess saturability, and uptake of the base should be examined in deenergized cells across which an artificial pH gradient is imposed. As with the weak acids used, the quenching effects of the buffer (which may vary with the pH of the buffer) must be determined. Binding controls are also needed here; we have employed killed cells and gramicidin-treated (1 μM) cells. Compounds that catalyze electroneutral exchange of OH^-/anion or H^+/cation could also be used. Weak acid protonophores at high concentration may collapse a pH gradient at moderately alkaline pH, but cannot be used reliably at very alkaline pH.

Many of our own studies and some of the protocols described below involve suspension of alkalophile cells at neutral or near-neutral pH values. If no methylamine uptake is observed with such suspensions, it is still premature to conclude that ΔpH = 0 and pH_{in} = pH_{out} unless the uptake of a weak acid (e.g., DMO) is also measured and found to be zero. Indeed, there are experimental conditions in which alkalophile cells are found to generate a ΔpH, acid out.[39]

Demonstration of the Na^+ Dependence of pH Homeostasis in Obligately Alkalophilic Bacteria

We have employed three experimental protocols to examine the requirement for specific ions for pH homeostasis in the obligate alkalophiles. As noted, each of the approaches has indicated a major role for Na^+. In the first, logarithmically grown cells of B. firmus RAB at pH 10.5 have been washed and resuspended in either 50 mM potassium carbonate or 50 mM sodium carbonate buffer at pH 10.5.[39,40] The cells resuspended in potassium carbonate show an immediate rise in the pH_{in} to 10.5 and a rapid loss of viability, whereas those in sodium carbonate (or control combinations of sodium and potassium carbonates) maintain a cytoplasmic pH below 9.0 and retain viability to a much greater extent over the first 30 to 60 min of incubation. The cells' ability to lower pH_{in} and retain viability are significantly enhanced when a nonmetabolizable solute whose inward translocation is coupled to Na^+ uptake is added in addition to Na^+. We have used 500 μM α-aminoisobutyric acid (AIB). The results support a role for Na^+ coupled symporters as a route of Na^+ entry that is relevant to pH homeostasis. It is not surprising, but should be noted, that the rate at which cells shifted to Na^+-free medium lose viability is highly dependent on the concentration of cells in the suspension and the number of times the cells are washed on harvesting. Use of less concentrated cell suspensions (e.g., $5-10 \times 10^7$ cells/ml) and more washes (2–3) result in much more rapid loss of viability than with thicker, less rigorously washed cell suspensions. This can be understood in terms of removal of bound and even, perhaps, some cytoplasmic Na^+ and adhering solutes which might facilitate Na^+ cycling.

The other two protocols involve pH shifts and are patterned after experimental designs used by Kroll and Booth,[48] Zilberstein et al.,[4] and McLaggan et al.[42] We have again used these protocols to demonstrate Na^+ dependence and AIB enhancement of pH homeostasis in B. firmus RAB. In one type of experiment, cells are equilibrated at pH 8.5 in the absence of Na^+ and then subjected to an alkaline shift in pH_{out} either with or without the simultaneous addition of Na^+ and/or AIB. Logarithmically grown cells are harvested and washed in 50 mM potassium carbonate buffer, pH 8.5. The cells are allowed to equilibrate for one hour at pH 8.5 with shaking at 30°; assays at that point confirm that $pH_{in} = 8.5$. In our use of this protocol, we then rapidly dilute (usually 1 : 10) the cells into potassium carbonate buffer at pH 10.5. The cell concentration after the pH shift is approximately 0.5 mg cell protein/ml. Some samples are shifted (diluted) into buffer containing Na^+ (in concentrations from 2 to 50 mM) and/ or 500 μM AIB. Controls include samples with similar amounts of added K^+ and a sample that is not shifted. Aliquots (usually 1 ml) are removed periodically to assay the pH. The first sampling is soon (15 to 30 sec) after the shift so that the "overshoot" phenomenon[4,42,48] can be observed. Using this kind of protocol, a rapid rise in pH_{in} up to 10.5 is observed in cells shifted to pH 10.5 with no Na^+ or low concentrations of Na^+. The presence of 50 mM Na^+ allows pH homeostasis with an initial overshoot acidification of the interior. Notably, in the presence of AIB, but not in its absence, concentrations of Na^+ as low as 2.5 mM become completely effective in facilitating pH regulation. These observations reinforce the caveat against ruling out the possible efficacy of low, contaminating Na^+ levels. Suppose, for example, that phosphate were a Na^+-coupled solute.

---

[48] R. G. Kroll and I. R. Booth, *Biochem. J.* **216**, 709 (1983).

Unless cells were kept phosphate free, the combined effects of residual Na^+ and residual phosphate might well facilitate the regulation of pH_{in}.

Finally, in a third experimental design, cells equilibrated at pH 8.5 as above are again shifted to pH 10.5, but only in potassium carbonate buffer. After 1 or 2 min, with pH_{in} already at 10.5, the suspensions are "treated" with various concentrations of Na^+, with added K^+ as a control, in some instances also adding AIB. While the cells under these conditions can still regulate pH_{in}, viability has been compromised, indicating that the most pH sensitive process(es) is other than an alkali-induced loss of the capacity for pH regulation.

While the protocols described will indicate Na^+ (or perhaps, with other species, different ion) involvement in pH homeostasis, they do not distinguish between the involvement of a secondary Na^+/H^+ antiporter and, for example, a primary Na^+ pump. Distinction between possible mechanisms of Na^+ function is generally made using inhibitors. Thus evidence for an electrogenic Na^+/H^+ antiporter in obligately alkalophilic bacilli includes: requirement of a $\Delta\psi$ per se for Na^+ fluxes—$^{22}Na^+$ efflux from cells at high pH[41] and $^{22}Na^+$ uptake by energized, everted vesicles[20,49]—and associated proton fluxes[38,49]; and inhibition of the Na^+-dependent acidification of the interior by protonophores, at least at the moderately alkaline pH that can be tested.[38,40]

Acknowledgment

The work in our laboratory was supported by Research Grant PCM 8121557 from the National Science Foundation.

[49] A. A. Guffanti, *FEMS Microbiol.* **17,** 307 (1983).

[29] Peptide Transport in Bacteria

By CHRISTOPHER F. HIGGINS and MARGARET M. GIBSON

Introduction

Three genetically distinct transport systems with overlapping substrate specificities serve to mediate the transport of small peptides across the cytoplasmic membrane and into the cytoplasm of both *E. coli* and *S. typhimurium*.[1] This multiplicity of systems makes measurement of trans-

[1] C. F. Higgins, *Microbiology* 17 (1984).

port difficult unless studied in strains genetically deficient in one or more of these permeases. This chapter is therefore divided into two sections: (1) the methods available for isolating and characterizing mutations in the genes encoding the various peptide transport systems, and (2) the various approaches which have been used to monitor peptide transport *in vivo* and to analyze the kinetic constants and substrate specificities of these systems. The methodology described will be based almost entirely on experience with the enteric bacteria *E. coli* and *S. typhimurium*. However, most of the principles and techniques can be readily adapted for the study of other organisms.

Methods for Isolating Peptide Transport-Deficient Mutations

There are three distinct peptide transport systems in *E. coli* and *S. typhimurium*. The most completely characterized of these is the oligopeptide permease (Opp). This is the major transport system under laboratory conditions and will transport almost any peptide containing from two to five amino acid residues. A second system, the tripeptide permease (Tpp), transports a range of di- and tripeptides but with a distinct preference for tripeptides containing hydrophobic amino acids. Tpp is a relatively minor system under normal laboratory conditions but its expression is strongly induced by anaerobiosis.[2] The third system, the dipeptide permease (Dpp), is rather poorly characterized. While being relatively specific for peptides containing only two amino acids, it will also handle a number of tripeptides. Bacteria deficient in all three of the peptide-specific permeases show no detectable transport of a wide variety of peptides.

Because of the relatively broad specificity of the peptide transport systems with respect to the amino acid side chains of the peptide substrates, a wide variety of naturally occurring and man-made toxic peptides can enter the cell via these permeases. Mutations which result in resistance to these peptides are relatively easy to select and are generally defective in the uptake of the toxic peptide. The following general procedures can be adopted for isolating and characterizing mutants resistant to toxic peptides. Specific details for each individual system are described separately below. Cells are grown to stationary phase (i.e., an overnight culture) in any suitable rich or minimal growth medium, sedimented by centrifugation (3000 rpm for 15 min) and resuspended in the original culture volume of minimal salts medium.[3] Of this washed culture 0.1 ml (2×10^8 cells) is spread onto a minimal agar plate[3] containing those supple-

[2] D. J. Jamieson and C. F. Higgins, *J. Bacteriol.* **160,** 131 (1984).

[3] J. R. Roth, this series, Vol. 17A, p. 3.

ments necessary to satisfy any auxotrophic requirement of the particular strain being used. Toxic peptides at the appropriate concentration can either be directly incorporated into the agar plate or can be applied to a filter paper disc placed in the center of the plate. Resistant mutants are isolated as colonies growing within the zone of growth inhibition around the disc. It is importnt to wash the cells in minimal medium and to avoid adding complex growth supplements (e.g., casamino acids, peptone) to the selective plates as these contain peptides which compete for uptake with the toxic peptides and thus reduce their toxicity.

Once putative peptide transport-deficient mutants have been isolated, a secondary screen should be used to confirm that the defect is indeed at the transport level. Peptide transport can be assayed directly (see below). Alternatively, a simpler approach for screening many putative mutants can be adopted. Peptide transport mutants selected by resistance to a toxic peptide simultaneously lose the ability to utilize peptides as sole source of a required amino acid. For example, although a proline auxo-troph can grow on the tripeptide prolylglycylglycine as sole source of proline, the ability to utilize this particular peptide is lost when an *opp* (oligopeptide permease) mutation is introduced. (Opp is the only one of the three peptide transport systems which can take up prolylglycylgly-cine.) Thus, it is generally most convenient to isolate peptide transport mutations in an appropriate auxotrophic background. The ability to utilize a particular peptide is then tested by radially streaking the mutant strain around a filter paper disc to which the peptide to be tested has been added. A reduced zone of growth, relative to that of the parental auxo-trophic strain, indicates a transport defect. Incidentally, the ability of peptide transport-proficient cells to utilize peptides as sole source of a required amino acid provides a convenient positive selection for cloning these genes.[1,4]

The Oligopeptide Permease

The most commonly used antibiotic for selecting mutations deficient in the oligopeptide permease is triornithine[5,6] which can be incorporated into minimal agar plates at a concentration of 400 μM or, alternatively, 1 μmol of the peptide can be added to a filter disc placed in the center of the plate. Triornithine-resistant colonies arise at a relatively high frequency (10^{-4} to 10^{-5}). Triornithine is specific to the oligopeptide permease and is

[4] M. M. Gibson, M. Price, and C. F. Higgins, *J. Bacteriol.* **160,** 122 (1984).
[5] Z. Barak and C. Gilvarg, *J. Biol. Chem.* **249,** 143 (1974).
[6] C. F. Higgins, M. M. Hardie, D. J. Jamieson, and L. M. Powell, *J. Bacteriol.* **153,** 830 (1983).

not taken up to any significant extent by Tpp or Dpp; all triornithine-resistant mutants which have been analyzed have been shown to be Opp⁻ and to map within the *opp* operon. No other locus has been identified which can be mutated to give triornithine resistance. Triornithine-resistant mutations can be shown to be in *opp* by their cotransductional linkage to the adjacent *trp* operon,[6] mapping at 35 min of the *S. typhimurium* chromosome and 27 min in *E. coli*. Triornithine-resistance mutations arise equally in all four of the genes which comprise the *opp* operon.[7] No method for specifically selecting mutations in any individual gene has been devised.

Several toxic peptides other than triornithine have also been successfully used to isolate *opp* mutations. Thus, *opp*⁻ strains have been selected from the zones of inhibition around filter paper discs impregnated with the tripeptides trilysine (2 μmol), norleucylglycylglycine (0.8 μmol), or glycylglycylhistidinolphosphate ester (0.5 μmol).[8] In addition, the synthetic toxic peptide alanylalanylaminoethylphosphonic acid has been used to isolate *opp* mutations (50 μg on a filter disc; Andrews and Short[9]). However, in addition to *opp* mutations this peptide also allows selection of mutations at least one additional locus which is not associated with the oligopeptide permease.

A few relevant observations about *opp* mutations should be mentioned. First, a high proportion of spontaneous *opp* mutations are deletions, frequently extending into adjacent genes. However, many others are leaky (missense) mutations, conferring resistance to triornithine but still permitting some uptake of other peptides. Second, many laboratory strains of *E. coli* are naturally Opp⁻; apparently long-term storage and subculturing in rich-medium provides a selection for loss of Opp function. This may be due to the toxicity of a number of peptides present in LB (C. G. Miller, personal communication). Thus, strains should be stored at −70° in 7% dimethyl sulfoxide. Third, it should be noted that different *E. coli* genetic backgrounds exhibit differing degrees of sensitivity to triornithine. The basis for this is not known.

The Tripeptide Permease

Most toxic peptides which enter the cell via Tpp are also substrates for Opp and can only be used to select *tpp* mutations in an *opp* background.

---

[7] B. G. Hogarth and C. F. Higgins, *J. Bacteriol.* **153,** 1548 (1983).

[8] B. N. Ames, G. F.-L. Ames, J. D. Young, D. Tsuchiya, and J. Lecocq, *Proc. Natl. Acad. Sci. U.S.A.* **70,** 456 (1973).

[9] J. C. Andrews and S. A. Short, *J. Bacteriol.* **161,** 484 (1985).

However, the synthetic peptide antibiotic alafosfalin is specifically transported by Tpp. This peptide exerts its toxic effects only after intracellular cleavage, releasing the toxic amino acid L-aminoethylphosphonic acid which inhibits D-alanine racemase and thus interferes with cell wall biosynthesis.

Alafosfalin-resistant mutants are selected by spreading cells onto minimal glucose plates supplemented with 80 μg/ml alafosfalin.[4] Mutations at three distinct genetic loci, *tppA (ompR)*, *tppB*, and *pepA*, can result in resistance to alafosfalin, although these loci can readily be distinguished when isolated in a leucine auxotroph. Mutations in *pepA* are alafosfalin resistant as this is the only one of the multiple cellular peptidases which is able to cleave alafosfalin. Strains which are *pepA leu* are able to utilize the peptide trileucine (3.3 μmol on a filter disc) as sole leucine source (as the cell contains several other peptidases able to cleave this peptide) while transport-deficient strains (*tppA* or *tppB*) are defective in their growth on this peptide. Mutations in *tppA* and *tppB* map at 74 and 28 min, respectively, on the *S. typhimurium* chromosome and can be distinguished in three ways. (1) By their relative resistance to alafosfalin when radially streaked around a filter disc containing 0.25 mg of alafosfalin. Strains carrying *tppB* mutations are totally resistant while *tppA* mutations confer only partial resistance. (2) By their ability to utilize trileucine. While *tppB* strains are totally unable to utilize trileucine when radially streaked around a disc containing 3.3 μmol of the peptide, *tppA* strains show a small amount of growth (although considerably less than the *tpp*+ parent). (3) When spread on a minimal glucose plate onto which is placed a disc containing 0.5 mg bacilysin (see section on dipeptide permease below), spontaneous bacilysin-resistant mutants arise at a frequency of 10^{-5} from *tppB* strains but at a frequency of less than 10^{-9} in *tppA* strains. These three phenotypic differences result from the fact that while *tppB* is the structural gene for the tripeptide permease and mutations in this gene totally disrupt Tpp function, *tppA* is a positive regulator of *tppB* and mutations at this locus still possess some residual transport function.[2,4] We have recently shown that *tppA* is identical with *ompR* which is known to regulate expression of the major outer membrane porins (Gibson and Higgins, unpublished results). *tppA* strains are therefore also OmpC⁻ and OmpF⁻, providing a further possible means of distinguishing *tppA* and *tppB* mutations from each other.

The toxic peptide norvalylaminoethylphosphonic acid can also be used to select *tpp* mutations when used at 40 μg/ml. This peptide has the advantage that peptidase mutations do not arise at a significant frequency, presumably because more than one of the intracellular peptidases are able to cleave this particular peptide.

The Dipeptide Permease

No peptide is known which is specifically transported by the dipeptide permease. *dpp* mutations must therefore be selected in an *opp tpp* genetic background. The most useful toxic peptide for isolating *dpp* mutations is bacilysin which is produced by certain *Bacillus* strains and inhibits glucosamine synthesis.[10] Mutants selected as resistant to this peptide incorporated into minimal agar plates at 100 μg/ml, map at 82 min on the *S. typhimurium* chromosome, and are defective in the uptake of a wide variety of dipeptides as well as some tripeptides.

Valine-containing peptides can also be used to select *dpp* mutations. *E. coli* K12 strains are defective in one of the isozymes of acetohydroxy acid synthetase and as a consequence are sensitive to valine and valine-containing peptides. *S. typhimurium* strains can also be rendered valine sensitive by introducing an *ilvG* mutation.[11] Thus, dipeptide transport-deficient mutations can be selected (in an *opp tpp* background) as resistant to 4 μM divaline, glycylvaline, or other valine-containing dipeptides in the presence of 0.8 mM leucine. The leucine competitively inhibits uptake of the free valine which is released as a result of peptide uptake, hydrolysis, and valine exodus. This valine is otherwise lethal to *dpp* strains and if leucine is omitted, results in the selection of valine-resitant mutants rather than mutants defective in the dipeptide permease.[12]

A variety of other toxic peptides can also be used to isolate peptide transport-deficient mutants (e.g., Payne *et al.*[13]) but as the mutations arising from such selections have not been characterized in detail their potential usefulness remains to be assessed.

Methods for Monitoring Peptide Uptake

Uptake of Radiolabeled Peptides

A number of studies of peptide uptake using radiolabeled peptides have been undertaken.[14–16] The methods employed are essentially those described elsewhere in this volume for measuring the uptake of radiola-

[10] M. Kenig and E. P. Abraham, *J. Gen. Microbiol.* **94,** 37 (1976).

[11] D. A. Primerano and R. O. Burns, *J. Bacteriol.* **150,** 1202 (1982).

[12] M. M. Gibson and C. F. Higgins, unpublished results (1984).

[13] J. W. Payne, J. S. Morley, P. Armitage, and G. M. Payne, *J. Gen. Microbiol.* **130,** 2253 (1984).

[14] J. L. Cowell, *J. Bacteriol.* **120,** 139 (1974).

[15] M. B. Jackson, J. M. Becker, A. S. Steinfield, and F. Naider, *J. Biol. Chem.* **251,** 5300 (1976).

[16] S. L. Yang, J. M. Becker, and F. Naider, *Biochim. Biophys. Acta* **471,** 135 (1977).

beled amino acids and sugars. However, there are several severe constraints encountered when attempting to monitor peptide uptake by such conventional procedures. First, once peptides enter the cytoplasm they are rapidly hydrolyzed to their constituent amino acids which rapidly undergo exodus from the cell. As peptide uptake is generally the rate-limiting step in this sequence of events this rapid exodus makes the accurate determination of initial rates of peptide uptake rather difficult. Second, very few suitable nonmetabolizable (nonhydrolyzable) substrates are known whose accumulation in the cytoplasm can be monitored. Third, and most importantly, very few radioactively labeled peptides are commercially available. Because of these constraints, and becuase the procedures used to study the uptake of radiolabeled peptides are essentially the same as those employed for studying sugar and amino acid uptake, these methods will not be detailed here. Rather, we will describe those methods which have been specifically developed for monitoring peptide uptake and discuss in some detail their relative advantages and disadvantages.

Measurement of Growth Rates

The simplest method for measuring peptide uptake relies upon the ablity of amino acid auxotrophs to utilize peptides as sole source of the required amino acid. Because *E. coli* lacks extracellular peptidases, yet contains a battery of highly active intracellular peptidases, the rate-limiting step in peptide utilization is their transport into the cytoplasm. Thus, growth rate is dependent upon the rate of peptide uptake. An overnight (saturated) culture of cells is washed twice in minimal salts medium and diluted 100-fold into minimal salts medium containing a carbon source (usually 0.4% glucose) and the peptide to be tested. In order to support growth of an auxotroph, peptide concentrations of at least 2 mM are generally required. Growth of the cells is followed by measuring the optical density of the culture at 600 nm over a period of 12–24 hr, depending on the rate of growth.

Such growth tests allow a gross assessment of the relative rates at which peptides and their analogs can enter the cell[17] and, in addition, provide a rapid screen to determine whether a transport system is functioning. However, the many limitations of this approach include (1) the impossibility of obtaining kinetic data. Growth curves are determined over an incubation period of 12–24 hr and generally require relatively high concentrations of substrate. (ii) The inability to measure transport under conditions which do not allow cell growth (i.e., in the presence of meta-

[17] J. W. Payne, *Adv. Microb. Physiol.* **13**, 55 (1976).

bolic inhibitors). (3) The insensitivity of the approach. There is a lower limit for the rate of peptide uptake below which the peptide will not support growth. Thus, the observation that a peptide fails to support growth does not necessarily mean that it cannot enter the cell but only that its rate of uptake is below the rather insensitive limit of detection (see, for example, Payne and Bell[18]). (4) An appropriate auxotroph is required, often a serious limitation when studying species other than the enteric bacteria. (5) Measuring optical density is relatively insensitive and a considerable length of time (12–24 hr) is required to perform such an assay.

The last of these problems has to some extent been overcome by measuring the synthesis of a specific enzyme (e.g., β-galactosidase) or the incorporation of radioactivity into protein, rather than optical density, as an estimate of the rate of cell growth. Although these methods do increase the rapidity and sensitivity of such assays they do not overcome the other inherent difficulties. These methods have been described in detail elsewhere[19-21] and will not be discussed further here.

Dansyl Chloride Method

Dansyl chloride reacts with primary amino groups of amino acids and peptides to form fluorescent derivatives which can be separated, identified, and quantified by two-dimensional thin-layer chromatography on polyamide sheets. Thus, if cells are incubated with a peptide the uptake of that peptide can be monitored using dansyl chloride to follow the disappearance of the peptide from the growth medium.[18] This procedure has several advantages over other techniques. Firstly, as the labeled peptides are separated from labeled amino acids by chromatography, the exodus of amino acids during incubation with the peptide does not interfere with the measurement of peptide uptake. Second, the uptake of more than one peptide can be monitored simultaneously. Third, the uptake of any peptide derivative can be studied as long as it has a primary amino group. Finally, this procedure can also provide considerable information on amino acid efflux from cells following peptide uptake.

Cells are grown at 37° in minimal glucose medium[3] to midexponential phase ($OD_{600} = 0.6$) and harvested by centrifugation at 3000 rpm for 15 min at room temperature. The cells are washed twice with equal volumes of 20 mM potassium phosphate buffer (pH 7.2) at 37° and finally resus-

[18] J. W. Payne and G. Bell, *J. Bacteriol.* **137,** 447 (1979).
[19] T. Cascieri and M. F. Mallette, *Appl. Microbiol.* **27,** 457 (1974).
[20] G. Bell, G. M. Payne, and J. W. Payne, *J. Gen. Microbiol.* **98,** 485 (1977).
[21] J. W. Payne and G. Bell, *FEMS Lett.* **1,** 91 (1977).

pended to about 10^9 cells/ml in 20 mM potassium phosphate buffer (pH 7.2), also at 37°. The optical density at 600 nm of the suspension is recorded and 1.0 ml of the cell suspension added to 1.3 ml of 20 mM glucose in 20 mM potassium phosphate buffer (pH 7.2) and equilibrated at 37° for 10 min. Transport is initiated by adding 0.2 ml of the appropriate peptide solution (10–500 μM in 20 mM potassium phosphate buffer). Samples (0.5 ml) are removed from the incubation mixture after appropriate time intervals and immediately passed through a 0.45-μm Millipore filter (13 mm diameter) to remove all cells. The most rapid sampling possible is about every 15 sec.

The filtered samples are placed on ice (or frozen at −20° for longer term storage) until analysis. Aliquots (50 μl) from each sample, together with 10 μl of a 0.5 mM solution of ornithine as an internal standard, are added to a Durham tube (30 × 6 mm) and freeze-dried *in vacuo*. Ten microliters of 200 mM sodium bicarbonate is added and again freeze dried. The sample is then dissolved by vigorous vortexing in 10 μl of a solution of dansyl chloride in acetone (2.5 mg/ml). In order for the dansylation reaction to proceed, the pH of this solution should be about 9.5 and should be checked by spotting a small sample on pH indicator paper. The tubes are sealed with a silicone stopper and incubated for 1 hr at 45°, the acetone is then evaporated *in vacuo* and the sample redissolved in 5 μl of 50% aqueous pyridine.

The dansylated samples are separated by thin-layer chromatography on 15 × 15 cm polyamide sheets. Samples (2 μl) are spotted onto one corner of the sheet and the chromatograms developed in the following solvent systems, ensuring thorough drying between solvents: (1) 50% aqueous formic acid (first dimension); (2) acetic acid : toluene (10 : 90 v/v) (second dimension); (3) methanol : butyl acetate : acetic acid (40 : 60 : 2 v/v/v) (second dimension). Resolution is frequently improved by rerunning in the first solvent. A mix of standard dansylated amino acids and/or peptides can be run on the reverse side of the sheets to facilitate identification of each spot. The chromatograms are examined under UV light and the concentration of each amino acid or peptide estimated by comparison of the fluorescence intensity with that of known standards of the same substance. The internal ornithine standard provides a check on the efficiency of the extraction and labeling procedures. A permanent record of each chromatogram can be made by photographing under UV light. Polyamide sheets can be reused after washing in acetone : water : ammonia (50 : 46 : 4). Figure 1 shows an example of data obtained by the dansyl chloride procedure.

Using this procedure the rate of peptide disappearance from the growth medium over a given period of time can be estimated to provide a

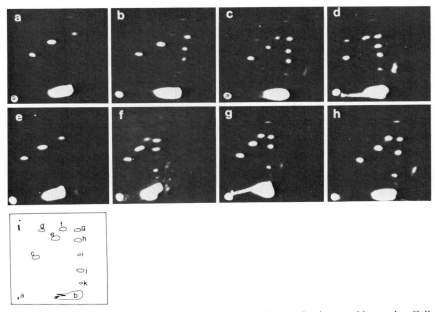

FIG. 1. Illustration of the dansyl chloride procedure for monitoring peptide uptake. Cells were incubated with the peptide glycylglycyl-L-leucine (a–d) or glycylglycyl-D-leucine (e–f) and samples removed from the incubation medium at time zero (a and e) and after increasing incubation periods (b–d and e–h). The samples were dansylated, separated by thin-layer chromatography, and the chromatogram photographed under UV light. Frame (i) is a diagram showing the identities of some of the dansylated derivatives: (a) origin, (b) dansyl hydroxide, (c) ornithine (internal standard), (d) leucine, (e) glycylglycyl-L-leucine (frames a–d) or glycylglycyl-D-leucine (frames e–h), (f) proline/valine, (g) dansyl ammonia, (h) alanine, (i) glycine, (j) glutamic acid. These data clearly show the rapid uptake of glycylglycyl-L-leucine and specific release of leucine and glycine while, over an identical time course, glycylglycyl-D-leucine is not transported to any significant extent. Certain amino acids (e.g., alanine and proline/valine) are released into the medium during incubation, even in the absence of peptide uptake.

measure of transport rate.[18] This method for studying peptide uptake has many advantages. (1) Because most peptides are rapidly cleaved by intracellular peptidases, exodus of the peptide following uptake is rarely a problem, facilitating the determination of initial rates of transport. In addition, one can study uptake over realtively short periods of time and at very low substrate concentrations, and therefore obtain reasonably meaningful kinetic data. (2) The inhibitory effect of one peptide on the uptake of another can be studied (assuming that the two peptides can be separated by the chromatographic techniques employed). (3) As the cells do not need to be growing, the effects of metabolic inhibitors, pH, temperature, etc. on peptide uptake can readily be assessed. (4) Very low rates of uptake (too low to be detected by growth rate) can be monitored. (5) Any peptide substrate with a primary amino group can be studied even if it is resistant to peptidase activity. (6) Peptide uptake is generally accompanied by exodus of specific amino acids, usually derived from the peptide. Because the dansyl chloride procedure detects amino acids which are released into the growth medium the procedure provides considerable information on peptidase activity, amino acid metabolism and exodus pathways. (7) The method does not require auxotrophs and can be applied to any species devoid of extracellular peptidase activity. The method has been successfully applied to *Streptococcus,* the yeast *Saccharomyces cerevisiae,*[22] and even to higher plants.[23]

The dansyl chloride method can also provide a means of examining intracellular accumulation of peptides which are resistant to peptidase cleavage. Cells are prepared and incubated with peptide as described above. Amino acids and peptides are quantitatively extracted from the cells by filtering 2 ml of the cell suspension through a 0.45-μm Millipore filter and washing twice with 2 ml of 20 mM potassium phosphate buffer (pH 7.2). The filter containing the cells is immediately added to 2 ml of boiling distilled water, 50 μl of toluene added, the cells mixed thoroughly by vortexing, and incubated at 100° for 15 min in a stoppered tube. The extract volume is made up to 2 ml with distilled water and debris removed by filtering through a 45-μm filter. Of this extract 500 μl is freeze-dried, dansylated, and chromatographed as described above. Accumulated unhydrolysed peptides within the cell, as well as free amino acid pools, can be quantitated from the fluorescence intensity of the dansylated compounds on the chromatograms.

Fluorescamine Procedure

Although the dansyl chloride method provides a very informative means of monitoring peptide uptake it does have certain drawbacks. Most fundamental is that it is a rather lengthy procedure, unsuitable for the analysis of a large number of different peptides or for comparison of uptake by a large number of different mutant strains. Thus, a fluorescence assay which can be carried out in solution, and not requiring the chromatography step, has been developed using fluorescamine.[24] This assay relies on the fact that under certain conditions fluorescamine will react with the primary amino groups of peptides but not of free amino acids. Thus, peptide disappearance from the growth medium can be monitored even in

[22] J. W. Payne and T. M. Nisbet, *FEBS Lett.* **119,** 73 (1980).
[23] C. F. Higgins and J. W. Payne, *Planta* **138,** 211 (1978).
[24] T. M. Nisbet and J. W. Payne, *J. Gen. Microbiol.* **115,** 127 (1979).

the presence of the free amino acids which are excreted into the medium during incubation.

Cells are grown, washed, and incubated with peptide as described above for the dansyl chloride procedure. At appropriate time points samples (0.5 ml) of the incubation medium are removed and filtered through a 0.45-μm Millipore filter to remove the cells. Aliquots of the filtered medium (50 μl; containing up to 10 nmol peptide) are added to 2.5 ml of 0.1 M sodium tetraborate in deionized water (pH adjusted to 6.2 with HCl). Then 0.5 ml of a fresly made solution of fluorescamine in acetone (0.15 mg/ml) is added and rapidly and thoroughly mixed. The fluorescence yield is read within 30 min of reaction using an excitation wavelength of 390 nm and an emission wavelength of 485 nm. Fluorescence yield is proportional to the amount of peptide in the medium. Under the assay conditions the fluorescence yield with free amino acids is only 2% of that obtained with peptides; thus exodus of amino acids from the cells during the period of the transport assay is not generally significant.

The fluorescamine procedure is the most sensitive means of measuring peptide uptake. An automated version of the assay has been described[25] which enables the continuous monitoring of peptide uptake and thus extremely accurate determination of transport kinetics. The fluorescamine procedure provides the most convenient method for monitoring peptide uptake. However, two limitations should be mentioned. First, peptides with an N-terminal proline residue cannot be assayed as they do not form a fluorescent adduct with fluorescamine. Second, competition experiments cannot be performed as the competing peptide will also react with fluorescamine. The fluorescamine procedure has been successfully used for measuring peptide uptake in *Streptococcus*, yeasts, and higher plants as well as in *E. coli*.

Concluding Remarks

Because of the broad specificity of the peptide transport systems, many toxic peptides can enter the cells via these systems and provide a simple means of selecting and characterizing peptide transport-deficient mutants. In addition, the difficulty in studying peptide uptake by conventional techniques employing radiolabeled substrates has led to the development of techniques using fluorescent labels. Not only are these techniques equally sensitive to those using radioisotopes but they can also provide a considerable amount of additional evidence about the fate of a transported peptide and its constituent amino acids. Indeed, the fluores-

[25] J. W. Payne and T. M. Nisbet, *J. Appl. Biochem.* **3,** 447 (1981).

camine procedure has clearly demonstrated some of the potential errors which can result from attempting to measure initial transport rates using isotopes when rapid exodus or metabolism of the transported substrate takes place.[22] These various methods are certainly applicable to many organisms and now allow the transport of peptides to be studied as readily as that of amino acids, sugars and other transported substrates.

[30] Methods for the Study of the Melibiose Carrier of *Escherichia coli*

By DOROTHY M. WILSON, TOMOFUSA TSUCHIYA, and T. HASTINGS WILSON

The melibiose carrier of *Escherichia coli* is a representative of the class of carriers that require a cation for cotransport.[1] An unusual feature of this transport system is that it is able to utilize Na^+, Li^+, or H^+ as the coupling cation depending on the conditions.[2] Another interesting property of the carrier is that different sugar substrates show different specificities for the cation requirement.[3] For example the natural substrate, melibiose, may be transported with either H^+ or Na^+ but not with Li^+, whereas the transport of thiomethyl-β-galactoside is effected with Li^+ or Na^+ but not with H^+. Another substrate, α-methylgalactoside, may be transported with any of the three cations. This chapter describes methods used for measuring the transport of both the sugar and cation substrates. Some of these methods such as the preparation of radioactive melibiose and the construction of the Li^+-selective electrode were devised specifically for studies of the melibiose transport system.

Bacterial Strains

A variety of strains of *E. coli* has been isolated for use in experiments on the melibiose carrier (Table I).[4-6] It is essential to start with a strain

[1] R. K. Crane, *Rev. Physiol. Biochem. Pharmacol.* **78**, 99 (1977).

[2] T. Tsuchiya and T. H. Wilson, *Membr. Biochem.* **2**, 63 (1978).

[3] T. H. Wilson, T. Tsuchiya, J. Lopilato, and K. Ottina, *Ann. N.Y. Acad. Sci.* **341**, 465 (1980).

[4] J. Lopilato, T. Tsuchiya, and T. H. Wilson, *J. Bacteriol.* **134**, 147 (1978).

[5] K. Ottina, J. Lopilato, and T. H. Wilson, *J. Membr. Biol.* **56**, 169 (1980).

[6] T. Tsuchiya, K. Ottina, Y. Moriyama, M. J. Newman, and T. H. Wilson, *J. Biol. Chem.* **257**, 50125 (1982).

METHODS IN ENZYMOLOGY, VOL. 125

TABLE I
Escherichia coli STRAINS

Strain[a]	Derived from	Melibiose genotype	Source
W3133	—	$melA^+B^{ts}$ [b]	Salvadore Luria
W3133-2	W3133	$melA^+B^+$	Lopilato et al.[4]
RA11	W3133-2	$melA^-B^+$	Lopilato et al.[4]
RE16	W3133-2	$melA^+B^-$	Ottina et al.[5]
RA11/pLC 25-33		$melA^-B^+/F'/pl\ melA^+B^+$	Tsuchiya et al.[6]
W3133-2S[c]	W3133-2	$melA^+B^{Li}$	Niiya et al.[7]

[a] All strains carry a *lacZY* deletion.
[b] ts, temperature sensitive.
[c] Lithium-resistant mutation.

from which the lactose carrier *(lacY)* has been deleted, as the *lac* carrier also is capable of transporting melibiose. All of the strains given in Table I were derived from W3133 (a K12 strain) which possesses a *lacZY* deletion. An interesting property of the carrier in K12 strains is the temperature sensitivity of the carrier protein. However, it has been found convenient for many studies to utilize mutants with transport systems that are functional at 37° (such as W3133-2 and its derivatives). An important mutant of W3133-2 is RA11. This strain lacks α-galactosidase and can be used when transport is required without metabolism of the sugar. It can be induced by growth on a rich medium in the presence of melibiose, although the sugar is not hydrolyzed. Another strain (RE16) lacks the transport carrier and serves as a control for sugar uptake by routes other than the melibiose carrier. It is preferable to induce this strain with α-methylgalactoside which enters the transport-negative cell more readily than does melibiose. Also included in Table I is a strain containing a plasmid from the Clarke-Carbon collection which possesses the genes *(melA, melB)* coding for α-galactosidase and for melibiose transport. The plasmid-containing strain produces larger amounts of transport protein than does the parent, which can be useful for purification studies.

In order to study the mechanism of cation cotransport it is helpful to isolate mutants with altered cation specificities. There is a convenient method of obtaining such mutants for the melibiose carrier by growth in the presence of Li$^+$. The growth of normal *E. coli* cells on melibiose minimal medium is strongly inhibited by 10 mM Li$^+$ and it is therefore possible to isolate Li$^+$-resistant mutants. Cells are grown either in minimal liquid medium containing 0.2% melibiose plus 10 mM LiCl or on agar plates containing the same ingredients. After 2–3 days cells will be found that are resistant to the inhibitory effects of Li$^+$. One of the most interesting mutants of this type was described by Niiya *et al.*[7] This organism (W3133-2S) is dependent on Na$^+$ or Li$^+$ for growth on melibiose as it has lost the ability to transport the disaccharide with protons.

Substrates and Transport Assays

Preparation of [³H]Melibiose. Since melibiose is the natural substrate for the melibiose carrier it is useful to obtain this sugar in radioactive form for studies of the transport. [³H]Melibiose is not available commercially but may be prepared from [³H]raffinose by the method described by Tanaka *et al.*[8] In this method the trisaccharide [³H]raffinose is treated with yeast invertase (β-fructofuranoside, EC 3.2.1.26) which splits the sugar to give the disaccharide melibiose as one of the products. The reaction mixture consists of 0.3 ml of a 1 mM solution of [³H]raffinose to which is added an equal volume of 0.2 M sodium acetate buffer, pH 4.9, containing 3 μg of invertase. The mixture is incubated at 50° for 90 min.[9] This is followed by the addition of 0.6 ml of 0.5 M potassium phosphate buffer, pH 7.0, and the mixture is heated in a boiling water bath for 3 min to inactivate the invertase. It is then cooled in an ice bath.

The solution which now contains [³H]raffinose, [³H]melibiose, and [³H]fructose is fractionated by column chromatography at room temperature. Samples of 1.2 ml are applied to a column (1.8 by 100 cm) of Sephadex G-15 which has been equilibrated with water. Elution is with water at a flow rate of 5.6 ml/hr and 1.6 ml fractions are collected. Alternatively paper chromatography[10] may be utilized to separate melibiose from raffinose and fructose.[5]

Transport Assay. In the assay of transport of [³H]melibiose by whole cells it is desirable to use a strain lacking α-galactosidase so that accumulation of the substrate without metabolism may be obtained. RA11 *(melA⁻ B⁺)* is a suitable strain for this purpose. Cells of RA11 are grown to late logarithmic stage in minimal medium supplemented with 1% tryptone (Difco) and also with 1 mM melibiose in order to induce the melibiose operon. The cells are harvested, washed twice, and resuspended at approximately 10⁹ cells/ml in an appropriate buffer such as potassium phosphate pH 7. The addition of 10 mM Na$^+$ increases the transport activity about 4-fold. In a typical experiment [³H]melibiose (60 μM, 0.4 μCi/ml) is

[7] S. Niiya, K. Yamasaki, T. H. Wilson, and T. Tsuchiya, *J. Biol. Chem.* **257**, 8902 (1982).
[8] K. Tanaka, S. Niiya, and T. Tsuchiya, *J. Bacteriol.* **141**, 1031 (1980).
[9] A. Goldstein and J. O. Lampen, this series, Vol. 42, p. 504.
[10] R. J. Block and G. Zweig, *in* "A Manual of Paper Chromatography and Paper Electrophoresis" (R. J. Block, E. L. Durran, and G. Zweig, eds.), p. 192. Academic Press, New York, 1958.

added to 1 ml of such a cell suspension at room temperature and 0.2 ml samples are removed at given time intervals. The samples are filtered through 0.65-μm (or 0.45-μm) Millipore filters followed by a 5 ml wash with buffer at room temperature. The filter containing the cells is placed in a vial, scintillation fluid is added and the vial is counted.

The transport of other radioactive sugar substrates may also be assayed by this method. In the case of nonmetabolizable sugars such as β-thiomethylgalactoside it is not necessary to use an α-galactosidase-negative strain except for purposes of comparison.

α-p-Nitrophenyl-D-galactoside (α-pNPG). Another substrate of the melibiose carrier is α-pNPG. A convenient method of measurement of transport of this sugar may be carried out with α-galactosidase-positive strains.[5] The sugar enters the cell via the melibiose carrier and is split by α-galactosidase into galactose and p-nitrophenol which gives a yellow color. Entry of the sugar is the rate-limiting step as the α-galactosidase activity is much greater than the rate of transport. The spectrophotometric assay is as follows. Induced cells are washed twice with 0.1 M potassium phosphate buffer, pH 7.0, and resuspended to a final concentration of 0.2 mg dry wt of cells per ml in the same buffer containing chloramphenicol (0.1 mg/ml) plus dithiothreitol (1 mM). After incubation at 30° for 10 min the reaction is initiated by addition of α-pNPG to give a final concentration of 0.5 mM, and the reaction vessels are shaken at 30°. At various intervals 3-ml samples are removed and placed in tubes containing 3 ml of 0.6 M Na$_2$CO$_3$. The tubes are then centrifuged to remove cells, and the yellow color of the p-nitrophenol in the supernatant is monitored with a Klett-Summerson colorimeter (No. 42 filter).

Other Sugar Substrates. Several additional α- and β-galactosides are substrates for the melibiose carrier. These are included in Table II which also shows the cations utilized for cotransport with these sugars. From this table it may be seen that the β-galactosides tested are cotransported with Na$^+$ and Li$^+$ but not with protons. The α-galactosides on the other hand can cotransport with H$^+$ as well.

Cation–Sugar Cotransport

H$^+$ Uptake with Sugars. Proton uptake associated with sugar transport is measured by the method of West,[11] using a pH electrode. Cells are washed twice with 120 mM choline chloride and resuspended in the same solution to 70–80 mg dry weight of cells/ml.[2,7] An aliquot (0.4 ml) of this cell suspension is diluted with 1.6 ml of a solution of 120 mM choline

[11] I. West, Biochem. Biophys. Res. Commun. 41, 655 (1970).

TABLE II
SUBSTRATES FOR THE MELIBIOSE CARRIER AND
THEIR CATION SPECIFICITIES FOR COTRANSPORT

Substrate	Cation for cotransport[a]		
	H^+	Na^+	Li^+
Melibiose	+	+	±
β-Thiomethylgalactoside	0	+	+
α-Methylgalactoside	+	+	+
β-Methylgalactoside	0	+	+
Galactose	n.d.	+	+
β-Thiodigalactoside	0	+	n.d.

[a] n.d., not determined; 0, no cation uptake.

chloride (or KCl) plus 30 mM KSCN. The incubation vessel consists of a vial with a plastic lid through which is passed a combined glass electrode. Nitrogen gas is passed (via a hypodermic needle) through a small hole in the lid, and another hole acts as a vent for the nitrogen as well as a port for introduction of solutions. The cell suspension is stirred with a very small magnetic stirring bar under anaerobic conditions at room temperature until it reaches a steady pH value (usually between pH 6.5 and 7.0) after 30–40 min. An anaerobic solution (20 μl) of 0.5 M sugar is then added and pH changes are continuously recorded. Calibration is carried out by the addition of small volumes of anaerobic solutions of 5 mM NaOH or 5 mM HCl. An example of H^+ uptake with different sugars is given in Fig. 1. In these experiments it is preferable to use a strain such as RA11 which lacks the hydrolytic enzyme α-galactosidase, otherwise the production of acid from metabolizable sugars masks the uptake of protons with the sugar after the first 30 sec.

Na$^+$ Uptake with Sugars. This method of measurement of Na$^+$ uptake involves the use of a Na$^+$-selective electrode. Cells are washed twice with 0.1 M morpholinopropanesulfonic acid (MOPS) which has been adjusted to pH 7.0 with tetramethylammonium hydroxide.[2] They are then resuspended in the same buffer to 40–50 mg dry weight of cells/ml. One milliliter of this suspension is diluted to 9 ml with either MOPS, pH 7.0 or 0.1 M N-tris(hydroxymethyl)methylglycine (Tricine) adjusted to pH 8.0 with tetramethylammonium hydroxide. NaCl is added to a final concentration of 25 μM. Cells are placed in a 15-ml glass vessel fitted with a plastic cap containing two 4-mm-diameter holes through which calomel and Na$^+$ electrodes are passed. Nitrogen gas is passed through a small (0.5 mm

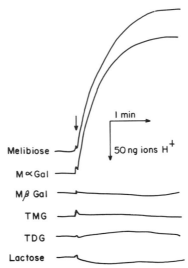

FIG. 1. Proton movement induced by the addition of galactosides. A 20 μl sample of 0.5 M galactoside was added to a 2 ml cell suspension of RA11 in 90 mM KCl plus 30 mM KSCN. An upward deflection represents a rise in the pH of the medium (taken from Tsuchiya and Wilson[2]).

diameter) opening in the cap, and a final 0.5-mm hole is used for the escape of N$_2$ and introducing solutions. The cell suspension is preincubated at room temperature and bubbled with N$_2$ gas until the background pNa of the medium reaches plateau level (approximately 30 min). Then 100 μl of an anaerobic 0.5 M solution of sugar is added with rapid stirring under anaerobic conditions. pNa was measured with a Na$^+$ electrode (Radiometer, Copenhagen). The records are calibrated by addition of known amounts of NaCl. An example of tracings obtained by this method is shown in Fig. 2.

    Li$^+$ Uptake with Sugars. 1. Construction of the Li$^+$-selective electrode.[12] To a small glass Petri dish (3 cm in diameter) add 3 ml of tetrahydrofuran. Then add very slowly 40 mg of poly(vinyl chloride) and dissolve completely. Three microliters of the Li$^+$ ionophore (N,N'-diheptyl-N,N'-5,5-tetramethyl-3,7-dioxanonandiamide) is added. The structure of the ionophore is given in Fig. 3. It may be purchased from Fluka AG, Chemische Fabrik, CH-9470 Buchs, Switzerland. Next, 0.2 mg of Na-tetraphenylboron (in 0.75 ml of tetrahydrofuran) is added. Finally, add 125 μl of dioctylphthalate.

[12] T. Tsuchiya, M. Oho, and S. Shiota-Niiya, *J. Biol. Chem.* **258**, 12765 (1983).

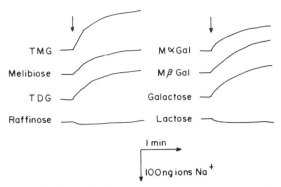

FIG. 2. pNa changes induced by the addition of various galactosides. Various sugars were added to RA11 at the time indicated by the arrow to give a final concentration of 5 mM. An upward deflection indicates a fall in Na$^+$ concentration in the incubation medium (taken from Tsuchiya and Wilson[2]).

The Petri dish is covered with a filter paper and allowed to stand overnight. The solvent evaporates and the membrane forms at the bottom of the Petri dish. A 6–8 mm square segment of the membrane is cut with a knife. The sheet is placed over the end of a polyvinyl transparent tube (approximately 12 cm/long and 4 mm in diameter). It is glued to the end of the tube with a very small amount of tetrahydrofuran (see Fig. 4). After 30 min the excess membrane is trimmed off and the inner compartment is filled with 10 mM LiCl solution. A silver wire coated with AgCl is prepared as follows. A 3 cm length of silver wire (0.5 to 1 mm in diameter) is attached to a length of copper wire. This silver electrode is placed in 0.1 M HCl and attached to the cathode of a power supply. A platinum electrode is placed in the same solution and attached to the anode. Two

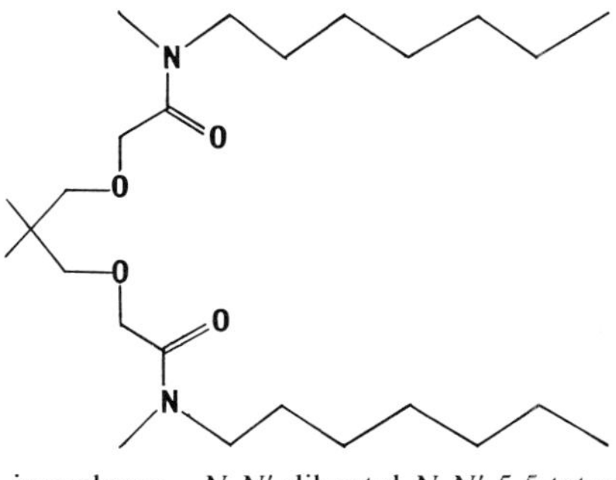

FIG. 3. Lithium ionophore: *N,N'*-diheptyl-*N,N'*-5,5-tetramethyl-3,7-dioxanonandiamide.

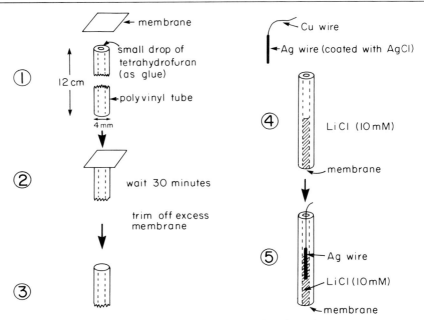

FIG. 4. Construction of Li⁺-selective electrode.

milliamps are passed through the solution for 5 min. The silver wire becomes black due to the deposit of AgCl.

The Ag–AgCl wire is then placed into the plastic tube containing LiCl solution (see Fig. 4). The upper 80% of the Li^+ electrode is then placed within a steel tube. When this metal tube is grounded it reduces electrical "noise." In addition the electrode becomes rigid and easier to manipulate. The metal support must not, however, touch the assay mixture.

2. *Measurement of Li^+ transport.*[12] An aliquot (0.3 ml) of the cell suspension is diluted with 2.7 ml of 0.1 M MOPS–Tris, pH 6.7. LiCl is added to a final concentration of 100 μM. Cells (15–20 mg of protein) are placed in an 8-ml plastic vessel fitted with a silicon cap containing two holes through which the Li^+ electrode and reference electrode (K801 of Radiometer, Copenhagen, Denmark) are passed. Nitrogen gas is introduced through a small opening (20-gauge needle) in the cap, and a final hole (18-gauge needle) is used for the escape of N_2 and for introducing solutions. An anaerobic solution of sugar (30 μl of 1 M) is added with rapid stirring under anaerobic conditions. The two electrodes are connected to a pH meter (PHM80, Radiometer), and the Li^+ concentration in the assay is recorded. The records are calibrated by addition of known amounts of LiCl (Fig. 5).

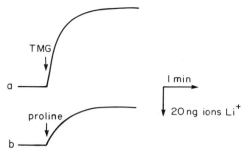

FIG. 5. Li^+ uptake by cells of *E. coli* in response to the addition of TMG or proline. Cells of W3133-2 were grown on minimal medium 63 plus 1% tryptone and (a) melibiose 0.2% to induce the melibiose carrier or (b) proline 5 mg/ml to induce the proline carrier. Cells were washed with 100 mM MOPS–Tris, pH 7. The assay mixture was 3 ml of 100 mM MOPS–Tris, pH 7 containing 100 μM LiCl. Cells were added to give a final concentration of 4 mg cell protein/ml. In the first experiment 30 μl of anaerobic TMG (1 M) was added at the arrow. Proline (10 μl of 30 mM) was added at the arrow in the second tracing.

Reconstitution of the Melibiose Carrier

The extraction of the carrier protein with detergent and the reconstitution of transport activity into proteoliposomes is a valuable technique. By this method one can simplify the experimental conditions by removing water-soluble molecules as well as many other membrane proteins. Complex metabolic interactions as well as other carriers may be avoided. A second value of the reconstitution procedure is the possibility of using it as an assay during the purification of the carrier protein.

Preparation of Bacterial Membranes. Cells are washed, resuspended in potassium phosphate (50 mM), pH 7.5 and disrupted by passage through an Aminco French pressure cell (Model 4-3398) at 19,000 psi. Centrifugation is carried out at 12,000 g for 10 min to remove intact cells. The supernatant is then centrifuged at 140,000 g for 45 min to collect the membranes. The pellet is washed once and suspended at a concentration of 10 mg protein/ml in 50 mM potassium phosphate, pH 7.5, and dithiothreitol (1 mM). An equal volume of 50% glycerol is added to the suspending buffer. Membranes are stored at $-80°$.

Solubilization and Reconstitution.[6,13,14] The general procedure is based on a method described by Newman and Wilson.[13] In a typical experiment 285 μl of 40 mM potassium phosphate, 10 mM sodium phosphate, pH 7.5, is added to a small centrifuge tube in an ice bath. To the tube are

[13] M. J. Newman and T. H. Wilson, *J. Biol. Chem.* **255**, 10583 (1980).
[14] D. M. Wilson, K. Ottina, M. J. Newman, T. Tsuchiya, S. Ito, and T. H. Wilson, *Membr. Biochem.* **5**, 269 (1985).

added 50 μl of membrane vesicles (10 mg protein/ml), 5 μl of 100 mM dithiothreitol, 30 μl of washed *E. coli* lipid (50 mg/ml), and 15 μl of 0.59 M melibiose. After blending on a vortex mixer for 5 sec, 35 μl of 15% octyl-β-D-glucopyranoside (w/v) in 50 mM potassium phosphate (pH 7.5) is added, and the tube is blended again. The suspension is incubated in an ice bath for 10 min, then centrifuged at 140,000 *g* for 1 hr. The supernatant fluid is carefully removed and 300 μl are mixed with 10 μl of bath-sonicated lipid (50 mg/ml) and with octylglucoside at a final concentration of 1.25%. The suspension is blended on a vortex mixer and incubated at 0°

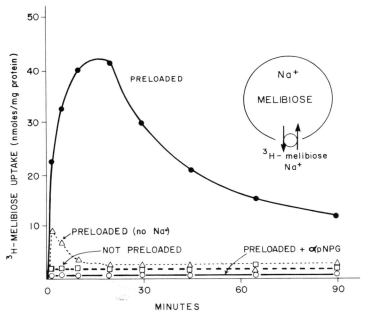

FIG. 6. Melibiose counterflow by reconstituted proteoliposomes. The proteoliposomes were centrifuged and resuspended in 50 μl of melibiose-free media with an ionic composition similar to that in which they had been prepared. The experiment was carried out as follows: (●) proteoliposomes containing 40 mM potassium phosphate, 10 mM sodium phosphate, and 20 mM melibiose were diluted into 40 mM potassium phosphate, 10 mM sodium phosphate, and 50 μM [³H]melibiose; (△) proteoliposomes containing 50 mM potassium phosphate and 20 mM melibiose were diluted into 50 mM potassium phosphate and 50 μM [³H]melibiose; (□) proteoliposomes containing 40 mM potassium phosphate and 10 mM sodium phosphate were diluted into 40 mM potassium phosphate, 10 mM sodium phosphate, and 50 μM [³H]melibiose; (○) proteoliposomes containing 40 mM potassium phosphate, 10 mM sodium phosphate, and 20 mM melibiose were diluted into 40 mM potassium phosphate, 10 mM sodium phosphate, 10 mM *p*-nitrophenyl-α-D-galactopyranoside, and 50 μM [³H]melibiose. In all cases the pH was 7.5 (redrawn from the data of Tsuchiya *et al.*[6]).

for 10 min. The mixture is then pipetted rapidly into 15 ml of the appropriate phosphate buffer, pH 7.5 (usually 40 mM potassium phosphate and 10 mM sodium phosphate), containing melibiose at a concentration of 20 mM and 1 mM DTT. The proteoliposomes which form are centrifuged in a type 42.1 (Beckman) rotor at 100,000 g for 90 min.

Measurement of Counterflow. The proteoliposome pellet is resuspended with a glass rod after addition of 100 μl of the appropriate buffer (usually 40 mM potassium phosphate, 10 mM sodium phosphate, pH 7.5) containing 1 mM dithiothreitol. Resuspension of the pellet is carried out in an ice bath to inhibit leakage of the preloaded sugar. Twenty microliters of reconstituted proteoliposomes is added to a solution containing 340 μl of the appropriate phosphate buffer, pH 7.5 plus 56 μM [3H]melibiose (2.8 μCi/ml). Samples (100 μl) are removed at intervals and filtered through the center of a 0.22-μm Millipore filter (Type GSTF). The filter is then washed with 3 ml of ice-cold buffer, placed in a vial with scintillation fluid and counted. Figure 6 shows an experiment in which good counterflow activity was obtained with melibiose-preloaded proteoliposomes in the presence of 10 mM Na+. In the absence of Na+ there was slight uptake of the [3H]melibiose. When the proteoliposomes were not preloaded, or when the competitive substrate, p-nitrophenyl-α-galactoside, was added to the system there was no uptake of [3H]melibiose.

[31] Assay, Genetics, Proteins, and Reconstitution of Proton-Linked Galactose, Arabinose, and Xylose Transport Systems of *Escherichia coli*

By Peter J. F. Henderson and Andrew J. S. Macpherson

It was established some time ago that separate systems transport each of the sugars D-galactose,[1,2] L-arabinose,[3] or D-xylose[4] into *Escherichia coli*. Subsequent work with mutants and structural analogs of the normal substrates showed that each individual sugar is transported by at least two

[1] B. L. Horecker, J. Thomas, and J. Monod, *J. Biol. Chem.* **235**, 1580 (1960).
[2] G. Buttin, *Adv. Enzymol.* **30**, 81 (1968).
[3] C. P. Novotny and E. Englesberg, *Biochim. Biophys. Acta* **117**, 217 (1966).
[4] J. D. David and H. Wiesmeyer, *Biochim. Biophys. Acta* **201**, 497 (1970).

routes[5-9] (galactose is a substrate for six transport systems,[6] but only two are of physiological significance). For each sugar at least one transport route operates by an H^+–sugar symport mechanism and probably involves one protein,[9-14] while the other is energized by a phosphorylated compound and involves at least two proteins, one being located in the periplasmic space.[7,8,12,15-20]

This chapter will be restricted to the former type of proton-linked transport system for galactose, arabinose, or xylose of designated phenotype GalP, AraE, and XylE, respectively[9,11,12]; the latter type containing a periplasmic binding protein is reviewed elsewhere.[21,22] A summary of their properties and nomenclature is given in Table I. The characterization of other sugar transport systems of E. coli is described in several recent reviews.[22-26]

Measurement of Sugar-Promoted Alkaline pH Changes with Intact Cells

In 1963 Mitchell[27] first proposed that β-galactoside transport into E. coli would be tightly coupled to proton translocation across the cytoplas-

[5] B. Rotman, A. K. Ganesan, and R. Guzman, J. Mol. Biol. 36, 247 (1968).

[6] H. L. Kornberg, Proc. R. Soc. London Ser. B 183, 105 (1973).

[7] R. Schleif, J. Mol. Biol. 46, 185 (1969).

[8] C. E. Brown and R. W. Hogg, J. Bacteriol. 111, 606 (1972).

[9] E. O. Davis, M. C. Jones-Mortimer, and P. J. F. Henderson, J. Biol. Chem. 259, 1520 (1984).

[10] V. M. S. Lam, K. R. Daruwalla, P. J. F. Henderson, and M. C. Jones-Mortimer, J. Bacteriol. 143, 396 (1980).

[11] P. J. F. Henderson, R. A. Giddens, and M. C. Jones-Mortimer, Biochem. J. 162, 309 (1977).

[12] K. R. Daruwalla, A. T. Paxton, and P. J. F. Henderson, Biochem. J. 200, 611 (1981).

[13] A. J. S. Macpherson, M. C. Jones-Mortimer, and P. J. F. Henderson, Biochem. J. 196, 269 (1981).

[14] A. J. S. Macpherson, M. C. Jones-Mortimer, P. Horne, and P. J. F. Henderson, J. Biol. Chem. 258, 4390 (1983).

[15] R. W. Hogg and E. Englesberg, J. Bacteriol. 100, 423 (1969).

[16] D. Kolodrubetz and R. Schleif, J. Bacteriol. 148, 472 (1981).

[17] W. Boos, Eur. J. Biochem. 10, 66 (1969).

[18] A. Robbins and B. Rotman, Proc. Natl. Acad. Sci. U.S.A. 72, 423 (1975).

[19] S. Harayama, J. Bollinger, T. Iino, and G. L. Hazelbauer, J. Bacteriol. 153, 408 (1983).

[20] C. Ahlem, W. Huisman, G. Neslund, and A. S. Dahms, J. Biol. Chem. 257, 2926 (1982).

[21] T. Silhavy, T. Ferenci, and W. Boos, in "Bacterial Transport," p. 127. Dekker, New York, 1978.

[22] C. Furlong, this series, Vol. 10 [98].

[23] R. Hengge and W. Boos, Biochim. Biophys. Acta 737, 443 (1983).

[24] H. R. Kaback, J. Membr. Biol. 76, 95 (1983).

[25] I. C. West, Biochim. Biophys. Acta 604, 91 (1980).

[26] S. S. Dills, A. Apperson, M. R. Schmidt, and M. H. Saier, Microbiol. Rev. 44, 385 (1980).

[27] P. Mitchell, Biochem. Soc. Symp. 22, 142 (1963).

PROPERTIES OF THE ARABINOSE, GALACTOSE, AND XYLOSE TRANSPORT SYSTEMS IN *E. coli*

System	Apparent M_r and location of gene product(s)		Gene[a,b] location (min)	K_m (μm)	Inhibition[a,c] by 5 mM arsenate (%)	Sugar–H⁺[a,c,e] symport	Mechanism of energization	Sensitivity to catabolite repression
	Membrane	Periplasm						
XylE	39,000[f,t]	—	91	70–170[f]	0	Yes	Chemiosmotic	?
AraE	37,000[g,t]	—	61	140–320[c]	8	Yes	Chemiosmotic	Resistant[c,h]
GalP	37,000[i]	—	64	50–450[j,k]	28	Yes	Chemiosmotic	Resistant[c,j]
XylF	Not known	37,000[l,m]	80	0.2–2.0[l,m]	72	No	Phosphate-bond?	?
AraF	37,000[n]	33,000[l,n-q]	45	4–6[c,o,p]	55	No	Phosphate-bond?	Sensitive[c,h]
MglP	51,000[r,u] 37,000	31,000[l,q-s]	45	0.4–0.7[s]	81	No	Phosphate-bond?	Sensitive[c,j]

[a] E. O. Davis, M. C. Jones-Mortimer, and P. J. F. Henderson, *J. Biol. Chem.* **259**, 1520 (1984).
[b] B. J. Bachmann, *Microbiol. Rev.* **47**, 180 (1983).
[c] K. R. Daruwalla, A. T. Paxton, and P. J. F. Henderson, *Biochem. J.* **200**, 611 (1981).
[d] V. M. S. Lam, K. R. Daruwalla, P. J. F. Henderson, and M. C. Jones-Mortimer, *J. Bacteriol.* **143**, 396 (1980).
[e] P. J. F. Henderson, R. A. Giddens, and M. C. Jones-Mortimer, *Biochem. J.* **162**, 309 (1977).
[f] P. J. F. Henderson, S. A. Bradley, and E. O. Davis, unpublished data.
[g] A. J. S. Macpherson, M. C. Jones-Mortimer, and P. J. F. Henderson, *Biochem. J.* **196**, 269 (1981).
[h] Kolodrubetz and R. Schleif, *J. Mol. Biol.* **151**, 215 (1981).
[i] A. J. S. Macpherson, M. C. Jones-Mortimer, P. Horne, and P. J. F. Henderson, *J. Biol. Chem.* **258**, 4390 (1983).
[j] D. B. Wilson, *J. Biol. Chem.* **249**, 553 (1974).
[k] P. J. F. Henderson and R. A. Giddens, *Biochem. J.* **168**, 15 (1977).
[l] Binding protein.
[m] C. Ahlem, W. Huisman, G. Neslund, and A. S. Dahms, *J. Biol. Chem.* **257**, 2926 (1982).
[n] D. Kolodrubetz and R. Schleif, *J. Bacteriol.* **148**, 472 (1981).
[o] R. Schleif, *J. Mol. Biol.* **46**, 185 (1969).
[p] R. W. Hogg and E. Englesberg, *J. Bacteriol.* **100**, 423 (1969).
[q] F. A. Quiocho and J. W. Pflugrath, *J. Biol. Chem.* **255**, 6559 (1980).
[r] S. Harayama, J. Bollinger, T. Iino, and G. L. Hazelbauer, *J. Bacteriol.* **153**, 408 (1983).
[s] W. Boos, *Eur. J. Biochem.* **10**, 66 (1969).
[t] From sequencing the DNA of the appropriate genes, the M_r predicted for XylE is 53,607 (E. O. Davis, unpublished) and for AraE is 51,683 (M. J. C. Maiden, unpublished).
[u] Recent characterization of *mgl* gene products in *S. typhimurium* indicate two membrane proteins of M_r 51,000 and 29,000, a binding protein of M_r 33,000, and a fourth protein of M_r 21,000 [N. Müller, M.-G. Heine, and W. Boos, *J. Bacteriol.* **163**, 37 (1985)].

mic membrane, but a direct experimental test was not described until 1970 by West.[28] The following assay procedure is based on that of West and Mitchell[29] with some modifications, including omission of iodoacetate, more rigorous substrate depletion, and operation at lower pH (5.8–6.8), which enable detection of proton-linked transport even when a second transport system is present.

Principle

In accordance with the predictions of the chemiosmotic theory[30,31] bacterial respiration or ATP hydrolysis generates an electrochemical gradient of protons across the cytoplasmic membrane (Fig. 1).[32–34] The polarity is inside negative and alkaline, compared with outside. This protonmotive force can be used to energize substrate transport, motility, or other processes.[35] Mitchell pointed out that an obligatory coupling of proton movement inward to sugar movement inward, i.e., symport (or the equivalent sugar–OH^- antiport) by the carrier protein(s) would effectively energize the sugar transport (Fig. 1), and explain the ability of the microorganisms to accumulate the nutrient even when the ambient sugar concentration is low.[27,35,36] In the steady state a cycling of protons would occur (Fig. 1), so to observe proton movement with sugar, West inhibited respiration (by removing O_2) and the glycolytic production of ATP (by adding 1 mM iodoacetate), and provided a counterion, SCN^-, or K^+, to prevent the development of a membrane potential, so that H^+–sugar symport could be measured as a net pH change in the medium.[28,29]

Materials and Methods

Growth of Bacteria. The exact growth conditions are probably not critical for subsequent observation of sugar-promoted alkaline pH changes, which have been detected in cells grown on either rich broth or minimal salts and harvested in either exponential or stationary phase. However, it is best to enhance the depletion of endogenous energy stores

[28] I. C. West, *Biochem. Biophys. Res. Commun.* **41**, 655 (1970).
[29] I. C. West and P. Mitchell, *J. Bioenerg.* **3**, 445 (1972).
[30] P. Mitchell, *Nature (London)* **191**, 144 (1961).
[31] D. G. Nicholls, "Bioenergetics—an Introduction to the Chemiosmotic Theory." Academic Press, London, 1982.
[32] B. A. Haddock and C. W. Jones, *Bacteriol. Rev.* **41**, 47 (1977).
[33] Y. Kagawa, *Biochim. Biophys. Acta* **505**, 451 (1978).
[34] M. Futai and H. Kanazawa, *Microbiol. Rev.* **47**, 285 (1983).
[35] P. Mitchell, *J. Bioenerg.* **4**, 63 (1973).
[36] P. Mitchell, *Symp. Soc. Gen. Microbiol.* **27**, 383 (1977).

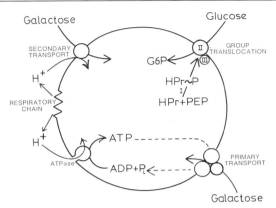

Fig. 1. Scheme for energization of galactose and glucose transport in *E. coli*. The large circle represents the cytoplasmic membrane. The H⁺–galactose symport (GalP, top left, see Table I) is energized by the protonmotive force which is generated by respiration and also used for ATP synthesis. The binding-protein galactose transport system (MglP, bottom right, see Table I) is energized by a phosphorylated compound, probably ATP. Both systems transport chemically unmodified galactose, in contrast to the glucose phosphotransferase (top right), which uses PEP to phosphorylate the sugar via the HPr protein, enzyme I, enzyme II, and factor III [H. L. Kornberg, *Proc. R. Soc. London Ser. B* **183**, 105 (1973); S. S. Dills, A. Apperman, M. R. Schmidt, and M. M. Saier, *Microbiol. Rev.* **44**, 385 (1980)].

of *E. coli* by harvesting in late log or stationary phase after oxygen supply becomes limiting for rate of growth.

E. coli cells are grown on minimal salts medium supplemented with 20 mM glycerol, 10 mM sugar, and any additions required by individual strains, e.g., amino acid, thymine. Culture volumes are usually 200, 400, or 1200 ml in 250 ml, 500 ml, or 2 liter growth flasks, respectively, incubated in an orbital incubator operating at 200 rpm and 37°. (30° may be necessary for some *E. coli* strains, e.g., those containing phage Mu-mediated DNA insertions.) The 10 mM sugar acts as inducer and growth substrate. Gratuitous inducer, such as 6-deoxy-D-galactose (D-fucose) for the galactose transport systems, or the natural sugar in mutants unable to metabolize it, can be used at lower concentrations (0.5–1 mM). Use of glycerol as growth substrate, instead of succinate for example, has the advantage of mildly repressing expression of the binding protein transport system without affecting the proton-linked transport.[4,12,17,37,38]

When their concentration reaches about 0.7 mg dry mass per ml, the bacteria are harvested by centrifugation (6000 g for 7.5 min in a 6 × 300 ml

[37] D. B. Wilson, *J. Biol. Chem.* **249**, 553 (1974).
[38] D. Kolodrubetz and R. Schleif, *J. Mol. Biol.* **151**, 215 (1981).

rotor for maximum recovery combined with ease of resuspension of the pellet), and resuspended in the same volume of 150 mM KCl, 5 mM glycylglycine, pH 6.6, 1 mM mercaptoethanol. This is incubated for 1 hr at the growth temperature with shaking at 200 rpm to promote removal of growth substrate and deplete energy reserves and ATP. The bacteria are again sedimented, resuspended in the same volume of 150 mM KCl/2 mM glycylglycine, pH 6.6, and immediately resedimented. Finally they are suspended in a much smaller volume of 150 mM KCl/2 mM glycylglycine, pH 6.6 to an A_{680} near 50, about 35 mg dry mass per ml. It should be noted that this procedure is by no means sufficient for complete deenergization, which may possibly be achieved by incubation with uncoupling agent plus methyl-α-D-glucoside.[39]

Apparatus. A pH electrode is inserted into a glass or perspex cell through an air-tight seal (Fig. 2). We routinely use a Pye Unicam M405 micro combination pH glass electrode. The temperature must be stabilized to $\pm 0.1°$ through a circulating water jacket, because the small pH changes observed, often less than 0.1 pH unit, are similar to changes in the standard electrode potentials, (i.e., of pH and reference electrode) that would be evoked by temperature fluctuations. An oxygen electrode and/or K$^+$ electrode can be inserted in the same suspension if desired. The glass cell we use is illustrated in Fig. 2, but a commercially available design made of perspex with our electrical measuring circuit (Fig. 2) has proved satisfactory provided that precautions to remove air are adequate.

Leakage of air into the bacterial suspension must be eliminated, its occurrence being indicated by respiration-dependent acidification of the medium. A stream of nitrogen or argon can be directed over the lid if necessary.

The pH electrode must be screened from external sources of electrical interference, since even static electricity on clothing can produce detectable currents in the high impedance electrode and its cable. Screening is most simply effected by sheathing the electrode and cable with metal foil connected to a common earth point.

If a K$^+$ electrode and medium of low K$^+$ concentration is included in the system it may be desirable to replace the KCl solution (usually >3 M) in the reference electrode with a lower concentration, e.g., 0.1 M, or connect a remote electrode to the bacterial suspension via a junction formed between the low potassium medium and the reference electrode solution in a capillary tube (Fig. 2). Details of a suitable system can be supplied.

[39] A. L. Koch, *J. Mol. Biol.* **59**, 447 (1971).

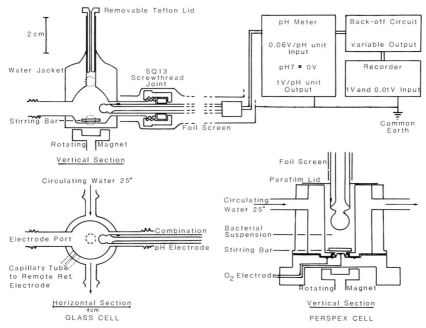

FIG. 2. Apparatus for the measurement of H+–sugar symport. On the left side is a diagram for a custom-built glass cell containing two SQ13 (Corning Quickfit) horizontal ports for a pH and a K+ electrode with a circulating water jacket for temperature control (25.0 ± 0.1°). An SQ18 larger size port will accommodate an oxygen electrode (Yellow Springs Instrument Company). The screened electrode output is measured by a high input impedance, low output noise pH meter connected to a recorder via a backing off circuit (top right). Alternatively, a perspex electrode cell (Rank Oxygen Electrode) can be used provided the lid adequately excludes air. The procedure is described in the text.

Sugar-promoted pH changes in *E. coli* generally exhibit $t_{1/2}$ values in the range 0.5–5 min at 25°; these are much slower than respiration-driven H+ expulsion.[31,32] The response time of the pH measuring system, which is usually 0.5–2 sec for 98% deflection, is not therefore a problem. Nevertheless, rapid mixing of additions should be ensured by vigorous stirring of the suspension with a rotating glass- or plastic-covered metal bar driven by an external magnet (Fig. 2); the stirrer speed must be constant, since variations will affect the electrode response.

When sugar is added to a suspension of such deenergized bacteria, it will diffuse into the cytoplasm provided an appropriate transport system is present in the membrane. If the mechanism is H+–sugar symport, such sugar movement will be accompanied by proton translocation, observed as an alkaline pH change of the medium outside the bacterial cells. The rate and total extent of the pH change depend upon the following factors: buffering capacity of the medium, i.e., $\Delta H^+/\Delta pH$ (nmol H^+ per unit), concentration of sugar added in relation to K_m and V_{max} of the transport system, whether the sugar is further metabolized, the intracellular volume of the bacteria, the size of any preexisting pH gradient across the membrane, the "leakiness" of the membrane itself to protons, and the availability of counterions to compensate for electrical imbalance. During the experimental period we routinely keep the stock bacterial suspension at room temperature, but storage in ice may be preferable. Some transport systems (GalP and XylE) appear to be more stable than others (LacY, AraE) which decline in activity over a period of hours.

Measurement of pH Changes. All solutions are kept anoxic by a slow stream of argon, care being taken to prevent evaporation. The measuring cell is thoroughly washed with distilled water, followed by at least three washes with anoxic 150 mM KCl, 2 mM glycylglycine, pH 6.6, kept at 25°. The empty cell is flushed with argon gas for about 1 min and then bacteria (0.5 ml of $A_{680} = 50$ equivalent to 17 mg dry mass) are added. The cell is immediately filled with anoxic 150 mM KCl/2 mM glycylglycine, pH 6.6 and the lid inserted. The recorder is turned on at a low sensitivity, corresponding to a full-scale deflection of about 1 pH unit. The pH usually falls by up to 0.3 units if glycerol was the sole growth substrate, followed by a slow alkaline drift toward a pH anywhere in the range 6.3–6.7. If a metabolizable sugar was present during growth the initial pH fall is usually more extensive and may not reverse, so the final pH is in the range 5.7–6.3 with, sometimes, an incurable acid drift. The pH value may be encouraged toward its natural equilibrium value by small additions of anoxic HCl or KOH, but a drift is often unavoidable if the experimenter sets his heart on some value other than that chosen by the cells! Equilibration usually takes 15–30 min; longer periods may indicate inadequate exclusion of air, insufficient depletion of substrate, or retention of metabolizable sugar from a previous experiment, i.e., failure to wash sufficiently.

When the drift rate is low, the full-scale recorder deflection is set to 0.05–0.10 pH units and anoxic standard acid or alkali equivalent to 30 nmol is added to calibrate the measurement. Then anoxic sugar (10–50 μl 0.5 M) is added (Fig. 3). After the subsequent pH change (Fig. 3) returns to the initial rate, the calibration is repeated. Provided the total excursion is less than 0.1 pH units there is a linear relationship between ΔpH and ΔH^+, so the repeated calibration should be the same as the first and the sugar-promoted ΔpH can be converted directly to ΔH^+.

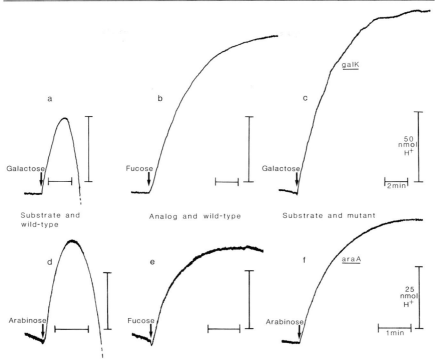

FIG. 3. Assay of H⁺–sugar symport using substrate analogs and *E. coli* mutants. The *lacY E. coli* strain ML35 (a, b) and *galK* strain W3092i (c) were grown on 20 m*M* glycerol plus 1 m*M* D-fucose to induce GalP activity. Each strain was harvested, depleted, washed, and resuspended under anaerobic conditions (see text) in a perspex cell (Fig. 2) in 7 ml 150 m*M* KCl, 2 m*M* glycylglycine. When galactose (25 μmol) was added to the *galK*⁺ strain the initial alkaline pH change (upward deflection) rapidly reversed as the sugar was metabolized (a), a problem circumvented by the use instead of a nonmetabolizable substrate analogue (b) or of a *galK* mutant (c). Similarly, *E. coli* strain K10 (d, e) and the *araA* strain SB5314 (f) were grown on 20 m*M* glycerol plus 1 m*M* L-arabinose to induce AraE activity. When arabinose (12.5 μmol) was added to the anaerobic suspension the initial alkaline pH change rapidly reversed (d), a problem circumvented by the use instead of a nonmetabolizable substrate analog (e) or of an *araA* mutant (f).

Factors Affecting the Sugar-Promoted pH Change. Commercially available sugars may be contaminated with compounds that, even at very low concentrations, produce a pH change. For example, recent batches of commercial methyl-β-D-thiogalactoside contain a contaminant producing alkaline pH changes. A control without cells present should be carried out to determine if this is a problem.

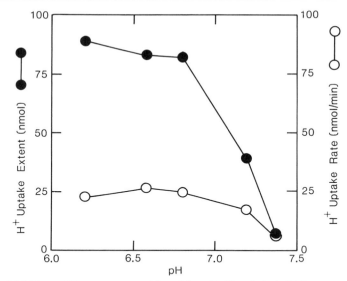

Fɪɢ. 4. Inhibition of H⁺–sugar symport by raising pH. *E. coli* strain K10 was grown on 15 m*M* galactose, harvested, and resuspended as described in the text. The alkaline pH change elicited by addition of 25 μmol D-fucose was measured as in Fig. 3, except that the pH was first adjusted to the values shown by addition of dilute HCl or NaOH in five separate experiments.

Contamination[40] of a sugar with D-glucose even at levels less than 0.1% is a severe problem with *E. coli,* because uptake through the phosphotransferase system[6,26] (a group translocation) is followed by glycolysis to lactate, acetate, and succinate, and rapid acidification results, obscuring any alkaline pH change resulting from transport of the main sugar.

Sugar-promoted pH changes become progressively inhibited at pH values above about 6.9 (Fig. 4). The effect is alleviated by KSCN or valinomycin + K⁺, indicating that above pH 6.9 indigenous ion movements compensating for the accumulation of positive charge during in-

[40] In all our experiments involving sugars it is important to remember that even the best commercial preparations are not absolutely pure. Disaccharides are often readily hydrolyzed in solution. Sugars are often difficult to recrystallize without substantial losses. The first method to try is to suspend the sugar in boiling ethanol and add water dropwise until the sugar dissolves. For the genetic experiments it is usually satisfactory to use commercial preparations at the lowest concentration that will yield good growth of colonies on plates (5–10 m*M* of a monosaccharide). For the growth of organisms in liquid culture the impurities may be preferentially utilized.

ward movement of H^+ become limiting. The pH dependence of H^+–sugar movement may have mechanistic significance.[24,25,41–43]

If the temperature of measurement is varied between 5 and 37° a reasonably linear Arrhenius plot of the relationship between log (initial rate of H^+ transport) and $1/T$ ($°K^{-1}$) is observed corresponding to energies of activation of 60–80 kJ/mol[12] This contrasts with nonlinear Arrhenius plots found when the temperature dependence of energized sugar transport is measured.

Value of Mutants

Elimination of Alternative Transport Routes. As shown above in Fig. 1 and Table I, galactose can enter *E. coli* by two transport routes, GalP and MglP.[2,5,17,37] Only the former is H^+ linked,[11] and galactose transport through GalP might be diminished substantially by transport through MglP. It is therefore desirable to assay H^+–galactose symport activity in a $GalP^+$ $MglP^-$ mutant to maximize uptake by the former route. For the same reason H^+–arabinose symport should preferably be measured in an $AraE^+$ $AraF^-$ strain (Table I) and H^+–xylose symport in a $XylE^+$ $XylF^-$ strain (Table I). It happens that galactose is also a good substrate for the H^+–lactose symporter LacY, but this is normally at a low level of activity in uninduced strains so use of a $LacY^-$ strain may not be obligatory for unambiguous measurements of GalP activity. In our experience arabinose and xylose are not significant substrates for H^+ symport systems other than AraE or XylE, respectively.

Elimination of Anaerobic Metabolism of the Physiological Substrate. Proton movement inward with a sugar substrate can quickly be obscured by acid excretion due to internal metabolism of the sugar to lactate, acetate, and succinate (Fig. 3). If a mutant lacking the *first* enzyme of the metabolic pathway is available, i.e., galactokinase-negative for galactose, arabinose isomerase-negative for arabinose, or xylose isomerase-negative for xylose, this problem is eliminated. Examples are shown in Fig. 3. Furthermore, such mutations decrease the concentration of sugar required for induction during growth, and may enhance the level of induction achieved (arabinose isomerase-negative mutants have their AraE activity enhanced by a factor $3–6\times$[12]).

Note that a mutation impairing activity of the *second* or subsequent metabolic steps will probably increase the extent of the pH change, be-

[41] E. Komor and W. Tanner, *J. Gen. Physiol.* **64,** 568 (1974).
[42] M. G. Page and I. C. West, *Biochem. J.* **196,** 721 (1981).
[43] G. J. Kaczorowski and H. R. Kaback, *Biochemistry* **18,** 3691 (1979).

cause functioning initial metabolic conversion(s) will reduce the accumulation of, and product inhibition by, internal free sugar.

Association of H⁺–Sugar Symport with a Particular Gene. Obviously, the coincident loss of the appropriate H⁺–sugar symport phenotype in a mutant lacking *galP*, *araE*, or *xylE* genes, and recovery of H⁺–sugar symport when the genetic lesion is cured, e.g., by deliberate transduction, transformation, conjugation, or spontaneous reversion, is a powerful method of associating the proton symport mechanism with the particular gene product. If the mutation is made by insertion of Mud(Ap^r*lac*) or λ*plac*Mu phage then coincident introduction of the LacY H⁺–lactose phenotype provides an elegant control; one example in which a Mud(Ap^r*lac*)I insertion impaired XylE activity is shown in Fig. 5.

Value of Substrate Analogs

For each of the sugar–proton symport systems of *E. coli* there is at least one substrate analog that is transported, but is not modified by the

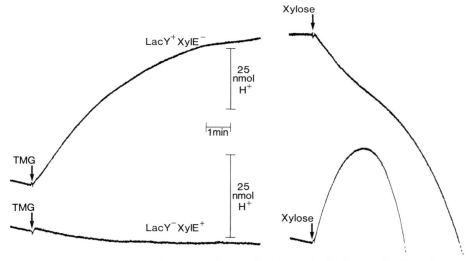

FIG. 5. Characterization of a Mud(Ap^r*lac*)-mediated mutation in the gene for H⁺–xylose symport. *E. coli* strains EJ18 and EJ15 (Table III) were grown on 10 mM xylose plus 20 mM glycerol, harvested, and resuspended as described in the text and by E. O. Davis, M. C. Jones-Mortimer, and P. J. F. Henderson, *J. Biol. Chem.* **259**, 1520 (1984). In separate experiments, 10 μmol methyl-β-D-thiogalactoside (TMG) or xylose was added to a suspension of each organism in the glass cell (Fig. 2) and the resulting pH change measured. The upper traces show that the Mud(Ap^r*lac*) mutant EJ18 had gained H⁺–TMG (LacY) activity and lost H⁺–xylose symport (XylE), compared to the parent strain, EJ15 (lower traces). Experiments performed by E. O. Davis and S. A. Bradley.

TABLE II
NONMETABOLIZED SUBSTRATE ANALOGS FOR H⁺–SUGAR SYMPORT SYSTEMS

Transport system	Substrate analogs	Inducers
GalP	D-Fucose[a]	D-Galactose[a]
	D-Talose[b]	D-Fucose[a]
	2-Deoxy-D-galactose[c]	6-Deoxy-6-fluoro-D-galactose[b]
	6-Deoxy-6-fluoro-D-galactose[b]	
AraE	D-Fucose[d,e]	L-Arabinose[d,e]
XylE	6-Deoxy-D-glucose[b]	D-Xylose[f,g]

[a] B. Rotman, A. K. Ganesan, and R. Guzman, *J. Mol. Biol.* **36**, 247 (1968).

[b] P. J. F. Henderson, P. Horne, E. O. Davis, and S. A. Bradley, unpublished observations.

[c] P. J. F. Henderson and R. A. Giddens, *Biochem. J.* **168**, 15 (1977).

[d] C. E. Brown and R. W. Hogg, *J. Bacteriol.* **111**, 606 (1972).

[e] K. R. Daruwalla, A. T. Paxton, and P. J. F. Henderson, *Biochem. J.* **200**, 611 (1981).

[f] J. D. David and H. Wiesmeyer, *Biochim. Biophys. Acta* **201**, 497 (1970).

[g] V. M. S. Lam, K. R. Daruwalla, P. J. F. Henderson, and M. C. Jones-Mortimer, *J. Bacteriol.* **143**, 396 (1980).

first enzyme of metabolism. They are listed in Table II. Each one can be used to measure the H⁺ symport activity even in wild-type strains, thus avoiding the acid excretion due to metabolism (see Fig. 3).

Measurement of Sugar Transport into Right-Side Out Subcellular Vesicles

The success of the following procedures for identifying membrane transport proteins with radioactive *N*-ethylmaleimide is critically dependent on the availability of right-side out subcellular vesicles prepared by the method of Kaback.[44] Such vesicles have played a central role in advancing our understanding of bacterial transport reactions.[45–47] The preparation method should remove cytoplasmic enzymes involved in sugar metabolism and periplasmic binding proteins involved in the activity of secondary transport systems (see above).[44–46] It is desirable, but not essential, to use a strain in which metabolic enzyme(s) and membrane-bound components of the second transport system have also been eliminated by appropriate mutations, and in which the H⁺–sugar symport

[44] H. R. Kaback, this series, Vol. 22, p. 99.

[45] H. R. Kaback, *Biochim. Biophys. Acta* **265**, 367 (1972).

[46] M. Futai, *in* "Bacterial Transport," p. 7. Dekker, New York, 1978.

[47] H. R. Kaback, *J. Cell. Physiol.* **89**, 575 (1977).

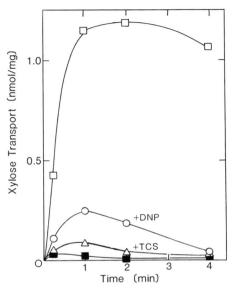

FIG. 6. Assay of proton-linked energized sugar transport using subcellular vesicles of *E. coli*. Right-side out subcellular vesicles were made from *E. coli* EJ15 grown on 10 m*M* xylose plus 20 m*M* glycerol and transport of 40 μ*M* [¹⁴C]xylose measured [see text; H. R. Kaback, this series, Vol. 22, p. 99; P. Horne and P. J. F. Henderson, *Biochem. J.* **210**, 699 (1983)]. (□) 20 m*M* ascorbate + 0.1 m*M* phenazine methosulfate (PMS); (■) no additions; (○) ascorbate + PMS + 1 m*M* 2,4-dinitrophenol; (△) ascorbate + PMS + 20 μ*M* tetra-chlorosalicylanilide. The observation of respiratory substrate-dependent, uncoupler-sensitive transport is diagnostic of an H⁺–sugar symport mechanism.

activity of interest has been amplified by genetic manipulation (see this volume [13]). H⁺–sugar transport activity in the vesicles is best measured by uptake of radioisotope-labeled sugar (0.04–0.2 m*M*) at 25° energized by 20 m*M* ascorbate plus 0.1 m*M* phenazine methosulfate, the respiration of which efficiently generates a protonmotive force[44,45] and is unimpaired by *N*-ethylmaleimide.[48] A detailed description is given below, and an example of energized xylose transport is shown in Fig. 6. Alternatively, sugar-promoted pH changes can be measured directly,[12,49,50] but this requires a higher concentration of vesicles derived from strains with amplified H⁺–sugar symport activity.

[48] H. R. Kaback and L. Patel, *Biochemistry* **17**, 1640 (1978).
[49] P. Horne and P. J. F. Henderson, *Biochem. J.* **210**, 699 (1983).
[50] L. Patel, M. L. Garcia, and H. R. Kaback, *Biochemistry* **21**, 5805 (1982).

In our laboratory, the preparation of right-side out vesicles was greatly facilitated by adoption of the procedure of Witholt et al.[51,52] for making spheroplasts from E. coli at any stage of growth. A protocol for the complete vesicle preparation will be provided on request. Some criteria of an acceptable vesicle preparation are as follows. There should be no contamination by intact cells or spheroplasts as judged by examination under a phase contrast light microscope at 800× magnification. There should be high respiration-dependent sugar transport activity reduced by a factor of at least 10-fold in the presence of uncoupling agent or in the absence of ascorbate plus phenazine methosulfate (Fig. 6). Neither glucose nor, in appropriately grown cells, glycerol should energize sugar transport, indicating the absence of glycolytic and Krebs cycle enzymes. There should be negligible activity of any binding protein-dependent transport system present originally in the intact cells. The vesicle suspension should be nonviscous, indicating the absence of DNA.

Genetics of the H⁺-Sugar Symport Systems

Unlike the contiguous lac operon the genes coding for transport, metabolism and regulation of galactose utilization are scattered around the chromosome of E. coli. A similar complexity exists for the genes of arabinose or xylose utilization. Only a summary of the genetic features of each H⁺–sugar symport system is included here, with a list of useful strains (Table III). However, there is much of interest to be learned about the operation, regulation, and evolution of each system, information that will almost certainly accrue when the DNA sequencing of their gene(s) is completed in the near future.

Galactose

The galP gene codes for H⁺–galactose symport in E. coli[11] and S. typhimurium.[53] Its approximate location is 64.2 min on the E. coli linkage map,[54] cotransducible with fda.[55] The regulatory gene galR at 62 min probably controls galP expression in a negative manner.[2,56] Constitutive mutants are available[57] (Table III), but the nature of their genetic lesion is

[51] B. Witholt, M. Boekhout, M. Brock, T. Kingma, H. van Heerikhuizen, and L. de Leij, Anal. Biochem. **74,** 160 (1976).

[52] B. Witholt and M. Boekhout, Biochim. Biophys. Acta **508,** 296 (1978).

[53] F. Nagelkerke and P. W. Postma, J. Bacteriol. **133,** 607 (1978).

[54] B. J. Bachmann, Microbiol. Rev. **47,** 180 (1983).

[55] C. Riordan and H. L. Kornberg, Proc. R. Soc. London Ser. B **198,** 401 (1977).

[56] B. von Wilcken-Bergmann and B. Müller-Hill, Proc. Natl. Acad. Sci. U.S.A. **79,** 2427 (1982).

[57] P. J. F. Henderson and R. A. Giddens, Biochem. J. **168,** 15 (1977).

TABLE III
Escherichia coli STRAINS

Strain	Relevant genotype	Relevant phenotype	Growth requirements	Reference
JM1424	$galP^+$ mgl	Enhanced inducible GalP$^+$	His, Thy, Ilv	a
JM1418	$galP$ mgl	GalP$^-$ amber mutant	His, Thy, Ilv	a
JM2053	$galP$ mgl	GalP$^-$ Tn10 insertion mutant	His, Thy, Ilv	a
JM1576	$galP^+$ mgl Δlac	Enhanced constitutive GalP$^+$	His, Ile, Leu	a, b
JM2071	$galP$ mgl Δlac	GalP$^-$ Tn10 insertion mutant	His, Ile, Leu	a, b
JM2531	mgl lac^+ $galP :: \lambda placMu1$	GalP$^-$ $\lambda placMu1$ insertion	His, Ile, Leu	c
S183 series	$galP^+$ mgl Δlac	Enhanced constitutive GalP$^+$	His, Thr, Leu, Met	d, e
W3092i	$galP^+$ $galK$ mgl lac	Inducible GalP$^+$, no metabolism		f
SB5314	$araE^+$ $araF^+$ $araA$	Enhanced inducible AraE$^+$		g, h
SB5313	$araE$ $araF^+$ $araA$	AraE$^-$		g, h
SB5159	$araE^+$ $araF^+$ $araC$	Enhanced constitutive AraE$^+$		g
JM1637	$\Delta(araE\ lysA\ galR)$ $araF$	AraE$^-$ deletion	Lys	i
JM1647	$\Delta(araE\ lysA\ galR)$ $araF$ $(\lambda d\ araE^+\ lysA^+)$ $(\lambda cI857\ s7)$	AraE$^+$ λ lysogen		i
JM2433	$araE$ $araA(B?)$ Δlac $araD$	Enhanced inducible AraE$^+$	His	j
JM2443	$araE :: \lambda placMu1$ $araA(B?)$ Δlac $araD$	AraE$^-$	His	j
RS1 E$^-$F$^-$	$araE$ $araF$	AraE$^-$ AraF$^-$		k
EJ15	$xylE^+$ $xylF^+$ Δlac	XylE$^+$	His	l
EJ18	$xylE :: Mud(ApRlac)I$ $xylF^+$	XylE$^-$ Mu insertion	His	l
JM2336	$xylE :: Mud(ApRlac)II$ $xylF^+$	XylE$^-$ Mu insertion	His	l
EJ68	$xylE^+$ $xylF^+$ $xylA$	Enhanced inducible XylE$^+$	His	m
EJ54	$xylE :: Mud(ApRlac)II$ $xylF^+$ $xylA$	XylE$^-$ Mu insertion	His	m
JM2390	$\Delta(xylE\ malB)$ $xylF^+$	XylE$^-$ deletion		c
EJ80	$xylF :: Mud(ApRlac)I$ $\Delta(xylE\ malB)$	XylE$^-$ XylF$^-$	His	m
EJ81	$xylF :: Mud(ApRlac)I$ $xylE^+$	XylE$^+$ XylF$^-$	His	m

a A. J. S. Macpherson, M. C. Jones-Mortimer, P. Horne, and P. J. F. Henderson, *J. Biol. Chem.* **258,** 4390 (1983).

not established (see also Buttin[2]). An amber mutation in *galP* has been characterized,[14] and Tn10[14] or λp*lac*Mu insertions in or very close to *galP* have been made (Table III). The biochemical evidence (below) is consistent with there being only one gene product, but genetical evidence as to the number of *galP* cistrons is not yet available. The location of *galP* is separate from that of the other galactose transport system, *mgl* (45 min[17,54]) and from the genes for enzymes of galactose metabolism, *galE,K,T* (17 min[54]) and *galU* (27 min[54]).

Arabinose

The *araE* gene codes for H⁺–arabinose symport in *E. coli*. Its location is near 61 min with the gene order[16,54,58] [*thyA*][58]*araE* orf[58] *lysR lysA galR*. The regulatory gene *araC* at 1 min controls expression by a positive (and negative?) interaction of the *araC* gene product and arabinose with the *araE* promoter region.[59] Constitutive mutants are available, and hyperinducible expression of *araE* seems to occur in *araA* mutants that are unable to metabolize the sugar[12] (Table III). Point, deletion, and insertion mutants of *araE* have been made (Table III). The gene has been cloned and its DNA sequence determined.[58] Biochemical and genetical evidence indicates that only one structural gene is involved in H⁺–arabinose transport activity.[13,16,58] Again, the location of *araE* is separate from the other arabinose transport system *araF,G* (45 min[16,54,60]) and the regulatory/structural genes *araC, araB,A,D* (1 min[54]).

[58] M. C. J. Maiden, unpublished data: in some *E. coli* strains the gene order may be *araE, lysA, thyA; orf* = open reading frame.

[59] C. Stoner and R. Schleif, *J. Mol. Biol.* **171,** 369 (1983).

[60] A. F. Clark and R. W. Hogg, *J. Bacteriol.* **147,** 920 (1981).

[b] P. J. F. Henderson, H. Hirata, P. Horne, M. C. Jones-Mortimer, T. Kaethner, and A. J. S. Macpherson, *in* "Biochemistry of Metabolic Processes," p. 339. Elsevier, Amsterdam, 1983.

[c] M. C. Jones-Mortimer, unpublished.

[d] A. Robbins and B. Rotman, *Proc. Natl. Acad. Sci. U.S.A.* **72,** 423 (1975).

[e] P. J. F. Henderson and R. A. Giddens, *Biochem. J.* **168,** 15 (1977).

[f] W. Boos, *Eur. J. Biochem.* **10,** 66 (1969).

[g] R. W. Hogg and E. Englesberg, *J. Bacteriol.* **100,** 423 (1969).

[h] K. R. Daruwalla, A. T. Paxton, and P. J. F. Henderson, *Biochem. J.* **200,** 611 (1981).

[i] A. J. S. Macpherson, M. C. Jones-Mortimer, and P. J. F. Henderson, *Biochem. J.* **162,** 309 (1977).

[j] M. C. Jones-Mortimer and M. C. Maiden, unpublished.

[k] R. W. Hogg, unpublished data.

[l] E. O. Davis, M. C. Jones-Mortimer, and P. J. F. Henderson, *J. Biol. Chem.* **159,** 1520 (1984).

[m] E. O. Davis, unpublished.

Xylose

The *xylE* gene codes for H^+–xylose symport in *E. coli*. Its location is near 91.4 min with the gene order[9] *aceA pgi xylE malB*. The regulatory gene *xylR* at 80 min probably controls *xylE* expression by a positive mechanism perhaps similar to the control of *araE* by *araC*.[59,61] Mutants constitutive for expression of *xylE* have not yet been characterized, though introduction of a *xylA* mutation seems to enhance expression of *xylE* (Table III). A series of mutants made by insertion of Mud(Ap*lac*)I, Mud(Ap*lac*)II, and λ*plac*Mu phage have been isolated[9] (Table III). There is probably one cistron in the *xylE* region, which has recently been cloned and its DNA sequence determined.[84] The location of *xylE* is again separate from the genes for the other xylose transport system *xylF* (80 min[9,61]) and from the regulatory/structural genes *xylR*, *xylA,B* (80 min[4,54,61]).

Identification of H^+–Sugar Transport Proteins by Labeling with Radioactive *N*-Ethylmaleimide

In a classical series of experiments[62-65] Kennedy and co-workers identified the LacY H^+–lactose protein of *E. coli* by relatively specific labeling with *N*-ethylmaleimide and by differential labeling with radioactive amino acids, exploiting strains containing inducible, constitutive, or impaired *lac*Y genes. In principle their strategy can also be applied to the identification of H^+–galactose,[14] H^+–arabinose,[13] and H^+–xylose transport proteins, but in practice refinements are necessary to overcome the following problems: wild-type *E. coli* strains contain a second transport system for each of these three sugars; available sugars and their analogues do not protect GalP, AraE, or XylE against *N*-ethylmaleimide as effectively as β-D-galactosyl-1-thio-β-D-galactoside (TDG) protects LacY against *N*-ethylmaleimide; the protecting agent of choice may be metabolized by intact cells; and available *E. coli* strains or mutants may contain a low proportion of GalP, AraE, or XylE proteins compared to the proportion of LacY protein expressed in available mutants.[66]

Subsequently, *p*-nitrophenyl-α-D-galactoside was shown to be a specific label for LacY,[24,67] but no specific affinity label for GalP, AraE, or XylE has yet been found.

[61] D. K. Shamanna and K. E. Sanderson, *J. Bacteriol.* **139,** 71 (1979).

[62] C. F. Fox and E. P. Kennedy, *Proc. Natl. Acad. Sci. U.S.A.* **54,** 891 (1965).

[63] C. F. Fox, J. R. Carter, and E. P. Kennedy, *Proc. Natl. Acad. Sci. U.S.A.* **57,** 698 (1967).

[64] J. R. Carter, C. F. Fox, and E. P. Kennedy, *Proc. Natl. Acad. Sci. U.S.A.* **60,** 725 (1968).

[65] T. H. D. Jones and E. P. Kennedy, *J. Biol. Chem.* **244,** 5981 (1969).

[66] J. K. Wright, R. M. Teather, and P. Overath, this series, Vol. 97, p. 158.

[67] G. J. Kaczorowski, G. Le Blanc, and H. R. Kaback, *Proc. Natl. Acad. Sci. U.S.A.* **77,** 6319 (1980).

Assessment of Efficiency with Which Sugars Protect Protein against Reaction with N-Ethylmaleimide

The factors affecting the ability of *N*-ethylmaleimide to react with —SH residues and to inactivate enzymes include *N*-ethylmaleimide concentration, reaction time, temperature, and pH.[64,68] Exposure of subcellular vesicles (1.5 mg protein per ml) to 1 m*M N*-ethylmaleimide for 15 min at 25° in 50 m*M* KP$_i$, pH 6.5, 10 m*M* MgSO$_4$ are effective conditions for 60–85% inactivation of the GalP, AraE, and XylE H$^+$–sugar transport systems.[13,14,69] A typical dose–response curve is shown in Fig. 7.

The efficiency of an individual sugar or structural analog as a protecting agent against *N*-ethylmaleimide is *not predictable* from its efficiency as a substrate or inhibitor of transport, because much higher concentrations of sugar may be required for equilibrium binding and protection compared with the K_m value for energized transport "K_T."[25,64,70,71] This is illustrated by the following data for the LacY H$^+$–lactose symport system, given by Wright *et al.*[66] The substrate analogue TDG exhibited a K_T value of 0.044 m*M*[66] and a K_D of 0.053 m*M* and gave protection efficiencies of 50–60% when used at a concentration of 20 m*M*, compared with a protective efficiency of only 3% for 20 m*M* lactose with a K_T value of 0.08 m*M* and K_D of 14 m*M*.

Samples of vesicles containing about 0.75 mg protein are suspended in 0.5 ml of 50 m*M* potassium phosphate, pH 6.6, 10 m*M* magnesium sulfate contained in a 1.5-ml Eppendorf microfuge tube. An appropriate sugar (10 μl M stock solution, 20 m*M* final concentration) is added to duplicate samples. Sugar is omitted from four samples. All the suspensions are incubated at 25° for 5 min, then 1 m*M N*-ethylmaleimide is added to all samples except for two of the controls. The *N*-ethylmaleimide is made up fresh from a desiccated store of the solid just before the experiment as a stock solution of 25 m*M* in H$_2$O or in 50 m*M* potassium phosphate, pH 6.6, 10 m*M* magnesium sulfate.

After 15 min the samples are cooled in ice for approximately 30 sec then sedimented for 5 min using an Eppendorf microfuge running in a cold room at 4°.[72] The supernatant fluid is carefully removed by aspiration and each pellet is resuspended in 0.5 ml potassium phosphate, 10 m*M* magne-

---

[68] J. L. Webb, "Enzyme and Metabolic Inhibitors," Vol. III, p. 337. Academic Press, New York, 1966.

[69] T. M. Kaethner and P. Horne, *FEBS Lett.* **113**, 258 (1980).

[70] E. P. Kennedy, M. K. Lumley, and J. B. Armstrong, *J. Biol. Chem.* **249**, 33 (1974).

[71] F. J. Lombardi, *Biochim. Biophys. Acta* **649**, 661 (1981).

[72] The reaction of *N*-ethylmaleimide with proteins can be stopped by addition of thiol reagents, e.g., mercaptoethanol, dithiothreitol, or cysteine,[64,66,68,69] rather than by cooling and rapid washing. However, we prefer to avoid any possible carryover of such a reagent, which would impair subsequent labeling with radioactive *N*-ethylmaleimide.

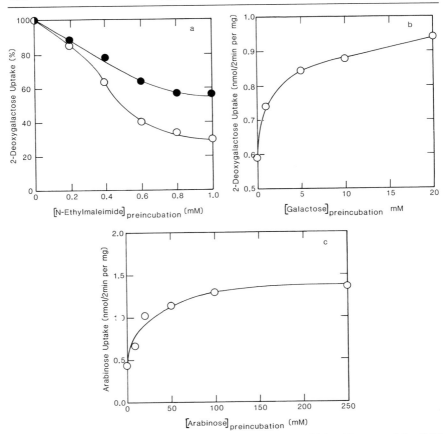

FIG. 7. Measurement of protection efficiency of sugars. (a) Right-side out vesicles (0.75 mg protein) derived from fucose-induced *E. coli* strain JM1424 (Table III) were preincubated at 25° in 0.5 ml of 50 mM potassium phosphate (pH 6.6), 10 mM MgSO$_4$, minus (○) or plus (●) 20 mM galactose added 5 min before the indicated concentration of N-ethylmaleimide. Subsequent incubation and washing were carried out as described in the text, and then ascorbate + PMS-energized 2-deoxygalactose (40 μM) transport was measured. Each value is the mean of duplicates expressed as a percentage of the control without N-ethylmaleimide (4.1 nmol/mg/min). (b) As (a) except that the concentration of galactose was varied as shown. (c) As (b) except that vesicles derived from arabinose-induced *E. coli* SB5314 (Table III) were used with the indicated concentrations of arabinose in the preincubation.

sium sulfate, by vigorous vibration on a vortex mixer, keeping the temperature at 0–4°. The resuspended vesicles must be homogeneous, which may require prolonged, determined, mixing.

The sedimentation and resuspension are repeated three more times. This extensive washing procedure is essential to remove sugar and N-

ethylmaleimide, residual amounts of which could affect subsequent measurements of transport activity. In our early experiments a control without N-ethylmaleimide but with sugar was included but was found to be unnecessary *provided all the washing steps were undertaken.*

Before the final resuspension any drops of residual liquid are removed from the walls of the tube with absorbent tissue, taking care not to disturb the pellet, which is resuspended to the original volume of 0.5 ml and stored in ice. Transport activity is then measured as follows.

Each sample is warmed and incubated for 3 min at 25°. After 2.5 and 2.75 min PMS (5 μl of 10 mM, final concentration 0.1 mM) and potassium ascorbate, pH 6.6 (10 μl of M, final concentration, 20 mM) are added to energize sugar uptake. At 3 min transport is initiated by the addition of a radioactively labeled sugar (10 μl 2 mM, specific activity 0.3–0.9 Ci/mol, final concentration 40 μM). Throughout the prior incubation and the sugar transport the suspension is bubbled with oxygen to maintain respiration. Samples (0.2 ml) are filtered with mild suction (800 mm Hg using a water pump) 15 sec and 2 min after the addition of labeled sugar. The filtration should be complete in less than 5 sec. If not, the protein concentration may be too high or the vesicles may be contaminated with DNA. More samples of smaller volume can be taken if desired. The filter is washed immediately with about 4 ml of sugar-free 0.1 M LiCl and transferred to 10 ml of scintillant [80% (v/v) toluene, 10% (v/v) 2-methoxyethanol containing 4 g/liter of 2,5-diphenyloxazole and 0.2 g/liter of 1,4-bis(5-phenyloxazol-2-yl)benzene] for detection of radioactivity.

Some results obtained for protection of GalP are illustrated in Table IV. In order to measure the degree of protection and to normalize data from different experiments a "protection index," *I*, is defined as follows

$$I(\%) = 100 \times \frac{\text{(sample with NEM and sugar)} - \text{(control with NEM)}}{\text{(control without NEM)} - \text{(control with NEM)}}$$

Either the transport after 15 sec (initial rate) or 2 min can be used for the calculation, the difference in *I* values not being greater than 2%, but the 2 min point is chosen to minimize the radioisotope counting error. For satisfactory comparison of different sugars, *duplicate or triplicate measurements* are necessary with one batch of vesicles. Absolute values of protection indices determined for the same sugar but with different batches of vesicles show a maximum variation of ±10% (Table IV). Nevertheless, the method shows that for the GalP transport system galactose, glucose, methyl-β-D-galactoside, and 6-deoxy-6-fluorogalactose are more effective protecting agents than any of the other test compounds (Table IV). Note that these four are all substrates also for the MglP transport system, so it is important to use a specific substrate of GalP, 2-deoxy-D-

TABLE IV
SUGARS PARTIALLY PROTECT GalP OR AraE AGAINST
INACTIVATION BY N-ETHYLMALEIMIDE

GalP[a,b]		AraE[a]	
Sugar added[c]	Protection index[d] (%)	Sugar added[e]	Protection index[d] (%)
Galactose	20.1 ± 0.9	Arabinose	14, 9
Methyl-β-D-galactoside	19.4 ± 0.5, 14.8 ± 0.4	Glucose	16
6-Fluorogalactose	18.8	Mannose	4
Glucose	18.6 ± 1.5	Fucose	5
Talose	15.0, 6.8 ± 2.8	Methyl-β-L-arabinoside	2
2-Deoxyglucose	7.4 ± 0.5	Inositol	0
Fucose	4.7 ± 0.2	Xylose	0
2-Deoxygalactose	0.4 ± 5.1	Galactose	−21

[a] A. J. S. Macpherson, Ph.D. thesis, University of Cambridge, 1982.
[b] A. J. S. Macpherson, M. C. Jones-Mortimer, P. Horne, and P. J. F. Henderson, *J. Biol. Chem.* **258**, 4390 (1983).
[c] 20 mM sugar, right-side out vesicles from fucose-induced *E. coli* JM1424, procedure as described in the text.
[d] Calculated as described in the text. Triplicate measurements ±SEM or duplicate measurements (no SEM).
[e] 20 mM sugar, right-side out vesicles from arabinose-induced *E. coli* SB5314, procedure as described in the text.

galactose,[57] for the transport measurements in subcellular vesicles of an *mgl* mutant to eliminate the risk that MglP activity is interfering. Glucose is not a desirable compound despite its effectiveness, because of its possible interaction with membrane components of the PtsG and PtsM transport systems, which react with sulfhydryl reagents.[6,26,73] Xylose and 6-deoxy-D-glucose effectively protect the XylE transport system, and arabinose is a weak protecting agent for AraE (Table IV).

Once an effective sugar or analog is chosen the concentration required for protection is ascertained by repeating the above procedure with different concentrations of the same sugar. In the cases of galactose protection of GalP, arabinose protection of AraE, or 6-deoxyglucose protection of XylE, an approximately hyberbolic dependence of protection efficiency on concentration is found (Fig. 7). Relatively high concentrations of arabinose are required to protect AraE (Table IV, Fig. 7) and, unexpectedly,

[73] G. T. Robillard and W. N. Konings, *Biochemistry* **20**, 5025 (1981).

glucose is also effective, while galactose has the reverse effect of increasing the sensitivity of AraE to N-ethylmaleimide (Table IV). These effects are being investigated further.

An alternative method of assessing protection efficiency is to measure the effect of the sugar on the rate at which N-ethylmaleimide inactivates transport activity.[64,74]

Differential Labeling with N[³H]- and [¹⁴C]Ethylmaleimide

A dual isotope labeling strategy for comparing protected with unprotected membrane proteins is desirable to correct for reaction of NEM with —SH groups in proteins other than the one(s) of interest.[74,75] It also enhances the detection of proteins that are only weakly protected and/or present as a small proportion, less than 1%, of the total membrane protein.[74] In our experience, repetition of the experiment with the labels reversed eliminates small but significant differences due mostly to the difficulty of achieving identical ¹⁴C- or ³H-labeled N-ethylmaleimide stock solutions, given the volatility of the n-pentane solvent and the difference in specific activities between commercially available N-[¹⁴C]- and [³H]ethylmaleimide. The procedure is as follows.

Prior Incubation with Unlabeled N-Ethylmaleimide ± Sugar. Four incubation mixtures are made in 1.5 ml Sarstedt microcentrifuge tubes. We generally use vesicles made the day before and kept overnight as the pellet in 100 mM K-phosphate, pH 6.6, at 0–4°, and diluted to 50 mM K-phosphate, pH 6.6, plus 10 mM MgSO₄ just before use (see tabulation below).

Addition	1 (Unprotected) (μl)	2 (Protected) (μl)	3 (Protected) (μl)	4 (Unprotected) (μl)
H₂O	20	0	0	20
0.5 M sugar	0	20	20	0
Vesicles (about 0.75 mg protein)	460	460	460	460
Mix and incubate for 10 min at 25°				
25 mM NEM	20	20	20	20

Mix and incubate for 15 min at 25°
Place all samples in ice for about 0.5 min.

---

[74] A. T. Phillips, this series, Vol. 46, p. 59.
[75] W. M. Hart and E. O. Titus, *J. Biol. Chem.* **248,** 1365 (1973).

Sediment in an Eppendorf microfuge in the cold room, 0–4°, at full speed for 5 min.[72]

All sedimentation and resuspension operations are performed at 0–4°.

First Wash Procedure. Resuspension of each pellet requires vigorous agitation on a vortex mixer followed by addition of ice-cold 0.1 ml 50 mM K-phosphate pH 6.6, 10 mM MgSO$_4$, further vigorous agitation, addition of 0.4 ml cold buffer, and final mixing.

This sedimentation/resuspension is repeated four times.

The final pellet is resuspended in 2 × 0.1 ml buffer.

Incubation with Radioactive N-Ethylmaleimide. The solution of radioactive *N*-ethylmaleimide in *n*-pentane received from the suppliers (New England Nuclear) is transferred to a small dry tube (about 2 ml maximum volume) with an ungreased *ground glass stopper.* This is kept inside another larger tube with a ground glass stopper containing *n*-pentane, which prevents evaporation of solvent from the inner tube. The stock is stored at −20°, when it remains reactive and without appreciable loss of solvent for at least a month. The following conditions were used with *N*-[*ethyl*-2-³H]maleimide of specific activity 48.9 Ci/mmol supplied as 250 μCi in 250 μl *n*-pentane and *N*-[*ethyl*-1-¹⁴C]maleimide of specific activity 42.9 mCi/mmol supplied as 50 μCi in 500 μl *n*-pentane. The additions were made into fresh 1.5 ml microcentrifuge tubes (see tabulation below).

Additions	1 (Unprotected)	2 (Protected)	3 (Protected)	4 (Unprotected)
H$_2$O	0	4.5 μl	0	4.5 μl
25 mM NEM	7.25 μl	2.75 μl	7.25 μl	2.75 μl
Radioactive NEM	25 μl [³H]NEM	50 μl [¹⁴C]NEM	25 μl [³H]NEM	50 μl [¹⁴C]NEM
	Incubated at room temperature for about 10 min with occasional vortex mixing so that the pentane evaporated			
Vesicles	200 μl	200 μl	200 μl	200 μl
	number 1	number 2	number 3	number 4

Mixed and incubated for 60 min at 25°.

Place all samples in ice for about 0.5 min.

Sediment in microfuge at full speed for 5 min.

Note that the total concentration of *N*-ethylmaleimide is below 1 mM to increase the specific activity, and the incubation period is longer to maximize the extent of reaction.

Second Wash Procedure. The pellet is resuspended in 0.1 ml buffer plus 0.4 ml buffer as before.

The sedimentation and resuspension is repeated three times.

A final sedimentation is made from K-phosphate buffer.

SDS–Polyacrylamide Gel Electrophoresis

The efficiency of solubilization in SDS and separation of membrane proteins by polyacrylamide gel electrophoresis is vital for the success of these protein identification methods. We use the buffers of Laemmli[76] with the modifications of Blattler et al.[77] in a slab gel system.[78] A 15% gel gives satisfactory separation of proteins in the M_r range 20,000–60,000 without too much accumulation of low M_r proteins at the gel front. The precise recipe and procedure is given in Table V. In our experience such gels can be stained, destained, and dried without cracking under the conditions described below; alternatively, all the proteins of M_r less than 120,000 can be electroblotted onto cellulose nitrate paper without appreciable loss.[79]

Preparation of Samples for Electrophoresis. Each pellet from above is resuspended in 0.1 ml sodium phosphate, pH 6.6, 10 mM MgSO$_4$.

Pellets 1 (^{3}H-unprotected) and 2 (^{14}C-protected) are mixed.

Pellets 3 (^{3}H-protected) and 4 (^{14}C-unprotected) are mixed.

Additions of 0.1 ml followed by 0.2 ml sodium phosphate, pH 6.6, 10 mM MgSO$_4$, buffer are used to wash any residual sample into the appropriate combined mixture.

Each tube is sedimented once and resuspended in 0.1 ml, followed by 0.15 ml, of sodium phosphate MgSO$_4$ buffer. A small measured volume (1–2 μl) is taken from each for determination of total ^{3}H/^{14}C content.

At this stage each sample can be frozen for storage at $-80°$.

Three parts by volume of each thawed sample containing approximately 120 μg protein (about 40 μl) is diluted with one part by volume of SDS dissolving buffer [4% (w/v) SDS, 40% (v/v) glycerol, 0.5 M mercaptoethanol, 40 mM Tris–HCl, pH 7.2, 0.005% (w/v) bromphenol blue],[76] and incubated at 37° for 15 min (or 60° for 10 min); *the samples must not be boiled,* a common practice for cytoplasmic proteins which may aggregate and precipitate hydrophobic transport proteins.[80] Each dissolved sample is applied to an SDS–PAGE 1-mm-thick slab gel (Table V) and electrophoresed at 20 mA constant current (70–250 V) for 5–8 hr at 4°, for a migration distance of 16–18 cm. Standards of known M_r are included in separate tracks (Table V): β-galactosidase, 116,000; bovine serum albumin, 68,000; ovalbumin, 43,000; carbonic anhydrase, 30,000; lysozyme, 14,400.

[76] U. K. Laemmli, *Nature (London)* **227**, 680 (1970).

[77] D. P. Blattler, F. Garner, K. Van Slyke, and A. Bradley, *J. Chromatog.* **64**, 147 (1972).

[78] F. W. Studier, *J. Mol. Biol.* **79**, 237 (1973).

[79] H. Towbin, T. Staehelin, and J. Gordon, *Proc. Natl. Acad. Sci. U.S.A.* **76**, 4350 (1979).

[80] R. M. Teather, B. Müller-Hill, V. Abrutsch, G. Aichele, and P. Overath, *Mol. Gen. Genet.* **159**, 239 (1978).

TABLE V

SDS–POLYACRYLAMIDE GEL CONDITIONS SUITABLE FOR SEPARATION OF H$^+$–SUGAR
SYMPORT PROTEINS

Running buffer
    28.8 g glycine, 6 g Tris base, 10 ml 10% SDS dissolved in 1 liter distilled H$_2$O, and kept at
    0–4°
Dissolving buffer (4 × final concentration)
    0.2 ml 1 M Tris–Cl, pH 7.2, 2 ml 10% SDS, 2.55 g glycerol, 0.2 ml mercaptoethanol, 0.6 ml
    distilled H$_2$O, 0.1 ml 0.5% bromphenol blue
Molecular mass standards
    Bovine serum albumin (M_r 68,000), ovalbumin (43,000), carbonate dehydratase (30,000),
    and lysozyme (14,400) are dissolved in H$_2$O so that the final concentration of each is 1
    mg/ml; 21 μl of the mixture plus 7 μl dissolving buffer is heated at 100° for 1 min and 20
    μl loaded per gel track
Samples
    20–75 μl of membrane preparation (maximum 120 μg protein) plus 1/3 vol of dissolving
    buffer is warmed at 37° for 15–30 min or 55–60° for 10 min
Running gel
    15 ml 30% acrylamide, 2.6 ml 1% bisacrylamide, 7.5 ml 1.5 M Tris–Cl, pH 8.7, 4.4 ml
    distilled H$_2$O are mixed in a 100 ml capacity Büchner flask and degassed under mild
    vacuum (water pump)
    0.3 ml 10% SDS, 10 μl TEMED, and 0.1 ml 10% ammonium persulfate are added to the
    degassed mixture
    The mixture is poured between the gel plates (1 mm spacer) until about 1 cm below the
    position of the comb
    Overlaid with degassed H$_2$O
    Left for 30 min, or until set
Stacking gel
    1.67 ml 30% acrylamide, 1.3 ml 1% bisacrylamide, 1.25 ml 1 M Tris–Cl, pH 6.8, 4.3 ml
    distilled H$_2$O are mixed in a 100 ml capacity Büchner flask and degassed under mild
    vacuum (water pump)
    0.1 ml 10% SDS, 5 μl TEMED, and 50 μl 10% ammonium persulfate are added to the
    degassed mixture
    The mixture is poured between the gel plates, the comb inserted, and left for about 30 min
    until set
    The comb is carefully removed, and the gel apparatus is placed in the cold room
Electrophoresis
    The samples are loaded into individual tracks from microsyringes, and electrophoresis
    carried out at 0–4°, 20 mA constant current, 60–250 V for 5–6 hr

---

Gel Slicing Procedures. We have used two methods. In the more
convenient one the gel is moved immediately to a Bio-Rad blotting apparatus and the proteins transferred to cellulose nitrate paper by passing a
current of 140–150 mA (constant 30 V) for 12–16 hr at 4°.[79] It is important
that the position of alternate tracks is marked with pyronin dye applied at
the beginning and toward the end of the SDS–PAGE. The outside tracks

contain molecular mass markers. The cellulose nitrate dries quickly in air. One standard track is removed and stained for protein for 15 min using 0.1% amido black, 45% (v/v) methanol, 10% (v/v) glacial acetic acid, 45% (v/v) H_2O, and destained for 1 min each in 3 changes of 90% (v/v) methanol, 2% acetic acid, 8% H_2O, followed by a brief exposure to H_2O which prevents subsequent shrinkage.[81] Each experimental track is cut into 1, 1.5, or 2 mm slices with sharp scissors using graph paper as supporting backing, noting the slice number(s) corresponding to the position of each M_r marker. The dry paper slices are transferred directly to 10 ml scintillant (toluene/methoxyethanol as above is suitable) in fresh scintillation vials which are tightly capped. It is most important that the paper is disintegrated by leaving for 12–24 hr with occasional vigorous shaking. They are then counted for radioactivity. This method is quick and simple, with the possible disadvantage that some proteins, particularly those of high M_r, may be lost in the blotting procedure, but we could not detect residual proteins in the running gel using Coomassie or silver staining.

In the second method[13,14] the gel is fixed and stained with about 250 ml [44% (v/v) methanol, 9% (v/v) acetic acid, 57% (v/v) H_2O, 0.97 g Coomassie Blue R per liter] at room temperature overnight. After destaining in 4–6 changes of [7% (v/v) methanol, 5% (v/v) acetic acid, 78% (v/v) water] shaken slowly at 37° for 5–12 hr, the gel is dried into Whatman No 1 paper on a Bio-Rad gel drier (2 hr at 80°). This temperature is critical if cracking of the gel is to be avoided. Using graph paper stapled to the side each track is cut into sequentially numbered 1 or 2 mm slices, noting the number(s) corresponding to the position of each M_r marker. Each slice is transferred to a new plastic vial (Zinsser polyvials 714), 0.4 ml 30% (v/v) hydrogen peroxide is added, the vial is tightly capped and then all are incubated at 60° for 12 hr or 55° for 24 hr (longer exposures risk loss of ^3H_2O and $^{14}CO_2$) during which each polyacrylamide slice swells and is solubilized. After cooling, scintillant, at least 10 ml of [67% (v/v) xylene, 33% (v/v) Triton X-114 containing 3 g of diphenyloxazole per liter] is added to each. The caps are screwed on tightly, the mixture shaken, and left for 12–24 hr to allow any fluorescence to decay, before counting for radioactivity.

Alternatively, each dried gel slice may be placed in scintillation vials with 25 μl water for 20 min followed by 10 ml Econo-fluor (New England Nuclear) containing 5% Protosol (New England Nuclear).[82] The vials are incubated at 45° with gentle shaking overnight, cooled, and counted.[82]

[81] W. Schaffner and C. Weissmann, *Anal. Biochem.* **56,** 502 (1973).
[82] V.A. Fried, *J. Biol. Chem.* **256,** 244 (1981).

Analysis of Data

The potentially tedious part of this labeling procedure is the determination of $^3H/^{14}C$ content of each slice and its expression as a percentage of the total, with ratios and differences. The calculations are trivial, but the use of a computer is highly advisable to minimize the time required for the calculations and to eliminate arithmetical errors.[83] The ^3H and ^{14}C content of each vial is determined by any of the established dual-isotope counting techniques,[83] compensating for variations in quenching which are generally small. It is important to minimize counting errors by collecting >5000 disintegrations per sample if possible. Two sets of six vials containing either [^3H]leucine or [^{14}C]leucine in the ranges of radioisotope found in the experimental samples are included for calculations of the spill of radioisotope between counter channels. To avoid errors due to differences in quenching, these, and four background vials, are made up and processed in parallel with the experimental samples, each containing a slice from the nonradioactive region of the gel.

The total of disintegrations due to ^3H in all the vials, i.e., for the entire gel track, is calculated, and so also is the total due to ^{14}C. Both the ^3H and ^{14}C in each vial are expressed as a percentage of the total for each isotope over the entire track.[13,14,82,83] If the specifically labeled peak is more than about 2% of the total the corresponding isotope should be normalized to $(100 + x)\%$ where x is the percentage due to the peak.[80,82] For proteins equally labeled by N-[^3H]ethylmaleimide and N-[^{14}C]ethylmaleimide the percentage values will be the same. The percentages are expressed as a difference ($\%^3H - \%^{14}C$) and as a ratio for each vial ($\%^3H/\%^{14}C$). The value of the ratio should be 1.0 for vials containing proteins that are not protected by sugar. The difference is useful because a protected protein present as a small proportion of the total protein may migrate in the same position as an unprotected protein present as a relatively high proportion. Then the change of ratio from a value of 1.0 will be small and obscured, but the absolute difference should still be large and detectable compared with that in other samples.

In addition, the program computes the linear relationship between log M_r and the distance of migration of the protein standards and calculates the apparent M_r of proteins in each slice.[83] A graph of % differences (or ratios) against M_r is plotted automatically to reveal peaks due to protected proteins (Fig. 8).

---

[83] We will supply full details of our counting procedure and the FORTRAN programs used for data analysis.

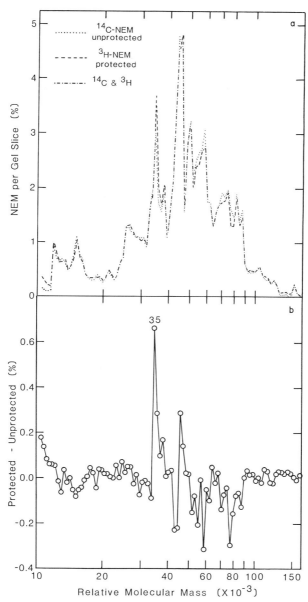

FIG. 8. Identification of arabinose-protected protein using differential labeling with ^{3}H/^{14}C N-ethylmaleimide. The experiments were performed as described in the text using vesicles from arabinose-induced SB5314 (Table III) and 100 mM arabinose for protection. (b) is the difference (^{3}H-protected $-$ ^{14}C-unprotected) of the percentage gel profiles shown in (a). The labels were reversed in the experiment for the percentage gel profiles of (c) so that (d) is the difference (^{14}C-protected $-$ ^{3}H-unprotected).

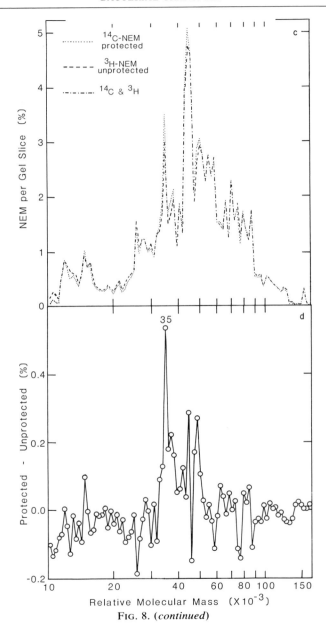

FIG. 8. (*continued*)

Results

One example of the technique is shown in Fig. 8 in which arabinose is used to protect AraE against *N*-ethylmaleimide. From Fig. 8a or 8c it is apparent that simple visual inspection of the ^{3}H or ^{14}C profile, i.e., of percentage per slice versus M_r value, is insufficient to detect a difference reliably. However, transformation to a plot of difference versus M_r reveals an obvious peak of M_r about 35,000 (Fig. 8b), *which occurs also when the labels are reversed* (Fig. 8d). Averaging the two sets, i.e., samples (1 + 2) versus (3 + 4), eliminates some systematic differences, probably due to differences between the labeled NEM solutions, and reinforces the significance of the peak of M_r 35,000.

By a similar dual-isotope technique a galactose-protectable peak of M_r 36,000 was identified in GalP$^+$ vesicles,[84] and a TDG-protectable peak of M_r 31,000 in LacY$^+$ vesicles[84] confirming previous results.[14,62,65,66,80] A xylose protectable peak of M_r 39,000 was found in XylE$^+$ vesicles.[84] The results may be ambiguous if a protein component of the other sugar transport system is present in the vesicles. Repetition of the experiment with vesicles from an appropriate *galP*, *araE*, or *xylE* mutant (this volume [13]) and consequent disappearance of the NEM-protected peak described above should confirm the identification.

Note. The *apparent* M_r determined by SDS–PAGE is less than the true M_r predicted from the nucleotide sequence for these very hydrophobic protein.[54,66,84,92]

Identification of Membrane Proteins by Comparing Wild-Type with Mutant Strains Using Differential Labeling with ^{3}H- and ^{14}C-Labeled Amino Acids

Kolber and Stein[85] devised a technique whereby the difference(s) in proteins between two cultures of *E. coli* could be detected by labeling one with ^{3}H-labeled amino acids and the other with ^{14}C-labeled amino acids. The method was used to identify the LacY protein,[65,66] the AraE and GalP proteins,[13,14] and a LacYLacA hybrid.[82] The procedures take time, but use of the blotting technique and computerized data analysis described above minimizes the inconveniences. It is very sensitive, capable of discriminating protein differences of 0.2% between membranes from different cultures.[13,14,82] Ideally, proteins from a parent strain are compared with proteins from a mutant in which only the gene of interest has been altered so that its protein product is absent or radically changed in M_r

[84] E. O. Davis, P. J. F. Henderson, and S. A. Bradley, unpublished data.
[85] A. R. Kolber and W. D. Stein, *Nature (London)* **209**, 691 (1966).

value, by methods described in this volume [13]. It is also possible to compare induced with uninduced strains.

Labeling Procedure. Two 50–100 ml *E. coli* cultures of the same initial cell density in the range 0.1 to 0.2 mg dry mass per ml are grown. The carbon source is identical with inducer in both, but one is a transport-positive strain, and the other is a mutant isogenic except for a transport-negative gene deletion (or transposon insertion, or Mu fusion, or amber base substitution i.e. a truncated or absent gene product, see this volume [13]) alternatively, the same strain and carbon source are in both cultures, with inducer only in one. After growth for one generation 100–150 μCi of ^{3}H-labeled amino acid mixture is added to one culture and 50–75 μCi of ^{14}C-labeled amino acid mixture to the other. If only one amino acid is used, e.g., leucine, the specific activity will have to be adjusted so that the rate of labeling is independent of the amino acid incorporation and the incorporation of label is directly proportional to the increase in cell mass.[82] Growth is continued for another generation to late exponential phase (about 0.6 mg dry mass/ml). The cultures are harvested separately by centrifugation, and then resuspended together at a concentration of 20–30 mg dry mass/ml in 0.2 *M* Tris–HCl, pH 8.0.

Preparation of Membrane Fractions. Spheroplasts are made by the method of Witholt[51,52] as follows. At zero time an equal volume of 0.2 *M* Tris–HCl, pH 8.0, 1 *M* sucrose, 1 m*M* EDTA is added to the suspension in 0.2 *M* Tris–HCl, pH 8.0, followed 90 sec later by sufficient lysozyme for a final concentration of 0.25 mg/ml, and 45 sec after that by 2 volumes of deionized water. The suspension is gently stirred at room temperature for 30 min when spheroplast formation is verified by phase-contrast microscopy.

Inclusion of 1 m*M* benzamidine and/or 0.1 m*M* phenylmethylsulfonyl fluoride throughout the following procedure may be desirable to inhibit proteases. The spheroplasts are sedimented at 47,000 *g* for 20 min at 4°. The supernatant can be retained as a source of labeled periplasmic proteins.

The spheroplasts are lysed by resuspension in 30 ml water with homogenization or by repeated freeze-thawing.[82] The lysate is centrifuged (47,000 *g*, 20 min, 4°). The supernatant may be retained as a source of labeled cytoplasmic proteins.

The following EDTA wash and freeze-thawing are designed to release extrinsic proteins[86] and may be omitted. The membrane precipitate is resuspended in 30 ml 1 m*M* EDTA (pH 8.0) and incubated at 37° for 30 min. This is centrifuged (47,000 *g*, 20 min, 4°) and the supernatant dis-

[86] A. B. Archer, A. Rodwell, and E. S. Rodwell, *Biochim. Biophys. Acta* **513**, 268 (1978).

carded. The pellet is placed in a freezing mixture of dry ice, 96% alcohol for 3 min, thawed at room temperature, and washed in H_2O. This freeze-thaw-wash is repeated once.

Alternatively, after the spheroplasts have been lysed, the suspension is centrifuged (47,000 g, 20 min, 4°) and the pellet is washed four times in 0.1 M sodium phosphate, pH 7.2, 1 mM 2-mercaptoethanol, finally resuspended to a protein concentration of 6–8 mg/ml in the same buffer and stored at −80°.

The method is preferable to making membranes as vesicles using osmotic lysis[44] or a French pressure cell[46] because the substantially higher yield minimizes wastage of expensive labeled amino acids.

Separation of Proteins and Data Analysis. Samples of radioactive amino acid labeled membrane equivalent to 100–150 μg protein are solubilized in SDS and treated exactly as described above for preparations labeled with N-ethylmaleimide.

Results

By this technique a mutant with a deletion in the *araE* gene was found to have lost a protein of M_r 36,000, which was regained when the deletion strain was reinfected with a λ phage carrying the *araE* gene[13]; also an arabinose-induced strain contained a labeled protein of M_r 36,000 absent from the uninduced strain.[13] Loss of a *galP* protein of M_r 36,000 from a Tn10 insertion mutant was also detected by dual-isotope labeling with amino acids (Fig. 9).[14,87]

Identification of Membrane Proteins by Comparing Wild-Type with Mutant Strains Using Two-Dimensional Gel Electrophoresis

The method is essentially the nonequilibrium pH gradient electrophoresis (NEPHGE) followed by SDS–PAGE of O'Farrell,[88,89] with modifications to avoid the use of special apparatus and to permit the loading of larger amounts of protein, thereby facilitating the detection of the protein spots by Coomassie blue staining.

Preparation of Membrane Samples. Membranes prepared as described above are sedimented (47,000 g, 20 min, 4°) and resuspended in 100 μl samples at 8–20 mg/ml in 9.5 M urea/2% (w/v) NP40/2% (v/v)

[87] P. J. F. Henderson, H. Hirata, P. Horne, M. C. Jones-Mortimer, T. Kaethner, and A. J. S. Macpherson, *in* "Biochemistry of Metabolic Processes" D. L. F. Lennon, F. W. Stratman, and R. N. Zahlten, eds.), p. 339. Elsevier, Amsterdam, 1983.

[88] P. Z. O'Farrell, H. M. Goodman, and P. H. O'Farrell, *Cell* 12, 1133 (1977).

[89] G. F-L. Ames and K. Nikaido, *Biochemistry* 15, 616 (1976).

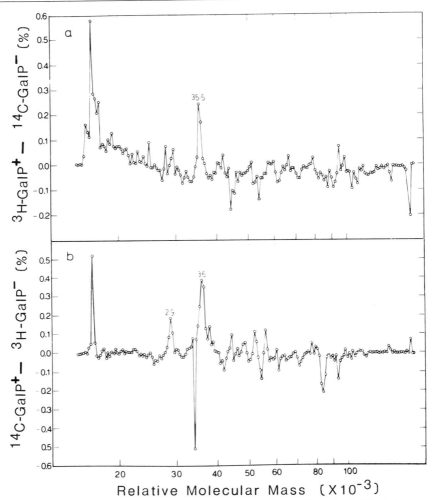

FIG. 9. Identification of a GalP protein using differential labeling of a GalP$^+$ strain and a GalP$^-$ Tn10 mutant with ^{3}H/^{14}C-labeled amino acids and one-dimensional gel electrophoresis. The procedure is described in the text and by A. Macpherson, M. C. Jones-Mortimer, P. Horne, and P. J. F. Henderson, *J. Biol. Chem.* **258,** 4390 (1983). (a) ^{3}H-labeled amino acids in the parent (JM1424) and ^{14}C-labeled amino acids in the mutant (JM2053). (b) Labels reversed.

ampholines (pH 3.5–10.0). To ensure that this "lysis buffer" is saturated with respect to urea, several crystals are added to each solubilized sample. These are incubated at 37° for 60 min, when 70 μl of each is loaded for nonequilibrium pH gradient electrophoresis.[88]

In some experiments one of the following strategies was used to reduce the amount of protein remaining as a precipitate at the top of the tube gel after the first dimension of electrophoresis.

Sodium dodecyl sulfate was included in the lysis buffer at a final concentration of 1–2% (w/v). This increased the number of proteins resolved by the technique (as reported by Ames and Nikaido[89]).

The protein solution in the lysis buffer was sonicated in a bath sonifier under N_2 gas for 10 min before the incubation at 37° for 1 hr. This procedure increased the solubilization of all proteins, but did not affect the number of species resolved.

The Nonidet P40 detergent in the gels and the lysis buffer was replaced by Lubrol PX at identical concentrations. This effects the resolution of a different spectrum of proteins.

To ensure that no inducible protein was left at the origin of the nonequilibrium pH gradient tube gel, the material precipitated from a dual-isotope labeled preparation during the first dimension of two-dimensional electrophoresis was collected and solubilized by one-dimensional SDS–polyacrylamide (15%) gel electrophoresis. The results of dual-isotope analysis of this gel indicated that no major inducible protein was present, although minor components were seen at M_r 18,000 and 61,000. These are probably insoluble in the NP-40 lysis buffer used in the first dimension of the two-dimensional separation.

Isoelectric Focusing in the First Dimension. Capillary tubes (length 13 cm) for the first dimension gels are cut from 0.5 ml pipets (of 5 mm diameter and 2 mm bore). Between each experiment every tube is acid washed, thoroughly rinsed in distilled water, and dried before use. The bottom of each is then *lightly* greased and sealed with a small piece of Parafilm.

For each batch of six tubes it is convenient to prepare 3 ml of 9.5 M urea/2% (w/v) NP40/2% (w/v) ampholines (pH 3.5–10.0)/4% (w/v) acrylamide/0.25% (w/v) bisacrylamide. To initiate polymerization, 24 μl of 10% (w/v) ammonium persulfate and 20 μl of 10% (v/v) tetramethyldiaminoethane are added to the gel solution just before pouring. Owing to the narrow diameter of the tubes, it is necessary to pour the gel using a 1-ml Brunswick 501-TB syringe fitted with a 20-gauge metal needle. To avoid air-locks one drop of gel solution is allowed to run slowly to the bottom of the tube, before filling rapidly to a premeasured distance of 9 cm. The gel solution is then overlaid with about 20 μl of 8 M urea/2% (w/v) NP40 and allowed to set for 2 hr.

After polymerization the overlay solution is discarded, the Parafilm removed, and each tube placed in a tube-gel electrophoresis tank. Each sample (2 mg total protein in a final volume of 60 μl, prepared as described above) is loaded, and overlaid with 20 μl of 8 M urea/2% (w/v) NP40/2%

(w/v) ampholines (pH 3.5–10.0). This is in turn overlaid with 0.01 M orthophosphoric acid until the tube is full. The upper reservoir is filled with 0.01 M orthophosphoric acid and the lower reservoir with 0.02 M sodium hydroxide, and the tank is connected (upper reservoir positive) to a high tension power supply. Electrophoresis is carried out at constant voltage at 250 V for 1 hr and subsequently at 450 V for a total of 2000 V-hr.

Upon completion of nonequilibrium pH gradient gel electrophoresis each gel is extruded from its capillary tube; to do this it is necessary to squirt distilled water between the gel and the capillary tube from both ends using a 100 μl Terumo syringe, then pressure is applied with a bulb-type pipet filler. The pH gradient formed during electrophoresis is measured using a surface electrode; this is essentially linear, ranging from pH 4.7 to 8.4. Each tube gel is then equilibrated in SDS buffer [10% (v/v) glycerol/5% (v/v) 2-mercaptoethanol/2.3% (w/v) sodium dodecyl sulfate/ 0.0625 M Tris–HCl (pH 6.8)] for 2 hr at room temperature with gentle agitation.

SDS–PAGE in the Second Dimension. Fifteen percent SDS–polyacrylamide gels are made up as described in Table V. The arrangement used consists of two slab gels run in parallel on the same apparatus, by staggering two notched glass plates with two sets of spacers (1.5 mm thick) on one back plate as shown in Fig. 10. The positions of the stacking and the separating gels are also staggered to correspond with this arrangement. It is not necessary to use plates specially cut or bevelled to form a well for the tube gel. Once the polyacrylamide gels have set, the gel plate assembly is clamped at an angle of 30° to the vertical, and using a length of Parafilm, the tube gels are laid along the edges of the notched plates. To form a well for the marker solution, a distance piece is inserted between the plates beside one of the tube gels. Molten agarose buffer [SDS buffer containing 1% (w/v) agarose] is then poured between each stacking gel and tube gel. The addition of agarose is continued until both gels are covered, as shown in Fig. 10. Once the agarose has set, the gel plates are clamped to the slab-gel tank, running buffer and markers are added (above), and electrophoresis is carried out at 40 mA constant current (60–250 V) for approximately 6 hr.

Results

Provided two gels are run exactly in parallel in both dimensions, one containing, for example, a GalP$^+$ strain and the other an otherwise isogenic GalP$^-$ strain, the patterns of spots are very similar (see Fig. 11). It is visually convenient to relate them to a line of spots as shown, which are

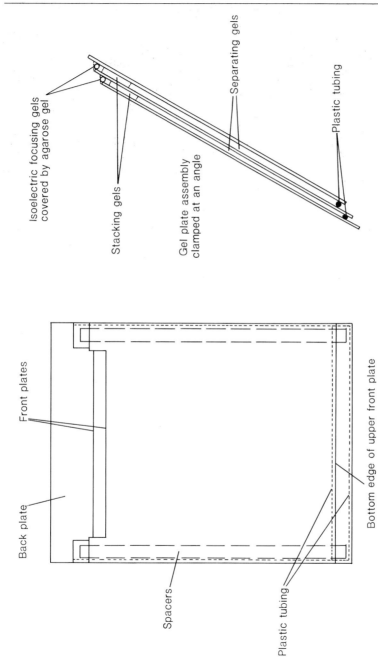

FIG. 10. Arrangement of second dimension gel plates to ensure reproducible migration of proteins. A normal vertical slab gel tank is used with the staggered sandwich arrangement shown from the front (left) and side (right). Details are described in the text.

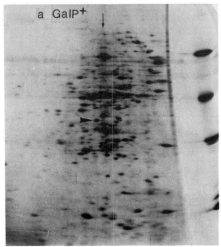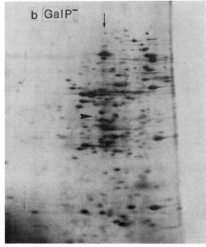

FIG. 11. Comparison of membrane proteins from GalP⁺ *E. coli* (a) with membrane proteins from a GalP⁻ Tn*10* insertion mutant (b) by two-dimensional gel electrophoresis. Experimental details are given in the text and by A. Macpherson, M. C. Jones-Mortimer, P. Horne, and P. J. F. Henderson, *J. Biol. Chem.* **258**, 4390 (1983). The position of the GalP protein is shown by a horizontal arrow; its apparent M_r is 37,000, comparable with the value of 36,000 deduced by one-dimensional gel electrophoresis of dual-isotope labeled proteins (Fig. 9). The vertical arrow indicates a line of spots which provide reference positions of known M_r common to all preparations.

also used as M_r markers. Thus a spot present only in the GalP⁺ strain was reproducibly detected, of M_r 36,000 and focusing at pH 5.5–5.8 (Fig. 11).[14] Similarly, an arabinose-inducible spot was detected, coincidentally of similar M_r to the *galP* gene product, but with a different isoelectric focusing position at a pH of about 6.4.[13]

Identification by Combining Dual-Isotope-Labeling of E. coli Membrane Proteins with Two-Dimensional Gel Electrophoresis

A powerful method of detecting nonabundant membrane protein is to unite the two approaches of dual-isotope labeling and two-dimensional gel electrophoresis. The procedure is illustrated by the following experiment used in the identification of AraE.

Dual-isotope-labeled membranes of succinate-grown (³H-labeled amino acid induced: ¹⁴C-labeled amino acid uninduced) *E. coli* strain SB5314 are solubilized, separated into discrete protein spots by two-dimensional NEPHGE, fixed, and stained as described above. The dried gel is cut into 350 (0.25 cm²) sections. These are dissolved and counted for radioactivity (above), and the results processed using a computer program written in FORTRAN; this is similar in principle to the program used to process one-dimensional dual-isotope data (above) but the data and results for the dual-isotope gel sections are held in a series of two-dimensional arrays. The radioactivity of the ³H and ¹⁴C isotopes in each section are expressed as percentages with respect to the entire gel, and printed out as contour diagrams (not shown) with the same orientation as the two-dimensional gel photograph in Fig. 11. The average percentage ratio (³H/¹⁴C) is also calculated for each section, and presented as another contour diagram (Fig. 12). This showed that the major arabinose-inducible protein was at the position deduced for AraE by inspection of Coomassie-stained gels (M_r 36,000, apparent pI 6.4), while other minor inducible components were observed at M_r 16,000, 18,000, 42,000, and 46,000, and a negatively induced protein at M_r 56,000 (Fig. 12). The identities of these proteins are unknown, but the M_r 42,000 and 46,000 proteins are in the approximate position of the *araG* gene product, judged from the relative positions of the major proteins resolved by nonequilibrium two-dimensional gel electrophoresis.[16] However, different techniques for solubilization of the membrane proteins prior to separation do result in qualitative differences between the species resolved[13,16] so the assignment of AraG remains equivocal in the absence of data from mutants with a lesion at the *araG* locus.

The inducibility of the section containing AraE (×5.26; Fig. 12), assessed by the dual-isotope technique, is less than expected from the inducibility of the AraE transport system (about ×50). This can be accounted for by the presence of other noninducible proteins which migrate close to AraE in this section.

Reconstitution of GalP or AraE Transport Activity of *E. coli* into Liposomes Made from Soybean Phospholipids

Principle. The principle of the method was reviewed by Racker *et al.*[90] It was modified by Newman *et al.*[91,92] who included *E. coli* lipid during the solubilization in octylglucoside, a critical step for reconstitution of LacY

[90] E. Racker, B. Violand, S. O'Neal, M. Alfonzo, and J. Telford, *Arch. Biochem. Biophys.* **198,** 470 (1979).

[91] M. J. Newman and T. H. Wilson, *J. Biol. Chem.* **255,** 10583 (1980).

[92] M. J. Newman, D. L. Foster, T. H. Wilson, and H. R. Kaback, *J. Biol. Chem.* **256,** 11804 (1981).

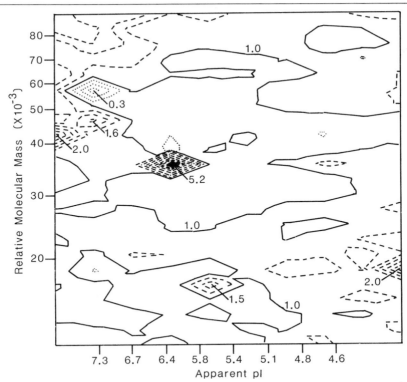

FIG. 12. Identification of an arabinose-inducible membrane protein by two-dimensional gel electrophoresis of a dual-isotope labeled membrane preparation of *E. coli* SB5314. An arabinose-induced and an uninduced culture of *E. coli* SB5314 were grown in parallel on 22.5 m*M* Na-succinate. After growth for 1 generation 100 μCi of [3H]-labeled amino acid mixture was added to the induced culture and 50 μCi of [14C]-labeled amino acid mixture to the uninduced culture; growth was continued for another generation to late exponential phase (to about 0.6 mg dry mass/ml). The bacteria from the two cultures were harvested, pooled, a single membrane preparation was made, and the proteins were subjected to two-dimensional gel electrophoresis (text). The dried gel was cut into 0.25 cm² sections each of which was dissolved and counted for radioactivity (text). Contours of [3H] radioactivity and [14C] radioactivity were plotted as percentage of the total for each isotope (A. J. S. Macpherson, Ph.D. thesis, University of Cambridge, 1982) and as [3H]/[14C] isotopic ratio in the figure above (normally plotted in different colours for clarity). —, ratio of 1.0; ---, ratios greater than 1.0 at intervals of 0.2; ···, ratios less than 1.0 at intervals of 0.2. The major induced peak of M_r 36,000 and apparent p*I* 6.4 corresponds to the position of AraE [A. J. S. Macpherson, M. C. Jones-Mortimer, and P. J. F. Henderson, *Biochem. J.* **196**, 269 (1981)].

activity into liposomes made of *E. coli* lipid (see also[24,93,94]). For reconstitution of galactose or arabinose transport activity soybean phospholipids are just as effective as *E. coli* lipid, and the liposomes can be conveniently made by probe, rather than bath, sonication.[95] Just like the LacY reconstitution procedure, the lipid must be purified by acetone precipitation and ether extraction,[91] it is essential to have added lipid present during solubilization, and a concentration of 1.25% octylglucoside is critical for the solubilization step.[95]

Sources of Membrane Protein. *E coli* strain JM1424 (Table III) has elevated levels of inducible GalP activity, and strain SB5314 has elevated levels of inducible AraE activity (Table III). Spheroplasts of each strain, prepared by the method of Witholt *et al.*[51,52] as described above, are made into right-side out vesicles by the method of Kaback,[44] frozen rapidly and stored at −80°. Inverted French pressure vesicles[46] would probably be equally effective, but it is convenient to be able to check the GalP or AraE activity of a sample of the right-side out vesicles by measuring ascorbate + PMS-energized uptake of the appropriate radioactive substrate (see above).

Preparation of Liposomes. A Branson sonifier 200 fitted with a microprobe is used instead of a bath-type sonicator.[95,96] The smallest practicable volume is about 0.15 ml, and a typical procedure is as follows: 140 μl of 50 mg/ml lipid (either purified asolectin or *E. coli* lipid), 7 μl 1 *M* potassium phosphate, pH 7.5, 3.5 μl 1 *M* galactose or arabinose, and 25 μl 50 m*M* potassium phosphate, pH 7.5, in a small test tube (0.8 cm i.d. × 4.5 cm) at 0° are sonicated for 3 min using 50% duty cycle pulses at a power setting of 2. The pulsed sonication, approximately 1 sec on and 1 sec off, and low power are critical features with small volumes. The opacity of the solution is visibly reduced by sonication. After a 1 min pause the pulsed sonication is repeated for 3 min, continued until the solution becomes clear. A further 3 min is sometimes necessary with *E. coli* lipid. A stream of argon must be directed over the mixture throughout.

Solubilization and Reconstitution of Galactose Transport Activity. The method is very similar to that described for the lactose and melibiose transport systems,[91,92,94,97] although for some experiments it is practicable

[93] C.-C. Chen and T. H. Wilson, *J. Biol. Chem.* **259**, 10150 (1984).

[94] D. L. Foster, M. L. Garcia, M. J. Newman, L. Patel, and H. R. Kaback, *Biochemistry* **21**, 5634 (1982).

[95] P. J. F. Henderson, Y. Kagawa, and H. Hirata, *Biochim. Biophys. Acta* **732**, 204 (1983).

[96] M. Kasahara and P. Hinkle, *J. Biol. Chem.* **252**, 7384 (1977).

[97] T. Tsuchiya, K. Ottina, Y. Moriyama, M. J. Newman, and T. H. Wilson, *J. Biol. Chem.* **257**, 5125 (1982).

to reduce the volumes of the components to the following. Membrane vesicles (0.25 mg protein in 34 μl), 15 μl 50 mg/ml lipid, 112.5 μl 50 mM potassium phosphate (pH 7.5), 3.7 μl 1 M galactose are thoroughly mixed, and 16.5 μl of 13.6% octylglucoside added on a vortex mixer. After 15 min the mixture is sedimented at 100,000 g for 1 hr. Liposomes (about 1.3 mg lipid in 33 μl) are treated with 3.25 μl of 13.6% octylglucoside and then mixed with 110 μl of the supernatant removed from the sedimentation above, taking care not to disturb the pellet. After 15 min the mixture is diluted with 5 ml of 50 mM potassium phosphate pH 7.5, 20 mM galactose, 1 mM dithiothreitol, and then centrifuged at 100,000 g for 1 hr. The centrifuge tube is carefully drained and residual fluid removed from the walls of the tube with tissue. The pellet of proteoliposomes is resuspended in 25 μl of 50 mM potassium phosphate, pH 7.5, 20 mM galactose, 1 mM dithiothreitol using a small glass rod and two passes into a 50 μl microsyringe. All steps are carried out at 0–4°. The diameters of the proteoliposomes are 20–70 nm when viewed by negative staining in the electron microscope.

Transport Assay. The method follows that of Newman and Wilson,[91] which was based on the entrance counterflow technique described by Wong and Wilson.[98] The volumes given below are convenient for assays with two or three timed samples, and the procedure is scaled up correspondingly for experiments in which more samples are required. At zero time proteoliposomes (0.6–1.5 μg protein in 5 μl) are diluted into a mixture of 0.235 ml potassium phosphate, pH 7.5, 2 mM dithiothreitol, with 0.01 ml [^{3}H]galactose (1 μCi) or a corresponding amount of labeled arabinose. Note that the proteoliposome suspension contains 20 mM sugar diluted to 0.4 mM at the start of the assay; at 15 sec a 0.1 ml sample is filtered (Millipore GSW filters, pore size 0.22 μm, prewashed in the medium below) and washed with about 2 × 2 ml ice cold 50 mM potassium phosphate, 1 mM dithiothreitol. A second sample is taken at 2 min. The filtration plus washing procedure takes about 45 sec. The radioactivity appearing in the proteoliposomes retained on the filter is determined by liquid scintillation counting. The assays are carried out at 15–18° to facilitate measurement of the initial rate of sugar uptake.

The method of Schaffner and Weissman[81] is used to measure the protein incorporated into the proteoliposomes. This is sufficiently sensitive and is immune to interference by lipid.

The sugar specificity can be investigated by including unlabeled other sugars at 4–20 mM final concentration before addition of proteoliposomes and measuring the extent of inhibition of galactose[95] or arabinose uptake.

[98] P. T. S. Wong and T. H. Wilson, *Biochim. Biophys. Acta* **196,** 336 (1970).

It corresponds with the specificity in intact cells, and for the GalP system, confirms that a component of MglP is not interfering. A strain with amplified XylE activity is not yet available, so its reconstitution has not been investigated.

Acknowledgments

We are grateful to the Science and Engineering Research Council for financial support. A. T. Paxton and S. A. Bradley provided expert technical assistance. We are grateful to Dr. M. C. Jones-Mortimer for help in mutagenesis of *E. coli* strains.

[32] Purification, Reconstitution, and Characterization of the *lac* Permease of *Escherichia coli*

By PAUL VIITANEN, MICHAEL J. NEWMAN, DAVID L. FOSTER, T. HASTINGS WILSON, and H. RONALD KABACK

Transport of β-galactosides in *Escherichia coli* is catalyzed by the product of the *lacY* gene,[1] the *lac* permease or *lac* carrier protein, which translocates substrate with H^+ in a symport (cotransport) reaction.[2] Accordingly, in the presence of a proton electrochemical gradient ($\Delta \bar{\mu}_{H^+}$), the permease utilizes the energy released from the downhill movement of hydrogen ion to drive uphill translocation of β-galactosides (see Kaback[3] for a review). Conversely, energy released from downhill movement of substrate along a concentration gradient is used to drive uphill movement of H^+ with generation of $\Delta \bar{\mu}_{H^+}$, the polarity of which depends on the direction of the substrate concentration gradient. The *lac* permease was identified as a membrane protein chemically in 1965[4] and functionally in 1970.[5] Eight years later, in rapid succession, the *lacY* gene was cloned into a recombinant plasmid, its product amplified[6] and synthesized *in vitro*,[7] and the sequence of the protein was deduced from the DNA se-

[1] G. N. Cohen and J. Monod, *Bacteriol. Rev.* **21**, 169 (1957).

[2] P. Mitchell, *Biochem. Soc. Symp.* **22**, 142 (1963).

[3] H. R. Kaback, *J. Membr. Biol.* **76**, 95 (1983).

[4] C. F. Fox and E. P. Kennedy, *Proc. Natl. Acad. Sci. U.S.A.* **54**, 891 (1965).

[5] E. M. Barnes and H. R. Kaback, *Proc. Natl. Acad. Sci. U.S.A.* **66**, 1190 (1970).

[6] R. M. Teather, B. Müller-Hill, V. Abrutsch, G. Aichele, and P. Overath, *Mol. Gen. Genet.* **159**, 239 (1978).

[7] R. Ehring, K. Beyreuther, J. K. Wright, and P. Overath, *Nature (London)* **283**, 537 (1980).

quence.[8] Shortly thereafter, it was demonstrated[9] that lactose transport activity can be solubilized and reconstituted into proteoliposomes, and a highly specific photoaffinity label for the permease was developed.[10] Subsequently, the use of these developments in concert led to the purification of a single protein, its identification as the product of the *lacY* gene, and the demonstration that it is the only polypeptide in the cytoplasmic membrane required for lactose/H^+ symport.[11,12]

The purpose of this article is to describe a detailed purification scheme that takes the *lac* permease from membrane to molecule in a fully functional state. As an overview, membranes from a strain of *E. coli* containing multiple copies of the *lacY* gene are first sequentially extracted with high concentrations of urea and cholate to affect about a 3-fold purification of the permease *in situ*. These steps are based on observations[13,14] demonstrating that treatment of right-side-out membrane vesicles with these reagents extracts considerable amounts of protein with little or no direct effect on β-galactoside transport activity. The membranes are then extracted with octyl-β-D-glucopyranoside (octylglucoside) in the presence of *E. coli* phospholipids which solubilizes most of the permease, but only about 15% of the remaining protein. In the final purification step, the permease is subjected to isocratic elution from DEAE-Sepharose. The overall strategy results in about a 30-fold purification of the permease relative to crude membranes in approximately 50% yield. Finally, reconstitution of purified *lac* permease into proteoliposomes and a variety of functional assays are described.

An alternative purification and reconstitution procedure for the *lac* permease has also been described.[15,16]

Materials. Valinomycin, carbonyl cyanide *m*-chlorophenylhydrazone (CCCP), sodium cholate, and octylglucoside were obtained from Calbiochem. *p*-Chloromercuribenzenesulfonate (*p*CMBS), dithiothreitol (DTT), lactose, deoxyribonuclease I (DNase I), phenylmethylsulfonyl fluoride

[8] D. E. Büchel, B. Groneborn, and B. Müller-Hill, *Nature (London)* **283**, 541 (1980).
[9] M. J. Newman and T. H. Wilson, *J. Biol. Chem.* **255**, 10583 (1979).
[10] G. J. Kaczorowski, G. LeBlanc, and H. R. Kaback, *Proc. Natl. Acad. Sci. U.S.A.* **77**, 6319 (1980).
[11] M. J. Newman, D. Foster, T. H. Wilson, and H. R. Kaback, *J. Biol. Chem.* **256**, 11804 (1981).
[12] D. L. Foster, M. L. Garcia, M. J. Newman, L. Patel, and H. R. Kaback, *Biochemistry* **21**, 5634 (1982).
[13] L. Patel, S. Schuldiner, and H. R. Kaback, *Proc. Natl. Acad. Sci. U.S.A.* **72**, 3387 (1975).
[14] E. Padan, S. Schuldiner, and H. R. Kaback, *Biochem. Biophys. Res. Commun.* **91**, 854 (1979).
[15] J. K. Wright, R. M. Teather, and P. Overath, this series, Vol. 97, p. 158.
[16] J. K. Wright and P. Overath, *Eur. J. Biochem.* **138**, 497 (1984).

(PMSF), β-D-galactopyranosyl-1-thio-β-D-galactopyranoside (TDG), and isopropyl-β-D-thiogalactopyranoside (IPTG) were purchased from Sigma. 2-Mercaptoethanol (BME) was from Eastman and [1-¹⁴C]lactose was obtained from Amersham. Diethyl ether, chloroform, acetone, benzene, and methanol (all distilled in glass) were purchased from Burdick and Jackson. DEAE-Sepharose CL-6B was from Pharmacia and urea (ultrapure) was from Bethesda Research Laboratories. All electrophoresis reagents and molecular weight standards were obtained from Bio-Rad. Membrane filters used for proteoliposome transport assays were type GSTF (02500; 0.22 μm) from Millipore. *p*-Nitro[2-³H]phenyl-α-D-galactopyranoside (NPG; 30 Ci/mmol) was synthesized by Yu-Ying Liu (Isotope Synthesis Group, Hoffmann-La Roche Inc.) under the direction of Arnold Liebman. Nigericin was provided by John Wesley, Hoffmann-La Roche Inc. *E. coli* T206 was contributed by Peter Overath, Max-Planck-Institut für Biologie, Tübingen, West Germany.

Determination of Protein. Protein is assayed throughout the purification and reconstitution as described by Schaffner and Weissmann[17] employing the modifications suggested by Newman et al.[11] For determination of protein in proteoliposomes, a background blank is routinely subtracted. The blank value is determined by processing liposomes "reconstituted" without added protein. It should be emphasized that the Lowry procedure[18] for measuring protein yields values that are about 1.5-fold higher than those obtained with the present method. A similar overestimation has been observed in another system.[19] As documented elsewhere[11] the accuracy of the modified Schaffner and Weissmann procedure has been confirmed by amino acid analysis of purified *lac* permease.

Growth of Bacteria. *E. coli* T206 which harbors the *lacY* gene in a multicopy recombinant plasmid is grown and induced with IPTG as described.[20] The cells are harvested by centrifugation, washed once in 100 m*M* potassium phosphate (K-phosphate, pH 7.5), frozen rapidly in liquid nitrogen, and stored at −180° until use.

Preparation of Membranes. Membranes are prepared essentially as described by Newman and Wilson.[9] Frozen cells are thawed at room temperature in 50 m*M* K-phosphate (pH 7.5) containing 10 m*M* MgSO₄ and collected by centrifugation. All subsequent steps are performed at

[17] W. Schaffner and C. Weissmann, *Anal. Biochem.* **56,** 502 (1973).
[18] O. H. Lowry, N. J. Rosebrough, A. L. Farr, and R. J. Randall, *J. Biol. Chem.* **193,** 265 (1951).
[19] W. H. M. Peters, H. G. P. Swarts, J. J. H. H. M. de Pont, F. M. A. H. Schuumans Stekhovan, and S. L. Bonting, *Nature (London)* **290,** 338 (1981).
[20] R. M. Teather, J. Bramhall, I. Riede, J. K. Wright, M. Fürst, G. Aichele, V. Wilhelm, and P. Overath, *Eur. J. Biochem.* **108,** 223 (1980).

0–4°. The cell pellet is resuspended (0.2 g wet weight/ml) by manual homogenization in 50 mM K-phosphate (pH 7.5) containing 1.0 mM DTT, 5.0 mM MgSO$_4$, 20 mM lactose, 30 μg/ml DNase I, and 0.5 mM PMSF. Disruption of cells is accomplished by subjecting the suspension to a single passage through a motor-driven French pressure cell (Travenol Laboratories, Inc.; Catalogue No. J5-598A) at 20,000 psi. Care is taken to maintain the cell lysate at 0–4° by use of a cooling coil connected to the outlet of the French pressure cell. Undisrupted cells are removed from the lysate by two low speed centrifugations (10,000 g for 20 min). The membrane fraction is then recovered from the supernatant by centrifugation at 140,000 g for 2 hr. Membranes are washed once by resuspension in 50 mM K-phosphate (pH 7.5) containing 1.0 mM DTT, 20 mM lactose, and 0.5 mM PMSF and centrifugation; the volume of buffer used for the wash is equivalent to the original volume prior to high-speed centrifugation. The washed membranes are then resuspended in the same solution to a protein concentration of 30–75 mg/ml. Aliquots of the suspension are frozen rapidly in liquid nitrogen and stored at −180° for subsequent use. The approximate yield of membrane protein is 1.5–2.0 g/100 g (wet weight) of cells.

Urea, Cholate, and Octylglucoside Extraction. The procedure described is essentially that of Newman et al.[11] incorporating the modifications suggested by Foster et al.[12] In order to follow the distribution and recovery of the *lac* permease throughout purification, a small portion of the membrane fraction prepared as described above is photoaffinity labeled with [³H]NPG as described by Kaczorowski et al.[10] and mixed with unlabeled membranes to yield a final specific activity of 10–50 nCi/mg of protein.

All steps are performed at 0–4° unless noted otherwise with 1.0 g total membrane protein as starting material. Membrane protein concentration is adjusted to 10 mg/ml by the addition of 50 mM K-phosphate (pH 7.5) containing 0.5 mM DTT and 10 mM lactose (final volume, 100 ml). An equal volume of freshly prepared 10 M urea containing 1.0 mM PMSF (at room temperature) is then slowly added to the ice-cold suspension while the latter is stirred vigorously on a magnetic stirrer. After addition of urea/PMSF, the suspension is stirred for an additional 10 min and centrifuged for 2 hr at 35,000 rpm in a Beckman Type 35 rotor. The supernatant is discarded, and the pellet is drained and resuspended by homogenization in 50 mM K-phosphate (pH 7.5) (final volume, 140 ml). While the suspension is mixed on a magnetic stirrer at 4°, 60 ml of 20% sodium cholate (w/v) (at room temperature) is slowly added. Stirring is continued for an additional 20 min, and the suspension is then centrifuged for 2 hr as described. The supernatant is discarded and the cholate-extracted residue is washed once by resuspension in about 150 ml of 10 mM K-phosphate (pH 5.9) at 4° and centrifugation for 2 hr. The washed pellet may be processed immediately or overlayed with a few milliliters of wash buffer and stored in an ice bucket overnight.

In the next step, urea/cholate extracted membranes are resuspended to a final volume of 81.3 ml in 10 mM K-phosphate (pH 5.9) at 4°. While the suspension is mixed on a magnetic stirrer, the following additions are made: 96.3 μl of 1 M DTT, 0.72 g of lactose, and 7.3 ml of acetone/ether-washed *E. coli* lipid at 50 mg/ml in 2.0 mM BME (cf. below for lipid preparation). When the ingredients are dissolved, 8.0 ml of 15% octylglucoside (w/v) in 10 mM K-phosphate (pH 5.9) is added slowly with continued stirring. Ten minutes after addition of octylglucoside, the suspension is centrifuged for 90 min at 40,000 rpm in a Beckman type 60 Ti rotor. The supernatant (~90 ml) contains 60–65% of the [^{3}H]NPG present in the original membrane fraction. If necessary, the pH of the extract is adjusted to 6.0 (at 4°) by the dropwise addition of 10 mM K$_2$HPO$_4$ containing 1.0 mM DTT, 20 mM lactose, 0.25 mg acetone/ether-washed *E. coli* lipid/ml, and 1.25% octylglucoside (w/v). The octylglucoside extract (0.5–0.6 mg of protein/ml) is rapidly frozen in liquid nitrogen and may be stored at $-180°$ for at least 6 months without significant loss of activity.

DEAE-Sepharose Chromatography. DEAE-Sepharose column chromatography is employed in order to isolate the *lac* permease from other protein contaminants present in the octylglucoside extract. Under the conditions described the *lac* permease is the only polypeptide to emerge from the column in a process that is essentially an isocratic elution. The procedure was first described by Newman *et al.*[11] for the isolation of small amounts of permease (<0.2 mg of permease) and was subsequently scaled-up and modified by Foster *et al.*[12] to yield larger quantities (4–5 mg of permease). The latter modifications are included in the description that follows.

DEAE-Sepharose CL-6B is extensively washed with deionized water and equilibrated overnight at room temperature in 10 volumes of 1.0 M K-phosphate (pH 6.0). The resin is then drained, and thoroughly equilibrated with 10 mM K-phosphate (pH 6.0) at ambient temperature. Equilibration is complete when the pH of the fluid above the resin is 6.00 at room temperature and may be facilitated by addition of small amounts of 10 mM K$_2$HPO$_4$. After equilibration, the volume of fluid is adjusted such that the settled resin volume comprises 71% of the total volume. The slurry is then degassed at room temperature and used to pour a 2.5 × 15 cm column (bed volume 75 ml; 106 ml of the degassed 71% slurry is required). The column is washed with degassed 10 mM K-phosphate (pH 6.0) and equilibrated to 4°C.

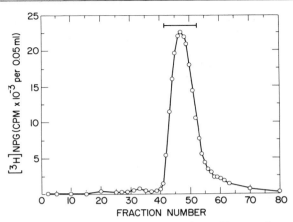

FIG. 1. Elution profile of the DEAE-Sepharose column. Twenty-nine milliliters of octylglucoside extract (0.55 mg protein/ml containing 1.19×10^7 cpm of [³H]NPG-photolabeled *lac* permease) was applied to a 2.5×15 cm DEAE-Sepharose column. The column was developed at 4° as described and fractions (2 ml) were collected. Aliquots (50 μl) of each fraction were assayed for radioactivity by liquid scintillation spectrometry. The fractions under the bar (41–53) contained 70% of the radioactivity applied to the column and yielded about 5 mg of purified *lac* permease.

Prior to sample application, 100 ml of degassed column buffer at 4° is passed through the column at a flow rate comparable to that employed during purification (45 to 60 ml/hr). Column buffer contains 10 mM K-phosphate (pH 6.0), 1.0 mM DTT, 20 mM lactose, 0.25 mg/ml of acetone/ether-washed *E. coli* lipid, and 1.25% octylglucoside (w/v).[21] The sample (25–30 ml of octylglucoside extract) is thawed rapidly, chilled on ice, and applied to the column at 4°. The column is then developed with column buffer and fractions of 2 to 3 ml are collected. Aliquots of each fraction are assayed by liquid scintillation spectrometry in order to locate the position of [³H]NPG-photolabeled permease added as a tracer. As shown in Fig. 1, most of the protein-associated radioactivity elutes in a symmet-

---

[21] Complete column buffer is made from the mixture of two stock solutions designated "monobasic" and "dibasic" column buffer. The former contains 10 mM KH₂PO₄ in addition to the other ingredients which are common to both: 1.0 mM DTT, 20 mM lactose, 0.25 mg/ml of acetone/ether-washed *E. coli* lipid, and 1.25% (w/v) octylglucoside. Complete column buffer is then prepared by the titration of degassed monobasic column buffer at 0–4° with degassed ice-cold dibasic column buffer until the final apparent pH of the mixture reads 6.00 (at 0–4°). A polymer body, gel-filled combination electrode is used for the measurement (Fisher Catalog No. 13-639-252) with no temperature compensation. It should be noted that glass electrodes under the same conditions yield slightly more acidic pH values, 5.7–5.8.

rical peak slightly behind the void volume of the column. In general, 70–80% of the radioactivity applied to the column is recovered in this peak (fractions 41 to 53). Importantly, about 30% of the protein applied also elutes as a single symmetrical peak that coincides with the profile observed for radioactivity (not shown). This suggests that the DEAE-Sepharose step affords a 2.5-fold enrichment of the permease relative to the octylglucoside extract, with a high degree of recovery (>70%). In order to assay column fractions for transport activity, a "rapid assay" may be used in which dilute proteoliposomes are trapped on membrane filters and assayed qualitatively for activity.[11,12] The assay circumvents the need for photolabeled *lac* permease as a tracer, since the column profile for activity coincides with the profile for [^{3}H]NPG-labeled permease.

Purified *lac* permease from small-scale preparations, containing 10–15 μg of permease protein/ml, is stable for at least a week at 4°.[11] Purified permease containing higher protein concentrations remains soluble for several hours in column buffer at 4°. Subsequently, however, the protein aggregates and precipitates out of solution. Since all attempts to prevent this phenomenon have been unsuccessful thus far, purified material from the DEAE-Sepharose column is reconstituted into proteoliposomes within 1–2 hr. Although column fractions containing purified permease can be reconstituted after freezing and storage in liquid nitrogen, the resultant proteoliposomes exhibit about a 50% loss in various transport activities.

Table I summarizes the recovery and yield of total protein and photolabeled *lac* permease at each step of purification. The overall procedure results in a 31-fold purification relative to the crude membranes used as starting material with a yield of 46% based on the recovery of photolabeled permease. Thus, 15 mg of purified *lac* permease is obtained from 1.0 g of crude membrane protein, a yield that is consistent with [^{3}H]NPG photolabeling studies indicating that at least 3% of the protein in *E. coli* T206 membranes is *lac* permease. In other words, a 31-fold enrichment of the [^{3}H]NPG label suggests that a high degree of purification is achieved, a contention supported by a number of criteria. Figure 2 shows the results from sodium dodecyl sulfate–polyacrylamide gel electrophoresis (SDS–PAGE) of urea/cholate-extracted membranes, the octylglucoside extract, and pooled DEAE fractions. The pooled DEAE fractions yield a single broad band when stained with a highly sensitive silver procedure.[22] The purified protein has an apparent M_r of 33,000, which is in close agreement

---

[22] B. R. Oakley, D. R. Kirsch, and N. R. Morris, *Anal. Biochem.* **105,** 361 (1980).

TABLE I

PURIFICATION OF THE *lac* PERMEASE

Fraction	Protein[a] (mg)	Recovery of [³H]NPG (%)	Purification[b] (fold)
Membrane vesicles	1000 (100)	100	1.0
Urea-extracted membrane	419 (42)	89	2.1
Urea/cholate extracted membrane	326 (33)	78	2.4
Octylglucoside extract	50 (5)	62	12.4
DEAE column eluent	15[c] (1.5)	46	31

[a] Total protein in each fraction. Values given in parentheses are the percentage of protein in each fraction relative to the starting material (1 g of membrane protein).

[b] Ratio of the specific activity of NPG (dpm/mg protein) in each fraction relative to that of the starting material.

[c] Yield from DEAE-Sepharose chromatography of 90 ml of octylglucoside extract, the volume obtained from 1 g of membrane protein as starting material.

with published values for the molecular weight of the permease as determined by SDS–PAGE.[6,23] Importantly, when membranes are prepared from cells that are not induced with IPTG, the band corresponding to purified permease is only a minor constituent of the octylglucoside extract of urea/cholate-treated membranes. Thus, the purified protein is induced by IPTG, a property expected of the product of the *lacY* gene in the recombinant plasmid.[20] Furthermore, the amino acid composition of the purified protein agrees closely with that predicted from the DNA sequence of the *lacY* gene,[8] and N-terminal sequence analysis demonstrates that the first 13 amino acids of the purified permease agree with the predictions derived from the DNA sequence.

As indicated from the DNA sequence of the *lacY* gene[8] and from the amino acid composition of the purified permease,[11] the molecular weight of the permease is 46,504, a value significantly higher than the M_r observed on SDS–PAGE (33,000). It is not known precisely why the permease exhibits a lower molecular weight on SDS–PAGE, although the highly hydrophobic amino acid composition of the protein suggests that the phenomenon may be due to unusually high binding of SDS. Moreover,

[23] T. H. D. Jones and E. P. Kennedy, *J. Biol. Chem.* **244**, 5981 (1969).

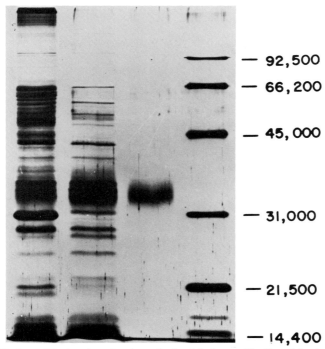

FIG. 2. SDS–PAGE of various fractions obtained during purification of the *lac* permease. From the left, lane 1: Urea/cholate-extracted membranes, 4 μg; lane 2: octylglucoside extract, 4 μg; lane 3: pooled DEAE fractions, 0.8 μg; lane 4: molecular weight standards, 2 μg. Gels were silver stained. (From Newman *et al.*[11])

circular dichroism measurements[24] indicate that purified permease maintains a high degree of secondary structure even in high concentrations of SDS. In any case, the electrophoretic mobility of the permease in SDS is highly dependent on the concentration of acrylamide used in the gel, and from the retardation coefficient,[25] the corrected M_r is about 46,000.

Preparation of E. coli Phospholipids. Purified *lac* permease functions best when reconstituted into native *E. coli* lipids apparently because of a specific requirement for phosphatidylethanolamine.[26] Thus, when the permease is reconstituted into asolectin (soybean phospholipids) as opposed to *E. coli* phospholipid, only about 15% of the entrance counterflow activity is observed.[9]

[24] D. L. Foster, M. Boublik, and H. R. Kaback, *J. Biol. Chem.* **258**, 31 (1983).
[25] D. M. Neville, Jr., *J. Biol. Chem.* **246**, 6328 (1971).
[26] C.-C. Chen and T. H. Wilson, *J. Biol. Chem.* **259**, 10150 (1984).

Two different methods for isolating *E. coli* phospholipids have been used and both are described below. The yield of lipid/gram wet weight of cells is similar for both, although Method 2 is faster and less cumbersome. Thus far, no major differences have been observed between proteoliposomes reconstituted with lipids prepared by the two procedures.

Method 1

The procedure described was modified from that of Ames.[27] *E. coli* B cells (300 g wet weight) are resuspended by homogenization in 347 ml of distilled water. From this step on, all operations are performed under an atmosphere of argon (or nitrogen) and protected from light. The homogenized cells are transferred to a glass-stoppered bottle (5 liters) and stirred with a magnetic bar. Chloroform (0.7 liters) and methanol (1.4 liters) are added, and stirring is continued at 4° for 12 hr. Particulate debris is then removed from the extract by centrifugation at 900 *g* for 15 min and the clear supernatant is retained. Lipids are separated from water-soluble material by the sequential addition of chloroform (645 ml) and distilled water (645 ml). The mixture is stirred for 3 hr at room temperature, and the magnetic stirrer is turned off in order to allow the extract to settle and partition into 3 distinct layers: an upper phase consisting of methanol–water, a brown particulate interphase, and a lower layer of chloroform.[28] The chloroform phase is carefully recovered and evaporated to dryness in a rotary evaporator at 30° (a small amount of benzene may be added to facilitate removal of trace amounts of water). The dried residue (about 3.7 g) is then subjected to extraction with acetone/ether (cf. below).

Method 2

The procedure was modified from that of Radin.[29] *E. coli* B cells (200 g wet weight) are added to a glass-stoppered bottle (5 liters) containing a magnetic stir bar. As in Method 1 care is taken to minimize exposure to oxygen and light. A mixture of 3 : 2 hexane/2-propanol (v/v) is prepared and 3.6 liters is added. The extract is stirred for 1.5 hr at room temperature until the cell paste is converted to finely dispersed debris. The debris is removed from the extract by filtration (under vacuum) through a scin-

[27] G. F. Ames, *J. Bacteriol.* **95,** 833 (1968).
[28] Alternatively, the rate of phase separation may be facilitated by subjecting the mixture to low speed centrifugation. Without centrifugation, complete separation may require 12 hr (at room temperature).
[29] N. S. Radin, this series, Vol. 72, p. 5 (1981).

tered glass filter funnel,[30] and the clear filtrate is evaporated to dryness in a rotary evaporator at 37° (as before, a small amount of benzene may be added to facilitate removal of water). The dry residue is dissolved in 20 ml of benzene, and the mixture is centrifuged at 2000 g for 15 min. The clear supernatant is removed and taken to dryness as before. The dried residue (about 3.0 g) is then subjected to extraction with acetone/ether (cf. below).

Acetone/Ether Extraction. The procedure was modified from that of Kagawa and Racker.[31] Dry lipid obtained from Method 1 or 2 is dissolved in 10 ml of chloroform/methanol (9:1, v/v). The mixture is then squirted forcefully into a glass flask containing a magnetic stir bar and 400 ml of acetone containing 2 m*M* BME. The acetone/BME solution should be at room temperature and thoroughly flushed with argon. Rapid stirring of the solution during addition of lipid produces smaller particles that are easier to wash. The vessel that contained the dried lipid is rinsed once with 10 ml chloroform/methanol (9:1, v/v) and the rinse is added to the rest of the preparation. The flask is then flushed with argon, made airtight and stirred at room temperature for 12 hr in the dark.

The waxy lipid pellet is recovered by centrifugation at 900 g for 15 min, and the drained pellet is washed once with an additional 400 ml of acetone/BME using a metal spatula to disperse the pellet. The lipid is recovered again by centrifugation, and the drained pellet is dried under a gentle stream of argon. The residue is then dissolved in 400 ml of diethyl ether containing 2 m*M* BME that was at room temperature and saturated with argon. The suspension is immediately centrifuged at 900 g for 15 min at 7°, the supernatant is removed carefully and evaporated to dryness in a rotary evaporator at room temperature. The residue is dissolved in a few milliliters of chloroform and transferred quantitatively to a tared glass vessel. The lipid is spread as a thin film on the sides of the vessel by slowly rotating the latter while evaporating the chloroform under a gentle stream of argon. In order to remove the last traces of solvent, the glass vessel is placed in a glass desiccator under vacuum; desiccation overnight at room temperature is sufficient if the lipid film is spread over a large surface area. The dried residue is weighed, resuspended to 50 mg/ml in argon-saturated 2 m*M* BME, and mechanically agitated until a viscous, homogeneous mixture is obtained. The lipid suspension is then divided into aliquots that are stored under argon at −180°. Acetone/ether extraction decreases the dried lipid residue weight by about 40%, and the material is greater than 95% phospholipid.

---

[30] To prevent clogging of the filter by particulate debris, the extract is first filtered through a "coarse" filter, and the resultant filtrate is then passed through a "medium" filter.

[31] Y. Kagawa and E. Racker, *J. Biol. Chem.* **246,** 5477 (1971).

Reconstitution. Purified *lac* permease is reconstituted into proteolipo-somes by using a modification[32] of the procedure described by Newman and Wilson.[9] In either case, the method is derived from the detergent dilution technique described originally by Racker *et al.*[33]

Liposomes are prepared for reconstitution by bath sonication[34] of ali-quots (1.0 ml) of acetone/ether-washed *E. coli* phospholipid prepared as described above in stoppered Pyrex test tubes under an atmosphere of argon (15 min at room temperature). The phospholipids are dispersed in 50 mM K-phosphate (pH 7.5) containing 2 mM BME to a final concentra-tion of 47.5 mg/ml. The sonicator bath contains 0.02% Triton X-100 and the level is adjusted to give maximal agitation.[35]

Bath sonicated liposomes (7.5 ml) are mixed with 0.65 ml of 15% octylglucoside (w/v in 50 mM K-phosphate, pH 7.5) and 25 ml of DEAE-Sepharose column fraction diluted with ice-cold column buffer to yield the desired protein concentration (generally 25–40 μg *lac* permease pro-tein/ml). The mixture is blended on a vortex mixer and incubated for 20 min on ice. Following incubation, the cloudy suspension is poured rapidly into a beaker containing 1135 ml of 50 mM K-phosphate (pH 7.5) contain-ing 1 mM DTT at room temperature. The suspension is stirred for 5 min and proteoliposomes are collected by centrifugation for 4 hr at 35,000 rpm in a Beckman Type 35 rotor (7°). The pellet is drained and resuspended in 10 ml of 50 mM K-phosphate (pH 7.5) containing 1.0 mM DTT (phospho-lipid concentration, 37.5 mg/ml). The proteoliposome suspension is then divided into aliquots (100 μl) which are frozen rapidly in liquid nitrogen and stored at −180°. As such, the preparation is stable for at least 2 years without a detectable loss of activity.

With each reconstitution, "control liposomes" are also prepared. The procedure used is identical to that described with the exception that an equivalent volume of fresh column buffer is substituted for the DEAE-Sepharose column fractions containing purified *lac* permease.

Freeze-Thaw/Sonication. For reasons that are not completely appar-ent, proteoliposomes prepared by octylglucoside dilution often exhibit variable transport rates unless they are subjected to a freeze-thaw cycle, followed by brief sonication.

Freeze-thaw/sonication offers the following advantages: (1) unilamel-lar vesicles devoid of internal structure (Figs. 3 and 5); (2) a relatively

[32] M. L. Garcia, P. Viitanen, D. L. Foster, and H. R. Kaback, *Biochemistry* **22,** 2524 (1983).
[33] E. Racker, B. Violand, S. O'Neal, M. Alfonzo, and J. Telford, *Arch. Biochem. Biophys.* **198,** 470 (1979).
[34] The bath sonicator used is 80 watts, 80 KHz, generator model G80-80-1, tank model T80-80-I-RS and may be obtained from Laboratory Supplies Co., Inc., Hicksville, NY.
[35] M. Kasahara and P. C. Hinkle, *J. Biol. Chem.* **252,** 7384 (1977).

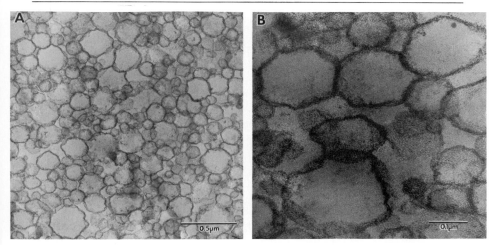

FIG. 3. (A, B) Electron micrographs of reconstituted proteoliposomes. Proteoliposomes containing purified *lac* permease were reconstituted by octylglucoside dilution followed by freeze-thaw/sonication. The preparation was fixed with 2% glutaraldehyde for 1 hr at 20°, postfixed with 1% OsO$_4$ in 50 mM cacodylate (pH 6.8), and embedded in Epon prior to sectioning. The study was performed by Miloslav Boublik and Frank Jenkins, Roche Institute of Molecular Biology. (From Garcia *et al.*[32])

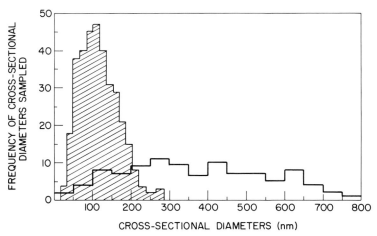

FIG. 4. Distribution of cross-sectional diameters of proteoliposomes. The number of structures exhibiting given cross-sectional diameters was measured from electron micrographs such as those shown in Fig. 3. The broad distribution profile corresponds to proteoliposomes subjected to a freeze-thaw step without sonication. The narrower peak (crosshatched) corresponds to the distribution of cross-sectional diameters observed for the same proteoliposomes following a brief period of sonication. (From Garcia *et al.*[32])

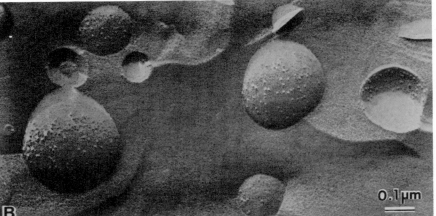

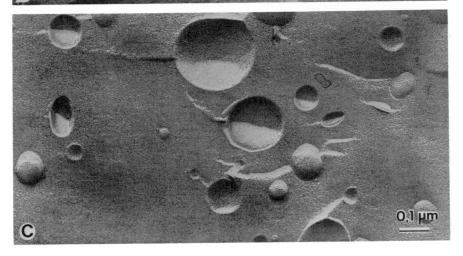

TABLE II
COMPARISON OF TURNOVER NUMBERS FOR *lac* PERMEASE IN ML 308-225
MEMBRANE VESICLES AND RECONSTITUTED PROTEOLIPOSOMES

| | Turnover numbers (sec^{-1}) | |
Reaction[a]	Membrane vesicles[b]	Proteoliposomes
$\Delta\Psi$-driven influx ($\Delta\Psi = 100$ mV)	16 ($K_m = 0.2$ mM)	16–21 ($K_m = 0.5$ mM)
Counterflow	16–39 ($K_m = 0.45$ mM)	28 ($K_m = 0.6$ mM)
Facilitated diffusion	8–15 ($K_m \approx 20$ mM)	8–9 ($K_m \approx 3.1$ mM)
Efflux	8 ($K_m = 2.1$ mM)	6–9 ($K_m = 2.5$ mM)

[a] All reactions were carried out at pH 7.5 and 25°. From Viitanen *et al.*[36]
[b] Determination of the amount of *lac* permease in ML 308-225 membrane vesicles is based on photolabeling experiments with [^{3}H]NPG which indicate that the permease represents about 0.5% of the membrane protein.

narrow size distribution with about 80% of the structures exhibiting cross-sectional diameters between 50 and 150 nm (Fig. 4); (3) reproducible turnover numbers for a variety of transport activities that are similar to those displayed by the permease in the bacterial membrane (Table II); and importantly (4) freeze-thaw/sonicated proteoliposomes are relatively impermeable to protons and other ions (e.g., see Fig. 3 of Garcia *et al.*[32]). The last property is not trivial, for it enables the imposition of long-lasting electrical potentials and pH gradients.

Freeze-thaw/sonication is carried out as follows: an aliquot of proteoliposomes (stored at −180°) is thawed at room temperature and the cloudy, viscous suspension is briefly blended on a vortex mixer. The bottom of the tube (1.5 ml conical; Eppendorf) is submerged to a depth of about 5 mm below the surface of the sonicator bath solution [0.02% Triton X-100 (v/v)].[34] As described in the preceding section, the level of the bath solution should be adjusted to yield maximum surface agitation and the tip of the microfuge tube should be placed in the center, the region of maxi-

FIG. 5. Freeze-fracture electron micrograph of proteoliposomes reconstituted with purified *lac* permease. Platinum/carbon replicas of freeze-fractured proteoliposomes (A and B) and liposomes (C) prepared by octylglucoside dilution followed by freeze-thaw/sonication are shown. Note the absence of particles on fracture faces of liposomes (C). The study was performed by Joseph Costello in the Department of Anatomy, Duke University Medical Center.

mal cavitation. Sonication is conducted in 3–4 sec bursts (8–15 sec, total time) with intermittent vortex mixing to recover material splashed onto the walls of the tube and is complete when the proteoliposome suspension appears only slightly opaque. Two different methods have been described for estimating the internal (trapped) volume of the proteoliposomes.[32] One microliter of the above suspension (37.5 μg lipid) yields an apparent internal volume of 0.056 $\pm$ 0.0002 μl using either $^{86}Rb^+$ or [1-^{14}C]lactose as a probe. In order to avoid fusion, freeze-thaw/sonicated proteoliposomes are kept at room temperature, and as such, remain stable at pH 7.5 for at least 8 h.

Relatively low magnification electron microscopy of platinum/carbon replicas of freeze-fractured, freeze-thaw/sonicated proteoliposomes containing purified *lac* permease confirms the unilamellar nature of the preparation (Fig. 5A). Higher magnification reveals that both the convex and concave fracture surfaces of the phospholipid bilayer exhibit a relatively uniform distribution of *lac* permease particles that are about 70 Å in diameter (Fig. 5B). Since particles, but no pits, are observed on both surfaces of the membrane, it seems likely that the purified permease has equal affinity for the phospholipids in each leaflet of the bilayer. Figure 5B also shows that there are no permease particles in the surrounding ice, suggesting that essentially all of the permease has been reconstituted into the phospholipid bilayer. Moreover, protein determinations on the final preparation demonstrate that about 90% of the purified permease is recovered with the proteoliposomes.

β-Galactoside Translocation: Overview. Studies with *E. coli* cells and cytoplasmic membrane vesicles show that *lac* permease catalyzes three basic translocation reactions—active transport, facilitated diffusion, and exchange.

Active transport is permease-mediated transport of β-galactosides against a concentration gradient, a process that requires $\Delta\bar{\mu}_{H^+}$ as a driving force. Physiologically, $\Delta\bar{\mu}_{H^+}$ is generated either via the respiratory chain or through hydrolysis of ATP by the H^+-ATPase, and is composed of electrical and chemical parameters according to the following relationship:

$$\Delta\bar{\mu}_{H^+}/F = \Delta\Psi - 2.3\ RT/F\ \Delta pH$$

where $\Delta\Psi$ represents the electrical potential across the membrane and ΔpH is the chemical difference in proton concentrations across the membrane (R is the gas constant, T is absolute temperature, F is the Faraday constant, 2.3 RT/F is equal to 58.8 at room temperature). Since proteoliposomes reconstituted with purified *lac* permease have neither a respiratory chain nor the H^+-ATPase, $\Delta\Psi$, and/or ΔpH must be generated nonphysiologically.

Proteoliposomes are prepared in 50 mM K-phosphate (pH 7.5) containing 1 mM DTT, and because they are relatively impermeable to protons and other ions,[32] ΔpH (interior alkaline) can be generated by diluting the preparation into identical buffer at lower pH. If ^{14}C-labeled lactose is included in the dilution medium, active transport of substrate is observed (cf. Fig. 1 of Foster *et al.*[12]). Alternatively, $\Delta\Psi$ (interior negative) is generated by means of an outwardly directed potassium diffusion potential ($K^+_{in} \rightarrow K^+_{out}$) when proteoliposomes are rendered permeable to the cation with valinomycin. The magnitude of $\Delta\Psi$ can be manipulated by diluting the ionophore-treated preparation into isoosmotic phosphate buffer containing known proportions of sodium and potassium. The magnitude of the $\Delta\Psi$ generated theoretically by a given potassium diffusion gradient in the presence of valinomycin is calculated from the following relationship:

$$\Delta\Psi \text{ (mV)} = -59 \log[K^+]_{in}/[K^+]_{out}$$

where $[K^+]_{in}$ and $[K^+]_{out}$ denote internal and external potassium concentrations, respectively, at the time of dilution.

The data presented in Fig. 6 show that proteoliposomes reconstituted with purified *lac* permease accumulate lactose in response to an artificially imposed $\Delta\Psi$ and that this phenomenon is abolished by the protonophore, CCCP. Note that when the proteoliposomes are diluted into

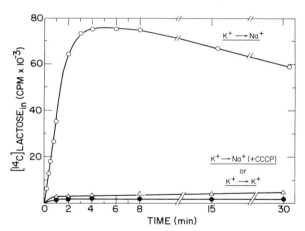

FIG. 6. $\Delta\Psi$-Driven uptake of [14C]lactose into proteoliposomes. Proteoliposomes were reconstituted with purified *lac* permease by octylglucoside dilution and freeze-thaw/sonication. The suspension contained [in 50 mM K-phosphate (pH 7.5)/1.0 mM DTT] 37.5 mg lipid/ml and 80 μg *lac* permease protein/ml. Valinomycin was added to 20 μM, final concentration, and aliquots (1 μl) were diluted 200-fold into ($\bullet$) 50 mM K-phosphate (pH 7.5) or ($\bigcirc$, $\triangle$) 50 mM sodium phosphate (pH 7.5) containing 0.3 mM [1-14C]lactose (19.7 mCi/mmol). Where indicated ($\triangle$), CCCP was added to the reaction mixture to a final concentration of 30 μM. At given times, reactions were terminated and samples were assayed as described.

equimolar K-phosphate (pH 7.5)—that is, in the absence of a potassium-diffusion potential—substrate accumulation is not observed. It is also clear from the data that the $\Delta\Psi$ generated by the potassium diffusion potential is stable for at least 8 min and only then starts to dissipate, resulting in loss of accumulated substrate. This behavior testifies to the high degree of passive impermeability exhibited by freeze-thaw/sonicated proteoliposomes.

In the absence of a driving force, *lac* permease catalyzes "facilitated diffusion," unidirectional net movement of β-galactosides resulting in equilibration of the sugar across the membrane. Since *lac* permease functions in a symmetrical fashion, it translocates substrate in either direction across the bilayer. Therefore, depending on the design of the experiment, one can study either facilitated influx or efflux. In either case, the term "facilitated diffusion" is technically a misnomer, since the permease simultaneously transports β-galactosides and protons. That is, although one substrate, the β-galactoside, moves downhill, the other substrate, hydrogen ion, moves uphill against its electrochemical gradient. This is not trivial, for it results in the formation of an opposing $\Delta\bar{\mu}_{H^+}$, due to a transmembrane net flux of protons, which acts to retard the rate of sugar equilibration. In proteoliposomes reconstituted with purified *lac* permease, for instance, initial rates of facilitated diffusion (influx[36] or efflux[32]) are increased 2- to 3-fold by the addition of appropriate ionophores.[37]

Lastly, *lac* permease catalyzes transmembrane β-galactoside exchange either in the absence (equilibrium exchange) or presence (entrance counterflow) of a substrate concentration gradient ($[\beta\text{-galactoside}]_{in} > [\beta\text{-galactoside}]_{out}$). The latter process is complex mechanistically since it represents a composite picture of the permease operating in two modes, efflux and exchange. In any event, counterflow may be used to assess the frequency with which the permease returns from the outer to the inner surface of the membrane in the loaded vs the unloaded form.[32,38–40]

---

[36] P. Viitanen, M. L. Garcia, and H. R. Kaback, *Proc. Natl. Acad. Sci. U.S.A.* **81**, 1629 (1984).

[37] As documented in Garcia *et al.*,[32] nigericin (an ionophore that promotes electroneutral exchange of protons for potassium or sodium) does not increase the initial rate of [^{14}C]lactose facilitated diffusion (efflux or influx) in proteoliposomes at pH 7.5. The apparent absence of ΔpH (interior alkaline for efflux or interior acid for influx) under these conditions is presumably due to the relatively high buffering capacity of potassium phosphate at pH values from 5.5 to 7.5. The presumption is supported by the observation that nigericin accelerates the rate of lactose efflux at more alkaline pH values.

[38] P. Viitanen, M. L. Garcia, D. L. Foster, G. J. Kaczorowski, and H. R. Kaback, *Biochemistry* **22**, 2531 (1983).

[39] G. J. Kaczorowski and H. R. Kaback, *Biochemistry* **18**, 3691 (1979).

[40] G. J. Kaczorowski, D. E. Robertson, and H. R. Kaback, *Biochemistry* **18**, 3697 (1979).

The transport assays just described are shown schematically in Fig. 7. A detailed experimental protocol for each case is included below. In order to measure true initial rates of translocation, transport assays are conducted on preparations that are reconstituted with relatively dilute solutions of purified *lac* permease. Generally 25–40 μg of protein per ml of column buffer is used during the reconstitution step. Nevertheless, as shown in Fig. 8, for $\Delta\Psi$-driven active transport of [^{14}C]lactose, initial velocities of uptake are linear over a relatively broad range of protein to lipid ratios. It is important to point out, however, that proteoliposomes reconstituted at high protein : lipid ratios are more permeable to ions and

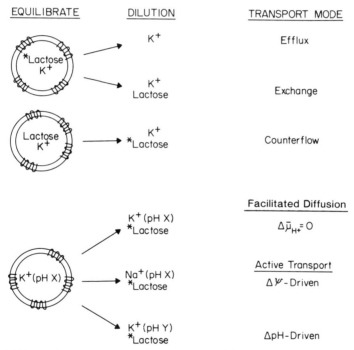

FIG. 7. Schematic representation of lac permease activity assays. Freeze-thaw/sonicated proteoliposomes containing purified *lac* permease are treated with valinomycin and equilibrated in 50 m*M* K-phosphate/1.0 m*M* DTT containing radioactive (*) lactose, unlabeled lactose, or no lactose (equilibrate). Aliquots of the suspension are then diluted into isoosmotic reaction mixtures containing phosphate buffer (potassium or sodium) with or without labeled or unlabeled lactose (dilution). For efflux and counterflow assays, [lactose]$_{in}$ > [lactose]$_{out}$ at the time of dilution, while the reverse is the case for facilitated diffusion, $\Delta\Psi$- and ΔpH-driven uptake. For equilibrium exchange, [lactose]$_{in}$ = [lactose]$_{out}$. Active transport of lactose is driven by $\Delta\Psi$ (interior negative) when [K$^+$]$_{in}$ > [K$^+$]$_{out}$ at the time of dilution or by a ΔpH (interior alkaline) when pH *x* > pH *y*. Details are given in the text.

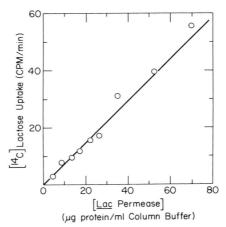

FIG. 8. Initial rate of $\Delta\Psi$-driven lactose transport by proteoliposomes reconstituted at various protein to lipid ratios. The *lac* permease was purified as described in the text and diluted with ice-cold column buffer to the protein concentrations indicated on the abscissa. An identical volume of each dilution was used for reconstitution into proteoliposomes, as outlined in the text, thereby varying the protein to lipid ratio of the preparations. All proteoliposomes were resuspended in 50 mM K-phosphate (pH 7.5)/1.0 mM DTT at 37.5 mg lipid/ml and assayed for $\Delta\Psi$-driven transport of [1-^{14}C]lactose (0.3 mM final concentration, 19.7 mCi/mmol) as described in Fig. 6. Initial velocities (V_i) of transport (cpm/min) were determined from the linear portion of the time courses.

cannot, therefore, maintain an artificially imposed $\Delta\bar{\mu}_{H^+}$ for prolonged periods of time.

Table II compares turnover numbers and apparent K_m values obtained from studies with purified *lac* permease and right-side-out cytoplasmic membrane vesicles. The values for proteoliposomes are derived from kinetic experiments[32,36,38] carried out using freeze-thaw/sonicated proteoliposomes prepared as described. Clearly, both the turnover numbers of the *lac* permease and the apparent K_m values for lactose are very similar in proteoliposomes and right-side-out membrane vesicles with respect to $\Delta\Psi$-driven lactose accumulation, counterflow, facilitated diffusion (influx) and efflux. Taken as a whole, the data argue strongly that the *lac* permease retains much of its activity during purification and reconstitution.

Detailed Transport Assays. Routinely, proteoliposome suspensions contain 50 mM K-phosphate (pH 7.5), 1 mM DTT, 37.5 mg phospholipid/ml, and 62.5–100 μg of purified *lac* permease/ml. When transport assays are conducted at pH 7.5, the proteoliposome suspension is used directly

as described below. For experiments at different pH values, proteoliposomes are first equilibrated to the desired pH. Thus, freeze-thaw/sonicated
proteoliposomes are added to at least 100 volumes of 50 mM K-phosphate
at the desired pH containing DTT. For pH values less than 7.5, 1 mM
DTT is sufficient, while higher concentrations of thiol (2–5 mM) are required at more alkaline pH values to stabilize the preparations. Proteoliposomes are then equilibrated for 30 min at room temperature and collected by centrifugation for 1 hr at 45,000 rpm in a Beckman Type 50 Ti
rotor. The pellet is resuspended in the buffer used for washing to a final
concentration of 37.5 mg of phospholipid/ml.

Transport assays are carried out individually in test tubes (12 × 75
mm), and temperature is maintained by incubation in a thermostatically
controlled water bath. Unless designated otherwise, all assays are conducted at 25°. Transport of [^{14}C]lactose by proteoliposomes is monitored
by rapid filtration. The filters are mixed cellulose nitrate/cellulose acetate
from Millipore (type GSTF 02500, 0.2 μm pore size) and other types of
filters yield variable results. Since the proteoliposomes are not retained
by cellulose acetate filters of the same pore size, it is likely that adsorption
phenomena, rather than entrapment, are involved. Prior to assay, the
filters are floated on distilled water (shiny side up) in a Petri dish. If the
filters do not wet evenly they are discarded.

Transport assays are terminated at desired times by quenching the
reactions with 3.5 ml of ice-cold 50 mM potassium (or sodium) phosphate
(pH 7.5). Permease-mediated loss of radioactive lactose is minimized by
the low temperature of the quench solution.[41] The samples are immediately filtered, and the filter is washed once with the same cold buffer.
Radioactivity retained on the filters is measured by liquid scintillation
spectrometry after dissolving the filters in 10 ml Bray's solution (New
England Nuclear).

Efflux. The proteoliposome suspension is treated with a small volume
of 10 mM valinomycin to yield a final concentration of 20 μM; submicroli-
ter volumes are accurately delivered using a 1-μl Hamilton syringe (Hamilton No. 7001 N). A small aliquot of radioactive lactose is then added to
the suspension to achieve the desired final concentration, generally 10
mM, and the suspension is incubated at room temperature for 1 hr in
order to allow lactose to equilibrate with the intravesicular space. Subsequently, an aliquot (1 μl) of the suspension is drawn into a 10-μl syringe
(Hamilton No. 801 N) and diluted 200-fold into 50 mM K-phosphate; the

---

[41] The lipid phase transition temperature of the acetone/ether-washed *E. coli* lipids used for
   reconstitution of the *lac* permease is 18.5° (as determined by differential scanning calorim-
   etry; unpublished observations).

pH of the dilution medium is identical to that of the proteoliposome suspension. Reactions are terminated at desired times and the samples assayed as described. If proteoliposomes are loaded with higher concentrations of lactose, efflux is initiated by dilution into a correspondingly larger volume of sugar-free buffer.

Equilibrium Exchange. Proteoliposomes are treated with valinomycin and equilibrated with 10 mM [^{14}C]lactose as described for efflux. Exchange is initiated by drawing an aliquot (1 μl) of the suspension into a 10-μl syringe and diluting 200-fold into 50 mM K-phosphate containing 10 mM unlabeled lactose. Reactions are terminated and assayed at desired times as described. Loss of intravesicular [^{14}C]lactose under exchange conditions is independent of pH and nearly 100 times faster than efflux at pH 5.5. On the other hand, exchange is only 25–30% faster than efflux at pH 9.5 and above.[38] The rate-determining step for efflux is either release of the symported proton from the permease at the external surface of the membrane or a reaction corresponding to return of the unloaded permease from the outer to the inner surface of the membrane; neither step is involved in exchange.[32,38-40]

Facilitated Diffusion (Influx). For measurements of facilitated diffusion proteoliposomes are concentrated 3- to 5-fold by centrifugation and resuspension in 50 mM K-phosphate containing DTT. Since the apparent K_m for facilitated influx is relatively high (see Table II), increasing the ratio of the intravesicular volume to external volume improves the signal-to-noise ratio of the assay. Concentrated proteoliposomes are treated with 20 μM valinomycin, and uptake is initiated by drawing a 1-μl aliquot of the suspension into a syringe and diluting into 50 mM K-phosphate (at the pH of the suspension) containing [^{14}C]lactose. Reactions are terminated and assayed at desired times as described.

Lactose-Induced Proton Influx. As a corollary to lactose facilitated diffusion measurements, influx of protons can also be measured in reconstituted proteoliposomes in response to an inwardly directed lactose concentration gradient.[12] Measurements of pH are performed in a closed electrode vessel that is jacketed and maintained at constant temperature. The vessel is continuously flushed with a stream of water-saturated nitrogen. By means of a lateral inlet, 2.5 ml of 150 mM KCl, 10 mM MgSO$_4$, and 60 μl of proteoliposomes reconstituted with *lac* permease are added to the vessel. The solution is adjusted to 5 μM in valinomycin in order to prevent generation of $\Delta\Psi$ (interior positive) during lactose-induced proton influx. Proton movements are initiated by addition of an aliquot of a freshly prepared stock solution of 0.5 M lactose to a final concentration of 10 mM. The lactose stock solution is first carefully adjusted to the pH of the proteoliposome suspension with a KOH solution. The suspension is stirred throughout the course of the experiment with a magnetic stir bar. A Radiometer pH meter (pHm84) connected to a Radiometer pH electrode (GK 2401 B) and a Radiometer chart recorder (REC 61 Servograph) are used to monitor pH continuously. Calibration of measured pH changes is performed at the conclusion of the experiment by addition of 10-μl aliquots of 1 mM HCl.

Entrance Counterflow. Proteoliposomes are treated with 20 μM valinomycin (final concentration) and equilibrated with 10 mM unlabeled lactose for 1 hr at room temperature. Counterflow is initiated by drawing a 1-μl aliquot into a syringe and diluting 200-fold into 50 mM K-phosphate (at the pH of the suspension) containing [^{14}C]lactose. The apparent K_m for external lactose in the counterflow assay is 0.6 mM at pH 7.5 (Table II). In calculating the external lactose concentration and its specific activity, it is important to take into account that 0.05 mM unlabeled sugar is contributed by addition of the proteoliposome suspension. Reactions are terminated and assayed as described above.

$\Delta\Psi$-*Driven Active Transport.* Proteoliposomes are treated with 20 μM valinomycin, and uptake is initiated by diluting a 1-μl aliquot 200-fold into 50 mM phosphate buffer (at the pH of the proteoliposome suspension). The dilution buffer also contains [^{14}C]lactose and different ratios of potassium and sodium to vary the magnitude of $\Delta\Psi$. When the dilution buffer is 50 mM sodium phosphate (pH 7.5), 200-fold dilution results in a $\Delta\Psi$ of -136 mV. In such theoretical calculations, the contribution of external potassium in the proteoliposome suspension to the total reaction mixture must be considered. Therefore, to achieve $\Delta\Psi$ values more negative than -136 mV, a dilution greater than 200-fold must be used. Despite the impermeability of proteoliposomes to protons, a large outwardly directed potassium diffusion potential leads to the formation of a ΔpH (interior acid) which decreases the overall driving force for lactose accumulation. The total driving force can be approximated, however, from the maximum level of lactose accumulation.[36]

Controls. All of the transport assays described include zero-time controls which are obtained by adding an aliquot of proteoliposomes to reaction mixtures that have already been diluted with ice-cold terminating buffer, followed by filtration and washing. For assays involving uptake of [^{14}C]lactose, the zero-time values are subtracted from each experimental time point. For efflux and exchange, the zero-time controls yield the initial internal concentration of radioactive lactose.

(N-Dansyl)aminoalkyl-1-thio-β-D-galactopyranoside (Dansylgalactoside) Fluoresence. Another convenient way to assay *lac* permease activity involves the use of any one of a series of dansylgalactosides in which the alkyl linkage between the galactosyl moiety and the dansyl group is

varied from 0 to 6 methyl groups.[42–46] Fluorescence emission at 500 nm is measured at an angle of 90° with excitation at 340 nm using a spectrofluorometer and 1 × 1 cm cuvettes. The sample chamber is maintained at 25° with a circulating water bath, and the light band pass for excitation and emission is 6 nm. Additions to the cuvette are made with Hamilton microsyringes and mixing is accomplished within 3–5 sec using a small plastic stick. Increases in dansylgalactoside fluorescence may be induced either by imposition of an outwardly directed potassium diffusion gradient in the presence of valinomycin (i.e., an artificially imposed $\Delta\Psi$) or by imposition of an outwardly directed β-galactoside concentration gradient (i.e., counterflow). Although evidence was presented initially to indicate that the increase in fluorescence is due specifically to binding of the dansylgalactosides to the permease, more recent experiments[47] suggest that fluorescence enhancement probably reflects transport of dansylgalactosides via the *lac* permease followed by nonspecific interaction with the membrane. In any event, the assay is highly specific for the *lac* permease, convenient, rapid, and highly sensitive.

For $\Delta\Psi$-driven uptake, the reaction mixtures (2 ml, total volume) contain various ratios of 50 mM potassium and sodium phosphate (at the pH of the proteoliposome suspension) and an appropriate concentration of dansylgalactoside. The reaction is initiated by rapid addition of proteoliposomes containing 5 μM valinomycin (10 to 20 μl) to the cuvette.

For counterflow, proteoliposomes are first equilibrated with 10 mM lactose for 1 hr in the presence of 5 μM valinomycin. Reactions are then initiated by rapid addition of proteoliposomes (10 to 20 μl) to 2 ml of 50 mM K-phosphate (at the pH of the proteoliposomes) containing an appropriate concentration of dansylgalactoside.

[42] J. P. Reeves, E. Shechter, R. Weil, and H. R. Kaback, *Proc. Natl. Acad. Sci. U.S.A.* **70**, 2722 (1973).

[43] S. Schuldiner, G. K. Kerwar, R. Weil, and H. R. Kaback, *J. Biol. Chem.* **250**, 1361 (1975).

[44] S. Schuldiner, H.-F. Kung, R. Weil, and H. R. Kaback, *J. Biol. Chem.* **250**, 3679 (1975).

[45] S. Schuldiner, R. Weil, and H. R. Kaback, *Proc. Natl. Acad. Sci. U.S.A.* **73**, 109 (1976).

[46] S. Schuldiner, R. Weil, D. E. Robertson, and H. R. Kaback, *Proc. Natl. Acad. Sci. U.S.A.* **74**, 1851 (1977).

[47] P. Overath, R. M. Teather, R. D. Simoni, G. Aichele, and V. Wilhelm, *Biochemistry* **18**, 1 (1979).

[33] Preparation of Monoclonal Antibodies and Site-Directed Polyclonal Antibodies against the *lac* Permease of *Escherichia coli*

By Nancy Carrasco, Doris Herzlinger, Waleed Danho, and H. Ronald Kaback

The *lac* permease (i.e., *lac* carrier protein) of *Escherichia coli* is an intrinsic membrane protein, the product of the *lacY* gene, that catalyzes the translocation of β-galactosides with hydrogen ion in a symport or cotransport reaction.[1] As described elsewhere in this volume,[2] the permease can be solubilized from the membrane, purified to homogeneity, reconstituted into phospholipid vesicles, and shown to catalyze all of the transport reactions typical of the native membrane with comparable activities and kinetic properties.

The permease is a 46.5-kDa polypeptide containing 417 amino acid residues of known sequence.[3] Based on circular dichroic measurements indicating that the protein has an exceptionally high helical content and on an analysis of the sequential hydropathic character of the protein, a secondary structure model has been proposed.[4] The model suggests that the permease consists of 12 hydrophobic α-helical segments that traverse the membrane in a zig-zag fashion connected by 11 shorter hydrophilic segments (Fig. 1). The model makes explicit predictions regarding those portions of the molecule that should be accessible to solvent at the surfaces of the membrane. In order to test these predictions, to determine the orientation of the permease in native and reconstituted membranes, and to obtain information regarding structure/function relationships, monoclonal antibodies (Mabs) against purified *lac* permease and polyclonal antibodies directed against synthetic peptides corresponding to portions of the permease (i.e., site-directed polyclonal antibodies) have been utilized.[5-9]

[1] H. R. Kaback, *J. Membr. Biol.* **76**, 95 (1983).
[2] P. V. Viitanen, M. J. Newman, D. L. Foster, T. H. Wilson, and H. R. Kaback, this volume [32].
[3] D. E. Büchel, B. Gronenborn, and B. Müller-Hill, *Nature (London)* **283**, 541 (1980).
[4] D. L. Foster, M. Boublik, and H. R. Kaback, *J. Biol. Chem.* **250**, 31 (1983).
[5] N. Carrasco, S. M. Tahara, L. Patel, T. Goldkorn, and H. R. Kaback, *Proc. Natl. Acad. Sci. U.S.A.* **79**, 6894 (1982).
[6] N. Carrasco, P. V. Viitanen, D. Herzlinger, and H. R. Kaback, *Biochemistry* **23**, 3681 (1984).
[7] D. Herzlinger, P. Viitanen, N. Carrasco, and H. R. Kaback, *Biochemistry* **23**, 3688 (1984).

METHODS IN ENZYMOLOGY, VOL. 125

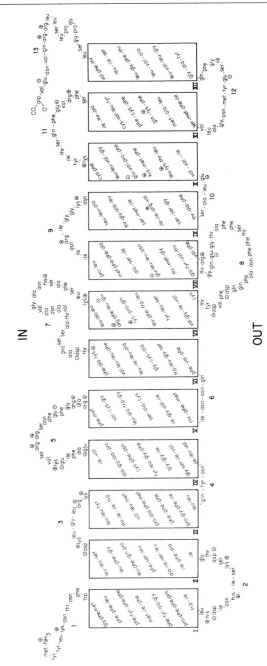

Fig. 1. Secondary structure model of the *lac* permease as predicted from the hydropathic profile of Foster *et al.*[4] Hydrophobic segments are shown in boxes as transmembrane, α-helical domains connected by hydrophilic segments.

Generation of Antibodies

Mabs

The sample of *lac* permease used for immunization is obtained directly from the DEAE-Sepharose column in the last step of purification,[2] dialyzed overnight against distilled water, lyophilized, and resuspended in phosphate-buffered saline (P_i/NaCl; 10 mM sodium phosphate, pH 7.2/ 150 mM NaCl) to a final concentration of 250 μg of protein and 415 μg of phospholipid per ml. BALB/c female mice (Charles River Breeding Laboratories) are immunized with 25 μg of purified *lac* permease. An aliquot of the immunogen (0.1 ml) is emulsified with 0.1 ml of complete Freund's adjuvant and injected subcutaneously. On day 24, serum samples obtained from the retroorbital sinus are tested for the presence of antibodies either by solid-phase radioimmunoassay (SP-RIA) or by immunoelectroblotting (cf. Screening Procedures). Mice with positive sera are given booster injections intraperitoneally on days 29 and 43 with 50 μg of antigen in 0.2 ml of P_i/NaCl. Three days after the last injection, the spleens are removed aseptically, and spleen cells are fused with P3X63Ag8.653 myeloma cells by using 50% polyethylene glycol 4000 (Merck).[10] The myeloma cells are maintained in RPMI 1640 growth medium supplemented with 15% heat-inactivated fetal calf serum (GIBCO), 2 mM glutamine, 1 mM sodium pyruvate, 50 mM 2-mercaptoethanol, and 100 units of penicillin and 100 μg of streptomycin per ml.

Hybridoma cells are selected with RPMI 1640 medium containing hypoxanthine, aminopterin, and thymidine.[11] Subcloning of positive wells is performed in two stages as described,[12] except that mouse macrophages are used as a feeder layer in place of mouse thymocytes. Selected hybridoma formal clones are expanded for ascites tumor production and frozen as reported.[11]

For large-scale antibody production, $\approx$5–7 $\times$ 10^6 hybridoma cells in 1.0 ml of P_i/NaCl are injected intraperitoneally in BALB/c mice primed at least 14 days earlier with 0.5 ml of pristane (Aldrich). Ascites fluid is collected after 10–15 days, clarified by centrifugation and stored at $-20°$.

[8] N. Carrasco, D. Herzlinger, R. Mitchell, S. DeChiara, W. Danho, T. F. Gabriel, and H. R. Kaback, *Proc. Natl. Acad. Sci. U.S.A.* **81**, 4672 (1984).

[9] N. Carrasco, D. Herzlinger, S. DeChiara, R. Mitchell, W. Danho, and H. R. Kaback, *Ann. N.Y. Acad. Sci.*, in press (1985).

[10] S. Frazekas de St. Groth and D. Scheidegger, *J. Immunol. Methods* **35**, 1–28 (1980).

[11] J. W. Littlefield, *Science* **145**, 709 (1964).

[12] R. C. Nowinski, M. E. Lostrom, M. R. Tam, and W. N. Burnette, *Virology* **93**, 111 (1979).

Site-Directed Polyclonal Antibodies

Peptide Synthesis. Peptides varying in size from 6 to 25 amino acid residues are synthesized in accordance with the amino acid sequence predicted from the nucleotide sequence of the *lacY* gene (Fig. 1). Peptide synthesis is performed by solid-phase methodology using *p*-hydroxymethyl polystyrene resin, the symmetrical anhydride procedure, and a Vega 250 chemistry module peptide synthesizer. The synthesizer is controlled by a Model 300 microprocessor from Vega Biochemicals. Deprotection and cleavage from the resin are achieved by treatment with anhydrous liquid HF using the modified procedure of Tam *et al.*[13] The peptides are purified by preparative high performance liquid chromatography using a μBondapak C_{18} column. The purity of the peptides is ascertained by analytical HPLC and amino acid analysis. The table lists the peptides synthesized with their corresponding sequences.

Conjugation of Peptides to Thyroglobulin. One micromole of peptide is dissolved in 500 μl of 0.1 *M* sodium phosphate (pH 7.2) containing 20 nmol of bovine thyroglobulin (Sigma); 22.5 μmol of glutaraldehyde (Sigma; grade I, 25%) is added and the mixture is incubated overnight at 4°. In order to conjugate hydrophobic peptides, dimethyl sulfoxide is added (up to 50%, v/v). Unreacted material is separated by gel filtration on Sephadex G-75. The final product has a molar ratio of peptide to thyroglobulin of about 50 : 1.

Antibody Production. Antibody is obtained by immunizing New Zealand White rabbits with 0.65 mg of peptide/thyroglobulin conjugate emulsified in 1.0 ml of complete Freund's adjuvant. The mixture is injected intradermally at $\approx$20 different sites, and the animals are given booster injections after 4–5 weeks by intradermal injection of the same amount of conjugate in incomplete Freund's adjuvant. Blood is drawn $\approx$10 days thereafter.

Purification of Antibodies

Mab

Antibody contained in ascites fluid is purified by affinity chromatography on protein A/Sepharose.[14] Clarified ascites fluid is applied to the column and washed with 0.1 *M* sodium phosphate (pH 8.1) until no protein is detected in the eluant. Bound immunoglobulins are eluted stepwise with 0.1 *M* sodium citrate buffer at pH 6.0, 4.5, and 3.5 to collect IgG_1,

[13] J. P. Tam, W. F. Heath, and R. B. Merrifield, *J. Am. Chem. Soc.* **105**, 6442 (1983).
[14] P. D. Ey, S. J. Prowse, and C. R. Jenkin, *Immunochemistry* **15**, 429 (1979).

CHEMICALLY SYNTHESIZED PEPTIDES OF THE *lac* CARRIER

Hydrophilic segment	Residues	Composition
1 (N-Terminus)	1–9	Met-Tyr-Tyr-Leu-Lys-Asn-Thr-Asn-Phe
2	35–47	Leu-His-Asp-Ile-Asn-His-Ile-Ser-Lys-Ser-Asp-Thr-Gly
3	56–79	Leu-Sr-Asp-Lys-Leu Gly-Leu-Arg-Lys-Tyr-Leu-Leu-Trp-Ile
4	94–107	Ile-Phe-Gly-Pro-Leu-Leu-Gln-Tyr-Asn-Ile-Leu-Val-Gly-Ser
5	130–145	Glu-Lys-Val-Ser-Arg-Arg-Ser-Asn-Phe-Glu-Phe-Gly-Arg-Ala-Arg-Met-Phe
6	163–168	Thr-Ile-Asn-Asn-Gln-Phe
7	185–198	Phe-Phe-Ala-Lys-Thr-Asp-Ala-Pro-Ser-Ser-Ala-Thr-Val-Ala
7a	205–221	His-Ser-Ala-Phe-Ser-Leu-Lys-Leu-Ala-Leu-Glu-Leu-Phe-Arg-Gln-Pro-Lys
8	233–258	Cys-Thr-Tyr-Asp-Val-Phe-Asp-Gln-Gln-Phe-Ala-Asn-Phe-Phe-Thr-Ser-Phe-Phe-Ala-Thr-Gly-Glu-Gln-Gly-Thr
10	302–316	Arg-Ile-Ile-Gly-Ser-Ser-Phe-Ala-Thr-Ser-Ala-Leu-Glu-Val-Val
11	334–345	Cys-Phe-Lys-Tyr-Ile-Thr-Ser-Gln-Phe-Glu-Val-Arg-Phe-Ser
12	370–383	Gly-Asn-Met-Tyr-Glu-Ser-Ile-Gly-Phe-Gln-Gly-Ala-Tyr-Leu
13 (C-Terminus)	407–417	Leu-Ser-Leu-Leu-Arg-Arg-Gln-Val-Asn-Glu-Val-Ala

IgG$_{2a}$, and IgG$_{2b}$, respectively, and each fraction is neutralized with 1.0 M Tris–HCl (pH 10).[15] Proteins eluted from the column are concentrated by evaporation under vacuum and dialyzed overnight against a 1000-fold volume of 50 mM potassium phosphate (pH 7.5) with two changes. Aliquots of the purified abs are frozen and stored in liquid nitrogen.

Site-Directed Polyclonal Antibodies

The IgG fraction is purified from rabbit serum by affinity chromatography on protein A/Sepharose in one step by eluting with 0.1 M sodium citrate (pH 3.5) and neutralizing with 1.0 M Tris–HCl (pH 10). Purified IgG is then subjected to a second purification on an affinity chromatogra-

[15] In order to reutilize the column, it is washed with 0.2 M glycine–HCl (pH 2.3) to strip remaining protein, equilibrated with 0.1 M sodium phosphate (pH 8.1) containing 0.03% sodium azide and stored at 4°.

phy column containing the appropriate synthetic peptide conjugated to Sepharose. To conjugate the peptides, 3.4 μmol of peptide is dissolved in 300 μl of P$_i$/NaCl (pH 12.5), mixed with 240 μl of epoxy-activated Sepharose (Pharmacia), and the mixture is incubated at 37° for 20 hr. Unreacted groups are blocked with 1.0 M ethanolamine (pH 8.0). The IgG fraction is recycled through the peptide affinity column for 2 hr, and the column is then washed with P$_i$/NaCl until no protein is detected in the eluant. Finally, bound IgG is eluted with 0.2 M glycine–HCl (pH 2.3) and adjusted to pH 7.0 immediately with 0.2 M Tris–HCl (pH 8.6). Aliquots are frozen and stored in liquid nitrogen.

Preparation of Fab Fragments

Monovalent Fab fragments are prepared from purified Mabs or site-directed polyclonal antibodies by papain digestion,[16] and the Fc portion is removed by chromatography on protein A/Sepharose.[14] Protein is determined as described[17] with bovine serum albumin (BSA) as standard, and the purity of the preparations is assessed by sodium dodecyl sulfate–polyacrylamide gel electrophoresis (SDS–PAGE).[18] Aliquots are frozen and stored in liquid nitrogen.

Screening Procedures

There are a variety of screening procedures for detecting Mabs and site-directed polyclonal abs two of which have been particularly useful with *lac* permease. Each procedure affords different advantages and limitations.

Solid-Phase Radioimmunoassay (SP-RIA)

This procedure is useful when many samples are to be assayed, and the antigen may be used in purified or unpurified form (e.g., membrane vesicles and detergent extracts). With unpurified material, however, false positives are sometimes observed which is not the case with purified *lac* permease. In any case, it is desirable to perform the assays in duplicate. Since most membrane proteins appear to adsorb to nitrocellulose (Millipore) or poly(vinyl chloride) (Dynatech), microtiter plates made of either

---

[16] B. B. Mischell and S. M. Shiigi, *in* "Selected Methods in Cellular Immunology." Freeman, San Francisco, 1980.

[17] O. H. Lowry, N. J. Rosebrough, A. J. Farr, and R. J. Randall, *J. Biol. Chem.* **193**, 265 (1951).

[18] U. K. Laemmli, *Nature (London)* **227**, 680 (1970).

material can be utilized, although the latter are easier to manipulate. All manipulations are carried out at room temperature.

1. Apply 10–40 μl of the antigen solution (5–20 pmol of *lac* permease) to each well of microtiter plate, except for those used as negative controls. Incubate the plate overnight in a desiccator.

2. Fill the wells with 5% BSA in 50 mM potassium phosphate (pH 7.5) and incubate for 1 hr to block nonspecific sites.

3. Remove BSA by inverting plate and shaking.

4. Add 50 μl of tissue culture supernatant, ascites fluid or purified ab diluted with 1% BSA in 50 mM potassium phosphate (pH 7.5) if necessary. Incubate 2 hr.

5. Remove antibody solution, and wash the wells 4 times with solution A (10 mM Tris–HCl, pH 7.4/0.9% NaCl), followed by 4 washes with solution B (solution A plus 0.05% NP-40).

6. Apply 2×10^5 cpm of ^{125}I-labeled protein A[12] to each well in 50 μl of 1% BSA in 50 mM potassium phosphate (pH 7.5) and incubate for 45 min.

7. Wash as in 5.

8. Detect bound radioactivity by autoradiography at −70° with Kodak XAR-5 film and an intensifier screen (Cronex, Lightening Plus; DuPont) or by excising the wells and counting in a gamma counter.

A typical example of a SP-RIA with purified *lac* permease and hybridoma cell tissue culture supernatants that was exposed overnight is shown in Fig. 2. Approximately 42% of the wells are positive, as judged from the film densities.

Immunoelectroblotting

This assay provides more specific information by direct visualization of the antigen and is particularly useful when the antigen is not purified. If different membrane preparations are to be tested with the same antibodies, they are electrophoresed on a single gel, electroblotted, and the entire nitrocellulose sheet is processed. If the same membrane preparation is to be tested with different antibodies, the nitrocellulose sheet is cut vertically into strips after electroblotting, and each strip is processed independently.

1. Purified *lac* permease or membranes are subjected to SDS–PAGE as described.[19]

2. Excise a small vertical strip of the gel and stain with 0.5% Coomassie brillant blue (w/v) dissolved in 10% acetic acid (v/v) and 50% metha-

[19] M. J. Newman, D. L. Foster, T. H. Wilson, and H. R. Kaback, *J. Biol. Chem.* **256,** 11804 (1981).

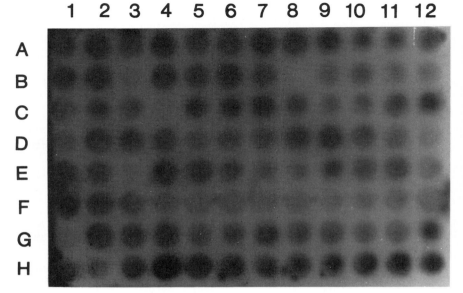

FIG. 2. SP-RIA of hybridoma tissue culture supernatants after primary cell plating. Purified *lac* permease (3.2 pmol) was applied in a 5 μl aliquot to each well and allowed to adsorb overnight. The plate was processed as indicated in the text and autoradiographed overnight. Control wells without antigen are located at positions H1 and H2 at the lower left.

nol (v/v). The stained gel is put aside and used later to assess the efficiency of electroblotting.

3. Electrotransfer the protein bands on to nitrocellulose[20] (BA85; Schleicher & Schuell) at 100 mA for 4 hr or overnight at 50 mA.

4. Stain the gel after electrotransfer to assess the efficiency of transfer (cf. 2).

5. Stain a narrow vertical strip of the nitrocellulose with 0.1% amido black (w/v) dissolved in 10% acetic acid (v/v), 43% methanol (v/v) for 20 min to visualize transferred material.

6. Block the main body of the nitrocellulose sheet with 5% BSA in 50 mM potassium phosphate (pH 7.5) for 1 hr.

7. Incubate the nitrocellulose sheet with crude or purified preparations of Mab or site-directed polyclonal antibody diluted as required in 1% BSA in 50 mM potassium phosphate (pH 7.5) for 2 hr.

[20] H. Towbin, T. A. Staehelin, and J. Gordon, *Proc. Natl. Acad. Sci. U.S.A.* **76**, 4350 (1979).

8. Wash the nitrocellulose sheet 4 times (5 min each) with 200 ml of solution A and 4 times with 200 ml of solution B.

9. Add 10^5 cpm of ^{125}I-labeled protein A in 1% BSA in 50 mM potassium phosphate (pH 7.5) for 50 min.

10. Wash as in 8.

11. Detect bound radioactivity by autoradiography at $-70°$ with Kodak XAR-5 film and a Cronex Lightening Plus intensifier screen (DuPont).

Results of an electroblotting experiment with culture supernatant from hybridoma 4B1[5] are shown in Fig. 3. Clearly, the hybridoma supernatant reacts with the major polypeptide in the purified *lac* permease that migrates at 33 kDa and with an aggregate of the permease at about 66 kDa.[21] Furthermore, essentially identical results are obtained with membrane vesicles containing the permease in addition to numerous other polypeptides.

If the efficiency of electroblotting is low, diffusion blotting[22] may be used. This procedure works well, and the volume of the transfer buffer is much smaller which conserves expensive materials such as octyl-β-D-glucopyranoside. However, longer times are required for transfer of the proteins from the gel to nitrocellulose (72 hr).

Binding Assays

In order to obtain information regarding the accessibility of different epitopes in the *lac* permease on the inner and outer surfaces of the membrane, direct binding assays are performed with ^{125}I-labeled antibodies in right-side-out and inside-out membrane vesicles and in proteoliposomes reconstituted with purified *lac* permease.[7]

Labeling of Antibodies—Glucose Oxidase/Lactoperoxidase-Catalyzed Iodination

1. Mix 100 μg of purified antibody [IgG or Fab fragments in 75 μl of 200 mM potassium phosphate (pH 7.5)] with 50 μl of Enzymobead reagent (Bio-Rad).

2. Add 1 mCi of Na^{125}I (Amersham; 300–600 mCi/ml diluted with water to a final concentration of 1 mCi/5 μl).

---

[21] Although the molecular weight of the *lac* permease is 46,500,[3,18,19] the protein migrates with an apparent M_r of $\approx$33,000 during SDS–PAGE in 12% gels.

[22] B. Bowen, J. Steinberg, U. K. Laemmli, and H. Weintraub, *Nucleic Acids Res.* **8,** 1 (1980).

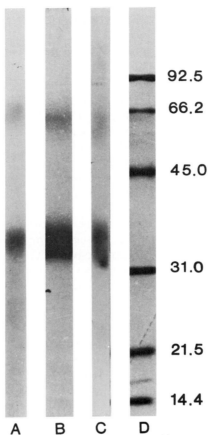

FIG. 3. Specificity of 4B1 hybridoma supernatant for the *lac* permease. Purified *lac* permease and T206 right-side-out membrane vesicles extracted with 5 *M* urea were subjected to SDS–PAGE (12% gels). Protein bands were then transferred to nitrocellulose by electroblotting as described. Individual strips were incubated with 4B1 hybridoma supernatant for 1 hr at room temperature, followed by washing and incubation with ^{125}I-labeled Protein A (~2 × 10[5] cpm/strip). Autoradiography was carried out at −70° for approximately 5 hr. (A) Purified *lac* permease after SDS–PAGE and staining with Coomassie brilliant blue; (B and C) autoradiograms of immunoblotted *lac* permease and T206 membrane vesicles, respectively; (D) molecular weight standards after SDS–PAGE and staining with Coomassie brilliant blue: phosphorylase (92.5 kDa), BSA (66.2 kDa), ovalbumin (45.0 kDa), carbonate dehydratase (31.0 kDa), soybean trypsin inhibitor (21.5 kDa), and lysozyme (14.4 kDa). From Carrasco *et al.*[5]

3. Start the reaction by adding 25 μl of 1% β-D-glucose and incubate for 7 min. Incubation time is critical with the antibody described here, as longer incubation periods modify the binding properties of the antibodies.

4. Terminate the reaction by separating free ^{125}I from ^{125}I-labeled protein by passage of the reaction mixture over an 8.0 ml Sephadex G-25 column previously equilibrated with 50 mM potassium phosphate (pH 7.5) containing 1% BSA.

5. Collect 0.5 ml fractions and determine radioactivity and protein in each fraction.

6. Pool fractions containing radiolabeled protein.

7. Determine the percentage of ^{125}I precipitated by trichloroacetic acid which is usually >90%. The specific activity of the ^{125}I-labeled protein ranges from 1 to 4 μCi/μg.

During iodination, certain antibodies are inactivated due presumably to modification of tyrosyl residues in the binding site. This can be avoided by performing the iodination reaction while the antibody is bound to antigen. Thus, abs directed against synthetic peptides are iodinated while bound to peptide/Sepharose affinity resin using chloramine-T. IgGs or Fab fragments (100 μg of protein) are incubated with 25 μl of resin in a Pasteur pipet and iodinated with 1 mCi of Na^{125}I (300–600 mCi/ml; Amersham).[23,24] Unbound material is washed through the resin with P$_i$/NaCl, and bound material is then eluted with 0.2 M glycine–HCl (pH 2.3) and adjusted immediately to pH 7.0 by addition of 0.2 M Tris–HCl (pH 8.6).

Binding of ^{125}I-Labeled IgG and ^{125}I-Labeled Fab Fragments

1. Right-side-out or inside out membrane vesicles from *E. coli* or proteoliposomes reconstituted with purified *lac* permease are incubated with 5% BSA in 50 mM potassium phosphate (pH 7.5) for 1 hr at room temperature.

2. Add a given amount of ^{125}I-labeled IgG or Fab fragments by mixing trace amounts of labeled with unlabeled material, and incubate at room temperature until maximum binding is achieved (usually <2 hr).

3. Preincubate 0.22-μm nitrocellulose filters (GSTF; Millipore) for 45 min with 5% BSA in 50 mM potassium phosphate (pH 7.5) to reduce nonspecific binding.

4. Flocculate vesicles by adding 1.0 ml of 50 mM potassium phosphate (pH 7.5) containing 250 μg of poly(L-lysine) (Sigma; molecular

---

[23] D. M. Weir, *in* "Handbook of Experimental Immunology." Blackwell, Oxford, 1978.
[24] E. C. Greenwood, W. M. Hunter, and G. S. Glober, *Biochem. J.* **89,** 114 (1963).

weight, 140,000), and after 10 sec, filter and wash twice with 3 ml of 50 mM potassium phosphate (pH 7.5) containing 5% BSA.

5. Radioactivity retained on the filters is assayed in a gamma counter.

Exemplary binding assays with ^{125}I-labeled Mab 4B1 (Fig. 4A) and 4B1 Fab fragments (Fig. 4B) show clearly that the epitope is accessible from the external surface of the membrane. Thus, both 4B1 and 4B1 Fab fragments bind almost exclusively to right-side-out vesicles relative to inside-out vesicles. Furthermore, with right-side-out vesicles, both immunological reagents exhibit a maximum at a concentration of IgG or Fab that is proportional to the amount of *lac* permease present (in the assays shown, 20 pmol of *lac* permease binds 10 pmol of 4B1 IgG and 20 pmol of 4B1 Fab, i.e., intact 4B1 binds bivalently and 4B1 Fab fragments bind monovalently) and then decreases as increasing amounts of reagent at constant specific activity are added.

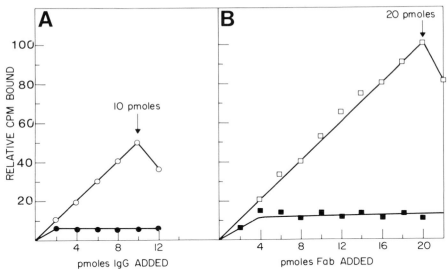

FIG. 4. Topological distribution of the 4B1 epitope on the periplasmic and cytoplasmic surfaces of the membrane. Right-side-out (open symbols) and inside-out (closed symbols) membrane vesicles prepared from *E. coli* ML 308-225 [46 µg of membrane protein/sample (closed symbols) containing approximately 20 pmol of *lac* carrier protein] were incubated in 50 mM potassium phosphate (pH 7.5) containing 5% BSA for 1 hr at 25°. Given amounts of ^{125}I-labeled 4B1 (A) or ^{125}I-labeled Fab fragments (B) were added, and the incubations were continued for 2 hr at 25°. Samples were then flocculated with poly(L-lysine), filtered, and assayed for bound radioactivity as described in the text. The data shown were corrected for nonspecific binding which was determined by incubation of vesicles with 10-fold molar excess unlabeled antibody (relative to *lac* permease) prior to incubation with ^{125}I-labeled antibody.

The efficacy of the technique and the unusual shape of the binding curves are related to two factors: (1) the high affinity of the antibody for its epitope and (2) the somewhat lower affinity of the iodinated antibody relative to that of the unmodified molecule. In the presence of excess antigen (i.e., *lac* permease), labeled antibody and unlabeled antibody bind independently until all of the binding sites are occupied. Further addition of antibody past saturation results in a "paradoxical" decrease in binding that is due to competition between labeled and unlabeled antibody molecules for a limited number of binding sites. The interpretation is verified by the study shown in Fig. 5 in which proteoliposomes reconstituted with 10.5 pmol of purified *lac* permease per sample were preincubated with increasing amounts of unlabeled Mab 4B1, and subsequently, a trace amount of ^{125}I-labeled 4B1 was added. As shown, binding of labeled 4B1 is not affected by preincubation with up to 4.6 pmol of unlabeled 4B1, but decreases precipitously between 4.6 and 5.8 pmol, approaching the level observed for nonspecific binding (i.e., liposomes prepared in the absence of *lac* permease). Therefore, exposure of 10.5 pmol of permease to 4.6–5.8 pmol of relatively high-affinity unlabeled 4B1 is sufficient to saturate the epitope, thus blocking subsequent binding of the labeled probe which

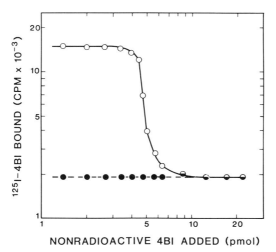

NONRADIOACTIVE 4BI ADDED (pmol)

Fig. 5. Effect of unlabeled 4B1 on binding of ^{125}I-labeled 4B1. Proteoliposomes containing 10.5 pmol of *lac* permease and 200 μg of *E. coli* phospholipid/sample (○) were incubated with given amounts of unlabeled antibody 4B1. After a 1 hr incubation at 25°, a trace amount of ^{125}I-labeled 4B1 was added (about 0.15 pmol) and incubation was continued for 1 hr, at which time the samples were filtered as described above. Identical experiments were carried out with liposomes (200 μg of phospholipid/sample) formed from *E. coli* phospholipids in the absence of protein (●). From Herzlinger *et al.*[7]

has a lower affinity. In this context, it is also noteworthy that the maximum level of ^{125}I-labeled 4B1 binding occurs at a concentration that consistently exhibits the same proportionality to the amount of *lac* permease (i.e., at the maximum, the stoichiometry of bound IgG: *lac* permease and Fab: *lac* permease is consistently 0.5 and 1.0, respectively), while the extent of paradoxical binding varies from one preparation of labeled IgG or Fab to another.

Concluding Remarks

Because of their affinity, high degree of specificity and inability to penetrate membranes, immunological probes such as those described here are particularly useful for structure/function studies of membrane proteins. In this regard, Mabs and site-directed polyclonal abs have been used as complementary approaches to the study of the *lac* permease, and it is apparent that each type of reagent has specific advantages and disadvantages.

Although laborious, the preparation of Mabs against purified immunogens such as *lac* permease is relatively straightforward. Furthermore, once the task is accomplished, these highly specific, molecularly uniform immunologic reagents can be prepared in almost unlimited quantity. On the other hand, it is not possible to predict a priori the molecular nature of the epitope within the immunogen, and identification may be extremely difficult once the Mab is in hand, as the epitope may be discontinuous (i.e., the antibody may not be directed against primary structure). This appears to be the situation with Mab 4B1. Thus, although studies with this Mab demonstrate that the 4B1 epitope is accessible on the exterior surface of the bacterial membrane and that this orientation is maintained in reconstituted proteoliposomes,[6,8] the immunoreactivity of the *lac* permease is drastically diminished when the protein is completely denatured, and it has not been possible to identify immunoreactive cyanogen bromide peptide fragments. In contrast, the epitope for 4A10R, another Mab obtained from the same fusion, is composed in part of the carboxy terminus of the permease which is on the cytoplasmic surface of the membrane.[25]

Another advantageous aspect of Mabs is their potential for inhibiting catalytic activity. Out of over 60 Mabs screened for inhibition of lactose: proton symport in right-side-out membrane vesicles and proteoliposomes, one Mab, 4B1, was found to inhibit lactose transport in a highly specific manner.[6] That is, Mab 4B1 inhibits all translocation reactions catalyzed by the permease that involve the coupled movement of lactose

---

[25] D. Herzlinger, N. Carrasco, and H. R. Kaback, *Biochemistry* **24**, 221 (1985).

and protons with no effect on exchange or binding of the high affinity substrate *p*-nitrophenyl-α-D-galactopyranoside.[5,26] No such inhibition has yet been observed with any of the site-directed polyclonal antibodies.

The most significant advantages of site-directed polyclonal antibodies is that they are directed against a predefined epitope and that their generation takes less time and less effort than Mabs. However, since site-directed polyclonal IgG usually represents a low percentage of the IgG fraction, availability is limited. Furthermore, site-directed polyclonal antibodies are produced by classical immunological methods, and they are not as uniform molecularly as Mabs. In any case, binding studies in right-side-out and inside-out membrane vesicles with radiolabeled site-directed polyclonal antibody directed against the carboxyl terminus of the permease demonstrate that this portion of the permease is on the cytoplasmic surface of the membrane.[8,27] Similar preliminary studies with site-directed polyclonal abs against hydrophilic segments 5 and 7a suggest that these segments are also more accessible from the cytoplasmic surface of the membrane. On the other hand, although antibodies directed against hydrophilic segments 1 (the amino terminus), 2, 6, 8, 10, 11, and 12 react with the appropriate peptides, as judged by SP-RIA, and with the *lac* permease on immunoblots, the antibodies do not react with the permease in the membrane. Thus, these segments either do not protrude sufficiently from the membrane or are buried within the tertiary structure of the polypeptide.

[26] G. Rudnick, S. Schuldiner, and H. R. Kaback, *Biochemistry* **15**, 5126 (1976).
[27] R. Seckler, J. K. Wright, and P. Overath, *J. Biol. Chem.* **258**, 10817 (1983).

[34] Glycerol Facilitator in *Escherichia coli*

By E. C. C. LIN

Unlike most carbohydrates, glycerol enters *E. coli* by facilitated diffusion rather than by active transport or vectorial phosphorylation. The facilitated diffusion is mediated by an inner-membrane carrier encoded by a gene (*glpF*) belonging to the same operon that contains the structural gene for glycerol kinase (*glpK*). This transport process, being independent of metabolic energy, can at best result in the equilibration of the substrate across the membrane barrier. It is a striking fact that glycerol is not known to be transported across any cell membranes. Perhaps nonspecific permeability of most biological membranes to this small uncharged compound prevented the evolution of an active transport system, which would effectually lead to the dissipation of energy.[1]

Principle of the Optical Assay

In theory, facilitated diffusion might be assayed by introducing a labeled substrate to a cell suspension and determining the initial rate of entry. In practice, this approach is feasible only if the rate of transport is slow relative to the time required for separating the cells from the incubation medium, or if the entering substrate is converted into a compound that cannot escape. Otherwise, the rate of substrate accumulation will rapidly diminish with the rise of its internal concentration. In the case of rat hepatocytes, it is possible to measure facilitated diffusion of [^{14}C]glycerol by following the accumulation of radioactivity in the cells, because the compound is trapped by internal phosphorylation and entry is the rate limiting step.[2] Unfortunately both mediated and nonspecific diffusions of glycerol across the cytoplasmic membrane of *E. coli* are too fast for this kind of measurement. However, a convenient optical assay for the glycerol facilitator was devised by taking advantage of the fact that a hypertonic solution of glycerol or its structural analog xylitol[3] causes transient shrinkage of the cytoplasmic volume.[4-7] As water exits, the cytoplasmic density increases, and this can be detected by decreases in transmitted light through the cell suspension (Fig. 1). In cases where the cytoplasmic membrane is impermeable to the solute, the plasmolysis (shrinkage) persists and consequently the decrease in optical density of the cell suspension is maintained. In cases where the solute penetrates the cytoplasmic membrane, the osmotic pressure across the membrane cannot be sustained; water reenters the cytoplasm, resulting in deplasmolysis (reswelling) and restoration of the original optical property of the cells. The time it takes to give an optical density of the cell suspension that is midway between that of maximal plasmolysis and that of full equilibration is the $t_{1/2}$ for the permeation constant of the solute. Figure 2 documents by

[1] E. C. C. Lin, *in* "The Cell Membrane" (E. Haber, ed.), p. 109. Plenum, New York, 1984.
[2] C.-C. Li and E. C. C. Lin, *J. Cell. Physiol.* **117,** 230 (1983).
[3] K. B. Heller, E. C. C. Lin, and T. H. Wilson, *J. Bacteriol.* **144,** 274 (1980).
[4] Y. Sanno, T. H. Wilson, and E. C. C. Lin, *Biochem. Biophys. Res. Commun.* **32,** 344 (1968).
[5] D. P. Richey and E. C. C. Lin, *J. Bacteriol.* **112,** 784 (1972).
[6] M. M. Alemohammad and C. J. Knowles, *J. Gen. Microbiol.* **82,** 125 (1974).
[7] M. O. Eze and R. N. McElhaney, *J. Gen. Microbiol.* **105,** 233 (1978).

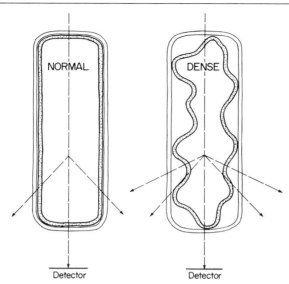

FIG. 1. Increased light scattering by a bacterial cell caused by plasmolysis.

electron micrographs the ability of a nonpenetrating compound (sucrose or sodium chloride) and the failure of a penetrating compound (glycerol) to sustain plasmolysis of *E. coli*.

Preparation of the Cells

To avoid catabolite repression and inducer exclusion,[8] L-alanine, L-aspartate, and L-glutamate (2 g/liter each) can be used as principal carbon and energy sources with the following amino acids as supplements (concentrations in mg/liter): L-arginine (hydrochloride), 60; L-cysteine (hydrochloride), 13; glycine, 30; L-histidine, 10; L-isoleucine, 30; L-phenylalanine, 30; DL-serine, 60; L-threonine, 30; L-tryptophan, 10; L-tyrosine, 20; L-valine, 30. Stock amino acid solutions are sterilized by filtration (450 nm pore size; Millipore Corp., Bedford, Mass). If accurate basal activity of glycerol facilitator need not be determined, 1% casein hydrolyzate or tryptone (both of which contain glycerol and *sn*-glycerol 3-phosphate as contaminants[9]) may be used instead of the synthetic amino acid mixture.[5] When induction is desired, glycerol or *sn*-glycerol 3-phosphate (the *glpKF* operon is a member of the *glp* regulon that responds to *sn*-glycerol 3-

[8] W. B. Freedberg and E. C. C. Lin, *J. Bacteriol.* **115,** 816 (1973).
[9] N. R. Cozarelli, W. B. Freedberg, and E. C. C. Lin, *J. Mol. Biol.* **31,** 371 (1968).

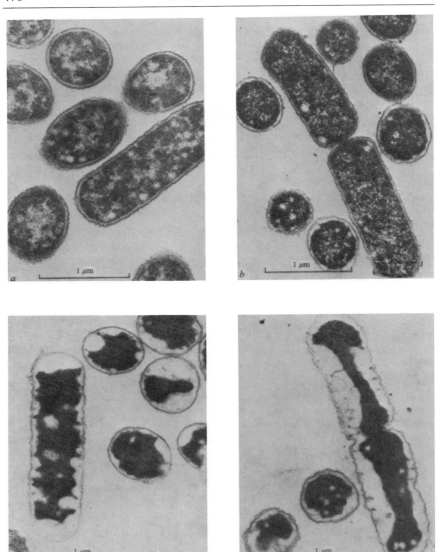

FIG. 2. Electron micrographs of stained sections of *E. coli* cells fixed under different osmotic conditions. Imidazole buffer (10 mM, pH 7) was used for all samples. (a) No additional solute added; (b) 0.34 M glycerol; (c) 0.2 M NaCl; and (d) 0.3 M sucrose. From Alemohammad and Knowles.[6]

phosphate as the true effector) is added at 5 mM. The lowest basal activity of glycerol facilitator is found when glucose is used as the carbon and energy source. However, the presence of glucose prevents induction of the *glp* system.

Cultures of *E. coli* K12 (25 ml in 300-ml flasks fitted with side arms for reading of turbidity in a Klett–Summerson colorimeter with a number 42 filter, or 250 ml in 2-liter flasks) are grown aerobically in mineral medium[10] with appropriate carbon and energy sources and are harvested during exponential phase by centrifugation. After being washed with a volume of 50 mM morpholinopropane sulfonate (MOPS), pH 7.0, equal to that of the culture medium, the cells are resuspended in that buffer to give a final density at 600 nm of about 3.0 (approximately 30 mg dry weight per ml).

Determination of the Half Time of Equilibration

The entry of a compound through the facilitator protein is detected by comparing the osmotic behaviors of cells with and without the membrane protein. The process is monitored photometrically at 25°. One milliliter of a 250 mM carbohydrate solution (hypertonic) in 50 mM MOPS (pH 7.0) is placed in a 1-ml cuvette (1 cm light path) and read in a spectrophotometer. A 0.1 ml sample of a concentrated cell suspension is added to the cuvette to give a final optical density at 600 nm of about 1.0. Rapid mixing is achieved by aspirating and expelling the contents several times with a Pasteur pipet. The optical density is continuously monitored with a recorder attached to the spectrophotometer. Figure 3 shows the responses of wild-type *E. coli* and a *glpF* mutant (lacking glycerol facilitator) when their *glp* regulons were induced. With glycerol as the solute, equilibration across the cytoplasmic membrane of induced wild-type cells occurred during the 10-sec period required for mixing. Consequently, no plasmolysis could be recorded. In contrast, *glpF* cells showed clear reswelling with a $t_{1/2}$ of approximately 0.5 min. With xylitol as the substrate analog, the $t_{1/2}$ for the facilitator-positive cells was about 0.7 min. For the facilitator-negative cells the $t_{1/2}$ was too long to be measured on the same time scale, as evidenced by the sustained plasmolysis.

Comments

Comparison of wild-type and *glpF* cells showed that glycerol facilitator is not highly specific. The table summarizes the $t_{1/2}$ of a number of compounds found to enter the cell through the *glpF* membrane protein.[3]

[10] S. Tanaka, S. A. Lerner, and E. C. C. Lin, *J. Bacteriol.* **93,** 642 (1967).

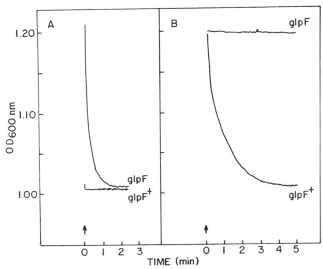

FIG. 3. Optical changes associated with shrinkage and swelling of *E. coli*. Cells were added (at arrow) to a solution of 250 m*M* glycerol (A) and xylitol (B). In the figure, a reading of 1.2 indicates shrinkage of the cell. The fall in optical density to 1.0 was due to reswelling of the cells as a result of penetration of glycerol or xylitol. From Heller *et al.*[3]

SPECIFICITY OF THE GLYCEROL FACILITATOR OF
E. coli K12

Compound admitted	$t_{1/2}$ of equilibration (min)	
	glpF[+]	*glpF*
Glycerol	<<0.1	0.4
DL-Glyceraldehyde	<<0.1	0.4
Glycine	5	∞
Urea	0.1	0.2
Erythritol	0.3	2.6
D-Arabitol	3.4	~10
L-Arabitol	4.6	~10
Ribitol	0.1	∞
Xylitol	0.7	∞
D-Galactitol (dulcitol)	7.0	∞
D-Mannitol	~25	∞
D-Sorbitol (glucitol)	8.8	∞

Compounds not transported by the glycerol facilitator include: sn-glycerol 1-phosphate, sn-glycerol 3-phosphate, glycylglycine, inositol (cyclohexitol), erythrose, D-arabinose, L-arabinose, D-ribose, D-xylose, D-galactose, D-mannose, and D-sorbose. Thus, it seems that 5-carbon and 6-carbon compounds can be admitted by the facilitator if they are in the form of straight chains, but not when they exist as rings. The glycerol phosphates are excluded, presumably because of the negative charges. The optical osmotic assay with xylitol as the substrate was validated by another assay in which the entry and exit rates of ^{14}C-labeled xylitol (100 mM) were measured in wild-type and glpF mutant cells. Because cells without glycerol facilitator are essentially impermeable to xylitol and cells fully induced in the transport protein have a convenient $t_{1/2}$ for the compound, this straight chain polyol is a substrate of choice. Xylitol is not a growth supporting compound for E. coli and is not metabolizable by it at a significant rate.

Permeation via the facilitator showed no competitive inhibition and substrate saturability. Furthermore, the process showed little temperature dependence. These features, together with the permeability of the substrate according to molecular size rather than molecular structure, suggest that the membrane protein provides an aqueous diffusion channel, estimated to have a diameter of about 0.4 nm.[3]

Note Added in Proof. Since this manuscript was submitted, a special case of accumulation of glycerol against a concentration gradient has been reported for the salt-tolerant yeast, *Debaryomyces hansenii* [L. Adler, A. Blomberg, and A. Nilsson, *J. Bacteriol.* **162,** 300 (1985)].

Acknowledgments

Work in the laboratory of the author was supported by Grant PCM83-14312 from the National Science Foundation and by Public Health Service grant 5 R01 GM11983 from the National Institute of General Medical Sciences.

[35] Polyhydric Alcohol Transport by Bacteria

By JOSEPH W. LENGELER

For most bacteria, the polyhydric alcohols, abbreviated as polyols hereafter, are excellent carbon sources. These include the three hexitols D-mannitol, D-glucitol (formerly D-sorbitol), or galactitol (formerly dulcitol) found abundantly in natural environments, and the pentitols D- and L-arabinitol (formerly arabitol), ribitol (formerly adonitol), and xylitol. In bacteria metabolism of these polyols has been shown[1-3] to be initiated either (1) by dehydrogenation of the free intracellular polyol to the corresponding ketose, or (2) by the vectorial phosphorylation of the polyol to the corresponding phosphate ester concomitant with its translocation across the membrane in a process called group translocation. In the former cases the polyols are taken up through active transport systems, and accumulated in a chemically unaltered form in the cells. This type of transport and metabolism is characteristic for pentitols in the enteric bacteria, for all polyols in most obligate aerobes, and for the heterofermentative lactobacilli. The transport process is fully reversible and most likely coupled to ion gradients. Hexitols, in contrast, are taken up and phosphorylated in most obligate and facultative anaerobes (including the Enterobacteriaceae) or in homofermentative lactobacilli by a series of phosphoenolpyruvate-dependent carbohydrate : phosphotransferase systems (abbreviated as PTS). In some of the latter bacteria, even the pentitols are taken up by phosphotransferase systems. These systems catalyze a solute transport by membrane-bound proteins, called Enzymes II.

Enzymes II are substrate-specific, and normally inducible. They catalyze the translocation of carbohydrates across the membrane with a concomitant phosphorylation of this carbohydrate to the corresponding phosphate ester at the expense of phosphoenolpyruvate and two or frequently three soluble phosphocarrier proteins (Enzyme I, HPr, and Enzyme III). Polyols are trapped in this process invariably as the 5-phosphate for pentitols and as the 6-phosphate for hexitols (note that for sterical reasons D-mannitol 1-phosphate and galactitol 1-phosphate are identical to the corresponding 6-phosphates). All polyol phosphates are the first intermediate in their subsequent metabolism. In wild-type strains, and under physiological conditions, group translocation is virtually irreversible. A general description of the PTS, of the purification of its major components and of their assays has been given recently in this series (Volume 90 [69–74]). Therefore, the present description will be restricted to the procedures for *in vivo* transport tests with intact cells and special mutants.

Principle of Assay. Suspensions of preinduced or constitutive mutant cells in an energized state are mixed in appropriate buffered media with radioactively labeled metabolizable or nonmetabolizable substrates and substrate analogs. Samples are withdrawn at short intervals and filtered through a bacteriological membrane filter. The filter is washed, dried, and

[1] J. B. Wolff and N. O. Kaplan, *J. Bacteriol.* **71,** 557 (1956).
[2] S. Tanaka, S. A. Lerner, and E. C. C. Lin, *J. Bacteriol.* **93,** 642 (1967).
[3] R. P. Mortlock, *Annu. Rev. Microbiol.* **36,** 259 (1982).

the radioactivity retained on it is measured. While this test itself is technically very simple, the preparation of the cells and the correct choice of the test conditions can be problematic.

Materials and Methods. The technical details as indicated have been worked out for enteric bacteria. The same methods can be applied to the study of other microorganisms after appropriated adaptations.

Choice of Strains and Isolation of Mutants. Most bacterial strains will take up each of the different polyols by more than one uptake system. Thus whenever possible mutant strains should be used in which all polyol transport systems except the one under study have been eliminated. Only kinetic data obtained from such strains can be expected to reflect the true substrate specificity and substrate affinity values of given transport system. Furthermore, mutant strains with a constitutive expression of this system should be used, since full induction of transport systems and the complete removal of the inducer before the tests can be difficult.

Media and Reagents. The standard minimal medium[2] contained 34 mM NaH$_2$PO$_4$ · H$_2$O, 64 mM K$_2$HPO$_4$, 20 mM (NH$_4$)$_2$SO$_4$, 1 mM HCl, 1 μM FeSO$_4$ · 7H$_2$O, 3 mM MgSO$_4$ · 7H$_2$O, 1 μM ZnCl$_2$, and 10 μM CaCl$_2$ · 2H$_2$O, at pH 7.0. Carbon sources were added to 10 mM, amino acids to 20 μg/ml, and vitamins to 1 μg/ml final concentration respectively in a filter sterilized form to the autoclaved liquid media.

Streptozotocin   [2-deoxy-2-(3-methyl-3-nitrosoureido)-D-glucopyranose was dissolved in water to 5 mg/ml, the pH adjusted to 5.0, and the solution kept frozen at −20°, not longer than 2 months. Since this drug is a strong mutagen and potential carcinogen, it must be handled and disposed of with adequate security measures. Destruction is best done under a hood with concentrated NaOH. MacConkey indicator plates contained 40 g of MacConkey Agar Base (Difco Laboratories, Detroit, MI) per liter. To this were added 10 g of polyols before autoclaving (15 min, 121°), or other sterilized carbohydrates after autoclaving to 1%: 2-Deoxy-*arabino*-hexitol (= 2-deoxymannitol = 2-deoxyglucitol), 2-deoxygalacitol, and 2-deoxypentitols (see below).

There are several convenient ways to isolate mutants with defects in the different polyol transport systems. Frequently, mutant selection procedures are preceded by mutagenesis of a culture according to any of various procedures described by J. R. Roth in this series (see Volume 17A [1]), or by transposon mutagenesis.[4]

---

[4] R. W. Davis, D. Botstein, and J. R. Roth, "Advanced Bacterial Genetics." Cold Spring Laboratory, Cold Spring Harbor, New York, 1980.

Streptozotocin Selection for Transport Negative Mutants

Principle.[5] The antibiotic streptozotocin is taken up and phosphorylated in sensitive bacteria by an Enzyme II^{Nag} of the PTS, specific for and inducible by N-acetylglucosamine. It is highly bactericidal but only after its uptake into the cells by generating intracellular diazomethane. Mutants unable to take up the drug due to the lack of the corresponding transport system, or due to the inability to energize uptake are resistant. Thus, if a mutagenized and starved bacterial culture is treated by the drug in the presence of a metabolizable carbon source, mutant cells lacking transport systems or a metabolic enzyme for this carbohydrate will survive preferentially. Since streptozotocin does not cause cell lysis and kills the cells irreversibly it is superior to penicillin or cycloserine in the selection of rare bacterial mutants.[6]

Procedure. Mutagenized cells are pregrown overnight in minimal medium containing N-acetylglucosamine as sole carbon source to eliminate unwanted auxotrophs or mutants with a defect in the general PTS proteins, and to preinduce the Enzyme II^{Nag}. The cells are centrifuged, washed once in minimal medium without supplements, and resuspended to 5×10^8 bacteria/ml in minimal medium containing all supplements, but no carbon source. They are starved under constant aeration at 37° during 60 min to deplete the internal energy pool. Next, the polyol, whose transport system is supposed to be absent in the wanted mutant, is added. Its concentration should be kept as low as possible (≤ 2 mM) since at high concentrations more than one transport system might be active. For high-affinity systems ($K_m \leq 10 \mu M$) 1 liter cultures with low cell titers (5×10^6 bacteria/ml) and low polyol concentration ($\leq 50 \mu M$) can be used, cell growth being monitored by plating appropriate dilutions on plates. As soon as the cell density increased by ~30%, streptozotocin is added by means of a mechanical pipet to 50 μg/ml. Any further increase of cell density will stop within the next 20 min. After 90 min, when the number of survivors dropped to 0.01%, the culture is centrifuged, the cells resuspended in minimal medium N-acetylglucosamine and grown out. This growth counterselects for unwanted PTS-negative and Enzyme II^{Nag}-negative mutants, and for pleiotropically carbohydrate-negative *cya* or *crp* mutants while the polyol transport-negative mutants sought after will grow out. Finally, the survivors are plated to 200 colonies per indicator plates. Among these from 0.01 to 1% are found to lack the transport

[5] J. Lengeler, *Arch. Microbiol.* **128**, 196 (1980).
[6] J. Lengeler, *FEMS Microbiol. Lett.* **5**, 417 (1979).

system or a metabolic enzyme for the polyol used during streptozotocin treatment.[7]

Suicide Mutant Selection

Principle.[8] Tritiated polyols, when added to preinduced and growing cells, will be incorporated into the DNA of wild-type cells and kill them, while barely affecting transport-negative mutants unable to take up the polyol.

Procedure. Mutagenized cells pregrown on the polyol under consideration are centrifuged, washed, and starved as described before. Then 2×10^7 bacteria/ml are resuspended in minimal medium containing all supplements in a screw cap tube and 30 μCi of tritiated polyol added per ml culture at a final concentration equivalent to approximately 10 times the K_m value of the polyol for its transport system. The tightly closed tubes (1 ml culture in a 25 ml tube) are incubated for 3 days at 37°, before the inactivation is ended by centrifugation of the culture. The cells are resuspended in minimal medium glycerol, grown out, and plated on appropriate indicator plates. The majority of these have defects in the corresponding polyol transport system.

Mutant Selection by Toxic Analogs or Intermediates

Principle.[8,9] Mutants with defects in polyol metabolic pathways frequently accumulate toxic phosphorylated intermediates if grown in the presence of this polyol. These include all mutants lacking the different polyol-phosphate dehydrogenases, the enzyme phosphofructokinase, or the enzyme fructose-bisphosphate aldolase. In strains of *E. coli* K12, the enzyme ketose-bisphosphate aldolase involved in galactitol metabolism is inactive at temperatures above 37°. If grown in the presence of the corresponding polyol, growth of such mutants on, e.g., glycerol is strongly inhibited except for mutant derivatives lacking the polyol transport system(s). These are selected in a positive way during such an incubation. This elegant selection procedure requires mutants with defects in a polyol catabolic pathway or alternatively in the glycolytic pathway. Such mutants are not needed if toxic nonmetabolizable polyol analogs are used. Instead mutants with a constitutive expression of the genes coding for a

[7] J. Lengeler, *Mol. Gen. Genet.* **179**, 49 (1980).
[8] J. Lengeler, *J. Bacteriol.* **124**, 26 (1975).
[9] E. Solomon and E. C. C. Lin, *J. Bacteriol.* **111**, 566 (1972).

polyol transport system are needed, since analogs do not induce such systems. Due to the broad substrate specificity of most carbohydrate transport systems, many analogs are substrates which in the case of the Enzymes II of the PTS are phosphorylated. Thus, 2-deoxy-*arabino*-hexitol, an analog of D-mannitol and of D-glucitol, or 2-deoxygalactitol are taken up and phosphorylated in the enteric bacteria, but for sterical reasons cannot be dehydrogenated to the corresponding ketose phosphate. If added to constitutive strains, these are inhibited rapidly, unless again the corresponding transport systems are lacking.[8,10,11] Finally, natural substrates taken up through unusual transport systems, e.g., D-arabinitol and the Enzyme II[Gat], can become toxic if toxic intermediates such as D-arabinitol 5-phosphate accumulate.[12,13]

Procedure. Mutant strains lacking a catabolic enzyme for a polyol and/or with a constitutive synthesis of such a system are grown exponentially in minimal medium glycerol to 2×10^8 bacteria/ml before the polyol or the toxic analog is added to 10 mM. The cultures are incubated until grown out completely. Alternatively, 5×10^8 bacteria are plated on a tryptone plate containing the selective polyol or polyol analog and grown out. Survivors are purified on indicator plates, there after negative and resistant colonies screened for defects in the corresponding transport systems.

Selection of Constitutive Mutants

Principle.[8] Many polyols can be taken up by a major transport system with a high affinity and induce this system. Most, however, can also be taken up by a minor system not inducible by that particular polyol. Such (a) minor system(s) for D-mannitol are the D-glucitol or the D-arabinitol transport system, for D-glucitol are the D-mannitol or the galacitol systems, for galacitol are the D-glucitol or the galacitol systems, for xylitol and probably L-arabinitol is the D-arabinitol system, for D-glucose and D-fructose is the D-glucitol transport system in the enteric bacteria. From mutants lacking the major transport system for a given polyol or carbohydrate, positive derivatives can be selected by plating out on minimal media plates containing this substrate as the only carbon source.[8,11,14,15]

[10] C. E. Delidakis, M. C. Jones-Mortimer, and H. L. Kornberg, *J. Gen. Microbiol.* **128**, 601 (1982).
[11] M. V. Sarno, L. G. Tenn, A. Desai, A. M. Chin, F. C. Grenier, and M. H. Saier, Jr., *J. Bacteriol.* **157**, 953 (1984).
[12] A. M. Reiner, *J. Bacteriol.* **132**, 166 (1977).
[13] G. A. Scangos and A. M. Reiner, *J. Bacteriol.* **134**, 501 (1978).
[14] M. C. Jones-Mortimer and H. L. Kornberg, *J. Gen. Microbiol.* **96**, 383 (1970).
[15] J. Lengeler and H. Steinberger, *Mol. Gen. Genet.* **164**, 163 (1978).

Except for true revertants, most derivatives carry mutations leading to the constitutive expression of a minor transport system. These suppressor mutations can normally be identified easily on MacConkey indicator plates. In contrast to the deep purple color of true revertant colonies, colonies of suppressor mutants are white with a red center.

Procedure. Few strains of *E. coli* are able to grow on pentitols. If a *pts* mutation is introduced into such a pentitol-positive strain (e.g., by strep-tozotocin selection), they lose their ability to grow on hexitols, while they remain able to grow on the pentitols D-arabinitol and ribitol. Mannitol-positive derivatives still lacking the PTS protein and thus negative for D-glucitol, galactitol, and D-fructose fermentation usually have a constitutive expression of the D-arabinitol transport system.[2] These mutants invariably become highly sensitive toward galactitol and galactitol-resistant colonies have usually lost the arabinitol transport system. In *pts* strains, selection for constitutivity can be done as indicated above.[2,8,11–15]

Complementation Assays for PTS Functions

Principle.[2,8] During the isolation of mutants with defects in different PTS proteins, it is often useful to have a convenient and rapid albeit semiquantitative complementation test. Crude cell extracts from a mutant strain lacking all substrate-specific and membrane-bound Enzymes II for polyols (e.g., the *E. coli* K12 strain JWL146 *mtlA gutA gatA*)[8] are used to supply an excess of the general PTS proteins Enzyme I and HPr. The Enzymes II to be tested are added in limiting amounts in the form of purified membrane vesicles. The assay is based on the separation of a radioactively labeled polyol from its phosphate formed by the complete PTS by binding the phosphate to a DEAE anion exchange filter paper and washing off the nonphosphorylated polyol. Besides being convenient and rapid, the test is very sensitive, reproducible, and specific. Because the exact amounts of PTS proteins in the test are unknown the test is only semiquantitative, but clearly sufficient for rapid screening programs. If done correctly, the kinetic values obtained from such tests are close to the value from reconstituted systems using purified components.

Procedure.[16] To obtain cell-free extracts containing the general PTS proteins Enzyme I and HPr, cells of strain JWL146 (triple negative for the three hexitol-specific Enzymes II) were grown to 2×10^9 bacteria/ml in minimal medium containing 1% casamino acids and 10 mM D-glucose under low aeration (500 ml culture in a 2-liter Erlenmeyer flask on a gyratory shaker at 150 rpm and 37°). To obtain Enzymes II, the cells were

[16] J. Lengeler, *J. Bacteriol.* **124,** 39 (1975).

grown under higher aeration (100 ml cultures in similar flasks) in minimal medium containing either the polyol to be tested for inducible strains, or 1% casamino acids for constitutive strains. The cultures were rapidly cooled, centrifuged, and washed twice in ice-cold NaCl (1%), before the cell pellets were frozen at −20° where they can be stored for months without losing appreciably any PTS activity. The frozen cells are resuspended to 200 mg of wet weight per ml in tris(hydroxymethyl)aminomethane (Tris) hydrochloride buffer (0.1 M, pH 7.6), or, if necessary, in other nonanionic buffers unable to bind to the DEAE filter. The cells are either treated sonically while being chilled in a −20° bath, or homogenates of the cells are passed twice through a French pressure cell (10,000 psi). The disrupted preparations are centrifuged at 30,000 g for 20 min and the supernatant used for enzyme assays. The Enzyme II activities in such extracts are stable to repeated freezing and thawing, while the [Enzyme I + HPr] activity drops after three consecutive freezings and thawings to below 5%, especially if diluted in addition. When membrane-free preparations are needed, cell lysis is best done by lysozyme treatment of frozen and rapidly thawed cells which are lysed more efficiently than nonfrozen cells. To 100 mg of frozen cells rapidly thawed in Tris (1 ml, 0.1 M, pH 7.6) containing 1 mM mercaptoethanol or DTT (DL-dithiothreitol) and 0.25 mM ethylene diaminetetraacetic acid and incubated for 10 min at 28° with gentle shaking are added 2 mg of crystalline lysozyme. After 15 min, 80 μl of a deoxyribonuclease solution (1 mg/ml in 0.5 M MgCl$_2$) is added, and incubated for 5 min before the extract is chilled and centrifuged (20 min at 30,000 g).

For Enzyme II assays the mixture contained in a total volume of 200 μl: 5 mM PEP, 50 μM MgCl$_2$, 100 μl extract of strain JWL146 (∼18 mg/ml of protein), 20 μl of ^{14}C-labeled polyol, and Tris (0.1 M, pH 7.6) to 200 μl. Final concentrations of the polyols in the test mixture were 30 μM D-[^{14}C]mannitol (2.5 Ci/mol), 200 μM D-[^{14}C]glucitol (5 Ci/mol), 100 μM [^{14}C]galactitol (5.5 Ci/mol). After a 5 min equilibration period of the extract at 25° to allow the formation of phospho-HPr, the reaction is started by the addition of 20 μl of the Enzyme II extract (1–5 mg/ml of protein) to be tested. At 1, 5, and 10 min, 50 μl samples are delivered on a DEAE filter disc (Whatman DE-81, 23 mm diameter) which is dropped immediately into 80% ethanol to stop any further reaction. The filters are washed three times in deionized water and the radioactivity on the filters is counted. To determine Enzyme I + HPr activities of a cell extract, increasing amounts of the latter are added to membrane preparations (eventually from a mutant lacking the general PTS proteins) containing high amounts of Enzyme II^{Mtl}. This Enzyme II is best suited for such tests since it does not require an Enzyme III and is very stable to freezing and thawing or dilution. All activities are expressed in nmol/min/mg of Enzyme II protein.

Purification of Polyols by Scavenging

Principle. Most commercially available polyol preparations, especially if labeled "for bacteriological use" contain up to 5% of contaminating isomers. These contaminations cannot be removed by recrystallization, nor for larger quantities by column chromatography. Cells of a mutant, however, able to take up the contaminating polyol through a high-affinity and constitutive transport system, but unable to ferment the polyol to be purified can be used to scavenge the contaminant bringing its concentration easily down to below 0.001%.

Procedure. Cells of the constitutive scavenger strain are pregrown exponentially on minimal medium glycerol to 5×10^8 bacteria/ml. The culture is centrifuged and washed carefully at room temperature, resuspended in minimal medium without a carbon source to 2.5×10^8 bacteria/ml, and starved for 30 min at 37° under constant aeration. The polyol to be purified, eventually after recrystallization, is added to 50–100 mM final concentration, and the cells incubated further until any increase of the culture density stops (normally after one doubling or 90 min). The culture is rapidly centrifuged, and the supernatant filtered immediately through a membrane filter (0.45 μm pore size) to eliminate all bacteria. The exact concentration of the purified polyol is determined by the periodate method according to Korn,[17] while the concentration of the contaminating polyol can be tested by isotope dilution assays. The inhibition of the uptake of low concentrations (close to the K_m value) of the contaminating polyol through its transport system by added purified polyol is used to calculate the concentration of the contaminant.

Preparations of Polyols or of Polyol Analogs

Principle.[1,18,19] Nonmetabolizable analogs or radioactively labeled polyols are not always available while the corresponding aldoses are. Aldoses and their analogs are reduced quantitatively by an excess of borohydride to the polyol form. This includes tritiated derivatives if tritium-labeled borohydride is used. The procedure is similarly suited for gram or milligram quantities. As an example, the preparation of 2-deoxy-*arabino*-hexitol from 2-deoxyglucopyranoside is described.

[17] E. D. Korn, *J. Biol. Chem.* **215**, 1 (1955).
[18] P. D. Bragg and L. Lough. *J. Am. Chem. Soc.* **79**, 4347 (1957).
[19] D. French, G. M. Wild, B. Young, and W. J. James, *J. Am. Chem. Soc.* **75**, 709 (1953).

Procedure. Ten grams of "glucose-free" 2-deoxyglucopyranoside is dissolved in 100 ml of distilled water. To this solution is added under constant stirring and dropwise a solution of sodium borohydride (2 g in 100 ml H_2O). The mixture is incubated at room temperature until gassing stops (from 30 to 90 min depending on the analog used[18]). The pH of the mixture is lowered to 5.0 by adding acetic acid (5 N) which destroys any excess of borohydride. The mixture is run over a small (1.3 × 23 cm) Dowex 50 WX-8 cation exchange column at a flow rate of 25 ml/hr to remove the sodium ions. For small preparations the resin is added batchwise and removed by filtration. The supernatant is brought to pH 6.0 before it is evaporated to dryness in a rotation evaporator at 50°. The syrup is dissolved in 150 ml of methanol, and the mixture evaporated again. This procedure is repeated at least three times until the volatile methyl borate disappears completely. The final material is recrystallized from methanol (150 ml) overnight by incubation at room temperature such that large crystals or a precipitate are formed. These are filtered off, dissolved in 15 ml of distilled water, and the amount of reducing activity left is determined, e.g., by the 2,3,5-triphenyltetrazolium method according to French *et al.*,[19] while the exact concentration of the polyol (-analog) is determined by the periodate method according to Korn.[17]

Growth of Cells and Preparation of Cell Suspensions for Transport Tests

The standard minimal medium used for all enteric bacteria is as described for the mutant selections. Maximal rates of polyol transport are observed in cells pregrown exponentially on the polyol to be tested, except for strains expressing the transport system in a constitutive way. These normally have the highest activities when growing exponentially on glycerol, on succinate, on one of the minor substrates for the transport system, or on other nonrepressing carbon sources such as casamino acids (1%). During growth on the major substrate a severe catabolite repression reduces the transport rates. Finally, when the cells have to be energized during the tests of active polyol transport systems, they must be pregrown on a mixture of the inducing polyol (5 mM) and of glycerol (5 mM). Cells from a fresh overnight culture, preferentially in the same medium, are diluted to 2.5 × 10⁷ bacteria/ml and grown to 2–5 × 10⁸ bacteria/ml with good aeration (not more than 15 ml culture in a 100 ml flask shaking at 200 rpm). The cells are harvested and washed by centrifugation at room temperature. One washing with minimal medium is sufficient unless the growth medium contained substrates for the transport system, e.g., as an inducer. In this case, two washings with careful resuspension of the cells are necessary. Washing at low temperature should be avoided since efflux rates for active transport systems can be considerable under these conditions. The cells are resuspended finally to 5×10^8 bacteria/ml in minimal medium lacking all supplements and carbon sources. They are used immediately for active transport system tests, or within 2 hr for PTS Enzyme II tests.

Reagents and Materials for Transport Assays

Labeled Polyols. For standard uptake assays, the final concentrations of the polyols during the tests were 5 μM D-[^{3}H]mannitol (20 Ci/mol), 25 μM D-[^{3}H]glucitol (10 Ci/mol), 25 μM [^{14}C]galactitol (5.5 Ci/mol), and 25 μM D-[^{14}C]arabinitol (3.4 Ci/mol). In general, final concentrations should not exceed the K_m value by more than 10-fold to minimize uptake through low-affinity and secondary transport systems.

Wash Medium. Standard minimal medium without any supplements is used as the wash medium at room temperature for active transport systems, and at 0° for all phosphotransferase systems.

Membrane Filters. Bacteriological membrane filters (cellulose acetate or nitrate; 25 mm in diameter, 0.65 μm pore size with a higher flow rate than the 0.45 μm filters used for sterilization) are used. If high bacterial densities ($\geq 5 \times 10^8$ bacteria/ml) are needed, glass fiber filters (e.g., Whatman GFL E) can be used on top of a membrane filter to prevent clogging of the pores.

Filtration Apparatus. An all-plastic filtration unit, preferentially six or more funnel manifold, similar to the one described by H. Rosenberg (this series, Volume 90 [33]) is best suited for the uptake assays as described.

Units. Rates should be expressed in nmol/min/mg of protein. For enteric bacteria growing exponentially in minimal media and treated as described, one absorbance unit measured at 420 nm corresponds roughly to 4.5–5.0 $\times$ 10^8 bacteria/ml, equivalent to 0.5 mg of wet weight or 0.25 mg of dry weight, and 0.125 mg of protein.

Transport Assays

Washed cells are normally resuspended to 1×10^8 bacteria/ml, but for high-affinity transport systems (K_m values below 10 μM) they should be even reduced to 2×10^7 bacteria/ml. To 0.9 ml of this suspension kept at 25° in a Wasserman tube, are added by means of a mechanical pipet 100 μl of the labeled polyol (10-fold concentrated). Three to four tubes are prepared for each test, adding the label to each of them individually under vigorous mixing. After 10, 20, 30 sec etc., transport is stopped. For PTS tests this is done by adding from a squeeze-bottle approximately 5 ml of ice-cold minimal medium to the tube, collecting the cells immediately on a membrane filter prepared before on the filter holder, and washing once with 5 ml of cold medium. Since the cells of enteric bacteria contain high amounts of intracellular PEP, no additional energy source is needed during uptake. Since, furthermore, the process is virtually irreversible under the test conditions used, cold wash medium which stops uptake efficiently can be used. With low cell densities and short incubation times aeration as the consequence of mixing the cells and the label is sufficient. If higher cell densities ($\geq 5 \times 10^8$ bacteria/ml) or larger volumes are used, the cells must be resuspended in small flasks under constant shaking, or air must be blown through the sample by means of capillary tubes. For each test series a 0 min sample is run in parallel by using cells incubated for at least 5 min before the test at 0°, adding the label, filtering, and washing at the low temperature. The amount of radioactivity on the dried filters is determined in a scintillation counter using 5 ml of the usual PPO-POPOP scintillant.

Assays of active transport systems are done in a similar way for the 0 min sample. For all others, however, the wash medium is used at 25°. If energized cells are needed, glycerol (1 mM) is added to 4.5 ml of glycerol-grown cells at 25° 5 min before the test under constant aeration. At time zero, 0.5 ml of labeled substrate (10-fold concentrated) is added and mixed rapidly. Samples are taken after 10, 20, 30 sec etc. and blown immediately into 5 ml of wash medium prepared in Wasserman tubes, filtered, and washed using an automated multisample pipet.

Comments. The sensitivity of the test depends on the specific activity of the labeled substrate and on the amount of bacteria filtered. At low substrate concentrations it is essential to lower the cell density, to keep incubation times as short as possible, and to use high specific radioactive activities. Normally, the PTS Enzymes II act virtually irreversibly and are insensitive to washing procedures. Efflux rates, however, can be considerable in the presence of external energy sources, or if metabolizable substrates are converted to labeled intermediates such as lactic acid which leak out rapidly from the cells. For active polyol transport systems, in contrast, leakage due to the reversibility of the systems can be very high if the temperature or the osmotic conditions are changed drastically during the test, or if the cells are exposed on the filters to a high air stream. Finally, Tris-buffered media should be avoided for growth and washings since the cells become osmotically fragile and leaky. Instead, nonchelating Good buffers should be used.

Energetization of the cells by metabolizable substrates during the uptake, and leakage of labeled intermediates can be minimized by using nonmetabolizable analogs. These, however, have other inconveniences.

When such analogs are taken up and phosphorylated by Enzymes II of the PTS, they can (1) cause drastic feedback inhibitions on Enzymes II and deplete the cells rapidly of PEP, and (2) be pumped out of the cells by a process called "inducer expulsion."[20] All these effects tend to falsify kinetic data. They can only be minimized by short incubation times, low substrate concentrations, the lack of an external energy source during the uptake tests, and careful controls. Deviations obtained from tests in intact cells, in crude cell extracts, and in reconstituted purified systems not only depend on variations of the ions or the transport components, but also on the presence or absence of cellular control mechanisms.[21] Substrates for the different polyol transport systems in the Enterobacteriaceae are (arranged according to decreasing affinities): for the D-mannitol system D-mannitol, 2-amino-2-deoxymannitol, 2-deoxymannitol, D-arabinitol, and D-glucitol; for the D-glucitol system D-glucitol, 2-deoxyglucitol (note that this compound is also 2-deoxymannitol, its correct name being 2-deoxy-*arabino*-hexitol), D-mannitol, xylitol, D-fructose, and D-glucose; for the galactitol system galactitol, D-glucitol, 2-deoxygalactitol, D-arabinitol, and L-fucitol; for the D-arabinitol system D-arabinitol, D-mannitol, galactitol, and xylitol; and for the ribitol system ribitol, D- and L-arabinitol, and xylitol.[1,8–16,22,23]

[20] J. Reizer and C. Panos, *Proc. Natl. Acad. Sci. U.S.A.* **77,** 5497 (1980).
[21] J. Lengeler and H. Steinberger, *Mol. Gen. Genet.* **167,** 75 (1978).
[22] G. R. Jacobson, C. A. Lee, J. E. Leonard, and M. H. Saier, Jr., *J. Biol. Chem.* **258,** 10748 (1983).
[23] G. R. Jacobson, L. E. Tanney, D. M. Kelly, K. B. Palman, and S. B. Corn, *J. Cell. Biochem.* **23,** 231 (1983).

[36] Alanine Carrier from Thermophilic Bacteria

By H. HIRATA

Introduction

The secondary active transport of nutrients across the bacterial membrane is catalyzed by a specific carrier protein(s) embedded in the membrane in a chemiosmotic manner as envisaged by Mitchell (see Refs. 1–5

[1] A. A. Eddy, *Curr. Top. Membr. Transp.* **10,** 279 (1978).
[2] F. M. Harold, *in* "The Bacteria" (L. N. Ornston and J. R. Sokatch, eds.), Vol. VI, p. 463. Academic Press, New York, 1978.

for recent reviews). In order to study the mechanism at the molecular level, the isolation of the carrier protein(s) from the membranes is an essential step. However, since the carrier protein(s) is expected to possess no catalytic activity other than mediating the transport of nutrients, the identification and purification of the protein can be performed only by following the activity facilitating the translocation of specific nutrient molecules into the reconstituted proteoliposomes unless a specific agent or substrate analog, which covalently binds to the active center of carrier protein(s), is available. The recent success in purifing a functional lactose carrier from membranes of *Escherichia coli* carrying the multicopy plasmid of the *lacY* gene, using octylglucoside and the 4-nitro[2-³H]phenyl-D-galactopyranoside labeling technique, provided a great advance in this context.[6]

General Principle

Membrane vesicles isolated from a thermophilic bacterium PS3 are extracted by ionic detergents and the detergent-soluble fraction is subjected to purification procedures including ion-exchange column chromatography and high-performance liquid chromatography (HPLC) with a hydroxylapatite column (HPHT column). At appropriate steps, proteins are reconstituted into proteoliposomes loaded with potassium salt. Then the activity of carrier protein(s) is determined as protein(s) catalyzing accumulation of radioactive substrate in response to a membrane potential created by K^+ diffusion via potassium ionophore, valinomycin.

Solubilization and Purification of Alanine Carrier

Definition of Activity. The alanine carrier activity is determined after reconstitution into proteoliposomes loaded with potassium salt. Specific activity of the alanine carrier is calculated from radioactivities accumulated in the reconstituted proteoliposomes 15 sec after the valinomycin addition subtracted by those 1 min prior to the addition on the basis of milligrams of protein in the proteoliposomes.

Bacterium. Thermophilic bacterium PS3, which was isolated from

[3] B. P. Rosen and E. R. Kashket, *in* "Bacterial Transport" (B. P. Rosen, ed.), p. 559. Dekker, New York, 1978.

[4] P. Mitchell, *Science* **206** (1979).

[5] I. C. West, *Biochim. Biophys. Acta* **604**, 91 (1980).

[6] M. J. Newman, D. L. Foster, Y. H. Wilson, and H. R. Kaback, *J. Biol. Chem.* **256**, 11804 (1981).

Mine Hot Spring in Shizuoka, Japan, by Dr. T. Oshima of Tokyo Institute of Technology, is cultivated with vigorous aeration in a medium containing a 0.8% polypeptone, 0.4% yeast extract, and 0.3% NaCl, pH 7.0, at 65°.

Preparation of Washed Membranes. The harvested cells (500 g, wet weight) are suspended in 4.5 liters of 50 mM Tris–sulfate, pH 8.0, at 36°, and 500 mg of lysozyme (EC 3.2.1.17) is added. The mixture is stirred for 30 min, and then 25 ml of 1 M MgCl$_2$ and 10 mg of DNase I (EC 3.1.21.1) are added. After 2 hr, the resulting lysate is centrifuged at 17,000 g for 20 min. The precipitate is homogenized and washed three times with 1-liter volumes of 50 mM Tris–sulfate, pH 8.0, and the final precipitate is collected and stored at −80° until use.

Purification of Alanine Carrier.[7,8] *Step 1. Solubilization.* Washed membranes (4 g of proteins) are suspended in a solution containing 2% Na-cholate, 1% Na-deoxycholate, 0.2 M Na$_2$SO$_4$, 30 mM Tris–sulfate (pH 8.0), and 0.5 mM dithiothreitol in a final volume of 400 ml. The mixture is sonicated in an ice bath for 5 min three times and then centrifuged at 140,000 g for 1 hr. This procedure is repeated again, and the resulting first and the second supernatant are combined and concentrated to approximately 50 ml in a Diaflo apparatus with a UM10 filter (Amicon Corp.). The concentrated extracts are dialyzed to remove salts and detergents against 2 liters of 50 mM Tris–sulfate (pH 8.0) containing 0.25 mM dithiothreitol for 17 hr at room temperature. The contents in the dialysis bag are then centrifuged and the pellets are suspended in 50 mM Tris–sulfate (pH 8.0). This preparation is designated as CDE-P.

Step 2. Triton Extraction. CDE-P is suspended in 20 ml of 2% Triton X-100 containing 0.2 M Na$_2$SO$_4$ and 0.5 mM dithiothreitol, sonicated for 1 min, three times, and centrifuged at 140,000 g for 1 hr.

Step 3. DEAE-Cellulose Column Chromatography. The supernatant is diluted 20-fold with distilled water and applied on a DEAE-cellulose column (300 ml of bed volume) equilibrated with 25 mM Tris–sulfate (pH 8.0) containing 0.25% Triton X-100 (Tris–Triton). The column is washed with 1 liter of Tris–Triton and then eluted with Tris–Triton containing 50 mM Na$_2$SO$_4$. Fractions with UV absorption are combined and mixed with Na-cholate at a final concentration of 1%. Then solid ammonium sulfate is added at 210 mg/ml. The precipitate is collected by centrifugation (20,000 g, 10 min) and suspended in 50 mM Tris–sulfate (pH 8.0). This ammonium sulfate fractionation is repeated once, and the preparation is designated as DE-1.

[7] H. Hirata, N. Sone, M. Yoshida, and Y. Kagawa, *J. Suramol. Struct.* **6**, 77 (1977).
[8] H. Hirata, T. Kambe, and Y. Kagawa, *J. Biol. Chem.* **259**, 10653 (1984).

Step 4. DEAE-TOYOPEARL Column Chromatography in the Presence of Triton and Urea. To the DE-1 preparation, Triton X-100 and solid urea are added to final concentration of 5% and 8 M, respectively. The solution is sonicated for 2 min (Branson Sonifier, model 200) and centrifuged at 140,000 g for 10 min. The supernatant is diluted 5-fold with distilled water and applied to a DEAE-TOYOPEARL 650M (Toyo Soda Mfg. Co., Tokyo) column (50 ml of bed volume) equilibrated with 5 mM Tris–sulfate (pH 8.0) containing 1% Triton X-100 and 4 M urea. The flow-through fractions are collected, combined, and diluted 4-fold with distilled water, and subjected to ammonium sulfate fractionation in the presence of 1% Na-cholate. After collecting by centrifugation (20,000 g, 10 min), the precipitate is washed twice and suspended with distilled water. The preparation is designated as UDETP-1.

Step 5. CM-Sepharose CL-6B Column Chromatography in the Presence of Triton and Urea. UDETP-1 is treated with Triton X-100 and urea as described for Step 4. After centrifugation, the supernatant is diluted 5-fold with distilled water and applied to a CM-Sepharose CL-6B (Pharmacia) column (20 ml of bed volume) equilibrated with 5 mM Tris–sulfate (pH 8.0) containing 1% Triton X-100, 4 M urea, and 20 mM Na$_2$SO$_4$. The flow through fractions are collected, combined, diluted, and fractionated by ammonium sulfate as in Step 3. This preparation is designated as UCM-1.

Step 6. HPLC with an HPHT Column. UCM-1 is treated as in Step 4, and after centrifugation, the supernatant is diluted 5-fold with distilled water and applied to an HPHT column (Bio-Rad) equilibrated with solvent A (5 mM Na-phosphate, pH 6.8, containing 1% Triton X-100 and 0.4 mM CaCl$_2$) using a Spectra Physics SP8700 HPLC System with a ISCO UA-5 UV monitor at 280 nm. An isocratic elution with solvent A for 10 min is followed by a linear gradient with solvent B (200 mM Na-phosphate, pH 6.8, containing 1% Triton X-100 and 0.01 mM CaCl$_2$) for 50 min. The flow rate is 1 ml/min. The eluted fractions are collected, pooled, diluted 4-fold with distilled water, and subjected to ammonium sulfate fractionation in the presence of Na-cholate as in Step 3. A typical elution profile is presented in Fig. 1. Most of the alanine carrier activity is recovered in peak II (HPHT-2).

Specific Activity and Purity of Preparation

Results of a typical purification are summarized in the table. In this particular case, the preparation after DEAE-cellulose column chromatography (DE-1) was used as the starting material. The specific activity of the final preparation (HPHT-2) was about 15-fold that of the DE-1. However, the specific activity of alanine transport by the original membranes and

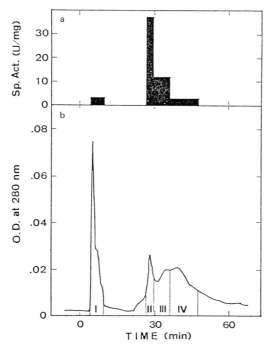

FIG. 1. Elution profile of HPLC with the HPHT column. The UCM-1 preparation (2.1 mg of protein) was treated and applied to the HPHT column as described in text. The elution was monitored at 280 nm (b). The eluted fractions were divided into four parts (I–IV) and the alanine carrier activity was assayed after reconstitution into proteoliposomes as described in the text (a). Data from Hirata et al.[8]

PURIFICATION OF ALANINE CARRIER[a]

Preparation	Protein (mg)	Alanine carrier activity		
		Specific (units/mg)	Total (units[b])	Yield (%)
DE-1	138.5	2.8	318.6	100
UDETP-1	20.0	12.3	246.0	77.2
UCM-1	4.2	31.5	132.3	41.5
HPHT-2[c]	1.0	42.8	42.8	13.4

[a] Data from Hirata et al.[8]

[b] One unit of alanine carrier activity is defined as the amount that transports 1 nmol of alanine under the standard assay conditions as described in text.

[c] The second peak of HPLC with a HPHT column as shown in Fig. 1a.

the proteoliposomes reconstituted with the cholate/deoxycholate extract (CDE-P) were around 1 and 2.4 nmol/min/mg of protein, respectively. Although it may be inadequate to compare directly the alanine transport activity by the original membranes with that by the proteoliposomes, a rough estimation indicates that the specific activity of the HPHT-2 preparation was more than 40-fold that of the original membranes. Polyacrylamide gel electrophoresis in the presence of sodium dodecyl sulfate revealed that the HPHT-2 consisted of a single polypeptide with an apparent M_r of 42,500.

Reconstitution of Proteoliposomes

Proteoliposomes are reconstituted from samples of the preparations obtained at each purification step and soy bean phospholipids according to the freezing and thawing method of Kasahara and Hinkle[9] with minor modifications. Partially purified soybean phospholipids, 40 mg in 1 ml of 0.25 M potassium phosphate (pH 8.0), are sonicated 5 times for 3 min at 1 min intervals under argon atmosphere at 0° with a probe type sonifier (Branson, model 200). To the resulting liposome suspension, samples of the preparations obtained at each purification step (0.1–0.5 mg of protein) is added. The mixture is then sonicated for 1 min, frozen quickly in a dry ice/ethanol bath, and thawed at room temperature, followed by a brief sonication (10 times of 0.5 sec pulse). This freeze-thaw sonication procedure is repeated three times. The resulting proteoliposomes are centrifuged at 140,000 g for 30 min, and suspended in 0.2 M sucrose containing 10 mM MgSO$_4$.

Transport Assay

Principle.[10] Addition of valinomycin to the reconstituted proteoliposomes loaded with potassium phosphate (pH 8.0) and suspended in a potassium free medium induces efflux of K$^+$ down the concentration gradient with concurrent generation of a membrane potential, interior negative. In the course of K$^+$ efflux, radioactive alanine accumulates in the proteoliposomes.

Reagents

Reconstitued proteoliposomes
Sucrose, 0.14 M

[9] M. Kasahara and P. C. Hinkle, *J. Biol. Chem.* **252**, 7383 (1977).
[10] H. Hirata, K. Altendorf, and F. M. Harold, *Proc. Natl. Acad. Sci. U.S.A.* **70**, 1804 (1973).

Magnesium sulfate, 0.013 M
Tris-maleate (pH 7.0), 0.1 M
L-[U-^{14}C]alanine, 10 μM (1 μCi/ml)
Valinomycin, 1 mg/ml of methanol
Sucrose–MgSO$_4$: 0.2 M sucrose containing 0.01 M MgSO$_4$

Procedure. The reconstituted proteoliposomes are placed in a test tube containing the reagents described above in total volume of 1 ml and incubated for 5 min at 40. Then 1 μl of valinomycin solution is added. At intervals before and after the valinomycin addition, 0.1-ml aliquots of the suspension are withdrawn and filtered through membrane filters (Sartorius, pore size 0.45 μm) that have previously soaked in sucrose–MgSO$_4$. The filters are washed with three portions of 5 ml of sucrose–MgSO$_4$, and then removed and placed on planchettes. Radioactivities are assayed by a gas-flow counter (Aloka LBC-451).

Comments

1. The final preparation, HPHT-2, consisting of a single polypeptide with an apparent M_r of 42,500 in functionally active form, can be obtained only by the use of a hydroxylapatite column for HPLC (HPHT column). Although the reason is unknown, conventional hydroxylapatite column chromatography provides no success at all.

2. Contaminants appeared in UCM-1 preparation (Step 5) distribute in the other fractions (peaks I, III, and IV, see Fig. 1). Particularly, a lower molecular weight polypeptide (an apparent M_r of 20,000), that is one of the major components of UCM-1, is recovered in peak I.

3. From the amino acid analysis of HPHT-2 preparation, the polarity index is calculated to be 33%, which indicates that the protein is slightly less hydrophobic than the lactose carrier of *Escherichia coli* whose polarity index is 29%.

4. An electrochemical potential difference of Na$^+$ also drives the alanine transport in proteoliposomes reconstituted with the HPHT-2 preparation. The Na$^+$-dependent transport of alanine is not sensitive to uncoupling agent, carbonyl cyanide *p*-trifluoromethoxyphenylhydrazone, while the valinomycin K$^+$-induced transport is completely inhibited by the agent.

[37] Generation of a Protonmotive Force in Anaerobic Bacteria by End-Product Efflux

By Bart ten Brink and Wil N. Konings

Introduction

Chemiosmotic phenomena, as postulated by Mitchell,[1,2] play a central role in bacterial metabolism. The chemiosmotic hypothesis postulates that proton translocation by primary proton pumps generates an electrochemical proton gradient, or protonmotive force (Δp) across the cytoplasmic membrane. This Δp consists of an electrical parameter, the membrane potential ($\Delta\psi$, inside negative) and a chemical parameter, the pH difference (ΔpH):

$$\Delta p = \Delta\psi - Z\Delta pH \qquad \text{(in mV)} \qquad (1)$$

where Z equals 2.3 RT/F and R, T, and F have their usual meaning.

The protonmotive force acts as a driving force or regulatory parameter in a large number of different metabolic processes.[3] Until recently only three primary proton pumps had been described in bacteria[4]: membrane-bound electron transfer chains,[5-7] the proton-translocating ATPase complex,[8,9] and, in halobacteria, the light-driven proton pump bacteriorhodopsin.[10] Strictly fermentative bacteria which do not possess electron transfer systems therefore seemed to be totally dependent on ATP hydrolysis for the generation and maintenance of a Δp. This generation of a Δp would consume a considerable fraction of the ATP formed by substrate level phosphorylation and as a consequence less ATP would be available for biosynthetic purposes.

[1] P. Mitchell, *Biol. Rev.* **41**, 445 (1966).

[2] P. Mitchell, *J. Bioenerg.* **4**, 63 (1973).

[3] K. J. Hellingwerf and W. N. Konings, *Adv. Microb. Physiol.*, in press (1986).

[4] W. N. Konings and P. A. M. Michels, *in* "Diversity of Bacterial Respiratory Systems" (C. J. Knowles, ed.), p. 33. CRC Press, Boca Raton, Florida, 1980.

[5] E. Padan, O. Zilberstein, and H. Rottenberg, *Eur. J. Biochem.* **63**, 533 (1976).

[6] S. Ramos and H. R. Kaback, *Biochemistry* **16**, 848 (1977).

[7] P. B. Garland, J. A. Downie, and B. A. Haddock, *Biochem. J.* **152**, 547 (1976).

[8] V. Liebeling, R. K. Thauer, and K. Jungermann, *Eur. J. Biochem.* **55**, 445 (1975).

[9] F. M. Harold, *Curr. Top. Bioenerg.* **6**, 83 (1977).

[10] W. Stoeckenius, R. H. Lozier, and R. A. Bogomolni, *Biochim. Biophys. Acta* **505**, 215 (1979).

In 1979, Michels *et al.*[11] proposed another mechanism for the generation of a protonmotive force in anaerobic bacteria. In their "energy-recycling model" it was postulated that metabolic end products are excreted via specific transport proteins in the cytoplasmic membrane and that together with the end products, protons are translocated from the cytoplasm to the external medium. As a result of this proton (and charge) translocation a protonmotive force is generated. Especially in fermentative bacteria, which continuously produce relatively large quantities of metabolic end products, this process could contribute significantly to the generation of a protonmotive force and consequently to the overall production of metabolic energy.

Solute Transport in Bacteria

Before describing the energy-recycling model in detail the most important solute transport systems of bacteria will be briefly reviewed. Solute transport across the cytoplasmic membrane of bacteria can occur by two major mechanisms. (1) Secondary transport systems: transport by these systems is driven by electrochemical gradients and will lead to the translocation of solute in unmodified form. (2) Group translocation: solute is substrate for a specific enzyme system in the membrane; the enzymatic reaction results in a chemical modification of the solute and release of the products at the cytoplasmic side. The only well-established group translocation system is the phosphoenolpyruvate phosphotransferase system (PTS).[4]

Most solutes are translocated across the cytoplasmic membrane by secondary transport either passively, without the involvement of specific membrane proteins or facilitated by specific carrier proteins. This latter mechanism is often termed "active transport."

Mitchell visualized three different systems for facilitated secondary transport. "Uniport": only one solute is translocated by the carrier protein. "Symport": two or more different solutes are translocated in the same direction by the carrier protein. "Antiport": two or more different solutes are translocated by one carrier in opposite directions.

The driving forces for solute transport will depend on the overall charge and the number of protons which are translocated, as well as on the solute gradient.[4] The driving force is composed of components of the protonmotive force and of the solute gradient, and translocation of solute will proceed until the total driving force is zero. At that stage a steady-

[11] P. A. M. Michels, J. P. J. Michels, J. Boonstra, and W. N. Konings, *FEMS Microbiol. Lett.* **5**, 357 (1979).

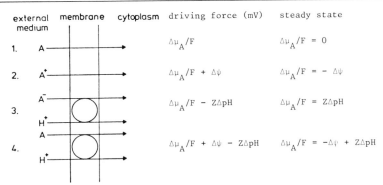

FIG. 1. Schematic presentation of four secondary transport processes. 1. Passive transport of a neutral solute; 2. passive transport of a cation; 3. facilitated transport of an anion in symport with one proton; 4. facilitated transport of a neutral solute in symport with one proton. For each process the driving force and the steady-state level of accumulation are given.

state level of accumulation is reached. Figure 1 shows schematically four different translocation processes. In example 1 transport of a neutral solute via passive secondary transport is shown. The driving force of this transport will be supplied only by the solute gradient and will be equal to $\Delta\mu_A/F = Z \log[A_{in}]/[A_{out}]$. A steady state will be reached when the solute gradient is dissipated ($\Delta\mu_A = 0$) and the internal solute concentration equals the external concentration. Example 2 shows transport of a monovalent positively charged solute by passive secondary transport. This transport will not only depend on the chemical concentration gradient of solute but also on the electrical potential. The driving force therefore will be $\Delta\mu_A/F = Z \log[A_{in}]/[A_{out}] + \Delta\psi$ and a steady state will be reached when $Z \log[A_{in}]/[A_{out}] = -\Delta\psi$. Unless the electrical potential is dissipated by movement of other ions across the membrane in the steady state, the internal concentration of solute will not be equal to the external concentration. In the absence of active primary transport systems net transport will already stop when the internal solute concentration is lower than the external concentration. However, when a $\Delta\psi$ (interior negative) is generated by other systems accumulation of solute can occur.

In a way similar to that shown above, the driving forces and steady-state levels of transport can be derived for other transport processes.[4] This is done in example 3 for a symport system by which a negatively charged solute is translocated together with one proton. The overall charge during this translocation process is zero and due to the translocation of one proton only the ΔpH will contribute to the driving force of this translocation process. In example 4 the driving force is determined for a neutral solute symported with one proton. In this process charge and protons are translocated and consequently the total protonmotive force ($\Delta\psi$ and ΔpH) contributes to the driving force.

A general equation for the driving force for transport of solute A with charge m in symport with n protons can be derived. This driving force is

$$n\Delta p + \Delta\bar{\mu}_A/F \qquad \text{(in mV)} \qquad (2)$$

Combination of Eqs. (1) and (2), and substitution of electrochemical gradient of A^m ($\Delta\bar{\mu}_A/F$) by the sum of the chemical gradient of A ($\Delta\bar{\mu}_A/F = Z$ $\log[A_{in}/A_{out}]$) plus the electrical component ($m\Delta\psi$) yields

$$\text{driving force} = \Delta\bar{\mu}_A/F + (n + m)\Delta\psi - nZ\Delta\text{pH} \qquad \text{(in mV)} \qquad (3)$$

The Energy-Recycling Model

In essence, the energy-recycling model describes the reversed process of solute uptake via a secondary transport system. The chemiosmotic hypothesis and the carrier model of Rottenberg[12] postulate that secondary transport processes can proceed in both directions across the membrane. The direction of transport is determined entirely by the directionality of its driving force. During solute uptake the energy present in Δp is used to drive the accumulation of solute into the bacterial cell, whereas during end-product excretion the energy in the product concentration gradient is used to generate a protonmotive force. Since metabolism of the energy source in fermentative bacteria leads to a continuous efflux of metabolic end products, a continuous generation of a protonmotive force takes place. The components of the protonmotive force which will be formed during the efflux process will vary with the translocated solute. If the efflux process leads to a net translocation of positive charges, a $\Delta\psi$ will be generated. A ΔpH will be formed if a net translocation of protons takes place. For the generation of both a $\Delta\psi$ and a ΔpH a net translocation of both protons and charges has to occur (see also Fig. 1).

In the detailed mathematical description of the energy-recycling model[11] Michels et al. took a homofermentative lactate-producing organism as a model system. The energy source glucose is accumulated by a PEP phosphotransferase system and during the translocation process converted to glucose 6-phosphate. The production of metabolic energy during glucose metabolism by such organisms can be divided in two distinctly different parts: substrate level phosphorylation and lactate excretion (see also Fig. 2).

[12] H. Rottenberg, *FEBS Lett.* **66**, 159 (1976).

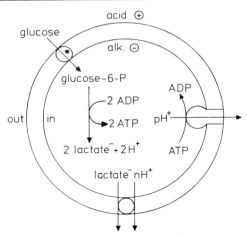

FIG. 2. Schematic presentation of the energy-recycling model in an organism with a homolactic glucose fermentation. Glucose is taken up by a phosphoenolpyruvate-dependent group translocation system (PTS).

Substrate level phosphorylation:

$$\text{glucose}_{out} + 2\ \text{ADP}_{in} + 2\ P_{in} \rightarrow 2\ \text{lactate}_{in}^- + 2\ H_{in}^+ + 2\ \text{ATP}_{in} + 2\ H_2O \qquad (4)$$

Lactate excretion:

$$\text{lactate}_{in}^- + n\ H_{in}^+ \rightarrow \text{lactate}_{out}^- + n\ H_{out}^+ \qquad (5)$$

where n equals the number of protons translocated together with lactate (the H^+/lactate stoichiometry) during carrier-mediated lactate excretion. The overall process is

$$\text{glucose}_{out} + 2\ \text{ADP}_{in} + 2\ P_{in} + 2(n-1)\ H_{in}^+ \rightarrow$$
$$2\ \text{lactate}_{out}^- + 2\ \text{ATP}_{in} + 2n\ H_{out}^+ + 2\ H_2O \qquad (6)$$

From the overall reaction it will be clear that if n equals 2, 4 protons and 2 positive charges are translocated across the cytoplasmic membrane per glucose molecule fermented. As a consequence both a $\Delta\psi$ and a ΔpH can be generated by the lactate excretion process, and less ATP has to be hydrolyzed for the generation of a protonmotive force. However, if n equals 1, only 2 protons and no charges are translocated. In that case lactate excretion will not result in the generation of a membrane potential and only a small pH difference will be created as a result of the acidification of the external medium by the produced lactic acid. This example clearly demonstrates that the H^+/solute stoichiometry during end-product excretion is a very important parameter in the energy-recycling process.

Experimental Approach

In the past 5 years a steadily increasing body of experimental evidence in favor of the energy-recycling model has been presented. These results were mainly obtained from studies on lactate excretion in streptococci. However, the excretion of other end products in other fermentative bacteria will most likely also lead to the generation of a protonmotive force. In this chapter the general approach and experimental techniques which can be used to study the generation of a protonmotive force by end-product efflux have been described. Some results obtained in *Escherichia coli* and *Streptococcus cremoris* will be given as experimental examples.

Role of Specific Transport Proteins in End-Product Translocation

It has been demonstrated above for lactate efflux that an effective generation of a protonmotive force by the excretion of neutral or negatively charged end products requires the net translocation of positive charges from the cytoplasm to the external medium. This can be achieved only if the end product is translocated in symport with protons via a specific transport protein. Evidence for the involvement of such a carrier in the end-product translocation can be obtained from studies on the translocation in the opposite direction, i.e., uptake studies of the end product. This can be achieved only if the driving force for end-product translocation is directed from the medium to the cytoplasm. In whole cells this can be realized by using starved cells with no internal end product present and imposing artificially a protonmotive force, inside negative and/or alkaline. For some end products it might be better to use isolated membrane vesicles instead of intact cells, in order to avoid effects of the metabolism. Isolation procedures of membrane vesicles have been described for a large number of organisms.[13–16] In cells and membrane vesicles a driving force for electrogenic secondary transport can be generated artificially by imposing a potassium diffusion potential across the membrane.[17] This is usually achieved by incubating concentrated cell or membrane vesicle preparations (30–50 mg protein/ml) with the K^+-ionophore valinomycin (1–2 nmol/mg protein) in the presence of 100 to 200 mM potassium ions and diluting small aliquots 100- to 200-fold in K^+-free buffer. The efflux of the K^+ ions will result in the generation of a $\Delta\psi$.

[13] H. R. Kaback, this series, Vol. 22, p. 99.
[14] W. N. Konings, *Adv. Microbiol. Physiol.* **15**, 175 (1977).
[15] R. Otto, R. G. Lageveen, H. Veldkamp, and W. N. Konings, *J. Bacteriol.* **149**, 733 (1982).
[16] W. N. Konings, this series, Vol. 56, p. 370.
[17] S. Schuldiner and H. R. Kaback, *Biochemistry* **14**, 5451 (1976).

A driving force for H^+-solute symport can also be created by the artificial formation of a ΔpH. This can be achieved by diluting cells or membrane vesicles, incubated in a buffer of high pH (7 to 8) into a buffer with a lower pH (4.5–6). Uptake of end products can be studied by adding radioactively labeled end products to the dilution media. Preferentially the end product with the highest radioactive label available should be used at different concentrations in the μM to mM range. The uptake is measured after filtration of the cells or vesicles over 0.45-μm pore size filters, followed by washing as described.[18]

In *Streptococcus cremoris* cells the $\Delta\psi$ generated by K^+ efflux can drive the uptake and accumulation of the end product lactate at pH 7.0,[19] indicating that under these conditions lactate uptake is a carrier mediated and electrogenic process (the H^+/lactate stoichiometry is higher than 1).

If the organism under study contains a proton-translocating electron transfer chain, a Δp can be generated by addition of the appropriate electron donor and acceptor. In membrane vesicles from anaerobically grown *Escherichia coli* lactate uptake could be driven by the addition of ascorbate/phenazine methosulfate under aerobic conditions (Fig. 3). The uptake of lactate has been studied in the pH range 5.5 to 8 in the presence of the ionophore nigericin. Under these conditions no ΔpH exists since nigericin exchanges H^+ for K^+ ions across the membrane but the $\Delta\psi$ is constant at about -100 mV in this pH range. Addition of nigericin (0.5 μM) to these vesicles at pH 5.5 almost completely inhibited lactate uptake, but had no effect at pH 8.0.[20] This result strongly indicates that at pH 5.5 lactate uptake is mainly ΔpH driven, while at pH 8.0 the $\Delta\psi$ is the main driving force. This implies that in *E. coli* membrane vesicles at pH 5.5 lactate uptake is an electroneutral process (1 H^+ per lactate symported), while at pH 8.0 the translocation is electrogenic (more than one H^+ per lactate symported). These observations supply strong evidence for a pH-dependent variation in the H^+/lactate stoichiometry during lactate transport in *E. coli* vesicles. The most likely explanation for this variation in stoichiometry is that the external pH affects the dissociation state of the carrier.[12,21]

Completely different techniques to study lactate uptake were used by Simpson *et al.*[22] with *Streptococcus faecalis*. They investigated the initial

[18] H. R. Kaback, this series, Vol. 31, p. 698.
[19] R. Otto, A. S. M. Sonnenberg, H. Veldkamp, and W. N. Konings, *Proc. Natl. Acad. Sci. U.S.A.* **77**, 5502 (1980).
[20] B. ten Brink and W. N. Konings, *Eur. J. Biochem.* **111**, 59 (1980).
[21] W. N. Konings, *Trends Biochem. Sci.* **6**, 257 (1981).
[22] S. J. Simpson, M. R. Bendall, A. F. Egan, and P. J. Rogers, *Eur. J. Biochem.* **136**, 63 (1983).

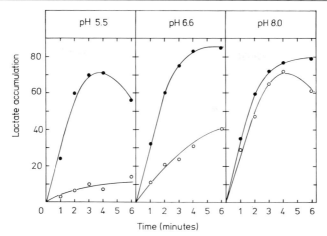

FIG. 3. Time course of L-lactate accumulation by membrane vesicles of *Escherichia coli* ML 308-225, under aerobic conditions at different pH values. In all experiments the membrane vesicles were energized by the oxidation of the electron donor potassium ascorbate (10 mM) plus phenazine methosulfate (100 μM). (●) No further additions; (○) plus nigericin (0.5 μM). Taken from ten Brink and Konings, *Eur. J. Biochem.* **111,** 59 (1980), with permission.

rates of uptake of protons and lactate anions resulting from the addition of high concentrations of lactate to resting intact cells. [^{14}C]Lactate was used to measure the rate of lactate influx. The proton influx was calculated from the changes in the external and cytoplasmic pH as determined with the ^{31}P NMR technique.[23,24] The initial rate of lactate uptake was independent of the external pH, while the rate of lactate influx induced proton influx strongly increased with the external pH. The ratio between proton influx and lactate influx rate, which equals the H$^+$/lactate stoichiometry (n), increased from about 1 at pH$_o$ 6.5 to about 2 at pH$_o$ 7.5.[22] This indicates that also in *S. faecalis* cells lactate uptake is a carrier mediated and electrogenic process at pH values near or above 7.0. This conclusion was confirmed by the observations that especially at high pH the addition of lactate strongly accelerated the rate of membrane depolarization in resting *S. faecalis* cells with an inside negative $\Delta\psi$.[22]

[23] G. Navon, S. Ogawa, R. G. Shulman, and T. Yamane, *Proc. Natl. Acad. Sci. U.S.A.* **74,** 888 (1977).

[24] K. Ugurbil, H. Rottenberg, P. Glynn, and R. G. Shulman, *Proc. Natl. Acad. Sci. U.S.A.* **75,** 2244 (1978).

End-Product Efflux under Nonphysiological Conditions

The main postulate of the energy-recycling model is that excretion of end products is carrier-mediated and that with the end product positive charges and/or protons are translocated from the cytoplasm to the external medium. To study this efflux process cells or membrane vesicles have to be loaded with high concentrations of end product. Usually this can be achieved by incubating deenergized cells or membrane vesicles for several hours at room temperature with the metabolic end product.[19,20] If the translocation of an acidic end product is investigated, incubation at low pH may shorten the incubation time needed, since then a large part of the product is present in the undissociated form, which is more membrane permeable. Membrane vesicles can also easily be loaded during the membrane vesicle isolation by performing the lysis of the protoplasts or spheroplasts in the presence of the end product.[20,25] The end product then will be entrapped in the membrane enclosed volume. Dilution of "loaded" cells or membrane vesicles into end product-free medium will create an outwardly directed end-product concentration gradient. Efflux of the end product then can lead to the generation of a protonmotive force. The generation of a membrane potential (inside negative) can be recorded by performing the dilution in a medium containing the radioactive labeled lipophilic cation tetraphenylphosphonium (TPP$^+$) at micromolar concentrations. For measurements of a ΔpH (inside alkaline) radioactively labeled weak acids like acetate or benzoate at micromolar concentrations have to be present in the dilution medium. The uptake by the cells or the membrane vesicles can be followed after filtration and counting of the radioactivity.

In a similar way the uptake of a solute driven by the protonmotive force generated by the efflux process can be followed. When lactate loaded starved cells of S. cremoris were diluted 100-fold into lactate-free buffer containing [^{14}C]leucine, a significant accumulation of leucine could be observed.[19] However, if a lactate gradient was not created, virtually no uptake of leucine occurred (Fig. 4A), indicating that lactate efflux and not ATP hydrolysis supplied the energy for leucine uptake. This could also be concluded from the observation that the ATPase inhibitor dicyclohexylcarbodiimide (DCCD) only slightly inhibited the leucine uptake in the presence of a lactate gradient (Fig. 4B). Figure 4C shows that lactate efflux induced leucine uptake is completely inhibited by the uncoupler carbonyl cyanide p-trifluoromethoxyphenylhydrazone (FCCP). This suggests that lactate efflux results in the generation of a protonmotive force,

[25] S. J. Simpson, R. Vink, A. F. Egan, and P. J. Rogers, FEMS Microbiol. Lett. 5, 85 (1983).

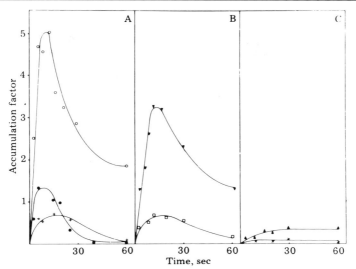

Fig. 4. Time course of L-lactate efflux-induced leucine uptake by *S. cremoris*. De-energized cells [100 mg/ml (dry wt)] were diluted 1 : 100 into choline/HEPES/KCl buffer at 25°. (A) Cells loaded with 50 mM choline L-lactate were diluted into choline/HEPES/KCl buffer (○) or into choline/HEPES/KCl buffer containing 50 mM choline L-lactate (●). Cells loaded with 50 mM choline chloride were diluted into choline/HEPES/KCl buffer (+). (B) Cells were preincubated with 25 μM DCCD for 30 min at room temperature. Cells loaded with 50 mM choline L-lactate were diluted into choline/HEPES/KCl buffer containing 25 μM DCCD (▼) or into the same medium supplemented with 50 mM choline L-lactate (□). (C) Cells loaded with 50 mM choline L-lactate were diluted into choline/HEPES/KCl buffer containing 10 μM FCCP (▲) or into the same medium containing 50 mM choline L-lactate (×). Taken from Otto *et al.*, *Proc. Natl. Acad. Sci. U.S.A.* **77**, 5502 (1980), with permission.

which subsequently drives the uptake of leucine. Similar results were obtained with membrane vesicles of *S. cremoris*[15] and *E. coli*[20]: lactate efflux resulted in uncoupler sensitive amino acid uptake. Direct evidence for the generation of a Δp by lactate efflux was obtained from experiments in which the uptake of the lipophilic cation tetraphenylphosphonium (TPP+) was studied. When *E. coli* membrane vesicles loaded with 50 mM lactate at pH 6.6 were diluted 100-fold into a buffer of pH 6.6 containing the same concentration of lactate, only a low level of TPP+ accumulation was observed.[20] On the other hand, when these loaded vesicles were diluted into lactate-free buffer, a high level of TPP+ accumulation (24-fold) was reached within 20 sec, followed by rapid efflux of the accumulated TPP+ (Fig. 5). Here, in analogy with the lactate efflux-induced amino acid uptake, the uncoupler FCCP also completely inhibited TPP+ uptake. These results clearly showed that in membrane vesicles of *E. coli*

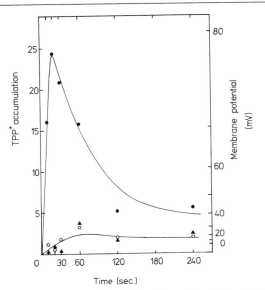

FIG. 5. Time course of L-lactate efflux induced tetraphenylphosphonium accumulation by *Escherichia coli* ML 308-225 membrane vesicles at pH 6.6 and 25°. Membrane vesicles loaded with 50 mM K L-lactate were diluted 100-fold into (●) buffer; (○) buffer + 50 mM K L-lactate; (▲) buffer + 10 μM FCCP. Taken from ten Brink and Konings, *Eur. J. Biochem.* **111,** 59 (1980), with permission.

a $\Delta\psi$ (inside negative) can be formed by efflux of lactate from the vesicles. The lactate efflux-induced $\Delta\psi$ was calculated with the Nernst equation from the increased level of TPP⁺ accumulation[20] to be about −55 mV. This means that at a pH near neutrality not only the uptake but also the efflux of the monovalent negatively charged lactate ion is carrier mediated and occurs in symport with more than one proton ($n > 1$). Similar results have been obtained with intact cells[19] and membrane vesicles[15] of *S. cremoris*. This phenomenon of solute efflux induced amino acid uptake and $\Delta\psi$ generation was not only observed for the end product lactate: efflux of thiomethylgalactoside or gluconate from *E. coli* cells[26,27] and lactose from *E. coli* vesicles[28] also resulted in Δp formation. The generation of a protonmotive force by carrier-mediated solute efflux therefore seems well established and justifies an examination of the transport mechanism of other metabolic end products.

[26] J. L. Flagg and T. H. Wilson, *Membr. Biochem.* **1,** 61 (1978).
[27] M. Bentaboulet, A. Robin, and A. Kepes, *Biochem. J.* **178,** 103 (1979).
[28] G. J. Kaczozowski and H. R. Kaback, *Biochemistry* **18,** 3691 (1979).

End-Product Efflux under Physiological Conditions

Experiments as described above can be used to investigate the basic assumptions of the energy-recycling model. Once these assumptions have been confirmed, it is of interest to study the end-product excretion process under physiological conditions, where an outwardly directed end-product gradient will be created by the bacterial metabolism and not artificially. This requires techniques to determine the magnitude of Δp, the end-product gradient and the H^+/end-product stoichiometry in growing or actively metabolizing bacteria. A variety of methods has been developed to measure the magnitude and composition of the protonmotive force in bacteria. In most cases the distribution of permeant ions and weak acids (or bases) is used in the determination of the $\Delta\psi$ and ΔpH, respectively. The distribution of the probe molecules can be determined chemically or by using radioactively labeled probes, either by measuring the internal or the external concentration of the probes. Usually a physical separation of both compartments by filtration or centrifugation is required which can result in leakage of the probe from the cytoplasm. To measure changes in the external concentration of the probe, either a specific electrode[29,30] or the flow dialysis technique[31,32] can be used.

Measurements of the components of the Δp and an end-product gradient under identical conditions, can best be performed with a procedure which is rapid and avoids washing steps or other manipulations that can change the actual gradients. A modified version of the silicon oil centrifugation technique[33] is such a procedure: the metabolizing cells are incubated with $\Delta\psi$ and ΔpH probes, separated from the external medium by centrifugation through a layer of silicon oil and collected in perchloric acid (PCA).

The ΔpH and $\Delta\psi$ values can be calculated from the accumulation of [^{14}C]benzoate and [^{3}H]TPP$^+$, respectively. Portions of 0.5 ml cell suspension (50–500 mg protein/ml) are incubated for 5 min with 0.5 μCi [^{14}C]benzoate (16 μM final concentration) and/or 2.5 μCi [^{3}H]TPP$^+$ (5 μM final concentration) and transferred to microfuge tubes containing 0.5 ml of silicon oil (density 1.02 g/ml) on top of 0.1 ml 11% (w/w) PCA (density 1.07 g/ml). After 3 min centrifugation at 12,000 g virtually all cells are present in the PCA fraction. In PCA, rapid disruption of the cells, inhibi-

[29] T. Shinbo, N. Kamo, K. Kurihara, and Y. Kobatake, *Arch. Biochem. Biophys.* **187**, 414 (1978).
[30] D. B. Kell, P. John, M. C. Sorgato, and S. J. Ferguson, *FEBS Lett.* **86**, 294 (1978).
[31] S. Ramos, S. Schuldiner, and H. R. Kaback, *Proc. Natl. Acad. Sci. U.S.A.* **73**, 1892 (1976).
[32] K. J. Hellingwerf and W. N. Konings, *Eur. J. Biochem.* **106**, 431 (1980).
[33] E. J. Harris and K. van Dam, *Biochem. J.* **106**, 759 (1968).

tion of further metabolism, and release of intracellular metabolites and probe molecules occur. Samples (50 μM) of the clear supernatant and the PCA fraction are taken and assayed for radioactivity. During the centrifugation also some extracellular fluid passes with the cells to the PCA fraction. In control experiments the volume of extracellular water can be estimated by adding membrane-impermeable ^{14}C compounds such as dextran, inulin, or taurine to the incubation medium. The total water volume that is transferred to the PCA fraction can be labeled with ^3H_2O. From the data on intracellular water and extracellular water in the PCA fraction and the amounts of radioactivity in the supernatants and PCA fractions, the accumulation ratios of the [^{14}C]benzoate and [^3H]TPP$^+$ can be calculated. However, before calculating the TPP$^+$ accumulation a correction for binding of TPP$^+$ to cell components has to be made.[34] This binding can be determined by incubating cells which are deenergized by a butanol treatment (7% n-butanol for 5 min) with [^3H]TPP$^+$. If also the concentrations of the metabolic end product under study are determined (i.e., enzymatically) in the supernatant and PCA fractions, the internal end-product concentration and therefore also the end-product gradient can be calculated.

From the data on $\Delta\psi$, ΔpH and the end-product concentration gradient ($\Delta\bar{\mu}_A/F$), the H$^+$/end-product stoichiometry during end-product excretion (n) can easily be calculated, if the assumption is made that during growth the driving force for the translocation process [Eq. (3)] is very close to zero. Rearrangement of Eq. (3) then yields

$$n = (m\Delta\psi - \Delta\bar{\mu}_A/F)\ \Delta p \tag{7}$$

We used the procedure described above to study the energy-recycling model in more detail in cells of *Streptococcus cremoris* growing in batch and chemostat culture[35,36] with lactose as sole energy source. Under these conditions lactate is the only end product of the energy metabolism. In these cells an outwardly directed lactate gradient is always present. Internal lactate concentrations between 50 and 200 mM were measured at external concentrations between 8 and 40 mM. Simultaneous measurements of $\Delta\psi$, ΔpH, and the lactate gradient in growing *S. cremoris* revealed that n was influenced by the external pH and the external lactate concentration.[35,36] During pH regulated growth in batch culture at pH 6.34 the magnitude and composition of Δp are essentially constant, but the

[34] J. S. Lolkema, K. J. Hellingwerf, and W. N. Konings, *Biochim. Biophys. Acta* **681**, 85 (1982).
[35] B. ten Brink and W. N. Konings, *J. Bacteriol.* **152**, 682 (1982).
[36] B. ten Brink, R. Otto, U.-P. Hansen, and W. N. Konings, *J. Bacteriol.* **162**, 383 (1985).

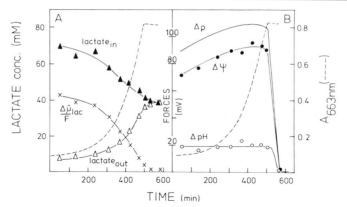

FIG. 6. Growth of *S. cremoris* Wg2 at 30° and pH 6.34 in batch culture on a rich medium supplemented with lactose (3 g/liter). The dotted line represents the time course of cell density measured at 663 nm. (A) Time course of the internal (▲) and external (△) lactate concentration and the lactate gradient (×) during growth. (B) Time course of the electrochemical proton gradient. The line for Δp was constructed by adding Δψ (●) and ΔpH (○). Taken from ten Brink, Otto, Hansen, and Konings.[36]

lactate gradient decreases sharply as the external lactate concentration increases (Fig. 6). As a consequence *n* [calculated from Eq. (7)] decreases with increasing external lactate concentration (Fig. 7) from 1.44 to about 0.9. If growth occurred at a higher pH, the H^+/lactate stoichiometry was higher over the entire lactate concentration range. The effects of the external pH and lactate concentration on *n* were determined in detailed studies with glycolyzing cells of *S. cremoris*[36]: increasing the external proton and lactate concentration results in lower values of *n*. *n* Values between about 2 (high pH, low external lactate) and 0.7 (low pH, high external lactate) could be determined.[36]

It should be evident that low *n* values result in a lower energy production from the lactate excretion process. Such a decrease of the *n* value is, however, essential since it prevents the cells from "self-poisoning." If the H^+/lactate stoichiometry would be fixed at 2.0, lactate excretion from cells with a Δp composed of a Δψ = −80 mV and ΔpH = −20 mV and 50 m*M* external lactate could occur only if the internal lactate concentration exceeded the unphysiologically high value of 5 *M*! [as can be calculated from Eq. (7)]. It should be noted that *n* values significantly lower than 1 indicate that less protons are excreted together with lactate than are formed internally during glycolysis. The remaining protons then have to be extruded by the action of the membrane bound ATPase, at the expense of ATP. H^+/lactate stoichiometries below 1 would therefore result in an energy loss.

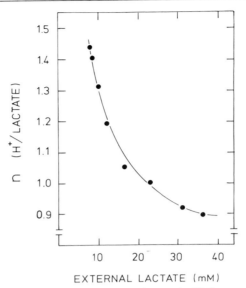

FIG. 7. Effect of the external lactate concentration on the H$^+$/lactate stoichiometry (n) in *S. cremoris* Wg2, grown in batch culture at pH 6.34. Taken from ten Brink, Otto, Hansen, and Konings.[36]

Metabolic Energy Yield by End-Product Efflux

The production of metabolic energy by substrate level phosphorylation processes during the breakdown of the energy source and by the excretion of the metabolic end products can be calculated if the metabolic pathways and the H$^+$/end-product stoichiometries are known. The energy present in excreted protons can be "translated" into ATP equivalents by dividing the number of protons excreted by p, the number of protons which have to flow back from the external medium to the cytoplasm per molecule of ATP synthesized via the membrane-bound ATPase. For instance, the overall production of metabolic energy by lactose fermentation and lactate excretion by *S. cremoris* can be calculated with the following equation:

$$\frac{\text{ATP equivalents produced}}{\text{mol of lactose consumed}} = 4 + 4(n - 1)/p \tag{8}$$

where n represents the H$^+$/lactate stoichiometry during lactate excretion. Per mol of lactose 4 molecules of ATP are synthesized by substrate level phosphorylation and 4 molecules of lactate are produced. Since per molecule lactate also one proton is formed intracellularly and lactate is excreted in symport with n protons, $4(n-1)$ represents the number of translocated protons which contribute to the generation of the protonmotive force. The calculated number of ATP equivalents produced per mol lactose consumed is shown in Fig. 8 for different values of n and p. If n and p both equal 2, then 6 ATP equivalents are formed, indicating that theoretically an energy gain of 50% can be the result of the lactate excretion process in *S. cremoris*.

For a number of well-known glucose fermentations the theoretical energy gain by end-product excretion has been calculated (Table I), assuming that per neutral end product (organic acids in undissociated form) one extra proton is translocated and that two protons are translocated per ATP synthesized. Table I shows that in theory between 17 and 100% extra metabolic energy can be produced by the energy-recycling process in these fermentations. A prerequisite for this energy gain is the presence of carriers for the alcohols and weak acids in the fermenting organism. Hardly any evidence for the existence of such carriers has been presented.

If the metabolic energy production of a microorganism is partly supplied by an end-product efflux process, changes in the H^+/end-product stoichiometry should be reflected in changes of the cell yield. In order to investigate this implication of the energy-recycling model, detailed yield

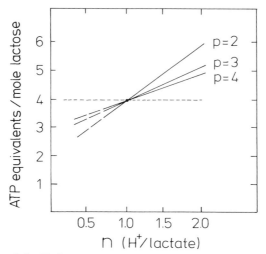

FIG. 8. Effect of the H^+/lactate stoichiometry (n) on the theoretical number of ATP equivalents produced per mol of lactose consumed by *S. cremoris* Wg2. p is the number of protons translocated per molecule of ATP synthesized. The lines were calculated from Eq. (8).

TABLE I
ENERGY GAIN BY END-PRODUCT EXCRETION IN A NUMBER OF GLUCOSE
FERMENTATIONS

Fermen-tation	Products (mol/mol glucose)	ATP produced by substrate level phosphorylation (mol/mol glucose)	ATP equivalents produced by energy-recycling Eq/mol (glucose)	Energy gain (%)
a	2 ethanol, 2 CO_2	2	1	50
b	2 lactate	2	1	50
c	1 lactate 1 ethanol, 1 CO_2	1	1	100
d	1 lactate 1.5 acetate	2.5	1.25	50
e	1 acetate 1 ethanol 2 formate	3	2	67
f	1 butyrate 2 CO_2, 2 H_2	3	0.5	17
g	0.5 butanol, 2 H_2 0.5 acetone, 2.5 CO_2	2	0.5	25

[a] Alcoholic fermentation (yeasts).
[b] Homolactic fermentation (lactate acid bacteria).
[c] Heterolactic bacteria (lactobacilli).
[d] Bifidum pathway (*Bifidobacterium bifidum*).
[e] Mixed acid fermentation (streptococci).
[f] Butyrate fermentation (clostridia).
[g] Butanol-acetone fermentation (clostridia). See text for the calculation of the ATP-equivalents production.

studies have to be performed, using chemostat cultures under energy limiting conditions. The molar growth yield corrected for maintenance requirement (Y^{max}, in g cell dry weight produced per mol energy source consumed) can be determined from a graph in which the specific energy source consumption rate (q, in mol energy source consumed per g dry weight per hr) is plotted as a function of the specific growth rate (μ, hr^{-1}), according to the following equation[37]:

$$q = \mu/Y^{max} + m_e \tag{9}$$

where m_e is the maintenance coefficient (in mol energy source consumed per g cell dry weight produced per hr). By varying the dilution rate of the

[37] K. L. Schulze and R. S. Lipe, *Arch. Mikrobiol.* **48**, 1 (1963).

TABLE II
EFFECT OF GROWTH pH AND EXTERNAL
LACTATE CONCENTRATION ON $Y_{lactose}$ OF
Streptococcus cremoris GROWN IN
CONTINUOUS CULTURE

pH	External lactate (mM)	$Y_{lactose}^{max}$ (g dry weight/ mol lactose)
7.0	27	61.2
6.4	27	55.2
5.7	27	51.2
6.3	25	56.3
6.3	85[a]	50.2

[a] 60 mM Na-lactate was added to the inflow medium. In the control experiment 60 mM NaCl was added.

chemostat μ can be varied. Yield studies performed in this way with lactose fermenting *S. cremoris*[19,36] showed that decreasing the growth pH[36] as well as increasing the external lactate concentration[19] resulted in lower $Y_{lactose}^{max}$ values, as could be expected on basis of the changes in n (Table II). A beneficial effect of lowering the external lactate concentration (and therefore increasing the lactate gradient and the value of n) was also demonstrated in continuous cultures of *S. cremoris* where the introduction of a lactate consuming *Pseudomonas* species increased the molar growth yield of the streptococci by over 50%.[38] This shows that detailed information on end-product excretion processes is necessary before bioenergetic parameters such as Y^{max} can be determined.

A protonmotive force generated by end-product excretion can in theory be used to drive the synthesis of ATP.[39] This ATP synthesis at the expense of a Δp, generated by end-product efflux, can be demonstrated in membrane vesicles or starved cells with low internal concentrations of ATP (low phosphorylation potential). A somewhat different approach is to study the effect of end-product influx on extravesicular ATP synthesis using inside-out membrane vesicles. Influx of lactate into inside-out vesicles of *S. faecalis* resulted in extravesicular ATP synthesis.[25] This ATP synthesis was determined from the incorporation of $^{32}P_i$ into ATP.[40] The

[38] R. Otto, J. Hugenholtz, W. N. Konings, and H. Veldkamp, *FEMS Microbiol. Lett.* **5**, 85 (1980).

[39] P. C. Maloney and T. H. Wilson, *J. Membr. Biol.* **25**, 285 (1975).

[40] T. Kagawa and N. Sone, this series, Vol. 55, p. 364.

lactate influx induced ^{32}P incorporation was almost completely inhibited by both the protonophore CCCP and DCCD, indicating that indeed a Δp (inside positive and acid) was generated by lactate influx.[25]

Conclusions

A number of experimental approaches can supply information about the generation of metabolic energy by end-product efflux:

1. Uptake experiments, to demonstrate carrier activity and to obtain information on the H^+/end-product stoichiometry.

2. Efflux experiments from preloaded membranes, to demonstrate end-product efflux driven Δp generation (and subsequent ATP synthesis).

3. Simultaneous measurements of the end-product gradient and Δp in growing or actively metabolizing cells, to obtain information on the H^+/end-product stoichiometry and possible contributions of energy recycling to the overall production of metabolic energy.

4. Yield experiments under various conditions.

Acknowledgments

The studies performed in the author's laboratory were supported by the Dutch Organization of Pure Scientific Research (ZWO). The help of Mrs. M. Pras and Mrs. M. Broens-Erenstein in the preparation of the manuscript is highly appreciated.

[38] Light-Driven Chloride Transport in Halorhodopsin-Containing Cell Envelope Vesicles

By Brigitte Schobert and Janos K. Lanyi

Halorhodopsin is the second retinal protein discovered in the plasma membrane of *Halobacterium halobium*. Like bacteriorhodopsin, it functions as a light-driven ion pump in the cytoplasmic membrane of this organism, but its substrate is chloride instead of protons, and it transports this ion into the cells, rather than outward. For unambiguous results the halobacterial strain[1] used in assays of halorhodopsin should be defective in bacteriorohodopsin, as is strain L-33, but such strains usually contain

[1] G. Wagner, D. Oesterhelt, G. Krippahl, and J. K. Lanyi, *FEBS Lett.* **131**, 341 (1981).

also slowly cycling rhodopsin, the third bacterial retinal protein. However, since the latter has no obvious transport function,[2] it does not interfere with most of the experiments and the results described can be attributed unequivocally to the functioning of halorhodopsin. The functional assays of halorhodopsin are either direct, i.e., measurement of the chloride transported, or more usually indirect, i.e., consist of following a secondary consequence of the chloride transport, such as membrane potential, passive proton uptake, or volume change.

Strain, Growth Conditions, and Preparation of the Vesicles

Halobacterium halobium strain L-33[1] is grown in a peptone medium with illumination, as described earlier.[3] Cell envelope vesicles are prepared by the sonication method, according to Lanyi and MacDonald.[3] This procedure results in a preparation with 85–90% right-side-out membrane orientation. The vesicle concentration can be measured by a protein determination,[4] with serum albumin as standard. The vesicles are stored in unbuffered 4 M NaCl (pH near 6.5) at 4° for months without significant loss of transport activity.

Measurement of Membrane Potential

The negative inside membrane potential, developed during illumination of the vesicles, is followed by determining a decrease in the fluorescence of the cyanine dye,[5] di-O-C$_5$.[3] As determined with K$^+$ diffusion potentials in the dark,[6] the fluorescence change is linear with potential up to about -120 mV.

For this experiment the vesicles are suspended in 4 M NaCl (protein concentration 0.13 mg/ml), with or without buffer but near neutral pH. The sample is in a 1 × 1 cm stirred and thermostatted (23°) fluorescence cuvette, total volume 3 ml. The dye is added with stirring to the vesicle suspension, from a 0.5 mM stock solution in methanol, so as to give a final concentration of 1.33 μM. The ratio between dye and protein concentration is important, because higher dye concentrations appear to result in less effective fluorescence quenching.

[2] R. A. Bogomolni and J. L. Spudich, *Proc. Natl. Acad. Sci. U.S.A.* **79,** 6250 (1982).
[3] J. K. Lanyi and R. E. MacDonald, this series, Vol. 56, p. 398.
[4] O. H. Lowry, N. J. Rosebrough, A. L. Fair, and R. J. Randall, *J. Biol. Chem.* **193,** 265 (1951).
[5] P. J. Sims, A. S. Waggoner, C. H. Wang, and J. F. Hoffman, *Biochemistry* **13,** 3315 (1974).
[6] R. Renthal and J. K. Lanyi, *Biochemistry* **15,** 2136 (1976).

The instrumentation constructed for this experiment is described schematically in Fig. 1. The light source for the measuring beam is a Kratos-Schoeffel 150 W xenon arc lamp. The excitation wavelength for fluorescence is isolated with a 450 nm interference filter (bandwidth 10 nm). A Princeton Applied Research Model 191 chopping motor, operating at 150 Hz, is placed between filter I and the cuvette. Filter II, placed after the cuvette and at right angle to the exciting beam, is a 510 nm interference filter (bandwidth 10 nm), which isolates the fluorescence emission. The latter is detected with a Centronix Q 4283B photomultiplier, cooled to $-20°$. The output is passed through a preamplifier and into a Princeton Applied Research Model 220 lock-in amplifier, tuned to the chopping frequency. The output is displayed on a strip-chart recorder. The actinic light source, a Sylvania tungsten-halogen project lamp (250 W, 120 V), is powered by regulated line voltage. The actinic beam is placed 180° from the fluorescence excitation beam, and is filtered through a 2 cm water layer, an additional heat-filter (hot-mirror) and a 530 nm cut-off filter. Virtually the entire volume of the sample is illuminated by the actinic light, and about 80% of the volume is sampled by the measuring light.

The relative changes in membrane potential are usually expressed as percentage quenching, rather than in mV. To calculate the latter, a calibration curve must be made for each vesicle batch, using KCl-loaded

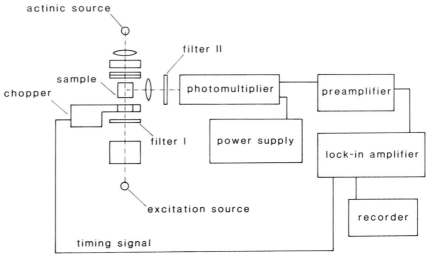

FIG. 1. Instrumentation constructed to measure illumination-dependent fluorescence and light-scattering changes in halorhodopsin-containing cell envelope vesicles. Details are explained in the text.

vesicles and valinomycin to produce K^+ diffusion potentials of known magnitude.[6,7]

Measurement of pH Changes

Light-dependent pH changes in the vesicle suspensions (1 mg/ml protein in unbuffered 4 M NaCl) are measured with a Beckman Altex semimicro combination electrode (Model 39822), connected to an Orion Research Model 701A pH meter, whose output is displayed on a strip-chart recorder. The sample is placed in a 1 × 1 cm cuvette in the thermostatted assembly of the spectrofluorimeter, and stirred throughout the measurement. During illumination L-33 vesicles alkalinize the medium. Typical pH changes are 0.5 and 0.7 units near neutral pH. Addition of a powerful uncoupler, such as SF 6847 (2 μM final concentration), results in an increase of both the rate and the extent of the light-dependent proton uptake, the latter resulting in pH changes near 1 unit. In order to correct for buffering of the pH change by protein and lipid in the vesicles, calibrations of buffering is always run after the experiments, adding several microliters of 10–20 mM HCl in 4 M NaCl. In the presence of uncoupler the light-dependent proton uptake amounts usually to about 50 nmol protons/mg protein, and the initial rate is 60 to 100 nmol H^+/min/mg protein.

The proton uptake will be abolished with membrane-permeable cations, such as tetraphenylphosphonium (final concentration up to 0.1 mM), or when gramicidin (final concentration of 0.1 μM) is added to increase the membrane permeability to Na^+. Under these conditions the light-dependent membrane potential is greatly diminished. These results, as well as those with the uncoupler, indicate that the proton uptake is not a primary process, but a secondary one which is passive and induced by the membrane potential. Thus, the protons detect membrane potential in this system, and the pH changes can be used to estimate both the extent of the potential (steady-state pH difference), and the rate of chloride uptake (initial rate of proton uptake).

It should be pointed out that the glass electrode shows an illumination artifact, corresponding to a decrease of about 0.025 pH units. With a large pH change this is negligible, but with small changes corrections must be applied.

Preparation of Vesicles in Chloride-Free Medium

In order to demonstrate an effect of chloride on the light-induced events in the vesicles, the NaCl in the suspension medium must be first

[7] B. Schobert and J. K. Lanyi, *J. Biol. Chem.* **257**, 10306 (1982).

replaced by a chloride-free solution of high ionic strength. The salt must be an antichaotrope to maintain the integrity of the vesicle membrane; 1.5 M Na$_2$SO$_4$ (unbuffered, at pH 6.5), or 3 M K phosphate (pH 6.5) are suitable. Because of the high specific gravity of these solutions, it is not possible to sediment the vesicles by centrifugation. Replacement is by dialysis overnight, using Spectrapor 2 (10 mm width) tubing filled with 2 ml of the vesicle suspension. We find it convenient to clamp the tubing to a plastic rod of the approximate length of a 500 ml measuring cylinder, with a magnetic sirring bar attached to its bottom. This arrangement ensures that the dialysis tubing is pressed flat during the dialysis, and that the dialyzate, which contains the sulfate or phosphate solution, is well stirred. Because of the large surface of the tubing, dialysis is complete after overnight stirring at room temperature. The final NaCl concentration in the vesicle suspensions is 16 mM (as calculated from the dilution of the chloride by the dialyzate), which is further lowered upon dilution into the assay mixtures to values no longer supporting significant chloride transport. Although this procedure was usually adequate for our measurements, more complete removal of chloride can be effected by a second dialysis. In all cases the dialysis must be performed in the dark because chloride-free halorhodopsin is very light sensitive.

Vesicles suspended in 1.5 M Na$_2$SO$_4$ must be stored at 23° because of the poor solubility of this salt at lower temperatures. For the same reason the temperature of the transport experiments should not be allowed to fall below this value. Even so, sulfate vesicles are stable under these conditions for a few days after their preparation. In contrast, phosphate vesicles can be stored at 4°, and maintain activity for several weeks.

Measurement of pH Change and Membrane Potential as Functions of Chloride Concentration

These measurements are done as described above. Vesicles suspended in 1.5 M sulfate, containing NaCl below 1 mM, show no significant light-dependent pH change. Upon addition of NaCl, using a 4 M stock solution, the light-dependent pH change, arising from the membrane potential created by the chloride transport, can be seen. The light-dependent pH response appears immediately after the addition of NaCl, in spite of the low permeability of these vesicles to chloride,[7] indicating that chloride on the vesicle exterior is effective. The K_m (apparent affinity) for this chloride effect is ~40 mM. The total magnitude of the proton uptake at saturating chloride concentration (200 mM and above) in the sulfate is only half of that found in 4 M NaCl.

The membrane potential created by chloride uptake in the sulfate or phosphate vesicles can be measured with the fluorescence dye method, as described above. No light-dependent fluorescence quenching is observed at chloride concentrations below 1 mM. Adding chloride allows the creation of membrane potential (negative inside) during the illumination, and the concentration dependency of this effect yields a K_m for chloride also about 40 mM.[7]

Measurement of Volume Changes

Relative volume increase due to chloride uptake can be conveniently measured by following light-scattering changes during the illumination of the vesicles. The instrumentation used is the same as for the fluorescence measurements (Fig. 1), with the only modification being that filter II is replaced with a 450 nm interference filter so that the measuring beam now detects Raleigh scattering at a 90° angle, rather than fluorescence. The protein concentration in the vesicle suspension is 0.5 mg/ml, and the total volume is 3 ml. The contents of the cuvette are stirred throughout the experiment. Care has to be taken to exclude dust, lint, and trapped air bubbles which greatly increase the noise and instability of the light-scattering signal. Since the strain L-33 vesicles take up chloride (and Na^+, particularly when gramicidin at 0.1 μM is added), they swell during illumination and the intensity of the scattered light at 90° decreases. It turns out that the vesicles have a much smaller volume in the sulfate or phosphate medium than in NaCl (amounting to 25–30% of the volume measured in the latter salt). Thus, these vesicles have a large swelling capacity, with linear light-scattering increase over at least 30 min of illumination. While a relative volume increase is easily measured from the slope of this change, it is difficult to determine absolute volume changes by the light-scattering method.

A chloride requirement for the transport activity can be demonstrated with this method also. At very low chloride concentrations no swelling is detected. After chloride is added, illumination causes very rapid increases in vesicle volume, and a K_m for chloride for this measurement is also about 40 mM.[7] In the presence of gramicidin, or valinomycin (5 μM) plus K^+, the rate of swelling is increased 2- to 3-fold.

Transport Activity under Different Conditions of Chloride Gradients

The ability to transport chloride against a concentration gradient is one of the features of a chloride pump, such as halorhodopsin. Vesicles can be loaded with different sulfate/chloride mixtures by dialyzing NaCl vesicles against 1.5 M Na$_2$SO$_4$ plus a concentration of NaCl calculated to produce the desired final chloride concentration after the dialysis is completed. In the first series of experiments the chloride concentration inside the vesicles is 10 mM, and the vesicles are suspended in sulfate plus chloride at outside concentrations from 10 to 100 mM, resulting in outside/inside ratios of 1 to 10. Transport activity is measured immediately after suspension of the vesicles, e.g., with the fluorescence method to detect membrane potential, since the half-life of the preformed chloride gradient is about 30 min.[7] In the second series of experiments the NaCl concentration inside the vesicles is set at 100 mM, and the outside concentration is varied between 10 and 100 mM, resulting in outside/inside ratios 0.1 to 1. The results of such experiments show[7] that the light-driven membrane potential, as well as other bioenergetic parameters of the transport, depend only on the external chloride concentration, and are influenced very little by a transmembrane chloride concentration gradient between 1/10 to 10.

Determination of a Chloride Gradient Created during Illumination

Since the accumulation of chloride during the illumination of the vesicles is accompanied by a volume increase, it is not entirely clear whether or not a chloride concentration gradient across the vesicle membrane is created. Such a gradient can be detected only indirectly. A convenient means of doing this is by the use of triphenyltin chloride (TPT), a compound which acts as an electroneutral chloride/hydroxyl antiporter,[8] thereby producing a pH gradient dependent on the existing chloride gradient. The pH change can be measured by conventional means in the suspension medium. Calibration of this method is with preformed chloride gradients. Vesicles (10 mg/ml), loaded with sulfate plus 100 mM chloride and suspended in the same salt mixture, are quickly diluted to 1 mg/ml with sulfate plus amounts of NaCl which give external chloride concentrations from 10 to 100 mM. Addition of TPT is after 10 min of incubation in the dark, anticipating this time delay in the subsequent illumination experiments. Acidification of the medium upon addition of the TPT in these calibration experiments is proportional to the log of the magnitude of the preformed chloride gradient up to about 5-fold, and levels off above this value.[7] Reverse preformed chloride gradients give TPT-dependent alkalinization under these conditions, as expected.

[8] M. J. Selwyn, A. P. Dawson, M. Stockdale, and N. Gains, *Eur. J. Biochem.* **14,** 120 (1970).

The creation of chloride gradient during the illumination of strain L-33 vesicles is measured by first suspending vesicles, loaded with sulfate plus 100 mM chloride, in the same salt mixture to 1 mg/ml, and incubating for 20–30 min to ensure equilibration of any small residual concentration difference. Absence of pH change upon adding TPT (or a minimal change due to artifact) confirms the absence of an initial chloride concentration gradient. The samples are then illuminated, in separate experiments, for different periods of time, from 2 to 10 min. Ten minutes after the illumination is terminated the light-dependent pH changes will have relaxed, and the TPT is added (final concentration 1 nM) in each experiment. From the magnitude of the observed acidification and the calibration curve, obtained as described above, the chloride gradient which existed when the light was turned off can be estimated. With increasing lengths of illumination the pH change upon addition of TPT increases. Figure 2 illustrates the results from such an experiment. This method yields a chloride gradient, inside/outside = 5- to 10-fold, after 10 min of illumination.

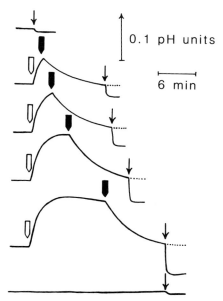

0.1 pH units

6 min

FIG. 2. pH traces obtained during the illumination of halorhodopsin-containing vesicles and the subsequent addition of triphenyltin. Each trace, including dark controls (top and bottom), is a separate experiment. Conditions are described in the text. Symbols: open wedges, light on; closed wedges, light off; arrows, addition of triphenyltin.

Chloride Transport Assay

Vesicles (10 mg/ml) are equilibrated with sulfate or phosphate plus 100 mM NaCl. Two 0.25 ml samples are withdrawn at 0 and 10 min, after which illumination is commenced in a stirred, thermostatted cuvette, and every 10 min another 0.25-ml aliquot is withdrawn. Each aliquot is placed immediately on top of one of a series of Agarose columns (BioGel A-50m, equilibrated with sulfate or phosphate and packed into 0.5 × 20 cm Econocolumns, Bio-Rad), and eluted with sulfate or phosphate. The first 2 ml of the eluants is discarded. The following 1 ml contains the vesicles (as indicated by turbidity), but no extravesicle chloride. The elution takes about 5 min. The collected vesicles are diluted to 3 ml with distilled water containing Triton X-100 (final concentration 0.03%), and the chloride content is measured with a Beckman chloride-sensitive electrode. The chloride content of the vesicles increases about 10-fold above its original value during 40 min of illumination.[7] Dark incubation causes insignificant increase in chloride content, but addition of gramicidin (0.1 μM) accelerates chloride uptake in the light by a factor of about 2.

Effects of Chloride upon the Absorption Spectrum of Halorhodopsin

The recording of good difference spectra of the highly turbid vesicle suspensions requires a lot of patience. We used a Gilson Model 2600 single-beam spectrophotometer, with digital scanning of the wavelength. Since this instrument (and many others) use a holographic grating, a spectral anomaly near 550 nm can arise, which is eliminated with a polarizing filter placed vertically in the beam after the light source. This arrangement completely eliminates the anomaly, but at the expense of reducing the light intensity by a factor of 2–2.5. Reduction of the signal-to-noise in these measurements is a problem, and a compromise must be reached between slit width (0.5 to 1.5 mm, corresponding to 0.9 to 2.7 nm bandwidth) and signal averaging (5 to 10 points averaged per wavelength reading).

The vesicles are loaded with sulfate and suspended in this salt solution to 5 mg/ml, and the pH adjusted to values between 5 and 9. The suspension is divided between the sample and reference cuvettes (1 cm pathlength, 3 ml total volume, stoppered), care being taken to avoid any difference between the two, resulting in mismatched spectra. Any trapped air, lint, temperature difference, or other source of mismatch will result in a slanted baseline and spectral features due to cytochromes, etc. Before any additions are made the baseline should be straight and at zero absorption.

Additions of NaCl to the sample cuvette are compensated by additions of Na_2SO_4 to the reference cuvette. The total absorption range is corrected for the dilution of the samples, particularly when repeated additions are made and a family of curves is produced. In the Model 2600 spectrophotometer this is conveniently done by rescaling the display. The chloride concentration can be increased in steps from 0 to 200 mM, with spectra recorded after each addition. Even by visual inspection, the vesicles appear less colored in the sulfate than in the NaCl. Addition of chloride causes increases absorption at 590 nm and decreases absorption at 410 nm.[9] At pH 5 the effect is very small, but it increases steeply with pH, up to 9. Between pH 5 and 8 the K_m for chloride for this effect is 40–50 mM, at higher pH the K_m is somewhat higher. The positive peak at 590 nm and the negative peak at 410 nm is most clearly observed at pH 9, with 200 mM chloride added to the sample cuvette.

From experiments with purified halorhodopsin, it seems now certain that these spectroscopic changes are caused by a pK increase for the halorhodopsin Schiff base upon chloride addition.[10] Hence, in the intermediate pH region chloride addition shifts the equilibrium between the protonated form (near 580 nm) and the deprotonated form (near 410 nm) in favor of protonation.

Estimation of Halorhodopsin Content of the Vesicles

Upon sustained illumination the halorhodopsin Schiff base is deprotonated to give a stable species with an absorption band at 410 nm,[11–13] particularly at alkaline pH.[9] This reaction can be used to estimate the amount of halorhodopsin in the cell envelope vesicles. For this test the envelopes are suspended in 4 M NaCl to 5 mg/ml protein, and the pH is adjusted to 9.0. The suspension is divided between a sample and a reference cuvette, as for the chloride-dependent difference spectra, and a baseline is recorded. The reference cuvette is then illuminated with yellow light. We find it most convenient to use the actinic illumination system in the spectrofluorimeter for this purpose. Difference spectra after the illumination yield a large absorption increase at 580 nm, corresponding to the halorhodopsin content of the membranes.[9] The length of illumination

[9] J. K. Lanyi and B. Schobert, *Biochemistry* **22**, 2763 (1982).
[10] B. Schobert, J. K. Lanyi, and D. Oesterhelt, *J. Biol. Chem.*, in press (1986).
[11] T. Ogurusu, A. Maeda, N. Sasaki, and T. Yoshizawa, *J. Biochem. (Tokyo)* **90**, 1267 (1981).
[12] M. Steiner and D. Oesterhelt, *EMBO J.* **2**, 1379 (1983).
[13] M. E. Taylor, R. A. Bogomolni, and H. J. Weber, *Proc. Natl. Acad. Sci. U.S.A.* **80**, 6172 (1984).

required depends on the light intensity; in our case 10 min is sufficient for maximal response. The difference spectra are stable for at least an hour. Care should be taken to avoid optical mismatch of the samples due to any heating during the illumination. An extinction coefficient of 50,000 M^{-1} cm^{-1} is used in calculating the halorhodopsin content. Typically, L-33 envelope vesicles contain 0.1 to 0.2 nmol halorhodopsin per mg protein.

[39] Sodium Translocation by NADH Oxidase of *Vibrio alginolyticus:* Isolation and Characterization of the Sodium Pump-Defective Mutants

By Hajime Tokuda

The marine bacterium *Vibrio alginolyticus* possesses a primary electrogenic Na^+ extrusion system (a Na^+ pump) which is coupled to respiration.[1,2] From the isolation of the Na^+ pump-defective mutants,[3] it became clear that the Na^+ pump is coupled to NADH oxidase which requires Na^+ for maximum activity and is highly sensitive to 2-heptyl-4-hydroxyquinoline *N*-oxide (HQNO), a specific inhibitor of the Na^+ pump.[4] Furthermore, it was recently shown that the NADH oxidase reconstituted into liposomes functions as the Na^+ pump.[5]

The Na^+ pump generates an electrochemical potential of Na^+ (a Na^+ motive force) which is an immediate driving force for various substrate transport systems[6] and flagella motility.[7] Since the activity of the Na^+ pump is maximum at pH about 8.5 and minimum at pH 6.0, generation of the Na^+ motive force shows striking difference in the sensitivity to a proton conductor, carbonyl cyanide *m*-chlorophenylhydrazone (CCCP), depending on external pH. At acidic pH, respiratory chain extrudes only H^+ leading to the generation of the protonmotive force (Δp). The generation of the Na^+ motive force at acidic pH is performed by a Δp-driven Na^+/H^+ antiport system.[1] Thus, collapse of Δp by CCCP concomitantly

[1] H. Tokuda and T. Unemoto, *Biochem. Biophys. Res. Commun.* **102**, 265 (1981).
[2] H. Tokuda and T. Unemoto, *J. Biol. Chem.* **257**, 10007 (1982).
[3] H. Tokuda, *Biochem. Biophys. Res. Commun.* **114**, 113 (1983).
[4] H. Tokuda and T. Unemoto, *J. Biol. Chem.* **259**, 7785 (1984).
[5] H. Tokuda, *FEBS Lett.* **176**, 125 (1984).
[6] H. Tokuda, M. Sugasawa, and T. Unemoto, *J. Biol. Chem.* **257**, 788 (1982).
[7] B. V. Chernyak, P. A. Dibrov, A. N. Glagolev, M. Yu. Sherman, and V. P. Skulachev, *FEBS Lett.* **164**, 38 (1983).

inhibits the generation of the Na^+ motive force. On the other hand, at alkaline pH, extrusion of Na^+ and generation of membrane potential ($\Delta\psi$) are primary processes performed by the Na^+ pump and take place in the presence of CCCP. The generation of $\Delta\psi$ in the presence of CCCP leads to the intracellular accumulation of H^+ until a steady state is reached. At the steady state, $\Delta\psi$ (inside negative) and ΔpH (inside acidic) of similar magnitude but in opposite polarity are generated, which results in the collapse of Δp. Since the generation of the inside acidic ΔpH is driven by CCCP-resistant $\Delta\psi$, the magnitude of the ΔpH never exceeds the magnitude of the $\Delta\psi$. This means that no net influx of H^+ occurs at the steady state, and that the $\Delta\psi$ at alkaline pH is stable even though the membrane is completely permeable to H^+. Therefore, the Na^+ pump activities in whole cells can be examined from CCCP-resistant Na^+ extrusion, CCCP-resistant $\Delta\psi$ generation, and CCCP-dependent H^+ accumulation. In this article, examinations of the Na^+ pump activity in whole cells and membranes are described.

Growth of *Vibrio alginolyticus*

A synthetic medium contains per liter: NaCl, 17.6 g; KCl, 0.75 g; K_2HPO_4, 0.35 g; $FeSO_4 \cdot 7H_2O$, 3 mg; $(NH_4)_2SO_4$, 2 g; $MgSO_4 \cdot 7H_2O$, 1.23 g; glycerol, 10 g; tris(hydroxymethyl)methylamine (Tris), 6 g. A complex medium contains per liter: polypeptone (Daigo Eiyo, Osaka, Japan), 5 g; yeast extract, 5 g; K_2HPO_4, 4 g; NaCl, 30 g. The pH of media is adjusted to 7.5. For growth at various pH values, 2-(N-morpholino) ethanesulfonic acid (MES), 4-(2-hydroxyethyl)-1-piperazineethanesulfonic acid (HEPES), or N-[tris(hydroxymethyl)methyl]glycine (Tricine) at 50 mM is supplemented in the complex medium and the pH is adjusted with NaOH. Cells are grown aerobically at 37°.

The stock culture of *V. alginolyticus* is maintained on slants of the complex medium supplemented with 0.2% glucose and 1.5% agar. This organism forms transparent and opaque variants. Sediment of the latter cell diffuses soon after a centrifugation. It is necessary to select the transparent colony on plates of the complex medium every 2 to 3 months. The opaque variant forms a shiny colony and is easily distinguished on the plate.

Isolation of the Na⁺ Pump-Defective Mutants

The growth of *V. alginolyticus* becomes remarkably resistant to CCCP when the Na^+ pump functions.[8] Although the growth at pH 6.5 on the

---

[8] H. Tokuda and T. Unemoto, *J. Bacteriol.* **156,** 636 (1983).

complex medium is completely inhibited by 5 μM CCCP, the growth at
pH 8.5 occurs even in the presence of 50 μM CCCP. These results indicate that the sodium motive force generated by the Na^+ pump is able to
support the growth of cells even though the generation of Δp is not possible. By selecting strains which are unable to grow at pH 8.5 in the presence of CCCP, the Na^+ pump-defective mutants can be isolated.[3]

Procedure. The wild-type cells harvested at the late logarithmic phase
of growth (10^8 cells/ml) on the complex medium are washed twice with
and resuspended in 0.1 M sodium citrate, pH 5.5, containing 0.3 M NaCl,
50 mM MgSO$_4$, 10 mM KCl. The cells are mutagenized at 37° for 10 min
with 50 μg/ml of N-methyl-N'-nitro-N-nitrosoguanidine and then washed
twice with 10 mM HEPES–NaOH, pH 7.0, 0.3 M NaCl, 50 mM MgSO$_4$,
10 mM KCl. The mutagenized cells are grown at 37° for two generations
on the complex medium containing 0.2% glucose and 50 mM HEPES–
NaOH, pH 7.0. The culture is then diluted with the complex medium
containing 50 mM Tricine–NaOH, pH 8.5, 0.2% glucose, 5 μM CCCP,
and 8×10^3 units/ml penicillin G. Growth and subsequent lysis of the cells
at 37° are monitored by measuring a turbidity at 600 nm. After the completion of lysis, unlysed cells are washed twice with and grown overnight on
the fresh complex medium at pH 7.0. The penicillin enrichment is repeated once more and the cells are grown on agar plates of the complex
medium at pH 7.0. Colonies on this plate are replicated on agar plates of
the complex medium containing 50 mM Tricine–NaOH, pH 8.5, 5 μM
CCCP. Strains unable to grow on the CCCP-containing plates are isolated
and purified by another replica plating. The isolated strains are next examined for the Na^+ pump activity as described in a later section.

Spontaneous revertants can be obtained by inoculating the heavy suspensions of the mutant strains on the CCCP-containing plates.

All the isolated strains should require Na^+ for growth and grow on
sucrose, but not on lactose, as a sole source of carbon.

Preparation of K^+-Depleted and Na^+-Loaded Cells

The cells growing on the synthetic medium maintain intracellular concentrations of K^+ and Na^+ at about 0.4 M and 80 mM, respectively.[9]
Although repeated washing of the cells with 0.4 M NaCl solution does not
change the internal concentration of these cations, treatment of the cells
with a membrane-permeable weak base, diethanolamine (DEA), and subsequent dilution of the DEA-treated cells into 0.4 M NaCl have been
found to cause the replacement of internal K^+ with external Na^+.[9] The

---

[9] H. Tokuda, T. Nakamura, and T. Unemoto, *Biochemistry* **20**, 4198 (1981).

release of internal K⁺ down the concentration gradient occurs via a K⁺/ H⁺ antiport system which functions at cytoplasmic pH value higher than about 8.0.[10] Since DEA enters the cells in its unprotonated form and alkalizes the cytoplasm by its protonation,[2] treatment of the cells with DEA induces the downhill efflux of K⁺. Although the efflux of K⁺ is coupled to the influx of H⁺, the cytoplasmic pH remains high because of the compensating influx of DEA. The release of K⁺ thus continues until almost all K⁺ is replaced with DEA. The dilution of DEA-loaded cells into 0.4 M NaCl solution causes the efflux of DEA with concomitant influx of Na⁺. Although this step is not characterized very well, the influx of Na⁺ presumably occurs via the Na⁺/H⁺ antiport system working in reverse.[10]

Procedure. The cells grown on the synthetic medium are harvested and treated twice with a half volume of 50 mM DEA–HCl, pH 8.5, 0.4 M NaCl at 25° for 10 min. K⁺ release from cells during the incubation can be followed by K⁺ electrode. The DEA-treated cells are collected and then treated twice with 50 mM Tricine–NaOH, pH 8.5, 0.4 M NaCl at 25° for 5 min. The cells treated in this manner retain essentially no K⁺ and about 0.4 M Na⁺.[10] The Na⁺-loaded cells are resuspended in 0.4 M NaCl at a concentration of about 50 mg protein/ml and kept on ice until use.

Na⁺ Extrusion by the K⁺-Depleted and Na⁺-Loaded Cells

Examination of Na⁺ extrusion by whole cells requires the K⁺-depleted and Na⁺-loaded cells. Since the Na⁺ extrusion by these cells requiers K⁺ as a countercation, the assay can be started by addition of K⁺.[6] The Na⁺ pump activity is examined from CCCP-resistant Na⁺ extrusion which is specifically inhibited by HQNO.[2]

Procedure. The Na⁺-loaded cells are resuspended in 50 mM Tricine–NaOH, pH 8.5, 0.4 M NaCl at a concentration of 2 to 3 mg protein/ml. The cell suspensions are equilibrated with ²²NaCl (about 10⁴ cpm/μl) on ice for 2 to 3 hr. Aliquots (50 μl) of the cell suspensions are transferred to a series of test tubes (12 × 75 mm) and preincubated at 25° for 5 min with 20 mM glycerol. Na⁺ extrusion is initiated by the addition of KCl at a final concentration of 10 mM. CCCP should be added prior to the start of assay. Microliter additions to the incubation mixture are made by using a Hamilton microsyringe (Hamilton Co., Reno, Nevada). The syringes are used in conjunction with a repeating dispenser (Hamilton Co.). The level of ²²Na⁺ retained by the cells is determined at the desired time by dilution of the cell suspension with 2 ml of 0.4 M NaCl at room temperature, immediate filtration through cellulose acetate filters (0.5 μm pore size,

[10] T. Nakamura, H. Tokuda, and T. Unemoto, *Biochim. Biophys. Acta* **692**, 389 (1982).

Millipore Corporation, Bedford, MA, or 0.45 μm pore size, Schleicher and Schull, Dassel, West Germany), and washing once with 2 ml of 0.4 M NaCl. The whole procedure of dilution and filtration should be performed in less than 10 sec. The filters are immediately removed from the suction apparatus and dissolved in 4 ml of Bray's scintillation liquid for the determination of radioactivity. The values obtained are corrected for background radioactivity which is obtained with samples boiled for 5 min prior to assay. Typical results of a Na⁺ extrusion experiment in cells of V. alginolyticus are shown in Fig. 1. If the wild-type cells are treated with 50 μM HQNO, CCCP-resistant Na⁺ extrusion by the Na⁺ pump is completely inhibited.[2]

Generation of Δψ and ΔpH

Although it is not strictly necessary, use of the K⁺-depleted and Na⁺-loaded cells is recommended for the determination of CCCP-resistant Δψ and CCCP-dependent ΔpH. In any case, it is necessary to keep K⁺ concentration minimum since K⁺ is a countercation for Na⁺ extrusion and collapses Δψ.[8]

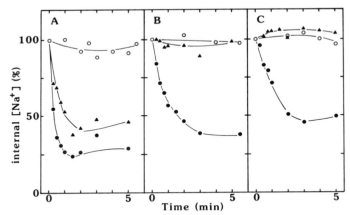

FIG. 1. Na⁺ extrusion by the K⁺-depleted and Na⁺-loaded cells of V. alginolyticus. Na⁺-loaded cells of the wild-type (A) and the Na⁺ pump-defective mutants, Nap 1 (B) and Nap 2 (C), were equilibrated with ²²Na⁺ (1 × 10⁴ cpm/μl) in 50 mM Tricine–NaOH, pH 8.5, 0.4 M NaCl. The cell suspensions were preincubated with 20 mM glycerol for 5 min at 25°. At 0 time, additions of nothing (○), 10 mM KCl (●), or 10 μM CCCP plus 10 mM KCl (▲) were made. The results are given as the percentage of radioactivity at 0 time. Na⁺ extrusion by the mutants cells were performed only by CCCP-sensitive Na⁺/H⁺ antiport system. Taken from Tokuda.[3]

Procedure. Generation of $\Delta\psi$ (negative inside) and ΔpH (acidic inside) is examined from the distribution of [³H]tetraphenyl phosphonium ion (TPP⁺) and [¹⁴C]methylamine, respectively. Level of steady-state accumulation of these probes can be determined by flow dialysis or filtration. Both methods give essentially the same values of $\Delta\psi$ and ΔpH. For the flow dialysis method, the reader is referred to some previous publications.[1,2,9] The filtration method is similar to that used for the assay of Na⁺ extrusion. The K⁺-depleted and Na⁺-loaded cells are resuspended at a concentration of about 1 mg protein/ml in 50 μl of 50 mM Tricine–NaOH, pH 8.5, 0.4 M NaCl, 20 mM glycerol. After 5 min preincubation at 25°, the assay is started by the addition of [³H]TPP⁺ (19 μM, 157 μCi/μmol) or [¹⁴C]methylamine (21 μM, 44 μCi/μmol). CCCP should be added at a final concentration of 10 to 20 μM prior to the start of assay. At time intervals, the uptake is terminated by addition of 2 ml of 0.4 M NaCl at room temperature and filtration with cellulose acetate filters. Cellulose nitrate filters give very high nonspecific binding of TPP⁺ and therefore can not be used for the determination of $\Delta\psi$.[11] The filters are washed once with 2 ml of 0.4 M NaCl and radioactivities are determined by liquid scintillation counting.

Steady-state concentration gradients are calculated by using a value of 3.3 μl of internal water space/mg of cell protein.[9] $\Delta\psi$ can be estimated according to the equation:

$$\Delta\psi = -\frac{RT}{F} \ln \frac{[\text{TPP}^+]_{\text{in}}}{[\text{TPP}^+]_{\text{out}}} \tag{1}$$

where $[\text{TPP}^+]_{\text{in}}$ and $[\text{TPP}^+]_{\text{out}}$ are concentrations of TPP⁺, inside and outside the cells, respectively, and the other symbols have their usual meaning. ΔpH is calculated from the distribution of methylamine (pK 10.62) as described.[12] Typical values of $\Delta\psi$ and ΔpH generated by various strains of *V. alginolyticus* are shown in the table.

Measurement of H⁺ Fluxes Induced by Oxygen Pulse

Procedure. The K⁺-depleted and Na⁺-loaded cells (about 1 mg protein) are resuspended in 2 ml of 0.2 mM Tricine–NaOH, pH 8.5, 0.4 M NaCl, 20 mM glycerol. The cell suspensions are kept anaerobic at 25° in a water-jacketed vessel (TTA 60 titration assembly, Radiometer, Copenhagen, Denmark) under the stream of water-saturated nitrogen gas until the medium pH becomes constant at pH 8.5. Readjustment of medium pH during the incubation should be done with small amount of 1 N NaOH.

[11] S. Schuldiner and H. R. Kaback, *Biochemistry* **14**, 5451 (1975).
[12] H. Rottenberg, this series, Vol. 55, p. 547.

EFFECTS OF CCCP ON THE GENERATION OF $\Delta\psi$ (NEGATIVE
INSIDE) AND ΔpH (ACIDIC INSIDE) AT pH 8.5 BY Na^+-LOADED
CELLS OF *V. alginolyticus*

Cells	$\Delta\psi$ (mV)		2.3RT/F ΔpH (mV)	
	−CCCP	+CCCP[a]	−CCCP	+CCCP[a]
Wild type	−159	−154	−24	−130
Nap 1	−160	−47	−19	ND[b]
Nap 2	−155	−40	−18	ND[b]
Nap 2R[c]	−157	−164	−31	−122

[a] At a final concentration of 20 μM.
[b] Not detectable.
[c] A spontaneous revertant of Nap 2.

The medium pH is monitored by a Radiometer galss electrode (G2040C)
attached to a pH meter (PHM84) with a calomel electrode (K701) as a
reference. The reference electrode contains a secondary salt bridge filled
with a saturated CH_3COOLi in order to minimize the release of K^+ from
the electrode. Output of the pH meter is supplied to an electrometer (HE-
101A, Hokuto Denko, Tokyo, Japan) which has a voltage-compensation
function. With appropriate setting of the voltage compensation, changes
in medium pH are recorded by a Hitachi 056 recorder. Oxygen pulse is
performed at pH 8.5 by addition of 100 μl air-saturated 0.4 M NaCl
solution at 25°. For the determination of H^+ pump activity, TPP^+ is added
to the anaerobic cell suspension at a final concentration of 2.5 mM in

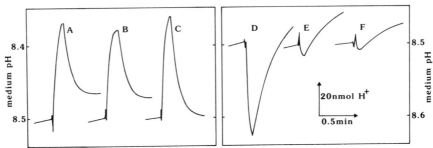

FIG. 2. H^+ extrusion and Na^+ pump-dependent H^+ uptake. Na^+-loaded cells of the wild
type (A and D), Nap 1 (B and E), and Nap 2 (C and F) were kept anaerobic at 25° in 2 ml of
0.2 mM Tricine–NaOH, pH 8.5, 0.4 M NaCl, 20 mM glycerol. TPP^+ (A to C) and CCCP (D
to F) were also included in the reaction mixture at final concentrations of 2.5 mM and 20 μM,
respectively. An oxygen pulse was made at pH 8.5 by the addition of 100 μl air-saturated 0.4
M NaCl. Changes in the medium pH were monitored by pH electrode. Taken from Tokuda.[3]

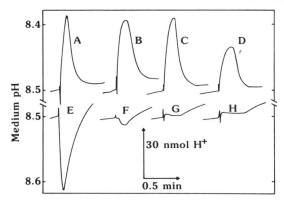

FIG. 3. Specific inhibition of the Na⁺ pump-dependent H⁺ uptake by HQNO. Na⁺-loaded cells of the wild type were treated with HQNO at final concentrations of 0 (A and E), 15 (B and F), 30 (C and G), and 50 μM (D and H). TPP⁺ (A to D) and CCCP (E to H) were also present in the assay mixture at the same concentrations as in Fig. 2. The cell suspensions were kept anaerobic and then pulsed with oxygen at pH 8.5 as described in Fig. 2. Taken from Tokuda and Unemoto.[4]

order to facilitate H⁺ extrusion. For the determination of the Na⁺ pump activity, oxygen pulse should be done in the presence of 20 μM CCCP. Under these conditions, $\Delta\psi$ generated by the Na⁺ pump drives the uptake of H⁺. Results shown in Fig. 2 indicate that the H⁺ pump activity of the mutant cells is comparable to that of the wild-type cells although the Na⁺ pump-dependent H⁺ uptake is observed only in the wild-type cells. Treatment of the wild-type cells with HQNO causes the specific inhibition of the Na⁺ pump with little effect on the H⁺ pump (Fig. 3). In the presence of TPP⁺ and CCCP, oxygen pulse does not cause any pH change and the buffering capacity of the reaction mixture should be titrated under these conditions.[2]

Preparation of Membranes for the Determination of Respiratory Activities

Like other marine bacteria, the cells of *V. alginolyticus* lyse when exposed to hypotonic medium.[13] In order to prepare membranes retaining high respiratory activities, complete lysis of the cells appears to be essential. The following procedure is a modification of the original method.[14] The membranes prepared by this method are not closed vesicles and

[13] T. Unemoto, T. Tsuruoka, and M. Hayashi, *Can. J. Microbiol.* **19**, 563 (1973).
[14] T. Unemoto, M. Hayashi, and M. Hayashi, *J. Biochem. (Tokyo)* **90**, 619 (1981).

permeable to dextran.[14] Therefore, these membranes are not feasible for transport assays but suitable for the examination of oxidation of membrane-impermeable substrates.

Procedure. Unless otherwise specified, all steps should be done at 4° and the volume of specified buffer solution is one-tenth of original culture. The cells harvested at the late logarithmic phase of growth are washed twice with 1 M NaCl, 20 mM Tris–HCl, pH 7.5, 5 mM EDTA. The washed cells are resuspended in 10 mM NaCl, 20 mM Tris–HCl, pH 7.5, 5 mM EDTA. The cell suspensions are incubated at 25° for 5 min and then centrifuged at 12,000 g for 15 min. Most cells lyse during the incubation and centrifugation. To complete the lysis, the precipitates are incubated again in the same buffer solution as above and centrifuged. The precipitates are resuspended in 10 mM NaCl, 20 mM Tris–HCl, pH 7.5, 0.5 mM EDTA. DNase and RNase at 5 μg/ml each and MgCl$_2$ at 2 mM are added to the suspensions. After 20 min incubation at 25°, EDTA is added at 5 mM and incubation of the suspension is continued for 5 min. Membrane fractions are then collected by centrifugation at 12,000 g for 20 min. If necessary, another cycle of the treatment with DNase and RNase should be done. After twice washing with 10 mM NaCl, 20 mM Tris–HCl, pH 7.5, 0.5 mM EDTA, membranes are resuspended at about 15 mg protein/ml in 10 mM NaCl, 20 mM Tris–HCl, pH 7.5, 10% glycerol and kept frozen at −70°.

Measurement of Respiratory Activities in Isolated Membranes

NADH Oxidase. NADH oxidase is examined at 30° from the decrease in absorbance at 340 nm by using a Hitachi 200-20 spectrophotometer. The standard assay mixture contains in 1 ml: 20 mM Tris–HCl, pH 7.5, 0.2 mM NADH (disodium salt), various concentrations of salt, and about 30 μg of membrane protein. The assay is started by the addition of membranes. Effects of HQNO are examined with membranes preincubated in the presence of various concentrations of HQNO. In this case, the assay should be started by the addition of NADH. As shown in Fig. 4, the NADH oxidase activity of membranes isolated from the wild type and revertant requires Na$^+$ for maximum activity. The stimulation by K$^+$ or other cations (not shown) is marginal. In contrast, the NADH oxidase activities of mutant membranes show no specific requirement for Na$^+$.

Inhibition by HQNO is specific to the Na$^+$-dependent NADH oxidase of the wild-type and revertant membranes (Fig. 5A). The NADH oxidase of the mutant membranes is much more resistant to HQNO even in the presence of Na$^+$. Moreover, the NADH oxidase activity of the wild type observed in the absence of Na$^+$ is also resistant to HQNO.

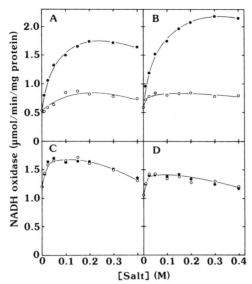

FIG. 4. Na⁺ requirement of NADH oxidase in *V. alginolyticus* membranes. The assay mixture in 1 ml contained 20 mM Tris–HCl, pH 7.5, 0.2 mM NADH, and various concentrations of KCl (○) or NaCl (●). The assay at 30° was started by addition of membrane suspensions containing about 30 μg of protein. Membranes examined were prepared from the wild type (A), Nap 2R (B), Nap 1 (C), and Nap 2 (D). Taken from Tokuda and Unemoto.[4]

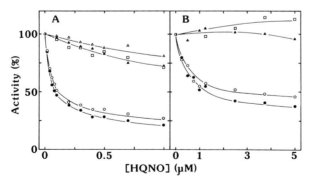

FIG. 5. Effect of HQNO on the NADH oxidase and QH$_2$ formation. The NADH oxidase activity (A) and the NADH-linked QH$_2$ formation (B) were determined in the presence of 0.2 M NaCl as a function of HQNO concentration. Where specified (□), the NADH oxidase activity in the absence of 0.2 M NaCl (A) and the G3P-linked QH$_2$ formation (B) were also examined with the wild type membranes. Membranes examined were prepared from the wild type (○,□), Nap 1 (△), Nap 2 (▲), and Nap 2R (●). Taken from Tokuda and Unemoto.[4]

Ubiquinol Formation. Membranes prepared from glycerol-grown cells retain inducible L-glycerol-3-phosphate (G3P) dehydrogenase which is linked to respiratory chain.[15] The activity of G3P-linked ubiquinol-1 (QH_2) formation is determined by measuring the decrease in absorbance at 275 nm taking a millimolar extinction coefficient (ε) of ubiquinone-1 (Q-1) as 12.4.[16] For the examination of NADH-linked QH_2 formation, the difference in absorbance between 248 and 267 nm should be determined taking ε as 7.8. Unemoto and Hayashi have found that absorbance changes due to the conversion of NADH to NAD can be neglected by using this wavelength pair.[17] The standard reaction mixture in 2 ml contains 2 mM Tris–HCl, pH 7.5, 25 μM diethyl dithiocarbamate, about 15 μg membrane protein, 5 μM Q-1 (Eisai Co., Tokyo, Japan), 10 mM KCN. Diethyl dithiocarbamate is a chelating agent added to protect enzyme activities from heavy metal cations. KCN is added to inhibit oxidation of QH_2. After preincubation of membranes at 30° for 3 min with various concentrations of salt and HQNO, the assay is initiated by the addition of disodium NADH or di(monocyclohexylammonium) G3P (Sigma) at final concentrations of 0.1 and 20 mM, respectively. Na^+-dependent QH_2 formation is observed only with the wild-type and revertant membranes in the presence of NADH. Neither the NADH-linked QH_2 formation of mutant membranes nor the G3P-linked QH_2 formation of all the membranes requires Na^+.[4] These results indicate that the Na^+ pump is coupled to the respiratory chain at the step of NADH:quinone oxidoreductase. Results shown in Fig. 5B indicate that only the Na^+-dependent NADH:quinone oxidoreductase is sensitive to HQNO which is a specific inhibitor of the Na^+ pump in whole cells.

[15] T. Unemoto, M. Hayashi, and M. Hayashi, *J. Biochem. (Tokyo)* **90**, 619 (1981).
[16] F. L. Crane and R. Barr, this series, Vol. 18, p. 137.
[17] T. Unemoto and M. Hayashi, *J. Biochem. (Tokyo)* **85**, 1461 (1979).

[40] Preparation, Characterization, and Reconstitution of Oxaloacetate Decarboxylase from *Klebsiella aerogenes,* a Sodium Pump

By PETER DIMROTH

The sodium transport decarboxylases have the unique property of being able to use decarboxylation energy as a deriving force for active Na^+ transport. Oxaloacetate decarboxylase (EC 4.1.1.2) from *Klebsiella aerogenes* was the first enzyme where this kind of energy transduction has been demonstrated.[1] Subsequently methylmalonyl-CoA decarboxylase (EC 4.1.1.41) from *Veillonella alcalescens*[2] and glutaconyl-CoA decarboxylase from *Acidaminococcus fermentans*[3] were also shown to be sodium transport decarboxylases. These enzymes have a number of properties in common, e.g., binding to the membrane, specific activation by Na⁺ ions, and the prosthetic group biotin. Oxaloacetate decarboxylase can be conveniently purified by affinity chromatography on monomeric avidin-Sepharose.[4] Incorporation of thus purified enzyme into phospholipid vesicles leads to the reconstitution of Na⁺ transport activity[5] which can therefore be studied with a clearly defined system in the absence of interfering contaminants.

Determination of Oxaloacetate Decarboxylase Activity

Assay a.[6] In this assay oxalacetate decarboxylation was followed by the decrease of enolic oxaloacetate at 265 nm (ε_{265} = 0.95 mM^{-1} cm^{-1}). The cuvette (d = 1 cm) contained in 1.0 ml at 25°: 100 mM Tris–HCl buffer, pH 7.5, 20 mM NaCl, and 1 mM oxaloacetate. After the keto–enol equilibrium was established as indicated by a constant absorbance reading the decarboxylation was initiated with the enzyme (0.01–0.1 U). The constant rate in the decrease of absorbance which was reached after a lag phase of about 0.5–2 min was used to calculate the activity of the decarboxylase.

Assay b.[6] For purified oxaloacetate decarboxylase samples (free of malate dehydrogenase) a coupled spectrophotometric assay with lactate dehydrogenase could be used. The cuvette (d = 1 cm) contained in 1.0 ml at 25°: 100 mM Tris–HCl buffer, pH 7.5, 20 mM NaCl, 0.3 mM NADH, 1 mM oxaloacetate, 6 U lactate dehydrogenase, and 0.005–0.03 U oxaloacetate decarboxylase to initiate the reaction which was followed continuously at 340 nm. In all determinations the rate of NADH oxidation before adding the enzyme was substracted. This consisted of the chemical decarboxylation of oxaloacetate and NADH oxidation by malate dehydrogenase which is present as a small contaminant in commercial lactate dehydrogenase.

[1] P. Dimroth, *Eur. J. Biochem.* **121,** 443 (1982).
[2] W. Hilpert and P. Dimroth, *Nature (London)* **296,** 584 (1982).
[3] W. Buckel and R. Semmler, *Eur. J. Biochem.* **136,** 427 (1983).
[4] P. Dimroth, *FEBS Lett.* **141,** 59 (1982).
[5] P. Dimroth, *J. Biol. Chem.* **256,** 11974 (1981).
[6] P. Dimroth, *Eur. J. Biochem.* **115,** 353 (1981).

Purification of Oxaloacetate Decarboxylase

Mass Production of Klebsiella aerogenes Cells

The cells were grown anaerobically at 35° in a medium containing 45 mM trisodium citrate, 1.6 mM MgSO$_4$, 30 mM KH$_2$PO$_4$, 11 mM (NH$_4$)$_2$SO$_4$, and 0.4% of a trace metal solution (20 g CaCl$_2$, 0.1 g CoCl$_2$, 0.1 g MnCl$_2$ · 4H$_2$O, 0.1 g Na$_2$MoO$_4$ · 2H$_2$O, 2 g FeSO$_4$ · 7H$_2$O per liter of 10 mM HCl). The pH was adjusted to 7.0 with solid NaOH. Inoculation was performed with 10% of a culture in the early stationary phase and several transfers were performed to reach the final stage where the bacteria were grown in a fermentor containing 200 liter growth medium. The culture was gently stirred for about 12 hr during which time the bacteria reached the stationary phase. After cooling to 15° the bacteria were collected by continuous centrifugation with a Westfalia Separator type KA-2. The cells (about 300 g wet weight) were washed twice with 2.5 liter 50 mM K-phosphate, pH 7.0, and stored at −20°.

Preparation of Membranes

Klebsiella aerogenes cells (30 g wet weight) were suspended in 120 ml buffer A (20 mM K-phosphate, pH 7.5, 0.5 M NaCl, 0.2 mM diisopropylfluorophosphate, 0.1 mM tosylphenylalanine chloromethyl ketone), and after adding 0.5 mg DNase I and 1 mM MgEDTA the cells were desintegrated by two successive passages through a French pressure chamber at 12,000 psi. Whole cells and large debris were removed by centrifugation at 8000 g for 10 min and the membrane vesicles were sedimented by centrifugation of the supernatant for 1 hr at 150,000 g. The pellet was washed with 30 ml buffer A and sedimented again as before.

Purification by Avidin-Sepharose Chromatography

The decarboxylase was solubilized by carefully homogenizing the membranes with 30 ml buffer A containing 2% Triton X-100 using a motor-driven plunger. The clear extract obtained after centrifugation at 150,000 g for 30 min was immediately applied to a monomeric avidin-Sepharose column (see below) and protein not retained was washed off the column with about 6 volumes of buffer A containing 0.1% Brij 58 (flow rate: 0.8 ml/min). The enzyme was eluted with buffer A containing 0.8 mM biotin and concentrated to about 1 ml by ultrafiltration over a PM10 membrane (Amicon). The enzyme could be stored for at least a year without loss of activity in liquid N$_2$. A summary of the purification procedure is shown in the table.

PURIFICATION OF OXALOACETATE DECARBOXYLASE FROM *Klebsiella aerogenes*

Step	Volume (ml)	Total protein (mg)	Total activity (U)	Specific activity (U/mg)	Recovery (%)
Crude cell extract[a]	122	3050	2568	0.84	100
Membrane vesicles	39	908	1396	1.5	54
Triton X-100 eluate	25	205	635	3.1	25
Eluate from avidin-Sepharose column	1.1	13.5	625	46.3	24

[a] Starting material: 30 g of wet packed cells.

Preparation of Monomeric Avidin-Sepharose

The avidin-Sepharose column was prepared by a modification of described procedures.[7,8] A suspension of 20 ml Sepharose CL-4B in 20 ml H_2O was activated with 0.56 g cyanogen bromide. The pH was maintained at about 11 for 10 min by the dropwise addition of 1 M NaOH. The gel was washed with 200 ml 0.01 M Na-phosphate, pH 7.0 and resuspended in 20 ml of the same buffer. Avidin (50 mg) was added to the suspension and this was gently shaken for 16 hr at 4°. The gel was filtered and resuspended in 20 ml 0.1 M 2-aminoethanol–HCl, pH 7.0 to block any unreacted sites on the resin. After washing the gel with 0.01 M Na-phosphate, pH 7.0 the tetrameric avidin was dissociated by five washes with 20 ml 6 M guanidine–HCl in 0.2 M KCl/HCl, pH 1.5. The gel was washed with 0.2 M K-phosphate, pH 7.5 and suspended in 20 ml of this buffer. Succinic anhydride (5 mg) was added and the suspension was gently shaken for 1 hr to succinylate the avidin. The gel was then equilibrated with 20 mM K-phosphate, pH 7.5 containing 0.5 M NaCl and was washed in a column with 0.8 mM biotin in this buffer to saturate the avidin with biotin. The biotin bound to low affinity binding sites was subsequently removed by washing the column with 0.2 M glycine–HCl, pH 2.0. After equilibration with 20 mM K-phosphate, pH 7.5 containing 0.5 M NaCl, 0.2 mM diisopropyl fluorophosphate, and 0.1% Brij 58 the column was ready for use. It was regenerated after each run by washing with 0.2 M glycine–HCl, pH 2.0 and reequilibration as above.

[7] K. P. Henrikson, S. H. G. Allen, and W. L. Maloy, *Anal. Biochem.* **94**, 366 (1979).
[8] R. A. Gravel, K. F. Lam, D. Mahuran, and A. Kronis, *Arch. Biochem. Biophys.* **201**, 669 (1980).

Properties

Catalytic Properties

Purified oxaloacetate decarboxylase was characterized by kinetic studies as follows[6]: pH optimum 6.5–7.5, K_m for oxaloacetate 150 μM, K_m for Na^+ 1 mM. If Na^+ ions were carefully excluded from all reagent solutions the residual oxaloacetate decarboxylase activity was negligible ($<0.3\%$ of V_{max}) even in the presence of 100 mM K-phosphate. The decarboxylase does not depend on a divalent metal ion since EDTA is without effect. The primary reaction product is CO_2 and not HCO_3^-.[9] The decarboxylase is completely inhibited by avidin but not by avidin saturated with biotin. L-Malate, 2-oxomalonate, glyoxylate, and oxlate inhibited the enzyme with increasing efficiency in that order. Oxalate is competitive for oxaloacetate.[10] Since the K_I (3.8 μM) is far below the K_m for oxaloacetate, oxalate probably acts as a transition state analog. The stereochemical course of oxaloacetate decarboxylation is with retention of configuration.[6]

Molecular Properties

Oxaloacetate decarboxylase from *Klebsiella aerogenes* is composed of three different polypeptide chains α, β, γ with M_r 65,000, 34,000, and 12,000, respectively, as determined by sodium dodecyl sulfate–gel electrophoresis.[9] For the detection of all three polypeptides on the gel, silver staining should be used since the β- and especially the γ-chain are very poorly stained with Coomassie brilliant blue. The α-chain contains about 1 mol covalently bound biotin but no biotin is present in the β- and γ-chain. The same pattern of three different polypeptides was obtained when the decarboxylase was analyzed by HPLC in a dodecyl sulfate containing buffer.[9] Oxaloacetate decarboxylase was incubated for 2 min with 1% sodium dodecyl sulfate and 5% 2-mercaptoethanol to dissociate the enzyme. The solution (20 μl, 5–20 μg protein) was applied to a TSK 250 column (30 × 0.75 cm) which was run at 20° with 0.2 M Na-phosphate, pH 7.0 containing 0.1% sodium dodecyl sulfate. The flow rate was kept constant at 0.5 ml/min with an HPLC pump and the polypeptides eluted from the column were detected by monitoring the absorbance at 280 or 220 nm. For preparative runs 0.5 ml (3.6 mg) of dissociated decarboxylase was fractionated on a TSK 250 column (60 × 2.15 cm) at a flow rate of 1.5 ml/min under otherwise identical conditions. The separation of oxaloacetate decarboxylase subunits by HPLC is shown in Fig. 1.

[9] P. Dimroth and A. Thomer, *Eur. J. Biochem.* **137,** 107 (1983).
[10] P. Dimroth, *Biosci. Rep.* **2,** 849 (1982).

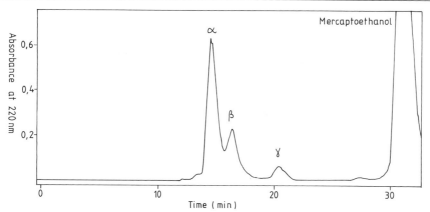

FIG. 1. Separation of oxaloacetate decarboxylase subunits by HPLC on a TSK 250 column (30 × 0.75 cm).

Partial Activities and Subunit Function

General

According to the mechanism of biotin-dependent carboxylases the mechanism of the biotin-dependent decarboxylases was expected to involve two distinct partial reactions, the transfer of the carboxyl group from the substrate to the biotin prosthetic group on the enzyme [Eq. (1)], and decarboxylation of the carboxybiotin enzyme to regenerate free biotin enzyme [Eq. (2)]. A third activity of the decarboxylases is the transport of Na$^+$ ions through the membrane [Eq. (3)]. It is conceivable that each of these activities is catalyzed by a distinct subunit.

$$
\begin{aligned}
\text{R-COO}^- + \text{E-biotin} &\rightleftharpoons \text{E-biotin-COO}^- + \text{RH} & (1)\\
\text{E biotin-COO}^- + \text{H}^+ &\rightleftharpoons \text{E-biotin} + \text{CO}_2 & (2)\\
\underline{n\ \text{Na}_{in}^+} &\rightleftharpoons \underline{n\ \text{Na}_{out}^+} & (3)\\
\text{R-COO}^- + \text{H}^+ + n\ \text{Na}_{in}^+ &\rightleftharpoons \text{RH} + \text{CO}_2 + n\ \text{Na}_{out}^+ &
\end{aligned}
$$

Determination of Carboxyltransferase[9,11]

The carboxyltransferase activity [Eq. (1)] was determined from the exchange of radioactivity between [1-^{14}C]pyruvate and oxaloacetate. The incubation mixtures contained in 1.2 ml at 25°: 100 mM K-phosphate, pH 7.5, 0.7 mM potassium-[1-^{14}C]pyruvate (343,000 cpm/μmol), 0.7 mM ox-

[11] P. Dimroth, *Eur. J. Biochem.* **121**, 435 (1982).

aloacetate, and the enzyme (about 10 μg of oxaloacetate decarboxylase or 10 μg of the isolated carboxyltransferase subunit), which was used to initiate the reaction. Samples (0.22 ml each) were transferred after 1–10 min into 10 μl 2 M HCl to terminate the exchange reaction. After neutralization of the reaction mixtures with 10 μl 2 M KOH and 20 μl 1 M Tris–HCl, pH 8.0, the oxaloacetate was reduced to malate with 2 μmol NADH and 20 U malate dehydrogenase. After 10 min at 25° the mixtures were acidified with 10 μl 2 M HCl and after adding 1 μmol each of pyruvate and malate, the acids were separated on Dowex 1 X-8 (Cl⁻) (1 × 10 cm) columns with a linear gradient from H$_2$O to 60 mM HCl (2 × 100 ml) and counted for radioactivity.

Isolation of the Carboxyltransferase Subunit (α)[9]

Isolation of the carboxyltransferase subunit was accomplished by dissociation of the α-chain from the membranes and purification by avidin-Sepharose chromatography. The membranes prepared from 30 g K. *aerogenes* cells as described above were suspended in 20 ml 5 mM K-phosphate, pH 7.5 containing 1 M LiCl and the suspension was frozen in liquid N$_2$. After thawing at room temperature the process was repeated three times which led to the destruction of more than 90% of the oxaloacetate decarboxylase activity. The particles were removed by centrifugation (20 min, 130,000 g), and the supernatant which did not contain any oxaloacetate decarboxylase activity was applied to the monomeric avidin-Sepharose column described above. The column was washed with about 150 ml 50 mM K-phosphate, pH 7.5 containing 0.3 M KCl and the carboxyltransferase was eluted in a yield of about 4.5 mg protein with 0.8 mM biotin in the same buffer. The carboxyltransferase was identified as the α-subunit by dodecyl sulfate gel electrophoresis (Fig. 2). It is devoid of β- and γ-subunits and ineffective in catalyzing oxaloacetate decarboxylation. The carboxyltransferase activity was about 9 U/mg protein and this was completely independent from the presence of Na⁺ ions. The α-chain is soluble in the absence of any detergents. This property and the ease with which it is released from the membrane characterize this subunit as a peripheral membrane protein.

The Sodium Pump

The sodium transport activity of oxaloacetate decarboxylase can be determined by the uptake of ²²Na⁺ into inverted bacterial membrane vesicles, prepared with a French press as described above.[1] However, leakiness of these vesicles is a major disadvantage for doing quantitative trans-

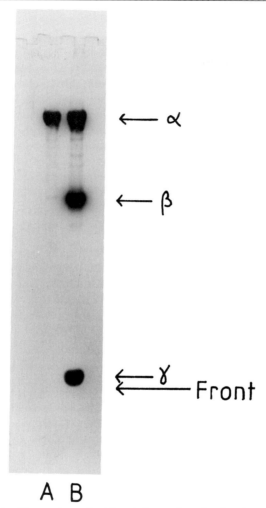

FIG. 2. Dodecyl sulfate–polyacrylamide gel electrophoresis of oxaloacetate decarboxylase and the isolated α-chain from this enzyme (Dimroth and Thomer[9]). Gel A contained 4 μg of the α-chain and gel B 9 μg oxaloacetate decarboxylase. The gels were stained with silver.

port studies. For these purposes the sodium transport activity had to be reconstituted by incorporating the purified oxaloacetate decarboxylase into membrane vesicles. Detergent dilution or detergent dialysis with octylglucoside as the detergent is an effective procedure to reach this goal.

Sodium Transport Assay

The incubation mixtures contained in 1.0 ml at 25°: 0.5 ml of reconstituted proteoliposomes, 30 mM K-phosphate, pH 7.0, 1 mM ^{22}Na$_2$SO$_4$ (500 cpm/nmol of Na$^+$), 0.5 mM dithioerythritol, and 1 mM Li-oxaloacetate to initiate the reaction. After 1/2–30 min samples (0.09 ml) were removed and passed within a few seconds over columns (2 × 0.5 cm) of Dowex 50 (K$^+$) to separate free ^{22}Na$^+$ from that entrapped inside the vesicles. The vesicular ^{22}Na$^+$ was eluted with 0.6 ml 30 mM K-phosphate, pH 7.0, and was determined by liquid scintillation counting. The specific radioactivity of ^{22}Na$^+$ was calculated from the radioactivity of the incubation mixture and from the actual Na$^+$ content as determined by atomic absorption spectrophotometry.

Reconstitution of Oxaloacetate Decarboxylase Catalyzing Na$^+$ Uptake[5,12]

The lipids used for reconstitution experiments were either soybean phosphatidylcholine (Sigma, type II-S) or lipids isolated from *K. aerogenes* when especially tight vesicles were required. For the preparation of these lipids 200 g *K. aerogenes* cells (wet weight) were suspended in 100 ml 2 M NaCl and extracted overnight with a mixture of 600 ml methanol and 300 ml chloroform. After centrifugation (2000 g, 5 min) the extraction of the pellet was repeated with half the volume of the two solvents. After washing the combined extracts by adding 300 ml each of water and chloroform the organic phase was evaporated under reduced pressure. The lipids were dissolved in a small volume of CHCl$_3$ and stored in sealed glass vials under reduced pressure at −20°. Prior to reconstitution, the solvent was removed with a stream of N$_2$ and the lipids were dissolved in peroxide-free ether. Removal of the solvent and dissolution in ether were repeated four times. Finally, the lipids were dried in a vacuum (<0.5 Torr) for 24 hr. The lipids (80 mg) were vigorously agitated on a vortex mixer for about 10 min under N$_2$ with 2.67 ml 3% octylglucoside in reconstitution buffer (30 mM K-phosphate, pH 7.0, 1 mM Na$_2$SO$_4$, 0.5 mM dithioerythritol, 1.5 mM NaN$_3$). Oxaloacetate decarboxylase (0.08 ml, 0.35 mg, 20 U) was then added to the lipid detergent mixture followed by dialysis against 1 liter reconstitution buffer for 24–48 hr at 4° with two changes of the dialysis buffer.[12] Alternatively, the transport can be reconstituted by a 20- to 40-fold dilution of the lipid–detergent–enzyme mixture with reconstitution buffer.[5] The proteoliposomes are concentrated by cen-

[12] P. Dimroth and W. Hilpert, *Biochemistry* **23**, 5360 (1984).

trifugation (200,000 g, 30 min) and resuspended in 2.7 ml reconstitution buffer.

Properties of the Na$^+$ Transport System

The reconstituted proteoliposomes performed a substrate-induced accumulation of Na$^+$ ions which was abolished by avidin but not by avidin saturated with biotin. In the presence of the Na$^+$ carrying ionophores monensin, nigericin or trinactin, the Na$^+$ uptake was abolished whereas the uncoupler carbonyl cyanide p-trifluoromethoxyphenylhydrazone had no significant effect on the transport. The transport was electrogenic creating a membrane potential of about 60 mV, a Na$^+$ concentration gradient equivalent to 50 mV, and a total sodium motive force of 110 mV.[1] The stoichiometry between Na$^+$ transport and oxaloacetate decarboxylation was 2:1 in the initial phase but decreased when steeper Na$^+$ concentration gradients had been developed.[12]

Reversal of Oxaloacetate Decarboxylase[12]

The decarboxylation of oxaloacetate is completely irreversible if catalyzed by the soluble enzyme. However, in a coupled vectorial transport system the decarboxylation can be reversed if a Na$^+$ gradient of proper direction and magnitude is applied. The most convenient assay of the reverse reaction is an isotopic exchange between $^{14}CO_2$ and oxaloacetate. Net carboxylation of pyruvate has been observed with a proteoliposomal system containing oxaloacetate decarboxylase and methylmalonyl-CoA decarboxylase. The Na$^+$ ion gradient established upon decarboxylation of malonyl-CoA served as the driving force for pyruvate carboxylation.

Oxaloacetate–$^{14}CO_2$ Exchange[12]

The incubation mixtures contained in 0.5 ml at 25°: 0.25 ml of oxaloacetate decarboxylase containing reconstituted proteoliposomes (7.5 mg lipid, 18 μg protein), 30 mM K-phosphate, pH 7.0, 1 mM Na$_2$SO$_4$, 0.5 mM dithioerythritol, 1.5 mM NaN$_3$, 6 mM KH^{14}CO$_3$ (3000 cpm/nmol), 10 U carbonate dehydratase, and 5 mM Li-oxaloacetate which initiated the reaction. Samples (0.12 ml) were transferred after appropriate incubation periods into 110 μl 0.2 M HCl to terminate the exchange reactions. After neutralizing the solutions with 0.2 ml 1 M Tris–HCl, pH 8.0, oxaloacetate was reduced to malate with 0.1 ml 10 mM NADH and 50 U malate dehydrogenase (25°, 10 min). The solutions were acidified with 0.1 ml 6 M HCl and evaporated to dryness at 95°. The residues were taken up in 0.2 ml 3 M HCl and evaporation was repeated. The acid-stable radioactivity ([^{14}C]malate) was determined by liquid scintillation counting.

Pyruvate Carboxylation with a Na$^+$ Gradient as Driving Force[12]

The reconstitution of proteoliposomes was performed as described above but in the additional presence of methylmalonyl-CoA decarboxylase[13] (0.05 ml, 0.26 mg, 7 U). The incubation mixtures contained in 1.0 ml at 25°: 0.5 ml of the reconstituted proteoliposomes (15 mg lipid, 60 μg protein), 30 mM K-phosphate, pH 7.0, 1 mM Na$_2$SO$_4$, 0.5 mM dithioery-thritol, 1.5 mM NaN$_3$, 20 mM KHCO$_3$, 2.7 mM K-pyruvate, 22 U citrate synthase, and 4.5 mM malonyl-CoA. Samples (0.16 ml) were added after appropriate times to 10 μl 2 M HCl to terminate the reactions. The citrate content of the centrifuged supernatants was determined in a coupled fluorometric assay with citrate lyase and malate dehydrogenase. NADH fluorescence was excited at 340 nm and the emission was recorded at 465 nm. The cuvette contained in a total volume of 2.0 ml: 100 mM Tris–HCl buffer, pH 8.0, 1 mM MgCl$_2$, 0.009 mM NADH, 1.2 U malate dehydrogenase, and the acidified sample (0.15 ml). The reactions were initiated by adding citrate lyase (2.5 U) and the change in fluorescence was determined and compared with that of a citrate standard. Under our conditions, the fluorescence change caused by 5 nmol citrate was 31%.

[13] W. Hilpert and P. Dimroth, this volume [41].

[41] Sodium Pump Methylmalonyl-CoA Decarboxylase from *Veillonella alcalescens*

By WILHELM HILPERT and PETER DIMROTH

Methylmalonyl-CoA decarboxylase from *Veillonella alcalescens* is one of the sodium pumps which are driven by decarboxylation energy.[1,2] Like the other sodium transport decarboxylases methylmalonyl-CoA decarboxylase is firmly bound to the cell membrane, specifically activated by Na$^+$ ions and contains the prosthetic group biotin. The enzyme is

[1] W. Hilpert and P. Dimroth, *Nature (London)* **296**, 584 (1982).
[2] W. Hilpert and P. Dimroth, *Eur. J. Biochem.* **132**, 579 (1983).

solubilized from the bacterial membrane with nonionic detergents and is extensively purified by monomeric avidin-Sepharose chromatography. If thus purified enzyme samples are incorporated into liposomes, the Na^+ pump activity is reconstituted.[3] With these reconstituted proteoliposomes the energetics and the stoichiometry of the Na^+ pump can be determined.

Determination of Methylmalonyl-CoA Decarboxylase Activity[2]

The activity of the decarboxylase was determined with a coupled spectrophotometric assay with propionyl-CoA carboxylase, pyruvate, kinase, and lactate dehydrogenase, as shown below:

(S)-Methylmalonyl-CoA + H^+ → propionyl-CoA + CO_2
Propionyl-CoA + ATP + HCO_3^- → (S)-methylmalonyl-CoA + ADP + P_i
ADP + phosphoenolpyruvate → ATP + pyruvate
Pyruvate + NADH + H^+ → lactate + NAD^+

The cuvette (d = 1 cm) contained in 1.0 ml at 25°: 100 mM K-phosphate, pH 7.0, 20 mM NaCl, 200 mM KHCO₃, 5 mM MgCl₂, 2 mM dithioerythritol, 1 mM phosphoenolpyruvate, 2 mM ATP, 0.3 mM NADH, 10 U pyruvate kinase, 40 U lactate dehydrogenase, 0.5 U propionyl-CoA carboxylase, and 0.08 mM (R,S)-methylmalonyl-CoA. The reaction was initiated with methylmalonyl-CoA decarboxylase (1–40 mU) and followed continuously at 340 nm. Under these conditions the initial velocity of NADH oxidation was proportional to the amount of the decarboxylase.

Another procedure to determine methylmalonyl-CoA decarboxylase activity is to hydrolyze the propionyl-CoA produced by phosphotransacetylase in presence of arsenate following the decrease of the thioester absorbance at 232 nm.

Decarboxylation of Malonyl-CoA by Methylmalonyl-CoA Decarboxylase[3]

In certain experiments malonyl-CoA was used as the substrate of the decarboxylase since the formation of acetyl-CoA can be more sensitively determined than that of propionyl-CoA. A highly sensitive assay of the decarboxylation reaction is required for reconstituted proteoliposomes which contain only low activities of the decarboxylase. The incubation mixtures contained in 0.5 ml at 25°: 30 mM K-phosphate, pH 7.0, 1 mM Na₂SO₄, 0.6 mM malonyl-CoA, and methylmalonyl-CoA decarboxylase

[3] W. Hilpert and P. Dimroth, *Eur. J. Biochem.* **138,** 579 (1984).

(about 10 mU). The reaction was terminated with 10 μl 4 M HClO$_4$ in samples (0.1 ml) taken after 1, 3, 6, and 10 min incubation. After 5 min at 0° the pH of these solutions was adjusted to pH 5–6 with 5 μl 5 M KOH, 25 μl 1 M KH$_2$PO$_4$, and 18 μl 1 M KHCO$_3$. The precipitate of KClO$_4$ was centrifuged after 20 min standing at 0° and washed with 100 μl H$_2$O. The acetyl-CoA content of the combined supernatants was determined in a coupled fluorometric assay with malate dehydrogenase and citrate synthase.[4] NADH fluorescence was excitated at 340 nm and the emission was followed at 465 nm. The acetyl-CoA content of the samples was calculated from a calibration curve with an acetyl-CoA standard. The sensitivity of the assay is about 1 nmol acetyl-CoA.

Purification of Methylmalonyl-CoA Decarboxylase

Growth of Veillonella alcalescens Cells

Veillonella alcalescens (ATCC 17745) cells were grown anaerobically at 37° in a medium of the following composition[5]: 0.5% trypton, 1% yeast extract, 2% sodium lactate, 0.05% KH$_2$PO$_4$, 0.001% MnSO$_4$, and 0.05% cysteine (separately sterilized). The pH was adjusted to 6.6 with solid KOH. The growth medium was inoculated with 10% of a culture in the early stationary phase and transfers were made after 24 hr when the bacteria had reached the stationary phase. The cultures were kept in volumetric flasks with a narrow neck which were sealed with pyrogallol soaked gauze plugs. Final growth was performed in 20 liter and 200 liter volumes in fermentors under gentle stirring. After 24 hr (early stationary phase) the suspension was cooled to 15° and the bacteria were harvested by continuous centrifugation with a Westfalia Separator type KA-2. The cells (200–400 g wet weight) were washed twice with 3 liter 50 mM K-phosphate, pH 7.0 and stored at −20°.

Isolation of Membranes

Diisopropyl fluorophosphate (0.1 mM) was included in all buffers used in the purification procedure. A suspension of 30 g (wet weight) V. alcalescens in 120 ml 50 mM K-phosphate, pH 7.0 containing 5 mM MgCl$_2$ and 3 mg deoxyribonuclease I was passed twice through an Aminco French pressure cell at 137.5 MPa (20,000 psi) to desintegrate the bacteria. Unbroken cells were removed by centrifugation at 31,000 g for 15

---

[4] P. B. Garland, in "Methoden der Enzymatischen Analyse" (H. U. Bergmeyer, ed.), p. 2042. Verlag Chemie, Weinheim, 1974.

[5] W. De Vries, T. R. M. Riedveld-Struijk, and A. H. Stouthamer, Antonie Leeuwenhoek J. Microbiol. **43**, 153 (1977).

PURIFICATION OF METHYLMALONYL-CoA DECARBOXYLASE FROM *Veillonella alcalescens*

Step	Volume (ml)	Total protein (mg)	Total activity (U)	Specific activity (U/mg)	Recovery (%)
Crude cell extract[a]	120	2250	680	0.3	100
Membrane vesicles	40	650	590	0.9	87
Triton X-100 eluate	60	270	453	1.7	67
Eluate from avidin-Sepharose column	14	4.7	155	33	23

[a] Starting material: 30 g bacteria (wet weight).

min. The membranes were sedimented in 90 min at 200,000 g and washed with 35 ml 2 mM K-phosphate, pH 7.0. Methylmalonyl-CoA decarboxylase was further purified from the membranes as described below. Membrane vesicles for Na⁺ transport studies were prepared similarly but at a pressure of 55 MPa (8000 psi).

Purification on Avidin-Sepharose

The membrane pellet was homogenized with 60 ml 2 mM K-phosphate, pH 7.0 containing 1% Triton X-100 using a motor-driven Teflon plunger and centrifuged for 30 min at 230,000 g. The supernatant was adjusted to 0.3 M KCl and applied to a monomeric avidin-Sepharose column (2.2 × 8 cm, biotin binding capacity: 35 nmol/ml) which was prepared as described.[6] The affinity column was washed with 10 mM K-phosphate, pH 7.0 containing 0.3 M KCl and 0.05% Brij 58 at a flow rate of 45 ml/hr until the absorption at 280 nm was below 0.1 (about 200 ml). The enzyme was eluted in about 14 ml volume with 10 mM K-phosphate, pH 7.0, 0.15 M KCl, and 1.5 mM biotin, concentrated to about 1 ml by ultrafiltration on a PM10 membrane (Amicon) and stored under liquid nitrogen. The results of the purification procedure are summarized in the table.

Properties

Catalytic Properties[2]

(S)-Methylmalonyl-CoA is the substrate for methylmalonyl-CoA decarboxylase. Apparent K_m for (S)-methylmalonyl-CoA (at 20 mM NaCl): 7 μM. Malonyl-CoA is also decarboxylated with an apparent K_m of 35 μM

[6] P. Dimroth, this volume [40].

at 20 mM NaCl. The V_{max} values for both substrates are the same. The pH optimum is in the range of pH 6.4–7.0. The decarboxylase is activated by Na$^+$ ions (apparent $K_m \sim 0.6$ mM at 0.1 mM (S)-methylmalonyl-CoA). If Na$^+$ ions were carefully excluded from all reagent solutions the residual methylmalonyl-CoA decarboxylase activity was less than 2% of V_{max}. The activation was specific for Na$^+$ ions since K$^+$, Li$^+$, Mg^{2+}, or Ca^{2+} were without effect at concentrations of 10 mM. The purified decarboxylase was completely inhibited by avidin but avidin saturated with biotin was without effect. The decarboxylation of methylmalonyl-CoA yielded CO$_2$ and not HCO$_3^-$ as a product.

Molecular Properties[2]

Methylmalonyl-CoA decarboxylase from $V.$ $alcalescens$ is composed of four different polypeptide chains: α, M_r 60,000; β, M_r 33,000; γ, M_r 18,500; δ, M_r 14,000. These molecular weights were determined by dodecyl sulfate–gel electrophoresis. The four subunits are also separated by HPLC on a TSK 250 column in a dodecyl sulfate containing buffer, analogous to the separation of oxaloacetate decarboxylase subunits.[6] The biotin is exclusively contained in the γ-subunit. A very convenient and sensitive method for the detection of biotin containing polypeptides is as follows. The polypeptides are separated by dodecyl sulfate–gel electrophoresis and transferred to nitrocellulose paper by Western-blot electrophoresis (15 hr, 30 V) in 20 mM Tris, 150 mM glycine, and 20% methanol. The paper is washed with 100 ml 10 mM Tris–HCl, pH 7.9 containing 0.15 M NaCl and 5% bovine serum albumin (30 min) and then incubated for 90 min with avidin labeled with fluorescein isothiocyanate (0.1 mg in 50 ml of the same buffer). After four washes with 500 ml 10 mM Tris–HCl, pH 7.9 containing 0.15 M NaCl, the biotin-containing polypeptides are detected by the fluorescence under an ultraviolet lamp. Protein bands containing 10 pmol biotin are clearly visible. The sensitivity of the method can be improved if avidin conjugated to peroxidase is used instead of the fluorescently labeled avidin. The biotin-containing polypeptides give rise to stained bands during incubation with H$_2$O$_2$ and a suitable dye, e.g., o-dianisidine. The separation of methylmalonyl-CoA decarboxylase subunits by dodecyl sulfate–gel electrophoresis and the detection of the biotin containing subunit with fluorescein isothiocyanate-labeled avidin is shown in Fig. 1.

Reconstitution of the Sodium Pump

The sodium transport activity of methylmalonyl-CoA decarboxylase can be determined with bacterial membrane vesicles which are prepared

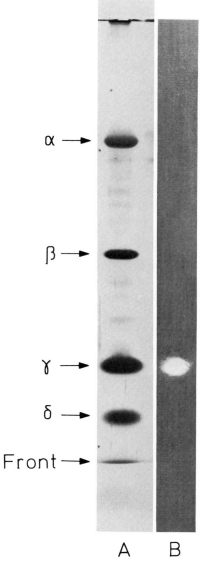

FIG. 1. Dodecyl sulfate–polyacrylamide gel electrophoresis of methylmalonyl-CoA de-carboxylase (Hilpert and Dimroth[2]). Gel A was stained with silver reagent and gel B was treated with fluorescein isothiocyanate-labeled avidin after Western-blot electrophoresis. The fluorescent spot visible under ultraviolet light was photographed.

with a French pressure cell.[1,2] These membrane vesicles, however, are very leaky for Na$^+$ ions, possibly due to other Na$^+$ conducting proteins in the crude membranes. Reconstitution of the sodium pump from purified methylmalonyl-CoA decarboxylase and phospholipids yields tightly sealed proteoliposomes which are suitable for quantitative Na$^+$ transport studies.[3] The reconstitution procedure is essentially the same as for oxaloacetate decarboxylase.[6] The incubation mixture contained in 2.8 ml reconstitution buffer (30 mM K-phosphate, pH 7.5, 1 mM Na$_2$SO$_4$) at 0°: 80 mg of homogenized phosphatidylcholine, 2.85% n-octylglucoside, and 0.1 mg methylmalonyl-CoA decarboxylase (28 U/mg). After 5 min, the incubation mixture was diluted with 60 ml ice-cold reconstitution buffer or dialyzed for 24 hr at 4° against 3 × 1 liter of this buffer. The proteoliposomes were collected by centrifugation (150,000 g, 40 min) and resuspended in 4 ml of reconstitution buffer. The transport experiments were performed immediately after reconstitution. Freezing and thawing partially destroyed the transport activity. The sodium ion transport was determined from the amount of ^{22}Na$^+$ passing through Dowex 50-K$^+$ columns as described.[6] The incubation mixtures contained in 1.15 ml at 25°: 1.0 ml of the reconstituted proteoliposomes, 30 mM K-phosphate, pH 7.0, 1 mM ^{22}Na$_2$SO$_4$ (500 cpm/nmol of Na$^+$), and 0.5 mM (R,S)-methylmalonyl-CoA to initiate the reaction. The uptake of radioactivity was determined from samples (0.1 ml) taken after 1/2–30 min incubation.

Properties of the Na$^+$ Transport System

The reconstitution process was highly asymmetric since about 80% of the decarboxylase was oriented in the proteoliposomes with the substrate binding site facing to the interior.[3] The reconstitution was dependent on the detergent and enzyme concentration with optimum results at 2.8% n-octylglucoside and 10 μg protein/mg phospholipid, respectively. The reconstituted proteoliposomes performed an active Na$^+$ uptake in response to the decarboxylation of methylmalonyl-CoA. The Na$^+$ uptake was fully inhibited by avidin but not by avidin saturated with biotin. The transport was electrogenic creating a membrane potential of about 58 mV, a Na$^+$ concentration gradient equivalent to 52 mV and a total sodium motive force of 110 mV. The stoichiometry between Na$^+$ transport and methylmalonyl-CoA decarboxylation was 2 : 1 in the initial phase and declined after an electrochemical Na$^+$ gradient had been developed.[7]

[7] P. Dimroth and W. Hilpert, *Biochemistry* **23**, 5360 (1984).

[42] Biotin-Dependent Decarboxylases as Bacterial Sodium Pumps: Purification and Reconstitution of Glutaconyl-CoA Decarboxylase from *Acidaminococcus fermentans*

By WOLFGANG BUCKEL

Introduction

Glutaconyl-CoA decarboxylase,[1,2] a certain type of oxaloacetate decarboxylase,[3–6] and methylmalonyl-CoA decarboxylase[7–9] form a closely related group of enzymes with three distinct properties: (1) they contain biotin, (2) their activities are specifically stimulated by Na^+ up to 300-fold but not by divalent cations, and (3) they are integral membrane proteins (Table I). Recently the three enzymes were characterized as sodium pumps.[1,2,4–6,8,9] They are able to convert the free energy of decarboxylation ($\Delta G° \approx -30$ kJ/mol) into an electrochemical Na^+ gradient.

The decarboxylases from six bacterial species have been purified by affinity chromatography on monomeric avidin-Sepharose and reconstituted to functional sodium pumps.[2,5,6,9] All enzymes consist of 3–4 different polypeptide chains. The size of those containing biotin vary considerably between the species (M_r 20–140K), whereas two polypeptides (~60K and ~35K) appear to be conserved in all analyzed organisms. The hydrophobic 35K peptide is able to bind Na specifically.[2,6] Possibly this part of the molecule is involved in the Na transport.

The three enzymes have been found exclusively in anaerobically grown heterotrophic eubacteria in which they catalyze essential decarboxylation steps of the fermentation pathways. Evidence for their physiological importance as energy converters has been obtained for *Propiogenium modestum*[10] and *Peptostreptococcus asaccharolyticus*.[11] In the

[1] W. Buckel and R. Semmler, *FEBS Lett.* **148**, 35 (1982).
[2] W. Buckel and R. Semmler, *Eur. J. Biochem.* **136**, 427 (1983).
[3] J. R. Stern, *Biochemistry* **6**, 3545 (1967).
[4] P. Dimroth, *FEBS Lett.* **122**, 234 (1980).
[5] P. Dimroth, *Biosci. Rep.* **2**, 849 (1982).
[6] P. Dimroth, this volume [40].
[7] J. H. Galivan and S. H. G. Allen, *J. Biol. Chem.* **243**, 1253 (1968).
[8] W. Hilpert and P. Dimroth, *Nature (London)* **296**, 584 (1982).
[9] W. Hilpert and P. Dimroth, this volume [41].
[10] W. Hilpert and B. Schink, and P. Dimroth, *EMBO J.* **3**, 1665 (1984).
[11] G. Wohlfarth and W. Buckel, *Arch. Microbiol.* **142**, 128 (1985).

TABLE I
BIOTIN-DEPENDENT SODIUM PUMPS

Type	Organism	Fermentation pathway
Glutaconyl-CoA decarboxylase: $$\text{CoAS}-\overset{\overset{\displaystyle O}{\|}}{C}-\overset{\overset{\displaystyle H}{\|}}{C}=C-CH_2-COO^- + H^+ \rightarrow \text{CoAS}-\overset{\overset{\displaystyle O}{\|}}{C}-C=\overset{\overset{\displaystyle H}{\|}}{C}-CH_3 + CO_2$$ (E)-Glutaconyl CoA crotonyl-CoA	*Acidaminococcus fermentans*[a,1,2]	Glutamate → butyrate[b]
	Peptostreptococcus asaccharolyticus[a,c,2]	Glutamate → butyrate[b]
	Clostridium symbiosum[a,2]	Glutamate → butyrate[b]
	Fusobacterium nucleatum[d,14]	Glutamate → butyrate[b]
Oxaloacetate decarboxylase: $$^-OOC-\overset{\overset{\displaystyle O}{\|}}{C}-CH_2-COO^- + H^+ \rightarrow {}^-OOC-\overset{\overset{\displaystyle O}{\|}}{C}-CH_3 + CO_2$$ Oxaloacetate pyruvate	*Klebsiella aerogenes*[a,4–6]	Citrate → acetate
	Salmonella thyphimurium[a,15]	Citrate → acetate
Methylmalonyl-CoA decarboxylase: $$\text{CoAS}-\overset{\overset{\displaystyle O}{\|}}{C}-\overset{\overset{\displaystyle CH_3}{\|}}{CH}-COO^- + H^+ \rightarrow \text{CoAS}-\overset{\overset{\displaystyle O}{\|}}{C}-CH_2-CH_3 + CO_2$$ (S)-Methylmalonyl-CoA propionyl-CoA	*Veillonella alcalescens*[a,7–9]	Lactate → propionate
	Propionigenium modestum[d,10]	Succinate → propionate

[a] The corresponding decarboxylase was purified and reconstituted to a sodium pump.

[b] Hydroxyglutarate pathway.[12]

[c] Formerly known as *Peptococcus aerogenes*.[13]

[d] The activity was detected in membrane preparations.

latter organism energy equivalent to 0.6 mol ATP/mol glutaconyl-CoA decarboxylated is conserved via a sodium gradient. The reason why sodium ions and not protons are pumped by these decarboxylases remains an intriguing question.

$$\text{Glutaconyl-CoA} + H^+ \xrightarrow{Na^+} \text{Crotonyl-CoA} + CO_2 \tag{1}$$

Glutaconyl-CoA [Eq. (1)] is an intermediate in the fermentation of glutamate via (R)-2-hydroxyglutarate to acetate and butyrate.[12,16] Crotonyl-CoA, the product of the decarboxylation, is partially reduced to butyryl-CoA from which butyrate is formed. The other part of crotonyl-CoA is hydrated to (3S)-3-hydroxybutyryl-CoA and oxidized to acetoacetyl-CoA. Thiolysis yields two acetyl-CoA which are used to synthesize acetylphosphate, a precursor of ATP, and to activate glutaconate to glutaconyl-CoA. This oxidative branch of crotonyl-CoA utilization is applied as the assay for the decarboxylase as shown in Scheme 1. The whole sequence of steps is initiated by a sparking amount of acetylphosphate which is necessary to avoid a lag phase. In summary 1 mol NADH is formed per mol of glutaconyl-CoA decarboxylated.[1]

The five auxiliary enzymes required for this assay are purified together from the soluble fraction of cell-free extracts from *Acidaminococcus fermentans* whereas the membrane fraction is used as starting material for the decarboxylase. Although this bacterium is a strict anaerobe, all the enzymes described are not inactivated by exposure to air.

[12] W. Buckel and H. A. Barker, *J. Bacteriol.* **117**, 1248 (1974).
[13] T. Ezaki, N. Yamamoto, K. Ninomiya, S. Suzuki, and E. Yabuuchi, *Int. J. Syst. Bacteriol.* **33**, 683 (1983).
[14] W. Buckel and H. Liedtke, unpublished.
[15] P. Dimroth, personal communication.
[16] W. Buckel, U. Dorn, and R. Semmler, *Eur. J. Biochem.* **118**, 315 (1981).

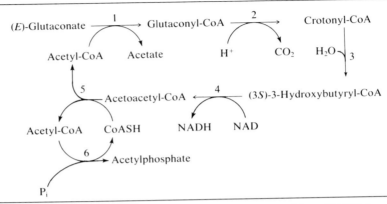

Sum: Glutaconate^{2-} + H_2O + NAD$^+$ + HPO$_4^{2-}$ → Acetate$^-$ + CO_2 + NADH + acetyl-phosphate^{2-}

SCHEME 1. Assay of glutaconyl-CoA decarboxylase. (1) Glutaconate CoA-transferase (EC 2.8.3.12); (2) glutaconyl-CoA decarboxylase (EC 4.1.1.-); (3) enoyl-CoA hydratase (crotonase) (EC 4.2.1.17); (4) (3S)-3-hydroxybutyryl-CoA dehydrogenase (EC 1.1.1.35); (5) acetyl-CoA acetyltransferase (thiolase) (EC 2.3.1.9); and (6) phosphate acetyltransferase (EC 2.3.1.8).

Assay of Glutaconyl-CoA Decarboxylase

Reagents

Triton phosphate buffer: 500 mM K-phosphate, pH 7.0 containing 1.0% Triton X-100

NaCl, 1 M

DTE/EDTA, 0.1 M each

NAD, 0.1 M (see below)

CoASH, Li salt, 0.01 M

Acetylphosphate, K-Li-salt, 0.01 M

Glutaconate, 0.1 M: glutaconic acid (EGA Chemie), recrystallized from ether, is dissolved in 0.2 M KOH and the pH is adjusted to 7

Auxiliary enzymes, 10 mg protein/ml (see below)

For routine assays ordinary reagent grade chemicals are sufficient. In cases when a low Na$^+$ concentration (≤0.1 mM) is required, KOH containing less than 0.002% Na (Merck, Darmstadt) is used for preparation of K$_2$EDTA and K$_2$-glutaconate solutions. NAD (grade I or II, Boehringer Mannheim) containing about 1% Na$^+$ is purified by a passage through AG 50 WX8, K$^+$. The eluate is concentrated *in vacuo* and its NAD content is determined enzymatically.[17] KH$_2$PO$_4$ and K$_2$HPO$_4$ supplied by Merck or Baker are used without further purification.

[17] M. Klingenberg, *in* "Methoden der enzymatischen Analyse" (H. U. Bergmeyer, ed.), 3rd Ed., p. 2098. Verlag Chemie, Weinheim, 1974.

Procedure. The reaction is measured at 340, 334, or 366 nm in a cuvette ($d = 1$ cm, final volume 1.0 ml) containing 0.1 ml Triton phosphate buffer, water up to 1.0 ml, DTE/EDTA and auxiliary enzymes, 0.02 ml each, CoASH, NAD, and acetylphosphate, 0.1 ml each, and not more than 100 mU decarboxylase. After at least 1 min at 25° the reaction is initiated by 0.01 ml glutaconate. The concentration of NADH increases linearly with time at least up to 0.1 mM. One unit of activity (1 U) is defined as 1.0 μmol NADH formed in 1.0 min at 25°. This assay is suitable for measuring glutaconyl-CoA decarboxylase activity in membrane fractions. In cell-free extracts NADH is slowly oxidized again by crotonyl-CoA and oxygen. Using the purified decarboxylase the assay may be applied for determination of glutaconate.[2]

Acidaminococcus fermentans

Growth. The strict anaerobe eubacterium *A. fermentans* ATCC 25085 is grown at 37° on media containing 0.1 M Na-glutamate, 0.4% yeast extract, 0.1% Na-thioglycolate (Merck, for large cultures a solution is prepared from the much cheaper acid and NaOH), VRB salts as indicated by Rogosa,[18] 500 nM biotin (10–100 mM stock solutions are prepared from the acid dissolved with $KHCO_3$ and stored at $-20°$), and 10 mM K-phosphate, pH 7.4. The media are autoclaved at 121° on the day of use.

Cultures are started in media dispensed in screw top tubes (10 ml portions) containing additional 0.2% agar (semisolid) and 2×10^{-4}% resazurin. The freeze-dried culture obtained from ATCC is suspended in 1 ml medium and transferred with a sterile Pasteur pipet into the solidified media at least 3 cm beneath the surface. Growth commences usually after 1–2 days. The cells spread slowly over the whole medium except on and about 5 mm below the surface where oxygen is present as indicated by the red color of resazurin. These cultures are used to inoculate 100 ml liquid media at 2-fold concentrations in 200-ml volumetric flasks. Then freshly sterilized water is added up to the neck in order to expose only a very small part of the medium surface to the air. The flasks are covered with two layers of aluminum foil and their content is mixed by gentle swinging, not by turning upside down. The next morning media in 2-liter volumetric flasks autoclaved during the night are inoculated in the same way using the 200 ml cultures. After growth for about 8 hr each of these 2-liter flasks is used to inoculate two 10-liter narrow necked bottles. Bacteria, concentrated medium, and water are mixed with a gentle N_2 stream. After standing overnight without stirring, the bacteria are harvested with a milk

---

[18] M. Rogosa, *J. Bacteriol.* **98**, 756 (1969).

centrifuge (Westphalia Separator). The cell paste (yield 3–4 g/liter) is stored at −80° up to several years without loss of activity.

Bacteria containing [^{14}C]biotin are grown in a 2-liter volumetric flask on a medium in which unlabeled biotin is replaced by 5 μCi [*carbonyl-*^{14}C]biotin (carrier-free, Amersham). The 200 ml culture used for the inoculum is grown on a medium to which no biotin is added. Under these conditions 80% of the radioactivity is incorporated into the bacteria.[2]

Maintenance of Cultures. A semisolid culture may be stored for several months at 4°. Bacteria below the red color of the oxygen indicator are used for further inocula. For longer periods bacteria are kept in stab cultures (2% agar) under a pyrogallol seal.[19] After each transfer cultures are checked by phase contrast microscopy for purity. In order to get single colonies, bacteria are plated and incubated anaerobically on the medium containing 2% agar (anaerobic chamber, Coy Laboratories, Ann Arbor, MI).

Purification of Glutaconyl-CoA Decarboxylase

Preparation of the Membranes. Frozen cells (120 g) are thawed with 240 ml 50 mM Na-phosphate, pH 7 containing 5 mM MgCl$_2$, 1 mM diisopropyl fluorophosphate (Serva, Heidelberg), and 2 mg DNase I (grade II, Boehringer Mannheim). The suspension is sonicated for 30 min at 90 W (Branson Sonifier) in a rosette vial immersed in an ice bath. Cell debris is removed by centrifugation at 12,000 g for 10 min. The light brown extract is again centrifuged at 200,000 g for 60 min. The supernatant is used for the preparation of the auxiliary enzymes (see below) and the pellet is suspended in 25 ml 50 mM Na-phosphate, pH 7.0 and homogenized using a plunger. The pellet obtained after a second centrifugation (200,000 g, 30 min) is designated as membrane fraction and homogenized again. The resulting suspension contains approximately 1 g protein as determined by the biuret method with bovine serum albumin as standard. It may be stored at −20° for several months without loss of activity.

Solubilization. Part of the membrane suspension (500 mg protein) is diluted to 400 ml with 80 ml 10% Triton X-100, 200 ml 1 M NaCl, 40 ml 0.1 M EDTA, 40 ml 0.2 M Na-phosphate, pH 7.0, 1 ml 40 mM diisopropyl fluorophosphate in ethanol, and water. The low protein concentration (≤1.5 mg/ml) and the presence of 0.5 M NaCl or KCl are essential for this step. The suspension is stirred at ambient temperature for 15 min during which time all particles dissolve. Centrifugation at 40,000 g for 30 min yields a clear supernatant which is used immediately for the next step.

---

[19] J. C. Rabinowitz, this series, Vol. 6, p. 703.

TABLE II
PURIFICATION OF GLUTACONYL-CoA DECARBOXYLASE

Step	Protein (mg)	Activity (U)	Yield (%)	Specific activity (U/mg protein)
Membrane fraction	500	350 (300–600)[a]	100	0.7 (0.6–1.2)
Triton extract	—	290	83 (40–100)	—
Avidin-Sepharose	8	221	63 (40–80)	28 (12–40)

[a] The data are obtained from a typical purification, those in parentheses indicate the variations among different runs. Modified from Buckel and Semmler.[2]

Affinity Chromatography. This step is performed at 4°. A column (inner diameter 2 cm) is filled with 20 ml monomeric avidin-Sepharose[6] equilibrated with 100 mM Na-phosphate, pH 7.0 containing 0.05% Triton X-100. The solubilized enzyme is pumped on the column (2 ml/min) followed by about 100 ml equilibration buffer until the absorbance at 280 nm (Uvicord, LKB) remains constant. About 90–100% of the decarboxylase is absorbed on the column. The enzyme is eluted with equilibration buffer to which 1.0 mM biotin is added. The enzyme containing fractions (A_{280} and enzymatic activity) are pooled and concentrated by ultrafiltration over a PM10 membrane (Amicon). The enzyme may be stored at $-20°$ for a year with little loss of activity. The whole purification procedure is summarized in Table II.

Auxiliary Enzymes

Principles of the Assay. Glutaconate CoA-transferase (a) is chosen as a representative for the five auxiliary enzymes (Scheme 1) which are purified together. Its activity is measured by a coupled assay in which glutaconyl-CoA is replaced by glutaryl-CoA which is readily available.[16]

$$\text{Glutaryl-CoA + acetate} \rightleftharpoons \text{glutarate + acetyl-CoA} \qquad \text{(a)}$$
$$\text{Acetyl-CoA + phosphate} \rightleftharpoons \text{acetylphosphate + CoASH} \qquad \text{(b)}$$

Phosphate acetyltransferase (b) also belonging to the auxiliary enzymes liberates CoASH which is determined by 5,5'-dithiobis(2-nitrobenzoic acid) (DTNB). The absorbance of the resulting thiol is measured at 405 nm, $\varepsilon = 14$ mM^{-1} cm^{-1}.[20]

[20] W. Buckel, K. Ziegert, and H. Eggerer, *Eur. J. Biochem.* **37**, 295 (1973).

Reagents

Sodium acetate and K-phosphate, pH 7.0, 1 M each

DTNB (0.05 M): 20 mg acid is dissolved in 1.0 ml water with a small amount of KHCO$_3$. Although the solution is kept in ice, it is not stable for more than 10 hr

Glutaryl-CoA (10 mM): from Sigma or prepared from CoASH, pH 5–6, stored at −20°.

Preparation of Glutaryl-CoA. CoASH, 10 mg in 1 ml 0.5 M KHCO$_3$ is acylated at ambient temperature with 10 μl portions of 1 M glutaric acid anhydride in peroxide-free *p*-dioxane. The reaction is followed by qualitative SH determination (1 μl sample is mixed with 1 μl DTNB solution on a filter paper). The pH, which should not drop below 8 during acylation, is adjusted to 5–6 after completion of the reaction by careful addition of 5 N HCl.

Glutaric acid anhydride is prepared by refluxing 50 g glutaric acid with 200 ml acetic anhydride for 3 hr. After evaporation to dryness, the residue is recrystallized from ether: needles, mp 55–56°.

Assay of Glutaconate CoA-Transferase. A cuvette (d = 1 cm, final volume 1.0 ml, λ = 405 nm) contains: 0.1 ml 1 M phosphate, 0.2 ml 1 M acetate, 0.02 ml DTNB, and the enzyme source (up to 30 mU). After at least 30 sec at 25° the reaction is started with 0.01 ml glutaryl-CoA. A blank is run without enzymes. In the absence of sufficient phosphate acetyltransferase 0.01 ml 0.1 M oxaloacetic acid (do not neutralize!) and 10 μg citrate synthase (Boehringer Mannheim) are added.[16]

Purification. To the first 200,000 g supernatant (~250 ml) from the membrane preparation ammonium sulfate is added (313 mg/ml, 50% saturation at 20°). After stirring for 30 min the suspension is centrifuged for 10 min at 40,000 g. The supernatant is brought to 80% saturation with ammonium sulfate (240 mg/ml) as above. The second pellet is dissolved in 20 mM phosphate, pH 6.5 and thoroughly dialyzed against this buffer over night at 4° (2 changes). The dialyzate is pumped (2 ml/min) on a DEAE-Sephacel column (5 × 10 cm, Pharmacia) previously equilibrated with dialysis buffer at 4°. The enzymes are eluted by applying a linear gradient from 20 to 400 mM K-phosphate, pH 6.5 (2 × 1 liter) and 25–30 ml fractions are collected. The auxiliary enzymes are eluted around 150–200 mM phosphate, after a slightly red peak (rubredoxin and butyrate CoA-transferase) and just before the brown to yellow main protein peak. The 55–80% ammonium sulfate fraction (351 and additional 179 mg/ml, respectively) of glutaconate CoA-transferase is chromatographed on a Sephacryl S-300 column (2.5 × 90 cm, equilibrated with 50 mM K-phosphate, pH 6.5 in 1 M NaCl at room temperature) in order to remove residual membrane vesicles with decarboxylase activity. The latter are well separated from the tailing main protein peak which is concentrated by ultrafiltration (PM10 membrane, Amicon) or by precipitation with ammonium sulfate (80% saturation). After extensive dialysis against 100 mM K-phosphate, pH 7.0, 1 mM DTE or DTT and 1 mM K$_2$EDTA the auxiliary enzymes are ready for use. They are stored at $-20°$ in 1.0 ml portions. Their activities are stable at least for 1 year.

Reconstitution of the Sodium Pump

Principle. The purified decarboxylase is reconstituted to a functional sodium pump by incorporation into phospholipid vesicles using the detergent dilution method.[2] In the sodium uptake assay ^{22}Na$^+$ entrapped into vesicles is determined by liquid scintillation counting. Free ^{22}Na$^+$ is removed by a passage through a cation exchanger.[1] The rate of decarboxylation is measured simultaneously at 366 (340) nm.

Materials

Buffer A: 50 mM K-phosphate, pH 7.0

Glutaconyl-CoA decarboxylase, at least 1 U/ml in buffer A containing 0.05% Triton X-100. Either the affinity chromatography is performed in this buffer or the Na buffer is exchanged by gel filtration on Sephadex G-25 (Pharmacia).

Phospholipid vesicles: A suspension of 100 mg crude phosphatidylcholine (Type II-S from soybeans, Sigma) in 2.5 ml buffer A at 0° is sonicated for 10 min, during which time clarity is obtained (Branson sonifier, small tip).

Octylglucoside, 20% (w/v)

NaCl, 0.1 M

^{22}NaCl, 200 μCi/ml, carrier free (Amersham)

NAD, DTE-EDTA, CoASH, acetyl phosphate, gluconate, and auxiliary enzymes as required for the decarboxylase assay with low Na content

Columns made from Pasteur pipets (tip length 5 cm) with a small cotton plug and filled with 2 cm cation exchanger AG 50 WX-8, 200–400 mesh. The fines from the resin (Bio-Rad) are removed by two decantations of the suspended material (20 liter water/500 g) after having settled for 20 min. The resin is washed on a large column with 2 volumes 4 N NaOH, water, 5 volumes 5 N HCl, water, and 5 volumes buffer A. The resin loaded with ^{22}Na is regenerated by the same procedure.

Scintillation vials, 20 ml

Scintillator for aqueous samples (e.g., Quickscint 212, Zinsser)

Monensin, Na-salt (Calbiochem-Behring) and valinomycin (Boehringer Mannheim), each 20 μM in ethanol

Procedure. Vesicles (1.25 ml) are mixed with 0.25 ml purified decarboxylase (1 U/ml), and 0.50 ml 20% octylglucoside (final concentration 2.5%). After 10 min at about 25° 48 ml cold buffer A (4°) is added. The pellet obtained by centrifugation of the diluted incubation (200,000 g; 30 min) is suspended in 2 ml buffer A.

Na^+ transport is assayed in a cuvette ($d = 1$ cm, $\lambda = 366$ nm) containing 0.4 ml reconstituted pump, 1.14 ml water, 0.06 ml buffer A; NAD, DTE/EDTA, and auxiliary enzymes, 20 μl each; CoASH, acetyl-phosphate, and 0.1 M NaCl, 10 μl each. In cases when no exact value of the specific activity of $^{22}Na^+$ is required, unlabeled NaCl may be omitted from the incubation. Thereby the sensitivity of the Na-uptake assay is increased. After addition of 5 μl ^{22}NaCl followed by 10 μl glutaconate, samples (0.05 ml) are withdrawn at times as indicated in Fig. 1 and transferred directly on top of a column. The column is washed with 200 μl buffer A followed by 750 μl buffer A. The whole eluate is collected in a scintillation vial, mixed with 9 ml scintillator and counted in a liquid scintillation counter at channel and gain settings as used for ^{32}P.

The formation of NADH (i.e., decarboxylation) is measured continuously in a photometer (Eppendorf) at 366 nm. Monensin may be added at a final concentration of 100 nM (5 μl/ml, Fig. 1, $t = 90$ min). The initial rate of Na^+ uptake is increased twofold if about 100 nM valinomycin (5 μl/ml) is present before addition of glutaconate. Thereby the initial rate of decarboxylation is accelerated by the same factor.

Properties of Purified Glutaconyl-CoA Decarboxylase

Polypeptide Composition. Gradient SDS–polyacrylamide gel electrophoresis[2,21] followed by silver staining[22] shows three polypeptides, α (M_r 140K), β (60K), and γ (35K). Additional bands α' (120K–100K) and below 60K at varying intensities are either impurities or proteolytic degradation products of larger chains. Whereas the α-bands contain the biotin, the γ-chain was characterized as Na-binding protein. It is resistant to tryptic digestion in 100 mM NaCl but not in 100 mM KCl. Since this polypeptide is very hydrophobic it might be involved in the Na^+ transport.[2]

[21] U. K. Laemmli, *Nature (London)* **227**, 680 (1970).
[22] J. Heukeshoven and R. Dernick, *Electrophoresis* **6**, 103–112 (1985).

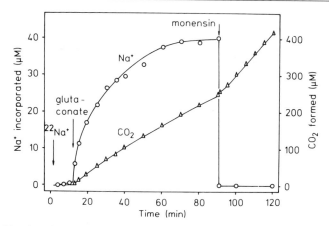

FIG. 1. Kinetics of the reconstituted sodium pump. For experimental details see text. In order to compare the rate of decarboxylation with that of sodium uptake, the concentration of incorporated $^{22}Na^+$ was based on the total volume of the cuvette. However, the amounts of the vesicles which were retained on the columns used for the sodium uptake assay (40%) were not considered. Reproduced from Buckel and Semmler.[2]

Molecular Weight. By sucrose gradient centrifugation (10–30%, w/v) the decarboxylase activity is recovered in two peaks of similar shape. The faster one comigrates with apoferritin (17.6 S, M_r 470K[23]) and the slower one with bovine catalase (11.3 S, M_r 250K[23]) indicating that the peaks represent the dimer and the monomer of the decarboxylase molecule. These data may lead to the hypothetical quaternary structure $\alpha_2\beta_2\gamma_2$. However, without information about shape and detergent content of the molecule, the data are very preliminary.

Catalytic Properties. The activity of the decarboxylase increases 6- to 8-fold by addition of 20 mM NaCl or 10 mM Na$_2$SO$_4$ to the "sodium-free" assay system (0.05–0.1 mM Na). By extrapolation to [Na] = 0 the overall activation is 10-fold with an apparent K_m = 1.0 mM. Li activates 5-fold with an apparent K_m = 100 mM. The activation by Na is inhibited by Li. The influence of other monovalent cations was not investigated. However, all the measurements were performed in the presence of 100 mM K.[1] K_m for glutaconyl-CoA was not determined. 3-Methylglutaconyl-CoA is not decarboxylated. As a biotin enzyme, the decarboxylase is inhibited by avidin but not by the avidin biotin complex. Only Trition X-100 and Brij 35 are able to keep the enzyme in an active state at 20–30°. Stronger

[23] M. H. Smith, *in* "Handbook of Biochemistry" (H. A. Sober, ed.), 2nd Ed., p. C-11. Chemical Rubber Co., Cleveland, 1970.

detergents such as Triton X-114, N-dodecyl-N-dimethylaminoxide, octylglucoside, deoxycholate, and SDS inactivate the enzyme at this temperature. Other inactivators are 10% ethanol at 37° and N-ethylmaleimide.[2] Dicyclohexylcarbodiimide (10^{-4} M) or tetrodotoxin (10^{-5} M) do not interfere with the enzyme nor influence its transport activity.

Acknowledgments

This work was supported by grants from the Deutsche Forschungsgemeinschaft und Fonds der Chemischen Industrie. The author thanks Roswitha Semmler and Henriette Liedtke for excellent technical assistance, Hannelore Trommer for typing the manuscript, and all the students who checked the described procedures.

[43] Anion Exchange in Bacteria: Reconstitution of Phosphate : Hexose 6-Phosphate Antiport from *Streptococcus lactis*

By SURESH V. AMBUDKAR and PETER C. MALONEY

Streptococcus lactis, a gram-positive anaerobe, expresses an unusual anion exchange that uses phosphate monoanion ($H_2PO_4^-$) to support an electroneutral antiport with certain mono- and divalent sugar 6-phosphates.[1,2,2a] Because substrates for exchange are natural products of a mannose phosphotransferase activity (*ptsM*) widely spread among bacteria,[3] we have suggested[2] that antiport could regulate metabolite levels by balancing a *pts*-mediated input with chemiosmotic exchange. There are also important similarities to phosphate shuttles in chloroplasts and mitochondria.[1,2] The bacterial example was first detected[1] by a characteristic phosphate self-exchange, and it is convenient to use this reaction in judging the success of reconstitution. The methods described below stem from

[1] P. C. Maloney, S. V. Ambudkar, J. Thomas, and L. Schiller, *J. Bacteriol.* **158**, 238 (1984).
[2] S. V. Ambudkar and P. C. Maloney, *J. Biol. Chem.* **259**, 12576 (1984).
[2a] S. V. Ambudkar, L. A. Sonna, and P. C. Maloney, *Proc. Natl. Acad. Sci. U.S.A.* (in press).
[3] N. D. Meadow, M. S. Kukurunzinska, and S. Roseman, *in* "Enzymes of Biological Membranes" (A. Martinosi, ed.), Vol. 3, p. 523. Plenum, New York, 1984.

those of Newman and Wilson[4] and Racker *et al.*[5] Our modifications may be of general use in the reconstitution of other transport systems.

Reconstitution

Materials

Octylglucoside (octyl-β-D-glucopyranoside, Calbiochem-Behring), 15% (w/v) in water; kept at $-20°$
Dithiothreitol, 100 mM; stored at $-20°$
Acetone/ether washed phospholipid (*E. coli* or *S. lactis*), 50 mg phospholipid/ml in 2 mM 2-mercaptoethanol; stored at $-70°$ under N_2
Membrane vesicles in 50 mM potassium phosphate (pH 7), 10 mM MgSO$_4$; at $-70°$
Stabilizing agent (glycerol, xylitol, sorbitol)
Potassium phosphate, 100 mM (pH 7); at room temperature
MOPS-K buffer [3-(N-morpholino)propanesulfonic acid][6] (pH 7.0). This has 20 mM MOPS-K, 75 mM K$_2$SO$_4$, and 2.5 mM MgSO$_4$; made fresh and stored at $4°$ until use.

Procedures

Preparation of Membrane Vesicles. Starting from a single colony, *S. lactis* (ATCC 7962) is grown 10–12 hr at 33–35° in batches of 4–8 liters, using a broth[1] supplemented with 1% (w/v) D-gluconate (Na or K salt)[1] to give full induction[1] of anion exchange. Right-side-out membrane vesicles, prepared as described,[2] are stored in 200–500 μl aliquots at 4–10 mg protein/ml in 50 mM potassium phosphate (pH 7.0), 10 mM MgSO$_4$.

Acetone/Ether Washed Phospholipid. Crude *S. lactis* lipid is extracted by the method of Ames[7]; crude lipid from *E. coli* is purchased from Avanti Polar Lipids Inc., Birmingham, AL. Phospholipids are purified by acetone precipitation and ether extraction in the presence of 2 mM 2-mercaptoethanol, as described.[4] The ether extract is brought to a small volume by evaporation under N_2 (20–23°), redissolved in 5–10 ml of ether, and after further drying under N_2, solvents are removed by a 3–5 hr lyophilization. Phospholipids are finally suspended at 50 mg/ml in 2 mM 2-mer-

[4] M. J. Newman and T. H. Wilson, *J. Biol. Chem.* **255**, 10583 (1980).
[5] E. Racker, B. Violand, S. O'Neal, M. Alfonzo, and J. Telford, *Arch. Biochem. Biophys.* **198**, 470 (1979).
[6] Solutions with MOPS were filter-sterilized (0.22 μm pore size) and stored in the cold.
[7] G. F. Ames, *J. Bacteriol.* **95**, 833 (1968).

captoethanol by vortex-dispersal under N_2; the lipid is stored under N_2 at $-70°$ in 1–3 ml portions for use in the next step.

Preparation of Liposomes. An aliquot of 0.2–1.2 ml phospholipid is placed in a pyrex test tube (1.5 × 15 cm) with potassium phosphate added to 10 mM. On gassing with N_2 (2–3 min) the tube is plugged with a rubber stopper, sealed with Teflon tape, and immersed about 0.5 cm at the center of a sonicator bath which has been filled with 0.02% Triton X-100 to give maximum visible agitation; the equipment is from Laboratory Supplies Co., Inc., Hickville, NY [G112 SP1G 80W/80KC (generator); G112 SP1T, 600V/80KC (bath)]. Liposomes are made by a simple sonication to clarity (15–25 min).

Solubilization with Octylglucoside. Solubilization occurs in a centrifuge tube (Beckman Ti50 rotor) with components brought to 20–23°. One adds 100 mM potassium phosphate (pH 7) (750–1000 μl, 75–100 μmol), acetone/ether washed phospholipid (90 μl, 4.5 mg), 100 mM dithiothreitol (12 μl, 1.2 μmol), and 15% (w/v) octylglucoside (88 μl, to 1.11%), in the presence or absence of stabilizing agent at 20% (v/v). After mixing, vesicles are added to give 1 mg protein (150–300 μl; controls have only vesicle suspension medium), followed by brief blending on a vortex mixer. The suspension is then placed on ice for 30 min, spun for 1 hr at 145,000 g (45,000 rpm; 4°), and the supernatant (0.3–0.6 mg protein) reserved for use in the next step.

Reconstitution. Reconstitution occurs on mixing of the 100 mM potassium phosphate (pH 7) (90 μl, 9 μmol), bath-sonicated phospholipid (110 μl, 5 mg), experimental or control extracts made above (788 μl, 175–400 μg protein), and enough 15% octylglucoside (~18.5 μl) to give 1.11%. After blending, the suspension is put on ice for 30 min before rapid injection into 25 ml of 100 mM potassium phosphate (pH 7), 1 mM dithiothreitol at 20–23°. This is left for 20 min, undisturbed, after which proteoliposomes are collected by a 1 hr centrifugation in the cold at 105,000 g (37,000 rpm, 42.1 rotor). On decanting the supernatant and swabbing the tube with cotton tipped applicators, the pellet is resuspended in 8 ml of MOPS-K buffer. Low speed centrifugation (7500 g for 15 min at 4°) removes a yellowish material. A second, high speed centrifugation (145,000 g for 1 hr at 4°) washes away most external phosphate. The final supernatant is discarded, the tube swabbed again, and the pellet taken up in 300 μl of MOPS-K buffer. If kept at 4°, this suspension is suitable for tests of phosphate exchange over the next 4 days. Protein assays (Schaffner and Weissman[8]) show 200–420 μg protein/ml in differ-

[8] W. Schaffner and C. Weissman, *Anal. Biochem.* **56,** 502 (1973).

TABLE I
RECONSTITUTION OF PHOSPHATE : HEXOSE 6-PHOSPHATE ANTIPORT

	$^{32}P_i$ incorporation			
	cpm/filter[a]		nmol/mg protein[b]	
Additions to assay	1 min	70 min	1 min	70 min
Proteoliposomes				
None	405	3325	9.8	95.2
2 mM glucose 1-phosphate	400	3120	10.3	89.7
2 mM 2-deoxyglucose 6-phosphate	35	180	0.3	1.2
Liposomes				
None	85	195	—	—
2 mM glucose 1-phosphate	60	170	—	—
2 mM 2-deoxyglucose 6-phosphate	27	147	—	—

[a] Each filter had 1.2 µg protein. Input radioactivity was about 100,000 cpm per filter, of which 120 cpm was retained by filter blanks; this blank has been subtracted.

[b] Calculated after subtracting counts found on liposomes. Direct assays showed a phospholipid : protein ratio of 55 : 1 for these proteoliposomes.

ent batches, while Hallen's assay[9] gives 8–13.5 mg phospholipid/ml. The usual phospholipid : protein ratio is near 30 : 1.

Assay of Anion Exchange

Materials

Proteoliposomes or liposomes, as above
MOPS-K buffer, as above
Na_3VO_4, 20 mM (Fisher Scientific Co.)
$KH_2{}^{32}PO_4$, 3 mM, 30–50 µCi/ml (stock from New England Nuclear Corp.)
Membrane filters, 0.22 µm pore size (GSWP 02500; Millipore Corp.; GSTF 02500 filters gave a 20% increased trapping)

Procedure

Assay of Phosphate Exchange. A 1 ml reaction volume contains proteoliposomes (40–100 µl, 20–40 µg protein), Na_3VO_4 (12 µl, 0.24 µmol), and MOPS-K buffer (873 µl) preincubated for 3 min at room temperature. To start an assay, the $KH_2{}^{32}PO_4$ (15 µl, 45 nmol) is added, and at timed

[9] R. M. Hallen, *J. Biochem. Biophys. Methods* **2,** 251 (1980).

TABLE II
THE EFFECT OF A PROTEIN STABILANT ADDED
DURING SOLUBILIZATION[a]

	$^{32}P_i$ incorporation (nmol/mg protein)	
Additions	1 min	70 min
None	4	30
Glycerol, 20% (v/v)	42	490

[a] *E. coli* phospholipid was used throughout.
Other conditions were as described in Table I.

intervals 50–100 μl aliquots are taken for vacuum filtration (20–22 lb/in.^{2}). Filters are washed (5–7 sec) with 5 ml of assay buffer and then placed in 6.5 ml ACS scintillation fluid for counting. Vanadate blocks hydrolysis[2] of sugar phosphates that might be added as inhibitors (or substrates); when used, these are added 30 sec before $^{32}P_i$. The assay of liposomes (no protein) gives nonspecific $^{32}P_i$ binding.

Table I gives an example of the assay of $^{32}P_i$ incorporation by proteoliposomes made with *S. lactis* phospholipids but no stabilizing agent. As expected from *in vivo* tests,[1,2] phosphate transport is blocked by sugar 6-phosphate, but not by sugar 1-phosphate. Table II shows that reconstitution is much improved when inactivation during solubilization[10,11] is prevented by a protein stabilant.[12]

Comments

It is essential to purify phospholipids under reducing conditions. The phospholipid used during solubilization is also important. Asolectin works poorly, but *E. coli* or *S. lactis* phospholipids give proteoliposomes in which steady state (70–90 min) $^{32}P_i$ content is increased 10- to 100-fold over liposomal blanks. *S. lactis* lipid is more effective than the *E. coli* product, probably because of a differential stabilization of the soluble protein. The phospholipid in a recipient bilayer seems less crucial, so that efforts to increase recovery are likely to be most efficient if they are directed at the solubilization step. This approach seems confirmed by the beneficial effect of glycerol when added at this stage (Table II). We have

[10] S. V. Ambudkar and P. C. Maloney, *Fed. Proc.* **44,** 1806 (Abstr. 8154) (1985).
[11] S. V. Ambudkar and P. C. Maloney, *Biochem. Biophys. Res. Commun.* **129,** 568 (1985).
[12] K. Gekko and S. N. Timasheff, *Biochemistry* **20,** 4677 (1981).

found the use of such stabilants essential to reconstitution of anion exchanges from other bacteria[13] and ATP-driven Ca^{2+} transport from streptococci.[14]

Acknowledgment

Supported by USPHS Grant GM24195 from the National Institutes of Health.

[13] S. V. Ambudkar, T. J. Larson, and P. C. Maloney (in preparation).
[14] S. V. Ambudkar, A. R. Lynn, B. P. Rosen, and P. C. Maloney (in preparation).

[44] Proton-Driven Bacterial Flagellar Motor

By ROBERT M. MACNAB

Introduction

Many bacterial species are motile, and the motility can be modulated by stimuli from the environment. These stimuli may be in various forms, such as chemicals, oxygen, light, pH, or temperature. In all cases where it has been examined, bacterial motility has been found to be energized, not by ATP hydrolysis, but by protonmotive force (PMF) or, in some species (see Imae *et al.,* this volume, [45][1]), sodiummotive force. This chapter considers methods for measuring PMF and the motility that is the consequence of its use. Methods for studying tactic responses will not be described explicitly. A number of other reviews may be consulted for further information regarding motility and taxis.[2-6] Chapter 45 by Imae *et al.*[1] in the present volume also contains pertinent information.

The best-studied type of bacterial motility derives from the use of external flagella. The external flagellar filaments are thin helical structures

[1] Y. Imae, H. Matsukura, and S. Kobayashi, this volume [45].
[2] H. C. Berg, *in* "Cell Motility" (R. Goldman, T. Pollard, and J. Rosenbaum, eds.), p. 47. Cold Spring Harbor Laboratory, Cold Spring Harbor, New York, 1976.
[3] H. C. Berg, M. D. Manson, and M. P. Conley, *in* "Prokaryotic and Eukaryotic Flagella" (W. B. Amos and J. G. Duckett, eds.), p. 1. Cambridge Univ. Press, London and New York, 1982.
[4] H. C. Berg, and S. Khan, *in* "Mobility and Recognition in Cell Biology" (H. Sund and C. Veeger, eds.), p. 486. De Gruyter, Berlin, 1983.
[5] R. M. Macnab and S.-I. Aizawa, *Annu. Rev. Biophys. Bioeng.* **13,** 51 (1984).
[6] J. S. Parkinson, and G. L. Hazelbauer, *in* "Gene Function in Prokaryotes" (J. Beckwith, J. E. Davies, and J. A. Gallant, eds.), p. 293. Cold Spring Harbor Laboratory, Cold Spring Harbor, New York, 1983.

that themselves perform no transduction of metabolic energy into mechanical work. They are rotated by motors at their base, and their rotation generates thrust. The motors are reversible, and it is this that permits control of behavior, since the reversal probability is controlled by the environment. The consequences of this reversal vary depending on the type of flagella the cell possesses and may be quite complicated. Thus characterization of motility is by no means trivial experimentally, and it will not be possible to provide simple prescriptions for carrying it out. A major simplification is possible in some cases where the cell can be tethered by a single flagellum and the output of a single motor monitored,[7] as is described below.

Initial evidence that bacterial motility lacked an obligatory relationship to ATP levels, and probably was driven by PMF, derived[8,9] from observations of whether motility was present or absent under a variety of conditions, such as presence or absence of respiratory or glycolytic energy sources, use of inhibitors and use of energy transduction mutants such as *uncA*. Subsequently, there have been extensive measurements of motility and its quantitative relationship to PMF.[10-14] Motility driven by an artificially imposed PMF has been demonstrated and characterized quantitatively.[10,11,13,15] It has also been shown that $\Delta\psi$ and ΔpH are equivalent in driving motility.[12-14] In cell envelope preparations (see below), the absence of a requirement for any organic metabolite, or for any ion other than proton (or hydroxyl), has provided final proof that the motor is driven by PMF.[16]

In certain alkalophilic bacteria that live at such high pH values that their PMF is small, sodium-motive force is used instead[17,18]; this is the subject of Chapter 45 by Imae *et al.*[1]

Finally, it should be noted that PMF can, under some circumstances at least, play a role in controlling the *switching* of the flagellar motor. Major

[7] M. Silverman, and M. Simon, *Nature (London)* **249**, 73 (1974).

[8] S. H. Larsen, J. Adler, J. J. Gargus, and R. W. Hogg, *Proc. Natl. Acad. Sci. U.S.A.* **71**, 1239 (1974).

[9] P. Thipayathasana and R. C. Valentine, *Biochim. Biophys. Acta* **347**, 464 (1974).

[10] M. D. Manson, P. Tedesco, H. C. Berg, F. M. Harold, and C. van der Drift, *Proc. Natl. Acad. Sci. U.S.A.* **74**, 3060 (1977).

[11] A. N. Glagolev and V. P. Skulachev, *Nature (London)* **272**, 280 (1978).

[12] S. Khan and R. M. Macnab, *J. Mol. Biol.* **138**, 563 (1980).

[13] M. D. Manson, P. M. Tedesco, and H. C. Berg, *J. Mol. Biol.* **138**, 541 (1980).

[14] J.-I. Shioi, S. Matsuura, and Y. Imae, *J. Bacteriol.* **144**, 891 (1980).

[15] S. Matsuura, J.-I. Shioi, Y. Imae, and S. Iida, *J. Bacteriol.* **140**, 28 (1979).

[16] S. Ravid and M. Eisenbach, *J. Bacteriol.* **158**, 1208 (1984).

[17] N. Hirota, M. Kitada, and Y. Imae, *FEBS Lett.* **132**, 278 (1981).

[18] N. Hirota and Y. Imae, *J. Biol. Chem.* **258**, 10577 (1983).

examples of this are aerotaxis and phototaxis, which appear to be more strictly "PMF taxis."[19-22] Another related aspect is pH taxis and certain classes of repellent-induced taxes, where perturbation of either internal or external pH is responsible for affecting motor switching.[23-25]

Measurement of PMF

PMF consists of two components, the membrane potential $\Delta\psi$ and the transmembrane pH difference ΔpH. Measurement of both components individually, as well as their sum, is generally necessary for a full understanding of any proton-driven device.

For many purposes, a simplification is possible with respect to PMF measurement because of the fact that bacteria display pH homeostasis; for example, *E. coli* under most conditions regulates its internal pH to around 7.6.[26,27] By working at an external pH of around the same value, the ΔpH is effectively zero, and its measurement can be omitted in all but the most rigorously quantitative experiments.

Because the methods used for measuring $\Delta\psi$ and ΔpH in the context of bacterial motility are essentially the same as those used in other contexts, they will only be outlined here. For further details see chapters in this series, Vol. 55 (Section III).

Measurement of $\Delta\psi$

Use of Microelectrodes. The use of conventional electrophysiological measurement of $\Delta\psi$ with microelectrodes is generally not feasible with bacterial cells on account of their small size (typical cell diameter ~1 μm). However, the normal process of cell division has been successfully manipulated in some cases by genetic and pharmacological means, in order to increase cell size to a usable value of around 5 to 10 μm.

The genetic approach involves, in *E. coli*, the use of mutants that carry lesions both in the *lon* gene (conferring failure of septation after UV

[19] B. L. Taylor, *TIBS* **8**, 438 (1983).

[20] D. J. Laszlo, B. L. Fandrich, A. Sivaram, B. Chance, and B. L. Taylor, *J. Bacteriol.* **159**, 663 (1984).

[21] D. J. Laszlo, M. Niwano, W. W. Goral, and B. L. Taylor, *J. Bacteriol.* **159**, 820 (1984).

[22] J. P. Armitage, C. Ingham, and M. C. W. Evans, *J. Bacteriol.* **161**, 967 (1985).

[23] M. Kihara and R. M. Macnab, *J. Bacteriol.* **145**, 1209 (1981).

[24] D. R. Repaske and J. Adler, *J. Bacteriol.* **145**, 1196 (1981).

[25] J. L. Slonczewski, R. M. Macnab, J. R. Alger, and A. M. Castle, *J. Bacteriol.* **152**, 384 (1982).

[26] E. Padan, D. Zilberstein, and S. Schuldiner, *Biochim. Biophys. Acta* **650**, 151 (1981).

[27] J. L. Slonczewski, B. P. Rosen, J. R. Alger, and R. M. Macnab, *Proc. Natl. Acad. Sci. U.S.A.* **78**, 6271 (1981).

irradiation) and *envB* or *mon* (conferring abnormal rounded morphology). The pharmacological approach involves a similar UV irradiation followed by incubation for ~3 hr with the antibiotic Mecillinam (6-amidinopenicillanic acid) at a concentration of ~1 μg/ml.[28,29]

Lipophilic Cation Selective Electrodes. Lipophilic cations such as tetraphenylphosphonium (TPP⁺) equilibrate across biological membranes and (used as a suitably low concentration, typically around 5 μM) achieve a chemical potential difference that is equal and opposite to the cell's $\Delta\psi$. The same principle has been applied to the design of an electrode selective for TPP⁺. The electrode is constructed using a polyvinyl chloride membrane that permits equilibration of the TPP⁺; the electrode then develops a potential that is equal and opposite to the chemical potential difference of TPP⁺ between the electrode buffer and the medium. An electrode concentration of TPP⁺ of 10 mM, and an initial external concentration of ~5 μM, are typically used. Storage of the electrode in a high concentration of TPP⁺ (10 mM) is necessary, the electrode must be preincubated overnight in buffer containing the desired external TPP⁺ concentration to minimize drift during the experiment, and calibration of the electrode must be performed for each experiment (for details regarding the construction and use of such electrodes see refs. 30, 31).

Changes in the external TPP⁺ concentration reflect redistribution across the cell membrane, and with a knowledge of the original amount of TPP⁺ that was added, the external concentration in the presence of the cells and the intracellular volume, the in/out ratio may be calculated (see ref. 30 for details) and the Nernst equation $\Delta\psi = -(RT/zF)\ln(TPP_{in}^+/TPP_{out}^+)$ used to calculate the membrane potential. Many strains of bacteria, especially of gram-negative species, require permeabilization with EDTA[32] (0.1 M Tris–HCl pH 8.1, 10 mM K⁺EDTA for 3 min at 36°, followed by 10-fold dilution into ice-cold 0.1 M phosphate, pH 6.6, and centrifugation and washing, and finally resuspension in the buffer for the experiment) in order for the TPP⁺ to cross the outer membrane and equilibrate across the inner membrane. In *E. coli,* certain mutants (*acrA*) defective in their outer membrane structure permit access of lipophilic ions without use of EDTA.[33]

[28] W. S. Long, C. L. Slayman, and K. B. Low, *J. Bacteriol.* **133**, 995 (1978).
[29] H. Felle, J. S. Porter, C. L. Slayman, and H. R. Kaback, *Biochemistry* **19**, 3585 (1980).
[30] N. Kamo, M. Muratsugu, R. Hongoh, and Y. Kobatake, *J. Membr. Biol.* **49**, 105 (1979).
[31] M. Eisenbach, *Biochemistry* **21**, 6818 (1982).
[32] S. Szmelcman and J. Adler, *Proc. Natl. Acad. Sci. U.S.A.* **73**, 4387 (1976).
[33] N. Hirota, S. Matsuura, N. Mochizuki, N. Mutoh, and Y. Imae, *J. Bacteriol.* **148**, 399 (1981).

Distribution of Radiolabeled Lipophilic Cations. Lipophilic cations such as TPP^+ may be used in ^3H-labeled form, typically at a concentration of ~5 μM, and their distribution measured by filtration assay,[34] flow dialysis assay,[35] or centrifugation assay.[36]

Correction for binding to the cell surface needs to be applied. The usual procedure involves measurement of the excess retention of label in deenergized cells; note that this assumes extent of binding is independent of energization, an assumption that may not always be valid.[37] For gram-negative bacteria the correction is not excessive. For gram-positive bacteria such as *Bacillus subtilis,* because of large errors associated with binding to the massive cell wall, and an anomalous dependence on TPP^+ concentration, the use of this and related lipophilic cations for $\Delta\psi$ measurement is problematical.[37]

Examples of the use of lipophilic cations for $\Delta\psi$ measurements in motility studies may be found in refs. 32, 38–41.

Distribution of $^{86}Rb^+$. An alternative approach is to measure the equilibrated distribution of $^{86}Rb^+$ (~8 μM) in the presence of valinomycin.[37,42] This technique, at least for gram-positive bacteria, appears to be less subject to anomalous binding artifacts than TPP^+.[37] However, while it gives a fairly reliable measurement of $\Delta\psi$, the value obtained is the one dictated by the external and internal K^+ concentrations, since this ion also equilibrates in the presence of valinomycin. Thus, the $\Delta\psi$ may or may not be at the value that existed prior to valinomycin addition, and is to a greater or lesser degree clamped thereafter.

Spectroscopic Techniques. Cyanine dyes such as $diS-C_3-(5)$[43] have proved convenient for qualitative measurement of $\Delta\psi$ in motility studies,[10,31,39,44] since they permit continuous monitoring during any desired manipulation such as energization or attractant addition. They provide an indication of changes in $\Delta\psi$, but not of its absolute value (unless very carefully calibrated against other techniques), and so their use for quantitative analysis of motility is limited. In certain photosynthetic bacteria a

[34] S. Schuldiner and H. R. Kaback, *Biochemistry* **14,** 5451 (1975).

[35] S. Ramos, S. Schuldiner, and H. R. Kaback, this series, Vol. 55, p. 680.

[36] R. G. Kroll and I. R. Booth, *Biochem. J.* **198,** 691 (1981).

[37] A. Zaritsky, M. Kihara, and R. M. Macnab, *J. Membr. Biol.* **63,** 215 (1981).

[38] S. Khan and R. M. Macnab *J. Mol. Biol.* **138,** 599 (1980).

[39] M. A. Snyder, J. B. Stock, and D. E. Koshland, Jr., *J. Mol. Biol.* **149,** 214 (1981).

[40] A. Zaritsky and R. M. Macnab, *J. Bacteriol.* **147,** 1054 (1981).

[41] Y. Margolin and M. Eisenbach, *J. Bacteriol.* **159,** 605 (1984).

[42] M. P. Conley and H. C. Berg, *J. Bacteriol.* **158,** 832 (1984).

[43] A. S. Waggoner, this series, Vol. 55, p. 689.

[44] J. B. Miller and D. E. Koshland, Jr., *Proc. Natl. Acad. Sci. U.S.A.* **74,** 4752 (1977).

$\Delta\psi$-dependent band shift of the absorption maximum of endogenous carotenoids provides a convenient internal method.[45]

Measurement of ΔpH

[31]P Nuclear Magnetic Resonance. The chemical shift of inorganic phosphate, which is an endogenous intracellular chemical, is a useful indicator of pH values between about 5.7 and 7.7 (a physiologically relevant range), and in phosphate-containing buffers permits simultaneous measurement of the external and internal values of pH, as well as their difference, ΔpH.[27,46] For studies under moderately alkaline conditions (up to pH 8.5), methyl phosphonate has been used successfully[27]; it can always give an indication of external pH[42] and, at least in some strains of *E. coli*, it is taken up (by an uncharacterized pathway) and therefore is able to give an indication of internal pH also. Using both phosphate and methyl phosphonate simultaneously, measurements of internal pH have been measured over a range of external pH from 5.5 to 8.5 in a single experiment.[27]

[31]P nuclear magnetic resonance has the added advantage that the values of ATP and ADP concentrations are available gratuitously, giving a good indication of the overall physiological well-being of the cells.

Distribution of Radiolabeled Weak Acids or Bases. Hydrophobic weak acids such as benzoate equilibrate across biological membranes in their uncharged, protonated form. Use of ^{14}C-labeled material at a low concentration ($\sim 5\ \mu M$) permits measurement of the total distribution and hence an estimate of the ΔpH from the equilibrium equation $[H_{in}^+][A_{in}^-]/[HA] = [H_{out}^+][A_{out}^-]/[HA]$, where $[HA]$ is the same in and out (HA being at equilibrium across the membrane), and where the concentration of A^- can be approximated to the total concentration of A-containing species (A^- + HA), provided the pH in both compartments is substantially (say at least 1.5 pH units) above the pK of the acid used. In those circumstances, the ΔpH is given simply by $\log([A_{in}]/[A_{out}])$. Where this approximation is not valid, the exact expression, $pH_{in} = \log\{([A_{in}]/[A_{out}])(10^{pK} + 10^{pH_{out}}) - 10^{pK}\}$, should be used. Distribution is usually measured by flow dialysis[35] or centrifugation assay.[36] Because of the rapid equilibration times of these weak acids, use of the filtration assay can cause errors associated with washing losses.

The choice of weak acid should take into account the metabolic capa-

[45] J. P. Armitage and M. C. W. Evans, *FEBS Lett.* **102**, 143 (1979).
[46] R. J. Gillies, J. R. Alger, J. A. den Hollander, and R. G. Shulman, *in* "Intracellular pH: Its Measurement, Regulation and Utilization in Cellular Functions" (R. Nuccitelli and D. W. Deamer, eds.), p. 79. Liss, New York, 1982.

bility of the bacterial species under study. Acetate is a poor choice for many species, being readily assimilated; benzoate is satisfactory for enteric bacteria, but not for some pseudomonads; the synthetic weak acid dimethyloxazolidinedione (DMO) is satisfactory for all species, but is rather insensitive at pH values below about 6 on account of its relatively high pK (6.3).

Illustrative examples of the use of radiolabeled weak acids for ΔpH measurements in connection with motility studies may be found in refs. 13, 24, 38, 41.

In situations where the cytoplasmic pH is lower than the external pH, the distribution of a radiolabeled weak base (such as [14C]methylamine[13] or ethanolamine[24,47]) may be used.

Spectroscopic Techniques. In situations where continuous monitoring of cytoplasmic pH is desired during motility studies, the lipophilic dye 9-aminoacridine, which displays pH-dependent fluorescence, has proved useful.[48]

Measurements of Proton Fluxes

Perturbation of External pH Measured by pH Electrodes. Provided the buffering capacity of the external medium is sufficiently low, net proton fluxes from cells to medium can be estimated from the change in external pH.[49] In the context of motility, where it is estimated that the motor contributes only around 0.1% of the total dissipative flux,[50] it becomes important to minimize the other contributions. Measurements at low temperature, such that the membrane ATPase but not the motor is kinetically inhibited, are in progress.[51]

31P Nuclear Magnetic Resonance. With a knowledge of the buffering capacity of the intracellular and extracellular compartments, it becomes possible to convert measured values of changes in internal and external pH, measured by the chemical shift values of inorganic phosphate, into net proton fluxes. It is convenient to use a high buffering capacity for the external medium so that its pH is not perturbed. This approach has not yet been applied to estimation of motility-associated fluxes, but has been used in connection with studies of pH taxis.[25,52]

[47] J.-I. Shioi and B. L. Taylor, *J. Biol. Chem.* **259**, 10983 (1984).
[48] E. A. Goulbourne, Jr. and E. P. Greenberg, *J. Bacteriol.* **143**, 1450 (1980).
[49] P. Mitchell and J. Moyle, *Eur. J. Biochem.* **7**, 471 (1969).
[50] R. M. Macnab, *Crit. Rev. Biochem.* **5**, 291 (1978).
[51] S. Khan and H. C. Berg, personal communication.
[52] A. Castle and R. M. Macnab, unpublished data.

Imposition of Artificial Protonmotive Force or Clamping of Naturally
Generated Protonmotive Force

For a detailed study of the relationship between PMF and motility, it is
often desirable for the experimenter to be able to impose the values of $\Delta\psi$,
ΔpH, or PMF. This requires either that the cell not be contributing actively to the generation of the PMF, or that the pharmacological agent
used have the capacity to defeat the cell's contribution.

A value of zero for PMF can be attained by the use of proton
ionophores such as CCCP (~100 μM).

A value of approximately zero for $\Delta\psi$ can be attained by employing an
external concentration of K^+ that is comparable to the internal value (in
E. coli, ~300 mM at 300 mosm,[53] but with the actual value depending on
osmolarity and other parameters), and then adding the K^+ ionophore
valinomycin (~5 μM). Because of the high conductance of valinomycin,
and the high and therefore not easily perturbed K^+ concentrations in both
compartments, the method is quite effective in clamping $\Delta\psi$ at close to
zero.[41] Pretreatment with EDTA is needed for gram-negative bacteria in
order to permit access of valinomycin to the cell membrane.

Similarly, a value of approximately zero for ΔpH can be attained by
the use of nigericin (which electroneutrally exchanges K^+ and H^+) at a
concentration of ~1 μM and at a K^+ comparable to the internal concentration.[41] Again, pretreatment with EDTA is necessary for gram-negative
bacteria.

Clamping of PMF, $\Delta\psi$, or ΔpH at nonzero values is more difficult,
especially in an actively metabolizing cell. In principle, the use of a low
concentration of CCCP should permit a reduced proton conductance,
such that the naturally generated PMF is not completely abolished, but in
practice this is difficult to achieve in a stable and reproducible fashion.

In the case of valinomycin or nigericin, the use of an external K^+
concentration lower than the internal value should impose a finite $\Delta\psi$ or
ΔpH, respectively.[14] For example, if the internal K^+ concentration is 300
mM, the use of an external K^+ concentration of 30 mM should give a $\Delta\psi$
of 60 mV (negative inside), provided the internal and external pools are
not appreciably perturbed. This is not always easy to achieve where the
cell is actively metabolizing. Most of the commonly studied bacteria, such
as E. coli or S. typhimurium, contain endogenous energy sources, and so
the problem cannot be circumvented by omitting an energy source from
the medium.

The use of species such as Streptococcus spp. has a great advantage

[53] W. Epstein and S. G. Schultz, J. Gen. Physiol. **49**, 221 (1965).

for bioenergetic studies, because in the absence of an exogenously supplied energy source, their PMF dissipates rapidly, the cells being unable to mobilize any endogenous reserves.[10,13] It then becomes possible to impose an artificial PMF, by K^+/valinomycin, K^+/nigericin, or by shift in external pH, without the complications inherent with species such as *E. coli*.

Parenthetically, it may be noted that for metabolizing cells, if the experiment is carried out at an external pH equal to the (regulated) internal value, the ΔpH is zero and the manipulation of $\Delta\psi$ is equivalent to a manipulation of PMF.

Use of Inhibitors

In addition to the ionophores, there are a number of other agents and inhibitors that have been useful in motility studies. These include inhibitors of respiration (e.g., cyanide[54]), ATP synthesis (e.g., arsenate[8]), and *S*-adenosylmethionine synthesis (cycloleucine[55]). There are also studies of inhibition of motility specifically. For example, lipophilic cations are found to inhibit motility of *Bacillus subtilis* at concentrations (e.g., 50 μM tetraphenylarsonium) where PMF is unaffected.[40] A recent study describes the effects of protein modification reagents (imidazole reagents, disulfide reagents, sulfhydryl reagents, amino reagents) on motor function.[42]

Reconstituted or Model Systems

In order to better understand flagellar motor function, one would like to be able to isolate and reconstitute it in defined ways. No such reconstitution experiments have yet been described. The closest approximation is a model system of cell envelopes where the flagellum is left intact throughout, but the cytoplasm can be replaced by any desired solution.

Envelopes of *E. coli* or *S. typhimurium* are prepared[56,57] by subjecting cells to limiting spheroplasting with penicillin (Penassay broth, 20% sucrose, 9 mM $MgSO_4$, 1000 units/ml penicillin, 35°) so that only small regions of cell membrane are unprotected by the peptidoglycan layer (incubation is continued until approximately 3% of the cells show detachment of the outer membrane; typically 30–40 min, but strain dependent).

[54] D. J. Laszlo and B. L. Taylor, *J. Bacteriol.* **145**, 990 (1981).
[55] J.-I. Shioi, R. J. Galloway, M. Niwano, R. E. Chinnock, and B. L. Taylor, *J. Biol. Chem.* **257**, 7969 (1982).
[56] M. Eisenbach and J. Adler, *J. Biol. Chem.* **256**, 8807 (1981).
[57] S. Ravid and M. Eisenbach, *J. Bacteriol.* **158**, 222 (1984).

Chloramphenicol (50 μg/ml) is added to prevent further cell growth, the cells pelleted and resuspended in a small volume of the high osmotic strength growth medium, and then subjected to osmotic shock by dilution into 50 mM potassium phosphate, pH 6.6, plus 0.1 mM EDTA (and any other desired components) and permitted to spontaneously reseal. In this way, it is possible to manipulate the presence or absence of particular ions, macromolecules, and small organic molecules. Under these conditions, the motors of a fraction of the envelopes are capable of rotation when provided with a PMF either by natural energization with a respiratory source, or by imposition of a gradient artificially.

Characterization of Bacterial Motility

Culture Preparation

Although the culture conditions of bacteria vary enormously, a few generalizations may be made. The first is that the degree of motility is often dependent on growth conditions, and the medium that provides optimal growth may or may not be the one that provides optimal motility; this must be tested empirically. For many purposes, nutrient broth is a good choice. Defined minimal media, such as Vogel–Bonner,[58] can also be used.

Also, motility varies with phase of growth; for enteric bacteria, exponential phase is generally best, and cells should have had the opportunity to grow for at least three generations after inoculation. In some species, including $E. coli,$ glucose and high temperatures cause repression of flagellar synthesis,[59,60] and therefore should be avoided.

It is desirable to periodically subject cells to positive selection for motility, since the growth cost of flagellar synthesis in a nonselecting environment (such as shaking liquid cultures) results in accumulation of spontaneous nonmotile (mostly nonflagellate) mutants. A convenient procedure for many species involves inoculation of about 5 μl of liquid culture at the center of a semisolid agar plate (0.35% agar, 0.65% tryptone works well) and incubating (in a closed box containing a beaker of water to prevent the agar surface from drying) for as long as is necessary (typically about 8 hr) to allow chemotactic swarming to occur toward nutrients in the plate. A loopful of cells is taken from the periphery of the spread and inoculated into liquid culture.

[58] H. J. Vogel and D. M. Bonner, J. Biol. Chem. 218, 97 (1956).
[59] J. Adler and B. Templeton, J. Gen. Microbiol. 46, 175 (1967).
[60] M. Silverman and M. Simon, J. Bacteriol. 120, 1196 (1974).

For the actual experiment, it is generally preferable to use a defined medium, which in some cases may be a simple buffer such as 10 mM phosphate. EDTA (0.1 mM), or an equivalent, is necessary to chelate trace amounts of Cu^{2+} and Ni^{2+} that inhibit motility.[59] In rich media, nutrients already present provide sufficient chelation capacity.

Types of Bacterial Motility

Bacterial motility exists in varied forms, depending on species.[61] Bacteria with external flagella may have one or many flagella, and these may be located at the pole or poles of the cell or at other random points around the cell. The cell body may or may not have appreciable helicity (as in spirilla) that contributes to the overall conversion of torque into thrust.

Spirochetes have flagella that are contained between the cell and the outer membranes, and that are believed to cause a rolling of the outer membrane around the helically shaped body to generate thrust.[62]

Gliding bacteria confine themselves to an interface, and reveal no obvious motor apparatus.[63,64]

With such a variety of forms of motility, only general advice can be given for the approaches that should be taken for its characterization.

Light Microscopy

Either phase-contrast or dark-field optics may be used. The optimal degree of magnification depends on the purpose of the experiment; for an initial estimate of the motility of a culture, around 250× (25× objective, 10× eyepiece) provides a large enough field of cells to be fairly certain that it is representative, while for more detailed observations, magnifications of around 640× (40× objective, 16× eyepiece) are preferable. A cell density of $\sim 3 \times 10^7$ ml^{-1} (OD$_{650}$ ~0.05) provides an image that is sufficiently occupied to be informative but not so crowded that it is confusing or that the cells are subject to anaerobiosis. A simple microscope slide and coverslip arrangement is suitable for quick inspection, but for respiring species is subject to oxygen limitation after a minute or two. An improved arrangement for longer viewing involves placing a coverslip across two flanking ones on a slide, to form a bridge, and then applying the sample (~7 μl) under the bridge. This arrangement has the added

[61] R. M. Macnab, in "Encyclopedia of Plant Physiology, New Series" (W. Haupt and M. E. Feinleib, eds.), Vol. 7, p. 207. Springer-Verlag, Berlin, 1979.

[62] H. C. Berg, J. Theor. Biol. **56**, 269 (1976).

[63] R. P. Burchard, Annu. Rev. Microbiol. **35**, 497 (1981).

[64] I. R. Lapidus and H. C. Berg, J. Bacteriol. **151**, 384 (1982).

advantage of providing a three-dimensional chamber, whereas the simple arrangement traps the cells into a more or less two-dimensional space.

Other, more specialized, arrangements have been designed, such as controlled atmosphere chambers for study of aerotoxis[47] and flow chambers that permit rapid exchange of medium.[65–67]

For quantitative study, temperature control is important. If experiments are to be carried out at ambient temperature, good room control is necessary. For experiments at other temperatures, a temperature-controlled microscope stage (e.g., Model TS-2, Bailey Instruments) should be used.

Measurement of Free-Swimming Motility

Visual Observation. A great deal of information may be gained from the use of simple visual observation, especially after some practice. While it is true that video recording may be used to document the results, much wasted recording can be avoided if the experimenter first spends time in direct study. Important parameters include the fraction of cells that are motile, the degree of homogeneity of those that are motile, the translational speed, and the character of the motility in terms of factors such as linear vs curved trajectories, frequency and appearance of direction changes (e.g., simple reversal vs tumbling vs flexing).

An improved visual estimate of swimming speed may be gained by the use of a network eyepiece reticle and a stop watch to time the interval between reticle marks.[68]

Photographic Recording. Dark-field illumination permits the generation of motility tracks that are a simple way of estimating swimming speed. The light source has to be fairly intense (for example, a 100 W mercury arc). A time exposure corresponding to about 10 body lengths displacement (for *S. typhimurium* or *E. coli*, ~1 sec) is convenient.[12] For measurement of track length, the negative may be projected on a screen or printed; an end correction for the cell body length should be made unless the tracks are long. Both for photographic and video records the image of an absolute-scale stage micrometer should be taken under identical optical conditions to those used for the motility images. The use of a stroboscopic light source (e.g., Strobex 136, Chadwick-Helmuth Co.)[69,70]

[65] H. C. Berg and P. M. Tedesco, *Proc. Natl. Acad. Sci. U.S.A.* **72,** 3235 (1975).

[66] J. L. Spudich and D. E. Koshland, Jr., *Nature (London)* **262,** 467 (1976).

[67] H. C. Berg and S. M. Block, *J. Gen. Microbiol.* **130,** 2915 (1984).

[68] E. P. Greenberg and E. Canale-Parola, *J. Bacteriol.* **130,** 485 (1977).

[69] R. M. Macnab and D. E. Koshland, Jr., *Proc. Natl. Acad. Sci. U.S.A.* **69,** 2509 (1972).

[70] J. L. Spudich and D. E. Koshland, Jr., *Proc. Natl. Acad. Sci. U.S.A.* **72,** 710 (1975).

provides punctate tracks that give detailed information on degree of body wobble, speed fluctuation etc.

Video Recording. The use of a moderately sensitive TV camera (e.g., Newvicon, Matsushita Corp.) provides a satisfactory image with conventional microscopic illumination. For recording, a unit with slow-playback and stop-frame capability is desirable; many of the video cassette recorders currently marketed for home use provide this at moderate cost. An accessory (e.g., Panasonic model WJ-810) providing superposition of a time-data record on the video image is almost a necessity for quantitative motion analysis, and provides useful record-keeping additionally; audio annotations regarding the experimental protocol are also useful, and are readily supplied via a microphone connected to the jack that is supplied on most recorders.

For estimation of speed, an initial frame may be marked on screen with a soft-tip pen, and the trajectory followed for a suitable number of frames and measured[38]; alternatively, the video screen may be subjected to a time exposure photograph as was described above for real-time analysis.[71]

For detailed characterization of motility, it should be realized that video recording provides images of only moderate quality, especially when single frames are examined; in visual observation, the eye performs much useful integration and can often extract more information than emerges from a video recording.

Automated Tracking. The image of a single cell can be made the target of an automatic tracking mechanism that permits the location of the cell, with respect to the chamber in which it is swimming, to be continously monitored, transferred to a recording device and subsequently analyzed.[2,72] Commercial instruments for motion analysis are also available (ExpertVision, Motion Analysis Corp.)

Population Assays. The migration behavior of a population of cells can be studied by monitoring the cell density as a function of position and time. An instrument of this sort, used for the study of bacterial response to defined chemical gradients, monitored the right-angle scattered light from a laser beam shone along the gradient axis.[73]

Quasielastic Light Scattering. Information about motion of a particle may be gained by the use of quasielastic light scattering and autocorrela-

[71] Y. Imae, T. Mizuno, and K. Maeda, *J. Bacteriol.* **159**, 368 (1984).
[72] H. C. Berg, *Rev. Sci. Instrum.* **42**, 868 (1971).
[73] F. W. Dahlquist, P. Lovely, and D. E. Koshland, Jr., *Nature (London) New Biol.* **236**, 120 (1972).

tion spectroscopy. Examples of the application of this technique to the motion of bacteria are presented in Chapter 45 of this volume.

Measurement of Motility of Tethered Cells

Except in the case of monoflagellate species, the motion of the cell is a complex result of the motion of the motors. An extremely useful technique, wherever it can be applied, is that of cell tethering.[7] The principle is simple: all of the flagella but one (and often that one also) are reduced in length to the point where they cannot perform significant work on the external medium, and where they cannot interfere with each other. The flagellum is then attached to a slide surface by antibody or other means, and the rotation of the cell monitored.

Preparation of Cells for Tethering. For monoflagellate strains, no deflagellation treatment is necessary, although shortening of the flagella might provide a tighter tether.

To reduce flagellar length in multiflagellate strains (and hence prevent simultaneous tethering by several flagella), a shearing procedure is employed (Waring blender, model 1120, with Eberbach 8580 semi-micro container, 30 sec; or multiple passages through a 23-gauge hypodermic needle), followed by a short period ($\sim$10 min) of regrowth that is then arrested by chloramphenicol (100 μg/ml). Where suitable mutants with an abnormally low degree of flagellation are available (mean number per cell <1), shearing is unnecessary.[12,66] Where mutants lacking the external flagellar filament are available, they may be tethered by their hook.[3]

Attachment of the flagellar stub or the hook to the glass surface occurs to some extent spontaneously, especially at higher ionic strength. However, to obtain a high yield of stably tethered cells, antiflagellar or antihook antibody is usually necessary. [Antiflagellar antibodies for certain species of enteric bacteria are available commercially, Lee Labs, Grayson, GA.] The coverslip is incubated in a humid chamber with a drop of antibody ($\sim$500$\times$ dilution from full strength) overnight, a drop of cells is added and allowed to sit for 5 min, the excess removed by blotting, another drop added and incubated, and the coverslip inverted onto the slot of a flow chamber that permits exchange of the medium to which the cells are exposed. [A bridged coverslip may be used for simple inspection, in which case the coverslip is first flushed with buffer to remove untethered cells.]

The flagella of some bacteria (notably *Streptococcus* spp.) show a sufficient affinity for a silanized surface (nitric acid cleaned coverslip, treated with Siliclad, Clay Adams) that antibody is unnecessary.[13] *Caulo-*

bacter crescentus, a polar monoflagellate, spontaneously tethers itself to an untreated glass surface.[74] These simple approaches are worth trying, before going to the trouble of raising antifilament antibodies.

Tethering of polarly multiflagellate strains by a single flagellum is problematical, and has not been described. An even greater problem arises in the case of the spirochetes, where release of the flagella from the constraint of the outer membrane is a prerequisite. Thus, for many species, analysis of the motility of free cells remains the only option at present.

Monitoring of the Behavior of the Tethered Cells. There are only two primary parameters associated with motion of a tethered cell—speed and direction—both of which need to be measured as a function of time.

For a simple characterization of switching behavior, visual inspection may be all that is necessary; for example, it is easy to make qualitative distinctions between highly clockwise- or counterclockwise-biased mutants and wild-type cells.

[By convention, counterclockwise and clockwise are defined from the point of view of an observer looking along the rotating object toward its rotating junction; therefore for a cell tethered to the underside of a coverslip, and viewed from above, the observed sense of rotation should be inverted. Also, it should be noted that in some microscopes, the observed image is the inverse of the real image (e.g., an R would appear as an Я); this can be readily checked by inspecting the letters on a stage micrometer, and the additional inversion between CCW and CW applied if necessary.]

For more quantitative work, either real-time automated motion analysis or video recording and subsequent analysis (manual or automatic) is indicated.

For analysis of images in real time, the use of a positionable pinhole in the image plane, and projection of the image onto a photomultiplier tube, permits speed analysis[2]; with two such pinholes, phase information and hence rotational sense could be obtained. Another approach has been the use of two linearly graded neutral density filters placed at right angles in the image planes of split beams from the microscope.[3,75] Each beam provides information about the motion of the center of "light-scattering mass," and the phase relationship provides the rotational sense.

Video recordings may be analysed by playing back in slow motion, and scoring the rotational speed and the duration of intervals in a given

[74] L. Shapiro, personal communication.
[75] S. Kobayashi, K. Maeda, and Y. Imae, *Rev. Sci. Instrum.* **48**, 407 (1977).

rotational sense; this can be done manually, or by entering the data to a computer in parallel with a time signal.[66] Automatic image analysis, of a similar sort to that employed in real time, may also be used.

Visualization of Flagella by High-Intensity Dark-Field Light Microscopy

Although bacterial flagella are very thin (~20 nm diameter), they can be visualized by high-intensity dark-field light microscopy.[76,77] A high-intensity lamp (500 W xenon arc), critical focusing using an oil-immersion dark-field condensor, and a medium-power (40×) dry objective (preferably fluorite) are needed. For video recording, the sensitivity of a silicon intensifying target vidicon (SIT) camera is essential. Use of this technique permits the individual flagella on a multiflagellate cell to be monitored, and has enabled a number of subtleties of bacterial motility to be described, such as the autonomous switching of individual flagella,[78] and the participation of polymorphic transitions in tumbling.[77]

Use of Markers

Physical markers can be of use in the study of motility. An important illustration of this is the attachment of antibody-coated beads [0.7 μm latex beads (Dow); 1 : 1000 diluted antifilament antibody in 10 mM Tris, pH 7.8 plus 0.1 M NaCl] to decorate an otherwise invisible flagellar filament with asymmetric clumps of beads and enable the rotation of these filaments to be detected.[7]

The motion of individual flagella on long aseptate cells of *E. coli* has been studied by attaching to them glutaraldehyde-fixed cells as markers, using antifilament antibody.[79] This enabled the conclusion to be reached that the motors switch autonomously in the absence of hydrodynamic or mechanical coupling.

Latex and other types of beads that spontaneously attach to the cell surface of gliding bacteria (*Cytophaga* and *Flexibacter* spp.) show linear and rotational motion that presumably reflect the activity of the as yet poorly understood motor apparatus of these species.[64,80]

[76] R. M. Macnab, *J. Clin. Microbiol.* **4**, 258 (1976).
[77] R. M. Macnab and M. K. Ornston, *J. Mol. Biol.* **112**, 1 (1977).
[78] R. M. Macnab and D. P. Han, *Cell* **32**, 109 (1983).
[79] A. Ishihara, J. E. Segall, S. M. Block, and H. C. Berg, *J. Bacteriol.* **155**, 228 (1983).
[80] J. L. Pate and L.-Y.E. Chang, *Curr. Microbiol.* **2**, 59 (1979).

Energetics of Motility

A detailed characterization of the flagellar motor output should include answers to the following questions: What is the relationship between the output torque and the PMF over a wide range, from zero to the maximum normally attained physiologically? Is there a linear range, where torque is proportional to PMF? Is there a threshold? Does the motor saturate above some particular value? Do a $\Delta\psi$ and a ΔpH of the same value produce the same output torque? Is the torque independent of load, and if so over what range? [The latter question is most easily answered by keeping the frictional geometry constant and varying the viscosity of the medium by addition of an agent such as Ficoll 400 (Pharmacia), which maintains Newtonian fluid characteristics.[13,81]] How many protons are used per revolution, and therefore—given the value of the PMF and the hydrodynamic work being performed—what is the efficiency?

Thus the most important parameters to be obtained are the hydrodynamic work, the PMF, and the stoichiometry of protons per revolution.

Estimation of Hydrodynamic Work Output

In a previous section, various approaches to measuring bacterial motility were described. This information by itself is of little value for energetic analyses unless it can be interpreted in terms of the amount of hydrodynamic work that it represents.

All of the hydrodynamic work that is performed by motile bacteria is in a regime of size, speed, fluid density, and viscosity (formally prescribed by the Reynolds number), such that there is no turbulence, no significant contribution of inertial forces, and where force–velocity and torque–angular velocity relationships are strictly linear.[82,83] This simplifies analysis greatly.

Estimation of the work output of tethered cells and of monoflagellate free cells is relatively straightforward.

A tethered cell may be approximated as an equivalent rotating sphere, with a scaling factor that has been estimated from macroscopic working models in a suitably viscous medium such that the Reynolds number is comparable to that of a tethered cell in water. Thus a cell of *E. coli* (rod-shaped with a length of ~3 μm and a cylindrical diameter of ~1 μm),

[81] H. C. Berg and L. Turner, *Nature (London)* **278,** 349 (1979).
[82] H. C. Berg, *Nature (London)* **249,** 77 (1974).
[83] E. M. Purcell, *Am. J. Phys.* **45,** 3 (1977).

tethered near its center, is approximately equivalent to a sphere of 1 μm radius.[2] The rotational frictional coefficient is then given by $f_{rot} = 8\pi\eta r^3$ (where η is the viscosity and r is the equivalent cell radius), the torque exerted on the surrounds by $\tau = f_{rot}\omega$ (where ω is the angular speed), the work done per revolution by $2\pi\tau$, and the power dissipation by $\tau\omega$. For a cell of *E. coli* rotating at 10 Hz, the work done per revolution is about 10^{-17} J and the energy dissipation is in the vicinity of 10^{-16} W.

For a monoflagellate free-swimming cell, the translational and rotational frictional coefficients of the cell are given approximately by $6\pi\eta r$ and $8\pi\eta r^3$, respectively, and for the filaments by more complex expressions.[76,84] The translational speed may be easily measured as was described above. Estimates of rotational speed are difficult at normal swimming speed, because the waveform is blurred. However, with the use of a high-viscosity medium (which should be Newtonian; Ficoll 400 generates such a medium, methylcellulose does not[81]), the speed may be slowed sufficiently that the rotational velocity of the flagellum may be estimated from the observed net phase velocity, after taking into account the phase velocity introduced by translation (if there were no slippage in the axial direction, the net phase velocity would be zero). The rotational velocity of the cell can be difficult to estimate if the cell is symmetrical and rotating precisely about its long axis, but often there is sufficient asymmetry for the rotational period to be recognized.

For multiflagellate cells, the issue is complicated by the unknown amount of energy that is being dissipated within the flagellar bundle, and in this situation the availability of information from tethered cells becomes especially valuable.

Estimation of Proton Stoichiometry per Revolution and of Efficiency

Subject to the caveats expressed above, the amount of hydrodynamic work performed by a flagellar motor can be estimated. Estimation of efficiency then requires a knowledge of the PMF and the number of protons that pass through the motor per revolution.

A lower bound to the number of protons can be estimated on the basis of the upper bound of efficiency, 100%. The energy of a proton at a PMF of, say, 150 mV is 2.4×10^{-20} J, and so at least $10^{-17}/(2.4 \times 10^{-20}) = 420$ protons are required to rotate a tethered cell at 10 Hz in water.

There are as yet no published measurements of the proton flux through the motor (see the earlier section on proton flux measurements).

[84] R. A. Anderson, *in* "Swimming and Flying in Nature" (T.Y.-T. Wu, C. J. Brokaw, and C. J. Brennen, eds.), Vol. 1, p. 45. Plenum, New York, 1975.

Mechanism of the Bacterial Motor

Ultimately, one would like to understand not only the energetics of bacterial motility, but also the mechanism.

Energetic studies can place important constraints on possible mechanisms: for example, the observation in *Streptococcus* that there are neither isotope nor temperature effects on the relationship between PMF and torque[85] suggests that proton association/dissociation events are not rate-limiting and that the device may be operating close to equilibrium, a hypothesis that is further reinforced by the observation that torque is independent of load over a wide range.[13]

However, energetic measurements alone cannot provide a complete answer. The isolation of the motor organelle, the determination of its chemical composition, and the genes responsible for its assembly and structure represent another approach.[86–88] So do the isolation and characterization of motility mutants[89]; the use of intergenic complementation as an indicator of protein/protein interactions[90]; the sequencing of motility genes in order to obtain protein sequence information[91]; and the use of molecular genetic techniques to study the consequences when particular motility genes are suddenly permitted to be expressed (for example, the MotB protein of *E. coli* restores torque in quanta[67]). Further discussion of these topics lies beyond the scope of this chapter; the original references should be consulted by interested readers.

[85] S. Khan and H. C. Berg, *Cell* **32**, 913 (1983).
[86] Y. Komeda, M. Silverman, P. Matsumura, and M. Simon, *J. Bacteriol.* **134**, 655 (1978).
[87] T. Suzuki and Y. Komeda, *J. Bacteriol.* **145**, 1036 (1981).
[88] S.-I. Aizawa, G. E. Dean, C. J. Jones, R. M. Macnab, and S. Yamaguchi, *J. Bacteriol.* **161**, 836 (1985).
[89] G. E. Dean, S.-I. Aizawa, and R. M. Macnab, *J. Bacteriol.* **154**, 84 (1983).
[90] J. S. Parkinson, S. R. Parker, P. B. Talbert, and S. E. Houts, *J. Bacteriol.* **155**, 265 (1983).
[91] G. E. Dean, R. M. Macnab, J. Stader, P. Matsumura, and C. Burks, *J. Bacteriol.* **159**, 991 (1984).

[45] Sodium-Driven Flagellar Motors of Alkalophilic *Bacillus*

By YASUO IMAE, HIROSHI MATSUKURA, and SYOYU KOBAYASI

Introduction

Bacteria swim by rotating their flagella, and each flagellum is driven at its base by a rotary motor embedded in the cell wall and cytoplasmic membrane. In the case of bacteria such as *Escherichia coli* and *Bacillus subtilis* which are living in moderate conditions, the energy for rotation of their flagellar motors is supplied by the electrochemical potential gradient of protons across the membrane, namely the protonmotive force (see review[1]). Furthermore, any ion other than H^+ is not required for the rotation of these flagellar motors.[2-4] Thus, these bacteria have the H^+-driven flagellar motors.

In addition to these neutrophilic bacteria, many bacteria live in extremely alkaline conditions where the available protonmotive force is considered to be quite small.[5] Among such bacteria, some strains of alkalophilic *Bacillus* grown optimally at pH 10–11 show best motility at the same pH range, although the protonmotive force of the cells under the condition is estimated to be smaller than -50 mV. Thus, it seems difficult to imagine that the flagellar motors of these bacteria are driven by the protonmotive force. Recent studies indicate that the presence of the electrochemical potential gradient of Na^+ is essential for rotation of these flagellar motors.[6,7] Thus, these alkalophilic *Bacillus* strains are considered to have the Na^+-driven flagellar motors.

In this chapter, we describe the method to measure the motility and bioenergetic parameters of alkalophilic *Bacillus*. Although the flagellar motors of these bacteria require Na^+ for their rotation and the rotation speed is optimal under alkaline conditions, the swimming pattern and swimming speed of these bacteria are indistinguishable from those of neutrophilic *Bacillus* such as *B. subtilis*. Hence, most methods for the

[1] H. C. Berg, M. D. Manson, and M. P. Conley, *Symp. Soc. Exp. Biol.* **35**, 1 (1982).

[2] M. D. Manson, P. Tedesco, H. C. Berg, F. M. Harold, and C. van der Drift, *Proc. Natl. Acad. Sci. U.S.A.* **74**, 3060 (1977).

[3] S. Matsuura, J. Shioi, Y. Imae, and S. Iida, *J. Bacteriol.* **140**, 28 (1979).

[4] S. Ravid and M. Eisenbach, *J. Bacteriol.* **158**, 1208 (1984).

[5] K. Horikoshi and T. Akiba, "Alkalophilic Microorganisms." Springer-Verlag, New York, 1982.

[6] N. Hirota, M. Kitada, and Y. Imae, *FEBS Lett.* **132**, 278 (1981).

[7] N. Hirota and Y. Imae, *J. Biol. Chem.* **258**, 10577 (1983).

motility measurements of neutrophilic bacteria described by R. M. Macnab[8] are also utilizable for alkalophilic *Bacillus*.

Cultivation of Bacteria

Strains. Motility of the following alkalophilic *Bacillus* strains is excellent: *Bacillus firmus* RAB,[9] *Bacillus* sp. YN-1,[7] *Bacillus* sp. 202-1,[10] and *Bacillus* sp. 8-1.[6] *Bacillus alkalophilus* ATCC27647 is filamentous and poorly motile.[11] These strains can be stocked as frozen cells at $-80°$ in the presence of 7% dimethyl sulfoxide in the growth medium.

The optimal growth of these strains is observed at pH 9–11, although their intracellular pH is below 9. These strains absolutely require Na$^+$ not only for swimming but also for their growth. The transport system for some amino acids requires Na$^+$ as a coupling ion.

Growth Medium. To obtain vigorously motile cells, a complex medium (AB-4 medium) of pH 9 to 10 is suitable for the growth of all the strains of alkalophilic *Bacillus*. The following solutions are separately autoclaved.

    AB-4 base
        Polypeptone, 10.0 g
        Yeast extract, 1.5 g
        Glucose, 10.0 g
        KH_2PO_4, 1.5 g
        H_2O, 1.0 liter
    1% $MgCl_2 \cdot 6H_2O$
    15% Na_2CO_3

To 100 ml of AB-4 base, 1 ml of 1% $MgCl_2$ solution and 5 ml of 15% Na_2CO_3 solution are added just before use. The final pH of the medium is about 9.5, and the final Na$^+$ concentration is about 150 mM. When 6.5 ml of 15% Na_2CO_3 is added, the medium pH is about 10. To make hard agar or semisolid plates, 1.5 or 0.35% of agar, respectively, is added to AB-4 base and autoclaved.

Selection of Vigorously Motile Cells. For motility experiments, it is important to select vigorously motile cells. A loopful of alkalophilic *Bacillus* cells is inoculated on the center of a semisolid plate, and the plate is incubated at 37° for 10 to 15 hr until the swarm front of the cells reaches to the plate edge. Nonmotile cells remain at the center of the plates and

---

[8] R. M. Macnab, this volume [44].
[9] M. Kitada, A. A. Guffanti, and T. A. Krulwich, *J. Bacteriol.* **152**, 1096 (1982).
[10] N. Nakamura, K. Watanabe, and K. Horikoshi, *Biochim. Biophys. Acta* **397**, 188 (1975).
[11] A. A. Guffanti, P. Susman, R. Blanco, and T. A. Krulwich, *J. Biol. Chem.* **253**, 708 (1978).

motile cells swarm out to the edge due to chemotaxis. Vigorously motile cells are obtained after several repeats of the swarming.

Cell Growth and Cell Preparation. Cells are grown in AB-4 medium at 37° with gentle aeration for 4–5 hr. For motility experiments, cells in early stationary phase of growth are suitable, because cells in the earlier growth stages are filamentous and poorly motile.

All the strains of alkalophilic *Bacillus* have a tendency to lyse in the absence of Na^+ in the medium, and, therefore, it is advisable to use a medium containing more than 5 mM of Na^+ for cell preparation. When the medium without Na^+ is used, the cell preparation should be carried out quickly to avoid the damage of the cells during the preparation.

TG medium consisting of 25 mM Tris–HCl buffer (pH 9.5), 5 mM glucose, and various concentrations of NaCl is routinely used as a cell preparation medium. As an exogenous energy source, glucose is suitable for many strains, because glucose uptake is mostly independent of Na^+ in the medium. For motility experiments, TG medium containing 0.1 mM EDTA (K^+-salt) is often used,[7] although the omission of EDTA has almost no effect on the swimming of most strains. Cells are harvested by centrifugation at 7000 g for 5 min at room temperature, washed twice with TG medium, and suspended in the same medium.

The cells stocked in the absence of Na^+ are nonmotile but the cells become motile by the addition of Na^+ to the medium. Only Na^+ can induce motility in all the strains. However, the recovery of motility by Na^+ is getting worse during the stock of the cells without Na^+. To test the effect of various cation species or of Na^+ concentrations on motility, therefore, the cells should be washed with TG medium containing 5 mM NaCl and stocked as a dense suspension in the same medium. Then, the cells are diluted 50-fold or more in various media. It is noteworthy that Na^+ concentrations lower than 0.1 mM are not enough to induce any translational swimming in most strains.

Measurement of Swimming Speed by Photographic Method

The photographic method is a simple but rough method to estimate the average swimming speed of bacteria. A drop of the cell suspension of alkalophilic *Bacillus* is transferred onto a microscope glass slide, and the movement of bacteria near the surface of the glass slide is immediately observed with a light microscope under dark-field illumination. The magnification of ×100 is enough for the motility measurements. A cover glass is not used to avoid the anexia. The movement of the cells, which are observed as moving white dots in dark, is recorded as swimming tracks by photographic method using Kodak Tri-X Pan film (Eastman Kodak Co.,

Rochester, NY) with an exposure time of 1–2 sec. The cell concentration is adjusted to give about 100 tracks on a picture. The swimming tracks on a film are cast on a paper by using a slide projector, and the length of each track is measured with a flexible measure. The average cell size is obtained from the nonmotile cells prepared by the addition of 10 μM gramicidin D and is subtracted from the average length of the swimming tracks. Then, the swimming speed of the cells is calculated based on the scale obtained from an objective micrometer. When a videotape recording system is available, the movement of bacteria under dark-field illumination is first recorded on videotapes with a TV camera attached to the microscope, and then the swimming tracks are photographed from the video scenes as above.

When a personal computer equipped with a tablet for graphics is available, the swimming tracks on the film are directly cast on the tablet and the length of each track can be measured by using an indicator pen. The average swimming speed and distribution of the swimming speed are directly calculated by the computer. As an example of the results obtained by this method, the relationship between the swimming speed of an alkalophilic *Bacillus* strain 202-1 and the medium pH is shown in Fig. 1.

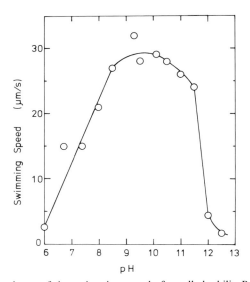

FIG. 1. pH dependence of the swimming speed of an alkalophilic *Bacillus* strain 202-1. The cells were suspended in TG medium containing 50 m*M* each of KCl and NaCl, and the medium pH was varied by the addition of HCl or KOH. The swimming speed at 30° was measured by the photographic method.

The data were analyzed by using an Apple II Plus computer equipped with Apple Graphics Tablet (Apple Computer Inc., Cupertino, CA).

In the case of the photographic method, data are collected only from the cells swimming nearly two-dimensionally on the surface of a microscope glass slide. It should be noted, therefore, that the average swimming speed estimated by this method might be slightly different from that of the cells swimming freely in three dimensions.

Motility Measurement by the Dynamic Light Scattering Method

The dynamic light scattering method is also useful for the quantitative determination of bacterial motility (see reviews[12–15]). Compared to the photographic method, the measurement of motility by this method can be carried out quite automatically on a considerably large number of the cells which are freely swimming in three dimensions. We describe here an application of this method to measure the swimming speed of alkalophilic *Bacillus*. Of course, this method is also applicable for the measurement of the swimming speed in various neutrophilic bacteria.

Principle. When actively swimming bacteria are irradiated by a laser light and the scattered light from the bacteria is observed through a small hole, the intensity of the scattered light fluctuates with time. The rapidity of the fluctuation is closely related to the swimming speed of the bacteria. The intensity, $I(t)$ is characterized by the normalized intensity autocorrelation function defined by

$$g^{(2)}(\tau) = \langle I(t)I(t + \tau)\rangle/\langle I(t)\rangle^2 \tag{1}$$

where τ is a delay time and the brackets denote time average. $g^{(2)}(\tau) - 1$ is usually plotted against τ, since $g^{(2)}(\tau)$ tends to 1 with increasing τ. When the scattering light comes from only bacteria (homodyne case), $g^{(2)}(\tau)$ is related with the normalized electric field autocorrelation function $g^{(1)}(\tau)$ by

$$g^{(2)}(\tau) = 1 + \beta|g^{(1)}(\tau)|^2 \qquad 0 < \beta \leq 1 \tag{2}$$

where β is a geometric parameter and is related to various factors of the experimental setup. When bacteria can be assumed to swim linear paths

[12] B. Chu, "Laser Light Scattering." Academic Press, New York, 1974.
[13] B. J. Berne and R. Pecora, "Dynamic Light Scattering." Wiley, New York, 1976.
[14] S. H. Chen and F. R. Hallett, *Q. Rev. Biophys.* **15**, 131 (1982).
[15] V. A. Bloomfield and T. K. Lim, this series, Vol. 48 [19].

as point scatterers in random directions, $g^{(1)}(\tau)$ is given by

$$g^{(1)}(\tau) = \int_0^\infty P_s(V) \frac{\sin qV\tau}{qV\tau} \, dV \tag{3}$$

where V is the swimming speed of bacteria, $P_s(V)$ is the swimming speed distribution and is normalized as $\int_0^\infty P_s(V) \, dV = 1$, and q is the value of the scattering vector; $q = (4\pi n/\lambda)\sin \theta/2$ in which n, λ, and θ are the refractive index of the medium, the wavelength of laser light *in vacuo*, and the scattering angle, respectively.

Measurement and Data Analysis. Cells in TG medium are put in a cuvette and incubated for a few minutes to equilibrate the temperature. Then, the cells are irradiated by a He–Ne laser ($\lambda = 6328$ Å), and the scattered light from the cells is observed at a scattering angle of 4°. Data are accumulated in a digital correlator for about 100 sec, and the correlation functions are calculated (for Instruments, see reviews[12,15]). Curve fitting for the data analysis is carried out by a personal computer. An example of the intensity autocorrelation function obtained from motile or nonmotile cells of alkalophilic *Bacillus* strain YN-1 is shown in Fig. 2. The $g^{(2)}(\tau) - 1$ for motile cells decays strikingly faster than that for nonmotile cells, indicating that the Brownian motion of the cells is mostly excluded from the data under the measurement conditions described above.

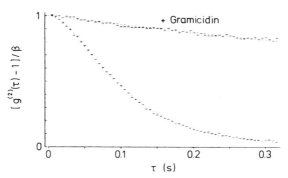

FIG. 2. Intensity autocorrelation functions obtained from motile and nonmotile cells of an alkalophilic *Bacillus* strain YN-1. Measurements were carried out at 30° by the cells in TG medium (pH 9.0) containing 20 mM NaCl. The light source was a He–Ne laser ($\lambda = 6328$ Å), and the scattering angle was 4°. Data were accumulated for 100 sec. Bars, experimental. Dots, calculated. Motile cells (lower lines): The dotted line was drawn according to Eq. (4). $\langle V \rangle_G = 16.1$ μm/sec. $[\langle V^2 \rangle_M]^{1/2} = 51.2$ μm/sec. $F_M = 0.072$. Nonmotile cells (upper lines): 1 μM gramicidin D was added to the cells to stop motility. The dotted line was drawn according to the equation in note 18. $D = 7.69 \times 10^{-9}$ sec/cm.

The decay process of the measured $g^{(2)}(\tau) - 1$ for motile alkalophilic *Bacillus* is mostly explained if the swimming speed has a Gamma distribution.[16] However, the process can be explained more perfectly if the data are assumed to contain an additional speed component which has a Maxwellian distribution.[17] In this case, the distribution of the swimming speed is described as a combination of two distributions with a factor of F_G for the Gamma distribution and of F_M for the Maxwellian distribution.

$$g^{(2)}(\tau) - 1 = \beta |F_G g_{s,G}^{(1)}(\tau) + F_M g_{s,M}^{(1)}(\tau)|^2 \qquad (4)$$

where $F_G + F_M = 1$. The observed $g^{(2)}(\tau) - 1$ can be fit with Eq. (4) by using the method of nonlinear least-squares fit (see the dotted line for motile cells in Fig. 2). The disturbance of the data by nonmotile cells[18] in a sample can be mostly neglected by using a fresh culture.

From the data shown in Fig. 2, the contribution of the speed component with the Maxwellian distribution to the decay curve is calculated to be only 7.2%. It is noteworthy that the average speed with this distribution is too high to assign it to that of the translational swimming of the cells, suggesting that this speed component may be derived from the motion such as rotation of flagella and some vibrational motion of cell body as reported by Stock.[19] Thus, the translational swimming of alkalophilic *Bacillus* strain YN-1 is considered to have the speed with the Gamma distribution.

Evaluation of the Method. By using the method described above, a linear relationship was obtained between the swimming speed of YN-1 cells and the logarithmic concentrations of Na^+ in the medium (Fig. 3).

---

[16] The speed distribution is given by

$$P_{s,G}(V) = [4/\langle V \rangle_G^2] V \exp(-2V/\langle V \rangle_G)$$

where $\langle V \rangle_G$ is an average swimming speed of the cells. Then

$$g_{s,G}^{(1)}(\tau) = [1 + (q\langle V \rangle_G \tau/2)^2]^{-1}.$$

[17] The speed distribution is given by

$$P_{s,M}(V) = (54/\pi)^{1/2}[\langle V^2 \rangle_M]^{-3/2} V^2 \exp[-(3/2)V^2/\langle V^2 \rangle_M]$$

where $\langle V^2 \rangle_M$ is mean-square speed of the distribution.[14] Then,

$$g_{s,M}^{(1)}(\tau) = \exp(-q^2 \langle V^2 \rangle_M \tau^2/6).$$

[18] When the cells are moving only by diffusion due to thermal agitation from the medium, $g^{(2)}(\tau) - 1$ can be written as

$$g^{(2)}(\tau) - 1 = \beta \exp(-2Dq^2\tau)$$

where D is the translational diffusion constant of the cells.

[19] G. B. Stock, *Biophys. J.* **22**, 79 (1978).

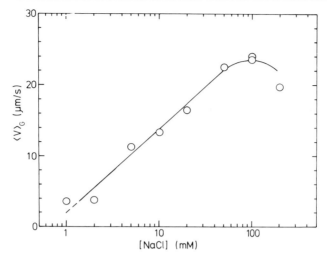

FIG. 3. Sodium concentration dependence of the swimming speed of an alkalophilic *Bacillus* strain YN-1. The experimental conditions were the same as described in the legend for Fig. 2, except that the NaCl concentrations were varied. $\langle V \rangle_G$ was calculated by Eq. (4).

The results are consistent with the data obtained by the photographic method.[7]

In application of the theoretical field autocorrelation functions given by Eq. (4) to real systems, however, the Bragg spacing, q^{-1} must be adjusted in a region suitable for bacterial swimming pattern and cell size. Thus, the straight path length of swimming, L should be

$$1/q \leqq L \tag{5}$$

and the shapes of bacteria usually deviate from a spherical symmetry so that the size of bacteria, d should be

$$1/q \geqq d \tag{6}$$

Wild-type bacteria usually have $L \geqq 10 \ \mu m$ and $d \approx 1 \ \mu m$. Therefore, when He–Ne laser ($\lambda = 6328$ Å) is used as a light source, the scattering angles must be set at between 0.5 and 4° to correctly measure the swimming speed of the cells. When argon-ion laser ($\lambda = 4880$ Å) is used, the scattering angle should be lowered.

Besides the translational swimming, bacteria have other types of motion such as tumbling, rotation, and wobbling. Most of these motions can be avoided by the measurement at low scattering angles. Figure 4 shows

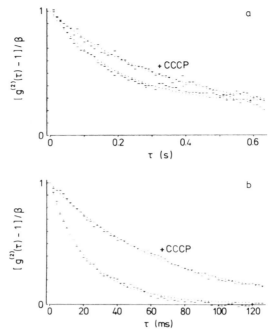

FIG. 4. Intensity autocorrelation functions obtained by an incessantly tumbling mutant of *E. coli* (*cheZ*292). Data were accumulated for 60 sec on the cells in 10 mM phosphate buffer (pH 7.0) containing 0.1 mM EDTA and 10 mM lactate. Motility was stopped by the addition of an uncoupler, 5 μM carbonylcyanide *m*-chlorophenylhydrazone. (a) Scattering angle, 4°. (b) Scattering angle, 20°. Dotted lines are drawn by the method of three-term cumulant expansion as described by Pusey *et al.*[20]

the $g^{(2)}(\tau) - 1$ obtained by using an incessant tumbling mutant of *E. coli* (*cheZ*292).[21] When the measurement was carried out at $\theta = 4°$, the $g^{(2)}(\tau) - 1$ obtained from either motile or nonmotile cells did not show much difference. Thus, the tumbling motion of the cells can be mostly neglected by the measurement at $\theta = 4°$. Of course, when $\theta = 20°$ was used, the $g^{(2)}(\tau) - 1$ for motile cells was more rapidly decreased than that for nonmotile cells. These data indicate that the use of different scattering angle for the measurement may provide the method to detect various types of motion in bacteria.

[20] P. N. Pusey, D. E. Koppel, D. W. Schaefer, R. D. Camerini-Otero, and S. H. Koenig, *Biochemistry* **13**, 952 (1974).
[21] J. S. Parkinson and S. R. Parker, *Proc. Natl. Acad. Sci. U.S.A.* **76**, 2390 (1979).

Preparation of Tethered Cells

Rotation of each flagellum of alkalophilic *Bacillus* is, though indirectly, observed by using tethered cells. Similar to the case of neutrophilic bacteria, antiflagellar antibodies prepared from a rabbit can be used for tethering the cells of alkalophilic *Bacillus* to a glass slide. After reducing the number of flagella on the cells by passing the cells through a thin needle or by using a small Potter-type cell homogenizer, a drop of the cells is placed on a glass slide which has spacers of 0.13 to 0.17 mm in thickness and mixed with a few drops of antiflagella antibody. Then, a cover glass is placed on it. Some rotating cells can be seen after 10 to 20 min. The medium held between the spacers can be easily changed by sucking the medium from one end and adding another medium to the other end. Tethered cells can rotate in either direction, clockwise or counterclockwise, and the rotation rate is decreased with decreasing Na⁺ concentrations in the medium.

Isolation of Flagella and Flagellar Basal Bodies

Flagella of alkalophilic *Bacillus* can be isolated and purified by the method described by Asakura and Iino.[22] The molecular weights of flagellins in SDS–polyacrylamide gel electrophoresis are 30,000 for strain YN-1, 35,000 for strain 202-1, and 78,000 for ATCC27647. Flagellar basal bodies, which are assumed to be an essential part of the flagellar motors, can be isolated by the method described by DePamphilis and Adler.[23] The electron micrographs of the flagella and flagellar basal bodies isolated from strain YN-1 look similar to those isolated from *B. subtilis*.[24]

Determination of Bioenergetic Parameters

The membrane potential of alkalophilic *Bacillus* strains can be measured by using radioactive and lipophilic cations such as [³H]triphenylmethyl phosphonium ion ([³H]TPMP⁺). The filtration method gives reasonable values of the membrane potential. The cells treated with 10 μM gramicidin D can be used for a zero membrane potential control. The membrane potential in most strains measured by using [³H]TPMP⁺ is -150 to -170 mV at pH 9.5 and slightly increases with increasing the medium pH up to 11. In TG medium at pH 9.5, the membrane potential of

[22] S. Asakura and T. Iino, *J. Mol. Biol.* **64,** 251 (1972).
[23] M. L. DePamphilis and J. Adler, *J. Bacteriol.* **105,** 396 (1971).
[24] N. Hirota, R. Kamiya, and Y. Imae, unpublished data.

the strains YN-1, 202-1, and RAB is not affected by the presence or absence of Na$^+$ in the medium.

A tetraphenylphosphonium ion-selective electrode[25] is also applicable to measure the time course of the changes in the membrane potential of the cells. In this case, the cells treated with some nonionic detergents such as octylglucoside and N-gluco-N-methylcaprylamide is suitable for a zero membrane potential control.

As reported by Hirota and Imae,[7] the membrane potential of the valinomycin-treated cells can be semiquantitatively varied by changing extracellular K$^+$ concentration. Under the condition, the membrane potential is roughly equal to the K$^+$ diffusion potential which is calculated by the concentration difference of K$^+$ between inside and outside of the cells. The intracellular K$^+$ concentration of YN-1 is about 500 mM.[7]

The ΔpH between inside and outside of the cells can be measured by the distribution of radioactive weak base such as [^{14}C]methylamine. The filtration method is also applicable. The use of glass-fiber filters (Whatman GF/C) is recommended for quick filtration, and the radioactivity trapped on the filter should be measured without washing and drying. The gramicidin-treated cells are used for a zero ΔpH control. The intracellular pH of RAB measured by this method is about 8.5 and stays nearly constant between the medium pH of 8.5–11.[9,26] Then, the protonmotive force of RAB cells is roughly estimated to be −30 to −50 mV at pH 11.

Chemotaxis in Alkalophilic *Bacillus*

All the alkalophilic *Bacillus* strains described above produce clear swarm in semisolid plates, indicating that these strains are chemotactically active. Consistent with this, the direction of rotation of tethered cells can be changed by the addition of nutrients. For example, the tethered cells of 202-1 in TG medium show either clockwise or counterclockwise rotation. However, when TG medium containing a 1/10 concentration of AB-4 medium is added, the direction of rotation of the tethered cells is fixed for a few minutes in counterclockwise. These results suggest that the information processing system of alkalophilic *Bacillus* is basically similar to that of neutrophilic bacteria.

[25] N. Kamo, M. Muratsugu, R. Hongoh, and Y. Kobatake, *J. Membr. Biol.* **49**, 105 (1979).
[26] S. Sugiyama, H. Matsukura, and Y. Imae, *FEBS Lett.* **182**, 265 (1985).

Section III

Transport in Mitochondria and Chloroplasts

A. Mitochondria
Articles 46 through 55

B. Chloroplasts
Articles 56 and 57

[46] Mitochondrial Outer Membrane Channel (VDAC, Porin) Two-Dimensional Crystals from *Neurospora*

By CARMEN A. MANNELLA

In the outer membranes of mitochondria isolated from plants and fungi, the principal protein components are integral polypeptides of molecular weight ~30,000.[1-4] X-Ray diffraction experiments on the plant membrane provided the first indication that these polypeptides comprise a fundamental in-plane structural unit in the mitochondrial outer membrane.[5] The subsequent discovery that the 30-kDa proteins form large passive diffusion channels when inserted into phospholipid membranes[6,3] implicates them as the basis of the extreme small-molecule permeability of the mitochondrial outer membrane.

The occurrence in mitochondria of a substance with potent *in vitro* pore-forming activity was first discovered in extracts of *Paramecium* mitochondria.[7] Colombini[8] was the first to assign the channel activity to the outer membrane in experiments with rat liver mitochondria, in which the 30-kDa polypeptide is present in smaller amounts relative to plant and fungal mitochondria. The functional entities were named VDAC (for Voltage-Dependent, Anion-selective Channels) on the basis of their electrical characteristics: the channels are more permeable to anions than cations and switch to a lower conductance state when a potential of 20 mV or more is imposed across the phospholipid bilayer.[7,8] The anion selectivity of the outer membrane channels may have physiological relevance, for example, in the ability of mitochondria to exclude large, multivalent cations.[9] Similarly, the gating of the channels by relatively low membrane potentials may occur *in situ*. A scheme whereby outer membrane permea-

[1] C. A. Mannella and W. D. Bonner, Jr., *Biochim. Biophys. Acta* **413**, 213 (1975).

[2] C. A. Mannella, *J. Cell Biol.* **94**, 680 (1982).

[3] H. Freitag, W. Neupert, and R. Benz, *Eur. J. Biochem.* **123**, 629 (1982).

[4] G. Daum, P. C. Bohni, and G. Schatz, *J. Biol. Chem.* **257**, 13028 (1982).

[5] C. A. Mannella and W. D. Bonner, Jr., *Biochim. Biophys. Acta* **413**, 226 (1975).

[6] L. S. Zalman, H. Nikaido, and Y. Kagawa, *J. Biol. Chem.* **255**, 1771 (1980).

[7] S. J. Schein, M. Colombini, and A. Finkelstein, *J. Membr. Biol.* **30**, 99 (1976).

[8] M. Colombini, *Nature (London)* **279**, 643 (1979).

[9] Polyamines (spermine, spermidine) are more effective inhibitors of rat liver mitochondrial respiration after the outer membranes are ruptured by hypoosmotic shock or digitonin [C. A. Mannella, R. Berkowitz, N. Capolongo, *Biophys. J.* **47**, 238a (1985)]. This suggests that diffusion of these polyvalent cations through the outer membrane (presumably via the VDAC channel) may be restricted.

bility is controlled by proximity of its channels to the inner membrane surface potential has been proposed.[10]

The 30-kDa VDAC polypeptides have also been referred to as mitochondrial porins, by analogy with the class of proteins in bacterial outer envelopes which also form large, passive diffusion channels in phospholipid bilayers.[11] Bacterial porins occur naturally in two-dimensional (2D) crystalline arrays which cover the outer envelope of certain gram-negative bacteria.[12] Close-packed but nonordered arrays of pore-like subunits were first reported to occur on the outer membranes of negatively stained plant mitochondria.[13] Periodic arrays of similar looking subunits were later discovered on outer membranes isolated from mitochondria of the fungus *Neurospora crassa*.[2] The evidence that the fungal arrays are composed of the 31-kDa pore protein has recently been provided by indirect immunoelectron microscopy.[14]

The occurrence of a biological macromolecular complex in planar crystalline form opens up a variety of approaches in analysis of electron microscopic images. Computer-based Fourier or correlation averaging techniques can be used to obtain projection maps with resolution on the order of 1.5 to 2 nm for negative stain embedded specimens[15] and better than 1 nm for glucose embedded specimens imaged under extreme low dose conditions.[16] Tilt-series projection data from different crystals can be combined in Fourier or direct space algorithms to reconstruct the three-dimensional (3D) density maps of the macromolecules.[17,18]

Another useful characteristic of macromolecules which form planar crystals is that they often can be induced to crystallize in three dimensions. In recent years, there have been several successful attempts at 3D crystallization of membrane proteins, among them one of the bacterial porins.[19] If such crystals prove suitable for X-ray diffraction studies, solu-

---

[10] C. A. Mannella, *in* "Encyclopedia of Plant Physiology, New Series" (D. A. Day and R. Douce, eds.), Vol. 18, p. 106. Springer-Verlag, Berlin and New York, 1985.

[11] H. Nikaido, this series, Vol. 97, p. 85.

[12] A. C. Steven, B. ten Heggeler, R. Muller, J. Kistler, and J. P. Rosenbusch, *J. Cell Biol.* **72**, 292 (1977).

[13] D. F. Parsons, W. D. Bonner, Jr., and J. G. Verboon, *Can. J. Bot.* **43**, 647 (1965).

[14] C. A. Mannella and M. Colombini, *Biochim. Biophys. Acta* **774**, 206 (1984).

[15] See the proceedings of the international workshop "Regular 2D Arrays of Biomacromolecules," published as "Electron Microscopy at Molecular Dimensions" (W. Baumeister and W. Vogell, eds.). Springer-Verlag, Berlin and New York, 1980.

[16] P. N. T. Unwin and R. Henderson, *J. Mol. Biol.* **94**, 425 (1975).

[17] See the review by L. A. Amos, R. Henderson, and P. N. T. Unwin, *Prog. Biophys. Mol. Biol.* **39**, 183 (1982).

[18] P. C. F. Gilbert, *Proc. R. Soc. London Ser. B* **182**, 89 (1972).

[19] R. M. Garavito and J. P. Rosenbusch, *J. Cell Biol.* **86**, 327 (1980).

tion of the structure of membrane proteins may be extended to atomic resolution.

Application of the above kinds of electron microscopic studies to a particular membrane protein is contingent on its forming at least 2D crystalline arrays. In general, if regular planar arrays of the protein are not observed in the native membrane, attempts at inducing crystallization usually are made after the protein has been extracted, purified, and functionally reconstituted. The *Neurospora* mitochondrial pore protein represents an intermediate case, the study of which may have more general application. Although arrays of this protein have not (as yet) been observed in outer membranes of intact mitochondria, they are seen in highly variable numbers in outer membrane isolates after dialysis against low salt buffer.[2] The observation, that inclusion of Ca^{2+} in the dialysis buffer stimulates array formation, implicated endogenous Ca^{2+}-activated phospholipase A_2 in the crystallization process.[20] Subsequent experiments demonstrated that the mitochondrial channel arrays could be reproducibly induced by digestion of the isolated outer membranes with exogenous phospholipase A_2.[21] The detailed procedure for inducing ordering of the *Neurospora* mitochondrial channel protein by dialysis in the presence of phospholipase A_2 is described below, along with a summary of the structural characteristics of the arrays.

Preparation of Crystalline Arrays of Mitochondrial Channels

Isolation of Neurospora Mitochondrial Outer Membranes

Principle. Outer mitochondrial membrane fractions are obtained by hyposmotic swelling of isolated mitochondria, followed by centrifugation on sucrose step gradients, the strategy first applied by Parsons and coworkers to liver mitochondria.[22] The percentage yield (based on marker enzyme recovery) of outer membranes by this procedure is not as good with *N. crassa* mitochondria ($\sim 10\%$) as it is with other types of mitochondria (e.g., typically 30% with higher plant mitochondria[23]). However, techniques which might increase outer membrane yield (sonication, digitonin treatment) are avoided because of undesired side effects (increased inner membrane contamination, modification of outer membrane composition, vesiculation of the outer membranes[10,23]).

[20] C. A. Mannella and J. Frank, *Biophys. J.* **37,** 3 (1982).
[21] C. A. Mannella, *Science* **224,** 165 (1984).
[22] D. F. Parsons and G. R. Williams, this series, Vol. 10, p. 443.
[23] R. Douce, C. A. Mannella, and W. D. Bonner, Jr., *Biochim. Biophys. Acta* **292,** 105 (1973).

Reagents

Isolation medium: 0.3 *M* sucrose, 1 m*M* EDTA, 0.3% bovine serum albumin, adjusted to pH 6.8 with NaOH.

Lysis medium: 0.25 m*M* EDTA, 0.25 m*M* EGTA, adjusted to pH 6.8 with NaOH.

Sucrose step solution: 0.80 *M* sucrose, 10 m*M* potassium phosphate buffer (pH 7.0), 0.25 m*M* EDTA, 0.25 m*M* EGTA.

Procedure. Mitochondria are isolated by differential centrifugation from *Neurospora crassa* ("slime" mutant, FGSC 326, grown in liquid culture) according to previously described procedures.[2] In the course of differential centrifugation, the high speed mitochondrial pellets have 3 layers, the dark-brown mitochondrial layer sandwiched between a viscous, tan overlay and a small, clear pellet at the bottom. The overlay (which contains fibrous cell debris) separates easily from the mitochondrial layer and can be removed during washes with a long-stem Pasteur pipet. As each mitochondrial layer is resuspended, care must be taken to avoid the clear pellet underneath, which contains polysomes and glycogen. (Glycogen particles will contaminate the final outer mitochondrial membrane fractions if the clear pellets are disturbed.)

The final mitochondrial pellets are gently resuspended with 1.0 ml isolation medium and the packed volume (= total volume − 1.0 ml) is noted. The mitochondrial suspension is then warmed to 35° for 5 min (to preswell the mitochondrial matrix space) and added rapidly to a beaker containing ice-cold lysis medium (40 ml for every 0.2 ml packed mitochondrial volume or fraction thereof). The lysate is centrifuged [3000 rpm (1000 *g*) 10 min, 4°, Sorvall SS34 rotor] and the supernatant is homogenized (5 strokes with a Teflon/glass homogenizer) and stirred briskly for an additional 25 min. Solid mannitol (2 g/40 ml lysate) is added to the suspension and stirred until dissolved. The suspension is then layered atop 0.80 *M* sucrose steps (40 ml suspension + 15 ml sucrose step per tube) and centrifuged in a swinging bucket rotor [Beckman SW25.2 rotor, 19,000 rpm (60,000 *g*) 90 min, 4°]. Following centrifugation, outer membranes which have lysed and separated from mitochondria are stopped by the sucrose step, appearing as a faint, colorless, light-scattering band. The mitochondrial inner membranes, along with the outer membranes which have not separated from them, pass through the 0.80 *M* sucrose step and form a dark-brown pellet at the bottom of the tube. Each outer membrane band is routinely collected in about 8 ml with a long-stem Pasteur pipet, although a fractionator (like the Isco Model 185) can be used for quantitative recovery of the bands.

Phospholipase A₂ Treatment: Induced Ordering of Membrane Proteins

Principle. Dilute suspensions of isolated mitochondrial outer membranes are incubated with low levels of soluble phospholipase A_2 during continuous dialysis against low salt buffer. The result is the formation of extended ordered arrays of the outer membrane channel protein.

Materials

Low salt buffer: 1 mM Tris–HCl, 0.1 mM EDTA, pH 7.5.

Phospholipase A_2 (phosphatide 2-acylhydrolase, EC 3.1.1.4): bee venom phospholipase A_2 (lyophilized, salt-free, Sigma Chemical Co.), 1 mg protein (~1500 units) dissolved in 2 ml glycerol/water (1 : 1 v/v). This solution is stored at −20° for periods up to 8 weeks, the glycerol preventing ice crystallization. One unit of the enzyme will hydrolyze 1 μmol of lecithin per min at 37° and pH 8.5. *Caution:* The lyophilized bee venom phospholipase A_2 is extremely allergenic.[24]

Dialysis tubing: standard cellulose tubing (MW cutoff ~10,000). Prior to use it is boiled for 20–30 min in 50 mM Na_2CO_3, 1 mM sodium EDTA (pH 7), then rinsed extensively with distilled, deionized water, in which it is stored.

Procedure. The outer mitochondrial membrane fractions collected following step gradient centrifugation are pooled and diluted with 3–4 volumes of low salt buffer. The final concentration of membrane protein is typically in the range 1–10 μg/ml. The outer membrane fraction is divided into several (2–4) equal aliquots and each is pipetted into a dialysis bag. A different activity of phospholipase A_2 is then added to each tube (e.g., 0, 0.3, 0.6, 1.2 units/ml). The tubes are sealed and stirred slowly in a large volume of low salt buffer at 4° for 15–20 hr. The contents of each tube are then transferred to a centrifuge tube and the membranes are pelleted (100,000 g, 90 min, 4°, SW27 rotor). The faintly yellow, transparent pellets are resuspended with a plastic-tipped pipettor in a small volume (100–250 μl total) of low salt buffer. These membrane suspensions can be used directly for making gel electrophoresis or negative-stain electron microscopic specimens.[2]

Comments. The soluble, bee venom phospholipase A_2 was chosen for these studies on the basis of low cost and ease of use (e.g., the activity of this enzyme is not dependent on the phospholipids being near their liquid-crystal/gel transition temperature, as with the pancreatic enzyme).[24] The

---

[24] R. C. Cottrell, this series, Vol. 71, p. 698.

bee venom enzyme is active on long chain phosphatidylcholines and phosphatidylethanolamines,[24] which comprise two-thirds of the phospholipids of the *N. crassa* mitochondrial outer membranes.[25] The bee venom enzyme is activated by Ca^{2+} and inhibited by other divalent cations, such as Mg^{2+}, Ba^{2+}, and Sr^{2+}.[24] In our procedure, Ca^{2+} is omitted from the dialysis buffer because, at recommended levels (0.1 to 1.0 mM), it causes aggregation of the outer mitochondrial membranes. The commercially available enzyme apparently contains tightly bound Ca^{2+} and is not significantly inhibited by the low level of EDTA (0.1 mM), which is included in the low salt buffer to prevent membrane aggregation and to chelate inhibitory divalent cations, which may be present at low levels.

Buffers besides Tris–HCl (1 and sometimes 10 mM, pH 7.5 at 4°) have occasionally been used during dialysis of the mitochondrial outer membranes in the presence of phospholipase A_2. Use of potassium phosphate (10 mM, pH 7.5) leads to aggregated membranes (perhaps due to the presence of divalent metal contaminants) and so it and other mineral buffers are avoided at this and subsequent steps. (In general, chemical or physical treatments which induce aggregation of the membranes inhibit phospholipase A_2-induced crystallization of the pore protein.) There are preliminary indications that substitution of zwitterionic Good buffers (10 mM MOPS or TES, pH 7.5) for Tris in the dialysis medium impedes the crystallization process. In addition, we have noted that prolonged preexposure of the membranes to MOPS (10 mM, pH 6.5) makes them somewhat resistant to the action of phospholipase A_2.

Precise determination of the starting membrane concentration is not routinely made each time the phospholipase reaction is run. When small specimens are required (as for electron microscopy), it is more convenient to run parallel reactions with varying enzyme-to-membrane ratios. Better control of the reaction conditions (for large-scale crystalline membrane preparations) would be achieved by maintaining a constant ratio of phospholipase A_2 activity to membrane concentration, which would require assaying (1) the specific activity of the particular enzyme preparation[24] and (2) the protein or lipid concentration of the membrane suspension.

Running the phospholipase reactions slowly, on dilute membrane suspensions, is essential for obtaining clean, high yields of the crystalline membranes. Increasing the temperature, increasing component concentrations, or omitting dialysis all result in membrane fractions with more amorphous vesicles and fewer crystalline membranes. These results point to the importance of allowing the products of phospholipid hydrolysis to

[25] G. Hallermeyer and W. Neupert, *Hoppe-Seylers Z. Physiol. Chem.* **355,** 279 (1974).

dialyze out of the reaction chamber before they can reach critical micellar concentration or can reinsert into other membranes.[26]

Characteristics of the Phospholipase A₂-Treated Membranes

Membrane Protein Content. The SDS–polyacrylamide gel electrophoretic patterns of outer mitochondrial membranes dialyzed overnight in the presence (lanes a–c) or absence (land d) of phospholipase A₂ are illustrated in Fig. 1. Following enzyme treatment, there is generally a higher recovery of the 31-kDa pore protein (60–70%) than of the other consistently observed polypeptides in the outer membrane fractions (20–40%). (Recoveries of each polypeptide are estimated by microdensitometry of bands in photographic negatives of silver-stained SDS–polyacrylamide gels.[21]) The recovery of VDAC channel-forming activity in the same mitochondrial outer membrane fractions following phospholipase A₂ treatment is ~50% (bilayer assays[7,8] performed by M. Colombini, University of Maryland), consistent with the recovery of the 31-kDa polypeptide.

Membrane Ultrastructure. Most of the structural characterization of *N. crassa* mitochondrial outer membranes, before and after phospholipase A₂ treatment, has been obtained by negative stain electron microscopy. So-called negative stains are aqueous solutions of heavy-metal salts which form thin, electron-dense, amorphous films when air dried. Biological particles embedded in these thin "glasses" are visualized primarily (but not exclusively, see below) by negative contrast in electron microscopic projection images. The stain tends to form casts of the biological particles, so that the electron intensity distribution in projection images is modulated primarily by the surface relief of the particles.

Several variations on the theme of negative staining techniques have been described in this series previously.[27,28] The procedure we use is based on that described by Munn.[28] A small aliquot (5 μl) of membrane suspension (0.2–1 mg protein/ml low salt buffer) is mixed with an equal volume of negative stain (1% uranyl acetate) on a carbon/Formvar-coated specimen grid, which has been glow discharged to render it hydrophilic.

---

[26] For the membrane concentration range defined (1–10 μg protein/ml = 1.1–11 μg phospholipid/ml), complete hydrolysis of the membrane phospholipids would release 10⁻⁵ to 10⁻⁶ M fatty acids and lysophospholipids. The range of critical micelle concentrations is 10⁻² to 10⁻⁴ M for the expected free fatty acids (P. Mukerjee and K. J. Mysels, *in* "Critical Micelle Concentration in Aqueous Surfactant Systems." National Bureau of Standards, Washington, D.C., 1971) and 10⁻³ to 10⁻⁵ M for the lysophospholipids (A. Helenius, D. R. McCaslin, E. Fries, and C. Tanford, this series, Vol. 56, p. 734).

[27] D. F. Parsons, this series, Vol. 10, p. 655.

[28] E. A. Munn, this series, Vol. 32, p. 20.

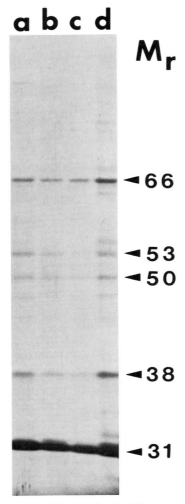

Fig. 1. Electrophoretic patterns (silver-stained) of *N. crassa* mitochondrial outer membrane proteins. The membranes in the specimens loaded in lanes a–c were treated with phospholipase A_2 by the procedure described in the text and contained 70–80% crystalline membranes. The specimen loaded in lane d was made from the same membrane isolate as that of lane c prior to phospholipase treatment. Its crystalline membrane content was less than 5%. The major outer membrane polypeptides are indicated by arrows at right, along with their apparent molecular weights ($\times 10^3$). The band at M_r 66,000 is bovine serum albumin, a component of the mitochondrial isolation medium. The intensity of this band can be greatly reduced by omitting the albumin from the final mitochondrial suspending buffer. The procedure for making the specimens and electrophoresing them on 12% polyacrylamide gels containing 0.1% sodium dodecyl sulfate have been described in detail previously.[2]

After 1–2 min, excess liquid is blotted from the grid, which is allowed to air dry for several minutes prior to examination in electron microscope.

There are striking changes in the shape and substructure of mitochondrial outer membranes after treatment with phospholipase A_2. The typical appearance of negatively stained membranes in control specimens, i.e., fractions dialyzed in the absence of added enzyme, is represented by the membrane of Fig. 2A. These are collapsed, round vesicles, 0.5–1.5 μm across, with irregular outlines and ripped and ruffled surfaces. There is no obvious periodic structure in the membrane plane, confirmed by the absence of periodic maxima in the optical diffraction from the image.

After exposure to phospholipase A_2, membranes like that of Fig. 2B can be seen. They are somewhat smaller than the original membranes and less irregular in outline. More significantly, the membranes begin to show signs of ordering of components, i.e., the appearance of small (100 nm diameter or less) patches of regular arrays of subunits in the membrane plane. This "polycrystalline" substructure is confirmed by the occurrence of broad maxima in the optical diffraction from these membranes.

After longer phospholipase A_2 treatment, membranes like those of Fig. 2C and D are the most common, in which most or all of the membrane surface is composed of one or a few periodic arrays. These arrays have previously been shown to be composed of the pore-forming polypeptide of the *N. crassa* mitochondrial outer membrane,[14] consistent with the selective recovery of this protein in sedimentable outer membranes after phospholipase A_2 treatment (Fig. 1). The crystalline membrane of Fig. 2C is a type often seen, a collapsed tube or cylinder, well ordered where the edges are straight and disordered at the rounded ends. The diffraction patterns from such membranes show discrete reflections on two superimposed and flipped parallelogram lattices, corresponding to the diffraction from an oblique array wrapped around a flattened tube. Average lattice parameters are $a = 12.6$ nm, $b = 11.1$ nm, $\theta = 109°$, with the short (b) lattice vector always parallel (or nearly so) to the straight, long axis of the membrane tube. With increased phospholipase A_2 treatment, a different array tends to predominate in the outer mitochondrial membrane fractions, a rectangular array with unit cell dimensions $a = 8.7$, $b = 5.1$ nm. The associated membranes have one or more straight edges and are usually open, folded sheets, unlike the closed tubes associated with the parallelogram lattice type. That the parallelogram (P) array is the precursor to the rectangular (R) array is strongly suggested by plotting the fractions of the two array types, observed in different outer mitochondrial membrane preparations, against the fraction of total crystalline membranes (Fig. 3). When the crystalline membrane content is below 50%, they are essentially all P arrays. However, as the crystalline membrane content in-

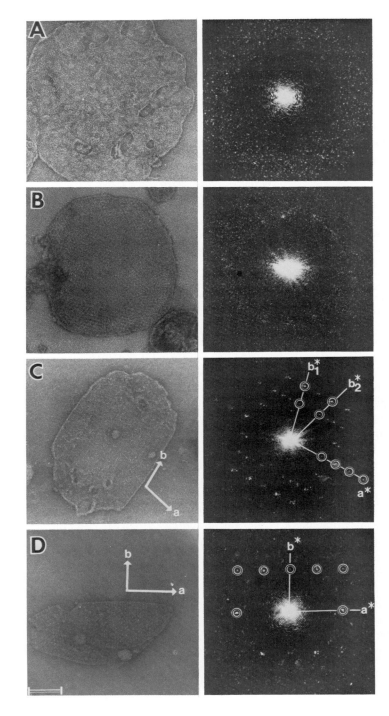

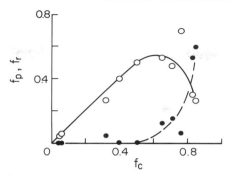

FIG. 3. Pore protein array type as a function of crystalline membrane content in *N. crassa* mitochondrial outer membrane specimens. Electron microscopic images were recorded from random fields of negatively stained membranes from 10 different preparations. In each case, 30–40 membranes were scored on the basis of their optical diffraction patterns as amorphous or crystalline and, if crystalline, as P or R arrays. f_c, f_p, and f_r are the contents of crystalline membranes, P arrays, and R arrays, respectively, measured as a fraction of the total number of membranes recorded. (Weakly diffracting, polycrystalline membranes were scored as amorphous.)

creases, the number of P arrays decline at the same rate at which the R arrays appear.

Stability of the Membranes. After isolation, native mitochondrial outer membranes (0.1–1 mg protein/ml) can be stored at 4° in low salt buffer for periods up to 1 week before they begin to aggregate. The phospholipase A₂ reaction can be run at any time within this period, although best crystalline membrane yields are generally obtained with freshly iso-

FIG. 2. *N. crassa* mitochondrial outer membranes in progressive stages of planar crystallization. Left: electron microscopic images of uranyl acetate embedded membranes. (Micrographs were recorded on Kodak S0163 film using a Philips EM301 operated at 100 kV. Bar at lower left is 100 nm.) Right: optical diffraction patterns recorded from the original electron image negatives. (Diffraction patterns were produced by illuminating a circular field on the negatives, 1–2 cm in diameter, with coherent radiation from a Jodon He-Ne laser. The patterns were recorded with a 35 mm camera using Kodak Pan-X film.) (A) Amorphous outer mitochondrial membrane. (B) Polycrystalline membrane. (C) Cylindrical outer membrane tube composed of pore protein in a single parallelogram, P, array. The principal axes of the two superimposed reciprocal lattices are indicated in the optical diffraction pattern. (The two a* directions coincide in this example, although this is not usually the case.) The directions of the real-space lattice vectors corresponding to reciprocal lattice 1 are shown at lower left in the micrograph. (D) Folded outer membrane sheet composed of rectangular, R, array. The lattices of the two overlapped membrane layers appear to exactly coincide, at least in the region on the right side of the membrane from which the pattern was recorded. Again, the principal reciprocal lattice axes and the directions of the real-space lattice vectors are indicated.

lated membranes. The crystalline membranes have a similar tendency to aggregate over a period of about a week when stored in concentrated suspensions in low salt buffer. Negatively stained specimens for serious electron microscopic studies are generally made from crystalline membrane fractions within the first 2 days after phospholipase treatment.

After the phopholipase-treated membranes have been pelleted and resuspended in low salt buffer, there is sometimes a slow conversion of the arrays from P to R geometry (over a period of several days). This indicates that low levels of phospholipase A_2 may be trapped in the pellets and resuspended with the membranes. Preliminary experiments indicate that the P to R transition may be prevented by centrifuging the phospholipase-treated membranes through a sucrose gradient prior to final pelleting.

Mechanism of Crystalline Array Formation

Slow Removal of Phospholipid. The number of crystalline membranes observed in outer mitochondrial membrane fractions after controlled phospholipase A_2 treatment is about the same as the number of large, amorphous vesicles present initially. This suggests that each array forms from the pore protein present in an individual native membrane. Simple incubation of the isolated outer mitochondrial membranes with either lysolecithin (0.1 mM) or glycerol monooleate (0.2–2 mM) under similar conditions does not induce crystallization of the membrane components. Thus, ordering of the protein components is due directly to the action of phospholipase A_2 on the membranes and not to indirect effects of the products of bulk phospholipid hydrolysis.

The simplest mechanism to account for the ordering of the pore protein after phospholipase treatment is that it occurs spontaneously as phospholipids are slowly released upon hydrolysis and the proteins gradually pack closer together in the membrane plane. Actual lipid analyses of the different types of mitochondrial outer membrane arrays are in progress. There are preliminary indications that phospholipase A_2 treatment of the native membranes causes loss of significant amounts of phospholipid. Not only do the membranes appear smaller after treatment, they are markedly denser. Native outer membranes from *N. crassa* mitochondria are composed of protein, phospholipid, and sterol in the ratios $11:12:2$ by weight[25] and display an apparent buoyant density of 1.081 g/cm^3 during sucrose gradient centrifugation.[2] After phospholipase treatment, the P and R arrays show apparent buoyant densities in the ranges 1.094–1.120 and 1.120–1.147 g/cm^3, respectively. Assuming the protein-to-cholesterol ratios of the membranes remain constant, the increased buoyant densities correspond to loss of 19–46% of the original phospholipid in forming the P arrays and 46–67% of the phospholipid in forming the R arrays.

The Role of Surface Charge. Experiments with phospholipases C and D (Mannella, unpublished) indicate that crystallization of pore protein in the mitochondrial outer membrane is effected by the polar headgroups of the phospholipids in the membranes. Zwitterionic phospholipids are hydrolyzed to neutral diacylglycerols by phospholipase C and to phosphatidic acid by phospholipase D. Dilute suspensions of crystalline outer membranes (prepared by the phospholipase A_2 treatment described above) were redialyzed (18 hr, 4°) against low salt buffer, after addition to the suspensions of phospholipase C (30 units of the *Clostridium* enzyme/ml) or phospholipase D (30 units of the peanut enzyme/ml). Optical diffraction patterns from electron microscopic images of negatively stained membranes indicate that both phospholipases C and D cause gradual loss of long-range order in the planar crystals. Since the disordering is induced by both phospholipases, the effect is probably related to decreased zwitterion concentration at the membrane surface. That electrostatic interactions among the membrane components is involved in the crystallization process is suggested by another observation, namely that dialysis of the pore protein arrays against 0.5 N NaCl causes loss of long-range order. This is expected if electrostatic interactions between fixed surface charges (screened in high ionic strength buffer) are important for maintaining crystalline order.

Structure of the Mitochondrial Channel Arrays

Array Geometries; Location of Phospholipid; Channel Triplets. In mitochondrial outer membrane preparations in which the crystalline membrane content exceeds 50%, parallelogram arrays occur with lattice angles smaller than 109°, the lattice angle of the usually observed P array. These novel parallelogram arrays appear to represent intermediate stages in the transition between the P and R arrays.

Fourier filtered projection images of the three kinds of mitochondrial channel arrays are shown in Fig. 4. The P array (Fig. 4A) has 6 dense stain loci per unit cell. The stain loci are arranged in hexagons with a 2-fold rotation axis in the center, giving the array p2 (or near p2) projection symmetry. There are stain-free regions at the four corners of each unit cell which decrease in size as the lattice angle decreases (going from the P array to the intermediate array of Fig. 4B) and which have essentially disappeared in the R array (Fig. 4C). These regions of the P array are likely composed of the phospholipids which are lost in the transition to the R array. The geometry of the R array is significantly different from the

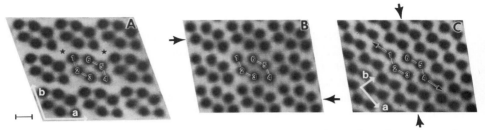

FIG. 4. Fourier filtered images of mitochondrial outer membrane channel arrays; transition from P to R arrays. Images recorded from phosphotungstate-stained membranes (as described in legend to Fig. 2) and Fourier filtered by the quasioptical procedure described in detail by Mannella and Frank.[37] (A) P array. Unit cell vectors defined by white arrows at lower left. Bar is 5 nm. The stain-filled channels are arranged into hexagonal groups of six, and these groups are organized on a parallelogram lattice. Large stain-free areas occur at the corners of the unit cells, two of which are marked by stars. (B) Intermediate array. The horizontal rows have slid 1 nm in the directions indicated by the large arrows, and the distance between adjacent rows has decreased by 0.3 nm, decreasing the size of the stain-free regions. (C) R array. The distance between the horizontal rows has decreased by another 0.4 nm (along the direction of the large arrows) and the two channel triplets in each hexagonal group have split apart by 0.8 nm (in the direction of the small arrows). The unit cell vectors of this array are at lower left. (Figure adapted from Mannella *et al.*[29])

P array. It has higher projection symmetry (p2mg) and a smaller unit cell, large enough to hold but two stain loci. However, there is a group of stain loci common to all the arrays: the triplet labeled 1–2–3 (or 4–5–6) in Fig 4A–C. Assuming each dense locus is a stain-filled channel (see below), this suggests that a channel triplet is the fundamental structural unit of the arrays. This is consistent with observations that functional channels insert from these membranes into phospholipid bilayers in groups of three and multiples of three.[29]

Size of the Channel. Low-resolution 3D reconstruction of the outer mitochondrial membrane P array indicates that the dense stain loci are stain-filled cylinders, normal to the membrane plane, consistent with transmembrane channels.[30] Unlike the channels formed by the *E. coli* porin protein OmpF, which fuse in the plane of the membrane,[31] the mitochondrial channels are independent. Projection maps of uranyl ace-

[29] C. A. Mannella, M. Colombini, and J. Frank, *Proc. Natl. Acad. Sci. U.S.A.* **80**, 2243 (1983).

[30] C. A. Mannella, M. Radermacher, and J. Frank *in* "Proceedings of the 42nd Annual Meeting of the Electron Microscopy Society of America" (G. W. Bailey, ed.), p. 644. San Francisco Press, San Francisco, 1984.

[31] A. Engel, D. L. Dorset, A. Massalski, and J. P. Rosenbusch *in* "Proceedings of the 41st Annual Meeting of the Electron Microscopy Society of America" (G. W. Bailey, ed.), p. 440. San Francisco Press, San Francisco, 1983.

tate-embedded mitochondrial pore protein arrays indicate a channel diameter of 2.0 to 2.5 nm, close to the bore of the best-fit channel model calculated from the X-ray scattering data from the plant mitochondrial outer membrane, 1.8–2.0.[32] This value for the channel diameter is consistent with that expected from the electrical step conductance of VDAC, 2 nm.[3] However, these values are considerably smaller than that based on liposome swelling experiments, 4 nm, obtained using reconstituted *N. crassa* VDAC with polyethylene glycols of varying molecular weights as the solutes.[33]

Number of 31-kDa Polypeptides per Channel. When isolated with Triton X-100, the rat liver VDAC channel protein displays an apparent molecular weight above 100,000 by gel filtration.[34] Sedimentation experiments indicate that only about 60-kDa of this protein–detergent complex is protein, suggesting the presence of a 31-kDa dimer.[35]

In the case of *N. crassa,* an estimate can be derived for the average protein mass associated with each stain-filled channel in the P array from the structural parameters of the array (obtained by electron microscopy) and its chemical composition (inferred from the composition of the native membranes and the buoyant density changes after phospholipase treatment). There is 1 channel per 22 nm² of surface area in the VDAC array. Given that the membrane is 5–7 nm thick,[5] that the channel bore is 2–2.5 nm, that the P array has about 40% less phospholipid than the native membrane, and that the partial specific volume (calculated from the amino acid composition[3]) of the *N. crassa* pore protein is 0.74 cm³/g, one can calculate that each channel in the array is associated with 35–63 kDa of protein. This result suggests that each *N. crassa* channel is no larger than a dimer of the 31-kDa polypeptide, and may even be a monomer.

Fixed Charges at the Channel. A systematic electron microscopic study of the staining characteristics of the *N. crassa* pore protein arrays suggests that there are numerous fixed charges which are located in the vicinity of the channel openings and which are positively stained by ionic "negative" stains.[36,37] This conclusion is consistent with a large percentage (25%) of acidic and basic amino acids in the *N. crassa* pore protein.[3] In addition, difference images calculated for Fourier filtered images of control and succinylated channel arrays suggest the presence of one prominent basic amino acid cluster per channel, in the immediate vicinity of the opening.[36] Succinylation of the rat liver pore protein reverses the

[32] C. A. Mannella, *Biochim. Biophys. Acta* **645**, 33 (1981).
[33] M. Colombini, *J. Membr. Biol.* **53**, 79 (1980).
[34] M. Colombini, *Ann. N.Y. Acad. Sci.* **341**, 552 (1980).
[35] M. Linden and P. Gellerfors, *Biochim. Biophys. Acta* **736**, 125 (1983).
[36] C. A. Mannella and J. Frank, *Biophys. J.* **45**, 139 (1984).
[37] C. A. Mannella and J. Frank, *Ultramicroscopy* **13**, 93 (1984).

ion selectivity of the channels and abolishes their voltage dependence.[38] Thus, the basic amino acid clusters detected in the difference images of the channel arrays may be involved in the mechanisms of ion selectivity and/or voltage gating of this channel.

[38] C. Doring and M. Colombini, *Biophys. J.* **45,** 44 (1984).

[47] Reconstitution of ADP/ATP Translocase in Phospholipid Vesicles

By REINHARD KRÄMER

The purposes for reconstituting a transport system can be 2-fold: (1) reconstitution may serve as the only assay for identification of the carrier protein, and (2) reconstitution is an ideal tool for elucidating the molecular function of a carrier protein. The last reason in particular is the one that really justifies the use of reconstitution methods. Several mitochondrial carriers have so far been incorporated into phospholipid vesicles. However, only in the case of the nucleotide carrier[1,2] and the phosphate carrier[3] have the requirements for reconstitution been fulfilled in the classical sense, which necessarily involves purification of the protein before incorporation into the membrane. The ADP/ATP translocator in particular is an example for the importance of functional reconstitution in elucidating the parameters which may modulate and regulate the transport reaction. In part due to the results obtained in the reconstituted system, this carrier is one of the best understood transport systems. Therefore it seems reasonable to describe in the following not merely the basic technical problems concerning the reconstitution of this membrane protein, but also some of the factors that have to be considered when the standard procedure is extended or changed for special purposes.

Isolation of Reconstitutively Active Adenine Nucleotide Carrier

Solubilization and purification of the ADP/ATP translocator as carboxyatractylate–protein complex has been described previously.[4] For

[1] R. Krämer and M. Klingenberg, *FEBS Lett.* **82,** 373 (1977).
[2] R. Krämer and M. Klingenberg, *Biochemistry* **18,** 4209 (1979).
[3] H. Wohlrab, H. V. J. Kolbe, and A. Collins, this volume [55].
[4] M. Klingenberg, H. Aquila, and P. Riccio, this series, Vol. 56, p. 407.

isolation of this protein in a reconstitutively active form, the nucleotide carrier must not be liganded with carboxyatractylate, since this tightly binding inhibitor cannot be easily removed. On the other hand, application of the basic procedure without added inhibitor leads to a protein, which—although highly pure—shows only low activity in the reconstituted system. The following procedure is thus designed to speed up purification in order to minimize denaturation of the carrier protein which is very labile in the solubilized form without bound inhibitors.

Solubilization. Beef heart mitochondria,[5] freshly prepared or from stock kept in liquid nitrogen, are rapidly added to the solubilizing buffer (final concentrations: 10 mg protein/ml, 3% Triton X-100, 150 mM Na$_2$SO$_4$, 0.5 mM EDTA, 20 mM Tricine–NaOH, pH 8.0). Effective solubilization is performed by repeated sucking through a pipet for 10 min at 0°.

Several classes of detergents have been tested for solubilization of the nucleotide carrier. In general, ionic detergents (e.g., bile salts), zwitterionic detergents (e.g., betaines, CHAPS), and nonionic detergents with high critical micellar concentration (octylglucoside, glucamides) are destructive to the carrier. Triton X-100 and laurylamidodimethylpropylamine oxide proved to be both effective and not too harmful to the protein[2,6]; recently, also dodecyloctaethylene glycol ether has been found to be suitable.[7] Low ionic strength results in poor yield of solubilized carrier protein and too high salt concentration leads to insufficient purification on hydroxyapatite.

Purification. As already described in detail,[4] advantage is taken of the specific nonadsorbance of this protein to hydroxyapatite under special conditions. Since the protein is very labile in the solubilized state, the purification procedure has to be fast. It is carried out in batch, thereby sacrificing yield but gaining high specific activity. We prepare hydroxyapatite ourselves,[2] but commercially available hydroxyapatite can also be used. If optimum purity of the protein is desired, the capacity of every batch of hydroxyapatite for adsorption of the ADP/ATP translocator has to be tested, as described below.

Hydroxyapatite is washed three times in distilled water and sedimented by centrifugation in reaction vessels (1.5 ml) to give a bed volume of 500–800 μl, depending on the capacity of the hydroxyapatite used for purification. The supernatant is removed; 500 μl of the solubilized mitochondria is then added to the reaction vessel at 0°. After thorough mixing

[5] H. Löw and I. Vallin, *Biochim. Biophys. Acta* **69**, 361 (1963).
[6] R. Krämer, H. Aquila, and M. Klingenberg, *Biochemistry* **16**, 4949 (1977).
[7] R. Krämer, *FEBS Lett.* **176**, 351 (1984).

with the hydroxyapatite by agitation with a small stirrer (by hand), an optically homogeneous light brown material is obtained. After 5 min incubation (0°) the reaction vessel is centrifuged for 5 min at 20,000 g in the cold. Higher centrifugal forces or longer times do not change the results. The supernatant contains about 5–6% of the applied solubilized protein which means, however, that the protein yield actually amounts to 10–12%. About 50% of solubilized carrier remains in the hydroxyapatite sediment due to the batch procedure. The supernatant is then immediately used for reconstitution (see below).

The purity of the protein can be tested by two methods. (1) The protein is isolated under identical conditions using [³H]carboxyatractylate-labeled mitochondria.[2] A pure carrier preparation would have to show binding of about 16 μmol [³H]carboxyatractylate/g protein. This method will not be described in detail here, since the labeled inhibitor is at present not commercially available. (2) The supernatant is analyzed by SDS–gel electrophoresis. The SDS–gel pattern should not contain significant protein bands with molecular weights higher than 32,000. If there are definite amounts of impurities in this region, the ratio of protein/hydroxyapatite should be lowered. Impurities in the lower molecular weight region cannot be avoided completely in this batch procedure. A protein band of about 28,000 molecular weight, however, which is routinely observed in this preparation in small amounts, has been identified as a proteolytic cleavage product of the adenine nucleotide carrier under these conditions.[8] On the other hand, a poor yield of the protein (i.e., below 4% of the applied material) indicates that the ratio of protein/hydroxyapatite should be increased.

The purity of the carrier protein obtained in our procedure, as measured by [³H]carboxyatractylate binding, is about 70–75%.[9] Using a modified and much more complicated procedure, purification can also be carried out with the ADP/ATP translocator stabilized by atractyloside (not carboxyatractyloside!). This leads to a virtually pure protein which is also reconstitutively active after removal of the inhibitor.[2] This method is important as proof for the identity of the ADP/ATP carrier; however, it does not lead to increased specific activity of the reconstituted carrier protein and is therefore not recommended.

Stability. Since the stability of the reconstituted protein is a crucial point, some further remarks are made on parameters found to influence the stability of the ADP/ATP translocator in the solubilized state. Four factors are of main importance. (1) The temperature must be kept at 0° until the protein is incorporated into the membrane. (2) High ionic

[8] H. Aquila, W. Eiermann, W. Babel, and M. Klingenberg, *Eur. J. Biochem.* **85**, 549 (1978).
[9] R. Krämer and M. Klingenberg, *Biochemistry* **16**, 4954 (1977).

strength is necessary for solubilization, however, not all types of salts are suitable. Na_2SO_4 proved to be the best of all salts tested, the addition of chlorides and phosphates should be avoided. (3) Although other detergents (see above) may be used for special purposes, Triton X-100 proved to be the best for isolation of active nucleotide carrier. (4) The protein is stabilized to a great extent by the presence of phospholipids. However, addition of phospholipids in amounts high enough to stabilize effectively against denaturation prevents further purification and can therefore not be used.

The relative instability of the solubilized protein also makes it impossible to remove the bound detergent at this stage. Removal of Triton X-100 (and phospholipids) leads to complete inactivation of the ADP/ATP translocator.[2]

Reconstitution

The freeze/thaw/sonication method for reconstitution is now a very common procedure and has already been described in this series.[10] Thus only the specific features of this method applied to the ADP/ATP translocator are dealt with in detail here.

Incorporation. The solubilized adenine nucleotide carrier is mixed rapidly with preformed sonicated liposomes (50–80 mg egg yolk phospholipids/ml water) at 0°.[11] The amount of protein that can be incorporated is restricted by the fact that the solubilizing detergent is still present and the concentration of Triton X-100 has to be kept below 5% of total lipids (w/w).[1] After incorporation the adenine nucleotide carrier is reasonably stable, i.e., it shows a half-life time of a few days as compared to about 1 hr in the solubilized state.[2]

If necessary, Triton X-100 can now be removed easily.[2] This definitely improves the stability and tightness of the liposomes for ions,[12] but does not increase the transport activity of the reconstituted carrier protein and is therefore not described in detail here.

Freeze/Thaw/Sonication Procedure. Upon incorporation the ADP/ATP translocator recovers its original affinity and specificity for inhibitor binding and substrate interaction, but does not catalyze adenine nucleo-

---

[10] E. Racker, this series, Vol. 55, p. 699.

[11] Phospholipids were extracted from egg yolks using chloroform/methanol according to M. A. Wells and D. I. Hanahan, this series, Vol. 14, p. 178. The crude egg yolk lipids which still contain some cholesterol (1–2%) are quite suitable for this reconstitution. If a well-defined phospholipid composition is desired, the lipids can be separated into pure phosphatidylcholine and phosphatidylethanolamine on silicic acid columns according to G. B. Ansell and J. N. Hawthorne, "Phospholipids." Elsevier, Amsterdam, 1964.

[12] R. Krämer, unpublished results.

tide exchange. For reconstitution of exchange activity a sonication step is necessary.[1] Although this leads to activation of the incorporated carrier molecules, the preparation obtained is not suitable for kinetic measurements, because of the small vesicles (25–35 nm diameter) produced by this method. Thus, before sonication an additional freeze/thaw step is carried out after incorporation of the protein. Freezing may be rapid (dry ice/acetone or liquid nitrogen), whereas thawing should take place in an ice/water bath, so that it will take about 15 min. Presence of sucrose should be avoided in this step. In contrast to reconstitution procedures published for several other proteins, the freeze/thaw cycle does not lead to improved protein incorporation in the case of the ADP/ATP translocator; it is, however, necessary to enlarge the size of the liposomes, thus making them more suitable for kinetic measurements. The sonication procedure is a crucial step. The optimum in sonication time is rather narrow, since the optimal curve is actually a superposition of the positive effect of sonication on the one hand, leading to reconstitutively active lipid vesicles, and the negative effect of this procedure on the other hand, causing inactivation of the protein.

Sonication is routinely performed with a sample volume of 1 ml in reaction vessels (1.5 ml) kept in an ice/water bath. We use a Branson sonifier B-30 with microtip (output power at position 4) in a pulse mode (20% duty). When sonicating liposomal suspensions containing 50–60 mg phospholipid/ml, the sonication optimum is between 40 and 60 sec total sonication time, i.e., 8–12 sec net sonication time. In the case of reduced volume or reduced phospholipid concentration, the sonication time should be shorter. It is recommended, when changing the reconstitution conditions, to optimize this critical step again.

At this stage of reconstitution, not only the size of the vesicles becomes defined, but also the composition of the internal space. After thawing, the vesicles break open and all ions and substrates which one wants to be present in the interior of the vesicles after sonication now have to be added. The activity to be reconstituted is a strict counterexchange of adenine nucleotides, i.e., it depends on the presence of internal substrate, thus 30 mM ADP or ATP is added together with buffer, usually 20 mM Tricine, pH 7.5.

Removal of External Substrates. In order to measure exchange kinetics with externally labeled nucleotides, the unlabeled external ADP or ATP present after the sonication step first has to be removed. The best way to do this is by gel filtration on Sephadex G-75 (Pharmacia). This procedure is superior to removal by anion exchangers since we always find significant unspecific adsorption of vesicles to the matrix of the exchanger, not to mention the difficulties with the pH in the eluant of ion-exchange columns. For a good separation of liposomes from external ions on Sephadex G-75, it is necessary to apply a high ratio of sample volume/ gel bed volume, i.e., 0.10–0.12. Lower ratios result in broad elution patterns of the applied vesicles. The turbid fractions eluted from the column, which contain the liposomes, are collected. When applying 1 ml sample volume on a Sephadex G-75 column of 10 ml bed volume (7 mm diameter), the first 150 μl of the eluted pool which mainly contain multilamellar vesicles is discarded. Then 1.5 ml is collected; the rest is discarded as well. The column is preequilibrated with a buffer containing all those solutes which are desired afterward to represent the outer hydrophilic space of the reconstituted system (usually 30 mM Na$_2$SO$_4$, 20 mM Tricine–NaOH, pH 7.5, 80 mM sucrose). Osmotic pressure of the internal space is balanced with sucrose or mannitol.

After separation on Sephadex G-75 columns there is a steep gradient of ions, at least of the substrates ADP or ATP, over the phospholipid membrane, therefore the reconstituted vesicles should now be used immediately for the transport assay.

Transport Assay and Calculation

Transport Assay. The transport assay is performed as inhibitor stop kinetics[13]; 100–200 μl of the vesicle suspension together with appropriate buffer and ions according to the design of the experiment (for standard experiments Sephadex G-75 column buffer is used) are thermostated in reaction vessels, taking 12 samples together in one aluminum block. The reaction is started by addition of 10 μl of radioactively labeled nucleotides on small metal "forks" (12 in one combined set) to every vessel while mixing rapidly. External nucleotides are usually added in 50–100 μM concentration, i.e., 3–10 times higher than the transport affinity K_m.[14] In order to avoid problems with membrane potential, only homologous nucleotide exchange is recommended for basic experiments (internal ADP/ external ADP or internal ATP/external ATP).[15] The exchange reaction is stopped by addition of inhibitors (see below) on another set of forks. Minimum reaction time is 5 sec. Blank values are obtained by addition of inhibitors before labeled nucleotides.

The samples are then applied to Dowex 1-X8 anion-exchange columns for separation of external and internal nucleotides. We use a 1 : 1 mixture of Dowex 1-X8 (Serva, Heidelberg, FRG), 50–100 mesh and 200–400

[13] E. Pfaff and M. Klingenberg, *Eur. J. Biochem.* **6**, 66 (1968).
[14] R. Krämer, *Biochim. Biophys. Acta* **735**, 145 (1983).
[15] R. Krämer and M. Klingenberg, *Biochemistry* **21**, 1082 (1982).

mesh in Cl⁻ form. At this stage ion exchanger can be used instead of Sephadex G-75 since the carrier is already blocked by inhibitors. In order to minimize unspecific adsorption, every column is preequilibrated with 500 μl of NaCl (50 mM), bovine serum albumine 5 mg/ml, egg yolk phospholipid liposomes 0.5%. A 200 μl sample of the proteoliposomes from the exchange assay is applied to a Pasteur pipet (0.5 cm diameter) which has been filled with the anion exchanger to a height of 4 cm. The eluate is collected after addition of first 100 μl, then 200 μl, then 500 μl of NaCl (50 mM). This combined eluate of 800 μl is analyzed for radioactivity in a liquid scintillation counter. The measured radioactivity represents the amount of labeled nucleotides in the internal volume of the proteoliposomes.

Inhibition of Transport. When performing the transport assay, a problem arises due to the random orientation of the carrier protein in the reconstituted system. The ADP/ATP translocator in the proteoliposomes is oriented both right-side-out and inside-out.[2,14] Since only right-side-out oriented carrier proteins are blocked by carboxyatractylate, the inside-out oriented carrier molecules have to be inhibited by other means. For this purpose we routinely use 100 μM carboxyatractylate together with 0.2 μM bongkrekate. Instead of bongkrekate, which unfortunately is not commercially available, the inhibitor palmitoyl-CoA (100 μM) can be used for inhibiting also the inside-out oriented carrier molecules.

Calculation of Exchange Kinetics. Only a few hints for the calculation of exchange kinetics in this system can be given here. In order to achieve a reasonably good evaluation of transport kinetics, a blank value and at least three samples at different exchange times should be measured for every experimental condition. In our experience, the blank values are much lower with ¹⁴C-labeled nucleotides (only about 0.1% of the amount of radioactivity applied to the columns) than with ³H-labeled nucleotides.

When evaluating transport kinetics it has to be considered that in fact not the uptake of nucleotides is measured, but the equilibration of external nucleotides (labeled) with internal nucleotides (unlabeled at the beginning of the experiment), due to the strict counterexchange catalyzed by the ADP/ATP translocator. From the time course of equilibration the true transport rates can be derived.[15] If the time resolution of the exchange kinetics appears to be insufficient, i.e., if the equilibration of external and internal label occurs too fast, there are several possibilities to increase the time required for equilibration. The optimum time resolution is obtained when the total amount of internal and external exchangeable nucleotides is of the same size (derivation cannot be given here). This can be achieved by changing the concentration of internal and/or external nucleotides or by dilution of the proteoliposomes. If this does not lead to an appropriate time resolution, the exchange has to be measured at lower temperature.

It should be mentioned here that the total amount of internal nucleotides available for the reconstituted carrier protein can be calculated from the exchange equilibration after long reaction times, usually 2–3 min. The ratio of internal and external radioactive label after equilibration, which is determined by separation on the Dowex columns, equals the ratio of internal and external amount of exchangeable nucleotides. Thus the amount of internal nucleotides can be calculated and since the concentration of internal nucleotides is known, the internal volume of reconstitutively active proteoliposomes can also be derived.

Factors Influencing the Activity of the Reconstituted ADP/ATP Translocator

In order to obtain optimum transport activity under varied conditions of reconstitution, the most important factors influencing the activity of the incorporated ADP/ATP translocator will be discussed below.

Carrier Protein. It is not easy to give a value for exchange activity in a "standard reconstitution experiment," since there is an extremely long list of parameters influencing the carrier activity (see below). In the case of purified egg yolk lipids and addition of 10% cholesterol, the true V_{max} for adenine nucleotide exchange is in the range of 2–4 mmol min^{-1} g^{-1} at room temperature.[16] If reconstitution is performed under standard conditions and the activity is found to be too low, this is normally due to inactivation of the carrier protein before it is incorporated into the liposomes. This can easily be tested by incorporation of a crude Triton X-100 solubilizate of mitochondria—thereby omitting the purification step—which should give a specific activity of about 150–300 μmol min^{-1} g^{-1} with the same liposomes as described above. Finally, it has to be mentioned that the activity of the reconstituted ADP/ATP exchange is significantly higher in the case of the adenine nucleotide carrier from beef heart mitochondria as compared to that from rate liver or neurospora mitochondria.

Hydrophilic Surroundings. The ion composition of the two surrounding water phases strongly influences the carrier activity. Especially if negatively charged membranes are used, the concentration of monovalent cations should be high enough (at least 50 mM) to minimize problems with surface potentials.[14] Divalent cations should be avoided since, due to the complexation, they lead to a drastic decrease of the apparent substrate affinity.[17] A sufficient amount of anions is necessary for specific activation of the carrier protein; however, these also compete with nucleotides at

[16] R. Krämer, *Biochim. Biophys. Acta* **693,** 296 (1982).
[17] R. Krämer, *Biochim. Biophys. Acta* **592,** 615 (1980).

the binding site.[18] The pH dependence is relatively broad, i.e., exchange measurements can be carried out from pH 6 to 8. We routinely use Na_2SO_4 (30 mM), Tricine–NaOH (20 mM) pH 7.5, and sucrose in a concentration that balances the internal osmolarity.

Hydrophobic Surroundings. There have been extensive studies on the phospholipid dependence of the reconstituted ADP/ATP carrier. The presence of phosphatidylethanolamine is essential for carrier activity;[19,20] negatively charged lipids activate the carrier protein,[14,20] but difficulties with the surface potential have to be considered under these conditions.[14] The ADP/ATP translocator cannot be reconstituted using phospholipids containing only saturated fatty acids.[12] Addition of cholesterol to the membrane enhances the carrier activity.[16] For standard experiments we use a mixture of purified egg yolk phospholipids[11] (ratio of phosphatidylcholine/phosphatidylethanolamine of about 6) with 10% (w/w) cholesterol.

Transmembrane Parameters. Membrane potential strongly modulates the carrier activity according to its physiological regulation mechanism, whereas a pH gradient does not significantly influence the exchange.[15,21] This, however, is only important for special investigations where the ADP/ATP carrier catalyzes a secondary active transport. In this article only the basic transport activity is described, which applies to the basic mode of this carrier, i.e., the passive equilibration of internal and external nucleotides.

[18] R. Krämer and G. Kürzinger, *Biochim. Biophys. Acta* **765,** 353 (1984).
[19] H. G. Shertzer and E. Racker, *J. Biol. Chem.* **251,** 2446 (1976).
[20] R. Krämer and M. Klingenberg, *FEBS Lett.* **119,** 257 (1980).
[21] R. Krämer and M. Klingenberg, *Biochemistry* **19,** 556 (1980).

[48] Fluorescent Nucleotide Analogs as Active Site Probes for the ADP/ATP Carrier and the Uncoupling Protein

By Martin Klingenberg

Introduction

The "natural" ligands of the mitochondrial ADP/ATP carrier (AAC), ADP and ATP, as well as the inhibitors of the atractylate group and the bongkrekate group, have no distinct optical signal which could be used for studying their interaction with the AAC. The same is true for the nucleotide ligands to the uncoupling protein (UCP), for ADP, ATP, GDP, GTP.

In 1974 great hopes were raised with the advent of the "formycin" nucleotides,[1,2] which differ from the adenine nucleotides only by exchange of C and N between positions 7 and 8 in the purine moiety. They could be expected to interact even with the very discriminating AAC binding site. As a result FoDP and FoTP were synthesized in our laboratory and found to be, relatively, the most active ADP or ATP analog.[3,4] The subsequent application as a fluorescent probe proved not to fulfill the expectations raised in the literature.[1,2] The fluorescence yield is low, the decay time very short (<1 nsec), and excitation and emission overlap with tryptophan fluorescence.[4] Only by energy transfer to vinylanthracene and the judicious selection of the wavelength a definite signal response on binding to the AAC could be elucidated.[4]

In 1973 dinitrophenyl- and trinitrophenyl-ATP were introduced with the fluorescent groups substituted at the 3'-O position of the ribose moiety.[5] These compounds proved to be strong ligands to the AAC but their poor fluorescence changes only little on binding and remains largely quenched.[6]

In 1979 Schäfer and his group introduced naphthoyl and its derivatives as a fluorescent substituent into the ribose moiety of nucleotides.[7] In particular the naphthoyl analog to the classical dansyl group, the 1,5-dimethylaminonaphthoyl (DAN) group proved to generate a most valuable fluorescent probe for AAC and for UCP when substituted as 3'-O esters into AMP, ADP, and ATP.[6,8–10]

The simpler naphthoyl (NAP)-3'-O esters of ADP and ATP were also fluorescent probes for the AAC; their fluorescence response was quite different and less informative.[6,11,12] Whereas the DAN-nucleotides are largely quenched in aqueous solution and emit a fluorescence signal on binding to the AAC, the NAP nucleotides fluoresce in H_2O and their

[1] D. C. Ward, E. Reich, and L. Stryer, *J. Biol. Chem.* **244,** 1228 (1969).
[2] D. C. Ward, A. Cerami, E. Reich, G. Acs, and L. Atlwerger, *J. Biol. Chem.* **244,** 3243 (1969).
[3] M. Klingenberg, *in* "The Enzymes of Biological Membranes: Membrane Transport" (A. N. Martonosi, ed.), p. 383. Plenum, New York, 1976.
[4] C. Graue and M. Klingenberg, *Biochim. Biophys. Acta* **546,** 539 (1979).
[5] T. Hiratsuka and K. Uchida, *Biochim. Biophys. Acta* **320,** 635 (1973).
[6] I. Mayer, A. S. Dahms, W. Riezler, and M. Klingenberg, *Biochemistry* **23,** 2436 (1984).
[7] G. Schäfer and G. Onur, *Eur. J. Biochem.* **97,** 415 (1979).
[8] M. Klingenberg, I. Mayer, and A. S. Dahms, *Biochemistry* **23** (1984).
[9] G. Schäfer and G. Onur, *FEBS Lett.* **109,** 197 (1980).
[10] M. Klingenberg, *in* "Structural and Functional Aspects of Enzyme Catalysis" (H. Eggerer and R. Huber, eds.), p. 202. Springer-Verlag, Berlin and New York, 1981.
[11] M. R. Block, G. J. M. Lauquin, and P. V. Vignais, *Biochemistry* **21,** 5451 (1982).
[12] M. R. Block, G. J. M. Lauquin, and P. V. Vignais, *Biochemistry* **22,** 2202 (1983).

fluorescence is decreased by about 20% on binding. Furthermore, the DAN-nucleotides discriminate strongly between the two functional states of AAC whereas the NAP-nucleotides seem not to differentiate.[11] Therefore in the past years use of fluorescent nucleotide derivatives in our hands has concentrated on the DAN-nucleotides because of their great advantages. Among the applications in studies on the AAC are visualization of the carrier sites in the membrane or in solubilized preparations, the distinction between functional states of the carrier binding site, the m- and c-state, determination of c/m-state distribution, and kinetic measurements of c/m-state transition as a function of ADP/ATP catalysis and numerous other factors. In general, the DAN-nucleotide fluorescence is a convenient signal for following influences at the binding site of the AAC, due to physical or to chemical factors.

Also to the UCP, DAN-ADP and DAN-ATP bind with a strong fluorescence enhancement. Here the fluorescence signal permits the visualization of UCP binding sites and the dependence of binding on various physical and chemical factors, as well as enabling us to follow the kinetics of UCP transition into other functional states.

Application of DAN-Adenine Nucleotides to the ADP/ATP Carrier

General Characteristics

The DAN derivatives DAN-AMP, DAN-ADP, and DAN-ATP have identical fluorescence properties. Their fluorescence is virtually quenched in aqueous solution, whereas a high fluorescence yield is obtained in nonaqueous media, such as dioxane and ethanol. The excitation (absorption) maximum in dioxane solution is at 344 nm and the emission peaks at 506 nm. On binding to the carrier in the mitochondrial membrane the uncorrected excitation is at 360 nm and the emission peak is at 505 nm. On binding to the solubilized AAC the emission peaks at 519 nm (Fig. 1). The fluorescence decay time in dioxane is around 8 nsec.

DAN-Nucleotides Applied to the ADP/ATP Carrier in the Mitochondrial Membrane

The DAN-nucleotides added to rat liver mitochondria do not give a marked fluorescence. In beef heart mitochondria they exhibit a relatively small signal which is fully enhanced in submitochondrial (sonic) particles. Also breaking the mitochondria by freezing-thawing or by detergents enhances the fluorescence response. Most remarkable is the virtually com-

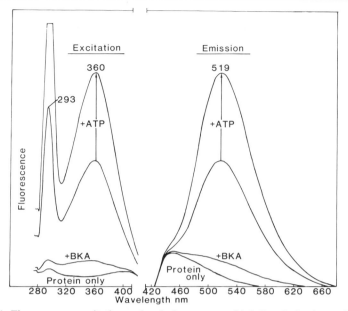

FIG. 1. Fluorescence excitation and emission spectra of 1,5-dimethylaminonaphthoyl 3'-O-AMP (DAN-AMP) bound to Triton-solubilized ADP/ATP carrier from beef heart mitochondria (from Klingenberg *et al.*[13]). Uncorrected spectra. DAN-AMP (20 μM) is added to an ADP/ATP carrier preparation, obtained by Triton X-100 extraction from beef heart mitochondria and fractionation on hydroxylapatite (see text). It contains 0.4 mg protein/ml and 0.6% Triton X-100, in Tris buffer of pH 7.2. Add 30 μM ADP and 10 μM bongkrekate (BKA) as indicated.

plete suppression of fluorescence by bongkrekate (BKA) addition. This indicates specific binding of the DAN-nucleotides to the AAC. The insensitivity to carboxyatractylate (CAT) addition shows that DAN-nucleotides bind only to the *m*-state of the carrier under fluorescence. This explains the required accessibility of the *m*-side of the membrane for the DAN-nucleotides. If the carrier is first liganded with CAT, DAN-nucleotide addition does not give fluorescence. Thus BKA is the most important tool for defining the AAC-linked fluorescence of DAN-nucleotides.

The fluorescence was shown to faithfully report physical binding to the carrier by a comparison with the binding of [³H]DAN-nucleotides. Although the fluorescence is completely suppressed by BKA there remains a definite BKA-insensitive [³H]DAN-nucleotide binding. The AAC-attributable binding is slightly decreased by Mg^{2+}, whereas the BKA-insensitive binding is increased to double the original amount in

SMP, without a fluorescence increase. Partially this represents binding of the DAN-nucleotides to the ATPase which is not fluorescent.[8] The fluorescent type of binding of the DAN-nucleotides is thus a peculiar property of the DAN-nucleotide interaction with the carrier. Furthermore, it is specific for the binding center in the m-state, since DAN-ADP and DAN-ATP appear to bind weakly also to the binding center in the c-state without fluorescence.[7]

The affinity of the DAN-nucleotide to the AAC in SMP, as determined with [³H]DAN derivatives, is quite high as compared to that of ADP or ATP: $K_D = 1.8 \ \mu M$ for DAN-AMP and 1.4 μM for DAN-ADP.[7] The maximum number of binding sites is 1.2 to 1.4 μmol/g protein. A direct comparison of the binding capacity for [³H]BKA and [³H]CAT with [³H]DAN-AMP binding shows good agreement between the number of binding sites for all three types of ligands.

The peculiar specificity to the m-state permits visualization of the kinetics of the transition from the c- to the m-state in SMP. For example, after adding CAT, the fluorescence of bound DAN-AMP is decreased only when ADP or ATP is also present. The slow fluorescence decrease demonstrates the ADP- or ATP-catalyzed transition from the m- to the c-state (see scheme, Fig. 2).

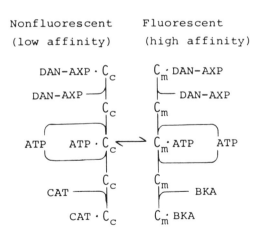

X = M, D, T

FIG. 2. Modes of DAN-nucleotide (DAN-N) binding to ADP/ATP carrier (from Klingenberg *et al.*[8]). Fluorescent and tight binding only to the m-state, nonfluorescent and loose binding to the c-state. The interaction of DAN-N with the m-state depends on the availability of ADP or ATP for facilitating the c- to m-state transition, since the binding site of the AAC can only make these transitions as the AAC–ADP or AAC–ATP complex.

DAN-Nucleotides Applied to the Solubilized ADP/ATP Carrier

The DAN-nucleotides bind also to the soluble and partially purified AAC with a fluorescence maximum at a slightly longer wavelength, 519 nm (Fig. 1).[13] The application is complicated by the detergent present to which DAN-nucleotides are adsorbed also under emission of fluorescence. Obviously the DAN moiety is attracted to the hydrophobic core of the detergent micelle where it is protected from quenching by H_2O. Fortunately, the affinity to Triton is not too high, so that, after removal of excess Triton, the fluorescence from the binding to the AAC stands out clearly. The partition β of DAN-AMP to Triton X-100 has been determined. It is defined as a dissociation constant β of DAN-nucleotides with Triton X-100 and given in mM. The constant is dependent on the salt concentration and varies, for example, from $\beta = 0.08$ mM at 10 mM Na_2SO_4 to $\beta = 0.25$ mM at 300 mM Na_2SO_4. Thus only half of the DAN-AMP is free in H_2O in a 1.2% (19 mM) Triton solution at 50 mM Na_2SO_4. In LAPAO, another useful detergent for isolating the AAC,[12] the partition is far more unfavorable, $\beta = 5$ mM at 50 mM Na_2SO_4.

The detergent-linked fluorescence of the DAN-nucleotides is easily differentiated from the AAC binding by BKA addition, as shown in Fig. 1. For studying the AAC with the DAN-nucleotide probes the following preparations are recommended.

Solubilized Preparation of Free AAC. One milliliter of frozen stored beef heart mitochondria containing 60 mg protein is thawed and added to 2.5 ml solution containing 8% Triton X-100, 10 mM tricine buffer pH 8, 0.2 mM EDTA, and 150 mM Na_2SO_4. After shaking for 10 min at 10°, the solubilizate is mixed with 4 ml of a pasty hydroxylapatite suspension prepared according to Bernardi,[14] followed by centrifugation at 800 g for 2 min.[14a] For partial removal of excess Triton the supernatant is treated with 4.5 g wet Amberlite beads (amberlite NAD-4). After shaking for 15 min the supernatant is removed by slow centrifugation and again treated with 2.5 g wet Amberlite. In a typical case the resulting preparation of 3 ml contains 1.2 mg protein/ml and 0.75% Triton X-100. The amount of AAC can be determined by binding of [³H]CAT. Typically it binds 4 μmol [³H]CAT/mg protein corresponding to 32% of intact AAC in this preparation. This unliganded AAC preparation is rather unstable and should be used within 4 hr.

A more stable and pure preparation of AAC useful for several applications of the DAN-nucleotide probe is the atractylate–AAC complex,

[13] M. Klingenberg, I. Mayer, and Appel, *Biochemistry* **24**, 3650 (1985).
[14] G. Bernardi, this series, Vol. 22, p. 325.
[14a] R. Krämer, H. Aquila, and M. Klingenberg, *Biochemistry* **16**, 4949 (1977).

ATR–AAC. It is essentially prepared according to the procedure described for the CAT–AAC complex.[15] The ATR–AAC can be obtained in a quite pure state (90%). Since ATR does not bind as tightly as CAT, the ATR–AAC is a relatively stable and pure source of AAC for functional studies which include the transition into the m-state. For example, with ATR–AAC in the presence of ADP or ATP, DAN-AMP also can bind extensively, permitting utilization of the fluorescence signal.[13]

Typical Incubation Conditions for Measurement of DAN-Nucleotide Fluorescence on Binding to the AAC

The basic medium contains 20 mM Tris–HCl, pH 7.2, 0.2 mM EDTA and, for mitochondria, additionally 0.25 M sucrose. The temperature should be between 0 and 10° with the labile soluble protein and can be up to 25° with submitochondrial particles and mitochondria. The concentrations for obtaining about 90% response are 10 μM for DAN-AMP and about 15 μM for DAN-ADP and DAN-ATP. The K_D for DAN-AMP is about 0.5 to 2 μM and for DAN-ADP and DAN-ATP 2 to 4 μM.[7,13] For facilitating m/c-state transition, addition of 10 to 20 μM ADP is required. ATP is less effective. For discriminating carrier binding add 5 μM BKA.

Application of DAN-Nucleotides as Probes for Determining c/m-State Distribution

In the solubilized AAC the full fluorescence response to addition of DAN-AMP (or DAN-ADP and DAN-ATP) is observed only in the presence of ADP or ATP (Fig. 3). This is interpreted to show that the AAC is, from the start, partially in the m-state and partially in the c-state. Only when the substrates for the ADP–ATP transport are provided the transition between the c- and m-state can be accomplished since the unloaded carrier cannot undergo this transition (see Fig. 2). Thus the carrier population in the c-state is largely converted into the m-state under the pulling force of the DAN-AMP binding. According to recordings in Fig. 3 the c/m distribution prior to addition of these proteins can be determined to be 40%/60%. The c/m distribution has been found to vary under the influence of various factors such as salts (anions) and SH reagents.

Starting with the ATR–AAC complex the fluorescence on DAN-AMP addition is nearly fully dependent on ADP or ATP, since with ATR the AAC is predominantly in the c-state.

---

[15] M. Klingenberg, H. Aquila, and P. Riccio, this series, Vol. 56, p. 229.

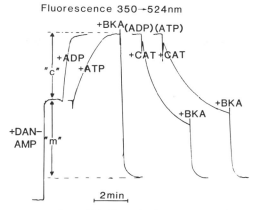

FIG. 3. Fluorescent recording of binding of DAN-AMP to Triton solubilized ADP/ATP carrier preparation (from Klingenberg et al.[13]). Fluorescence increases on addition of 1.5 μM DAN-AMP to 0.5 mg protein/ml. Add 20 μM Tris buffer at pH 7.2. Further additions are of 30 μM ADP, 30 μM ATP, 10 μM CAT, and 10 μM BKA as indicated.

Measurement of the Kinetics of Transition between c- and m-State

Figure 3 indicates that DAN-nucleotides can also be applied to study the kinetics of the c/m-state transitions. For example, the increase of fluorescence is faster with ADP than with ATP and the decrease is slower with CAT than with BKA. The increase reflects the rate of c- to m-state transition and the decrease that of m- to c-state transition. These rates can be measured using any type of mixing device. Measurement of the binding rate on addition of DAN-nucleotides requires stopped-flow devices with a high time resolution.

One application is aimed at determining the influence of the ADP and ATP concentration on the rate of c- to m-state transition (Fig. 4). The limited rate of the fluorescence increase can be covered over a wide range and shows strong differences between ADP and ATP. The maximum rates are more than twice as high with ADP than with ATP and at 20 μM the difference is even greater, $v^{ADP}/v^{ATP} = 4$. The fluorescence increment is nearly independent of the ADP or ATP concentration in accordance with their catalytic influence. It slightly decreases with the ADP and ATP concentration as a result of competition with DAN-AMP at the binding site. For reciprocal plots one evaluates $K_M^{ADP} = 1.8$ μM and $K_M^{ATP} = 28$ μM. These values are remarkable since no related affinity data are available so far for the soluble protein because of the low affinity of the nucleotides ($K_D > 30$ μM). Since $K_M \geq K_D$ the actual K_D can be much higher.

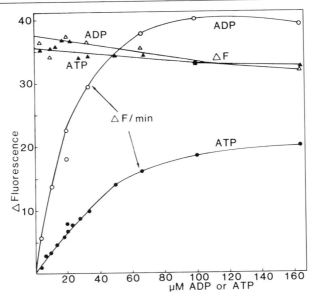

Fig. 4. The dependence of the rate of fluorescence increases on the concentration of ADP and ATP. Conditions as in Fig. 3. ΔF corresponds to the difference of fluorescence $\Delta F = F(-\text{BKA}) - F(+\text{BKA})$ (Klingenberg, unpublished).

Also the reversed reaction m- to c-state can be followed by way of the DAN-AMP fluorescence decrease. Here CAT serves as a trap for the c-state (follow the pathway in Fig. 2). The rates are much slower and limited by the DAN-AMP dissociation. In this direction, too, ADP is several times more effective than ATP in stimulating the transition.

In general, the rate of the ADP- or ATP-dependent DAN-AMP fluorescence changes has been found to be much more sensitive to environmental factors than the extent. This holds for the influence of pH, anions, phospholipids, temperature, etc. It is clear that the transition rate between the c- and m-state must involve important conformation changes in the AAC which require high activation energy. Therefore, it is quite reasonable that such a delicate catalytic process is highly sensitive to various factors.

The fluorescent DAN-nucleotides provide an up to now unique tool for determining and analyzing the translocational transition of the AAC. They are superior to fluorescent SH reagents which can also be used as probes for the transition to the m-state (Klingenberg, unpublished results). However, the signal of the fluorescent SH reagents is generally rate limited by the reaction of the probe with the SH group rather than by the transition rate to the *m*-state, whereas the binding of the DAN-nucleotides is so rapid that it is not a rate-limiting step.

Application of DAN-Nucleotides to the Uncoupling Protein of Brown Fat Mitochondria

The uncoupling protein (UCP) in brown fat mitochondria has been shown to have some striking structural similarities to the AAC (see Aquila *et al.*[16]). Functionally the UCP contrasts with AAC, since UCP transports the "smallest" solute, H^+, and AAC about the "largest," ADP and ATP. The UCP, too, interacts with ADP and ATP, however, in contrast to their function with the AAC, these nucleotides are now inhibitors. Also, in UCP the interaction is less specific, since GDP and GTP are good ligands too.

DAN-nucleotides bind also to UCP, exhibiting fluorescence, similar to the AAC.[17] This has been shown both for membrane-bound and for solubilized UCP. In contrast to the AAC, the UCP does not accept DAN-AMP but only DAN-ADP and DAN-ATP. A further difference is that in UCP the DAN-nucleotides can only bind from the *c*-side, since there is no nucleotide binding site in the *m*-state.

Interaction with the Membrane Bound and Soluble UCP and Differentiation from Binding to AAC

DAN-ATP or DAN-ADP bind to brown fat mitochondria with a strong fluorescence signal. This interaction is reversed by subsequent addition of ATP, but is insensitive to CAT addition. This difference, compared to other mitochondria, shows that DAN-ATP has direct access to the *c*-side of the UCP. The interaction of DAN-ATP with UCP can be differentiated from that with the AAC by adding low amounts of ATP ($<5 \ \mu M$), which do not affect DAN-ATP binding to the AAC. On breaking the brown fat mitochondria by sonication the fluorescence is further enhanced. This increase is abolished by BKA and is obviously due to DAN-ATP binding to AAC from the *m*-side. On the other hand DAN-ATP has full accessibility to the UCP in sonic particles. This adds to the evidence that most sonic particles are permeant and not closed vesicles.

Interestingly, ATP and DAN-ATP have a different influence on the recoupling of respiration in brown fat mitochondria. Whereas ATP and the other purine nucleotides inhibit the uncoupling typical for brown fat

---

[16] H. Aquila, T. Link, and M. Klingenberg, *EMBO J.* **4,** 2369 (1985).
[17] M. Klingenberg, *Biochem. Soc. Trans.* **12,** 390 (1984).

mitochondria, DAN-ATP and DAN-ADP do not recouple electron transport despite their binding to UCP. This can be rationalized, as elucidated below, by assuming that DAN-ATP binding does not induce a conformational transition of UCP into the "closed" states, as ATP does.

Since the isolated UCP is quite stable in Triton X-100 even without inhibitor protection—in contrast to AAC—DAN-nucleotides can be easily applied to purified preparations of UCP. Here again the fluorescent adsorption of DAN-ATP to the Triton micelles has to be accounted for and excess Triton has to be removed by Amberlite treatment (see method for AAC). Differentiation to the Triton binding is again achieved easily using the strong competition by ATP of DAN-ATP binding.

Typical Application Conditions for Binding of Fluorescent DAN-Nucleotides to UCP

Because of strong pH dependency of nucleotide binding to UCP, the medium should be kept at 7.0. It should have low ionic strength to avoid strong anion competition with nucleotide binding. In particular multivalent buffer anions compete with nucleotides.

Isolated brown adipose tissue mitochondria are incubated in an isotonic medium containing 0.25 M sucrose and 20 mM HEPES buffer, pH 6.5, at 1 to 2 mg protein/0.3 ml, 12°, in a 5 × 5 mm fluorescence cuvette. Saturation (90%) of fluorescent binding is obtained at 15 μM DAN-ADP or 10 μM DAN-ATP. By addition of 100 μM ATP the fluorescence is slowly reversed at $t_{1/2}$ = 2 min, with 1 mM ATP at $t_{1/2}$ = 1.0 sec.

For the fluorescence binding of DAN-nucleotides to the isolated protein the usual solubilized UCP preparation can be used.[18] Add 0.1 mg UCP protein to 0.3 ml incubation medium consisting of 20 mM HEPES buffer, pH 6.5, 12°. Add DAN-ADP and DAN-ATP in the same concentration as with mitochondria.

Kinetics of DAN-ATP and ATP Interaction with UCP

Whereas the rate of ATP interaction with UCP was found to be slow—order of seconds to minutes dependent on pH and temperature—the binding of DAN-ATP is very fast. Measured by stopped flow it has a half time of 50 msec (Klingenberg, unpublished). This peculiar difference is impressingly demonstrated when following the fluorescence signals given in Fig. 4, showing the effect when both DAN-ATP and ATP are added simultaneously.[17] After an almost instantaneous full rise of the fluorescence it takes several minutes until the tightly binding ATP has taken over

---

[18] M. Klingenberg and C.-S. Lin, this series, Vol. 126, p. 490.

at the binding site and displaced DAN-ATP. This experiment demonstrates that by applying the fluorescence of DAN-ATP the binding of ATP can be kinetically analyzed. This rate is not limited by the dissociation of DAN-ATP but by the transition of the ATP–UCP complex into the conformation of an inhibited ATP–UCP complex with tight ATP binding (Klingenberg, unpublished). DAN-ATP permits the observation of the active conformation prior to the inhibited state.

An outstanding feature of the nucleotide binding to UCP is the strong pH dependency. Extensive binding studies have been performed with the various nucleotides, ADP, ATP, GDP, GTP and analogs such as dATP or AMP-PNP to analyze the H^+ dissociating group involved (see Klingenberg[17] and Lin and Klingenberg[19]). The fluorescence of DAN-ATP gives another convenient method to measure the pH influence on the concentration dependence, for determining the K_D. Particularly in the range of pH < 5.5, binding determinations by equilibrium methods are difficult because the $K_D < 10^{-6} M$ is so small that errors occur in determining the very small free nucleotide concentration. Here the rate of DAN-ATP binding followed with stopped flow techniques can extend the range by making use of the relation $K_D = v_{off}/v_{on}$. With the determination of the binding rate (v_{on}) of DAN-ATP by rapid mixing equipment the measurement can be extended into a range where the $K_D < 10^{-5} M$.

Preparation of 1,5-Dimethylaminonaphthoyladenine Nucleotides (see also Mayer et al.[6] and Onur et al.[20])

The starting material is commercially available 1-naphthoic acid. For the synthesis of 1,5-nitronaphthoic acid spread 30 g 1-naphthoic acid in a beaker and slowly add 60 g of 65% HNO_3. Then heat the paste slightly for about 15 min while stirring until no nitrous gases are developed any more. Then remove HNO_3 by washing with H_2O and dissolve the residue in 240 ml 2 M Na_2CO_3 solution. From the resulting dark brown solution the remaining residue is removed and the nitronaphthoic acid precipitates on addition of 6 M HCl until a pH 4.0 is reached. The yellow precipitate is a mixture of 1,5- and 1,8-nitronaphthoic acid. The dried powder is dissolved in 300 ml methanol, and on cooling 1,5-nitronaphthoic acid precipitates with a smp 239°.

For synthesizing 1,5-aminonaphthoic acid (see Ekstrand[21]), 4 g of 1,5-nitronaphthoic acid is dissolved in 60 ml 100% acetic acid and heated to 55°. Then add in portions 16 ml of powdered Fe and keep the temperature

[19] C.-S. Lin and M. Klingenberg, *Biochemistry* **21,** 2950 (1982).

[20] G. Onur, G. Schäfer, and H. Strotmann, *Z. Naturforsch.* **38c,** 49 (1983).

[21] A. G. Ekstrand, *J. Prakt. Chem.* **42,** 273 (1890).

below 60° by cooling for 20 min. Add 60 ml H_2O. Separation from sediments and extraction from the supernatant is achieved by several applications of diethyl ether. Dry the ether extracts with Na_2SO_4 and precipitate the 5-aminonaphthoic acid with HCl gas. Recrystallize from methanol and ether.

For synthesis of 1,5-dimethylaminonaphthoic acid dissolve 500 mg 1,5-aminonaphthoic acid in 2 ml 4 M NaOH, heat to 50°, and add 4 × 200 μl dimethyl sulfate plus 200 μl 4 M NaOH at 2 min intervals. Extract the unreacted aminonaphthoic acid twice with 10 ml ether. From the residual solution precipitate with 6 M HCl the 1,5-dimethylaminonaphthoic acid (1,5-DAN) and recrystallize in methanol plus H_2O.

For synthesis of 1,5-DAN-adenine nucleotides (see Gottiks et al.[22]) dissolve 103 mg (480 μmol) of 1,5-DAN in freshly distilled and dried dimethylformamide (DMF). Add 182 mg carbodiimidazole. After 15 min add 62 mg AMP (or ADP, ATP, respectively) dissolved in 2.5 ml H_2O dropwise to the carbodiimidazole solution. Allow to react for 4 hr at room temperature while stirring in the dark, lyophilize, take up with 0.5 ml H_2O, adjust to pH 3.5 with HCl, and extract 4 times with ether. Neutralize to pH 7, extract again with ether to remove excess free 1,5-DAN and carbodiimidazole.

Purify the resulting DAN-AMP on LH 20 dextran gel columns (2 × 35 cm) by elution with H_2O. Collect fractions with absorption at 259 nm and fluorescence. The pooled fractions are further purified on DEAE-cellulose column (1 × 15 cm) which is activated with 0.5 M NH_4HCO_3, pH 8.0 and then washed with H_2O. The column is eluted with a linear H_2O/0.6 M triethylamine bicarbonate buffer gradient. Test the fractions for absorption and fluorescence. Assay on thin layer chromatography plates coated with silica gel in 2-butanol–propanal–H_2O (1 : 1 : 1). The lyophilized substance is taken up and stored at −180° in ethanol. Slow hydrolysis occurs in aqueous solution.

Acknowledgments

This work was supported by grants from the Deutsche Forschungsgemeinschaft (Kl 134/21-23).

[22] B. P. Gottiks, A. A. Kraevskii, P. P. Purygin, T. L. Tsilevich, Z. S. Belova, and L. N. Rudzite, *Bull. Acad. Sci. USSR, Div. Chem. Sci.* (Engl. Transl.) 2453 (1967).

[49] Techniques for NMR and ESR Studies on the ADP/ATP Carrier

By KLAUS BEYER and ANTON MUNDING

The applicability of magnetic resonance techniques to proteins of the inner mitochondrial membrane is generally limited. Our investigation of the mitochondrial ADP/ATP carrier benefits from the large abundance of this protein in heart mitochondria. In addition, this carrier protein can be stabilized by the specific transport inhibitors carboxyatractylate (CAT), atractyloside (ATR), and bongkrekic acid (BKA). Thus standard techniques of membrane solubilization and protein purification can be applied to the protein–inhibitor complexes.[1] The stability of the CAT–carrier complex is of great importance with regard to the inevitably long data acquisition times needed in NMR experiments.[2]

The ADP/ATP carrier molecule itself does not provide suitable intrinsic labels for NMR or ESR studies such as heteronuclei or transition metal atoms. Proton NMR on the solubilized carrier protein is not feasible owing to the large excess of detergent necessary to avoid protein aggregation. In membrane-bound as well as in detergent-solubilized protein preparations, the only nucleus to be easily observed by NMR is ^{31}P from phospholipids associated with the protein surface. Thus the interaction of the carrier with surrounding phospholipids was studied in phospholipid membranes and in detergent micelles by ^{31}P NMR spectroscopy.[2,3]

The perturbation caused by the carrier protein in the headgroup region of phospholipid bilayers may be examined after reincorporation of the purified protein into membranes of optional phospholipid composition.[2] In order to obtain maximal contact between the phospholipids and the protein surface, the protein/phospholipid ratio must be adjusted to the highest possible value whereas residual detergent should be essentially absent. Unfortunately these goals are not attainable simultaneously. The very effective method of detergent dialysis as applied successfully in many reconstitution studies with other membrane proteins is not applicable to the ADP/ATP carrier. Dialyzable detergents of high critical micelle concentration (CMC) either fail to solubilize or tend to denature this protein. Thus stepwise procedures for protein reincorporation and Triton

[1] M. Klingenberg, H. Aquila, and P. Riccio, this series, Vol. 56, p. 407.
[2] K. Beyer and M. Klingenberg, *Biochemistry* **22,** 639 (1983).
[3] K. Beyer and M. Klingenberg, *Biochemistry* **24,** 3821 (1985).

removal were developed, providing a compromise between the requirements of high protein and low detergent content in the final sample. Under certain conditions the protein-loaded membranes can be further enriched by a centrifugation step using D_2O buffer.

Strong binding of a single phospholipid species to the protein surface cannot be observed in the reincorporated membrane preparations by ^{31}P NMR. In the Triton-solubilized protein, however, ^{31}P signals of residual phospholipids are easily resolvable.[3] Specific lipid–protein interactions may be evident from differences in line width of the phospholipid signals and line narrowing upon denaturation of the protein, as described shortly.

In spin label ESR studies the amino acid sequence and three-dimensional structure of the protein dictates the choice of appropriate spin probes. Ideally, spin labeling should introduce a single spin probe molecule into each protein subunit. It has been shown that in the m-state of the ADP/ATP carrier one SH group per protein subunit is accessible to radioactive N-ethylmaleimide.[4,5] These SH groups are masked in the c-state of the protein. Thus spin-labeled maleimides can be bound specifically to the ADP/ATP carrier by rearranging the protein from the c-state to the m-state. Spin labeling of the carrier in mitochondria requires blocking of the large background of foreign SH groups before rearranging the protein. The paramagnetism of the protein-bound nitroxyl groups may be abolished by chemical reduction. Addition of membrane-impermeable reducing agents can afford information about the sidedness of SH groups in the spin-labeled protein.[6] The combination of both specific labeling and spin reduction will be described in detail.

Indirect spin labeling of the ADP/ATP carrier can be achieved by binding a spin probe to the inhibitor CAT. The carrier can be labeled with this CAT derivative (CATSL) without the need of a troublesome prelabeling procedure.[7] In an earlier study[8] spin-labeled fatty acids have been attached to the inhibitor via the fatty acid anhydrides. In the approach described here the CAT molecule is labeled by 2,2,5,5-tetramethyl-1-oxy-3-pyrroline-3-carboxylic acid or the corresponding pyrrolidine analog using carbonyldiimidazole as a coupling reagent.[7] In CATSL the spin label is located much closer to the sugar moiety of the inhibitor molecule than in the fatty acid derivatives. The resulting restriction in motional freedom of

[4] H. Aquila, W. Eiermann, and M. Klingenberg, *Eur. J. Biochem.* **122**, 133 (1982).

[5] H. Aquila and M. Klingenberg, *Eur. J. Biochem.* **122**, 141 (1982).

[6] A. Munding, K. Beyer, and M. Klingenberg, unpublished experiments.

[7] A. Munding, K. Beyer, and M. Klingenberg, *Biochemistry* **22**, 1941 (1983).

[8] G. J. M. Lauquin, P. F. Devaux, A. Bienvenue, C. Villiers, and P. V. Vignais, *Biochemistry* **16**, 1202 (1977).

carrier-bound CATSL allows the overall mobility of the protein molecule to be evaluated by conventional and saturation transfer ESR.

Reincorporation of the ADP/ATP Carrier into Phospholipid Membranes

One-Step Method

Egg yolk phospholipids (80 mg) of the desired composition (see below) are dissolved in 2 ml of chloroform in a 100 ml round-bottomed flask. The solvent is removed on a rotatory evaporator and the resulting phospholipid film is thoroughly dried at 10 Pa for at least 6 hr. Eighteen milliliters of the CAT-carrier complex, typically containing 0.7 to 1.0 mg protein per ml and 0.5 to 0.6% (w/v) of Triton X-100 in 200 mM NaCl and 10 mM 4-morpholinopropanesulfonic acid (MOPS), pH 7.2, is added. The suspension is homogenized by gentle shaking at room temperature under an atmosphere of nitrogen, followed by incubation at 4° for 1 hr. The large amount of Triton added along with the protein leads to partial transformation of the phospholipid film into mixed micelles. Excess detergent is removed by addition of 70 g of wet Amberlite XAD-2 beads per g of Triton X-100. After gentle shaking under nitrogen at room temperature for 1 hr the lipid–protein mixture is separated from the beads by sucking it through a narrow plastic tip. The complex is centrifuged at 136,000 g for 1 hr. For ^{31}P NMR the pellet[9] is resuspended in 1 ml of a buffer containing 200 mM NaCl, 10 mM MOPS, pH 7.2, and 20% (v/v) D$_2$O. In the final complex 70–80% of the added protein is recovered. The ratio phospholipid/protein is 200 to 300 mol/mol and phospholipid/Triton is 15 to 20 mol/mol.

Stepwise Incorporation

Reincorporation may be achieved by stepwise protein addition to a phospholipid suspension. In this case the ratio of phospholipid to protein in the starting mixture is higher than in the one-step method. When the carrier is incorporated in this way, the protein is contained largely in unilamellar vesicles roughly 100 to 200 nm in diameter as shown by freeze fracture electron microscopy. These vesicles are easily separated by centrifugation in D$_2$O.

A film of 150 mg dry egg phospholipids in a 100-ml flask is suspended in 6 ml buffer (200 mM NaCl, 10 mM MOPS, pH 7.2). Twelve milliliters of

---

[9] A small translucent patch frequently appears at the bottom of the fluffy pellet consisting of almost exclusively of phospholipid.

solubilized CAT protein is added and the mixture is incubated for 0.5 hr at 4° under nitrogen. The detergent concentration is reduced by adding Amberlite beads as described above. After centrifugation of the phospholipid–protein complex the pellet is resuspended in 6 ml buffer and the protein incorporation is repeated with another 12 ml of solubilized CAT protein. After two incorporation steps about 40 to 50% of the added protein is found in the final pellet. The phospholipid/protein molar ratio is similar to that in the first incorporation method. However, the final ratio of phospholipid/Triton of 5 to 8 mol/mol indicates that detergent removal is less effective when this procedure is used.

The protein–phospholipid mixture can be fractionated by centrifugation in D_2O. The pellet is resuspended in 5 ml of D_2O containing 200 mM NaCl, 10 mM MOPS, pH 7.2, and centrifuged at 136,000 g for 1 hr. The turbid subphase may be separated from the floating layer by a Pasteur pipet. In the subphase containing about 80 to 90% of the protein the protein/phospholipid ratio is enhanced approximately 2-fold.

Remarks

1. When the one-step method is used, protein particles are found in unilamellar as well as in multilamellar structures. D_2O centrifugation fails to separate protein-loaded and -unloaded membranes.

2. The protein may be further enriched in the membranes by repeating the stepwise protein addition. Alternatively, higher amounts of protein may be added to a dried lipid film using appropriately higher amounts of Amberlite beads. This leads always to a limiting ratio of phospholipid/protein of 100 ± 20 mol/mol and to somewhat elevated Triton concentrations in the final membranes.

Protein Enrichment for ^{31}P NMR Studies in Detergent Micelles

The protein concentration as obtained after the standard purification procedure[1] should be enhanced at least 10-fold before ^{31}P spectroscopy on the residual phospholipids[10] in the solubilized ADP/ATP carrier can be performed.

Forty milliliters of the carrier solution containing 0.6 to 1 mg of protein per ml and 0.5–0.6% (w/v) Triton X-100 is pressure dialyzed in a 50 ml chamber using an Amicon PM-10 membrane filter to a final volume of 1 ml. Depending on the starting solution, the protein and Triton concentrations in the final sample are 15 to 30 mg/ml and 120 to 200 mg/ml, respectively.[11] At 9° and a spectrometer frequency of 36 MHz the ^{31}P NMR

[10] E. London and G. W. Feigenson, *J. Lipid Res.* **20,** 408 (1979).
[11] The time required for 40-fold volume reduction is about 3 hr.

signals of phosphatidylethanolamine and phosphatidylcholine appearing at -0.4 and -1.1 ppm, respectively, from 85% phosphoric acid as an external standard exhibit a line width of about 3 Hz. At Triton concentrations >200 mg/ml the signal line widths increase, presumably due to the formation of large nonspherical micelles. As described elsewhere,[3] the ^{31}P signal from cardiolipin is strongly broadened due to specific binding of this lipid to the carrier protein.

The CAT–protein–cardiolipin complex may be further purified by centrifugation in a sucrose density gradient. One milliliter of the concentrated protein solution as obtained after pressure dialysis, followed by 1 ml of the gradient buffer, is layered on the top of the 5–20% (w/v) sucrose gradient (40 ml) and then centrifuged 24 hr at 48,000 rpm in a Beckman 50 Ti vertical rotor.[12] The gradient contains 0.5% (w/v) Triton X-100, 100 mM NaCl, 0.05 mM EDTA, and 10 mM MOPS, pH 7.2. Fractions of 1 ml are collected. The protein-containing fractions are pooled and dialyzed three times against 700 ml buffer (0.5% Triton X-100, 200 mM NaCl, 10 mM MOPS, pH 7.2) for 3 hr. After pressure dialysis as described above only trace amounts of phosphatidylethanolamine and phosphatidylcholine are detectable by ^{31}P NMR spectroscopy. After extensive signal averaging the protein-bound cardiolipin appears as a broad signal of 30 to 40 Hz line width. The cardiolipin–protein interaction may be disrupted by denaturing the protein as reported elsewhere.[3]

Spin Labeling of the ADP/ATP Carrier

Synthesis of Spin-Labeled CAT (CATSL)

Fifty milligrams of carbonyldiimidazole is dissolved rapidly in 400 μl of essentially amine-free dimethylformamide (DMF).[13] The solution is added at once to 25 mg of powdered 2,2,5,5-tetramethyl-3-pyrroline-1-oxy-3-carboxylic acid. CO_2 development ceases after a few seconds. The mixture is allowed to stand for 10 min at room temperature. Twenty milligrams CAT in 400 μl H_2O is added and the reaction mixture is left overnight at room temperature. The white precipitate is discarded by centrifugation and the solvent is removed by lyophilization. The yellow oily residue is treated with 1.5 ml of carefully dried acetone. The insoluble precipitate is centrifuged, washed with another portion of acetone, and

---

[12] H. Hackenberg, Doctoral thesis, University of Munich, 1979.

[13] Traces of amine may be removed by stirring 5 ml of analytical grade DMF with 1.5 g of Dowex 50W ion-exchange resin in the H^+ form for 0.5 hr. After filtration the procedure is repeated. Then the solvent is treated twice with an equal volume of molecular sieve for 2 hr. Alternatively, DMF may be purified by fractional distillation of a mixture of 60 ml DMF, 8 ml benzene, and 2.4 ml water.

dried with a stream of nitrogen. The remaining white powder is dissolved in 200 μl of 50% (v/v) ethanol and applied to two 20 × 20 × 0.1 cm silica gel thin layer plates. The plates are developed in chloroform–methanol–water–acetic acid (55 : 22 : 4 : 8, v/v/v/v). Small strips are cut from both sides of the plates and stained with vanillin reagent (0.5% vanillin in 50% phosphoric acid). The 1 cm broad band at a R_f value of 0.38 containing the spin-labeled CAT is scraped off and the product is eluted by 50% (v/v) ethanol. Purification of thin-layer chromatography should be repeated in order to remove byproducts. The overall yield as determined by the spin density of the purified CATSL is 2 to 4%.

Spin Labeling with CATSL

Mitochondria are isolated from beef heart and stored at −176° as described previously.[14] The mitochondria are suspended at a concentration of 40 mg of protein per ml in 250 mM sucrose, 2 mM CaCl$_2$, 0.5 mM EGTA, and 10 mM MOPS, pH 7.2. In order to prevent spin reduction in the bound CATSL molecule, electron transport is inhibited by adding 0.5 mM KCN. The desired amount of the spin-labeled inhibitor in 10% (v/v) ethanol is added along with 50 μM ADP. After incubation at 10° for 10 min, the mitochondria are spun down at 12,000 g. Unbound CATSL is removed by washing the mitochondria three times in the same buffer. For the ESR experiment the pellet is sucked into a quartz flat cell or transferred to a quartz capillary. The capillary may be centrifuged for a few minutes at 12,000 g to enhance the spin density by further condensing the mitochondrial suspension.

A minor portion of nonspecific binding to the mitochondrial membrane becomes visible when the carrier sites are blocked by preincubation of the mitochondria with 0.6 mM CAT for 5 min before CATSL addition. This portion of the spin labeled inhibitor is weakly immobilized, in contrast to the extremely immobile carrier-bound CATSL. In order to obtain the pure spectral component from specifically bound CATSL, spectra recorded with and without CAT pretreatment may be computer subtracted. The various additions and spectral subtraction may be represented in short as

(mitochondria + CATSL) − (mitochondria + CAT + CATSL)

where the incubation sequence is put in parentheses and the minus sign means spectral subtraction.[15]

For spin labeling of the detergent-solubilized carrier, one has to start from the protein, prelabeled with [³H]ATR. The radioactive analog of

14 P. V. Blair, this series, Vol. 10, p. 78.
15 To simplify the scheme, the addition of ADP has been omitted.

ATR[5] serves as a probe for protein intactness, since ATR stabilizes the carrier protein somewhat less than CAT. ATR, in contrast to CAT, can be easily displaced by CATSL. The [^{3}H]ATR-loaded ADP/ATP carrier is isolated in 100 mM Na$_2$SO$_4$, 0.05 mM EDTA, 0.05 mM NaN$_3$, and 10 mM MOPS, pH 7.2, as detailed previously.[5] Before spin labeling the [^{3}H]ATR protein is pressure dialyzed to a final concentration of about 6 mg protein per ml. Binding of CATSL to Triton micelles can interfere with binding to the carrier protein. Thus the Triton concentration in the resulting solution should be adjusted to 0.4–0.5 mg of Triton per mg of protein by incubation with an appropriate amount of Amberlite XAD-2 beads for 0.5 hr at 0°. The Triton-binding capacity of the beads is about 70 mg of detergent per g of wet beads.[16] Excess CATSL may be eventually removed by chromatography of the mixture on a 0.8 × 10 cm Sephadex G-75 column eluted with the isolation buffer containing 0.2% Triton X-100.

Binding of Spin-Labeled Maleimides

Freshly prepared beef heart mitochondria (10 mg protein/ml) in the mitochondria buffer described in the preceding section are incubated with 0.8 mM of the desired maleimide spin label (MSL)[17] and 0.2 mM ADP at 0° for 5 min. Excess spin label is removed by washing the mitochondria twice with the same buffer. After this treatment the bulk of the mitochondrial SH groups is MSL labeled. The corresponding ESR spectrum exhibits strongly and weakly immobilized spin label portions. In a parallel experiment the mitochondria are incubated for 5 min with 0.2 mM CAT at 0° prior to spin labeling by MSL as described. Computer subtraction of the ESR spectra obtained in the absence and in the presence of CAT yields the spectrum of carrier-bound MSL[15]:

$$(\text{mitochondria} + \text{MSL}) - (\text{mitochondria} + \text{CAT} + \text{MSL})$$

SH groups in the detergent-solubilized ADP/ATP carrier can be labeled by MSL again starting from the protein preloaded with ATR or [^{3}H]ATR. Unlike spin labeling with CATSL, in the present procedure a conformational transition from the c- to the m-state of the carrier is induced, thereby unmasking two SH groups in the dimeric protein. This rearrangement is accompanied by the release of ATR.

The [^{3}H]ATR carrier complex is isolated by Triton X-100 and pressure dialyzed as described in the preceding section. Protein rearrangement and MSL binding are initiated by addition of 0.4 mM ADP, 0.3 μmol BKA/mg

---

[16] P. W. Holloway, *Anal. Biochem.* **53**, 304 (1973).

[17] A collection of various maleimide spin labels may be obtained from Aldrich, Milwaukee WI 53201.

of protein, and 0.3 μmol MSL/mg of protein. The mixture is incubated for 10 min at 4°. Unbound MSL is removed on a 0.8 × 10 cm Sephadex G-75 column. After pressure dialysis to a final protein concentration of about 2 to 3 mg/ml, the spin-labeled ADP/ATP carrier is transferred to a quartz capillary for ESR measurement.

As will be described in detail elsewhere,[6] considerable CAT-insensitive MSL binding can be observed when MSL molecules with long aliphatic chain between the maleimide and the nitroxyl moieties are employed. This binding portion can be visualized by replacing the inhibitor BKA in the above incubation mixture by 0.15 μmol CAT/mg of carrier protein. Computer subtraction then furnishes the ESR spectrum due to spin labeling of the SH groups unmasked in the conformational transition from the m- to the c-state. The procedure reads[15]:

$$(\text{Protein} + \text{BKA} + \text{MSL}) - (\text{Protein} + \text{CAT} + \text{MSL})$$

Probing the Membrane Sidedness of Spin-Labeled SH Groups

The membrane sidedness of carrier-bound MSL in mitochondria can be determined by pretreating the mitochondrial membrane with impermeable SH reagents or by chemical reduction of bound MSL from the cytosolic side, again making use of the potential of CAT to mask SH groups in the ADP/ATP carrier.

The mitochondrial suspension (10 mg protein/ml; either preincubated with CAT or without CAT pretreatment) is incubated for 2 min at 0° with 0.2 mM 2,2'-dinitro-5,5'-dithiodibenzoic acid (DTNB). Spin labeling of the mitochondria with MSL is performed as described above. Subtraction of ESR spectra from DTNB-treated mitochondria with and without CAT yields the portion of the carrier-specific SH-groups, which is inaccessible to DTNB:

$$(\text{mitochondria} + \text{DTNB} + \text{MSL}) -$$
$$(\text{mitochondria} + \text{CAT} + \text{DTNB} + \text{MSL})$$

Alternatively, 5 min after spin labeling by MSL, 1.6 mM glutathione and 16 mM of $(NH_4)_2Fe(SO_4)_2 \cdot 6H_2O$ is added to an aliquot of the mitochondrial suspension. The mitochondria are centrifuged at 12,000 g and the pellet is rapidly transferred to a quartz capillary.[18] The residual spin signal is equivalent to the total number of SH groups in the matrix side of the mitochondria that have been labeled by MSL. A second aliquot is

---

[18] The interval between Fe^{2+} addition and determination of spin density by ESR should not exceed 1.5 min, since slow leakage of the reducing agent into the matrix space leads to progressive reduction of the spin signal from matrix-bound MSL.

pretreated with CAT as described above prior to MSL addition and spin reduction. The difference of the ESR spectra from the two samples then allows the evaluation of the number of SH groups in the ADP/ATP carrier facing the matrix side of the mitochondrial membrane. The total labeling and spectral subtraction scheme is[15]

$$(\text{mitochondria} + \text{MSL} + \text{Fe}^{2+}) -$$
$$(\text{mitochondria} + \text{CAT} + \text{MSL} + \text{Fe}^{2+})$$

[50] Fluorescent Probes of the Mitochondrial ADP/ATP Carrier Protein

By Marc R. Block, François Boulay, Gérard Brandolin, Yves Dupont, Guy J. M. Lauquin, and Pierre V. Vignais

Extrinsic and intrinsic fluorescence probes have been used to study conformational changes and nucleotide binding sites in the mitochondrial ADP/ATP carrier protein. This carrier catalyzes the exchange between external and internal ADP and ATP across the inner membrane of mitochondria. The extrinsic fluorescence probes to be described below include representative fluorescent analogs of ADP or ATP, namely, naphthoyl-ADP and naphthoyl-ATP, two nontransportable nucleotides which however bind with high affinity to the ADP/ATP carrier, and formycin triphosphate, a transportable nucleotide. They also include derivatives of atractyloside (ATR) a specific inhibitor of ADP/ATP transport, namely dansyl-ATR, dansyl-4-aminobutyryl-ATR, and naphthoyl-ATR. The presence of tryptophanyl residues in the carrier molecule (five tryptophan per subunit of M_r 32,000) confers to the isolated protein a fluorescence which is sensitive to environmental changes. The fluorescence changes monitored by extrinsic and intrinsic fluorescent probes has led to the conclusion that the ADP/ATP carrier can adopt two conformations probably associated with the transport process.[1,2] The CATR conformation is recognized by carboxyatractyloside (CATR) and its derivative atractyloside (ATR), two inhibitors which attack the ADP/ATP carrier in mitochondria from the outside, whereas the BA conformation is recognized by bongkrekic acid (BA), an inhibitor which attacks the carrier from the inside, after penetration into the mitochondrial matrix.

[1] G. Brandolin, Y. Dupont, and P. V. Vignais, *Biochem. Biophys. Res. Commun.* **98,** 28 (1981).
[2] M. R. Block, G. J.-M. Lauquin, and P. V. Vignais, *Biochemisry* **22,** 2202 (1983).

In the experiments described, corrected fluorescence spectra are recorded on an SLM 8000 fluorimeter, and kinetic analysis of fluorescence changes are performed either with a Perkin-Elmer MPF 2A fluorimeter in the case of the membrane-bound ADP/ATP carrier (mitochondria), or with a high sensitivity spectrofluorimeter (BioLogic Co, ZIRST 38240 Meylan, France) equipped with a 75 W Xe (Hg) lamp. In all cases, a 1 × 1-cm thermostatted fluorescence quartz cuvette is used with the standard 90° geometry for the excitation and emission radiation; the content of the cuvette is continuously stirred by means of a small propeller rotating at 2500 rpm. Reagents are injected into the medium in volumes of 2 to 10 μl, using Hamilton automatic syringes. Control additions are made to correct the signal for dilution and quenching effects. The emission fluorescence light is collected by a photomultiplier through a focusing quartz lens and appropriate filters whose characteristics will be given later. The amplified and filtered signals are recorded on a standard potentiometric recorder.

Beef heart mitochondria are prepared by the method of Smith,[3] and rat heart mitochondria by the method of Chance and Hagihara.[4] The soluble ADP/ATP carrier protein is obtained from beef heart mitochondria after extraction with laurylamido-N,N'-dimethylpropylamine oxide (LAPAO)[5] and chromatography on hydroxyapatite.[6] Desalting of the carrier protein is performed by chromatography on AcA 202 (IBF) equilibrated in a medium containing 50 mM MOPS, 0.1 mM EDTA, and 0.5% (w/v) LAPAO at the final pH of 7.0.

Extrinsic Fluorescent Probes

Fluorescent Nucleotides

Synthesis of Naphthoyl-ADP and Naphthoyl-ATP (N-ADP and N-ATP). The method used[7] is based on the principle of esterification of nucleotides at the level of the ribose moiety as described by Gottikh *et al.*[8] and adapted by Guillory and Jeng.[9] It includes an activation step of naphthoic acid by a carboxyl reagent: 1,1'-carbonylbisimidazole. 1-Naph-

[3] A. L. Smith, this series, Vol. 10, p. 81.

[4] B. Chance and B. Hagihara, *Proc. Int. Congr. Biochem., 5th, Moscow* **5**, 3 (1963).

[5] G. Brandolin, J. Doussière, A. Gulik, T. Gulik-Krzywicki, G. J.-M. Lauquin, and P. V. Vignais, *Biochim. Biophys. Acta* **592**, 592 (1980).

[6] P. Riccio. H. Aquila, and M. Klingenberg, *FEBS Lett.* **56**, 133 (1975).

[7] M. R. Block, G. J.-M. Lauquin, and P. V. Vignais, *Biochemistry* **21**, 5451 (1982).

[8] B. P. Gottikh, A. A. Kraysesky, N. B. Tarussova, P. P. Purygin, and T. L. Tsilevich, *Tetrahedron* **26**, 4419 (1985).

[9] R. J. Guillory and S. J. Jeng, this series, Vol. 46, 259.

thoic acid (0.1 mmol) and 1,1'-carbonylbisimidazole (0.3 mmol) are dissolved in 0.1 ml of dimethylformamide dried over molecular sieves and calcium hydride. The mixture is stirred for 15 min prior to introducing 50 μmol of ADP or ATP dissolved in 0.5 ml of water. The reaction is allowed to proceed for 2 hr at room temperature. The solvents are evaporated to dryness, and the gummy residue is washed three times with acetone followed by centrifugation, to eliminate the excess of naphthoic acid and imidazole. The acetone washed residue is finally dissolved in a mixture of ethanol/water (1/1, v/v) and applied to a sheet of Whatman paper No. 3. The chromatogram is developed with *n*-butanol, acetic acid, and water (5/2/3, v/v/v). N-ADP and N-ATP can be readily identified by their deep violet fluorescence under ultraviolet light; they have R_f values of 0.62 and 0.51, respectively. The fluorescent strips corresponding to the derivatives are cut out and eluted with water. The molecular extinction coefficients of N-ADP and N-ATP at pH 7 are 6200 M^{-1} cm^{-1} at 300 nm and 15,400 M^{-1} cm^{-1} at 260 nm. For both derivatives in solution in a buffered medium at a pH between 6.5 and 7.5, the fluorescence spectrum is characterized by an excitation peak at 310 nm and by an emission peak at 395 nm.

Use of N-ADP and N-ATP as Fluorescent Reporters of the Conformational State of the ADP/ATP Carrier. Principle. N-ADP and N-ATP bind to mitochondria, but they are not transported. They inhibit competitively ADP/ATP transport. When [^{14}C]N-ADP or [^{14}C]N-ATP is incubated with mitochondria, a large amount of radioactivity is recovered in the pellet after centrifugation. The bound [^{14}C]N-ADP and [^{14}C]N-ATP which are displaced by CATR are considered to be bound to the nucleotide sites of the ADP/ATP carrier; they are referred to as specifically bound N-ADP and N-ATP. N-ADP or N-ATP binds to apparently homogeneous sites; the number of sites is 1.4–1.6 nmol/mg protein and the K_d value 3 μM.[7] In fluorescence experiments, the release of the specifically bound N-ADP by either CATR or BA is accompanied by a fluorescence increase.[2] N-ADP can therefore be used as a reporter of bound CATR or BA on the ADP/ATP carrier.

Fluorescent assay. Beef heart mitochondria (2 mg protein) are suspended in 2.5 ml of a saline medium made of 120 mM KCl, 1 mM EDTA, 10 mM Mes, pH 6.5 with 5 μM N-ADP in a 1 × 1-cm fluorescence cuvette. The temperature is maintained at 10°. After 3 min, i.e., when fluorescence has attained a steady level, 10 nmol CATR is added (this amount of CATR corresponds to a saturating concentration). The fluorescence level increases rapidly; this rapid phase which lasts for less than 0.5 sec is followed by a slow one which reaches completion in about 20 min (Fig. 1, trace 2). Addition of 10 nmol BA, a saturating concentration (trace 1) or 10 nmol ADP (trace 3) shortens the slow phase period to a few

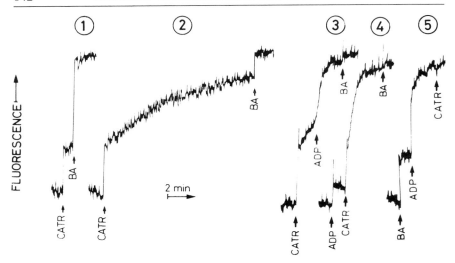

Fig. 1. ADP-induced stimulation of release of bound N-ADP from beef heart mitochondria upon addition of CATR or BA. Details are given in the text.

seconds. A final addition of BA does not modify the plateau of fluorescence attained after addition of ADP. Similar results are obtained when CATR is replaced by BA (10 nmol) (Fig. 1, trace 5). When the same amount of ADP (10 nmol) is added first in the absence of CATR or BA (Fig. 1, trace 4), only a very small increase of the fluorescence level results; this small effect of ADP in the absence of CATR and BA is attributable to a competitive release of bound N-ADP by ADP. Consequently, the large effect of ADP observed in the presence of CATR and BA cannot be due to a competitive effect. It is known that CATR and BA cannot bind simultaneously to the same carrier unit.[10] To explain the large effect of ADP on bound N-ADP, it is therefore proposed that in the mitochondrial membrane, the carrier units are distributed in two categories, depending on their accessibility and reactivity to CATR and BA; those reacting with CATR are identified by the CATR-induced release of bound N-ADP, and those reacting with BA by the BA-induced release of bound N-ADP. For convenience, the carriers of the first type are said to be in the CATR conformation and those of the second type in the BA conformation. In the absence of the external ADP or ATP, the spontaneous transition between the two conformations is slow, as reflected by the slowness of the second phase of the CATR- or BA-induced release of

[10] M. R. Block, R. Pougeois, and G. J.-M. Lauquin, *FEBS Lett.* **117**, 335 (1980).

bound N-ADP. It is considerably accelerated by the addition of small amounts of ADP or ATP; the concentration of ADP required for the half-maximal stimulation of N-ADP release is 3 μM, a value similar to the K_m found for ADP transport.[11]

Assay of Nucleotide Binding to a Soluble Purified Preparation of the ADP/ATP Carrier Protein. Two fluorescent nucleotides, N-ATP and for-mycin triphosphate (FTP), will be considered.[12,13] In contrast to N-ATP which binds to the ADP/ATP carrier, but is not transported, FTP is transported in rat heart mitochondria at a rate which is about half that measured for ATP; the K_m is in the range of 20–50 μM.[13] In experiments with N-ATP, the emitted light is recovered through a K_1 (410 nm) Balzers filter. In the case of FTP (λ_{ex} 305 nm, λ_{em} 370 nm), the emitted light is recovered through a 0-52 Corning filter coupled to a 7-54 Corning filter, the resulting band-pass being centered at 370 nm.

In routine assays, the fluorescence cuvette is filled with 2.0 ml of a medium consisting of 80 mM glycerol, 40 mM MOPS, pH 7.0, 4×10^{-5} M EDTA, and 0.2% laurylamido-N,N-dimethylpropylamine oxide (w/v). The purified carrier protein is added at a concentration of 0.04–0.05 mg/ml. Binding of N-ATP results in a decrease of fluorescence, and that of FTP in an increase of fluorescence. Conversely, the CATR-induced release of bound N-ATP results in an increase of fluorescence[12] and the CATR-induced release of bound FTP in a fluorescence decrease.[13]

The specific binding of N-ATP or FTP is assessed as follows. To the cuvette containing the medium and 0.08 mg of the ADP/ATP carrier protein, N-ATP and FTP are added at final concentrations of 2 and 5 μM, respectively (final volume 2 ml). After 1 min, when the fluorescence has reached a stable level, CATR is added. The fluorescence level immediately increases in the case of bound N-ATP, and on the contrary decreases in the case of bound FTP, corresponding to the release of N-ATP or FTP, respectively (Fig. 2). Note that 50 μM CATR are required for the full release of bound FTP. Based on the CATR-induced fluorescence changes, the titration of N-ATP and FTP binding sites in the soluble ADP/ATP carrier has been performed.[12,13] The nucleotides are added at increasing concentrations up to 10 μM. For each concentration of N-ATP, when the fluorescence has reached a steady level, the specifically bound N-ATP is determined from the relative fluorescence change $\Delta F/\Delta F_{max}$ induced by addition of 2 μM CATR; the same procedure is followed for FTP, except that in this case, CATR is used at a much higher concentra-

[11] E. D. Duée and P. V. Vignais, *J. Biol. Chem.* **244**, 3920 (1969).
[12] Y. Dupont, G. Brandolin, and P. V. Vignais, *Biochemistry* **21**, 6343 (1982).
[13] G. Brandolin, Y. Dupont, and P. V. Vignais, *Biochemistry* **21**, 6348 (1982).

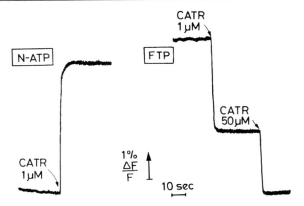

FIG. 2. CATR-induced fluroescence changes of bound N-ATP and FTP in the soluble ADP/ATP carrier protein. Details are given in the text.

tion, $\simeq 50\ \mu M$. The binding parameters of N-ATP and FTP with respect to the soluble carrier in detergent have been assessed by this procedure; they can be summarized as follows. Two different affinity classes of N-ATP binding sites ($K_{d_1} < 10$ nM and $K_{d_2} \simeq 0.45\ \mu M$) are present in equal amounts on the ADP/ATP carrier. In the case of FTP, four classes of sites differing by affinity ($K_{d_1} < 10$ nM, and K_{d_2}, K_{d_3}, K_{d_4} ranging between 0.5 and 2 μM) are demonstrated although the total number of FTP sites per mg of carrier protein is the same as the total number of N-ATP sites; the differences in the binding properties of N-ATP and FTP have been discussed in terms of specific interactions of the ADP/ATP carrier with nontransportable and transportable nucleotides. Only the transportable nucleotides would induce a conformational change of the carrier responsible for the splitting of two classes of binding sites into four classes of sites.[12,13]

Fluorescent Derivatives of ATR

The formulas of the three fluorescent derivatives described thereafter, 6'-O-dansyl-ATR, 6'-O-dansyl-4-aminobutyryl-ATR, and naphthoyl-ATR[14] are listed in Fig. 3.

6'-O-Dansyl-ATR. ATR (12.5 μmol) and dansyl chloride (37 μmol) are dissolved in pyridine freshly distilled over KOH-ninhydrin and dried over calcium hydride. The mixture is stirred at room temperature in the dark for 16 hr. Pyridine is evaporated *in vacuo* between 35 and 40°. The gummy

[14] F. Boulay, G. Brandolin, G. J.-M. Lauquin, and P. V. Vignais, *Anal. Biochem.* **128**, 323 (1983).

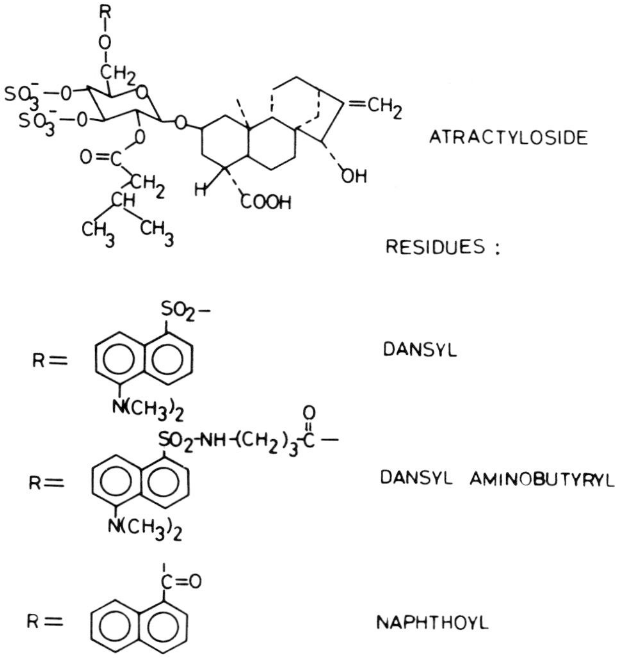

FIG. 3. Adducts used in synthesis of fluorescent derivatives of atractyloside.

residue is dissolved in 2 ml of methanol/water (50/50, v/v) and filtered through Millex FG. The filtrate is then subjected to HPLC using a μBondapak C_{18} column (300 × 7.8 mm, 10 μm) and a linear methanol gradient in acetic acid and NH_4 acetate obtained with two pumps, one delivering a mixture of acetic acid, 1 M NH_4 acetate, and water 1/1/98 (v/v/v) and the other, a mixture of acetic acid, 1 M NH_4 acetate, and methanol (1/1/98, v/v/v). Dansyl-ATR is eluted at a concentration of methanol of 90%.

The purity of dansyl-ATR can be checked by ascending chromatography on a Whatman K6 silica gel plate. The chromatogram is developed in chloroform/methanol/acetic acid/water (60/20/0.5/0.5, v/v/v). Dansyl-ATR is identified as a single fluorescent spot with an R_f value of 0.40. In 0.02 M phosphate buffer, pH 7.2, dansyl-ATR shows two peaks of absorption, one at 250 nm, the other at 332 nm. The concentration of dansyl-ATR is calculated from the absorption peak at 332 nm, using the extinction coefficient given for dansyl amino acids, i.e., 3400 M^{-1} cm^{-1}.

Using an excitation light centered at 340 nm, the fluorescence emission spectrum of dansyl-ATR in dimethylformamide, ethanol, or pyridine exhibits a peak at 522 nm. The fluorescence of dansyl-ATR is much higher in apolar solvents compared to aqueous solutions; for example in 0.02 M phosphate buffer, pH 7.2, the emission peak is 40 times lower than in dimethylformamide, ethanol, or pyridine.

6'-O-Dansyl-4-aminobutyryl-ATR. Dansyl chloride (600 μmol) in 2 ml of anhydrous acetone is added dropwise with stirring to a solution of 4-aminobutyric acid (300 μmol dissolved in 6 ml of pyridine and water in equal proportions). A pH of 9 is maintained by addition of triethylamine. Stirring is continued for 12 hr at room temperature. Pyridine is removed under reduced pressure below 30°. The residual solid is dissolved in 5 ml of water and the aqueous solution brought to pH 2 by addition of 2 N HCl. Dansyl-4-aminobutyric acid is extracted from the acid solution into 10 ml of ethylacetate. This extraction is repeated three times. The pooled extracts are washed with 20 ml of saturated NaCl solution and evaporated under reduced pressure below 30°. The residue is solubilized in 0.01 M NaOH; after a final extraction by diethyl ether, the alkaline solution containing the purified dansyl-4-aminobutyric acid is neutralized. An amount of dansyl-4-aminobutyric acid corresponding to about 10 μmol is dried under nitrogen and treated with diethyl ether to remove all traces of water. The residue is solubilized in 1 ml of anhydrous pyridine, and a stoichiometric amount of thionyl chloride is added and left to react for 3 min to form the chloride derivative of dansyl-4-aminobutyric acid. Then ATR (10 μmol) in 2 ml of anhydrous pyridine is added and the coupling reaction is left to proceed overnight at 4°. Pyridine is removed by rotary evaporation at 20°. The unreacted dansyl-4-aminobutyric acid is removed by repeated washings with anhydrous acetone. Dansyl-4-aminobutyryl-ATR is finally purified by HPLC using the same procedure as that described for the purification of dansyl-ATR. Both derivatives have the same retention time with the methanol gradient used. The absorption and fluorescent spectra of dansyl-4-aminobutyryl-ATR and dansyl-ATR are the same. In thin-layer chromatography under the same conditions as those used for dansyl-ATR, a single fluorescent spot with an R_f value of 0.45 is found.

6-O'-Naphthoyl-ATR. 1-Naphthoic acid (100 μmol) and carbonyl-diimidazole (300 μmol) are mixed in 100 μl of dry dimethylformamide. After 30 min at room temperature, ATR (12 μmol in 0.4 ml of water) is added with mixing. The suspension immediately becomes opaque and then clarifies after stirring for 2 hr. Unreacted 1-naphthoic acid is removed by repeated extractions with 5 ml of diethyl ether each time. The water phase is evaporated to dryness *in vacuo*. The residue is dissolved in 0.5 ml of a mixture of methanol/water (50/50, v/v). The solution is processed by HPLC with a μBondapak C_{18} column (300 × 7.8 mm, 10 μm) using the same methanol gradient as that described for the preparation of dansyl-ATR. Naphthoyl-ATR is eluted with 72% methanol. The purity of naphthoyl-ATR can be checked by chromatography on TLC with the same solvent system as that used for dansyl-ATR; naphthoyl-ATR migrates with an R_f of 0.35.

The fluorescence excitation peak of naphthoyl-ATR is at 312 nm. In contrast to dansyl-ATR, the relative intensity of the fluorescent emission of naphthoyl-ATR is much higher in aqueous solutions than in organic solvents. The peak of fluorescence emission is at 390 nm in 10 mM phosphate buffer, pH 7.2 and displaced to 360 nm in dimethylformamide.

Use of the Fluorescent Derivatives of ATR to Probe the CATR and BA Conformations. Dansyl-ATR, dansyl-4-aminobutyryl-ATR, and naphthoyl-ATR behave as competitive inhibitors of ADP transport in mitochondria with K_i values of about 50 nM. A typical experiment bearing on fluorescence changes of bound dansyl-4-aminobutyryl-ATR (DG-ATR) upon addition of the specific ligands ADP, CATR, and BA is described. The fluorescence emission of DG-ATR is measured at 520 nm, with an excitation light centered at 340 nm. The thermostated fluorescence cuvette ($t = 25°$) is filled with 2.5 ml of a medium consisting of 250 mM sucrose, 1 mM EDTA, and 10 mM HEPES, pH 7.0. An aliquot of 0.1 ml of a suspension of beef heart mitochondria in 250 mM sucrose corresponding to 0.6 mg protein is added and thoroughly mixed. DG-ATR is added to the final concentration of 1.5 μM (Fig. 4). Fluorescence increases abruptly to reach a plateau (trace 1). Addition of CATR (20 μM) results in a fluorescence decrease (ΔF) to a stable value in about 6 min. A control experiment (trace 2) is carried out with CATR added prior to DG-ATR; the fluorescence increase levels to the same value as that attained in the first experiment after addition of CATR. These two experiments taken together indicate that the change of fluorescence ΔF is linked to the occu-

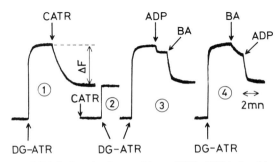

Fig. 4. CATR- and BA-induced release of bound DG-ATR in beef heart mitochondria. For details, see text.

pancy of a specific site on the ADP/ATP carrier by DG-ATR; the increase of fluorescence corresponds to the binding of DG-ATR, and conversely the decrease of fluorescence corresponds to the release of DG-ATR, resulting from the occupancy of the DG-ATR site by CATR.

The following two experiments concern the effect of ADP on the binding of BA and the displacement by BA of the bound DG-ATR. In trace 3, ADP (50 μM) is added after the maximal fluorescence has been reached following addition of DG-ATR; a small decrease of fluorescence is observed, probably due to competition with DG-ATR. Addition of BA (5 μM) results in a larger decrease of fluorescence to a stable value equal to that obtained after addition of CATR (trace 1). The reverse order of additions, i.e., BA added prior to ADP (trace 4) shows a limited effect of BA on the fluorescence decrease; this effect is amplified by ADP, and the fluorescence decrease attains the same stable value as in the preceding traces. ADP can be replaced by ATP with similar effects. These results can be interpreted on the same basis as those of the experiment carried out with naphthoyl-ADP, i.e., the ADP/ATP carrier in the mitochondrial membrane exists in two conformations, the CATR conformation and the BA conformation. The CATR conformation binds DG-ATR, which results in a maximal fluorescent signal. The bound DG-ATR is removed by CATR by an effect of direct competition, with the concomitant fluorescence change, ΔF. At the temperature of the experiment, 25°, all the carrier units rapidly reach the CATR conformation upon addition of DG-ATR. On the other hand the transition from the CATR conformation to the BA conformation is accelerated by addition of ADP (or ATP).

Tryptophanyl Residues of the ADP/ATP Carrier as Intrinsic Fluorescence Probes

When irradiated with a light beam centered at 295 nm, the ADP/ATP carrier protein in solution shows a fluorescence emission spectrum characteristic of tryptophan with a broak peak centered at 330 nm. Upon addition of ADP or ATP, the emission fluorescence peak is shifted to the red and centered at 340 nm. This modification in the fluorescence emission spectrum is prevented by CATR. The kinetics of the ADP- or ATP-induced fluorescence changes have been investigated with the BioLogic spectrofluorimeter, using an excitation wavelength at 295 nm and an optical filter system with a band-pass centered at 355 nm. This filter system consists of a 0-54 Corning filter coupled to an ultraviolet light 7-54 Corning filter. Due to the displacement of the spectrum to the red when the carrier protein is incubated with ADP or ATP, the fluorescent signal centered at 355 nm is increased when ADP or ATP is present in the medium.

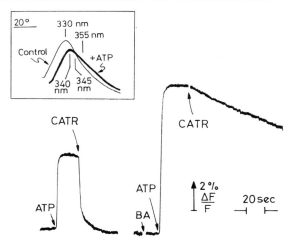

FIG. 5. ATP-induced change in the intrinsic fluorescence of the soluble ADP/ATP carrier protein. For details, see text.

Typical kinetics of fluorescence changes induced by ATR, CATR, or BA are presented in Fig. 5. A sample of freshly prepared ADP/ATP carrier corresponding to 0.3–0.4 mg protein is mixed with a medium consisting of 80 mM glycerol, 20 mM MOPS, 0.4 mM EDTA, and 0.5% (w/v) laurylamido-N,N-dimethylpropylamine oxide (final volume 2 ml). The cuvette is thermostated at 20°. After stabilization of the fluorescence base line, 5 μl of 1 mM ATP is added. Fluorescence increases rapidly, the increase corresponding to 5% of the initial fluorescence (Fig. 5). Upon addition of 5 μl of 1 mM CATR within a 2- to 3-min period, the ADP or ATP induced fluorescence signal is completely reversed. When added prior to ADP or ATP, CATR prevents the fluorescence rise. The fluorescent response obtained with ATP is virtually doubled when 5 μl of 1 mM BA is added prior to ATP. Here again, when the signal had reached its maximum, the addition of CATR results in full reversal. A full description of the experiments performed on the ADP(ATP)-, CATR-, and BA-induced changes of the intrinsic fluorescence of the soluble ADP/ATP carrier protein can be found in a recent paper by Brandolin et al.[15]

[15] G. Brandolin, Y. Dupont, and P. V. Vignais, *Biochemistry* **24,** 1991 (1985).

[51] ADP/ATP Carrier: Analysis of Transmembrane Folding Using Pyridoxal Phosphate

By Werner Bogner and Heinrich Aquila

Chemical modification is an approach to study the disposition of components at the membrane surface. It can be extended to evaluate the path of a polypeptide chain through the membrane if several modifications with impermeable reagents from the two sides of the membrane are performed and if the positions of these modifications within the sequence of the protein are determined.

The primary structure of the ADP/ATP carrier from beef heart mitochondria has been determined by sequencing the protein.[1] The lysines are the most frequently occurring residues for which specific reagents exist. Twenty-three such residues are distributed rather homogeneously over the polypeptide chain. The criteria which a surface-labeling reagent should meet have already been discussed in general by Tinberg and Packer.[2] The substrates and the specific inhibitors of the carrier, CAT and BKA,[3] are negatively charged molecules demanding positively charged residues at the protein surface for binding. Therefore modification of the lysine residues could, in addition, show residues involved in binding of the inhibitors. We selected pyridoxal phosphate for this purpose as it reacts faster with the carrier protein than trinitrobenzene sulfonate, as monitored by the remaining CAT-binding capacity. The reagent has also been described to be a kind of affinity reagent for phosphate binding centers.[4]

Since the carrier protein undergoes extensive conformational changes in the translocation step,[5] the protein has to be fixed in a defined conformation by inhibitor binding for the studies described here.[6,7] On the other hand, the specific inhibitors are relatively large molecules, which could

---

[1] H. Aquila, D. Misra, M. Eulitz, and M. Klingenberg, *Hoppe-Seyler's Z. Physiol. Chem.* **363**, 345 (1982).

[2] H. M. Tinberg and L. Packer, this series, Vol. 56 [55].

[3] Abbreviations: ATR, atractylate; BKA, bongkrekate; CAT, carboxyatractylate; DABTC, dimethylaminoazobenzene thiocarbonyl-; HPLC, high-pressure liquid chromatography; MOPS, morpholinopropanesulfonic acid; PIPES, piperazine-*N*,*N*'-bis(2-ethanesulfonic acid); PLP, pyridoxal phosphate or phosphopyridoxyl-; SMP, submitochondrial particles.

[4] G. E. Means and R. E. Feeney, *J. Biol. Chem.* **246**, 5532 (1971).

[5] M. Klingenberg, *in* "Dynamics of Energy-Transducing Membranes" (L. Ernster *et al.*, eds.), p. 511. Elsevier, Amsterdam, 1974.

prevent the access of a reagent by sterical hindrance. Therefore either of two different inhibitors was used: (1) atractylate or its carboxy derivative, bound from the cytosolic side to the protein in the c-state and (2) bongkrekate, bound from the matrix side to the carrier in the m-state.[8,9] Labeling of the protein from the c-side with the impermeable reagent pyridoxal phosphate (PLP) was performed in mitochondria (right-side-out), while labeling from the m-side was obtained with submitochondrial particles (inside-out).

Preparations

Beef heart mitochondria were isolated in Tris–sucrose buffer as described by Blair.[10] Loading of the mitochondria with the inhibitors atractylate, carboxyatractylate, or bongkrekate has been described.[6,11] Submitochondrial particles (SMP) were prepared from a portion of these mitochondria by freezing and thawing followed by sonication at 0° with a Branson Sonifier Model J17V, maximal output, for 10 × 5 sec at a protein concentration of 20 mg/ml in a buffer containing 50 mM sucrose, 15 mM MgCl$_2$, 10 mM PIPES, pH 7.4.[12] The particles of a 18,000 g supernatant were collected at 140,000 g for 30 min, yield 30–35%. In order to label pyridoxal phosphate, it was reduced with NaB^{3}H$_4$ and reoxidized with MnO$_2$, as described by Stock et $al.$[13] (see also Ref. 14). The yield of [^{3}H]PLP was 30–40% with a specific radioactivity of 1–3 × 10^6 dpm/ nmol. It was stored in portions at −32°.

Methods

Incubation with PLP. Mitochondria or submitochondrial particles loaded with the different inhibitors are suspended in 250 mM sucrose, 50

[6] H. Aquila, W. Eiermann, W. Babel, and M. Klingenberg, *Eur. J. Biochem.* **85,** 549 (1978).

[7] B. B. Buchanan, W. Eiermann, P. Riccio, H. Aquila, and M. Klingenberg, *Proc. Natl. Acad. Sci. U.S.A.* **73,** 2280 (1976).

[8] M. Klingenberg, P. Riccio, H. Aquila, B. Schmidt, K. Grebe, and P. Topitsch, *in* "Membrane Proteins in Transport and Phosphorylation" (G. F. Azzone *et al.* eds.), p. 229. North-Holland Publ., Amsterdam, 1974.

[9] M. Klingenberg, *Enzymes Biol. Membr.* **3,** 383 (1976).

[10] P. V. Blair, this series, Vol. 10 [12].

[11] H. Aquila and M. Klingenberg, *Eur. J. Biochem.* **122,** 141 (1982).

[12] M. Klingenberg, *Eur. J. Biochem.* **76,** 553 (1977).

[13] A. Stock, F. Ortanderl, and G. Pfleiderer, *Biochem. Z.* **344,** 353 (1966).

[14] O. Ribaud and M. F. Goldberg, *FEBS Lett.* **40,** 41 (1974).

mM triethanolamine–HCl, 2.5 mM MgCl$_2$, 0.5 mM EDTA, pH 8.0, to a final concentration of 20 mg/ml and allowed to react at room temperature with 6 mM [³H]PLP for 15 min in the dark. The temperature is lowered to 5° and the Schiff's bases are reduced with a slight excess of NaBH$_4$ until the yellow color disappears. Ten minutes later the suspension is diluted with 2 volumes of cold buffer and centrifuged at 20,000 g for 20 min. The loose pellet is resuspended in 250 mM sucrose, 10 mM MOPS, pH 8.2, and centrifuged again.

Isolation of the Labeled Carrier Protein

The PLP-modified carrier is conventionally isolated as CAT complex.[15,16] For the modified ATR[11] or BKA[6] complexes a revised procedure is used. The isolation procedure follows mainly that for the BKA protein[6,16] with the following modifications: 100 μM p-chloromercuribenzoate is present during Brij 58 extraction to prevent the proteolytic breakdown,[6] and 3 min after Triton X-100 addition 1 ml hydroxylapatite suspension per 30 mg protein is added. The suspension is stirred for 3 min and then centrifuged for 1 min at 10,000 g. The supernatant is applied to the hydroxylapatite column as described in Ref. 16.

Alternatively the ATR or CAT protein is isolated first and then labeled at a protein concentration of 1 mg/ml with PLP as described above for mitochondria or SMP. Addition of 0.6 ml of cold acetone and 0.1 ml of 50% trichloroacetic acid to 1 ml solution of the isolated PLP-labeled inhibitor complexes will precipitate the protein which is collected by centrifugation and washed twice with water and acetone.

Delipidation, Carboxymethylation, Citraconylation, and Cleavage with Thermolysin

Delipidation. Fresh lower phase and fresh upper phase are prepared according to Folch as described in Ref. 17. Twenty milligrams of protein is dissolved in 0.5 ml formic acid at 0° and 15 ml of chloroform/methanol (2 : 1, v/v) is then added. After 10 min, 3 ml water is added and the milky mixture is stirred for 30 min. The phases are separated by centrifugation for 20 min at 3000 g. The lipid-containing lower phase is removed and the upper and interphase washed with fresh lower phase. The pooled lower

[15] P. Riccio, H. Aquila, and M. Klingenberg, *FEBS Lett.* **56,** 133 (1975).
[16] M. Klingenberg, H. Aquila, and P. Riccio, this series, Vol. 56 [36].
[17] N. S. Radin, this series, Vol. 14 [44].

phases are washed with fresh upper phase. The upper phases and the interphase are combined and concentrated in a rotatory evaporator and finally lyophilized.

Carboxymethylation.[18] Ten milligrams of delipidated protein is dissolved in 1 ml of a solution containing 8 M guanidinium chloride, 500 mM Tris–chloride, 0.2 mM EDTA, pH 8.2, and reduced with 3 μmol dithioerythritol at 50° for 2 hr in a N_2 atmosphere. Sodium iodoacetate (6.6 μmol) is added at 20° and the reaction is stopped after 30 min with 6% thioglycolic acid. The protein is dialyzed extensively against water and the precipitate is collected by centrifugation and dried by lyophilization.

Citraconylation.[19] Ten milligrams of carboxymethylated protein is dissolved in 1 ml 6 M guanidinium–chloride at 0° and 60 μl citraconic anhydride is added within 1 hr in 10 μl portions. The pH measured with a glass electrode was kept between 8.6 and 8.8 by means of 40% NaOH. *O*-Methylmaleyl ethers, probably formed as by-products, are destroyed with 1 M hydroxylamine at pH 8.6 for 1 hr, then the mixture is dialyzed against water containing 50 μl ammonia per liter for at least 30 hr. During this period the initially precipitating protein dissolves again. The solution is lyophilized.

Cleavage. The citraconylated protein is dissolved in 4 ml 100 mM N-methylmorpholine acetate, pH 8.2, at 52° and cleaved with 100 μg thermolysin (Serva) for 2 hr. After lyophilization the peptides are decitraconylated with 2 ml 20% acetic acid overnight. The acid is removed by lyophilization.

Separation of the Peptides

The lyophilized sample, dissolved in 100 μl 80% formic acid, is diluted with 700 μl water and immediately applied to HPLC. A reversed-phase column (Merck, Lichrosorb RP 18, 4.6 × 250 mm, 10 μm) thermostatted at 50° and equilibrated with 50 mM ammonium acetate, pH 4.0, is used. The peptides are eluted at a flow rate of 1 ml/min by means of 100 ml of a linear gradient ranging from 0 to 50% (v/v) of acetonitrile in equilibration buffer. The absorbance at 215 nm is recorded and the ^{3}H radioactivity of 0.8 ml fractions is determined by counting 10-μl aliquots. After concentration to about 20 μl by means of a Speed Vac Concentrator (Savant), the fractions are pooled according to the radioactivity profile and lyophilized.

The different pools are further purified by thin-layer peptide mapping

---

[18] A. M. Crestfield, S. Moore, and W. H. Stein, *J. Biol. Chem.* **238**, 622 (1963).
[19] P. J. G. Butler and B. S. C. Hartley, this series, Vol. 25 [14].

following the description of Heiland *et al.*[20] A sample of the pool (5–10 nmol) in 20% acetic acid is applied to the origin on a 20 × 20-cm plastic-backed cellulose thin-layer plate (Macherey and Nagel, Polygram Cell 300) and electrophoresis is performed for 2 to 3 hr at 400 V and 15–20° in a buffer containing pyridine/acetic acid/acetone/water (20 : 40 : 150 : 790, v/v), pH 4.4. After drying of the plates under a well-ventilated hood, ascending chromatography in the second dimension with the solvent *n*-butanol/acetic acid/water/pyridine (15 : 3 : 12 : 10, v/v) is performed (6–7 hr). The plates are dried under a hood again. Pyridoxyllysine-containing peptides can be detected by blue fluorescence on illumination at 350 nm. The peptides of stronger spots are recovered by scraping and eluting 3 times with 200 μl 20% acetic acid. The acid is removed by lyophilization and the peptide dissolved in 100 μl 50% pyridine. The peptides obtained are almost pure. Since thermolysin, a protein with a broad specificity, is used, several overlapping peptides are generated.

A 2-μl aliquot is used for liquid scintillation counting, 20 μl is hydrolyzed, and the amino acid composition determined. The rest is subjected to sequence determination.

Sequencing of the Peptides

The labeled peptides are sequenced by manual Edman degradation, applying the 4-*N*,*N*-dimethylaminoazobenzene 4′-isothiocyanate/phenyl isothiocyanate double coupling method of Chang *et al.*[21] The procedure as described in full detail by Chang[22] is modified in the following way. The conversion is carried out in 50% trifluoroacetic acid for 50 min at 50°. For the primary identification of the thiohydantoin derivatives, 3 × 3-cm polyamide sheets (Schleicher and Schuell, F 1700) are used with DABTC-diethylamine and DABTC-ethanolamine as internal references. Leucine is separated from isoleucine by thin-layer chromatography of the thiohydantoins on silica plates (Merck, HPTLC precoated plates Silica gel 60, without fluorescent indicator) with the solvent chloroform/ethyl acetate (90 : 10, v/v). After the acid treatment, the product of the *N*-ε-pyridoxyl derivative cannot be identified directly since it is not soluble enough in butylacetate, the extractant after the cleavage step (see also ref. 23). However, the position of the modified lysine residue can be disclosed by liquid scintillation counting of the extract when [^{3}H]pyridoxal phosphate

[20] I. Heiland, D. Brauer, and B. Wittmann-Liebold, *Hoppe-Seyler's Z. Physiol. Chem.* **357**, 1751 (1976).
[21] J. Y. Chang, D. Brauer, and B. Wittmann-Liebold, *FEBS Lett.* **93**, 205 (1978).
[22] J. Y. Chang, this series, Vol. 91 [41].
[23] S. Tanase, H. Kojima, and Y. Morino, *Biochemistry* **18**, 3002 (1979).

is used. There is good agreement of the maxima of radioactivity with the lysine positions known from the amino acid sequence.

Examples of Results and Comments

General Remarks

Initially the influence of PLP on functional parameters was studied. It was found that PLP (5 mM) inhibits ADP/ATP exchange and also binding of specific inhibitor ligands such as [³H]CAT, [³H]ATR, and [³H]BKA in mitochondria to about 80%. This indicates that one or more lysine groups are involved in the binding of inhibitors and perhaps of substrates as well. If the carrier in the mitochondria was first loaded with these ligands, the same exposure to PLP did not markedly remove [³H]CAT or [³H]BKA, however, it released [³H]ATR to about 70%. This is explained by the weaker affinity of ATR to the carrier. If inhibitor complexes of the isolated carrier protein were treated with PLP, [³H]CAT remained bound whereas both [³H]ATR and [³H]BKA were released to about 80–90%, according to the marked weakening of BKA binding in the isolated carrier.[6] Since after removal of BKA the isolated protein in the m-state tends to denature, no further studies were performed with this material. In contrast, the isolated protein in the c-state, after removal of ATR, seems to stay in the native state, as indicated by CD spectra.

Incorporation of PLP into the Isolated Complexes

In order to localize lysine residues at the surface of the protein, the isolated ATR and CAT complexes were treated with [³H]PLP and the modified lysine positions determined. The result is shown in Fig. 1. Apart from position 22, which will be discussed later, the patterns of the two proteins are rather similar. A strong labeling of lysine at positions 48, 106, 146, and 262 is found. Positions 42, 62, 205, and 267 are less reactive, and the amount of labeling in positions 9, 91, 93, 95, 162, 165, 198, and 244 is below one-tenth. No labeling at all is found in positions 32, 259, 271, 294, and 295; these residues may be buried. The lysine in position 51 is trimethylated and thus cannot react.

Inside/Outside Distribution

The procedure outlined above is applied to examine the inside/outside distribution of the exposed lysine residues. For this purpose the labeling pattern of CAT or BKA loaded mitochondria is compared with that observed in submitochondrial particles derived from the loaded mitochon-

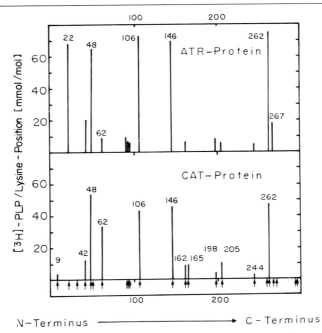

FIG. 1. [³H]PLP incorporation into the isolated ATR and CAT protein complexes. The abscissa represents the amino acid sequence; the 23 lysine positions are marked by arrows. The ordinate shows the quantity of labeling per position.

dria. The data are summarized in Fig. 2. In general, more lysine groups are accessible to PLP in mitochondria, such as positions 22, 91, 93, 95, 106, 162, 198, 205, 259, 262, and 267. In SMP, only lysines 42, 48, and 146 are distinctly labeled. The labeling of positions 42 and 48 also in mitochondria may be an artefact. Especially in the case of BKA loading, the mitochondria may have become leaky to some extent.

The labeling pattern of some positions strongly depends on the loading with CAT or BKA. Especially remarkable is lysine-22, which seems to be involved in the binding of the inhibitors ATR and CAT, as indicated by the different labeling pattern of the isolated ATR and CAT complexes, as well as by the patterns of ATR (not shown) and CAT loaded mitochondria.

Additionally, more PLP is incorporated into positions 42 and 48 if CAT is replaced by BKA. Similarly, one cysteine group reacts with *N*-ethylmaleimide when the protein is brought into the *m*-state.[11,24] This

24 H. Aquila, W. Eiermann, and M. Klingenberg, *Eur. J. Biochem.* **122,** 133 (1982).

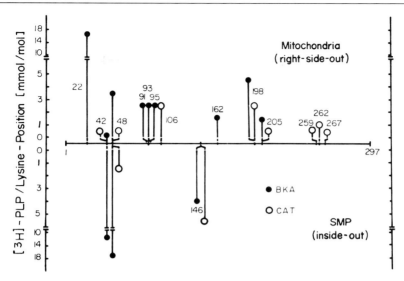

FIG. 2. Comparison of [³H]PLP incorporation into the ADP/ATP carrier in sonic particles and mitochondria, in the CAT-bound c-state (○) and the BKA-bound m-state (●). The abscissa again represents the amino acid sequence; the ordinate shows the quantity of labeling per position.

cysteine residue has now been identified to be cysteine-56 (unpublished results). Thus the higher reactivity of lysine-42 and -48 and of cysteine-56 may reflect a conformational change in this region of the protein. In contrast to lysine-42 and -48 (Fig. 1), cysteine-56 is unreactive in the isolated CAT protein.

Positions 91, 93, 95, and 162 show a small amount of labeling when BKA-loaded mitochondria are treated with PLP, but none if CAT was bound. Since CAT is attached to the same side, a shielding effect of CAT or small conformational changes could explain this fact. Position 106 is only labeled in CAT-loaded mitochondria, therefore a conformational change must be assumed.

A number of lysine groups are more accessible to PLP when the carrier is within the detergent micelle (compare Figs. 1 and 2). This can be explained by a removal of the phospholipid headgroups which may protect the protein in the membrane. Moreover, negatively charged headgroups may form salt bridges causing an inactivation of the lysines located in this area. Taking into consideration both the sidedness of the lysine groups and the differences in reactivity with PLP observed in the two conformational states, the folding pattern depicted in Fig. 3 is proposed. The folding pattern takes into account the existence of one β-strand and

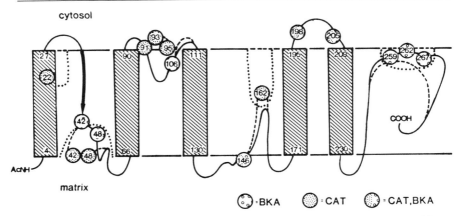

FIG. 3. Possible spanning of the ADP/ATP carrier across the membrane as derived from lysine labeling in the c- and m-states in sonic particles and mitochondria. The dotted circles represent lysines labeled in the c-state, the hatched circles indicate lysines labeled in the m-state. Dotted lines show possible pore-forming regions. Cylinders symbolize α-helices and the arrow points out a β-strand.

possibly six largely hydrophobic amino acid stretches of 18 to 23 residues (α-helices) which are possible candidates for membrane-spanning segments. The model is based on the versions published by us recently.[25,26]

[25] W. Bogner, H. Aquila, and M. Klingenberg, *FEBS Lett.* **146,** 259 (1982).
[26] W. Bogner, H. Aquila, and M. Klingenberg, *in* "Structure and Function of Membrane Proteins" (W. Quagliariello and F. Palmieri, eds.). Elsevier, Amsterdam, 1983.

[52] Chemical Modifications and Active Site Labeling of the Mitochondrial ADP/ATP Carrier

By Marc R. Block, François Boulay, Gérard Brandolin, Guy J. M. Lauquin, and Pierre V. Vignais

A peculiarity of the mitochondrial ADP/ATP carrier resides in its binding asymmetry for the specific inhibitors: atractyloside (ATR), carboxyatractyloside (CATR), and bongkrekic acid (BA). When added to mitochondria, ATR and CATR bind to the outer face of the carrier (face exposed to the cytosol in the cell), and BA binds to its inner face (face exposed to the matrix space).[1] ATR and CATR on one hand, and BA on the other compete for binding to the carrier, and they cannot bind at the same time to the same carrier unit.[2] A likely explanation is that the ADP/ATP carrier can take two conformations referred to as CATR and BA conformations; ATR and CATR bind to the CATR conformation and BA to the BA conformation.

The aim of this chapter is to summarize the effects of chemical modifications on the binding of ATR, CATR, and BA to the membrane-bound ADP/ATP carrier and to discuss these effects in terms of specific sites and conformations. We also report experiments on the mapping of the ATR binding site; for these mapping studies, the membrane-bound ADP/ATP carrier is first photolabeled by azido derivatives of ATR; it is then cleaved by chemical reagents and the localization of the covalently bound ATR in the peptide fragments is investigated.

Chemical Modifications of the ADP/ATP Carrier

Two typical cases of inactivation of inhibitor binding sites are discussed: (1) inactivation of both ATR (CATR) and BA binding sites and (2) selective inactivation of the ATR (CATR) binding site. For these binding studies, beef heart mitochondria[3] and [3H]ATR, [14C]acetyl-CATR, and [3H]BA are used.

Preparation of [3H]ATR, [14C]Acetyl-CATR, and [3H]BA

Radiolabeling of ATR and BA by 3H has been reported in this series.[4]

Since 6'-O-acetyl-CATR mimics CATR by its inhibitory and binding properties, 6'-O-[14C]acetyl-CATR is currently used as a substitute for radiolabeled CATR. Its preparation is briefly described.[2] Triethylamine (160 μl) is added to 600 μl of an aqueous solution of CATR (30 mg/ml), and 12 successive additions of 2 μl of [14C]acetic anhydride are made at 4-min intervals with vigorous stirring at room temperature. The reaction mixture is then dried under vacuum and washed 5 times with methanol. The residue is dissolved in 1 ml of methanol. [14C]Acetyl-CATR is purified by HPLC with a C_{18} μBondapak column (300 × 7.8 mm, 10 μM) (Waters). The solvent consists of water, methanol, acetic acid, and 1 M ammonium

[1] P. V. Vignais, M. R. Block, F. Boulay, and G. Brandolin, *in* "Membranes and Transport" (A. N. Martonosi, ed.), Vol. 1, p. 405. Plenum, New York, 1982.
[2] M. R. Block, R. Pougeois, and P. V. Vignais, *FEBS Lett.* **117**, 335 (1980).
[3] A. L. Smith, this series, Vol. 10, p. 81.
[4] P. M. Vignais, G. Brandolin, G. J.-M. Lauquin, and J. Chabert, this series, Vol. 55, p. 518.

acetate (59/41/1/2, v/v/v/v). The column is eluted at the rate of 2 ml/min. The retention times for the eluted products are as follows: CATR 14 min, acetyl-CATR 25 min, ATR 39 min. Acetyl-CATR is characterized by TLC on a silica gel analytical plate (K_6 Whatman) with a mixture of chloroform, methanol, acetic acid, and water (55/25/6/4, v/v/v/v) as developing solvent. Acetyl-CATR has an R_f of 0.46; it is stable for at least 6 months at $-20°$ in methanol.

The binding capacity of beef heart mitochondria for [³H]ATR, [¹⁴C]-acetyl-CATR, and [³H]BA is roughly 1.3–1.5 nmol/mg protein; the K_d values are 20–30, 2–5, and 20–40 nM, respectively. Because of the high affinity of acetyl-CATR for the ADP/ATP carrier, the acetyl-CATR–carrier complex can be isolated, and the binding stoichiometry of [¹⁴C]acetyl-CATR can be readily assessed. Beef heart mitochondria (1 mg/ml) are incubated for 30 min at 0° in a medium consisting of 120 mM KCl, 10 mM Mes, and 1 mM EDTA, pH 6.5. After centrifugation, the mitochondrial pellet is solubilized by 4% Triton X-100, 0.5 M NaCl, and 10 mM MOPS, pH 7.2. The [¹⁴C]acetyl-CATR–carrier complex is purified on a hydroxyapatite column equilibrated with a buffer consisting of 0.5% Triton X-100, 0.1 M NaCl, and 10 mM MOPS, pH 7.2. The excluded peak is concentrated by pressure dialysis on a PM10 Amicon membrane and applied to an Ultrogel type AcA 34 column equilibrated and eluted with the same buffer as that used for the hydroxyapatite chromatography. The elution pattern (Fig. 1) shows that the peak of radioactivity is superimposed on

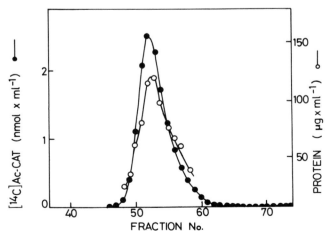

FIG. 1. Chromatography of the [¹⁴C]acetyl-CATR–ADP/ATP carrier complex on an AcA 34 column. Note the superimposition of the protein peak and the peak of bound [¹⁴C]acetyl-CATR. If present, free [¹⁴C]acetyl-CATR would be eluted after bound [¹⁴C]acetyl-CATR.

the peak of protein, and that there is no free [^{14}C]acetyl-CATR arising from dissociation of the complex. From the size of the peaks, it can be calculated that 1 mol [^{14}C]acetyl-CATR binds to 1 mol carrier dimer ($M_r \simeq 64,000$), a typical case of half-site reactivity.

Chemical Modifications Resulting in the Simultaneous Inactivation of ATR and BA Binding Capacity

Inactivation of binding of ATR, acetyl-CATR, and BA to the membrane-bound ADP/ATP carrier can be achieved by means of the α-diketonic reagents, phenylglyoxal and 2,3-butanedione.[5] The chemical reaction is performed in a borate buffer. Borate probably interacts with the condensation products of the butanedione and the guanidinium group[6]; it also interacts with phenylglyoxal itself, lowering the concentrations of the reactive species[7] and thereby decreasing the rate of the chemical modification. Solutions of 0.1 M butanedione in 1 M borate buffer, pH 7.8, or 0.5 M phenylglyoxal in dimethylformamide are prepared just before use. As an example, we describe the chemical modification of the membrane-bound ADP/ATP carrier by phenylglyoxal and the resulting inactivation of [^{3}H]ATR and [^{3}H]BA binding.[5] A suspension of beef heart mitochondria is preincubated for 5 min at 25° at the concentration of 1 mg protein/ml in 50 mM borate, pH 7.8 and 200 mM sucrose, and an aliquot sample of the phenylglyoxal solution is added at an appropriate concentration (ranging from 1 to 10 mM). The reaction is stopped at various intervals of time between 5 and 30 min by cooling to 0° and by lowering the pH by a 5-fold dilution in 50 mM KCl and 70 mM Mes, pH 6.5. Then [^{3}H]ATR or [^{3}H]BA is added at the final concentration of 0.4 and 0.6 μM, respectively; these concentrations correspond to the high affinity saturation plateau for binding of ATR and BA. Incubation is for 40 min at 0°. After centrifugation, the pellet of mitochondria is treated with 1 ml of 5% Triton X-100 and 0.5 M NaCl at 20°. Radioactivity is counted by liquid scintillation. The plot of the log of the reciprocal of the half-time of inactivation against the log of the phenylglyoxal concentration[8] gives straight lines with slopes close to 1 for ATR binding and to 2 for BA binding (Fig. 2). This suggests that two distinct arginyl residues are critical for the binding of ATR and BA. It is noteworthy that the number of ATR and BA sites is decreased without modification of the K_d value of the remaining sites. Control assays are carried out to check whether ATR binding and BA binding are protected

[5] M. R. Block, G. J.-M. Lauquin, and P. V. Vignais, *Biochemistry* **20**, 2692 (1980).
[6] J. F. Riordan, *Biochemistry* **12**, 3915 (1973).
[7] S.-T. Cheung and M. L. Fonda, *Biochem. Biophys. Res. Commun.* **90**, 940 (1979).
[8] F. Marcus, S. M. Schuster, and H. A. Lardy, *J. Biol. Chem.* **251**, 1175 (1976).

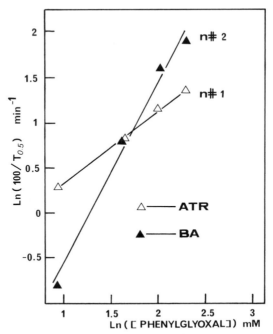

FIG. 2. Inactivation of [³H]ATR and [³H]BA binding to beef heart mitochondria by phenylglyoxal. For details, see text.

against inactivation by preincubation of mitochondria with BA and ATR, respectively. Indeed BA protects against inactivation of BA binding and ATR protects against inactivation of ATR binding, corroborating the presence of two specific binding sites for ATR and BA.[5]

Chemical Modifications Resulting in the Selective Inactivation of [³H]ATR or [¹⁴C]Acetyl-CATR Binding

ATR binding to mitochondria is inactivated with only a minor effect on BA binding after the following treatments: UV light irradiation,[9] incubation with 2-hydroxy-5-nitrobenzyl bromide (HNB),[5] a reagent for tryptophanyl residues and also thiol groups,[10] incubation with *N*-ethylmaleimide (NEM),[11] a well known thiol-alkylating agent, and

[9] M. R. Block, G. J.-M. Lauquin, and P. V. Vignais, *FEBS Lett.* **104,** 425 (1979).
[10] R. H. Horton and D. E. Koshland, Jr., *J. Am. Chem. Soc.* **87,** 1126 (1965).
[11] M. R. Block, G. J.-M. Lauquin, and P. V. Vignais, *FEBS Lett.* **131,** 213 (1981).

N-ethoxycarbonyl-2-ethoxy-1,2-dihydroquinoline (EEDQ),[11] a carboxyl-activating agent.[12]

Chemical Modification with HNB. A solution of 0.1 M HNB is prepared in anhydrous dimethylformamide and used immediately. Aliquots of the HNB solution are added with vigorous stirring to 5 ml of the suspension of beef heart mitochondria (0.2 mg protein per ml) in a saline medium consisting of 120 mM KCl, 20 mM Mes buffer, pH 6.5, and 2 mM 2-mercaptoethanol. Rapid mixing is essential because HNB is quickly hydrolyzed in water (the half time of hydrolysis is less than 1 min). After 30 min of incubation, the mitochondria are recovered by centrifugation and resuspended in the same saline medium as above for binding assays with radiolabeled ATR and BA. The incubation is terminated by centrifugation. The pellet of sedimented mitochondria is solubilized by 1 ml of 5% Triton X-100 and 0.5 M NaCl at 20°. Prior to radioactivity counting by liquid scintillation, the mixture is acidified with HCl (final concentration $\simeq$ 1 N) to avoid quenching due to the strong yellow color of the nitrophenol moiety of HNB bound to protein at neutral or basic pH.

UV Light Irradiation. Beef heart mitochondria are suspended in a saline medium consisting of 0.12 M KCl, 10 mM Mes, and 1 mM EDTA, pH 6.5 at a final concentration of 5 mg of protein/ml. The suspension is pipeted into a 5-cm-diameter Petri dish placed horizontally on crushed ice at about 10 cm from a 15 W Philips T-UV germicid lamp equipped with a reflector. The maximum emission is at 254 nm. After irradiation for different periods of time, aliquot fractions (routinely 0.2 ml equivalent to 1 mg of protein) are withdrawn and incubated at 0° for 40 min with 5 ml of a buffer made of 0.12 M KCl, 10 mM Mes, pH 6.5, and 1 mM EDTA, supplemented with [³H]ATR or [³H]BA. The binding of [³H]ATR and [³H]BA is estimated as described under the preceding section, except that the acidification step is omitted.

Chemical Modification with EEDQ or NEM. Beef heart mitochondria (final concentration 1 mg protein/ml) are incubated with different concentrations of EEDQ at 25° in a saline medium made of 120 mM KCl, 10 mM Mes, pH 6.5, and 1 mM EDTA. At time zero, an aliquot fraction of a freshly made solution of EEDQ (1 M dimethylformamide) is added with vigorous stirring. The reaction is stopped by a 5-fold dilution with the same medium cooled at 0° and supplemented with radioactive ATR, acetyl-CATR, or BA for binding assays.

A similar procedure is adopted for inactivation by NEM except that the medium is buffered with MOPS, pH 7.2 and that the reaction is stopped with a 5-fold dilution with the cold MOPS medium supplemented

¹² B. Belleau and G. Malek, *J. Am. Chem. Soc.* **90,** 1651 (1968).

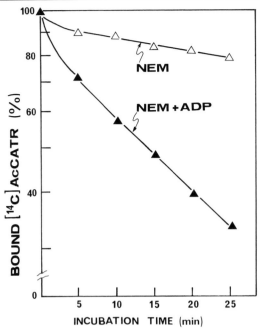

Fig. 3. Inactivation of [^{14}C]acetyl-CATR binding to beef heart mitochondria by NEM. Enhancing effect of ADP. The medium used consists of 120 mM KCl, 10 mM MOPS, pH 6.5, and 1 mM EDTA. Beef heart mitochondria are used at the concentration of 1 mg/ml, NEM at the concentration of 200 μM, ADP at the concentration of 10 μM. After different periods of reaction with NEM and NEM plus ADP at 25°, aliquot fractions are withdrawn, diluted 5-fold with cold medium adjusted at pH 6.5, and supplemented with cysteine to stop the NEM reaction. [^{14}C]Acetyl-CATR binding is assayed by addition of a saturating concentration of [^{14}C]acetyl-CATR (2 μM).

with 1 mM cysteine. The rate of inactivation of ATR or acetyl-CATR binding is markedly accelerated by micromolar amount of ADP. This is illustrated by the data in Fig. 3 which show the inhibition of [^{14}C]acetyl-CATR binding by NEM and by NEM plus ADP.

Interpretation of the Differential Inactivation Data. Treatment of mitochondria by UV light, HNB, NEM, or EEDQ results in the decrease of the total number of the ATR binding sites without alteration of the K_d value of the remaining ATR sites. A possible explanation for the much higher sensitivity of ATR binding compared to BA binding is that the ADP/ATP carrier units are distributed between two conformational states in the mitochondrial membrane, the CATR conformation and the BA conformation[13]; all the above modifications would stabilize the BA con-

---

[13] M. R. Block, G. J.-M. Lauquin, and P. V. Vignais, *Biochemistry* **22**, 2202 (1983).

formation and would therefore prevent ATR from binding to the carrier. If this hypothesis is correct, the following properties are predicted. (1) Inhibition of ATR binding must be stimulated by any ligand that favors the BA conformation, or by transportable nucleotides which favor the transition between the ATR and BA conformations. (2) Like BA binding, binding of inner nucleotides to the matrix side of the carrier (or externally added nucleotides in the case of inside-out submitochondrial particles) must escape inactivation. These predictions are verified for NEM.[11,14] The acceleration of inactivation of ATR binding by NEM caused by ADP (or ATP) (Fig. 3) indicates that the NEM-alkylated thiol group does not belong to the ATR binding site, but rather controls the reversible transition of the CATR conformation to the BA conformation. By means of chemical cleavage of the [^{14}C]NEM-labeled ADP/ATP carrier, one thiol group only has been found to be alkylated by NEM; this thiol group belongs to Cys-56.[15]

In contrast to inactivation by NEM, inactivation by UV light, HNB or EEDQ is not enhanced by ADP or ATP.[11] Furthermore, with N-ADP as a probe of the nucleotide binding, it can be shown that nucleotide binding to the matrix side of the carrier (tested by binding of externally added N-ADP to inside-out submitochondrial particles) is quickly inactivated by UV light or butanedione whereas the number and the affinity of the BA binding sites remain unaltered.[16] These results are interpreted to mean that ATR (CATR) and BA bind to different sets of amino acid residues, which is not incompatible with the two postulated CATR and BA conformational states of the ADP/ATP carrier.[13]

Photolabeling of the ATR Binding Site in the ADP/ATP Carrier

Photolabeling of the membrane-bound ADP/ATP carrier by azido derivatives of ATR has been used for mapping studies of the ATR binding site.

Synthesis of Radiolabeled Azido Derivatives of ATR and ADP

Principle

Short- and long-arm azido compounds, such as *p*-azido[^{14}C]benzoic acid, arylazido[^{3}H]aminopropionic, and aminobutyric acids, are coupled to the primary alcohol group of the glucose moiety of ATR and to the

[14] P. V. Vignais and P. M. Vignais, *FEBS Lett.* **26**, 27 (1972).
[15] F. Boulay and P. V. Vignais, *Biochemistry* **23**, 4807 (1984).
[16] R. Block, G. J.-M. Lauquin, and P. V. Vignais, *Biochemistry* **21**, 5451 (1982).

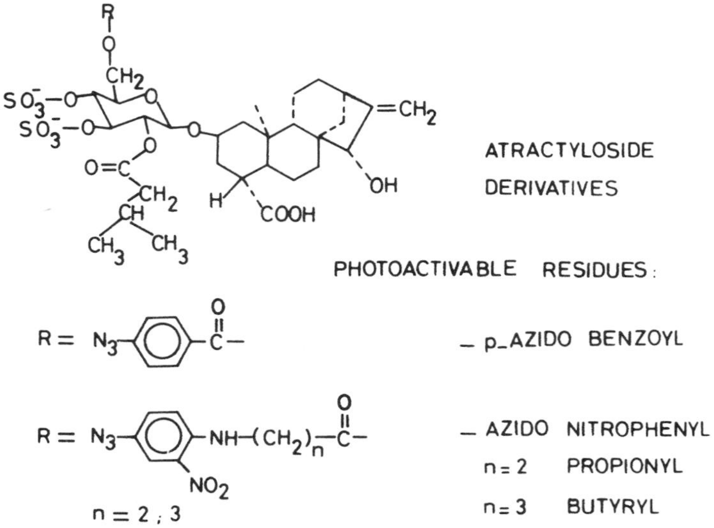

FIG. 4. Photoactivable derivatives of atractyloside (azido-ATR derivatives).

secondary alcohol group in the 3' position of the ribose moiety of adenosine di- or triphosphate. The resulting p-azidobenzoyl, azidonitrophenylpropionyl, and azidonitrophenylbutyryl derivatives of ATR[17] are listed in Fig. 4.

Methods

Preparation of 6'-O-p-Azido[14C]benzoyl ATR. First p-azido[14C]benzoic acid is synthesized from p-amino[14C]benzoic acid as described by Hixson and Hixson.[18] p-Amino[14C]benzoic acid (20 μmol, 10^8 dpm/μmol) is dissolved in 900 μl of 4 N sulfuric acid. The solution is cooled in an ice bath; 100 μl of 0.3 M sodium nitrite is added dropwise, and the mixture is kept in the ice bath with mechanical stirring for 50 min. Then, 2 ml of diethyl ether is added, followed by 300 μl of 0.2 M sodium azide. The solution is stirred in the dark for another 30 min and is left to stand. On standing, the ether layer separates rapidly; it is withdrawn and the remaining aqueous phase is washed 4 times with 2 ml of ether. The ether fractions are pooled, washed with 2 ml of an aqueous solution of NaCl, poured into a 100-ml flask, and the ether is evaporated under a stream of nitrogen. The dry residue consisting of p-azido[14C]benzoic acid is ready

[17] F. Boulay, G. J.-M. Lauquin, A. Tsugita, and P. V. Vignais, *Biochemistry* **22,** 477 (1983).
[18] S. H. Hixson and S. S. Hixson, *Biochemistry* **14,** 4251 (1975).

for coupling to ATR. The coupling step is carried out after activation of the carboxylic group of the *p*-azido[^{14}C]benzoic acid by *N*,*N'*-carbonyl-diimidazole.[19] Carbonyldiimidazole (70 μmol) is dissolved in 330 μl of dimethylformamide dried over a 3 Å molecular sieve and calcium hydride; the solution is added to the *p*-azido[^{14}C]benzoic acid in the flask. After stirring in the dark for 15 min at room temperature, a mixture of ATR (20 μmol) and triethylamine (20 μmol) in 1.2 ml of water is added dropwise within 2 min, and stirring is continued for 30 min. A longer contact does not improve the yield of coupling. The mixture is then evaporated *in vacuo* below 40°. The gummy residue is dispersed in 1 ml of distilled water and evaporated to dryness to remove the residual dimethylformamide. This procedure is repeated at least 3 times. The unreacted *p*-azido[^{14}C]-benzoic acid is removed by several washings of the dry residue with chloroform. Finally, the residue is dissolved in 1 ml of methanol, and the methanolic solution is purified by HPLC using a μBonkapak C$_{18}$ column (300 × 7.8 mm, 10 μm) with isocratic elution at 2 ml/min at 25°. The solvent used is a mixture of methanol, 1 *M* NH$_4$-acetate, acetic acid, water (58/1/1/40, v/v/v/v). The main product, identified by mass spectrometry as the mono-*p*-azidobenzoyl ester of the glucose disulfate moiety of ATR (6'-*O*-*p*-azidobenzoyl-ATR), is eluted with a retention time of 25 min. The unreacted *p*-azidobenzoic acid is eluted with a retention time of 15 min. The yield of recovery of 6'-*O*-*p*-azidobenzoyl-ATR is about 25%. The purity is assessed by TLC on silica gel with a solvent system made of chloroform, methanol, acetic acid, water (55/20/3/3, v/v/v/v); its R_f value is 0.34. The UV spectrum of 6'-*O*-*p*-azidobenzoyl-ATR in ethanol exhibits an maximum absorption at 272 nm (ε_{272} = 14600 M^{-1} cm^{-1}) and a shoulder at 290 nm.

Synthesis of 6'-O-[3-N(4-Azido-2-[^{3}H]nitrophenyl)amino]propionyl-atractyloside ([^{3}H]NAP$_3$-ATR) and of 6'-O-[4-N(4-Azido-2-[^{3}H]nitrophe-nyl)amino]butyryl-atractyloside ([^{3}H]NAP$_4$-ATR). These preparations have already been described in this series.[4] The method has been recently improved.[17] The starting materials, namely, [^{3}H]NAP$_3$ and [^{3}H]NAP$_4$, and the final products, [^{3}H]NAP$_3$-ATR and [^{3}H]NAP$_4$-ATR, are purified by HPLC. In brief, β-[^{3}H]alanine and 4-amino[^{3}H]butyric acid in aqueous solution are diluted with the respective unlabeled compound to a specific radioactivity of about 10^9 dpm/μmol. To 100 μl containing 100 μmol of ^{3}H-labeled compound is added an equimolar amount of (4-fluoro-3-ni-trophenyl)azide in 400 μl of dimethyl sulfoxide. The mixture is made alkaline with 50 μl of triethylamine and stirred overnight at 60°. After

[19] B. P. Gottikh, A. A. Krayseski, N. B. Tarussova, P. P. Purygin, and T. L. Tsilevich, *Tetrahedron* **26**, 4419 (1970).

acidification with 50 μl of acetic acid, the reaction products, i.e., [^{3}H]NAP$_3$ and [^{3}H]NAP$_4$ are separated by isocratic elution at a flow rate of 2 ml/min from a μBondapak C$_{18}$ column (300 × 7.8 mm, 10 μm) at 25°. The solvent system is made of methanol, acetic acid, water (50/2/48, v/v/v). The retention times for [^{3}H]NAP$_3$ and [^{3}H]NAP$_4$ are 36 and 49 min, respectively. The coupling step on ATR is carried out as previously described for p-azido[^{14}C]benzoic acid. The final residue is dissolved in 1 ml of methanol and this solution is subjected to HPLC on μBondapak C$_{18}$ column (300 × 7.8 mm, 10 μm). The isocratic elution is performed at a flow rate of 2 ml/min in methanol/1 M NH$_4^+$ acetate, acetic acid, water (60/1/1/38, v/v/v/v) at 25°. The retention times for [^{3}H]NAP$_3$-ATR and [^{3}H]NAP$_4$-ATR are 28 and 40 min, respectively. The final yield is around 15%. Purity is checked by TLC on silica gel plate with a solvent system made of chloroform, methanol, acetic acid, water (55/20/2/1, v/v/v/v). The R_f value of [^{3}H]NAP$_3$-ATR is 0.43 and that of [^{3}H]NAP$_4$-ATR is 0.46. The UV spectrum in ethanol exhibits peaks of absorption at 260 nm (ε_{260} = 27,200 M^{-1} cm^{-1}) and 460 nm (ε_{460} = 5900 M^{-1} cm^{-1}).

Application to the Mapping of the ATR and ATP Binding Sites of the ADP/ATP Carrier

All of the above photoactivable derivatives of ATR have virtually the same affinity and specificity as ATR when the binding is assayed under reversible conditions, i.e., in the absence of light. Upon photoirradiation covalent binding to the membrane bound ADP/ATP carrier occurs.

Covalent Photolabeling of Beef Heart ADP/ATP Carrier Protein. With [^{3}H]NAP derivatives of ATR, photoirradiation is performed with a 250 W Osram halogen lamp for 30 min or with a 1000 W xenon lamp equipped with a parabolic reflector (Müller GmbH, Freisinger Str 1, 8059 Moosinning, Germany) for 1 min through a glass filter to avoid UV irradiation. On the other hand, photolabeling by p-azido[^{3}H]benzoyl ATR requires UV light that can be provided by a 15 W Phillips UV germicidal lamp for 10 min or by the xenon XBO lamp for 1 min without any glass filter under an atmosphere of argon. The radioactive derivatives of ATR (40 nmol) are preincubated with the suspension of beef heart mitochondria (20 mg of protein) for 30 min in the dark in a standard medium consisting of 120 mM KCl, 5 mM MOPS, and 1 mM EDTA, pH 6.8; the final volume is adjusted to 10 ml. The mixture is introduced in a 50 ml flask that is rotated horizontally in an ice bath. After irradiation, the mitochondria are sedimented by centrifugation for 10 min at 20,000 g and washed once with 10 ml of the standard medium.

Purification of the Photolabeled ADP/ATP Carrier Protein. The photolabeled carrier protein is extracted by Triton X-100 and purified by
hydroxyapatite chromatography as described by Riccio *et al.*[20] with some
modifications including washing with organic solvent to remove bound
lipids.[17] The hydroxylapatite pass-through fraction is concentrated and
dialyzed on a PM10 Amicon membrane. The concentrated solution is
treated with 5 volumes of acetone at $-20°$ overnight. The resulting precipitate, around 1 mg, is recovered by centrifugation and washed with 5 ml of
acetone at $-20°$ to remove the residual Triton X-100. Then the pellet is
dissolved in 500 μl of 80% formic acid at 25°. After solubilization, 1 ml of
ethanol is added with 5 μl of 5 M NaCl, followed by addition of diethyl
ether, and the temperature is brought down to $-20°$ to precipitate the
protein. The precipitate is freeze-dried. The same method can be used for
large scale preparation of photolabeled carrier.

Fragmentation of the Photolabeled ADP/ATP Carrier Protein. The
beef heart ADP/ATP carrier protein contains seven methionyl residues
and four cysteinyl residues which are potential sites for chemical cleavage. Before cleavage at methionyl residues with cyanogen bromide, cysteinyl residues are modified with DTNB, and lysyl residues are succinylated according to the following procedure. The protein (8 mg) is dissolved
in 1 ml of 7 M guanidinium chloride, 1 mM EDTA, and 0.1 M sodium
phosphate, pH 7.2, under a nitrogen stream and left in contact for 1 hr at
37° with 5 mM DTT for full reduction. A 6-fold excess of DTNB with
respect to the total amount of sulfhydryl groups is added. The pH is
maintained at 7.2 by addition of 0.1 N NaOH and the mixture is left to
incubate at 37° for 30 min. The thionitrobenzoylated protein is succinylated at room temperature following the general procedure described by
Klotz.[21] The succinylated protein is desalted by passing through a column
of Ultrogel AcA 202 (10 × 2.5 cm) equilibrated in 50 mM ammonium
bicarbonate, pH 7.8, and then freeze-dried. The cleavage of the peptide
bonds with cyanogen bromide is performed in 70% formic acid as described by Gross.[22] Tryptamine at the final concentration of 1 mM is
added to protect tryptophan against oxidation. A large peptide of M_r
23,000 (CB1) together with a number of small peptides are found to accumulate. Radioactivity is present only in the CB1 peptide. The detailed
procedure is described in the paper by Boulay *et al.*[17] The TNB-modified
and succinylated CB1 fragment is isolated by gel filtration through a

---

[20] P. Riccio, H. Aquila, and M. Klingenberg, *FEBS Lett.* **56**, 133 (1975).
[21] I. M. Klotz, this series, Vol. 11, p. 576.
[22] E. Gross, this series, Vol. 11, p. 238.

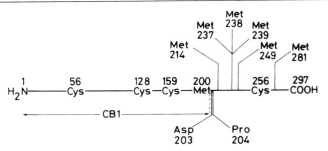

FIG. 5. Alignment of peptide fragments obtained after cleavage of the ADP/ATP carrier protein at methionyl residues by cyanogen bromide, and at cysteinyl residues by cyanylation followed by incubation at alkaline pH. In the carrier photolabeled by azido-ATR, only the fragment extending from Cys-159 to Met-200 is labeled.

column of Sephadex G-50 fine (1.5 × 90 cm) equilibrated in 50 mM ammonium bicarbonate, pH 7.8, and then freeze-dried.

Cyanylation of sulfhydryl groups, followed by cleavage of alkaline pH is carried out as follows.[17] A sample (5 mg) of the TNB modified and succinylated CB1 fragment is dissolved in 1 ml of 7 M guanidinium chloride, 1 mM EDTA, and 0.1 M sodium phosphate buffer, and 50 μl of 0.1 M NaCN or 0.1 M Na^{14}CN (2 × 10^7 dpm/μmol) is added. A yellow color develops rapidly. After 30 min at 35°, the cyanylated CB1 fragment is desalted by gel filtration on a 10 × 1 cm column of AcA 202 equilibrated in 0.015 M NaHCO$_3$/Na$_2$CO$_3$, pH 9.8. After a 24 hr incubation at 37°, the solution is neutralized with 6 N HCl and freeze-dried. The freeze-dried residue is dissolved in 70% formic acid and applied on a BioGel P-10 (60 × 1 cm) equilibrated in 70% formic acid. The column is eluted at a flow rate of 5 ml/hr. The peptides obtained by cyanide cleavage are separated by BioGel chromatography. Most of the covalently bound ^{3}H radioactivity is recovered in a peptide of $M_r \simeq 4500$. The amino acid composition and the C-terminal amino acid sequence of the four peptides are reported by Boulay et al.[17] Based on these data and on the amino acid sequence reported by Aquila et al.,[23] the four peptides have been aligned. The labeling by the [^{3}H]NAP$_4$-ATR is located in the peptide fragment extending from Cys-159 to Met-200 (Fig. 5).

[23] H. Aquila, D. Misra, E. Eulitz, and M. Klingenberg, Hoppe Seyler's Z. Physiol. Chem. 363, 345 (1982).

[53] Measurement of Citrate Transport in Tumor Mitochondria[1]

By RONALD S. KAPLAN, RISA A. PARLO, and PETER S. COLEMAN

Introduction

It is now well established that the inner membrane of mitochondria from rat liver maintains a highly selective anion permeability through the action of numerous substrate-specific exchange carrier systems.[2–4] One such carrier, the tricarboxylate or citrate carrier, catalyzes an electroneutral exchange of a tricarboxylate anion (e.g., citrate, *threo*-D_s-isocitrate, *cis*-aconitate) with either another tricarboxylate anion, a dicarboxylate (e.g., malate or succinate), or with phosphoenolpyruvate.[4–6] While the kinetics of citrate transport has been studied in detail with isolated normal rat liver mitochondria,[5,6] our work with mitochondria isolated from two different Morris hepatomas was the first in-depth study of this transporter in tumor mitochondria,[7] and has led to a better understanding of the role of this vital organelle in the bioenergetics and intermediary metabolism of neoplasia.

Among the most consistent phenotypic alterations of tumor cells relative to their normal counterparts is an elevation in the cholesterol content of their various cellular membranes.[8,9] Interestingly, this membrane cholesterol enrichment coincides with an observed deregulation (e.g., a loss of dietary feedback inhibition) of the cholesterol biosynthetic pathway *de novo* in the tumor cell itself,[8,10,11] a logical outcome of which might be a continuous rate of cholesterogenesis during tumor growth. These phe-

[1] The work with Morris hepatoma mitochondria was supported by a grant from the USPHS, CA28677.

[2] A Fonyo, F. Palmieri, and E. Quagliariello, *in* "Horizons in Biochemistry and Biophysics" (E. Quagliariello, F. Palmieri, and T. P. Singer, eds.), Vol. 2, pp. 60–105. Addison-Wesley, Reading, Massachusetts, 1976.

[3] M. Klingenberg, *Essays Biochem.* **6,** 119 (1970).

[4] K. F. LaNoue and A. C. Schoolwerth, *Annu. Rev. Biochem.* **48,** 871 (1979).

[5] F. Palmieri, I. Stipani, E. Quagliariello, and M. Klingenberg, *Eur. J. Biochem.* **26,** 587 (1972).

[6] B. H. Robinson, G. R. Williams, M. L. Halperin, and C. C. Leznoff, *J. Biol. Chem.* **246,** 5280 (1971).

[7] R. S. Kaplan, H. P. Morris, and P. S. Coleman, *Cancer Res.* **42,** 4399 (1982).

[8] H. W. Chen, A. A. Kandutsch, and H.-J. Heiniger, *Prog. Exp. Tumor Res.* **22,** 275 (1978).

[9] R. Van Hoeven and P. Emmelot, *J. Membr. Biol.* **9,** 105 (1972).

[10] P. Pedersen, *Prog. Exp. Tumor Res.* **22,** 190 (1978).

[11] J. R. Sabine, *Prog. Biochem. Pharmacol.* **10,** 269 (1975).

nomena are especially true of primary and transplanted hepatomas, but have also been documented for many other tumors, both animal as well as human.[12-14] A fundamental metabolic requirement for the decontrolled synthesis of cholesterol is the maintenance of a sufficiently large (and continuously supplied) pool of its precursor, acetyl-CoA, in the cytoplasm where the sterol is synthesized. It is known that the two-carbon acetate units, required as precursor for both fatty acid and cholesterol biosynthesis, arise from the enzymatic cleavage of cytoplasmic citrate. In turn, the supply of cytoplasmic citrate is furnished by carrier-mediated citrate efflux from its primary site of formation within the mitochondrial matrix. These facts highlight the conspicuous importance of the citrate carrier to tumor cell metabolism, in that the extent to which tumor mitochondria are capable of (perhaps accelerated) citrate export might dramatically affect the supply of cytosolic acetyl-CoA available to the proliferating tumor. Recent work from our laboratory suggests that indeed, a direct relationship exists between the operation of the citrate carrier in tumor mitochondria and the observation that tumor mitochondrial membranes are cholesterol rich.[15,16]

The methods that follow illustrate two approaches that we have used in our studies of citrate transport with isolated Morris hepatoma mitochondria. The first method takes advantage of detailed procedures suitable for the measurement of initial velocities, and thereby allows for the determination of apparent K_m and V_{max} values for the citrate carrier. This approach isolates, mechanistically speaking, the discrete kinetics of the citrate carrier from complications due to concurrent mitochondrial metabolism. The second approach outlines procedures for measuring the metabolic flux of pyruvate carbons as precursors to the carbons of extramitochondrial citrate (and thus, of cytosolic acetyl-CoA), according to the metabolic sequence:

$$\text{pyruvate}_o \rightarrow \text{pyruvate}_i \rightarrow \text{acetyl-CoA}_i \nearrow \text{citrate}_i \rightarrow \text{citrate}_o$$
$$\text{oxaloacetate}_i$$

The subscripts o and i refer, respectively, to outside and inside of the mitochondrial matrix compartment. Both of our experimental approaches share some common strategic manipulations (such as inhibitor-stop meth-

[12] P. S. Coleman and B. Lavietes, *CRC Crit. Rev. Biochem.* **11**, 341 (1981).

[13] H.-J. Heiniger, H. W. Chen, O. L. Applegate, Jr., L. P. Schacter, B. Z. Schacter, and P. N. Anderson, *J. Mol. Med.* **1**, 109 (1976).

[14] P. Pani, R. A. Canuto, R. Garcea, and F. Feo, *Biochem. Soc. Trans.* **1**, 972 (1973).

[15] P. S. Coleman and R. A. Parlo, *Ann. N.Y. Acad. Sci.* **435**, 129 (1984).

[16] R. A. Parlo and P. S. Coleman, *J. Biol. Chem.* **259**, 9997 (1984).

odology), and some major differences (such as concurrent mitochondrial substrate oxidation and the temperature at which incubations are conducted). The important subjects of inquiry that are illustrated by applications of the following methods and procedures include (1) a full characterization of the kinetic parameters of the citrate carrier in tumor versus host and normal tissue mitochondria; (2) the effect of temperature on these kinetic parameters and implications for tumor mitochondrial membrane structure/function; and (3) the operation of the carrier under conditions that allow mitochondrial metabolism and substrate flux to proceed.

Animals and Tumor Tissue Sources

ACI strain male rats can be purchased from Harlan Sprague–Dawley (Walkersville, MD) while Buffalo strain male rats are obtainable from Simonsen Laboratories (Gilroy, CA). Rats are fed and watered *ad libitum* and housed under a 12-hr light–dark cycle. Morris hepatomas 3924A and 16 were originally obtained from the laboratories of Dr. H. P. Morris, Howard University College of Medicine, Washington, D.C., as tumor-bearing animals.[17] Hepatoma 3924A is a rapidly growing, poorly differentiated tumor, whereas hepatoma 16 is a slowly growing, well-differentiated tumor.[18,19] Both tumor lines have been carried in our laboratory at New York University, by serial intramuscular transplantation into the hindlimbs of recipient (host) rats. The transplantation procedure employs methods similar to those described previously,[20] and is performed with a #21 Silverman biopsy needle (trochar). Hepatoma 3924A is transplanted every 11–17 days into ACI rats, whereas tumor 16 is transplanted every 6–9 months into Buffalo rats. These tumors represent two extremes in the rate-of-growth spectrum of the Morris hepatoma lines, and have demonstrated biological stability over many generations. Furthermore, they differ significantly from each other in their mitochondrial content per tumor

---

[17] The Morris hepatomas are transplantable hepatocellular carcinomas in rats which were originally induced by the chronic feeding of known chemical carcinogens in their diet. Further compiled information on the Morris hepatoma series may be obtained from (1) H. P. Morris and W. E. Criss, eds., *Adv. Exp. Med. Biol.*, **92** (1978); and (2) H. P. Morris and L. J. Slaughter, in "Liver Carcinogenesis" (K. Lapis and J. Johannesen, eds.). Hemisphere Publ., New York, 1979. There is some question as to the continuing availability of the Morris hepatoma series of tumors directly from Howard University. Investigators interested in working with these tumors would be best advised to contact Dr. L. J. Slaughter at the Howard University College of Medicine, Washington, D.C., 20059, in order to inquire about the status of the particular tumor line of interest before proceeding with their research plans.

[18] H. P. Morris, *Adv. Cancer Res.* **9,** 227 (1965).

[19] H. P. Morris and L. J. Slaughter, *Adv. Exp. Med. Biol.* **92,** 1 (1978).

[20] H. P. Morris and B. P. Wagner, *Methods Cancer Res.* **4,** 125 (1968).

cell,[21,22] and have been found to possess remarkably different levels of mitochondrial membrane cholesterol, not only with respect to each other, but to normal liver mitochondria as well (see Table V).

Methods for Determination of Citrate Carrier Kinetics

Basic Principles

The kinetics of the citrate carrier may be determined *in vitro* by measuring the exchange of extramitochondrial substrate with either (1) endogenous substrates already present in the mitochondrial matrix, or (2) with a substrate (e.g., citrate) preloaded into the matrix. It has been suggested that exchange with endogenous substrates may more accurately reflect mitochondrial transport *in vivo*.[6] However, it is also possible that the endogenous substrate levels are significantly altered during the mitochondrial isolation procedure. Furthermore, kinetic measurements obtained under these conditions reflect the combined effect of the properties of the tricarboxylate carrier and the availability of endogenous exchangeable substrates. Clearly, the presence of nonsaturating levels of matrix substrates would substantially alter the observed transport kinetics.

When the aim is to examine the kinetic properties of the citrate carrier itself, free of the constraints of limiting and undefined internal substrate levels,[5,23] mitochondria are first preloaded with a saturating concentration of a transportable substrate (e.g., citrate). The exchange reaction is then carried out under such explicitly defined conditions. Unless otherwise indicated, the procedures described below are performed at a controlled temperature of 10°. Incubations are carried out at this temperature due to the rapidity of the exchange process and the difficulty in measuring initial transport velocities at higher temperatures.

Stock Solutions of Buffers and Inhibitors[24]

1. Buffer A: 0.6 M KCl, 120 mM HEPES, 6 mM EGTA, pH 7.2.
2. Isotonic buffer: 0.1 M KCl, 20 mM HEPES, 1 mM EGTA, pH 7.0.

[21] K. F. LaNoue, J. G. Hemington, T. Ohnishi, H. P. Morris, and J. R. Williamson, *in* "Hormones and Cancer" (K. W. McKerns, ed.), p. 131. Academic Press, New York, 1974.

[22] J. R. Schreiber, W. X. Balcavage, H. P. Morris, and P. L. Pedersen, *Cancer Res.* 30, 2497 (1970).

[23] F. Palmieri and M. Klingenberg, this series, Vol. 56, p. 279.

[24] Abbreviations: BTC, 1,2,3-benzenetricarboxylate; BSA, bovine serum albumin (Cohn fraction V); PPO, 2,5-diphenyloxazole; POPOP, 1,4-di-[2-(5-phenyloxazolyl]benzene; EGTA, ethylene glycol bis(β-aminoethyl ether)-N,N,N',N'-tetraacetic acid; HEPES, 4-(2-hydroxyethyl)-1-piperazineethanesulfonic acid; PCA, perchloric acid.

3. Antimycin (Sigma): 3 mM in absolute ethanol, diluted with absolute ethanol prior to use.
4. Rotenone (Pfaltz and Bauer): various concentrations, 50–500 μg/ml in absolute ethanol, prepared fresh before use.
5. Sodium arsenite: 100 mM.
6. Benzenetricarboxylic acid hydrate (Aldrich or ICN/K & K). BTC: 400 mM, initially dissolved in H_2O, made basic with KOH. The pH is brought to 7.0 with HCl, and the solution is stored at 0–4°.
7. Mersalyl acid (Sigma): 20 mM, prepared and stored as with BTC.
8. Phenylsuccinic acid (Aldrich): 500 mM, prepared in 50% (by volume) ethanol, pH adjusted to 8.0 with KOH, and stored at 0–4°.

Where indicated the transport inhibitors BTC, mersalyl, and phenylsuccinate are added to mitochondrial incubations as isotonic buffer solutions. Accordingly, the stock solution of a given inhibitor is diluted with H_2O and Buffer A to yield the desired final inhibitor concentration in isotonic buffer.

Isolation and Functional Characterization of Tumor and Control Mitochondria

For the transport kinetics experiments described below, mitochondria are isolated from tumor, host, and normal rat liver via the Nagarse method of Kaschnitz *et al.*,[25] employing the modifications previously described.[7] Protein concentration is determined by a modified biuret procedure.[26] Two types of controls are employed in these transport kinetics studies. The host liver (i.e., the liver of the tumor-bearing rat) serves as a control since it represents nontransformed liver tissue from the same animal as the tumor and thus accounts for variations between individual animals. However, host liver characteristics also reflect a tumor-imposed stress. Therefore, normal liver is used as a control as well.[27] A total of 6 different types of mitochondrial preparations are examined in the following kinetics protocols.

Before undertaking transport studies it is essential to demonstrate that mitochondria from tumor and control tissues are not substantially damaged by the isolation procedure. This can be achieved by measuring ADP/O and respiratory (acceptor) control ratios (RCR) employing standard methods.[28] ADP/O ratios approaching 3 with NAD-linked respiratory

[25] R. M. Kaschnitz, Y. Hatefi, P. L. Pedersen, and H. P. Morris, this series, Vol. 55, p. 79.
[26] R. S. Kaplan and P. S. Coleman, *FEBS Lett.* **63,** 179 (1976).
[27] G. Weber, *in* "The Molecular Biology of Cancer" (H. Busch, ed.), p. 487. Academic Press, New York, 1974.
[28] R. W. Estabrook, this series, Vol. 10, p. 41.

chain substrates (e.g., glutamate) indicate that the three energy-coupling sites are operative in the particular mitochondrial isolate. Similarly, we have considered RCR values above 3 to mean that the mitochondrial preparation is functionally and physically intact,[29] and therefore suitable for transport studies. These measurements should be performed with each type of mitochondrial preparation before beginning a transport investigation.

Determination of Sample Radioactivity by Liquid Scintillation Counting

NCS tissue solubilizer (Amersham) is used to digest mitochondrial pellets and suspensions for at least several hours at 50°. Samples are cooled and neutralized with 30 μl glacial acetic acid/ml NCS. A toluene-based scintillation cocktail containing 6 g PPO and 75 mg POPOP per liter is used to count the ^{14}C-labeled samples. Aqueous (nonmitochondrial) samples are counted in ACS (Amersham), a xylene surfactant-based cocktail. Counting efficiency is determined by the internal standard method using aliquots of [^{14}C]toluene (7.08 $\times$ 10^5 dpm/ml, Amersham).

Kinetics of Citrate Influx in Exchange for Endogenous Substrates

Incubation Medium. The reaction medium used is similar to that of Palmieri *et al.,*[5] and contains the isotonic buffer (0.1 M KCl, 20 mM HEPES, 1 mM EGTA, pH 7.0) *plus* 0.25 nmol antimycin/mg total protein, 0.4 μg rotenone/mg total protein, and 0.3 mg BSA/ml (which *includes* the BSA used to resuspend the mitochondrial pellet). Subsequently, 2.0 mg mitochondrial protein/ml and various concentrations of [1,5-^{14}C]citrate (Amersham) are added (specific radioactivity = 4.1–8.1 nCi/nmol). Reactions are stirred with a magnetic "flea" and are performed in a final volume of 1.0 ml.

Reaction Procedures. Mitochondria are preincubated in the transport reaction mixture for 60 sec. The reaction is then initiated by the addition of [^{14}C]citrate and is quenched after various time intervals by the addition of 0.05 M BTC in isotonic buffer. BTC is a competitive inhibitor of the citrate carrier.[5] Twenty seconds later, 0.5 μmol mersalyl/mg total protein in isotonic buffer is added, thus blocking the dicarboxylate carrier[5] and partially inhibiting the α-ketoglutarate carrier as well.[30] The latter two anion transporters have low affinities for citrate. Due to the possibly slow

---

[29] J. B. Chappell and R. G. Hansford, *in* "Subcellular Components, Preparation and Fractionation" (G. D. Birnie, ed.), Ed. 2, p. 77. Univ. Park Press, Baltimore, 1972.

[30] "Cell Biology" (P. L. Altman and D. D. Katz, eds.), Table 50, p. 195. Federation of American Societies for Experimental Biology, Bethesda, Maryland, 1976.

reaction of mersalyl with accessible sulfhydryl groups,[31] the incubation is stirred for 2 min before continuing the protocol. Then the reaction mixture is diluted approximately 3-fold with isotonic buffer containing 1 mM BTC and 10 mM phenylsuccinate. The phenylsuccinate serves to more completely block the α-ketoglutarate carrier.[32] The diluted reaction mix is then centrifuged (16,000 g, 5 min, 4°; HB-4 rotor, RC-5 centrifuge, Dupont-Sorvall), the supernatant is discarded, and the top of the pellet is rinsed with 10 ml isotonic buffer containing 1 mM BTC and 10 mM phenylsuccinate. The radioactivity in the pellet is determined as previously described.

Transport Controls. Control incubations contain 0.05 M BTC (no additional isotonic buffer) in the original reaction mix prior to addition of mitochondria. The BTC concentration in the control reactions is maintained at 0.05 M until the addition of mersalyl. In all other ways the reaction sequences for the control and experimental incubations are identical. The radioactivity appearing in the final control pellet is due to the sum of citrate influx by means other than passage via the citrate carrier (e.g., on the dicarboxylate and/or α-ketoglutarate carriers, by leak, etc.) and the citrate that is nonspecifically trapped within or adsorbed to the surface of the pellet. The control values are subtracted from the experimental values.

Precautions. Our preliminary work indicated that altering the BSA concentration in the transport incubation mix changed the measured rate of citrate influx. Therefore, the BSA level is held constant and includes the BSA that is added as part of the mitochondrial suspension (i.e., the relative portion of the total amount of BSA added during the suspension of the final mitochondrial pellet in the isolation procedure). Also, to account for inevitable protein loss during the subsequent handling of the incubation, it is necessary to perform a protein determination (employing an occasional parallel incubation) on the centrifugal pellet following the transport protocol given above. Data are normalized to this protein value. Finally, we allowed citrate influx reactions to proceed for 15 sec with normal ACI liver and hepatoma 3924A mitochondrial preparations, but only for 5 sec with all other preparations. In all cases, data collected during these time intervals corresponded to initial velocities of citrate

[31] F. Palmieri, S. Passarella, I. Stipani, and E. Quagliariello, *Biochim. Biophys. Acta* **333,** 195 (1974).

[32] F. Palmieri, E. Quagliariello, and M. Klingenberg, *Eur. J. Biochem.* **29,** 408 (1972); we included mersalyl and phenylsuccinate additions in our procedures because initial experiments indicated that even in the presence of BTC, a slow efflux of intramitochondrial [14C]citrate occurred during the subsequent centrifugations to separate mitochondria for citrate uptake analysis. This slow efflux was inhibited by mersalyl *plus* phenylsuccinate.

TABLE I
KINETIC CHARACTERISTICS OF [^{14}C]CITRATE INFLUX INTO
UNLOADED MITOCHONDRIA[a]

Mitochondrial source	$K_m(\mu M)$	V_{max}(nmol/min/mg)
1. Normal liver (ACI rat)	28.6 ± 4.3	8.0 ± 2.4
2. Host liver (3924A)	24.4 ± 4.8	7.9 ± 1.2
3. Hepatoma 3924A	25.5 ± 10.4	6.6 ± 2.7
4. Normal liver (Buffalo rat)	25.4 ± 2.3	3.2 ± 1.3
5. Host liver (16)	28.6 ± 2.7	4.4 ± 1.6
6. Hepatoma 16	23.5 ± 10.2	6.1 ± 0.8

[a] Five to seven different [^{14}C]citrate concentrations between 7.5 and 400 μM were tested per experiment. Lineweaver–Burk plots were subjected to linear regression by least-squares analysis. Data are from three separate mitochondrial preparations from each of the six types of tissue. From Kaplan et al.[7] with permission of Cancer Research.

influx under the given experimental conditions, with the possible exception of host liver (3924A) mitochondria, where the 5 sec reaction rate approached, but may have slightly underestimated, the initial velocity.

Results. Table I depicts the apparent K_m and V_{max} values which were determined from Lineweaver–Burk plots. Our data show that all six mitochondrial preparations display similar K_m values. The V_{max} values also are quite similar for hepatoma 3924A mitochondria and its controls, but with tumor 16 we found the V_{max} value to be somewhat increased over the normal liver value.

Loading Mitochondria with [^{14}C]Citrate

For accurate determination of the transport properties of the citrate carrier itself it is necessary to load a high concentration of a transportable substrate into the mitochondrial matrix. Most of the transport experiments we shall describe deal with citrate vs citrate exchange since this dynamic antiport process does not result in the buildup of a membrane potential or pH gradient, either of which would impose other more complex restrictions on the transport process.[33] Thus, in all subsequent experiments, both tumor and normal mitochondria are preloaded with high levels of citrate.

Reaction Procedures. Mitochondria (8 mg protein/ml final concentration) are added to a reaction mix that already contains isotonic buffer, 0.3 nmol antimycin/mg mitochondrial protein, 0.4 μg rotenone/mg mitochon-

[33] M. L. Halperin, S. Cheema-Dhadli, W. M. Taylor, and I. B. Fritz, Adv. Enzyme Regul. 13, 435 (1975).

drial protein, and 2 mM (final concentration) sodium arsenite. This mixture is preincubated for 60 sec at 10°. Loading is then initiated by the addition of 0.75 mM [^{14}C]citrate (specific radioactivity = 4.5 nCi/nmol) in isotonic buffer. The final volume varies depending on the quantity of mitochondria to be loaded. After 10 min under these conditions the reaction mixture is diluted with 6 volumes of ice cold isotonic buffer, and is centrifuged at 16,000 g for 5 min at 4° (SS-34 rotor, RC-5 centrifuge, Dupont-Sorvall). The top of the pellet is rinsed twice with the isotonic buffer, and the pellet is resuspended carefully by hand in a glass-Teflon homogenizer in 97 mM KCl, 19.5 mM HEPES, 1 mM EGTA, pH 7.0, and 2 mM sodium arsenite, *plus* (per mg mitochondrial protein) 0.25 nmol antimycin and 0.4 μg rotenone. For our studies,[7] the final mitochondrial protein concentrations in the loaded suspension were 12.5 mg/ml (tumor 16 and its controls), 2.12 mg/ml (normal and host (3924A) liver preparations), and 4.25 mg/ml (tumor 3924A).

Precautions. Arsenite (an inhibitor of α-ketoglutarate dehydrogenase[34]) as well as antimycin and rotenone (electron transport chain inhibitors) are included in this loading procedure in order to block citrate metabolism. In fact, our studies with normal liver mitochondria indicated that even when high levels of the electron transport inhibitors are included in both the citrate loading incubation and the final loaded mitochondrial suspension, a slow conversion of [^{14}C]citrate to $^{14}CO_2$ occurs.[35] This oxidation is largely abolished by the inclusion of the arsenite in both the loading mix and the final loaded mitochondrial suspension, in which case less than 8% of the loaded [^{14}C]citrate is metabolized to $^{14}CO_2$ within 3 hr.

Results. With this loading procedure, an average internal [^{14}C]citrate concentration $\simeq$20 mM is attained,[36] whereas the average external citrate concentration remains <5 μM when a suspension of freshly loaded mitochondria is placed in a typical "efflux" incubation mix (see below). We found that internal citrate, loaded this way, slowly leaks out of mitochondria. Therefore, all transport reactions with loaded mitochondria are performed within 2–3 hr of loading, during which time the external citrate level remains low.

Kinetics of [^{14}C]Citrate Efflux from [^{14}C]Citrate-Loaded Mitochondria

Incubation Medium. The reaction mix contains isotonic buffer, 0.25 nmol antimycin and 0.4 μg rotenone (both per mg total protein), mito-

[34] D. R. Sanadi, M. Langley, and F. White, *J. Biol. Chem.* **234,** 183 (1959).

[35] R. S. Kaplan, Ph.D. thesis, New York University, 1980.

[36] The intramitochondrial citrate concentration after loading is based on a matrix volume assumed to be 1 μl/mg mitochondrial protein.

chondria (see below), BSA (see below), and various concentrations in isotonic buffer of external [^{12}C]citrate (final volume = 3.0 ml, 10°).

The concentration of mitochondria that permits accurate determinations of initial transport rates is found empirically and varies depending on the type of mitochondria being examined. Thus, we found that between approximately 2 and 3 mg mitochondrial protein/incubation gave initial rate data that could be satisfactorily measured and reproduced with all six types of mitochondria. In all reactions, the mitochondrial protein/BSA (w/w) ratio is kept constant at 6.5 ± 0.1. Except where noted, all [^{14}C]citrate transport inhibitors are added as isotonic buffer solutions.

Reaction Procedures. [^{14}C]Citrate-loaded mitochondria are preincubated in the reaction mix for 60 sec, at which time [^{14}C]citrate efflux is initiated by the addition of [^{12}C]citrate to the incubation.

Before continuing with these detailed methods, it is necessary to make a distinction between *experimental* (E-aliquot) and *control* (C-aliquot) samples that will be removed from the incubation for analysis. E-aliquots are removed at various times after [^{12}C]citrate addition and placed into ice cold 0.05 *M* BTC in order to quench the transport reaction. The concentration of the [^{12}C]citrate in the E-aliquot at this point is designated as *Y* μM, and varies depending on the concentration of [^{12}C]citrate added to trigger the efflux reaction.

Ten seconds *prior* to the transport-triggering [^{12}C]citrate addition (at *t* = 50 sec), a C-aliquot is transferred from the incubation to a solution of 0.05 *M* BTC *plus Y* μM [^{12}C]citrate.

With these manipulations it is clear that throughout the remainder of the procedure, the tubes containing the E-aliquot and C-aliquot will contain identical concentrations of BTC, mitochondria, and [^{12}C]citrate. Consequently, the control tube data will account for all extramitochondrial [^{14}C]citrate that does *not* arise as a result of the timed exchange transport reaction.

Twenty seconds after an E- or C-aliquot is placed into BTC, 0.5 μmol mersalyl/mg total protein is added. Two minutes later, 10 m*M* phenylsuccinate is added. The transport-quenched reaction aliquots are then centrifuged (SS-34 rotor) at 16,000 *g* for 2 min (4°), and the supernatants are counted in the scintillation spectrometer. Typically, several pellets are analyzed for protein, and all data are normalized to this amount of protein. The C-aliquot tube is corrected with respect to leak rate (see below), and then the radioactivity of the corrected C-aliquot tube is subtracted from that of the E-aliquot tubes.

Measuring the Leak Rate. We found that some [^{14}C]citrate efflux occurs as a leak during the 10 sec time interval between the removal of the C-aliquot (at *t* = 50 sec) and the transport-triggering addition of

[^{12}C]citrate (at t = 60 sec). In order to quantify this efflux, leak incubations should be routinely performed, in which isotonic buffer is added in place of [^{12}C]citrate. Any ensuing efflux of [^{14}C]citrate occurs in the absence of externally added counteranion, and therefore under conditions that are identical to those which exist between the t = 50 and t = 60 sec time points (i.e., 10 sec interval) in a typical efflux incubation. From these leak detection incubations, [^{14}C]citrate leak rates are calculated, and used to correct for citrate movement during the 10 sec interval in the standard efflux incubations.

Finally, control incubations are also performed in which 0.05 M BTC (without additional isotonic buffer) is initially present in the transport reaction mix (see Table II). The subsequent addition of a saturating [^{12}C]citrate level (0.4 mM) resulted in negligible amounts of [^{14}C]citrate efflux on other carriers (Table II).

Results. Employing the above procedures, initial [^{14}C]citrate efflux rates can be determined with tumor mitochondria and their control preparations. We found efflux to be linear with respect to time for 15–50 sec, depending on the mitochondrial source and the external citrate concentration used to trigger the transport. The data presented in Fig. 1 indicate that citrate efflux kinetics with tumor and normal mitochondria are similar under the experimental conditions employed. Only with tumor 3924A

TABLE II

V_{max} OF [^{14}C]CITRATE EFFLUX FROM PRELOADED MITOCHONDRIA IN EXCHANGE FOR EITHER EXTERNAL CITRATE OF MALATE[a]

Mitochondrial source	V_{max}: citrate vs citrate (nmol/min/mg protein)		V_{max}: citrate vs malate (nmol/min/mg protein)	
	−BTC	+BTC	−BTC	+BTC
1. Normal liver (ACI rat)	8.67 ± 1.85	0.27 ± 1.12	8.13 ± 1.86	1.93 ± 3.91
2. Host liver (3924A)	11.85 ± 2.59	0.53 ± 0.19	11.48 ± 2.07	1.38 ± 1.79
3. Hepatoma 3924A	7.37 ± 3.21	0.47 ± 1.91	7.79 ± 5.58	2.86 ± 1.73
4. Normal liver (Buffalo rat)	6.21 ± 0.97	0.54 ± 0.47	6.62 ± 1.19	1.46 ± 1.54
5. Host liver (16)	6.45 ± 0.77	0.25 ± 0.33	7.63 ± 2.04	0.28 ± 2.40
6. Hepatoma 16	7.25 ± 0.46	0.33 ± 0.69	8.69 ± 0.53	2.96 ± 1.71

[a] Efflux was triggered by addition of either 0.4 mM [^{12}C]citrate or 5.0 mM [^{12}C]malate. Prior to initiating the exchange reaction, a C-aliquot has removed from the incubation and placed immediately into a solution containing 0.05 M BTC, *plus* the same [^{12}C]citrate or malate concentration present in the E-aliquot-BTC mixtures. Leak incubations were not required since the observed transport rates were considerably greater than leak rates. For further details see Kaplan *et al.*[7] Reproduced with permission of *Cancer Research.*

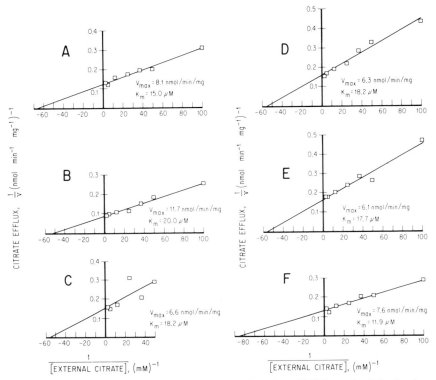

FIG. 1. Lineweaver–Burk plots of [^{14}C]citrate efflux from [^{14}C]citrate-loaded mitochondria. Data were obtained from 3 individual mitochondrial preparations for each type of tissue, with the exception of hepatoma 3924A mitochondria, the data for which were derived from an average of 8 preparations. Best fit lines were constructed via linear regression least squares analysis. (A) Normal ACI rat liver; (B) host liver (3924A); (C) hepatoma 3924A; (D) normal Buffalo rat liver; (E) host liver (16); (F) hepatoma 16. From Kaplan *et al.*[7] with permission of *Cancer Research.*

mitochondria is there some scatter in the Linewaver–Burk plot (Fig. 1C, correlation, coefficient = 0.744, linear probability = 90%), which may be due to their known fragility.[25] This, in turn, could be a consequence of the documented altered lipid composition of their membranes.[12,37–39]

With minor modification (see Table II), the above procedures can also be used to measure the efflux of citrate triggered by added malate. This

[37] F. Feo, R. A. Canuto, R. Garcea, and L. Gabriel, *Biochim. Biophys. Acta* **413,** 116 (1975).
[38] R. Morton, C. Cunningham, R. Jester, M. Waite, N. Miller, and H. P. Morris, *Cancer Res.* **36,** 3246 (1976).
[39] R. C. Reitz, J. A. Thompson, and H. P. Morris, *Cancer Res.* **37,** 561 (1977).

measurement may be especially important, since it is thought that the efflux of citrate from the matrix is coupled, primarily, to the influx of malate *in vivo*.[4] Comparative V_{max} data shown in Table II indicate similar values for [14C]citrate efflux regardless of whether the exchange is triggered by external citrate or malate. Additionally, BTC inhibits both exchange reactions in all preparations studied.

Kinetics of [14C]Citrate Influx into [12C]Citrate-Loaded Mitochondria

Reaction Procedures. By incorporating the following minor changes, the standard citrate efflux procedure can be used to study the kinetics of [14C]citrate *influx* into [12C]citrate preloaded mitochondria. The transport reaction is initiated after the 60 sec preincubation by the addition of a given concentration of [14C]citrate (~4.5 nCi/nmol) to the reaction mix. Similar to the strategy given above, the C-aliquot is removed at $t = 50$ sec and placed into a solution containing 0.05 M BTC and the appropriate concentration (i.e., $Y \mu M$) of [14C]citrate. Other conditions are identical to those previously described. Following the centrifugation step, the supernatants are discarded. The tops of the pellets are rinsed twice with isotonic buffer containing 1 mM BTC and 10 mM phenylsuccinate, and the radioactivity of the pellets is then determined.

Results. Table III indicates similar V_{max} values for citrate influx into preloaded mitochondria for all preparations studied. BTC effectively in-

TABLE III
V_{max} OF [14C]CITRATE INFLUX INTO [12C]CITRATE-LOADED
MITOCHONDRIA[a]

Mitochondrial source	V_{max} (nmol/min/mg protein)	
	−BTC	+BTC
1. Normal liver (ACI rat)	12.08 ± 1.96	0.12 ± 2.06
2. Host liver (3924A)	12.15 ± 3.41	0.23 ± 1.11
3. Hepatoma 3924A	9.40 ± 2.72	−3.30 ± 4.35
4. Normal liver (Buffalo rat)	7.27 ± 1.11	−0.69 ± 0.82
5. Host liver (16)	10.26 ± 3.11	0.25 ± 0.36
6. Hepatoma 16	8.65 ± 0.93	1.16 ± 1.09

[a] Data were obtained according to the methods given in the text. Influx was triggered by addition of 0.4 mM [14C]citrate. When added, 0.05 M BTC (not in isotonic buffer) was present prior to the addition of mitochondria. For further details, see Kaplan *et al.*[7] Reproduced with permission of *Cancer Research*.

hibits this influx. Furthermore, comparing the data of Table II with those of Table III illustrates the fact that, within experimental error, the stoichiometry of the exchange transport is most likely 1 : 1 for all mitochondrial preparations. Finally, a comparison of the data from Tables I and III supports the contention that preloading the mitochondria yields a higher V_{max} for citrate influx, due, no doubt, to the fact that the concentration of endogenous exchangeable substrate upon mitochondrial isolation is not saturating for the citrate carrier.

The Effect of Temperature

One of the difficulties in conducting studies on mitochondrial transport kinetics is the requirement to perform the experiments at well below physiological temperature. Unless the temperature is reduced, carrier-mediated transport occurs too rapidly to allow for the accurate measurement of initial rates required for V_{max} and K_m determinations. In spite of these difficulties, it is important to conduct transport studies at several different temperatures, when comparing tumor and normal mitochondria (see below). In fact, an altered temperature dependency has been observed by other laboratories for the activities of several tumor mitochondrial membrane functions, including anion transport.[40,41] Such changes may be due to the well documented alterations in the lipid composition (particularly with respect to cholesterol) of the tumor mitochondrial membranes.[12,42-45]

Reaction Procedures. The rate of [14C]citrate efflux as a function of temperature is measured with hepatoma 3924A mitochondria and its controls according to the standard protocol given above, modified as follows. First, aliquots of isotonic buffer are pH adjusted at different temperatures to remain constant at 7.0 over the range of temperatures studied. Second, 0.4 mM [12C]citrate is added to the reaction mix to initiate [14C]citrate efflux. Depending on the incubation temperature, transport reactions are carried out over varying time intervals (5–120 sec) to obtain initial export velocities. At high temperatures (e.g., 21°) the efflux reaction is linear for only 5–10 sec, whereas at low temperatures (e.g., 3°) it is necessary to

[40] F. Feo, R. A. Canuto, R. Garcea, A. Avogadro, M. Villa, and M. Celasco, *FEBS Lett.* **72**, 262 (1976).
[41] H. S. Sul, E. Shrago, S. Goldfarb, and F. Rose, *Biochim. Biophys. Acta* **551**, 148 (1979).
[42] M. Esfahani, A. R. Limbrick, S. Knutton, T. Oka, and S. J. Wakil, *Proc. Natl. Acad. Sci. U.S.A.* **68**, 3180 (1971).
[43] P. V. Vignais, *Biochim. Biophys. Acta* **456**, 1 (1976).
[44] K. Y. Hostetler, *in* "Membranes in Tumor Growth" (T. Galleotti, A. Cittadini, G. Neri, and S. Papa, eds.), p. 481. Elsevier, Amsterdam, 1982.
[45] F. Feo, R. A. Canuto, G. Bertone, R. Garcea, and P. Pani, *FEBS Lett.* **33**, 229 (1973).

TABLE IV
CHARACTERISTICS OF THE TEMPERATURE DEPENDENCY OF CITRATE
EFFLUX KINETICS[a]

Mitochondrial source	E_a (kcal/mol)	Q_{10}	V_{max} (37°) (nmol/min/mg)	T_c (°)
1. Normal liver (ACI rat)				
High temperature	21.5	3.66	283.1	
Low temperature	34.4	9.11	—	12.5
2. Host liver (3924A)				
High temperature	21.0	3.56	330.0	
Low temperature	41.4	14.43	—	11.0
3. Hepatoma 3924A	28.9	5.95	591.7	NONE

[a] Values for E_a, Q_{10}, and the V_{max} extrapolated at 37° were calculated from the data of Fig. 2. E_a, activation energy; Q_{10}, factor by which the reaction velocity changes for a 10° increase in temperature; T_c, phase transition temperature. From Kaplan et al.[7] with permission of Cancer Research.

allow the reaction to continue for 2 min in order that a measurable initial rate be obtained. Initial citrate transport rates cannot be determined accurately at temperatures greater than 21° due to the rapidity with which the limited quantity of loaded [^{14}C]citrate is exported from the mitochondrial matrix.

Results. Arrhenius plots are shown in Fig. 2, and the information derived from them is summarized in Table IV. Control mitochondrial preparations display biphasic plots, with phase transition temperature breaks in the vicinity of 12°. However, tumor 3924A mitochondria exhibited linearity over the entire temperature range studied (5–21°). Extrapolating the data to 37° indicates that citrate export from the tumor 3924A mitochondria may occur more than 2-fold faster than from normal control liver mitochondria (Table IV), a result that is clearly relevant to the proposed relationship between deregulated cholesterogenesis and cholesterol enrichment of membranes in tumor tissue (cf. Introduction).

Methods for Determination of Metabolite Flux

Basic Principles

The Arrhenius plot studies described above revealed that as physiological temperature is approached, tumor mitochondria appear to have the capacity to export citrate considerably faster than do normal mito-

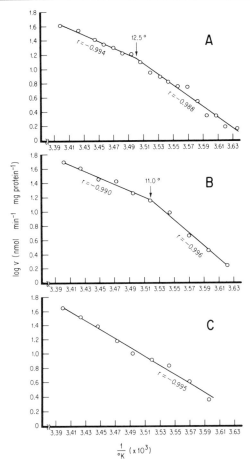

FIG. 2. Arrhenius plots of [¹⁴C]citrate efflux rates from [¹⁴C]citrate-loaded mitochondria in exchange for [¹²C]citrate with tumor 3924A, its host and normal mitochondrial preparations. Efflux was triggered with 0.4 mM externally added [¹²C]citrate. Leak incubations were not performed. Linear regression least-squares analysis was performed and correlation coefficients (r) are shown. (A) Normal ACI rat liver (6 preparations; temperature range = 2–21°); (B) host liver (3924A) (3 preparations; temperature range = 3–21°); (C) tumor 3924A (3 preparations; temperature range = 5–21°). Points shown are the means of preparations studied. From Kaplan et al.[7] with permission of *Cancer Research*.

chondria. In order to test the premise that tumor mitochondria do, in fact, export citrate from the matrix more rapidly than normal organelles do, the following metabolic flux studies are performed. Since these studies do not attempt to measure initial transport rates, experiments can be carried out at higher temperatures, and under incubation conditions that can explicitly permit mitochondrial metabolism to proceed.

Determination of Mitochondrial State 3 Respiratory Rates

Incubation Medium. The basic reaction medium is similar to that of Kaschnitz *et al.*[25] and contains 250 mM mannitol, 20 mM KCl, 10 mM KH$_2$PO$_4$, and 5 mM MgCl$_2$, pH 7.4. We used a mitochondrial protein concentration that ranged from 1.3 mg/ml (normal liver) to 2.0 mg/ml (tumor) in a final incubation volume of 3.0 ml. Furthermore, we ensured that the mitochondrial protein/BSA ratio was kept constant at 2:1 for all tumor and normal mitochondrial incubations. Respiratory chain substrates are added to a final concentration of 5 mM each. With particular substrates (e.g., pyruvate, citrate, and isocitrate), malate (1 mM) should also be added as counter exchange anion.

Reaction Procedures. Respiratory (acceptor) control ratios (RCR) and ADP/O ratios are determined according to the graphic procedure of Estabrook.[28] Oxygen consumption is assayed polarographically at 30° with a Clark-type oxygen electrode apparatus (e.g., Model 53, Yellow Spring Instruments).

State 3 oxygen uptake rates are triggered with 250 μM ADP after the reaction medium has been allowed to achieve temperature equilibration (about 2 min). Sequential ADP additions (up to three/incubation) should be made to permit calculation of the mean state 3 rate for each substrate/ mitochondrial type examined. In specific incubations with pyruvate or citrate as the primary respiratory chain substrate, the citrate carrier inhibitor BTC is also present. In such incubations, BTC is added 30 sec *after* the addition of substrate, and 30 sec *prior* to the addition of ADP.

Results. Table V shows state 3 rates for both tumor and normal liver mitochondria fueled with a variety of respiratory chain substrates, in both the absence and presence of BTC. In the absence of BTC, normal liver mitochondria and the mitochondria of "minimally deviated" tumor 16 display fairly high (that is, "normal") rates of oxygen consumption with all substrates tested. However, when respiration is fueled with pyruvate or citrate, the mitochondria from the rapidly proliferating tumor 3924A yield very slow state 3 rates ($\leq$1 ng atom O/min/mg). Yet, Table V also indicates that the presence of BTC appears to "restore" this otherwise negligible state 3 respiration with pyruvate, and particularly citrate, and tumor 3924A mitochondria. A likely interpretation of this phenomenon is that by inhibiting citrate export from tumor 3924A mitochondria, BTC permits the *metabolic* conversion of citrate in the matrix (whether the citrate is generated there via the enzymatic decarboxylation of externally

TABLE V

STATE 3 RATES OF OXYGEN CONSUMPTION BY TUMOR AND NORMAL RAT LIVER MITOCHONDRIA[a]

Substrate	Tumor 3924A		Normal ACI liver		Tumor 16		Normal Buffalo liver	
	−BTC	+BTC	−BTC	+BTC	−BTC	+BTC	−BTC	+BTC
Pyruvate	Negligible	11.9 ± 2.5	38.8 ± 5.1	29.9 ± 7.4	14.4 ± 1.3	12.1 ± 2.2	23.2 ± 3.2	13.4 ± 2.5
Citrate	Negligible	47.9 ± 9.5	43.9 ± 4.2	48.7 ± 8.3	36.5 ± 8.0	32.5 ± 6.5	28.6 ± 2.7	38.9 ± 3.4
Isocitrate	82.1 ± 11.4	80.1 ± 9.8	41.9 ± 1.1	39.8 ± 5.3	29.0 ± 1.1	28.8 ± 1.1	24.6 ± 2.0	26.9 ± 2.8
Glutamate	66.6 ± 8.7	—	59.8 ± 2.1	—	47.0 ± 1.4	—	28.7 ± 3.4	—
α-Ketoglutarate	92.0 ± 13.3	—	112.8 ± 14.9	—	41.3 ± 8.7	—	87.3 ± 9.3	—
Succinate	78.8 ± 7.6	—	113.0 ± 13.0	—	88.4 ± 16.6	—	78.8 ± 5.2	—
Mitochondrial membrane total cholesterol (μg/mg protein)	25.7 ± 6.7		4.9 ± 0.9		13.7 ± 4.1		6.8 ± 1.6	

[a] For O_2 uptake experiments, duplicate incubations from 4 individual preparations from each tissue source were performed as described in the text. Data are given as the mean ±SD. For membrane cholesterol determinations, aliquots of freshly isolated mitochondria (20 mg protein) were added to 0.5 ml isopropyl alcohol, the mixture vortexed, and stored at −70° until assayed. Thawed samples, containing about 2.0 mg mitochondrial protein, were analyzed for total cholesterol according to J. G. Heider and R. L. Boyett. J. Lipid Res. **19**, 514 (1978). Values are expressed as the mean ±SD of triplicate assays from more than 8 individual mitochondrial preparations from each tissue source.

added pyruvate, or originally arises via the influx of the citrate added to the incubation 30 sec prior to the addition of BTC). The negligible state 3 rates with pyruvate or citrate and tumor 3924A mitochondria could thus be due to a rapid citrate efflux from matrix to external medium, as implied by the data from the Arrhenius plot studies (cf. Fig. 2, Table IV).

Table V also depicts the membrane cholesterol levels found in mitochondria from both the slow and rapidly growing hepatomas and their normal control livers. These data suggest that a correlation may exist between the extent of cholesterol enrichment of tumor mitochondrial membranes, and the abberrant respiratory rates with pyruvate or citrate. This correlation is extended to include the rate of citrate efflux via the following methods.

Mitochondrial Export of Pyruvate-Derived Citrate

Incubation Medium. Freshly isolated mitochondria (1–2 mg/ml protein) are incubated at 25° in a reaction medium identical to that employed for oxygen consumption studies. Pyruvate (0.5 mM) *plus* malate (0.1 mM) provide the carbons for the intramitochondrial formation of citrate. Reactions are triggered with 5 mM ADP, and the sequential incubations are performed both in the absence and the presence of 150 μM BTC. In the latter case, the mitochondria are preincubated in the reaction medium *plus* BTC for 1 min prior to addition of pyruvate/malate or ADP. Three control incubations are run in parallel reactions, and consist of (1) mitochondria alone, (2) mitochondria *plus* pyruvate/malate, and (3) mitochondria *plus* ADP.

Reaction Procedures. The basic method is that of Williamson and Corkey[46] in which mitochondria are incubated in the reaction medium described above for various periods of time between 5 and 80 sec. At the end of each time interval a 1.0-ml aliquot is removed from the incubation and layered above a band (3 mm wide) of silicone oil (sp. grav. = 1.05, General Electric Co.) which serves as an interface on top of 0.3 ml of 14% (w/v) perchloric acid (PCA) in 1.5-ml conical polypropylene centrifuge tubes. Immediately, the tube is spun (12,000 g, 2 min, room temperature) in the rapid acceleration Eppendorf microfuge (model 5412, Brinkmann). To help assure the proper timing of such a procedure, the incubation aliquots are taken up in an automatic pipettor (e.g., the Gilson Pipetman) and layered onto the silicone oil barrier about 2–3 sec prior to the end of the defined time interval, then centrifuged immediately. Only the mitochondria penetrate the oil barrier and are precipitated in the PCA layer as

---

[46] J. R. Williamson and B. E. Corkey, this series, Vol. 55, p. 200.

a centrifugal pellet. Any mitochondrially generated citrate that was exported to the external medium during the timed incubation interval remains above the silicone oil. Intramatrix citrate, on the other hand, comes down together with the mitochondria and is released into the PCA layer. With this protocol, the earliest incubation time interval capable of being reliably assayed by us was 5 sec after addition of ADP to the incubation medium in order to trigger mitochondrial metabolism.

Extramitochondrial (i.e., exported) citrate is assayed on 0.8 ml of the postcentrifugal supernatant above the silicone oil layer. To this aliquot, 0.2 ml of cold 14% PCA is added and this mixture is then centrifuged (12,000 g, 2 min) to remove residual protein. The supernatant is neutralized with KOH, recentrifuged as before to remove $KClO_4$, and this last supernatant is used for the determination of citrate (see below). The steady-state intramitochondrial citrate level can be determined by assaying the contents of the PCA layer after removal of the silicone oil layer by aspiration, in which case a 0.2-ml aliquot of the PCA layer is neutralized, centrifuged, and then analyzed for citrate. Experimental data for the levels of intra- and extramitochondrial citrate are corrected for endogenous citrate that is present within the mitochondria prior to the addition of pyruvate/malate to the incubation.

The coupled enzyme reaction assay of Moellering and Gruber[47] is quite satisfactory for measuring the citrate levels in this system. The starting, as-yet incomplete, assay mix contains 160 mM glycylglycine buffer (pH 7.8), 0.19 μM NADH, 3.8 units/ml malate dehydrogenase (MDH), 8.8 units/ml lactate dehydrogenase (LDH), and the citrate-containing sample. Ultimately, the reactions that will occur in this assay are

$$\text{citrate} \xrightarrow{\text{citrate lyase}} \text{oxaloacetate} + \text{acetate} \tag{1}$$

$$\text{oxaloacetate} + \text{NADH} + \text{H}^+ \xrightarrow{\text{MDH}} \text{L-malate} + \text{NAD}^+ \tag{2}$$

$$\text{CO}_2 \downarrow \text{(spontaneous)}$$

$$\text{pyruvate} + \text{NADH} + \text{H}^+ \xrightarrow{\text{LDH}} \text{L-lactate} + \text{NAD}^+ \tag{3}$$

It is important to note that an aliquot of sample, derived as given above, contains in addition to citrate, a certain amount of unconverted pyruvate that had been added at the start to the incubation medium, *plus* any endogenous mitochondrial pyruvate. Thus, prior to the addition of the citrate lyase enzyme, reactions (2) and (3) are allowed to proceed until all of the excess pyruvate is converted to lactate (at least 5 min at 23°). This is verified when a stable A_{340} value is recorded for 1 min, and only then are 27 units/ml citrate lyase added to make the assay reaction mix complete,

[47] H. Moellering and W. Gruber, *Anal. Biochem.* **17**, 369 (1966).

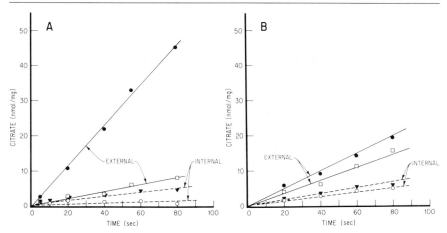

FIG. 3. Extra- and intramitochondrial citrate levels over time with tumor and normal mitochondria. Mitochondria were incubated for various times in the presence of 0.5 mM pyruvate *plus* 0.1 mM malate *plus* ADP, then subjected to rapid centrifugation through a silicone oil barrier (see text). Each point represents the mean of duplicate incubations performed on 4 individual mitochondrial preparations. (A) ●, external; ○, internal; tumor 3924A. □, external; ▼, internal; normal ACI. (B) ●, external; ○, internal; tumor 16. □, external; ▼, internal; normal Buffalo.

thereby triggering reaction (1). The final assay volume is 3.14 ml and the sample is read against air. In our hands this assay is sensitive to 1.9 ± 0.4 nmol citrate/mg.

Results. Mitochondria from the rapidly growing tumor 3924A display a dramatically enchanced rate of pyruvate-derived citrate efflux (4.2-fold) compared with normal control (ACI) mitochondria (Fig. 3A), while the matrix citrate level is reduced to 14% that of the control mitochondria. These tumor 3924A mitochondria also possess fivefold enrichment in the cholesterol content of their membrane (Table V). In contrast, mitochondria from the slowly growing tumor 16, whose membranes are cholesterol-enriched only 2-fold (Table V), exhibit a correspondingly smaller 1.3-fold increase in the pyruvate-to-citrate flux. The steady-state matrix citrate level for tumor 16 mitochondria is 82% of normal (Buffalo) controls. These data (see also ref. 16) corroborate those of Table IV and help sustain the proposal that a direct functional correlation exists between the rate of the carrier-mediated efflux of citrate and the extent of tumor mitochondrial membrane enrichment with cholesterol.

[54] Partial Purification and Reconstitution of the Tricarboxylate Carrier from Rat Liver Mitochondria*

By FERDINANDO PALMIERI, ITALO STIPANI, GIROLAMO PREZIOSO, and REINHARD KRÄMER

In addition to many other carriers, the inner mitochondrial membrane contains a specific exchange system for the transport of tricarboxylates, malate and phosphoenolpyruvate.[1,2] This carrier which is present in liver but absent in heart and brain[3,4] has important functions in the pathways of fatty acid synthesis and gluconeogenesis and in the transfer of reducing equivalents across the inner mitochondrial membrane.

A 94-fold purification of a tricarboxylate-binding protein from Triton-extracted submitochondrial particles was obtained in our laboratory by affinity chromatography on a Sepharose-aminobenzene 1,2,3-tricarboxylate column.[5] Unfortunately, the protein fraction isolated by this method was not able to catalyze citrate/citrate exchange when incorporated into phospholipid vesicles.

Later on, we succeeded in reconstituting citrate transport activity using detergent-solubilized mitochondrial particles[6] and, recently, we reported the functional reconstitution of a partially purified tricarboxylate carrier from rat liver mitochondria.[7] The citrate carrier is about 25-fold enriched and—when reconstituted into phospholipid vesicles—closely resembles in its properties the transport system observed in mitochondria.

Isolation and Reconstitution of the Tricarboxylate Carrier

Solubilization. Fresh rat liver mitochondria[8] or submitochondrial particles[9] were solubilized in a buffer containing 4% Triton X-100, 50 mM

---

* This paper is dedicated to the memory of Professor Enzo Leone.
[1] K. F. LaNoue and A. C. Schoolwerth, *Annu. Rev. Biochem.* **48,** 871 (1979).
[2] A. J. Meijer and K. Van Dam, in "Membrane Transport" (S. Bonting and J. De Pont, eds.), p. 235. Elsevier, Amsterdam, 1981.
[3] J. B. Chappell, *Br. Med. Bull.* **24,** 150 (1968).
[4] F. E. Sluse, A. J. Meijer, and J. M. Tager, *FEBS Lett.* **18,** 149 (1971).
[5] F. Palmieri, G. Genchi, I. Stipani, and E. Quagliariello, in "Structure and Function of Energy-Transducing Membranes" (K. Van Dam and B. F. Van Gelder, eds.), p. 251. Elsevier, Amsterdam, 1977.
[6] I. Stipani, R. Krämer, F. Palmieri, and M. Klingenberg, *Biochem. Biophys. Res. Commun.* **97,** 1206 (1980).
[7] I. Stipani and F. Palmieri, *FEBS Lett.* **161,** 269 (1983).

NaCl, 2.5 mM benzene 1,2,3-tricarboxylate (Merck), 3 mg/ml cardiolipin (Serdary), and 10 mM morpholinopropanesulfonic acid, pH 7.0 at a concentration of 10 mg protein/ml. Frozen mitochondria can also be used which, however, leads to about 50% loss of activity. Addition of the phospholipid cardiolipin and of benzene 1,2,3-tricarboxylate, a specific inhibitor of the carrier, to the solubilization buffer is important to protect the transport protein during the procedure. It has been found that the presence of cardiolipin, in particular, protects the citrate carrier against denaturation by the detergent.[7,10]

Effective solubilization is achieved by vortexing the mitochondria with 1 ml of solubilization buffer for a few seconds at 0°. After 20 min at 0° the solubilizate is centrifuged at 147,000 g for 45 min; the sediment is discarded.

Solubilization has been attempted also with several other detergents. Triton X-100 proved to be the best for isolation of an active citrate carrier protein. Other nonionic detergents like Genapol X-80 (Hoechst), Emulphogen BC 720 (GAF), and octaethylene glycol mono-n-dodecyl ether (Nikkol) were considerably less effective. It was not possible to solubilize a reconstitutively active carrier with the nonionic detergents laurylamidodimethylpropylamine oxide or octylglucoside, or with the polyoxyethylene glycol homologs of Triton, Brij 58, Lubrol WX, and Tween 20, or with the ionic detergent cholate.

Partial Purification. As in the case of other mitochondrial carrier proteins, the tricarboxylate carrier can be partially purified by chromatography on hydroxyapatite. Of the supernatant after ultracentrifugation (3.5–5 mg protein) 600 μl is applied to 0.6 g dry hydroxyapatite (Bio-Rad) in Pasteur pipets at 4°. The protein is eluted by solubilization buffer. The first 1.5 ml, which contains about 4% of the applied protein, is collected. Table I reports the results obtained by applying the Triton X-100 extract of rat liver mitochondria to hydroxyapatite in the presence and absence of cardiolipin. By this procedure, in the presence of cardiolipin, the specific activity of the reconstituted citrate/citrate exchange increases 22-fold and the total activity in the eluate after hydroxyapatite accounts for 73% of that applied to the column. Thus the citrate transport protein passes through hydroxyapatite almost quantitatively whereas the bulk of the protein is retained.

---

[8] M. Klingenberg and N. Slenczka, *Biochem. Z.* **331**, 486 (1959).

[9] G. L. Scottocasa, *in* "Biochemical Analysis of Membranes" (A. M. Maddy, ed.), p. 55. Chapman & Hall, London, Wiley, New York, 1976.

[10] I. Stipani, G. Prezioso, V. Zara, V. Iacobazzi, and G. Genchi, *Bull. Mol. Biol. Med.* **9**, 193 (1984).

| | | Citrate exchange[a] | | |
Step	Protein applied or eluted (mg)	Specific activity	Total activity	Recovery (%)
Mitochondrial extract (−DPG)	4.32	34	146	
HTP eluate from mitochondrial extract (−DPG)	0.20	460	92	63
Mitochondrial extract (+DPG)	4.72	85	401	
HTP eluate from mitochondrial extract (+DPG)	0.16	1840	294	73

[a] The activity of the reconstituted citrate/citrate exchange is expressed as μmol 10 min^{-1} g protein^{-1} (specific activity) and μmol 10 min^{-1} (total activity). DPG, Cardiolipin; HTP, hydroxyapatite.

The SDS–PAGE of the pass-through fraction shows at least 6 protein bands in the region of 28–35 kDa.[7] Three out of these 6 components have clearly been identified as the adenine nucleotide carrier, the phosphate carrier,[11] and the mitochondrial porin.[12]

Incorporation into Preformed Liposomes. Phospholipids from turkey egg yolk (Sigma) with addition of 25% (w/w) mitochondrial phospholipids[13] are suspended at a concentration of 125 mg/ml in 12 ml of 50 mM NaCl, 20 mM citric acid, and 20 mM morpholinopropanesulfonic acid, pH 7.5 and pulse-sonicated (20 sec sonication, 20 sec interval) with a Branson sonifier under nitrogen in an ice/water bath for about 1 hr.

For incorporation, 1.3 ml liposomes is rapidly mixed with 50 μl of the extracts from rat liver mitochondria or submitochondrial particles or the hydroxyapatite pass-through fractions. The mixture is frozen in liquid nitrogen, thawed in a water bath at 15–20°, and pulse-sonicated (0.2 sec sonication, 0.8 sec interval, output power 35 W) for 60 sec at 0°.

Assay of Reconstituted Citrate Transport. The citrate outside the proteoliposomes has to be removed in order to measure citrate/citrate exchange using externally added labeled substrate. This is achieved by passing the proteoliposomes through an anion exchange column (AG1 X-8, 50 mesh, Bio-Rad) in acetate form which has been preequilibrated with 150 mM sucrose.

[11] F. Bisaccia and F. Palmieri, *Biochim. Biophys. Acta* **766,** 386 (1984).
[12] V. De Pinto, M. Tommasino, R. Benz, and F. Palmieri, *Biochim. Biophys. Acta* **813,** 230 (1985).
[13] G. Rouser and S. Fleischer, this series, Vol. 10 [69].

Proteoliposomes (1.25 ml) are applied to a 0.5 × 8 cm column. 1.1 ml of the pass-through fraction is collected, diluted with 0.25 ml of a buffer consisting of 50 mM NaCl, 20 mM morpholinopropanesulfonic acid, pH 7.0, distributed into reaction vessels, and incubated at 30° for 4 min. Addition of [14C]citrate (Amersham) at a final concentration of 0.3 mM starts the exchange reaction. The transport is stopped after appropriate time intervals (1–10 min) by adding benzene 1,2,3-tricarboxylate (6 mM) and instant cooling to 0°. In order to obtain blank values the inhibitor is added before the substrate. The samples are then applied to AG1-X8 (100–200 mesh) columns in chloride form (0.5 × 5 cm) which have been preequilibrated with 150 mM sucrose. The eluate after addition of first 1 × 100 μl, then 2 × 200 μl, then 3 × 400 μl of 150 mM sucrose is directly collected in scintillation vials and counted in a scintillation counter.

Some Properties of the Reconstituted Tricarboxylate Carrier

The basic properties of the reconstituted citrate carrier, e.g., substrate affinity, inhibitor specificity, tissue specificity, and the absolute requirement of an appropriate counteranion, generally resemble those known for this transport system from mitochondria.[14] The K_m and the V_{max} values for citrate uptake in the reconstituted system are 0.28 mM and 338 μmol citrate min^{-1} g protein^{-1}, respectively, at 30° and pH 7.0. Benzene-1,2,3-tricarboxylate is a competitive inhibitor with respect to citrate, with a K_i of 0.13 mM. Table II shows that [14C]citrate exchanges not only with internal citrate but also with cis-aconitate, phosphoenolpyruvate, and malate. Externally added cis-aconitate, isocitrate, phosphoenolpyruvate, and malate strongly inhibit the reconstituted citrate/citrate exchange activity. Furthermore, besides the specific inhibitor benzene 1,2,3-tricarboxylate, also p-iodobenzyl malonate and p-hydroxymercuribenzoate strongly inhibit the exchange of citrate. In contrast, data not shown demonstrate that virtually no [14C]citrate is taken up by the proteoliposomes when they contain no anion, or anions which are not substrates of the tricarboxylate carrier. Externally added trans-aconitate, ADP, oxoglutarate, malonate, fumarate, phosphate, pyruvate, and aspartate do not compete with citrate for transport, and 2 mM NEM has only little effect on citrate exchange. Furthermore, in agreement with the observations that the tricarboxylate carrier is virtually absent in heart and brain mitochondria, the reconstituted [14C]citrate/citrate exchange activity of the hydroxyapatite eluate is very low (<26 μmol 10 min^{-1} g protein^{-1}) when

[14] F. Palmieri, I. Stipani, E. Quagliariello, and M. Klingenberg, Eur. J. Biochem. 26, 587 (1972).

TABLE II

Substrate Specificity and Inhibitor Sensitivity of the Reconstituted
Tricarboxylate Carrier[a]

Internal anion	External anions and inhibitors	Citrate exchange (μmol 10 min^{-1} g protein^{-1})
cis-Aconitate	[^{14}C]Citrate	1680
Phosphoenolpyruvate	[^{14}C]Citrate	1596
Malate	[^{14}C]Citrate	1463
Citrate	[^{14}C]Citrate	2025
Citrate	[^{14}C]Citrate + cis-aconitate	394
Citrate	[^{14}C]Citrate + threo-D$_s$-Isocitrate	380
Citrate	[^{14}C]Citrate + Phosphoenolpyruvate	508
Citrate	[^{14}C]Citrate + Malate	602
Citrate	[^{14}C]Citrate + Benzene 1,2,3-tricarboxylate	107
Citrate	[^{14}C]Citrate + p-iodobenzyl malonate	344
Citrate	[^{14}C]Citrate + p-hydroxymercuribenzoate	162

[a] The proteoliposomes contained 20 mM of the indicated anions; 0.3 mM [^{14}C]citrate was added either after 2 min preincubation with p-hydroxymercuribenzoate (0.2 mM) or simultaneously with the indicated external anions (2.7 mM).

crude extracts of mitochondria or submitochondrial particles from heart and brain (instead of liver) are applied to hydroxyapatite.

There is a 2-fold effect of phospholipids on the activity of the reconstituted citrate carrier. First, the phospholipid composition of the liposomes used for reconstitution is important. Addition of 25% (w/w) mitochondrial phospholipids, extracted from beef heart mitochondria[13] or alternatively, addition of 5% cardiolipin increases the activity of the reconstituted citrate/citrate exchange 3- to 4-fold as compared to pure egg yolk lipids. Furthermore, addition of cardiolipin to the solubilization buffer (3 mg/ml) enhances the transport activity 3- to 4-fold, presumably due to the protecting influence against inactivation by detergents.[10]

[55] Isolation and Reconstitution of the Phosphate Transport Protein from Mitochondria

By Hartmut Wohlrab, Hanno V. J. Kolbe, and Anne Collins

Mitochondrial inorganic phosphate transport is a process essential for steady-state oxidative phosphorylation as carried out by mitochondria in the cell. The transport occurs across the inner mitochondrial membrane via (1) the highly active phosphate carrier (phosphate transport protein, PTP) which catalyzes, as is generally believed, electroneutral transport of inorganic phosphate (P_i^- exchange against OH^-, or P_i^- cotransport with H^+), or (2) the low activity dicarboxylate carrier which has been shown to catalyze electroneutral exchange of divalent anions (P_i^{2-} or dicarboxylate^{2-}). This methodological note concerns itself only with the beef heart mitochondrial phosphate transport protein (PTP). Procedures for the protein from other tissues are referenced.

Three PTP preparations, that differ in purity, will be described. The preparation of highest purity requires the most effort. Preparation I consists of PTP and ADP/ATP carrier with very small amounts of contaminating proteins (M_r of 35,000–38,000 and less than 15,000).[1] Preparation II has all contaminating proteins of 35,000–38,000 M_r and some of lower M_r removed.[2] PTP Preparation III is highly purified.[3] These preparations can be incorporated into liposomes and have been shown to catalyze transmembrane pH gradient-dependent and N-ethylmaleimide (NEM)-sensitive net phosphate transport.

PTP Preparation I

Reagents and Solutions

Medium A: 10 mM sodium phosphate, 0.1 mM EDTA, 130 mM sodium chloride, adjust to pH 7.2 with sodium hydroxide.

Medium B: 10 mM sodium phosphate, 0.1 mM EDTA, 130 mM sodium sulfate, adjust to pH 7.2 with sodium hydroxide.

[1] H. Wohlrab, A. Collins, and D. Costello, *Biochemistry* **23**, 1057 (1984).
[2] H. Wohlrab, H. V. J. Kolbe, U. B. Rasmussen, and A. Collins, *in* "Epithelial Calcium: Phosphate Transport" (F. Bronner and M. Peterlik, eds.), pp. 211–216. Liss, New York, 1984.
[3] H. V. J. Kolbe, D. Costello, A. Wong, R. C. Lu, and H. Wohlrab, *J. Biol. Chem.* **259**, 9115 (1984).

Swelling medium (SM): 10 mM sodium phosphate, 0.33 mM EDTA, adjust to pH 7.2 with sodium hydroxide.

A/DTT: Medium A made to 5 mM dithiothreitol.

B/DTT/X-100: Medium B made to 5 mM dithiothreitol and 0.5% (v/v) Triton X-100 [Triton X-100 is permitted to flow gravimetrically out of 2-ml glass pipet, thus yielding less than 0.5% (v/v)].).

16% X-100/A/DTT: 4.2 ml A/DTT plus 0.8 ml Triton X-100.

0.5% X-100/A/DTT: 19.4 ml A/DTT plus 0.62 ml 16% X-100/A/DTT.

Mersalyl: 64 mM mersalylic acid in water with enough sodium hydroxide added to dissolve it.

Medium C$_i$: 10 mM Tris base, 10 mM 1,4-piperazinediethanesulfonic acid, pH$_i$ (intraproteoliposomal pH usually 8.0) adjusted with potassium hydroxide.

Medium C$_i$/DTT: Medium C$_i$ made to 5 mM dithiothreitol.

BioBeads SM-2 (BB): Washed[4] and stored in water in amber glass bottle at 4°; BB (Bio-Rad) are weighed out for use after placing them briefly on filter paper to remove the water.

Lipids: Plant phosphatidylcholine, plant phosphatidylethanolamine, beef heart cardiolipin (calcium salt) (Avanti); sodium phosphatidate (egg) (Calbiochem) was converted to calcium salt.[1]

Beef heart mitochondria: Prepared according to standard procedures.[5]

Lipids

Add 80 μl phosphatidylcholine (50 mg/ml chloroform) into amber (1 ml) ampule, dry with ultrahigh purity argon (keep dried lipid near bottom of ampule to facilitate complete sonication later with 250 μl medium). Place under high vacuum for at least 2 hr or overnight. Then add 250 μl medium A, sonicate in sonicator cup (Heat Systems, Plainview, NY) for 1 min (5–10 °) or until dispersion appears optically uniform; transfer dispersed lipids to small (3.5 ml) plastic test tube, flush with argon, stopper the tube, and store on ice for same day use.

Hydroxylapatite Column

Add 3.5 g of hydroxylapatite (fast flow powder, Calbiochem) gently into 21 ml B/DTT/X-100. Mount column (Bio-Rad, 1 × 13 cm, 15 cm long tubing attached to outflow to increase hydrostatic head) on ring stand, add 10 ml B/DTT/X-100 to column and let 2 ml flow through. Then add the

[4] P. W. Holloway, *Anal. Biochem.* **53**, 304 (1975).
[5] S. Joshi and D. R. Sanadi, this series, Vol. 55F, p. 384.

hydroxylapatite/B/DTT/X-100 slurry, let hydroxylapatite settle briefly and open valve; wash column with 35 ml B/DTT/X-100 (4°).

Purification Method

All operations, unless noted otherwise, are carried out at 4°. Thaw one test tube [6 ml frozen ($-20°$) beef heart mitochondria (13 nmol cytochrome b/ml[6])]. Keep as cold as possible during thawing. Centrifuge 10 min (17,000 g). Disperse pellet with glass/Teflon homogenizer in 60 ml SM and centrifuge 10 min (17,000 g). Wash mitochondrial pellets two times in A/DTT medium. Drain final pellets well and suspend in A/DTT to a final volume of 5 ml, then add 980 μl 16% X-100/A/DTT while vortexing. Centrifuge solubilized mitochondria 10 min (166,000 g, Beckman 50Ti rotor).

Add supernatant to large (75 ml) glass test tube with 1.6 g BB and a small magnetic flea; flush with argon, cover with parafilm, and stir for 30 min. Add solution to hydroxylapatite column, followed by 0.5% X-100/A/DTT. Collect 1 min fractions. OD (275 nm) of fractions (use 1 ml water plus 25 μl from fraction) should increase with the fifth or sixth fraction. Pool high OD fractions (start pooling when OD (275 nm) is 50% of maximum until it is decreasing to 50% of maximum). Should yield 6.2 ml total pool. Add pooled fractions to large (14 ml) plastic test tube, add sonicated phosphatidylcholine (250 μl), and vortex. Add to large (75 ml) glass test tube with 3.2 g BB and small magnetic flea; flush with argon, cover with parafilm, and stir 45 min.

Add solution to large (75 ml) glass test tube with 1.3 ml glycerol (volume of glycerol should be 20% of sample volume); disperse glycerol with Pasteur pipet. Parcel this final mixture as 200-μl aliquots each into 1.5 ml cryotubes (Vangard International), flush with argon, close, and store in liquid nitrogen freezer.

Preparation of Catalytically Active PTP Proteoliposomes

Lipids/Liposomes

Add 126 μl phosphatidylethanolamine (50 mg/ml chloroform), 99 μl phosphatidylcholine (50 mg/ml chloroform), and 37.5 μl calcium phosphatidate (20 mg/ml chloroform) into an amber (1 ml) ampule; blow off the solvent with ultrahigh purity argon. Ethyl ether (100 μl) is added to redissolve the lipids and dried with argon to near the bottom of the ampule.

---

[6] H. Wohlrab, *Biochemistry* **18,** 2098 (1979).

Place lipids under high vacuum for at least 2 hr at room temperature. Remove ampule from vacuum and place onto ice; flush with argon. Add 250 μl of medium C_i and cover with parafilm. Sonicate ampule in sonicator cup for 6 min (5–10°, cooled with water bath) or until lipid dispersion is visually homogeneous and clear. Transfer lipids to small (3.5 ml) plastic test tube. Adjust pH of lipids to that of medium C_i with 2 μl increments of 0.35 N potassium hydroxide while vortexing. Estimate pH with pH paper. This solution is named RL (Reconstitution Lipids).

PTP Proteoliposomes

PTP (200 μl, frozen Preparation I) is thawed and vortexed for 1 sec before being transferred to a small (3.5 ml) plastic test tube. A small gel (P-6DG) filtration column is used to eliminate the glycerol and to exchange PTP into medium C_i/DTT. The column (0.7 × 3.8 cm bed volume, Bio-Rad) is equilibrated with 6 ml medium C_i/DTT. PTP (200 μl) is added and eluted with 450 μl medium C_i/DTT and the eluate discarded. Then 330 μl medium C_i/DTT is added and eluate (P6/PTP) saved and vortexed (2 sec).

Prepare reconstitution mixture in a 3.5-ml plastic test tube: 110 μl RL, 60 μl P6/PTP, and 25 μl medium C_i made to 0.6% (v/v) glycerol. Flush test tube with argon and cap it. Mount plastic test tube with reconstitution mixture in drill press and spin fast enough so that mixture, submerged slowly in liquid nitrogen, freezes only on wall of test tube. Thaw mixture at room temperature; then vortex at low speed for 8 sec.

This reconstituted mixture is ready for transport assays. Centrifugation (12,000 g, 15 min) pellets protein and lipid (about 50% of total), that do not contribute to the transport activity. This step is essential before measuring net efflux of phosphate from the proteoliposomes since transport-inactive proteoliposomes contribute dramatically to the $^{32}P_i$ background. The transport activity catalyzed by these proteoliposomes (stored at 4°) is stable for at least 24 hr.

Anion-Exchange Column

These columns are used to separate the extraproteoliposomal inorganic phosphate from the intraproteoliposomal phosphate. The commercial anion exchange resin (AG1-X8, chloride, 50–100 mesh, Bio-Rad) has to be changed to the formate form since formate is much more readily exchanged off the resin by inorganic phosphate. To convert 150 g of chloride resin, wash it 12 times with 4 liters 1 N sodium hydroxide (brief gentle stirring in a large beaker) or until chloride is no longer detectable in the washing medium. To 1 ml of washing medium (after washing) are added 4 to 5 drops of concentrated nitric acid and the mixture is vortexed; then 100 μl of 1% (w/v) silver nitrate is added and if the solution does not become turbid, sufficient chloride has been washed off. The resin is then washed 3 times with 4 liters water, one time with 4 liters 1 N formic acid, and 3 times with 4 liters water to yield a pH equal to that of water. The resin is stored in water (4°).

The columns (0.7 × 11.5 cm, bed volume) are prepared and kept from running dry. Pass 500 μl of bovine serum albumin (30 mg/ml water) through the column, followed by 4 ml GNN [5% glycerol (v/v), 0.1 mM NaN$_3$]. Add 30 μl RL to 1 ml medium C$_i$ and pass mixture through anion exchange column, followed by 3 ml GNN. Keep anion exchange column in 4° cold box to facilitate movement from room temperature transport experiment to cold anion-exchange column.

Transport Experiment

Prepare medium C$_{ie}$ (medium C$_i$ plus 1 mM potassium phosphate, generally at pH 6.8) with ^{32}P$_i$ (12.5 KBq/ml) and add 500-μl aliquots into separate plastic test tubes. Carry out these operations at room temperature (22°) with ^{32}P$_i$ behind a leucite shield and wear two pairs of disposable rubber gloves to minimize exposure to radiation.

Time points for phosphate uptake are obtained as follows. For $t = 0$ sec, add 5 μl mersalyl and then 25 μl proteoliposomes to the 500 μl ^{32}P$_i$ medium C$_{ie}$. Then wait the longest phosphate uptake point planned before adding the mixture onto the anion exchange column. In this manner a maximum estimate for mersalyl-insensitive phosphate uptake is obtained. Thus when phosphate uptake up to 150 sec (Fig. 1) is planned, let inhibited sample ($t = 0$ sec) sit for 150 sec, then add to anion exchange column, followed by 3 ml GNN. Collect the liquid that comes off the column during sample application and GNN elution into one scintillation vial, add 6.5 ml Liquiscint (National Diagnostics) and count in liquid scintillation counter. For $t = 4$ sec, add 25 μl proteoliposomes at $t = 0$ sec and 5 μl mersalyl at $t = 4$ sec; wait 146 sec (150 sec minus 4 sec) before placing sample on anion-exchange column.

Proteoliposomes do not completely wash off the column with 3 ml GNN. About 10 to 15% of the radioactivity remains on the column but elutes with the next 3 ml GNN. To circumvent this problem and to monitor the capacity of the column to bind phosphate, add ^{32}P$_i$ medium C$_{ie}$ samples with mersalyl but without proteoliposomes (m point). In practice, start with two of these m points, then alternate sample and m point; after having applied up to 30 samples per column, add two more m points to make sure the column has not reduced its ^{32}P$_i$ binding capacity.

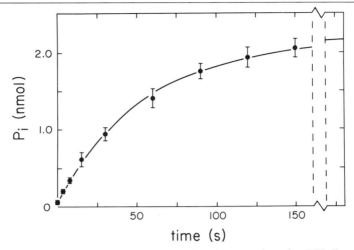

FIG. 1. Computer fit for net (zero-trans) phosphate uptake using PTP Preparation I proteoliposomes with pH_e 6.2, pH_i 8.0, $(P_i)_i = 0$ mM, $(P_i)_e = 0.35$ mM. The vertical bars are the standard deviation of the experimental points from four separate experiments. Mersalyl-insensitive transport has not been subtracted.

In Fig. 1 are uncorrected results from a typical P_i uptake experiment; note that the mersalyl-insensitive transport ($t = 0$ sec point) is very small. Specific transport rates can be calculated by determining the concentration of PTP from optical scans of Coomassie Blue R250 stained PTP in sodium dodecylsulfate polyacrylamide gels,[7] which have recently been quantitated with human carbonate dehydratase as reference.[3]

PTP Preparation II

This preparation is obtained after one additional purification step beyond those for Preparation I. It removes contaminating proteins of M_r between 35,000 and 38,000 and some with M_r less than 15,000.[2] This additional step was based on a report[8] that the ADP/ATP carrier interacts strongly with Celite (diatomaceous earth). Pasteur pipets, plugged with a small amount of glass wool, were filled with 500 mg of dry Celite (Celite 535, Roth, Karlsruhe, Federal Republic of Germany) that had been treated with 2-mercaptoethanol and cooled to 4°. The 2-mercaptoethanol treatment of Celite was carried out as follows. Celite (100 g) was mixed at

[7] H. Wohlrab and N. Flowers, *J. Biol. Chem.* **257**, 28 (1982).
[8] H. V. J. Kolbe, J. Bottrich, G. Genchi, F. Palmieri, and B. Kadenbach, *FEBS Lett.* **124**, 265 (1981).

50° for 10 hr with 650 ml of ,1 M 2-mercaptoethanol at pH 7.0. This treatment was repeated and the Celite was then mixed (10 hr) at room temperature with 650 ml distilled water and dried. Fines were removed to increase the flow rate of the column and to reduce the surface area of Celite. Part of the pooled hydroxylapatite column eluate (400 μl) (fresh Preparation I, before second BB treatment, phosphatidylcholine and glycerol addition) was added to the celite column. Medium A/Triton X-100 (1.5%, v/v) (700 μl) was added to wash off proteins with no affinity for celite. The phosphate transport protein was then eluted with medium A/Triton X-100 (1.5%, v/v)/sodium dodecyl sulfate (0.3 or 0.6%, w/v). To the eluted proteins was added 0.1 volume of phospholipids (46 mg/ml medium A) [phosphatidylethanolamine (PE)/phosphatidylcholine (PC)/cardiolipin (CL), 15/4/4, w/w] and mixed with a small magnetic flea for 45 min with 0.5 g BB/ml mixture under high purity argon. This preparation can now be handled as Preparation I, i.e., pass through P-6DG column and reconstitute.

PTP Preparation III

This highly purified preparation is of interest, first because it is very clean and second because it has been exposed to urea and SDS and still, after reconstitution, is able to catalyze phosphate transport. The purification procedure that we report here is a scaled down version of that published[3] and requires less time to carry out.

The eluate from the hydroxylapatite column (4°) (see Preparation I) is mixed with 3.2 g BB/6.4 ml eluate and stirred for 30 min at 4°. The BB are removed and an equal volume of medium UM (22°) [6.2 M urea, 240 mM sodium phosphate, 5 mM dithiothreitol, 0.8% (w/v) sodium dodecyl sulfate (Fluka), made to volume with water and adjusted then to pH 7.0 with 10 N sodium hydroxide] is added (PTP/UM). The following steps are carried out at room temperature (22°). Add 500 mg of hydroxylapatite to 3 ml UM and pour slurry into Pasteur pipet plugged with silanized glass wool, rinse with 2 ml UM; add 1.6 ml PTP/UM, then 1.6 ml UM, followed by 1.2 ml SUM (medium UM, except 340 mM sodium phosphate plus 7.5 mg bromphenol blue/100 ml). At this time the bromphenol blue front should be at bottom of column. Collect protein (SUM/PTP) by adding 0.4 ml SUM to the column.

Dried lipids (6 mg; see the table) were sonicated with 180 μl CHAPS/DM [25 mg CHAPS(Calbiochem)/ml DM; dialysis medium (DM) is medium C_i with 2 mM dithiothreitol] (10°) until dispersion appears visually homogeneous. To 340 μl dispersed lipids add 40 μl SUM (control) or 40 μl of SUM/PTP. About 100 μl of this mixture is added into a microdialysis

RECONSTITUTION OF BEEF HEART PTP PREPARATION III

Proteoliposome lipids	Protein[a]	Phosphate uptake (nmol/30 sec)
PE : PC : CaPA[b] (25 : 20 : 3)	—[c]	−0.010
	Preparation III	0.227
	Preparation I	0.806
PE : PC : CaPA : CaCL[b] (25 : 20 : 3 : 3)		
	—	0.003
	Preparation III	0.288
	Preparation I	0.690

[a] The same amount of PTP, determined from Coomassie Blue stained SDS gels, was present in the various reconstitution mixes.

[b] Phosphatidylethanolamine (PE), phosphatidylcholine (PC), calcium phosphatidate (CaPA), calcium cardiolipin (CaCL); lipid ratios are (w/w).

[c] Only buffer and no PTP was added to the reconstitution mixture. (HW21984)

well (BRL) and dialyzed for 18 hr against 1 liter DM with two dialysis medium changes.

The dialysate is assayed directly for transport without the freeze-thaw-vortexing step. Freeze-thaw-vortexing generates high mersalyl-insensitive phosphate uptake. The phosphate uptake within 30 sec is not as high per PTP protein from Preparation III than Preparation I. The purification and reconstitution methodologies involved however are quite different and the transport activities may not be directly comparable.

Comments

Identification and purification of the phosphate transport protein, like any other enzymatically active protein, depends on SDS–polyacrylamide gel electrophoresis and quantitative N-terminal amino acid analysis, in addition to reasonable activity measurements. Thus, while Preparation I consists of PTP and ADP/ATP carrier, traces of protein ($<1/10$ the amount of PTP) of M_r close to PTP copurify in addition to very low levels of lower M_r proteins. One of the proteins of M_r close to that of PTP also reacts with N-ethylmaleimide, a necessary but insufficient identifying criterion for PTP. The question remained, could this band be PTP with a 10- to 20-fold higher turnover number. This protein as well as others of this M_r range will not adhere to celite, while PTP and ADP/ATP carrier do. The fraction that passes through the celite shows no transport activity after incorporation into liposomes. The PTP and ADP/ATP carrier can be eluted from the celite with SDS containing buffer. These eluted proteins

(Preparation II) account for all the transport activity that can be obtained from the fraction incorporated into liposomes before and after the Celite column. It is clear that Preparation II contains only PTP and ADP/ATP carrier without any other proteins in this or higher M_r region. Traces of lower M_r proteins, however, are still present and these are removed during the steps leading to Preparation III.

[56] Transport of Metabolites across the Chloroplast Envelope

By H. W. HELDT and U. I. FLÜGGE

Introduction

A chloroplast is surrounded by an envelope consisting of two membranes, the inner and the outer envelope membrane. Due to a porin,[1] the outer envelope membrane is unspecifically permeable to solutes up to a molecular weight of 10,000, whereas the inner envelope membrane is the osmotic barrier and thus the site of specific metabolite transport.[2]

Chloroplasts are the site of photosynthesis in a plant. By utilizing the energy from the sun's radiation, the chloroplasts fix CO_2 in order to supply the plant cell with substrates required for maintenance and growth. The main substrate delivered from the chloroplasts is dihydroxyacetone phosphate (DHAP). Transport of the fixed carbon into other plant cells, e.g., for the growth of other parts of the plants, for filling of seeds or the supply of roots, requires that the DHAP is transformed into another transport metabolite, often sucrose. Sucrose is synthesized in the cytosol. The fixed carbon can also be temporarily stored in the chloroplasts as starch, this storage often being mobilized in the subsequent night by phosphorolytic as well as hydrolytic starch breakdown.

The provision of DHAP by the chloroplasts requires a transfer of water, CO_2 and inorganic phosphate (P_i) across the envelope into the chloroplasts, and an outward transfer of DHAP. Whereas the transfer of CO_2 is due to unspecific permeation of the inner envelope membrane,[3] transport of P_i and DHAP is catalyzed by a specific translocator, also

[1] U. I. Flügge and R. Benz, *FEBS Lett.* **169**, 85 (1984).
[2] H. W. Heldt and F. Sauer, *Biochim. Biophys. Acta* **234**, 83 (1971).
[3] K. Werdan, H. W. Heldt, and G. Geller, *Biochim. Biophys. Acta* **283**, 430 (1972).

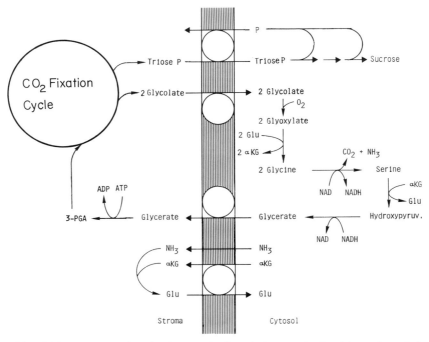

FIG. 1. Transport reactions in photosynthesis and photorespiration. For review of photorespiration see Ogren.[5] Glycolate and serine are transported by the same translocator (p. 713).

transporting 3-phosphoglycerate (3-PGA) in a counterexchange mode. This translocator of the inner envelope membrane, named phosphate-triose phosphate-phosphoglycerate translocator, or in short, phosphate translocator,[4] catalyzes the export of DHAP for different purposes. First, by exporting DHAP in exchange with P_i, the fixed carbon is delivered to the cytosol as outlined above (Fig. 1). Second, by an export of DHAP in exchange with 3-PGA, a shuttle is introduced, by which the chloroplast provides the cytosol with ATP and NAD(P)H generated by photosynthetic electron transport (Fig. 2). There are two ways in which the DHAP exported to the cytosol can be converted to 3-PGA. The conversion via NAD-glyceraldehyde dehydrogenase and phosphoglycerate kinase results in the generation of NADH and ATP,[6] whereas the conversion by the nonphosphorylating and nonreversible NADP glyceraldehyde phos-

[4] H. W. Heldt and L. Rapley, *FEBS Lett.* **10,** 143 (1970).
[5] W. L. Ogren, *Annu. Rev. Plant Physiol.* **33,** 415 (1984).
[6] U. Heber, *Proc. Int. Congr. Photosynth. Res., 3rd, 1974* p. 1335 (1975).

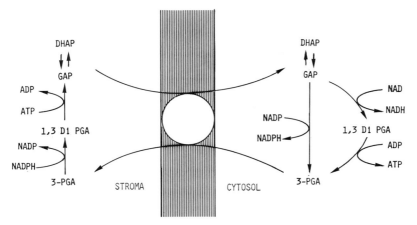

FIG. 2. Metabolite shuttle between the chloroplast stroma and the cytosol.

phate dehydrogenase leads to the formation of NADPH.[7] Another way to transfer reducing equivalents across the envelope is by a malate–oxaloacetate shuttle as discussed later. Two more translocators are needed in order to convert glycolate, the byproduct of photosynthesis, into 3-PGA, which can be fed into the Calvin cycle again (Fig. 1). The photorespiratory cycle enabling this conversion involves a glycolate–glycerate translocator[8,8a,9] for the release of glycolate from the chloroplasts and for the uptake of the produced glycerate into the chloroplast again, where it is then phosphorylated to 3-PGA. Furthermore, a dicarboxylate translocator catalyzing a 2-oxoglutarate–glutamate exchange is required for the refixation of the NH$_3$ that has been released during the oxidation of glycine, a step in the photorespiratory cycle.[10,11] This refixation of NH$_3$ taking place in the chloroplast involves the conversion of 2-oxoglutarate to glutamate, as catalyzed by glutamine and glutamate synthases.

The release of the products of starch degradation from the chloroplasts involves two translocators, the phosphate translocator for the products of phosphorolytic degradation (DHAP and 3-PGA) and a glucose translocator for the products of hydrolytic degradation.[12]

[7] G. J. Kelly and M. Gibbs, *Plant Physiol.* **52**, 111 (1973).
[8] K. T. Howitz and R. E. McCarty, *Biochemistry* **24**, 3645 (1985).
[8a] K. T. Howitz and R. E. McCarty, *Biochemistry* **24**, 2645 (1985).
[9] S. P. Robinson, *Plant Physiol.* **70**, 1032 (1982).
[10] K. C. Woo, *Plant Physiol.* **71**, 112 (1982).
[11] S. C. Sommerville and W. L. Ogren, *Proc. Natl. Acad. Sci. U.S.A.* **80**, 1290 (1983).
[12] G. Schäfer, U. Heber, and H. W. Heldt, *Plant Physiol.* **60**, 286 (1977).

The various chloroplast transporters have been mainly characterized by using silicone layer filtering centrifugation.[13]

Phosphate–Triose Phosphate–Phosphoglycerate Translocator

This translocator, dealt with in more detail in a following report (Flügge and Heldt, this volume [57]), accepts at its binding site P_i (or arsenate), 3-PGA, DHAP and glyceraldehyde phosphate, the apparent K_m values for the transport of these substances being between 0.1 and 0.4 mM. 2-Phosphoglycerate, phosphoenolpyruvate, glycerol 1-phosphate, and erythrose 4-phosphate are also transported, but the K_m values are at least one order of magnitude higher.[14] Sulfate and sulfite were shown to be transported by this carrier,[15] although a separate sulfate carrier in chloroplasts has been postulated.[16] As mentioned in the introduction, transport by the phosphate translocator proceeds in a strict counterexchange mode. For each molecule transported inward, another one is transported outward[14] so that the total amount of P_i (and phosphate esters) in the stroma is kept constant. The rate of unidirectional phosphate transport in spinach chloroplasts was found to be more than three orders of magnitude lower than that of counterexchange.[17,18] The activation energy of phosphate transport as determined from the temperature dependence has been evaluated to be 67 kJ (0–12°).[14] Studies of the pH dependence of transport suggest that not only P_i and DHAP but also 3-PGA are transported as divalent anions.[14] Since in contrast to the divalent P_i and DHAP anions, 3-PGA is predominantly a trivalent anion at physiological pH that must equilibrate with the divalent form in a pH dependent reaction, a proton gradient across the envelope would be expected to influence transport. During CO_2 assimilation of isolated chloroplasts mainly DHAP is found to be released to the medium, although the stromal level of 3-PGA is generally much higher than that of DHAP.[19] This apparent restriction of 3-PGA release is only observed in the light and thus appears to be due to a pH gradient which is found between the external space and the stroma of

[13] H. W. Heldt, this series, Vol. 69, p. 604.
[14] R. Fliege, U. I. Flügge, K. Werdan, and H. W. Heldt, *Biochim. Biophys. Acta* **502,** 232 (1978).
[15] R. Hampp and J. Ziegler, *Planta* **137,** 309 (1977).
[16] G. Mourioux and R. Douce, *Biochemie* **61,** 1283 (1979).
[17] K. Lehner and H. W. Heldt, *Biochim. Biophys. Acta* **501,** 531 (1978).
[18] G. Mourioux and R. Douce, *Plant Physiol.* **67,** 470 (1981).
[19] H. W. Heldt, U. I. Flügge, and R. Fliege, in "The Proton and Calcium Pumps" (G. F. Azzone, M. Avron, J. C. Metcalfe, E. Quagliariello, and N. Siliprandi, eds.), p. 105. Elsevier, Amsterdam, 1978.

illuminated chloroplasts. This proposed role of a pH gradient in the regulation of the chloroplast phosphate translocator could be verified, when the isolated phosphate translocator protein was incorporated into liposomes and the influence of pH gradients on the transport into liposomes was investigated (see Flügge and Heldt, this volume [57]).

Pyrophosphate and citrate, which bind to the carrier without being transported to any significant extent, are suitable competitive inhibitors of the transport of P$_i$, DHAP, and 3-PGA.[14] The phosphate translocator is also inhibited by reagents which covalently modify specific amino acids such as pyridoxal 5'-phosphate, trinitrobenzene sulfonate, p-diazobenzosulfonate,[20] and diisothiocyanodisulfonic acid stilbene (M. Rumpho and G. E. Edwards, personal communication) reacting with lysine residues, or butanedione and phenylglyoxal reacting with arginyl residues.[20] Reagents reacting with sulfhydryl groups, such as p-chloromercuribenzene sulfonate or mersalyl, are also strong inhibitors of the chloroplast phosphate translocator.[14,21] Linolate has been reported as being inhibitory to the chloroplast phosphate translocator.[22]

Dicarboxylate Transport

In spinach chloroplasts, a number of dicarboxylates such as L-malate, succinate, 2-oxoglutarate, L-aspartate, and L-glutamate were found to be specifically transported across the envelope[4,17], and very similar results were also obtained with chloroplasts of pea leaves[23] and maize mesophyll cells.[24] Maleate is not transported in contrast to its trans isomer fumarate, and the C$_3$ compound malonate does not react with the carrier.[17] Dicarboxylate transport is also an exchange process, but is less strictly coupled to the simultaneous countertransport of another dicarboxylate ion[17] than the substrate exchange catalyzed by the phosphate translocator. A specific inhibitor of dicarboxylate transport has not yet been reported. Mercurials, e.g., p-chloromercuribenzoate, cause some inhibition, but the inhibitory effect of these substances on dicarboxylate transport is less pronounced than on phosphate transport.[25]

Dicarboxylate transport was initially attributed to a single dicarboxyl-

[20] U. I. Flügge and H. W. Heldt, in "Function and Molecular Aspects of Biomembrane Transport" (E. Quagliariello et al., eds.), p. 373. Elsevier, Amsterdam, 1979.
[21] S. P. Robinson and J. T. Wiskich, FEBS Lett. **78**, 203 (1977).
[22] L. M. Akamba and P. A. Siegenthaler, Plant Cell Physiol. **20**, 405 (1979).
[23] M. O. Proudlove and D. A. Thurman, New Phytol. **88**, 255 (1981).
[24] D. A. Day and M. D. Hatch, Arch. Biochem. Biophys. **211**, 738 (1981).
[25] K. Werdan and H. W. Heldt, Proc. Int. Congr. Photosynth. Res., 2nd, 1971 p. 1337 (1972).

ate translocator of the inner envelope membrane.[4] More detailed investigations suggest that the transport might be mediated by more than one carrier, and that these carriers have overlapping specificity.[17] Studies of dicarboxylate transport by chloroplasts isolated from a photorespiratory mutant of *Arabidopsis thaliana* indeed indicate that two carriers are involved in dicarboxylate transport, one of these transporting dicarboxylates but not glutamine, and the other one transporting dicarboxylates at a lower rate and also glutamine.[11] A transport of glutamine, which was being competitively inhibited by asparagine and also by glutamate, malate, and succinate, has been reported.[26] The experiments with the photorespiratory mutant of *Arabidopsis thaliana* also showed that dicarboxylate transport, by catalyzing a 2-oxoglutarate–glutamate exchange, is an essential step in the photorespiratory nitrogen metabolism. Subsequent studies on the transport of 2-oxoglutarate revealed that although 2-oxoglutarate was a powerful competitive inhibitor of malate transport, the binding of 2-oxoglutarate for being transported, at least in part, occurs on a different site from that of the other dicarboxylates transported.[27,28]

Transport of Oxaloacetate

In spinach chloroplasts, dicarboxylate transport is also competitively inhibited by oxaloacetate with a K_i value of 0.3 mM which is very similar to the K_m for transport of malate (0.4 mM),[17] as also being observed with chloroplasts from maize mesophyll cells.[24] Chloroplasts, which had been loaded with radioactively labeled dicarboxylates like malate, show a release of the label on the addition of oxaloacetate, like on the addition of malate or succinate.[17] Although direct measurements of oxaloacetate transport had not been done at the time, as radioactively labeled oxaloacetate was not available, these results strongly indicate that in chloroplasts from spinach as well as from maize mesophyll, dicarboxylate transport is also able to catalyze a malate–oxaloacetate exchange. A malate–oxaloacetate exchange by this dicarboxylate transport system, however, appears to be unsuitable for the function of a malate–oxaloacetate shuttle *in vivo*. Due to the equilibrium constant of malate dehydrogenase (3×10^{-5}),[29] millimolar concentrations of malate are in equilibrium with micromolar concentrations of oxaloacetate, making a transport of

[26] D. J. Barber and D. A. Thurman, *Plant Cell Environ.* **1**, 297 (1978).
[27] M. O. Proudlove, D. A. Thurman and J. Salisbury, *New Phytol.* **96**, 1 (1984).
[28] K. C. Woo, U. I. Flügge, and H. W. Heldt, *Proc. Int. Congr. Photosynth. Res., 6th* p. 685 (1984).
[29] J. R. Stern, S. Ochoa, and F. Lynen, *J. Biol. Chem.* **198**, 313 (1952).

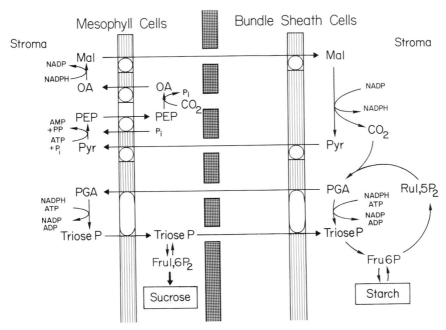

FIG. 3. Transport reactions in photosynthesis by maize leaves. For review Hatch and Osmond.[31]

oxaloacetate by the same translocator as for transport of malate virtually impossible.[30]

In a C_4 plant, such as maize, there is an absolute requirement for a malate-oxaloacetate shuttle, as the oxaloacetate formed by carboxylation of the PEP in the cytosol of the mesophyll cells has to be transported into the chloroplasts for being reduced to malate, which is then released from the chloroplasts, and subsequently diffuses into the bundle sheath cells (Fig. 3).[31] Polarographic measurements of oxaloacetate reduction by intact chloroplasts suggest that a malate–oxaloacetate shuttle is also functioning in chloroplasts from C_3 plants.[10,32] Recent measurements of [^{14}C]oxaloacetate uptake into chloroplasts from spinach and maize show that the high rate of oxaloacetate uptake is saturated at very low concentrations, and that the competition by malate is very low [e.g., maize, K_m

[30] C. Giersch, *Arch. Biochem. Biophys.* **219**, 379 (1982).
[31] M. D. Hatch and B. Osmond, *in* "Encyclopedia of Plant Physiology" (C. R. Stocking and U. Heber, eds.), New Ser., Vol. 3, p. 144. Springer-Verlag, Berlin and New York, 1976.
[32] W. A. Anderson and C. M. House, *Plant Physiol.* **64**, 1064 (1979).

(oxaloacetate) 53 μM, K_i (malate) 7.5 mM].[33] It remains to be elucidated in which way the transport of oxaloacetate interacts with a countertransport of malate.

Transport of Glycolate

Glycolate is formed in the chloroplasts by oxygenation of ribulose 1,5-bisphosphate, followed by hydrolysis of the phosphoglycolate produced. In order to be oxidized in the peroxysomes, the glycolate has to cross the envelope. In swelling studies the chloroplast envelope was found to be permeable to the protonated form of glycolate,[34] and subsequent studies by silicone layer filtering centrifugation indicated that rapid uptake of glycolate into chloroplasts occurred by penetration of the inner envelope membrane by the undissociated glycolic acid.[35] More recently, when the glycolate uptake was measured within the first 3 sec after the addition of the glycolate, the uptake rates were found to be saturated with respect to glycolate concentration, and to be inhibited by sulfhydryl reagents such as N-ethylmaleimide and by structurally related compounds such as glyoxylate.[8] These data indicate that transport of glycolate is a carrier-mediated process with maximum transport rates being high enough to sustain the export of glycolate during photorespiratory carbon fluxes.[8a] From the dependence of the kinetic properties of the translocator on stromal and medium pH it is concluded that transport occurs together with a proton or as a hydroxyl antiport.[8a]

Transport of Glycerate

The glycerate, formed in the photorespiratory cycle from glycolate, has to be taken up into the chloroplasts, in order to be phosphorylated by the glycerate kinase located in the stroma and then fed back again into the Calvin cycle. The apparent K_m of glycerate consumption by intact chloroplasts (1 mM)[36] turned out to be the apparent K_m of specific glycerate transport.[9] Glycerate transport is also inhibited by p-chloromercuriphenyl sulfonate.[9] Light increases both the initial rate and extent of glycerate uptake. Studies with uncoupler reagents indicate that glycerate uptake is

[33] M. D. Hatch, L. Dröscher, U. I. Flügge, and H. W. Heldt, *FEBS Lett.* **178**, 15 (1984).
[34] U. Enser and U. Heber, *Biochim. Biophys. Acta* **592**, 577 (1980).
[35] T. Takabe and T. Akazawa, *Plant Physiol.* **68**, 1093 (1981).
[36] U. Heber, M. R. Kirk, H. Gimmler, and G. Schäfer, *Planta* **120**, 31 (1974).

driven by a proton gradient established across the envelope in the light.[37] After the completion of this report, evidence has been presented showing that glycerate and glycolate are transported by a single translocator also transporting D-lactate and glyoxylate.[8]

Transport of Pyruvate

In the pathway of C_4 photosynthesis, pyruvate has to be transported into the mesophyll chloroplasts to be converted into phosphoenolpyruvate, which is released again to the cytosol to be carboxylated to oxaloacetate. In the transport of both pyruvate and PEP specific translocators are involved.

Evidence for a carrier-mediated pyruvate transport with an apparent K_m (Pyr) of 0.6–1 mM comes from work with mesophyll chloroplasts from the C_4 plant *Digitaria sanguinalis*.[38] Transport is inhibited by the sulfhydryl reagent mersalyl and by pyruvate analogs such as phenyl pyruvate (K_i 5.7 mM) and α-cyano-4-hydroxycinnamic acid (K_i 1.2 mM).[38] It may be noted, however, that the inhibitory effects of these substances on the chloroplast transport are much less pronounced than the corresponding effects on mitochondrial pyruvate transport. Measurements of the intracellular pyruvate distribution in maize leaves indicate, that in the steady state of photosynthesis the pyruvate concentration in the mesophyll chloroplasts is considerably higher than in the cytosol.[39] These findings suggest that transport of pyruvate into the chloroplasts is an active process. Transport measurements with isolated mesophyll chloroplasts indeed show that during illumination pyruvate is transported against a concentration gradient.[40] Preliminary results indicate that this active transport is driven by a light-dependent cation gradient across the envelope, perhaps similar to the transport of glycerate dealt with in the preceding section. Whereas in spinach chloroplasts, being almost impermeable to pyruvate,[38,41] such a pyruvate transporter has not been found, in pea chloroplasts a pyruvate transporter with an apparent K_m of 0.33 mM[23] has been reported.

In mesophyll chloroplasts from C_4 plants, such as maize, phosphoenolpyruvate (PEP) is rapidly transported in counterexchange with P_i,

[37] S. P. Robinson, *Plant Physiol.* **75,** 425 (1984).
[38] S. C. Huber and G. E. Edwards, *Biochim. Biophys. Acta* **462,** 583 (1977).
[39] M. Stitt and H. W. Heldt, *Biochim. Biophys. Acta* **808,** 400 (1985).
[40] U. I. Flügge, M. Stitt, and H. W. Heldt, *FEBS Lett.* **183,** 335 (1985).
[41] U. Heber and G. H. Krause, *in* "Photosynthesis and Photorespiration" (M. D. Hatch, C. Osmond, and R. O. Slatyer, eds.), p. 218. Wiley (Interscience), New York, 1971.

DHAP, and 3-PGA.[42,43] The apparent K_m of PEP transport is about one order of magnitude lower than in spinach.[43] It remains to be elucidated whether this PEP translocator is identical with the phosphate–triose phosphate–3-phosphoglycerate translocator.

Transport of Amino Acids

Early measurements of the osmotic response of pea chloroplasts to various neutral amino acids have led to the proposition that two carriers with apparent K_m values in the order of 100 mM transport glycine and serine across the chloroplast envelope.[44] In contrast, with spinach chloroplasts neutral amino acids were found to penetrate the envelope only very slowly[45,46] and the final concentrations reached in the stroma were similar to the concentrations in the medium. Since the rate of amino acid uptake was linearly dependent on the concentration[46] and also on the hydrophobicity of the amino acids studied, it was concluded that these amino acids permeate the envelope by simple diffusion.[45,46] In isolated pea leaf chloroplasts, a carrier-mediated transport of leucine has been reported to exist.[47] For transport of glutamate, aspartate, glutamine, and asparagine see the section on Dicarboxylate Transport.

Glucose Transport

Several pentoses and hexoses, such as D-xylose, D-ribose, D-glucose, D-mannose, and D-fructose, but essentially not L-glucose, are transported across the envelope. For D-glucose the apparent K_m is 20 mM.[48] When millimolar concentrations of these sugars are added to isolated chloroplasts, the internal concentration reaches about the external one. With high external glucose concentrations, however, glucose transport into the chloroplasts appears to be inhibited.[48] This explains why glucose can be used as an osmoticum for chloroplasts, although it is specifically transported across the envelope. The glucose transporter also appears to transport maltose.[49] This transporter plays a role in exporting glucose and maltose, arising from hydrolytic brakedown of chloroplast starch.

[42] S. C. Huber and G. E. Edwards, *Biochim. Biophys. Acta* **462,** 603 (1977).
[43] D. A. Day and M. D. Hatch, *Arch. Biochem. Biophys.* **211,** 743 (1981).
[44] P. S. Nobel and Y. S. Cheung, *Nature (London)* **237,** 207 (1972).
[45] H. Gimmler, G. Schäfer, H. Kraminer, and U. Heber, *Planta* **120,** 47 (1974).
[46] H. W. Heldt, *in* "The Intact Chloroplast" (J. Barber, ed.), p. 215. Elsevier, Amsterdam, 1976.
[47] J. S. McLaren and D. J. Barber, *Planta* **136,** 147 (1977).
[48] G. Schäfer, U. Heber, and H. W. Heldt, *Plant Physiol.* **60,** 286 (1977).
[49] A. Herold, R. C. Leegood, Ph. H. McNeil, and P. R. Simon, *Plant Physiol.* **67,** 85 (1981).

It has been reported recently that ascorbate is taken up into intact chloroplasts with an apparent K_m between 18 and 40 mM.[50,51] Interference of ascorbate transport with substrates of dicarboxylate and phosphate transport could not be detected. Transport of ascorbate resulted in an equilibrium of the ascorbate concentrations between the stroma and the medium.[51] Ascorbate transport was competitively inhibited by dehydroascorbate, and blocked by the sulfhydryl reagent p-chloromercuriphenyl sulfonate.[51] These findings indicate that the ascorbate uptake into chloroplasts is a carrier-mediated process. The relation between this transport and the glucose transporter (see above) remains to be elucidated.

ATP and ADP Transport

Transport of adenine nucleotides across the chloroplast envelope proceeds by counterexchange and is usually slow.[52] It is highly specific for external ATP; transport of ADP into the stroma is about one order of magnitude slower. It appears that this transport does not contribute to an export of ATP from the stroma to any considerable extent. The high specificity for external ATP suggests that the chloroplast ATP translocator mainly acts in the opposite direction, transporting ATP from the cytosol to the stroma. This could play a role in supplying the chloroplast during the night with ATP generated by glycolysis or respiration. In chloroplasts isolated from peas, the ATP transporter also transports pyrophosphate, although with a much lower rate.[53] When pyrophosphate is present in the medium, exchange with internal adenylates may deplete stromal adenylate pools leading to an inhibition of photosynthesis of isolated chloroplasts.[53,54] The activity of the ATP translocator depends on the developmental state of the chloroplasts. Inhibition of CO_2 fixation by pyrophosphate in pea chloroplasts indicated that young leaves transport ATP faster than mature leaves.[53] For mesophyll chloroplasts from the C_4 plant *Digitaria sanguinalis*, the ATP translocator was found to be very active.[55] In pea and maize mesophyll chloroplasts, a rapid counterexchange between ATP and phosphoenolpyruvate has been demonstrated.[56] The authors suggest that this counterexchange may be dependent upon

[50] J. W. Anderson, D. A. Walker, and C. H. Foyer, *Planta* **158**, 442 (1983).
[51] E. Beck, A. Burkert, and M. Hofmann, *Plant Physiol.* **73**, 41 (1983).
[52] H. W. Heldt, *FEBS Lett.* **5**, 11 (1969).
[53] S. P. Robinson and J. T. Wiskich, *Plant Physiol.* **59**, 422 (1977).
[54] Z. S. Stankovic and D. A. Walker, *Plant Physiol.* **59**, 428 (1977).
[55] S. C. Huber and G. E. Edwards, *Biochim. Biophys. Acta* **440**, 675 (1976).
[56] G. Woldegiorgis, S. Voss, E. Schrago, M. Werner-Washburne, and K. Keegstra, *Biochem. Biophys. Res. Commun.* **116**, 945 (1983).

the interaction of both ATP translocator and phosphate translocator. Further studies will be required to verify this suggestion.

From competition experiments a transport of purines and pyrimidines into isolated pea chloroplasts has been suggested.[57]

Concluding Remarks

As shown in this chapter, our present knowledge about transport processes across the inner envelope membrane is still quite fragmentary. This may be due in part to the fact that only a comparatively small number of researchers have been working in this area up to now. It was our aim to summarize the present knowledge on chloroplast transport processes in order to make it easier for new workers entering the field.

[57] D. J. Barber and D. A. Thurman, *Plant Cell Environ.* **1**, 305 (1978).

[57] Chloroplast Phosphate–Triose Phosphate–Phosphoglycerate Translocator: Its Identification, Isolation, and Reconstitution

By U. I. FLÜGGE and H. W. HELDT

The phosphate–triose phosphate–phosphoglycerate translocator (the phosphate translocator) of the chloroplast is located in the inner envelope membrane and catalyzes a strict counterexchange of inorganic phosphate (P_i), triose phosphate [dihydroxyacetone phosphate (DHAP) and glyceraldehyde phosphate], and 3-phosphoglycerate (3-PGA).[1] The main function of this translocator is to enable the export of fixed carbon in the form of triose phosphate from the chloroplasts to the cell in exchange with inorganic phosphate. The phosphate translocator also catalyzes the export of DHAP in exchange with 3-PGA, by which the cytosol is provided with the products of the light reaction, i.e., ATP and reducing equivalents in the form of NAD(P)H. This transport can therefore be regarded as a crucial step in the overall reaction of photosynthesis.

[1] H. W. Heldt and L. Rapley, *FEBS Lett.* **10**, 143 (1970).

The kinetic properties of this transport have been described[2] (see also Heldt and Flügge, this volume [56]). A more detailed understanding of this transport system, however, comprises (1) the identification of a distinct membrane polypeptide component involved in this transport, (2) the isolation of this membrane polypeptide, and (3) the reconstitution and characterization of the isolated phosphate translocator into artificial membranes under strictly defined conditions.

Identification of the Translocator Protein

For isolating the phosphate translocator protein, the envelope membrane polypeptides are solubilized by the use of detergents. The isolation of the translocator protein requires a means of detecting it. In principle this could be achieved by reconstitution of the solubilized protein into artificial membranes and measurement of transport activity. This approach, however, is not suitable here, as the phosphate translocator gradually loses activity when being exposed to detergents.[3] Alternatively, the membrane protein can be labeled by radioactive ligands. Unfortunately there are no inhibitors known, which would bind specifically and with high affinity to the chloroplast phosphate translocator and thus could be used as a label, like carboxyatractyloside for the mitochondrial ATP/ADP translocator[4] or oubain for the Na^+-K^+ pump.[5] Instead, for labeling of the chloroplast phosphate translocator, reagents, reacting selectively with specific amino acid residues of proteins, have been employed.

Since the substrates transported by the phosphate translocator are bound as divalent anions,[2] positively charged groups, such as lysine and arginine residues, were expected to be present at the active site of the translocator. Reagents employed and reacting selectively with lysine groups are pyridoxal 5'-phosphate (PLP) or 2,4,6-trinitrobenzene sulfonate (TNBS), and with sulfhydryl groups, p-chloromercuribenzene sulfonate (pCMS) and N-ethylmaleimide (NEM) are used. Although such reagents might bind to several membrane proteins, these can be used for the identification when a correlation between the concentration and time dependence of the incorporation into a particular membrane polypeptide and the inhibition of transport is established.

[2] R. Fliege, U. I. Flügge, K. Werdan, and H. W. Heldt, *Biochim. Biophys. Acta* **502,** 232 (1978).
[3] U. I. Flügge and H. W. Heldt, *Biochim. Biophys. Acta* **638,** 296 (1981).
[4] Klingenberg, M., *in* "Membrane Transport" (A. W. Martonosi, ed.), Vol. 4, p. 511. Plenum, New York, 1985.
[5] H. J. Schatzmann, *Helv. Phys. Acta* **11,** 346 (1953).

Measurement of the Inhibitory Effect of Amino Acid Reagents on
   Substrate Transport

   The transport of phosphate, 3-phosphoglycerate, and dihydroxyacetone phosphate into intact chloroplasts is measured by silicone layer filtering centrifugation.[6] Intact chloroplasts are prepared as described by
Heldt and Sauer.[7] The incubations are normally carried out in a medium
containing 0.33 M sorbitol, 50 mM N-2-hydroxyethylpiperazine-N-2-
ethanesulfonic acid (HEPES)-KOH, 1 mM MgCl$_2$, 1 mM MnCl$_2$, and 2
mM EDTA, pH 7.6 (medium 1). Uptake is initiated by adding 5–10 μl
labeled substrate to 200 μl chloroplast suspension (0.1 mg chl/ml) which
has been preincubated with given concentrations of the inhibitors for the
times indicated and is terminated by centrifugation through the silicone oil
into 20 μl 1 M HClO$_4$.

General Procedure for Labeling of the Phosphate Translocator Protein
   by Amino Acid Reagents

   For the labeling procedure intact chloroplasts equivalent to 3 mg chlorophyll are suspended in medium 1 and treated with the inhibitor for the
times indicated. Labeling with [^{3}H]NEM is terminated by the addition of
an excess of DTT, and labeling with PLP and TNBS by the subsequent
addition of a 3-fold molar excess of NaB^{3}H$_4$ (30 mCi/mmol). The chloroplast are then washed three times with medium 1 and envelope membranes are isolated as described by Douce et al.[8] The purified envelope
membranes are analyzed by SDS–polyacrylamide gel electrophoresis
(SDS–PAGE) as described by Neville[9] using 12.5% acrylamide and 0.1%
N,N'-methylenebisacrylamide. Gels are either stained with Coomassie
Blue[10] and scanned at 578 nm or analyzed for radioactivity by slicing and
digestion with NCS (Amersham and Searle)[11] followed by liquid scintillation counting. All steps following the incubation with PLP or TNBS are
done with minimal exposure to light.

Identification of the Phosphate Translocator with Sulfhydryl Reagents

   The phosphate translocator is strongly inhibited by low concentrations
of the sulfhydryl reagent p-chloromercuribenzene sulfonate (pCMS).[12]

[6] H. W. Heldt, this series, Vol. 69, p. 604.
[7] H. W. Heldt and F. Sauer, Biochim. Biophys. Acta 234, 83 (1971).
[8] R. Douce, B. R. Holtz, and A. A. Benson, J. Biol. Chem. 248, 7215 (1973).
[9] D. M. Neville, J. Biol. Chem. 246, 6328 (1971).
[10] J. K. Hoober, J. Biol. Chem. 245, 4327 (1970).
[11] R. S. Basch, Anal. Biochem. 26, 185 (1968).
[12] K. Werdan and H. W. Heldt, Int. Congr. Photosynth., 2nd, Stresa p. 1337 (1979).

This inhibition can be released by subsequent incubation with DTT.[13] *N*-Ethylmaleimide (NEM), another SH reagent, also inhibits phosphate transport, but reacts irreversibly. Incubation of intact chloroplasts (0.1 mg chl/ml) with 1 mM [³H]NEM (30 min at 20°, pH 8.4; specific activity 45 Ci/mol) and subsequent isolation of envelope membranes and analysis of the membrane polypeptide by SDS–PAGE show that almost all the membrane polypeptides have become labeled.[13] This is to be expected, since NEM is known as an unspecific sulfhydryl reagent. To increase the specificity of the NEM labeling, the chloroplasts are first incubated with pCMS (0.25 mM for 2 min) to inhibit phosphate transport. NEM (5 mM, 30 min) is then added in order to react with other remaining membrane polypeptides not related to phosphate transport. After the addition of DTT (4 mM, 5 min) the activity of phosphate transport is almost totally restored, which shows that the phosphate translocator is protected by pCMS from reacting irreversibly with NEM. By subsequent incubation of the chloroplasts with [³H]NEM (see above), only 2–3 polypeptides become labeled, one of which is a membrane polypeptide with an apparent molecular weight of 29,000.[13] When the chloroplasts are treated with increasing concentrations of NEM (0.2–2.0 mM) and the inhibition of phosphate transport activity and the incorporation of the [³H]NEM label into the membrane polypeptides is measured in parallel experiments, correlation between the inhibition of transport activity and incorporation of the radioactive label is only observed for the 29-kDa polypeptide. Of the other membrane polypeptides the labeling is nearly independent of the NEM concentration employed. It appears from the data that the 29-kDa membrane polypeptide is involved in phosphate transport across the inner envelope membrane.

Identification of the Phosphate Translocator with Lysine- and Arginine-Specific Reagents

The reactions between a lysine residue of a protein and PLP and TNBS are illustrated in Fig. 1A and B. The reversible binding of PLP to the lysine residue (the rate constants were calculated to be 160 M^{-1} min^{-1} for the forward reaction and 0.027 min^{-1} for the back reaction, respectively[14]) can be rendered irreversible by a reduction of the Schiff base formed by NaBH$_4$. Using NaB³H$_4$ a stable radioactive label is incorporated into the protein. As shown in Fig. 1B, the product between a lysine residue and TNBS can be reduced by NaBH$_4$ (or NaB³H$_4$) leading to a

---

[13] U. I. Flügge and H. W. Heldt, *in* "Function and Molecular Aspects of Biomembrane Transport" (E. Quagliariello *et al.*, eds.), p. 373. Elsevier, Amsterdam, 1979.

[14] U. I. Flügge and H. W. Heldt, *FEBS Lett.* **82,** 29 (1977).

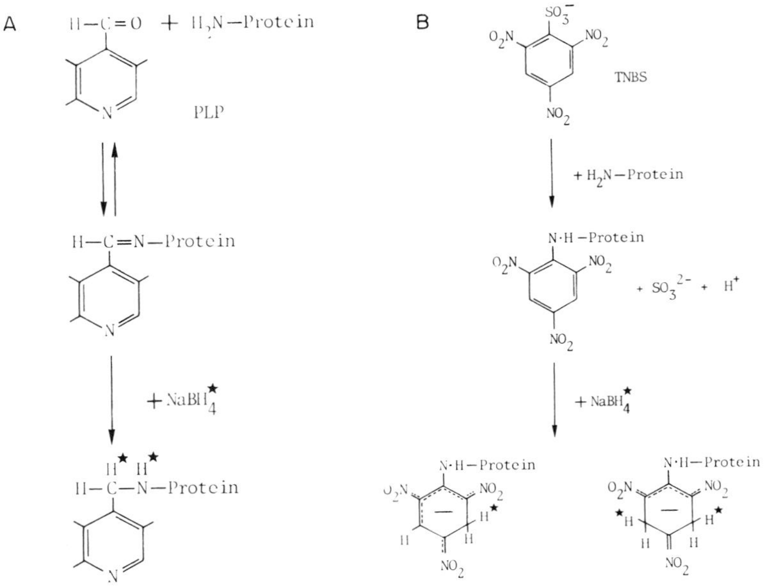

FIG. 1. Reaction between the ε-amino group of a lysine residue and PLP (A) and TNBS (B).

irreversible formation of a 1:1 or 1:2 adduct between the trinitrophenyl moiety and hydride ions.[15] Both PLP and TNBS as well as phenylglyoxal (PGO), reacting with arginine residues, turned out to be suitable and strong inhibitors of the chloroplast phosphate translocator.[13,14,16] The inhibition of the transport activity by these inhibitors, however, is substantially reduced if phosphate or 3-PGA, both substrates transported by the translocator, are added prior to the addition of the inhibitors (Table I). These findings show that the substrates bound to the active site of the phosphate translocator have a protective effect against the inhibition of the transport. In order to exclude that the observed protection of the transport activity is due to an unspecific anion effect, the experiment can also be performed using 2-phosphoglycerate (2-PGA) instead of 3-PGA. 2-PGA, which has a 10–20 times lower affinity to the phosphate translocator, also has no marked effect against the inhibition.

As mentioned above the adducts between PLP and TNBS, respectively, and the lysine group which is probably located at the active site of the phosphate translocator can be labeled by reduction with NaB^3H_4. Figure 2A shows the resulting pattern after labeling of the intact chloro-

[15] C. L. Parrot and S. Shifrin, *Biochim. Biophys. Acta* **491**, 114 (1977).
[16] U. I. Flügge and H. W. Heldt, *Biochem. Biophys. Res. Commun.* **84**, 37 (1978).

TABLE I

INHIBITION OF PHOSPHATE TRANSPORT BY
LYSINE- AND ARGININE-SPECIFIC INHIBITORS
AND THE EFFECT OF SUBSTRATES[a]

Additions	Inhibition % of control		
	PLP	TNBS	PGO
None	0	0	0
Inhibitor	41	64	68
Phosphate, then inhibitor	4	33	33
3-PGA, then inhibitor	2	36	34
2-PGA, then inhibitor	25	64	63

[a] Chloroplast concentration 0.10–0.15 mg chl/ ml. The inhibitor concentrations were 0.4 mM PLP, 10 min; 0.2 mM TNBS, 5 min; 10 mM PGO, 30 min. Substrates (10 mM) were added to the chloroplasts 10 min prior to the inhibitors. Data from Flügge and Heldt.[13]

plasts with TNBS/NaB^{3}H$_4$ [0.2 mM TNBS for 5 min followed by the addition of 0.6 mM NaB^{3}H$_4$ (30 mCi/mol) for 10 min; 4°] and analyzing the membrane polypeptides by SDS–PAGE. The radioactivity almost exclusively appears together with the 29-kDa polypeptide. About the same result is obtained by labeling of the membranes with PLP/NaB^{3}H$_4$ (0.5 mM PLP for 20 min followed by the addition of 2 mM NaB^{3}H$_4$ for 10 min; 4°)[14] Table II shows that the preincubation of the chloroplasts with phosphate or 3-PGA prior to the addition of the inhibitor leads to a large

TABLE II

INCORPORATION OF PLP AND TNBS INTO THE 29-kDa
MEMBRANE POLYPEPTIDE AND THE EFFECT OF
SUBSTRATES[a]

Additions before incubation with inhibitor	Incorporation of radioactivity % of control	
	PLP/NaB^{3}H$_4$	TNBS/NaB^{3}H$_4$
None	100	100
Phosphate	64	45
3-PGA	59	58
2-PGA	83	101

[a] Data from Flügge and Heldt.[13]

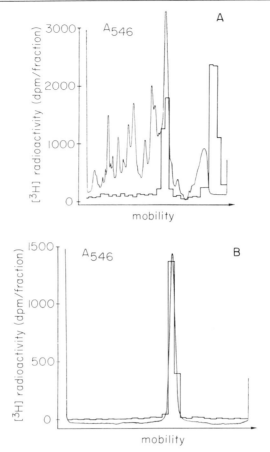

FIG. 2. SDS–PAGE of envelope membrane proteins (A) and of the purified phosphate translocator (B). Separate gels were used for the absorbance scan (continuous curve) and for the radioactivity (discrete lines). Data from Flügge and Heldt.[3]

decrease of the incorporation of the label into the 29-kDa polypeptide whereas 2-PGA has nearly no effect. Apparently, the protective effect of substrates against the incorporation of the radioactive label into the 29-kDa polypeptide has the same specificity as the protection against the inhibition of the transport as shown in Table I. From these results the 29-kDa polypeptide is identified as the chloroplast phosphate translocator containing a lysine and an arginine residue at the substrate binding site. Moreover, with TNBS/NaB^{3}H$_4$ a specific label for the phosphate translocator is available for the purification procedure of this translocator.

Isolation of the Phosphate Translocator

Envelope membranes, isolated from intact chloroplasts as described by Douce et al.,[8] are used as starting material for the purification of the phosphate translocator. The specific labeling procedures by TNBS/NaB³H₄ as described above represents the basis for the identification and isolation of the protein. All the following procedures are carried out at 4° and in the dark. Thirty milligrams of envelope membrane protein (about 0.1 μCi³H/mg protein) is suspended in 6 ml of 20 mM 4-morpholinopropanesulfonic acid (MOPS)-KOH, pH 7.4 and 0.02% NaN₃ (buffer A) and preextracted with 1.5% of the mild polyglycol detergent Brij 58 (30 min, 4°) in order to remove loosely bound proteins. By this procedure about 30–40% of the total membrane proteins are released from the membrane whereas the phosphate translocator as an intrinsic membrane polypeptide is only slightly affected. After centrifugation (120,000 g for 1 hr) the Brij residue is treated for 1 hr at 4° with buffer A containing 6% of the nonionic detergent Triton X-100 which is very effective in the solubilization of the phosphate translocator protein as well as the total membrane protein. Residual undissolved material is subsequently removed by centrifugation (120,000 g for 1 hr) and the clear yellow supernatant is applied to a DEAE-Sepharose CL-6B column (1 × 30 cm) which has been preequilibrated with a medium containing buffer A, 0.5% Triton X-100, and 0.2 M NaCl and is eluted with the same buffer. In contrast to most other membrane proteins, the phosphate translocator is not bound to the column and appears in the column pass through. This fraction is brought to 0.25 M NaCl and 5 mM DTT and concentrated by ultrafiltration or by polyethylene glycol dialysis. The addition of NaCl and DTT after this step is absolutely essential to prevent formation of high aggregation products of the translocator. Further purification is achieved by ultracentrifugation on a sucrose density gradient [10–20% (w/w), 30 ml] in a medium containing 0.25 M NaCl, 0.02% NaN₃, 0.1 mM EDTA, 0.2% Triton X-100, 5 mM DTT, 5 mM KH₂PO₄, and 20 mM MOPS, pH 7.4 for 14 hr at 420,000 g at 5°.

With the chosen centrifugation time, the non-protein-bound Triton X-100, lipids, and carotenoids remain at the top of the gradient, whereas the translocator is recovered half way down the gradient and thereby separated from aggregated forms of the translocator found at the bottom and other impurities. After this stage of purification, analysis of the fractions containing the phosphate translocator by SDS–PAGE yields a single polypeptide and all the radioactivity is recovered in this band (Fig. 2B³). The specific content of the preparation cannot be increased by a second sucrose density gradient centrifugation or gel chromatography. Table III summarizes the results of the various purification steps by which the

TABLE III
ISOLATION OF THE PHOSPHATE TRANSLOCATOR FROM TNBS/NaB³H₄-LABELED
ENVELOPE MEMBRANE[a]

Preparation	Protein (mg)	Specific content (dpm/mg protein) $\times 10^{-2}$	yield (%)
Starting material	27	410	100
Sediment after Brij 58 extraction	16.4	545	78
Supernatant of the Triton X-100 extraction	9.2	665	63
DEAE-Sepharose CL-GB pass through	1.0	1550	16
sucrose gradient centrifugation	0.4	2140	8

[a] Data from Flügge and Heldt.[3]

phosphate translocator is purified to apparent homogeneity. The increase of the specific content is 5- to 6-fold, indicating that the phosphate translocator represents 15–20% of the total envelope membrane protein which is in good agreement with earlier results.[13] This purified membrane protein is suitable for studying protein chemistry and the hydrodynamic properties of the translocator-Triton X-100 micelle. Part of these results is included in Table VI. For reconstitution experiments, however, this highly purified protein is not suitable, since its functional integrity is decreased during the long isolation procedure due to the presence of detergent. To obtain a functionally active transport protein, a batch procedure using hydroxyapatite has to be applied which was originally described for the rapid isolation of the mitochondrial ATP/ADP translocator.[17] For this purpose, envelope membranes (5 mg/ml) are solubilized with 3% Triton X-100 for 2 min and then applied to a hydroxyapatite (Bio-Rad, HTP) column of 300 mg dry gel per 0.5 ml Triton X-100 extract. The elution is performed with 20 mM MOPS, pH 7.4 and 0.5% Triton X-100 and the pass through of the column containing the phosphate translocator of 70–80% purity is immediately used for reconstitution experiments.

Reconstitution of the Phosphate Translocator

For the preparation of liposomes acetone-washed soybean phospholipids (50 mg/ml) are sonicated for 10 min in a solution containing 120 mM

[17] R. Krämer, H. Aquila, and M. Klingenberg, *Biochemistry* **16**, 4949 (1977).

P$_i$ or 3-PGA, 50 mM potassium gluconate, and 100 mM Tricine–NaOH, pH 7.8, or 100 mM PIPES–NaOH, pH 6.8. The high internal metabolite concentration is employed in order to avoid a limitation of the counterexchange by outward transport. Incorporation of the phosphate translocator is achieved by the freeze-thaw technique originally described by Kasahara and Hinkle.[18] Of the hydroxyapatite pass through, which are obtained within 15 min, 0.3 ml is combined with the sonicated phospholipids, mixed, and frozen in liquid nitrogen. The phospholipid/detergent ratio is kept above 15 in order to avoid the damaging effect of Triton X-100 on the stability of the liposomes. After being thawed at 0°, the liposomes are sonicated again for 18 sec while being kept in an ice bath. To remove the external medium, the liposomes with the incorporated phosphate translocator are passed over a Sephadex G-75 column which has been equilibrated with buffer containing 300 mM sodium gluconate, 50 mM potassium gluconate, and 20 mM Tricine–NaOH, pH 7.8 or 20 mM PIPES–NaOH, pH 6.8. The liposomes in the eluant are ready for the uptake experiments.

Uptake is measured by adding radioactively labeled substrate, [^{32}P]phosphate, 3-[^{14}C]PGA, or [^{14}C]DHAP (prepared as described by Flügge *et al.*[19]) to 200–300 μl of the liposomes (20°). The uptake is terminated by the addition of the inhibitor PLP (final concentration 30 mM). Incubation with PLP before the addition of the labeled substrates yields the corresponding blank. Immediately afterward, the mixture is placed on a Dowex AG 1-X8 (acetate form, 100 mesh) anion-exchange column (0.5 × 5 cm) and eluted with H$_2$O. Aliquots of the eluted liposomes are counted for radioactivity.

Uptake of radioactively labeled substrates into the liposomes is only observed when the liposomes have been preloaded with an exchangeable anion such as P$_i$ or 3PGA, but not when preloaded with salts such as potassium chloride or 2-PGA which has a 10- to 20-fold lower affinity toward the phosphate translocator.[2] Thus, like in intact chloroplasts, the reconstituted transport facilitates an obligatory and specific counterexchange of anions.[3] The apparent K_m values (substrate concentration causing half-maximal rate of transport) for P$_i$ and 3-PGA of this reconstituted system are about 2 and 1.5 mM respectively, and are therefore higher than the corresponding values found in intact chloroplasts. The efficiency of the reconstitution as expressed in terms of the turnover number (300–500 min^{-1}, 20°) is relatively high and accounts for 6–10% of the original activ-

[18] M. Kasahara and P. C. Hinkle, *J. Biol. Chem.* **257,** 7384 (1977).
[19] U. I. Flügge, J. Gerber, and H. W. Heldt, *Biochim. Biophys. Acta* **725,** 229 (1983).

ity. This reconstituted system is well suited to elucidate how an H^+ gradient can influence the preferential direction of the transport of P_i, DHAP, and 3-PGA across the chloroplast envelope.

As outlined in the introduction, the functioning of the phosphate translocator during illumination implies a preferential export of DHAP from the chloroplast into the cytosol (although this level is much lower than that of 3-PGA[20]) and a preferential import of 3-PGA into the chloroplasts whereas in the dark these substrates are transported equally well into either direction. It was suggested that the influence of light on the transport is related to the proton gradient across the envelope, resulting from an alkalinization of the stroma due to the light-driven proton transport into the thylakoid space.[21] This hypothesis can be checked by examining the influence of proton gradients across the liposome membrane on the substrate transport. A proton gradient between the liposomal compartment with a prefixed pH and the external medium can be adjusted by addition of buffer solution in order to change the external pH. If an inward directed proton gradient across the liposome membrane is to be applied, liposomes prepared as described above with an internal pH of 7.9 are used and buffer is added to adjust the external pH to 6.8. The gradient is maintained for several minutes and can be dissipated by the addition of the ionophore nigericine which induces an H^+/K^+ exchange allowing internal pH of the liposome to equilibrate with the pH of the external medium. For monitoring the proton gradient, liposomes are prepared in the presence of the impermeable dye, phenol red, and the absorbancy is measured using a dual wavelength spectrophotometer (Fig. 3). When an outward directed proton gradient is to be applied, the liposomes are prepared in a medium of pH 6.8 and the proton gradient is adjusted by adding alkaline buffer solution to change the external pH to 7.8.

For the sake of simplicity, here only the effects of proton gradients on the transport of DHAP and 3-PGA in the heteroexchange mode, i.e., the exchange of DHAP with 3-PGA in the internal liposome volume or vice versa will be dealt with. The concentration dependence was studied by varying the external substrate concentrations, and the resulting K_m and V_{max} values were evaluated from double reciprocal plots.

From the results in Table IV, the following conclusions can be drawn:

1. In the case of DHAP transport an inward directed proton gradient results in an inhibition due to an increase in the apparent K_m (DHAP) with

[20] R. McC. Lilley, C. J. Chon, A. Mosbach, and H. W. Heldt, *Biochim. Biophys. Acta* **460**, 259 (1977).
[21] H. W. Heldt, U. I. Flügge, and R. Fliege, *in* "Mechanism of Proton and Calcium Pumps" (M. Avron *et al.*, eds.), p. 105. Elsevier, Amsterdam, 1978.

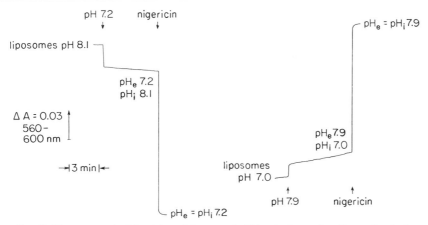

FIG. 3. Measurement of the internal liposomal pH by dual-wavelength spectrophotometry. Liposomes (4 mg/ml) with internal phenol red (0.3 mM) at pH 8.1 (left) or 7.0 (right) were used. The pH pump was adjusted by addition of acidic or basic buffer solution. When indicated 5 μM nigericin was added. Downward (upward) deflection indicates acidification (alkalinization) of the internal liposome space. Data from Flügge et al.[19]

the V_{max} (DHAP) being unaltered, an outward directed proton gradient having an opposite effect in decreasing the apparent K_m (DHAP).

2. In the case of 3-PGA transport an inward directed proton gradient leads to a stimulation associated with an increase of V_{max} (3-PGA). An outward directed proton gradient also has the opposite effect in decreasing the V_{max} (3-PGA). Under both conditions the apparent K_m (3-PGA) are not affected.

TABLE IV

INFLUENCE OF THE H$^+$ GRADIENT ON K_m (mM) AND V_{max} (μmol/mg PROTEIN PER min) AS MEASURED IN THE HETEROEXCHANGE MODE[a]

Gradient	External pH	Without H$^+$ gradient		With H$^+$ gradient	
		K_m	V_{max}	K_m	V_{max}
Transport of DHAP					
Inward directed H$^+$	6.8	1.30	9.7	1.93	9.7
Outward directed H$^+$	7.8	1.18	9.7	0.74	9.7
Transport of 3-PGA					
Inward directed H$^+$	6.7	0.72	10.0	0.72	14.9
Outward directed H$^+$	7.8	2.17	10.0	2.17	7.5

[a] Data from Flügge et al.[19]

TABLE V

MODEL CALCULATION OF THE INFLUENCE OF
THE LIGHT-INDUCED H^+ GRADIENT ON THE
FLUXES OF DHAP AND 3-PGA

Conditions	Transport DHAP/ 3-PGA		Ratio outward/inward
	Outward	Inward	
Light	1.2	0.21	5.9
Dark	0.15	0.62	0.24

Thus, in the reconstituted system the transport of DHAP and 3-PGA is influenced by a proton gradient in a different manner. In the case of DHAP (and also P_i) the observed effects are associated with changes in the apparent K_m values whereas in the case of 3-PGA transport the application of a proton gradient is linked to a change of V_{max}. These observations suggest that DHAP and 3-PGA are recognized and bound to the translocator differently. The fact that both substrates are transported as twice negatively charged anions implies that the phosphate moiety in the case of DHAP and P_i contains both negative charges. In the case of 3-PGA, however, it contains only one negative charge whereas the other one is contributed by the carboxylic group. These differences may partly account for the different responses to an applied proton gradient. For a more detailed discussion see Flügge et al.[19] The question arises, whether the effects of the proton gradient observed in the reconstituted system can explain the considerable changes in the transport of DHAP and 3-PGA between illumination and darkness as observed with protoplasts or intact chloroplasts. Using the cytosolic and stromal concentrations of DHAP, P_i, and 3-PGA as measured in illuminated protoplasts[22] in combination with K_m and V_{max} values presented in Table V and assuming a simple Michaelis–Menten characteristic for the transport, the transport rates for each metabolite can be calculated. The model calculation presented in Table V shows that in the light the ratio of DHAP/3-PGA transported is more than 8-fold higher for the outward transport and 3-fold lower for the inward transport than the corresponding rates obtained in the dark. Thus, under conditions with an applied proton gradient (light conditions) the species transported out is indeed mainly DHAP with 3-PGA transported mainly into the chloroplasts. These fluxes are reversed under conditions

[22] M. Stitt, W. Wirtz, and H. W. Heldt, *Biochim. Biophys. Acta* **593**, 85 (1980).

without a proton gradient (dark conditions). However, the presented effects of a proton gradient on the selectivity of the substrate transport by the phosphate translocator cannot fully explain the DHAP/3-PGA gradients between illumination and darkness as observed in intact chloroplasts.[21] One additional factor might be the increasing stromal Mg^{2+} concentration in the light, which results in an inhibition of 3-PGA transport and stimulation of DHAP transport (unpublished results).

Concluding Remarks

The properties of the chloroplast phosphate translocator are summarized in Table VI. The inhibition studies reveal that two cationic amino residues, a lysine and an arginine, are involved in the binding of the anionic substrates to the active site of the translocator protein. Furthermore, the translocator protein contains a sulfhydryl group essential for transport activity. The phosphate translocator is the largest portion of the total envelope membrane protein. Parallels exist between the chloroplast phosphate translocator and the mitochondrial ATP/ADP translocator.

TABLE VI

SOME CHARACTERISTICS OF THE CHLOROPLAST PHOSPHATE–TRIOSE
PHOSPHATE–PHOSPHOGLYCERATE TRANSLOCATOR

Specificity	Inorganic phosphate and phosphate in 3′, not in 2′ position of C_3 compounds; translocator catalyzes an 1 : 1 exchange
Specific inhibitors	Pyridoxal 5′-phosphate
	2,4,6-Trinitrobenzene sulfonate
Translocation activity	0.25 μmol/min mg protein (20°)
NEM binding sites	7.2 μmol/g membrane protein
Content	53 μmol/g chloroplast protein
Molecules/chloroplast	1×10^6
Molecular activity (20°)	
Intact chloroplast	5000 min^{-1}
Reconstituted system	300 min^{-1}
Translocator density of the envelope	2.5 $pmol/cm^2$
Flux rate per membrane area	7.3 $nmol/min/cm^2$
Stokes radius of the	
Translocator–Triton X-100 micelle	6.32 nm
Sedimentation coefficient	3.50×10^{-13} (sec)
Diffusion coefficient	3.39×10^{-7} cm^2/sec
Molecular weight of the protein–Triton X-100 micelle	177,500
Protein moiety	61,000
Minimal molecular weight (SDS–PAGE)	29,000

Both translocators have the function of transporting the main product of each organelle into the cytosol. Both translocators catalyze a strict counterexchange and both translocator proteins are dimers with subunits of almost identical size (phosphate translocator 29 kDa, ATP/ADP translocator 30 kDa[4]). It may be noted that the molecular activity of the chloroplast translocator is about one order of magnitude higher than that of the mitochondrial ATP/ADP translocator. For the ATP/ADP translocator, a gated pore mechanism has been postulated.[4] One may speculate that the chloroplast phosphate translocator also functions in this way.

Author Index

Numbers in parentheses are footnote reference numbers and indicate that an author's work is referred to although the name is not cited in the text.

Subject Index

A

Acetic acid, in measurement of bacterial intracellular pH, 340
Acetyl-carboxyatractyloside
radiolabeled, binding to beef heart mitochondria, 660–661
radiolabeling, 659–660
Acetyl-CoA acetyltransferase (thiolase), 550
Acetylsalicylic acid, in measurement of bacterial intracellular pH, 340, 356
Acheloplasma, 260
growth conditions, 262, 263
Acheloplasma laidlawii
growth medium, 261
transport processes, 261
transport studies, 264
Acidaminococcus fermentans
growth, 551–552
maintenance of cultures, 552
membrane preparation, 552
Acidophiles
cytoplasmic buffering capacity, determination, 358–361
passive proton permeability, determination, 361
ΔpH, determination, 355–358
pH homeostasis, 354–361
study of, 353
resting cells, intracellular pH, 346
Acridine derivatives, in measurement of bacterial intracellular pH, 340
Active transport, 46, 493. *See also* Binding protein-dependent active transport
bacterial
binding-protein dependent, 281
facilitated diffusion, 279–280
osmotic shock-sensitive, 279–280
phosphotransferase, 279–280
types, 279–280
cotransport carriers, 9

energy conversion in, efficiency, 14–15
exchange-transport carriers, 9–10
nonequilibrium thermodynamics, 12–14
permease-mediated, in *E. coli,* 444
primary, 3
energetics, 5–6
secondary, 3
electrogenic, 9–10, 11
energetics, 9–11
two flow system, 12–13
types, 9
Adenosine diphosphate
assay, in heart mitochondria, 19
radiolabeled azido derivatives, 665–666
ADP/ATP carrier or translocator
ATR and ATP binding sites, mapping, 668–670
ATR binding site, photolabeling, 665–670
binding of spin-labeled maleimides, 632, 637–638
binding to inhibitors, 658–659
calculation of exchange kinetics, 616–617
chemical modifications, 659–665
by 2,3-butanedione, 661
with N-ethoxycarbonyl-2-ethoxy-1,2-dihydroquinoline, 663–665
with N-ethylmaleimide, 662, 663–665
with 2-hydroxy-5-nitrobenzyl bromide, 662, 663–665
by phenylglyoxal, 661–662
resulting in selective inactivation of radiolabeled ATR or acetyl-CATR binding, 662–665
resulting in simultaneous inactivation of ATR and BA binding capacity, 661–662
with ultraviolet light, 662, 663–665
and chloroplast phosphate translocator, 729–730

P

Partition coefficient
  effect of water on, 129
  measurement, 120
    by RP–HPLC, 120
Pea, chloroplast. *See* Chloroplast
Pentitols, metabolism, in bacteria, 473–474
Peptostreptococcus asaccharolyticus,
    biotin-dependent decarboxylase, 547–
    549
Periplasmic protein, processing, 133
pH
  external, perturbation measured by pH
    electrodes, 569
  homeostasis
    in bacteria, 565
      K^+ transport in, 349–351
      Na^+ transport in, 351–352
      mechanism, 347–352
    in neutrophiles, 353
    role in cell physiology, 337–338
  intracellular
    calculation, 338–343
    in growing bacterial cells, under
      various conditions, 346–347
    measurement in bacteria
      by colorimetry, 345
      determination of probe distribu-
        tion, 342–343
      by fluorimetry, 345
      by NMR spectroscopy, 343–345
      probes for, 356
    probes used in measurement, 339–
      340
    regulation in bacterial cells, 337–365
    in resting bacterial cells, under
      various conditions, 345–346
  regulation, in *E. coli,* 329
  transmembrane difference, measure-
    ment, 568–569
    by distribution of radiolabeled weak
      acids and bases, 568–569
    by ^{31}P NMR, 568
    spectroscopic techniques, 569
Phenylglyoxal, identification of chloroplast
    phosphate translocator using, 720–722
Phosphate acetyltransferase, 550
Phosphate : hexose 6-phosphate antiport, of
    S. lactis, 558–563
  assay of anion exchange, 561–562

reconstitution, 559–561
  effect of stabilant, 562
Phospho-2-dehydro-3-deoxygluconate
    aldolase, 166
Phosphoenolpyruvate carboxylase, 167
  isolation of mutants, by targeted muta-
    genesis, 167–168
Phosphoenolpyruvate-dependent carbohy-
    drate : phosphotransferase systems,
    474
  functions, complementation assays for,
    479–481
Phosphoenolpyruvate-dependent phospho-
    transferase system, of mycoplasmas,
    260
Phosphoenolpyruvate-phosphotransferase
    membrane transport system, 280, 493,
    496
Phosphofructokinase, 164
Phosphonium, probes, for measurement of
    membrane potential, 48–51
Photorespiration, transport reactions, 706,
    707
Photosynthesis, transport reactions, 706
  in maize, 711
Photosystem I
  reaction center chlorophyll P700, re-
    duction by cytochrome c, pH
    dependence, 71–73
  reaction center chlorophyll P700, and
    ferrocyanide, effect of salt concen-
    tration on reduction, 68–69
Photosystem II
  from blue-green algae, purification, 34
  primary acceptor Q, reaction with
    ferricyanide, phenazine methosul-
    fate, and benzoquinone, effect of
    salt concentration, 69–70
  quinones, redox forms, 110–111
Phototaxis, 565
pIN III-C, 143
  cloning staphylococcal nuclease A into,
    148–149
  linker polynucleotides, partial DNA
    sequences, 143, 144
pIN III-*ompA*, 140–143
  cloning β-lactamase into, 143–146
  cloning staphylococcal nuclease A into,
    146–148
  construction, 141